Aquatic Chemistry

An Introduction Emphasizing
Chemical Equilibria in Natural Waters

Aquatic Chemistry

An Introduction Emphasizing Chemical Equilibria in Natural Waters

WERNER STUMM

Professor, Swiss Federal Institute of Technology, Zurich

JAMES J. MORGAN

Professor of Environmental Engineering Science
California Institute of Technology

A WILEY-INTERSCIENCE PUBLICATION

JOHN WILEY & SONS
New York • Chichester • Brisbane • Toronto • Singapore

Copyright © 1981 by John Wiley & Sons, Inc.

All rights reserved. Published simultaneously in Canada.

Reproduction or translation of any part of this work
beyond that permitted by Sections 107 or 108 of the
1976 United States Copyright Act without the permission
of the copyright owner is unlawful. Requests for
permission or further information should be addressed to
the Permissions Department, John Wiley & Sons, Inc.

Library of Congress Cataloging in Publication Data:

Stumm, Werner, 1924–
Aquatic chemistry.

"A Wiley-Interscience publication."
Includes bibliographical references and indexes.
1. Water chemistry. I. Morgan, James J.,
1932-   joint author. II. Title.
GB855.S78 1981   551.4   80-25333
ISBN 0-471-04831-3

Printed in the United States of America

10 9 8 7 6 5 4

Preface to the Second Edition

In the ten years since the first edition of *Aquatic Chemistry* was published, the field has grown and matured considerably in terms of data, unifying concepts, techniques, and applications. A new edition is now in order. Our aim in this substantially revised edition remains the same: to present a quantitative treatment of the variables that determine the composition of natural waters by drawing upon basic chemical principles. Chemical equilibrium continues to be the central theme in our treatment, but steady-state and dynamic models using mass-balance approaches and kinetic information have been given more attention. Those readers who find the formalism of thermodynamics in Chapter Two too abstract at first should not hesitate to proceed to subsequent chapters where they can gain a thorough acquaintance with applications of equilibrium concepts to aquatic chemistry. They may then wish to return to Chapter Two for detailed review of key concepts. *Aquatic Chemistry* emphasizes a teaching approach and is designed to assist the reader in applying general principles and methodologies to systems of particular interest. More than 100 numerical examples illustrate the most important aspects of natural water chemistry.

This book is written for those who share a concern and sense of responsibility for the environment. Because humankind has become a more powerful participant in the biosphere, a chapter giving a chemical and ecological perspective on pollution and its control has been added (Chapter 11). We attempt to show how pollution affects water quality and how the aquatic ecosystem responds to human impact, especially to stress caused by chemical perturbation. The distribution of pollutants in air, water, and soil is discussed and the environmental movement assessed. It is clear that pollution is no longer a local or regional problem, but that people can influence and change global chemical cycles of the environment.

Among the other more significant improved features of this new edition we should mention the following changes. A new chapter (Chapter 8) introduces the subject of organic carbon compounds in natural water systems. There is a more detailed consideration of seawater chemistry, including activity conventions in seawater (Chapters 4 and 6). Major revision of the treatment of the solid–solution interface (Chapter 10) reflects significant advances in concepts and experimental developments during the past decade. Inclusion of material on stable and radioactive isotopes (Chapters 4 and 5) illustrates their value in characterizing physical and chemical processes in natural water systems. Greater emphasis on kinetics in a number of chapters (Chapters 2, 4, 5, 7, 9, 10,

11) shows the interplay between kinetic considerations and equilibria. The consideration of metal–ligand interactions (Chapter 6) including ion pairs and complexes, inorganic speciation, complexing by organics, alkylation and multicomponent, multispecies models is expanded. There is a more extensive compilation of thermodynamic data (free energies, enthalpies, entropies, and equilibrium constants) for some of the more important species and reactions in natural water systems.

Zurich, Switzerland   Werner Stumm
Pasadena, California   James J. Morgan
February 1981

Acknowledgments

We thank Drs. Charles R. O'Melia, George Jackson, Phil Singer, J. Lyklema, James Pankow, Russ McDuff and Elisabeth Stumm-Zollinger; and our students Dieter Diem, Robert Kummert, Laura Sigg, Christine Matter, Neal Handly, and Robert Kaiser for the review of individual chapters. We are grateful also to the colleagues who provided us with valuable suggestions for corrections and improvements to the first edition. Appreciation is expressed to Mrs. W. Kumpfe and Mrs. Elaine Granger for typing the manuscripts.

The hospitality of Caltech during 1977–78 and the awarding of a Fairchild Distinguished Scholarship to Werner Stumm is gratefully acknowledged, as is the hospitality of EAWAG to James Morgan in 1973 and 1978.

W.S.
J.J.M.

Contents

Aquatic Chemistry

An Introduction Emphasizing
Chemical Equilibria in Natural Waters

1

Introduction: Scope of Aquatic Chemistry

Aquatic chemistry is concerned with the chemical processes affecting the distribution and circulation of chemical compounds in natural waters; its aims include the formulation of an adequate theoretical basis for the chemical behavior of ocean waters, estuaries, rivers, lakes, groundwaters, and soil water systems, as well as the description of the processes involved in water treatment. Obviously, aquatic chemistry draws primarily on the fundamentals of chemistry, but it is also influenced by other sciences, especially geology and biology.

This book does not cover all aspects of aquatic chemistry; a large part is devoted to the demonstration that elementary principles of physical chemistry can be used to identify some of the pertinent variables determining the composition of natural water systems. The student of chemistry is not always fully aware that the well-known laws of physical chemistry not only apply in the laboratory but also regulate the course of reactions taking place in nature. During the hydrological cycle water interacts continuously with matter. Thus a progressive differentiation of matter is achieved by processes of weathering, soil erosion, and soil and sediment formation. These separation processes accomplished by nature on a large scale have been likened [1] to the sequence of separations carried out during the course of a chemical analysis. The processes—dissolution and precipitation, oxidation and reduction, acid–base and coordinative interactions—are the same in nature as in the laboratory; although this book contains treatment of topics similar to those found in an analytical chemistry text, it endeavors to consider the spatial and temporal scales of the reactions in nature as entirely different from those in the laboratory; for example, in analytical chemistry precipitates (of frequently metastable and active compounds) are formed from strongly oversaturated solutions, whereas in natural water systems the solid phase is usually formed under conditions of slight supersaturation; often crystal growth and aging continues for geological time spans. Interfacial phenomena are particularly important because most chemical processes of significance in nature occur at phase discontinuities.

The actual natural water systems usually consist of numerous mineral assemblages and often of a gas phase in addition to the aqueous phase; they nearly always include a portion of the biosphere. Hence natural aquatic habitats are characterized by a complexity seldom encountered in the laboratory. In order to understand the pertinent variables out of a bewildering number of possible ones, it is advantageous to compare the real systems with their idealized counterparts.

Models to Obviate Nature's Complexity. Simplified and manageable models may be used to illustrate the principal regulatory factors that control the chemical composition of natural waters and in turn the composition of the atmosphere. To be useful a model need not be realistic as long as it produces fruitful generalizations and valuable insight into the nature of aquatic chemical processes and improves our ability to describe and to measure natural water systems.

The theory of *thermodynamic equilibrium* appears to be the most expedient concept to facilitate identification of many variables relevant in determining the mineral relations and in establishing chemical boundaries of aquatic environments. Free energy concepts describe the thermodynamically stable state and characterize the direction and extent of processes that are approaching equilibrium. Discrepancies between predicted equilibrium calculations and the available data of real systems give valuable insight into those cases in which chemical reactions are not understood sufficiently, in which non-equilibrium conditions prevail, or in which the analytical data are not sufficiently accurate or specific. Such discrepancies thus provide an incentive for future research and the development of more refined models.

By comparing the actual composition of seawater (sediments + sea + air) with a model in which the pertinent components (minerals, volatiles) with which water has come into contact are allowed to reach true equilibrium, Sillén [2] in 1959 epitomized the application of equilibrium models for portraying the prominent features of the chemical composition of this system. His analysis, for example, has indicated that, contrary to the traditional view, the pH of the ocean is not buffered primarily by the carbonate system; his results suggest that heterogeneous equilibria of silicate minerals comprise the principal pH buffer systems in oceanic waters. This approach and its expansion have provided a more quantitative basis for Forchhammer's suggestion of 100 years ago that the quantity of the different elements in seawater is not proportional to the quantity of elements which river water pours into the sea but is inversely proportional to the facility with which the elements in seawater are made insoluble by general chemical actions in the sea. Although inland waters represent more transitory systems than the sea, equilibrium models are also useful here for interpreting observed facts. We can obtain some limits on the variational trends of chemical composition even in highly dynamic systems, and we can speculate on the type of dissolved species and solid phases one may expect.

Natural waters indeed are open and dynamic systems with variable inputs and outputs of mass and energy for which the state of equilibrium is a construct. *Steady-state models* reflecting the time-invariant condition of a reaction system may frequently serve as an idealized counterpart of an open natural water system. The concept of free energy is not less important in dynamic systems than in equilibrium systems. The flow of energy from a higher to a lower potential or energy "drives" the hydrological and the geochemical cycles [3, 4]. The ultimate source of the energy flow is the sun's radiation.

Ecological System. In natural waters organisms and their abiotic environments are interrelated and interact on each other. Because of the continuous input of solar energy (photosynthesis) necessary to maintain life, ecological systems are never in equilibrium. The ecological system, or ecosystem, may be considered a unit of the environment that contains a biological organization made up of all the organisms interacting reciprocally with the chemical and physical environment. In an ecosystem the flow of energy and of negative entropy is reflected by the characteristic trophic structure and leads to material cycles within the system (cf. E. P. Odum [5]). In a balanced ecological system a steady state of production and destruction of organic material, as well as of production and consumption of O_2, is maintained.

The distribution of chemical species in waters and sediments is strongly influenced by an interaction of mixing cycles and biological cycles. Radioisotope measurements may often be used to establish the time scale of some of these processes. Similarly, evaluation of the fractionation of stable isotopes aids in the quantitative interpretation of biogeochemical and environmental processes and cycles.

Kinetics. Our understanding of nature is seriously hampered by a lack of kinetic information on reactions typically encountered in natural waters. Often, mass balance models may be used to describe the dynamics of some processes.

While the emphasis is on chemical equilibria, kinetic information can often be interwoven with equilibrium information in order to gain understanding of the major factors that influence the chemistry of the aquatic environment.

Figure 1.1 gives in an abbreviated periodic table the important elements typically found in fresh water and seawater, their dominant forms of occurrence, and their representative concentrations.

Water as a Resource and as a Life Preservation System. Aquatic chemistry is of practical importance because water is a necessary resource for humans. We are not concerned with the quantity—water as a substance is abundant—but with the quality of the water and its distribution. Humans in their social and cultural evolution continue to be successful in diverting energy for the advancement of their own civilization. In the Northern Hemisphere, anthropogenic energy use already exceeds biotic energy flux (photosynthesis). Despite the ecosphere's remarkable buffering and feedback mechanisms, humans have now become sufficiently powerful to influence some global chemical cycles [6]. Thus we are altering the natural pattern of our environment at an accelerating rate; we can hardly circumvent influencing nature and with it the hydrological cycle.

Conservation of aquatic resources cannot be achieved by avoiding human interference with the aquatic habitat. Water pollution control cannot solely consist of waste treatment. To what extent can the ocean be used as a sink of wastes? How can we improve the fertility of the ocean and exploit its production

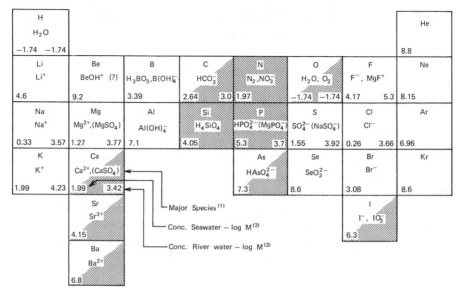

Figure 1.1 Some of the more important elements in natural waters, their form of occurrence, and their concentration. Elements whose distribution is significantly affected by biota are shaded. P, N, and Si (fully shaded) are often depleted in surface waters. (1) Species in parentheses are major ion pairs in seawater. (2) Concentrations (M = mol liter^{-1}) valid for seawater from P. G. Brewer, in *Chemical Oceanography*, Vol. 1, J. P. Riley and G. Skirrow, Eds., Academic, New York, 1975. (3) From A. D. Livingstone, *Chemical Components of Rivers and Lakes*, U.S. Geological Survey Paper No. 440G, 1963.

of harvestable food for the increasing world population? How can we restore the ecological balance between photosynthetic and respiratory activity in nutritionally enriched (polluted) receiving waters?

By evaluating the strength of significant global emission sources and by comparing natural and pollutant fluxes, we have to assess more quantitatively human influences on the natural metabolism of fresh and marine waters. We have to develop the ability to modify and manipulate our aquatic environment in order to maintain its quality as a life preservation system and as a reservoir of genetic diversity. It is hoped that the following discussion will enhance the learning process toward a better understanding of the aquatic environment. "Man masters nature not by force but by understanding" [7].

REFERENCES

1   K. Rankama and Th. G. Sahama, *Geochemistry*, University of Chicago Press, Chicago, 1950.

2   L. G. Sillén, in *Oceanography*, M. Sears, Ed., American Association for Advancement of Science, Washington, D.C., 1961.

3   B. Mason, *Principles of Geochemistry*, Wiley, New York, 1966.

4   H. J. Morowitz, *Energy Flow in Biology*, Academic, New York, 1968.

5   E. P. Odum, *Science*, 164, 262 (1969).

6 R. M. Garrels, F. T. Mackenzie, and C. Hunt, *Chemical Cycles and the Global Environment*, W. Kaufman, Los Altos, Calif., 1975.

7 J. Bronowski, *Science and Human Values*, Harper and Row, New York, 1965.

APPENDIX: THE INTERNATIONAL UNITS, SOME USEFUL CONVERSION FACTORS, AND NUMERICAL CONSTANTS

The Système International (SI) units, based on the metric system, were designed to achieve maximum internal consistency.† The SI system is based on the following set of defined units:

Physical quantity	Unit	Symbol
Length	meter	m
Mass	kilogram	kg
Time	second	s
Electric current	ampère	A
Temperature	kelvin	K
Luminous intensity	candela	cd
Amount of material	mole	mol

The main derived units are:

Force	newton	$N = kg\ m\ s^{-2}$
Energy, work, heat	joule	$J = N\ m$
Pressure	pascal	$1\ Pa = Nm^{-2}$
Power	watt	$W = J\ s^{-1}$
Electric charge	coulomb	$C = A\ s$
Electric potential	volt	$V = W\ A^{-1}$
Electric capacitance	farad	$F = A\ s\ V^{-1}$
Electric resistance	ohm	$\Omega = V\ A^{-1}$
Frequency	hertz	$Hz = s^{-1}$
Conductance	siemens	$S = AV^{-1}$

Useful Conversion Factors

Energy, Work, Heat

$$1\ \text{joule} = 1\ \text{volt-coulomb} = 1\ \text{newton meter}$$
$$= 1\ \text{watt-second} = 2.7778 \times 10^{-7}\ \text{kilowatt hours}$$
$$= 10^7\ \text{erg}$$
$$= 9.9 \times 10^{-3}\ \text{liter atmospheres}$$
$$= 0.239\ \text{calorie}$$
$$= 1.0365 \times 10^{-5}\ \text{volt-faraday}$$
$$= 6.242 \times 10^{18}\ e\ V$$
$$= 5.035 \times 10^{22}\ cm^{-1}\ \text{(wave number)}$$
$$= 9.484 \times 10^{-4}\ \text{BTU (British thermal unit)}$$
$$\approx 3 \times 10^{-8}\ \text{kg coal equivalent}$$

† The SI system has been criticized, however, as being neither convenient for nor relevant to physical chemistry [A. W. Adamson, *J. Chem. Educ.*, **55**, 634 (1978)].

Power

$$1 \text{ watt} = 1 \text{ kg m}^2 \text{ s}^{-3}$$

$$= 2.39 \times 10^{-4} \text{ kcal s}^{-1} = 0.860 \text{ kcal h}^{-1}$$

Entropy (S)

$$1 \text{ entropy unit, cal mol}^{-1} \text{ K}^{-1} = 4.184 \text{ J mol}^{-1} \text{ K}^{-1}$$

Pressure

$$1 \text{ atm} = 760 \text{ torr} = 760 \text{ mm Hg}$$
$$= 1.013 \times 10^5 \text{ N m}^{-2} = 1.013 \times 10^5 \text{ Pa (Pascal)}$$
$$= 1.013 \text{ bar}$$

Coulombic Force.   Coulomb's law of electrostatic force is written, in SI units, as

$$F = \frac{q_1 \times q_2}{4\pi\varepsilon\varepsilon_0 d^2} \tag{1}\dagger$$

The charges q_1 and q_2 are expressed in coulombs (C), the distance in meters (m), and the force F in newtons (N). The dielectric constant ε is dimensionless. The permittivity in vacuum is $\varepsilon_0 = 8.854 \times 10^{-12} \text{ C}^2 \text{ m}^{-1} \text{ J}^{-1}$. Thus, to calculate a coulombic energy E we have

$$E(\text{joules}) = \frac{q_1 \times q_2}{4\pi\varepsilon\varepsilon_0 d} \tag{2}$$

Important Constants

Avogadro's number ($^{12}C = 12.000\ldots$) $N_A = 6.022 \times 10^{23} \text{ mol}^{-1}$

Electron charge, e $= 4.803 \times 10^{-10}$ abs esu

(= charge of a proton) $= 1.602 \times 10^{-19} \text{ C}$

1 Faraday $= 96,490 \text{ C mol}^{-1}$ (= electric charge of 1 mol of electrons)

Electron mass, m $= 9.1091 \times 10^{-31} \text{ kg}$

Permittivity of a vacuum, ε_0 $= 8.854 \times 10^{-12} \text{ C}^2 \text{ m}^{-1} \text{ J}^{-1}$

† In the old cgs system of units, equation 1 was written as $F = q_1 \times q_Y/\varepsilon d^2$ in which units were so defined that ε was dimensionless; with ε in vacuum, $\varepsilon = 1$.

Speed of light in a vacuum, c	$= 2.998 \times 10^8 \text{ m s}^{-1}$
Gas constant, R	$= 8.314 \text{ J mol}^{-1} \text{ K}^{-1}$
	$= 0.082057 \text{ liter atm deg}^{-1} \text{ mol}^{-1}$
	$= 1.987 \text{ cal deg}^{-1} \text{ mol}^{-1}$
Molar volume (ideal gas, 0°C, 1 atm)	$= 22.414 \times 10^3 \text{ cm}^3 \text{ mol}^{-1}$
Planck's constant, h	$= 6.626 \times 10^{-34} \text{ J s}$
Boltzmann's constant, k	$= 1.3805 \times 10^{-23} \text{ J K}^{-1}$
Ice point	$= 273.15 \text{ K}$
$R \ln 10$	$= 19.14 \text{ J mol}^{-1} \text{ K}^{-1}$
$RT_{298.15} \ln x$	$= 5706.6 \log x \text{ J mol}^{-1}$ or $1364.1 \log x \text{ cal mol}^{-1}$
$RTF^{-1} \ln 10$	$= 59.16 \text{ mV at 298.2 K}$
$\dfrac{RT_{298.2}}{\text{volt mol}} \ln x$	$= 0.05916 \log x, \text{ volt mol}^{-1}$

The Earth—Hydrosphere System

Earth area	$5.1 \times 10^{18} \text{ cm}^2$
Oceans area	$3.6 \times 10^{18} \text{ cm}^2$
Land area	$1.5 \times 10^{18} \text{ cm}^2$
Atmosphere mass	$52 \times 10^{17} \text{ kg}$
Ocean mass	$13{,}700 \times 10^{17} \text{ kg}$
Pore waters in rocks	$3{,}200 \times 10^{17} \text{ kg}$
Water locked in ice	$165 \times 10^{17} \text{ kg}$
Water in lakes, rivers	$0.34 \times 10^{17} \text{ kg}$
Water in atmosphere	$0.105 \times 10^{17} \text{ kg}$
Total stream discharge	$0.32 \times 10^{17} \text{ kg year}^{-1}$
Evaporation = precipitation	$4.5 \times 10^{17} \text{ kg year}^{-1}$

2

Chemical Thermodynamics and Kinetics

2.1 INTRODUCTION

Natural waters attain their chemical composition through a variety of chemical reactions and physicochemical processes. Among these are acid–base reactions, gas-solution processes, precipitation and dissolution of solid phases, complexation reaction of metals and ligands, oxidation–reduction reactions, adsorption processes at interfaces, and distribution of solutes between aqueous and non-aqueous phases. These different classes of reactions and processes will be examined in detail in succeeding chapters. In this chapter we consider some idealized models for natural waters and natural water systems, and we review the principles of two alternative kinds of models for actual systems: *thermodynamic* models and *kinetic* models. Thermodynamic, or *equilibrium*, models for natural systems have been more extensively developed than kinetic models. They are simpler in that they require less information, but they are nonetheless powerful when applied within their proper restrictions. Equilibrium models for aquatic chemical systems are given the greatest emphasis in this book. However, kinetic information needs to be brought to bear in the analysis of natural water systems when the assumptions of equilibrium models no longer apply. Since rates of different chemical reactions in water can differ enormously, kinetic and equilibrium descriptions will often be needed for the same system.

The composition of a natural water, symbolized by the concentration of a constituent A, C_A, results from chemical reactions in the water itself, from processes that transfer constituents between the water and other parts of the system (atmosphere, solid matter in suspended or sedimentary form, the biota, other liquid phases) and from fluxes into and out of the system. Figure 2.1 is a schematic representation of a general natural water system. The concentration $C_A = n_A/V$, where n_A is the mole number and V is the volume of water, can be altered by variations in n_A, i.e. dn_A, brought about by fluxes, transfers, and reactions. The time-invariant, or stationary-state value of the chemical composition of the water, C_A, is given by the condition $dC_A/dt = 0$, but this state has different origins in models for closed and continuous, open systems.

Equilibrium and Kinetic Models

The basic differences between thermodynamic models and kinetic models for a stationary state of the system can be exemplified by considering a single hy-

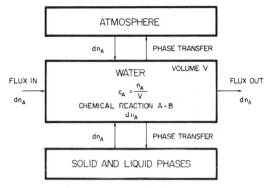

Figure 2.1. General representation of a natural water system treated as an open system. The system receives fluxes of matter from the surroundings and undergoes chemical changes symbolized by the reaction A = B. The time-invariant condition is represented by $dC_A/dt = 0$.

pothetical chemical reaction

$$A - B$$

taking place in the water. We assume no transfer of A or B between the water and other phases.

A stationary-state *thermodynamic* model for this simple system requires that the system be *closed*, that is, that no matter be exchanged with the surroundings. This means, strictly, that the material flows of A and B between the system and the surroundings must be zero: $\sum dn_{A, flux} = 0$, $\sum dn_{B, flux} = 0$. Then, application of equilibrium thermodynamic principles, to be developed below, yields an *equilibrium distribution* for C_A and C_B in the water. The equilibrium model requires fixed temperature and pressure, homogeneous distribution of A and B throughout the water, a specified total mole number for constituent A or B, and the volume of the water. In practical terms, the information needed to find C_A and C_B for the equilibrium state is the effective *equilibrium constant* K_{AB} (a function of pressure, temperature, and composition) and the *total* concentration of A (or B) in the system, $C_{A, 0} = n_{A, 0}/V$. The defining relationships are

$$C_{A, 0} = C_A + C_B \tag{1}$$

$$\frac{C_B}{C_A} = K_{AB} \tag{2}$$

Equations 1 and 2 can be solved to yield the equilibrium values of C_A and C_B. The results are

$$C_A = \frac{C_{A, 0}}{1 + K_{AB}} \tag{3}$$

$$C_B = C_{A, 0} - C_A \tag{4}$$

A stationary-state kinetic model for the corresponding homogeneous (well-mixed) *open* system, that is, one that can exchange matter with the surroundings, requires kinetic information about the transformations of A and B. Needed are the values of the forward and backward *rate constants*, k_f and k_b, represented in

$$A \; \underset{k_b}{\overset{k_f}{\rightleftharpoons}} \; B \tag{5}$$

The elementary rate laws for this reaction are assumed to be given by the simple velocity expressions

$$v_f = k_f C_A \tag{6}$$

$$v_b = k_b C_B \tag{7}$$

and the time rates of change of C_A and C_B from the *reaction* by

$$\frac{dC_A}{dt} = -k_f C_A + k_b C_B \tag{8a}$$

$$\frac{dC_B}{dt} = k_f C_A - k_b C_B \tag{8b}$$

The effective rate constants are functions of temperature, pressure, and solution composition, as is the equilibrium constant K_{AB}. For the open system, the material balance information needed consists of the steady mole fluxes per unit volume of A and B to the system, given by $r\bar{C}_{A,0}$ and $r\bar{C}_{B,0}$, where $r = Q/V$ is the fluid transfer rate constant for the system (the rate of flow Q divided by the volume V) and $\bar{C}_{A,0}$ and $\bar{C}_{B,0}$ are the *inflow* concentrations of A and B, respectively. For the steady state, $dC_A/dt = dC_B/dt = 0$, and assuming the inflow and outflow rates the same, we have

$$\frac{dC_A}{dt} = r\bar{C}_{A,0} - k_f C_A + k_b C_B - rC_A = 0 \tag{9}$$

$$\frac{dC_B}{dt} = r\bar{C}_{B,0} + k_f C_A - k_b C_B - rC_B = 0 \tag{10}$$

The stoichiometry of the reaction leads to $C_A + C_B = \bar{C}_{A,0} + \bar{C}_{B,0}$. These equations can be solved for the steady-state values C_A and C_B. The results are

$$C_A = \frac{r\bar{C}_{A,0} + k_b(\bar{C}_{A,0} + \bar{C}_{B,0})}{k_f + k_b + r} \tag{11}$$

$$C_B = \frac{r\bar{C}_{B,0} + k_f(\bar{C}_{A,0} + \bar{C}_{B,0})}{k_f + k_b + r} \tag{12}$$

Comparison of equations 3 and 11 shows the essential points of difference between the stationary states of closed and open systems. For the closed system *equilibrium* is the time-invariant condition. The total mole number of each independently variable constituent (A or B in the example), the volume, and the equilibrium constant (a function of temperature, pressure, and composition) for each independent reaction (K_{AB} in the example) are required to define the equilibrium composition C_A. For the open system the *steady state* is the time-invariant condition. The mass transfer rate constant, the inflow mole number of each independently variable constituent, and the rate constants (functions of temperature, pressure, and composition) for each independent reaction are required to define the steady-state composition C_A. It is thus clear that open-system models of natural waters require more information than closed-system models to define stationary-state compositions. Furthermore, the basic data required for equilibrium models, namely, equilibrium constants, Gibbs free energies, entropies, enthalpies, and heat capacities, are at present more abundant than the corresponding information required for kinetic models, namely, experimental rate laws, rate constants, and activation energies. We will turn later in this chapter to comparisons of equilibrium and steady-state conditions for simple types of chemical reactions.

When the general system of Figure 2.1 is considered once more, it is clear that introducing additional chemical reactions and including pertinent transfer processes between the water and atmosphere on the water and solid or liquid phases will increase the mathematical complexity of a closed-system or open-system model. Additional equilibrium constants for chemical reactions and distribution of constituents between phases are required for the closed system; additional rate constants are required for the kinetic processes in the open system, and more material transfer rate constants ($r = Q/V$) may be required for chemical fluxes into other phases. Apart from the increased complexity, the basic differences between the idealized models are those already illustrated for the simple case of a single reaction.

Rates, Equilibrium, and Reaction Types

It should not be inferred that kinetic information is irrelevant for closed systems. The equilibrium state is the stationary state in a closed system, but it may be attained slowly for some reactions. Thermodynamic information enables us to identify reactions that are *possible*, to calculate the equilibrium composition of a solution, and to find the maximum useful work done or minimum energy needed for a process, but definite conclusions concerning the time required to reach equilibrium require kinetic information. Some solution reactions are so rapid that equilibrium can be attained essentially instantaneously (characteristic times of microseconds). Other reactions are sufficiently slow (characteristic times of minutes to years) that equilibrium is attained extremely slowly. For similar reasons, it cannot be assumed that the steady state is *rapidly* attained in an open-system model. Kinetic information must be used to describe the rate of approach to the appropriate stationary state for closed and open systems.

It may prove useful to keep some simple classes of reactions in mind as we turn to developing first equilibrium and then kinetic frameworks for natural water systems. Consider first the following simple representation of a reaction:

$$A + B = C + D$$

This conveys *stoichiometric* information. It is a statement of *material balance*. If at least one of the constituents in the reaction is a component of a homogeneous mixture, that is, a solution (aqueous, gaseous, solid, or other), the reaction has an equilibrium constant:

$$A + B \xrightleftharpoons{\quad K \quad} C + D$$

If A, B, C, and D are all pure solids or liquids, there is in general no equilibrium condition involving all these constituents and no defined equilibrium constant.

If K for a solution reaction is of sufficient magnitude that essentially only products C and D exist in appreciable concentrations at equilibrium, it is sometimes said that the reaction "goes to completion" and the representation

$$A + B \longrightarrow C + D$$

is used. This case is the basis of so-called stoichiometric models in closed systems.

A kinetic description of a reversible reaction is

$$A + B \xrightleftharpoons[k_b]{k_f} C + D$$

accompanied by expressions for the experimentally observed rate law in each direction with rate constants k_f and k_b. For the case where the reaction is practically irreversible, the description is

$$A + B \xrightarrow{\quad k \quad} C + D$$

and only one rate constant is needed to describe the kinetics of the reaction.

Each of these reactions can be described in either thermodynamic terms or in kinetic terms by drawing upon available energetic and dynamic information, or by carrying out appropriate experiments to obtain required data. A useful set of energetic data for a reaction would include molar values of the entropy, enthalpy, Gibbs free energy, heat capacity, and volume of each species in the reaction at 25°C and a pressure of 1 atm, plus additional information describing the variation of enthalpy, entropy, Gibbs free energy, and heat capacity with temperature. Kinetic data for a reversible reaction would include the rate laws and rate constants in both forward and reverse directions and energetic information for the effect of temperature on the rate law parameters. For both equilibrium and kinetic models it is usually necessary to deal with nonideality effects in equilibrium expressions and rate expressions, either by developing data for constant ionic media, for example, seawater, or by using models for activity coefficients, which describe departures from ideality.

In summary, it is proposed that complex natural water systems can be investigated quantitatively by means of idealized models. The models are basically of two kinds:

Open systems, which exchange material with their surroundings, that is, can vary their mole numbers both by flows and reactions

Closed systems, which do not exchange material with their surroundings, that is, contain a fixed total mass of components but which can vary their mole number by reactions and internal processes

For each of these idealized models there is a stationary state. For an *open system*, this is the steady state. Rate laws and steady material flows are required to define the steady state. For a *closed system*, equilibrium is the stationary state. Equilibrium may be viewed as simply the limiting case of the stationary state when the flows from the surroundings approach zero. The simplicity of closed-system models at equilibrium is in the rather small body of information required to describe the time-invariant composition. We now turn our attention to the principles of chemical thermodynamics and the development of tools for the description of equilibrium states and energetics of chemical change in closed systems.

2.2   THERMODYNAMICS AND CHEMICAL EQUILIBRIUM

Objectives

Among the objectives of our treatment of chemical thermodynamics are:

1   To review the principles for determining the composition of chemical reaction systems at equilibrium.
2   To identify useful criteria for determining the direction of spontaneous chemical reactions and processes in any system, but particularly in systems at constant temperature and pressure.
3   To determine the minimum energy required to carry out phase transfers and chemical processes at constant temperature and pressure.
4   To establish useful relationships between the equilibrium constant and thermodynamic functions such as standard Gibbs free energies, enthalpies, and entropies.
5   To describe the effects of temperature and pressure on the equilibrium state of a system.

The key thermodynamic functions for describing reaction equilibrium and phase equilibrium at constant temperature and pressure are found to be G, the Gibbs free energy, and μ, the chemical potential, both thermodynamic functions of the state of a system. Most of our attention therefore will be given to identifying the important characteristics of the Gibbs free energy and the chemical potential and pointing out their relationship to the properties of a system and to other useful thermodynamic functions.

Thermodynamic Systems

Thermodynamic *systems* are parts of the world selected for study. The region around the system is called the *surroundings* or *environment*. A *boundary* (real or imaginary) separates the system from the surroundings. An *isolated system* cannot exchange matter, heat, or work with its environment. An *adiabatic system* cannot exchange heat or matter. A *closed system*, already discussed in Section 2.1, can exchange heat and work, but not matter.

Systems consist of one or more parts called *phases.* A phase is a region of the system that is physically and chemically homogeneous, that is, spatially uniform in its properties. Figure 2.2 shows a simplified version of a thermodynamic system of the sort often met with in aquatic chemistry. Aqueous, gaseous, and various solid and liquid phases may be included, and the gases and solid and liquid phases may be pure phases or homogeneous mixtures, that is, solutions.

The phases are characterized by uniform temperature and pressure and by compositions defined by the set of mole numbers $n_1, n_2, \ldots, n_k$, where k is the number of substances in the system. An extensive property such as the volume completes the specification of the state of each phase.

Thermodynamic Properties

Thermodynamic systems are characterized by a small number of variables or properties. Familiar macroscopic properties of a system are the temperature, pressure, volume, density, concentration of dissolved components, viscosity, surface tension, and so on. Properties are functions of the state of a system. Properties can be classified as either *extensive* or *intensive*. Extensive properties are additive; that is, they depend upon the number of moles. The value of an

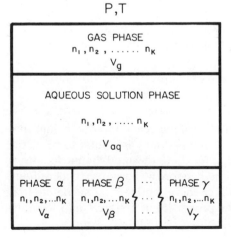

Figure 2.2. Simplified model for the thermodynamic description of natural water systems. P and T are intensive variables, and the mole numbers n_i in each phase are extensive variables which together determine the volume, mass, composition, and other properties of the system.

extensive property of a system, for example, the volume, is the sum of the values for the various parts. Intensive properties are not additive; they do not require specification of the mass of the phase to which they refer. Examples of *extensive* properties are number of moles, mass, volume, energy, and entropy. Examples of *intensive* properties are pressure, temperature, density, concentration of dissolved components, molar volume, and chemical potential.

State of a System

The thermodynamic state of a system is defined by specifying the values of a sufficient number of properties. The *phase rule* of Gibbs [1] gives the number of intensive properties that must be specified in order to define the state of a system. In general, an extensive property and two intensive properties are sufficient to specify completely the state of a system consisting of a pure phase, for example, liquid water. A mole of liquid water at 1 atm pressure and a temperature of 25°C is in a defined state. All other intensive properties such as refractive index, density, viscosity, and surface tension are fixed by specifying the two intensive properties T and P. Similarly, the remaining extensive properties, mass, volume, internal energy, heat capacity, entropy, and enthalpy, are also fixed.

Suppose that sodium chloride (NaCl) is dissolved in the liquid water phase. Now, *three* intensive properties, say T, P, and the mole fraction of NaCl, $X_{NaCl} = n_{NaCl}/(n_{H_2O} + n_{NaCl})$, in the solution must be specified along with an extensive property to define the state of the one-phase system. The meaning of the term *extensive property* is that the property depends directly on the number of moles. The set of mole numbers determines the *composition* and *size* of a system. Consider the pure liquid water phase. Increasing the number of moles of water while holding the *two* intensive properties, P and T, constant results in a proportional increase in mass, volume, energy, and entropy. Consider the NaCl–H_2O solution. Increasing the total number of moles in the solution while holding the *three* intensive properties P, T, and X_{NaCl} constant results in a proportional increase in mass, volume, energy, entropy, and all other extensive properties. Applications of the Gibbs phase rule to systems of several phases are discussed in Chapter 5.

Thermodynamic Processes

A process is a change in some property of a system. Familiar examples are changes in temperature (heating, cooling), changes in pressure (expansion, compression), increasing the surface area of a phase, changes in the mole numbers of substances through chemical reactions, phase transfer, and so on. Changes that take place spontaneously are called *natural processes* or *irreversible processes*. A *reversible process* in a system is one that can be reversed without causing an appreciable change in any other system. Reversible processes are approached in real systems by allowing only infinitesimally small differences between the internal forces within a system and the external forces acting on the system, for example, an electrochemical galvanic cell in which an external electric force balances almost perfectly the chemical energy of a system (there may be a negligibly small difference), allowing a small current to flow. Reversible processes involve a succession of equilibrium states.

Processes represent changes in functions of state. Hence the change in a property between one state and another is independent of path. Variations in system properties or state functions in a process, for example, dV, dP, dT, and dE (the energy) are exact differentials.

Modes of Energy Transfer

A system receives energy from its surroundings in the form of heat q and work w. Heat and work are not system properties. Their values are dependent on path, and their differentials dq and dw are therefore inexact.

A number of ways of transferring energy to a system as work are possible. Mechanical, or pressure–volume work is most familiar. The mechanical work done on a system is given by the expression $-PdV$, where P is the pressure and dV the change in volume. This expression is of the form *intensive property* times *extensive variation*. Other forms of thermodynamic work transferred to a system and their expressions are given in Table 2.1.

TABLE 2.1   EXPRESSIONS FOR THERMODYNAMIC WORK dw DONE ON A SYSTEM

Type	Intensive Property	Extensive Variation	Expression
Expansion	Pressure, P	Volume, dV	$-PdV$
Electrical	Potential, E	Charge, de	$-Ede$
Gravitational	Force, mg	Height, dh	$mg\,dh$
Chemical	Chemical potential, μ	Moles, dn	μdn
Surface	Interfacial tension, γ	Area, dA	γdA

2.3   BASIC THERMODYNAMIC PRINCIPLES AND RELATIONSHIPS

What follows is a summary of essential definitions, principles, and fundamental relationships of chemical thermodynamics. It is not intended to be a full development of the topic. For extensive, critical treatments the reader may consult a number of authors, among them Guggenheim [1], Denbigh [2], Lewis and Randall [3], Wall [4], and Prigogine and Defay [5].

Fundamental Variables and Characteristic Functions

The whole of the fundamental theory of chemical thermodynamics can be represented in terms of five fundamental variables, two modes of energy transfer and one characteristic state function. The theory is developed on the basis of the three principles or laws of thermodynamics. The *fundamental variables of a system are*:

T, absolute temperature, an intensive property

S, entropy, an extensive property

P, pressure, an intensive property

V, volume, an extensive property

n_i, mole number of component i, an extensive property

The *modes of energy transfer for a system are*:

q, heat transferred to a system from surroundings

w, work done on a system by surroundings

The *characteristic state function* is:

E, the internal energy, an extensive property

In addition, three other characteristic state functions have been defined:

H, the enthalpy, an extensive property

A, the Helmholtz free energy, an extensive property

G, the Gibbs free energy, an extensive property

The Zeroth Law of Thermodynamics

Systems in thermal equilibrium have the same *temperature*. In effect, this principle introduces temperature T as a thermodynamic property.

The First Law of Thermodynamics

$$dE = dq + dw \tag{13}$$

The change in the internal energy, dE, is the sum of the heat transferred *to* the system, dq, and the work done *on* the system, dw. This is the energy conservation principle for a system. For a finite change we can write

$$\Delta E = q + w \tag{14}$$

The Second Law of Thermodynamics

Two alternative but equivalent statements are found useful. Both are based on the definition of the entropy as an extensive function of the state of the system. Figure 2.3 represents a *closed system* and its *surroundings*. The *closed system plus surroundings* constitutes an *isolated system*. The system is characterized by temperature T, pressure P, and a set of mole numbers n_i. The system has volume V. It receives heat dq from the surroundings.

Statement One. The total entropy change of the closed system is dS_{sys} and is the sum of changes *inside* the system, dS_{int}, and entropy *transferred* to the system from its surroundings, dS_{sur}:

$$dS_{sys} = dS_{int} + dS_{sur} \tag{15}$$

The entropy transferred to the system is defined by

$$dS_{sur} = \frac{dq}{T} \tag{16}$$

For a reversible process or equilibrium state of the system, the second law states that the internal entropy change is zero:

$$dS_{int} = 0 \tag{17}$$

For a spontaneous or natural process in the system,

$$dS_{int} > 0 \tag{18}$$

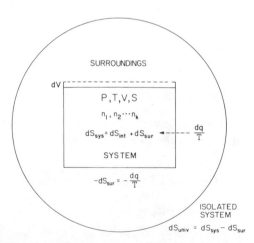

Figure 2.3. Illustration of a closed system, surroundings, and an isolated system or "universe" of a system plus surroundings. Heat transferred to the system, q, is positive, and that lost from the surroundings is $-q$. The entropy change of the system, dS_{sys}, is the *sum* of an internal change dS_{int} and a flow from the surroundings dS_{sur}.

Therefore, for any possible process, either reversible or natural,

$$dS_{int} \geqslant 0 \tag{19}$$

or, stated another way,

$$dS_{sys} \geqslant \frac{dq}{T} \tag{20}$$

The internal entropy change for a closed system is zero at equilibrium and positive for a spontaneous process. The heat transferred to a closed system, divided by T, is equal to or less than the entropy increase for any possible process.

Statement Two.   The total entropy change of the "universe," that is, the isolated system made up of the closed *plus* its surroundings, dS_{univ}, is the sum of the entropy changes inside the system and in the surroundings. The change in the entropy of the surroundings is $-dS_{sur}$. Therefore

$$dS_{univ} = dS_{sys} - dS_{sur} \tag{21}$$

$$dS_{univ} = dS_{sys} - \frac{dq}{T} \tag{22}$$

For any possible process in the *isolated system,*

$$dS_{univ} \geqslant 0 \tag{23}$$

or

$$dS_{sys} \geqslant \frac{dq}{T} \tag{24}$$

The entropy of an isolated system increases in any spontaneous process and attains a maximum for any reversible process.

   Comparing the two alternative statements we note that the internal entropy change of a system is precisely the entropy change of the universe. The two statements are equivalent. We will emphasize the internal entropy production, pointing out its relationship to changes in characteristic state functions such as the Gibbs free energy.

Combined First and Second Laws

For a closed system on which only expansion work is done,

$$dw = -PdV \tag{25}$$

For equilibrium and fixed composition, that is, $dn_i = 0$ for all conceivable processes, equation 24 gives

$$dq = TdS_{sys} \tag{26}$$

therefore the first and second laws combined give

$$dE = TdS_{sys} - PdV \tag{27}$$

We now omit the system subscript on the entropy for simplicity:

$$dE = TdS - PdV \tag{28}$$

This is a fundamental relationship for a closed system of fixed composition. It shows that the internal energy is a function of entropy and volume. For given S and V in a closed system at equilibrium, $dE = 0$.

An equilibrium state represents a minimum internal energy. To obtain useful state functions of fundamental variables other than S and V, Legendre transformations of the variables result in the following characteristic functions. The enthalpy

$$H = E + PV \tag{29}$$

The Helmholtz free energy

$$A = E - TS \tag{30}$$

The Gibbs free energy

$$G = H - TS = A + PV \tag{31}$$

The differentials are

$$dH = TdS + VdP \tag{32}$$

$$dA = -SdT - PdV \tag{33}$$

$$dG = -SdT + VdP \tag{34}$$

In addition, for the entropy of the system, the differentials are

$$dS = \frac{1}{T} dE + \frac{P}{T} dV \tag{35}$$

and

$$dS = \frac{1}{T} dH - \frac{V}{T} dP \tag{36}$$

All the foregoing relations apply to closed systems at equilibrium.

The Gibbs free energy is of particular use in chemistry because its natural variables are P and T. Many chemical processes of interest in chemistry take place at constant temperature and pressure (e.g., atmospheric pressure). Thus the conditions

$$dP = 0 \qquad dT = 0 \qquad dG = 0 \tag{37}$$

define the conditions of reversible processes or equilibrium states for conditions of practical interest. Some processes take place under conditions of constant temperature and volume (isochoric), for example, a rigid system. The Helmholtz energy is then the natural state function, with

$$dV = 0 \qquad dT = 0 \qquad dA = 0 \tag{38}$$

describing the condition of equilibrium.

For *spontaneous* processes in a closed system, the second law gives the criterion $dS > dq/T$. Corresponding criteria in terms of the characteristic state functions E, H, A, and G are obtained by substituting the *inequality* of the second principle into the statement of energy conservation. We obtain

$$dE < TdS - PdV \tag{39}$$

For a spontaneous process at constant entropy and volume, $dE < 0$; that is, the internal energy must decrease. By corresponding substitutions for dH, dA, and dG the following criteria are obtained for *spontaneous* processes:

$$dP = 0 \qquad dS = 0 \qquad dH < 0 \tag{40}$$

$$dT = 0 \qquad dV = 0 \qquad dA < 0 \tag{41}$$

$$dT = 0 \qquad dP = 0 \qquad dG < 0 \tag{42}$$

A parallel development is possible in terms of the change in internal entropy dS_{int}. The heat transferred is

$$dq = TdS - TdS_{int} \tag{43}$$

and the internal energy change is

$$dE = TdS - TdS_{int} - PdV \tag{44}$$

The corresponding changes in H, A, and G are

$$dH = TdS - TdS_{int} + VdP \tag{45}$$

$$dA = -SdT - TdS_{int} - PdV \tag{46}$$

$$dG = -SdT - TdS_{int} + VdP \tag{47}$$

and, in terms of the system entropy S,

$$dS = \frac{1}{T}dE + \frac{P}{T}dV + TdS_{int} \tag{48}$$

and

$$dS = \frac{1}{T}dH - \frac{V}{T}dP + TdS_{int} \tag{49}$$

Now, for all possible processes in *any* system, the second principle states

$$dS_{int} \geq 0 \tag{19}$$

or, since $T \geq 0$,

$$TdS_{int} \geq 0 \tag{50}$$

Therefore

at constant S, V: $dE = -T dS_{int} \leqslant 0$ (51)

at constant E, V: $dS = T dS_{int} \geqslant 0$ (52)

at constant S, P: $dH = -T dS_{int} \leqslant 0$ (53)

at constant H, P: $dS = T dS_{int} \geqslant 0$ (54)

at constant T, V: $dA = -T dS_{int} \leqslant 0$ (55)

at constant T, P: $dG = -T dS_{int} \leqslant 0$ (56)

Note that at constant temperature and pressure (an isothermal, isobaric process)

$$dG = -T dS_{int}$$ (57)

or

$$dS_{int} = -\frac{1}{T} dG$$ (58)

The increase in internal entropy is directly proportional to the decrease in the Gibbs free energy. The usefulness of E, H, A, S, and G is that they are state functions and thus their *changes* are independent of path, unlike the heat and work that occur in the first and second laws.

We can now assemble the inequalities and equalities to summarize the criteria for spontaneous processes and equilibrium processes in closed systems. Table 2.2 contains the result.

TABLE 2.2   CRITERIA FOR PROCESSES IN CLOSED SYSTEMS

State Function	Fundamental Variables	Criteria	
		Spontaneous	Reversible
S	E, V	$dS > 0$	$dS = 0$
S	H, P	$dS > 0$	$dS = 0$
E	S, V	$dE < 0$	$dE = 0$
H	S, P	$dH < 0$	$dH = 0$
A	T, V	$dA < 0$	$dA = 0$
G	T, P	$dG < 0$	$dG = 0$

2.4   PROCESSES AT CONSTANT PRESSURE AND TEMPERATURE

For any equilibrium process in a closed system at constant temperature and pressure the Gibbs free energy is a minimum and the criterion of equilibrium is

$$dG = 0$$ (59)

The criterion for a spontaneous process is

$$dG < 0 \tag{60}$$

It will be useful to establish the relationship of changes in the Gibbs energy to changes in other state functions, particularly the enthalpy and the entropy, and relationships of these functions to measurable quantities such as heat and heat capacity. For processes in systems of variable composition, relationships between dG, the changes in numbers of moles, and the equilibrium constant can be developed fully after we have obtained expressions for the chemical potential in a homogeneous mixture.

From the first law for a closed system of constant composition, equation 13 becomes, if only PV work is involved,

$$dE = dq + dw = dq - PdV \tag{61}$$

in which PdV is the reversible expansion work done on the system by the surroundings. For a state change at constant pressure, equation 61 can be written

$$dq_p = dE + PdV \tag{62}$$

where dq_p is the heat adsorbed by the system at constant pressure. Since P, V, and E are state properties, dq_p must represent a change in a state property. The property is the enthalpy. The definition of the enthalpy, $H = E + PV$, yields $dH = dE + PdV + VdP$, which is $dH = dE + PdV$ at constant P. Therefore enthalpy changes can be obtained directly from measurement of the heat absorbed at constant pressure. For a finite change in a system from state 1 to state 2 for which $H_2 - H_1 = \Delta H$,

$$\Delta H = q_p \tag{63}$$

It is important to emphasize that this relationship applies *only* if PV work is the sole form of work done on the system.

The heat capacity is defined by

$$C = \frac{dq}{dT} \tag{64}$$

which for constant P becomes

$$C_p = \frac{dq_p}{dT} = \left(\frac{dH}{dT}\right)_p \tag{65}$$

Calorimetry at constant pressure thus can provide information on C_p and ΔH for closed-system processes.

From the definition of G as a state function, $G = E + PV - TS = H - TS$, the change in G is

$$dG = dH - TdS - SdT \tag{66}$$

At constant temperature, therefore

$$dG = dH - TdS \tag{67}$$

For a finite change in a system at constant T from state 1 to state 2 for which $H_2 - H_1 = \Delta H, S_2 - S_1 = \Delta S,$ and $G_2 - G_1 = \Delta G,$ equation 67 takes the form

$$\Delta G = \Delta H - T\Delta S \tag{68}$$

This is one of the most important relationships in chemical thermodynamics.

If the surroundings of a system which is maintained at constant temperature and pressure consist of large heat bath (reservoir) at constant T, then the heat absorbed from the bath at constant pressure q_p is ΔH for a process within the system. The entropy change of the bath is simply $-q_p/T = -\Delta H/T$. The entropy change for the universe of the system plus the bath, ΔS_{univ}, is

$$\Delta S_{univ} = \Delta S_{sys} + \Delta S_{sur} \tag{69}$$

or

$$\Delta S_{univ} = \Delta S_{sys} - \frac{\Delta H}{T} \tag{70}$$

or, multiplying by T,

$$T\Delta S_{univ} = T\Delta S_{sys} - \Delta H \tag{71}$$

Recalling that $T\Delta S_{univ} \geqslant 0$ for any possible process, we have

$$\Delta H - T\Delta S_{sys} \leqslant 0 \tag{72}$$

for any possible process. The expression $\Delta H - T\Delta S_{sys}$ is of course ΔG (compare equation 68). In terms of the second law, we see that ΔG is the difference between the heat absorbed, $q_p = \Delta H$, and the product of the temperature and the entropy change of the system, $T\Delta S_{sys}$. Summarizing

$$\Delta G = -T\Delta S_{univ} \tag{73}$$

$$\Delta H = q_p \tag{74}$$

$$\Delta G = \Delta H - T\Delta S \tag{68}$$

Because S, H, and G are functions of the state of the system, the differences for any process, ΔS, ΔH, or ΔG, depend only on the properties of the two states, not on the path connecting them. However, the heat and work transferred *are* dependent on path. Thus, at constant temperature and pressure, the entropy difference between two states for a system of specified composition is well-defined. The heat and work accompanying the change depend on the path, that is, whether it is reversible or irreversible. The content of the second law for a finite change in state at constant pressure and temperature,

$$q_p \leqslant T\Delta S = T(S_2 - S_1) \tag{75}$$

is that the heat absorbed is equal to $T(S_2 - S_1)$ for a reversible path and less than $T(S_2 - S_1)$ for an irreversible path. The quantity $T(S_2 - S_1)$ is independent of the path, depending only on T, P, and the composition of the system.

The Third Law of Thermodynamics

The third law states that the entropy of a perfectly ordered substance (e.g., a perfect crystal) approaches zero as the absolute temperature approaches zero:

$$S \to 0 \quad \text{as} \quad T \to 0$$

The basis of the third law in statistical mechanics is Boltzmann's equation, $S = k \ln \Omega$, where Ω is the number of ways of achieving a given energy state and k is Boltzmann's constant (or R/N, where N is Avogadro's number). Equation 32, for constant pressure, yields

$$dS = \frac{dH}{T} \tag{76}$$

From the definition of heat capacity C_p (equation 65),

$$dH = C_p \, dT \tag{77}$$

Therefore

$$dS = C_p \frac{dT}{T} \tag{78}$$

Integrating between $T = 0$ and $T = T$,

$$S = \int_0^T C_p \frac{dT}{T} \tag{79}$$

This equation is the basis for calculating entropies when heat capacities are known as a function of temperature. If phase transitions occur between absolute zero temperature and the temperature, then the entropy of the transitions, q_{trans}/T_{trans}, must be included in the calculation of S.

Temperature Dependence of ΔG

The basic equation is

$$dG = -S dT + V dP \tag{34}$$

which for constant pressure can be written

$$\left(\frac{\partial G}{\partial T}\right)_P = -S \tag{80a}$$

For a finite change in state with $\Delta G = G_2 - G_1$ and $\Delta S = S_2 - S_1$,

$$\left(\frac{\partial \Delta G}{\partial T}\right)_P = -\Delta S \tag{80b}$$

This is one form of the Gibbs–Helmholtz equation.

A useful transformation is obtained by differentiating $(\Delta G/T)$:

$$\frac{\partial(\Delta G/T)}{\partial T} = -\frac{\Delta G}{T^2} + \frac{1}{T}\cdot\left(\frac{\partial\Delta G}{\partial T}\right)$$

Substituting equation 80b yields

$$\frac{\partial(\Delta G/T)}{\partial T} = -\frac{\Delta G}{T^2} - \frac{\Delta S}{T} \tag{81}$$

combining with $\Delta G = \Delta H - T\Delta S$ we obtain

$$\left[\frac{\partial(\Delta G/T)}{\partial T}\right]_P = -\frac{\Delta H}{T^2} \tag{82}$$

Equation 82 is an alternative form of the Gibbs–Hclmholtz equation and is useful for describing the influence of temperature on energetics and equilibrium.

Enthalpies of Reaction and Standard Enthalpies

The condensation of water vapor to liquid water can be represented by

$$H_2O(g) = H_2O(l) \tag{83}$$

where (g) signifies the gaseous or vapor state and (l) signifies the liquid state. If P and T are fixed, the *change* in enthalpy for equation 83 is a definite quantity. If the pressure is 1 atm and the temperature 25°C, or 298.15 K, the change is referred to as a *standard enthalpy change*, that is, a change under arbitrarily chosen *standard conditions*. The standard enthalpy change is denoted by $\Delta H°$. To specify individual numerical values of the enthalpies for $H_2O(g)$ and $H_2O(l)$ an additional convention is required, one for zero enthalpy. For the reaction

$$\tfrac{1}{2}O_2(g) + H_2(g) = H_2O(g) \tag{84}$$

the standard enthalpies of $O_2(g)$ and $H_2(g)$ are taken to be zero. The enthalpy change for reaction 84 is thus the change in enthalpy for forming water vapor, $H_2O(g)$, from its elements in their standard states. We denote this change as $\Delta H_f°$ (1 atm, 298.15 K). The standard state of the elements is chosen as their most stable form, solid, liquid, or gas, under the standard conditions of temperature and pressure.

The standard enthalpy of formation of water vapor is found from experiment to be -57.798 kcal/mol, or -241.8 kJ/mol. The standard enthalpy change for the condensation of water vapor (reaction 83) $\Delta H°_{cond}$, is found from experiment to be -10.519 kcal/mol. Therefore, because H is a function of state,

$$\Delta H°_{f,\,H_2O(l)} = \Delta H°_{f,\,H_2O(g)} + \Delta H°_{cond} \tag{85}$$

or $\Delta H°_{f,\,H_2O(l)} = -68.317$ kcal/mol. Noting that the standard enthalpies of formation are *molar quantities*, we will adopt the equivalent notation

$$\Delta H°_{f,i} \equiv \overline{H}°_{f,i} \tag{86}$$

where $\bar{H}^\circ_{f,i}$ is the standard partial molar enthalpy of formation of substance or species i and pressure (the general concept of *partial molar quantities* for homogeneous mixtures is discussed later).

Other examples will serve to illustrate the standard enthalpy conventions:

$$
\begin{array}{ll}
C(graphite) + O_2(g) = CO_2(g) & \bar{H}^\circ_{f,CO_2(g)} \\
CO_2(g) = CO_2(aq) & \Delta H^\circ \\
\hline
C(graphite) + O_2(g) = CO_2(aq) & \bar{H}^\circ_{f,CO_2(aq)} = \bar{H}^\circ_{f,CO_2(g)} + \Delta H^\circ
\end{array}
$$

$$
\begin{array}{ll}
Ca(s) + \tfrac{3}{2}O_2(g) + C(graphite) = CaCO_3(s) & \bar{H}^\circ_{f,CaCO\ (s)} \\
CaCO_3(s) = Ca^{2+}(aq) + CO_3^{2-}(aq) & \Delta H^\circ \\
\hline
Ca(s) + \tfrac{3}{2}O_2(g) + C(graphite) = Ca^{2+}(aq) + CO_3^{2-}(aq) &
\end{array}
$$
$$
\bar{H}^\circ_{f,Ca^{2+}(aq)} + \bar{H}^\circ_{f,CO_3^{2-}(aq)} = \bar{H}^\circ_{f,CaCO_3(s)} + \Delta H^\circ
$$

$$
\begin{array}{ll}
\tfrac{1}{2}O_2(g) + H_2 = H_2O(l) & \bar{H}^\circ_{f,H_2O(l)} \\
H_2O(l) = H^+(aq) + OH^-(aq) & \Delta H^\circ \\
\hline
\tfrac{1}{2}O_2(g) + H_2(g) = H^+(aq) + OH^-(aq) & \bar{H}^\circ_{f,H^+(aq)} + \bar{H}^\circ_{f,OH^-(aq)} =
\end{array}
$$
$$
\bar{H}^\circ_{f,H_2O(l)} + \Delta H^\circ
$$

2.5   FUNDAMENTAL EQUATION FOR SYSTEMS OF VARIABLE COMPOSITION

A phase consisting of one component, for example, pure calcium carbonate crystal or liquid water, can change its energy by an increase in the number of moles. Thus the fundamental equation for a closed system at equilibrium,

$$dE = TdS - PdV \tag{28}$$

must be modified to reflect the effect of variations in mole number for a pure phase. In a similar fashion, a variation in the mole number of a component in a solution due to a phase transfer or chemical reaction causes a variation in the energy. Thus, for the energy function E, the variables S and V are not sufficient to specify the state of an open phase or a system of variable composition. Instead of

$$E = E(S, V) \tag{87}$$

with the total differential

$$dE = \left(\frac{\partial E}{\partial S}\right)_V + \left(\frac{\partial E}{\partial V}\right)_S \tag{88}$$

we need

$$E = E(S, V, n_1, n_2, \ldots, n_k) \tag{89}$$

where n_i is the mole number of component i, with the total differential

$$dE = \left(\frac{\partial E}{\partial S}\right)_{V,n_i} dS + \left(\frac{\partial E}{\partial V}\right)_{S,n_i} dV + \sum_i \left(\frac{\partial E}{\partial n_i}\right)_{S,V,n_j} dn_i \qquad (90)$$

for $n_i \neq n_j$. The mole numbers of all chemical species are held constant for the variations with entropy and volume, whereas all but one of the mole numbers, n_j, are held constant for the variation of n_i. The variations of n_i may be caused by an increase in the *amount* of a pure phase, by *transfer* of a component to a phase, and by *diffusion, mixing, chemical reaction*, and so on. Comparing equations 28 and 88, for expansion work only, we see that

$$\left(\frac{\partial E}{\partial S}\right)_{V,n_i} - T \qquad \left(\frac{\partial E}{\partial V}\right)_{S,n_i} = -P$$

In a similar fashion for the Gibbs free energy G we write

$$G = G(P, T, n_1, n_2, \ldots, n_k) \qquad (91)$$

and the total differential

$$dG = \left(\frac{\partial G}{\partial P}\right)_{T,n_i} dP + \left(\frac{\partial G}{\partial T}\right)_{P,n_i} dT + \sum_i \left(\frac{\partial G}{\partial n_i}\right)_{P,T,n_j} dn_i \qquad (92)$$

Comparing equations 34 and 92 we note

$$\left(\frac{\partial G}{\partial P}\right)_{T,n_i} = V \qquad \left(\frac{\partial G}{\partial T}\right)_{P,n_i} = -S$$

The term $(\partial G/\partial n)_{P,T,n_j}$ is the *partial molar Gibbs* free energy, an *intensive* property. It was defined by Gibbs as the chemical potential of component i:

$$\mu_i \equiv \left(\frac{\partial G}{\partial n_i}\right)_{P,T,n_j} \qquad (93)$$

Recalling the expression for chemical work done on a system (Table 2.1) we can interpret

$$\left(\frac{\partial G}{\partial n_i}\right)_{P,T,n_j} dn_i = \mu_i dn_i$$

as the increased capacity of a phase to work as a result of an infinitesimal increase in the content of component i in the phase.

For a *pure substance* (one-component) system we may write

$$dG = -SdT + VdP + \left(\frac{\partial G}{\partial n}\right)_{T,P} dn \qquad (94)$$

and note that $(\partial G/\partial n)_{T,P}$ is the *molar* free energy. The molar free energy of a pure substance is the chemical potential.

Comparing equations 90 and 92 it can be readily shown that

$$\left(\frac{\partial G}{\partial n_i}\right)_{P,T,n_j} = \left(\frac{\partial E}{\partial n_i}\right)_{S,V,n_j} \tag{95}$$

Therefore

$$\mu_i \equiv \left(\frac{\partial E}{\partial n_i}\right)_{S,V,n_j} \tag{96}$$

By writing the total differentials for the enthalpy H and the Helmholtz free energy A, where

$$H = H(S, P, n_1, n_2, \ldots, n_k) \tag{97}$$

and

$$A = A(T, V, n_1, n_2, \ldots, n_k) \tag{98}$$

we find

$$\mu_i = \left(\frac{\partial E}{\partial n_i}\right)_{S,V,n_j} = \left(\frac{\partial H}{\partial n_i}\right)_{S,P,n_j} = \left(\frac{\partial A}{\partial n_i}\right)_{T,V,n_j} = \left(\frac{\partial G}{\partial n_i}\right)_{T,P,n_j} \tag{99}$$

The set of fundamental equations for systems of variable composition is then

$$dE = TdS - PdV + \sum_i \mu_i dn_i \tag{100a}$$

$$dS = \frac{1}{T} dE + \frac{P}{T} dV - \frac{1}{T} \sum_i \mu_i dn_i \tag{100b}$$

$$dH = TdS + VdP + \sum_i \mu_i dn_i \tag{100c}$$

$$dS = \frac{1}{T} dH - \frac{V}{T} dP - \frac{1}{T} \sum_i \mu_i dn_i \tag{100d}$$

$$dA = -SdT - PdV + \sum_i \mu_i dn_i \tag{100e}$$

$$dG = -SdT + VdP + \sum_i \mu_i dn_i \tag{100f}$$

Comparison of equations 10a–10f with the corresponding expressions for the total differentials in terms of partial derivatives (i.e., equations 90 and 92 yields

$$T = \left(\frac{\partial E}{\partial S}\right)_{V,n_i} = \left(\frac{\partial H}{\partial S}\right)_{P,n_i} \tag{101}$$

$$P = -\left(\frac{\partial E}{\partial V}\right)_{S,n_i} = -\left(\frac{\partial A}{\partial V}\right)_{T,n_i} \tag{102}$$

$$V = \left(\frac{\partial H}{\partial P}\right)_{S,n_i} = \left(\frac{\partial G}{\partial P}\right)_{T,n_i} \tag{103}$$

$$S = -\left(\frac{\partial A}{\partial T}\right)_{V,n_i} = -\left(\frac{\partial G}{\partial T}\right)_{P,n_i} \tag{104}$$

Criteria for Change and Equilibria in Systems of Variable Composition

A link between the quantity $\sum_i \mu_i dn_i$ for an open phase and the internal entropy production of the second law, TdS_{int}, is made evident by comparing equations $100a$–$100f$ and equations 44–49.

For example, we have

$$dG = -SdT + VdP - TdS_{int} \tag{47}$$

and

$$dG = -SdT + VdP + \sum_i \mu_i dn_i \tag{100f}$$

It is clear that the internal entropy production at constant T and P is

$$dS_{int} = -\frac{1}{T} \sum_i \mu_i dn_i \tag{105}$$

TABLE 2.3 CRITERIA FOR SPONTANEOUS AND REVERSIBLE PROCESSES IN SYSTEMS OF VARIABLE COMPOSITION

State Function	Constant Fundamental Variables	Process		
		Spontaneous	Reversible	Impossible
S	E, V	$dS > 0$	$dS = 0$	$dS < 0$
		$\sum_i \mu_i dn_i < 0$	$\sum_i \mu_i dn_i = 0$	$\sum_i \mu_i dn_i > 0$
S	H, P	$dS > 0$	$dS = 0$	$dS < 0$
		$\sum_i \mu_i dn_i < 0$	$\sum_i \mu_i dn_i = 0$	$\sum_i \mu_i dn_i > 0$
E	S, V	$dE < 0$	$dE = 0$	$dE > 0$
		$\sum_i \mu_i dn_i < 0$	$\sum_i \mu_i dn_i = 0$	$\sum_i \mu_i dn_i > 0$
H	S, P	$dH < 0$	$dH = 0$	$dH > 0$
		$\sum_i \mu_i dn_i < 0$	$\sum_i \mu_i dn_i = 0$	$\sum_i \mu_i dn_i > 0$
A	T, V	$dA < 0$	$dA = 0$	$dA > 0$
		$\sum_i \mu_i dn_i < 0$	$\sum_i \mu_i dn_i = 0$	$\sum_i \mu_i dn_i > 0$
G	T, P	$dG < 0$	$dG = 0$	$dG > 0$
		$\sum_i \mu_i dn_i < 0$	$\sum_i \mu_i dn_i = 0$	$\sum_i \mu_i dn_i > 0$

For a *spontaneous* process, $dS_{int} > 0$ and $\sum_i \mu_i dn_i < 0$. At *equilibrium*, $dS_{int} = 0$ and $\sum_i \mu_i dn_i = 0$. Thus an increase in internal entropy for a spontaneous process is associated with changes in mole numbers driven by differences in chemical potentials. For *any* system, with its defining constraints (isolated, adiabatic, etc.), the direction of spontaneous change is that for which $\sum_i \mu_i dn_i < 0$. For isobaric, isothermal systems, G is the characteristic function and

$$-TdS_{int} = dG = \sum_i \mu_i dn_i \tag{106}$$

Table 2.3 summarizes the criteria for spontaneous processes and equilibrium processes for systems of variable composition. For all closed systems, we note that for

a spontaneous process: $\sum_i \mu_i dn_i < 0$

a reversible process: $\sum_i \mu_i dn_i = 0$

an impossible process: $\sum_i \mu_i dn_i > 0$

Thus, for a reversible or equilibrium process, we have $dG - \sum_i \mu_i dn_i = TdS_{int} - 0$, and G is a *minimum*.

2.6   THE MEANING OF THE GIBBS ENERGY: USEFUL WORK AT CONSTANT TEMPERATURE AND PRESSURE

Up to now we have assumed that no work other than PV work is transferred to or from a system. Table 2.1 lists a number of kinds of thermodynamic work other than PV work. If we denote net or other-than-PV work done on the system as w', then the total work done on the system is

$$dw = -PdV + dw' \tag{107}$$

and the first law takes the form

$$dE = dq + dw = dq - PdV + dw' \tag{108}$$

From the second law, equation 20,

$$dE \leqslant TdS - PdV + dw' \tag{109}$$

where the equality applies to a reversible transfer ($dq = TdS$) and the inequality to an irreversible transfer ($dq < TdS$).

From the definitions of G and E,

$$dG = dE + VdP + PdV - TdS - SdT \tag{110}$$

From which

$$dG \leqslant VdP - SdT + dw' \tag{111}$$

At constant T and P,

$$dG \leqslant dw' \tag{112}$$

Thus, for any possible process

$$dw' \geqslant dG \tag{113}$$

where the equality refers to a *reversible* process and the inequality to an irreversible process, the negative of w' is work done *by* the system. Therefore

$$-dw' \leqslant -dG$$

For a finite change in state variables, $\Delta G = G_2 - G_1$, and the work done is $-w'$. Then

$$-w' \leqslant -\Delta G \tag{114}$$

The maximum (reversible) work that can be done by a closed system at constant T and P is given by the negative of the change in the Gibbs energy for a spontaneous process within the system. If $dG = 0$ (the system itself is at equilibrium), the system can of course do no work. If the transfer of work from the system to another system is carried out irreversibly, the net work done will be less than the decrease in Gibbs free energy.

By similar arguments, the minimum non-PV work that must be provided *to* a system (chemical, electrical, gravitational work, etc.) to *drive* a nonspontaneous chemical process is ΔG, the change in free energy for the process. Under irreversible conditions of driving the process ($T\Delta S' > 0$) the work needed w' is greater than ΔG.

For processes carried out at constant temperature and volume the Helmholtz free energy is the pertinent function. The change in the Helmholtz free energy dA is related to the total work dw, both PV *and* net, done on the system. Thus

$$-dw \leqslant -dA \tag{115}$$

or

$$-dw' + PdV \leqslant -dA \tag{116}$$

Therefore the relationship between the variations in the Gibbs and Helmholtz free energies is

$$dG = dA + PdV \tag{117}$$

or, for finite changes in state functions, ΔG, ΔA, and ΔV

$$\Delta G = \Delta A + P\Delta V$$

The Gibbs variation does not include the work in expanding the system against a constant pressure; that is, it gives the maximum *net* work.

Case of an Electrochemical Cell

If the net work is electrical work of an electrochemical cell, the work done by the chemical system is

$$-dw' = E\,de \tag{118}$$

where E is electric potential and de is the variation in charge associated with the variation in mole numbers in the system. For a reversible electrochemical process at constant T and P,

$$-dw' = -dG \tag{113}$$

Therefore

$$-dG = E\,de \tag{119}$$

If the variation in the ion mole number is dn, the charge per ion Z, and F the Faraday (charge per mole),

$$-dG = ZFE\,dn \tag{120}$$

For a spontaneous cell reaction the electric potential is positive. If the process involves a finite change of one mole number in the cell, we may represent the change in free energy as ΔG, or

$$-\Delta G = ZFE \tag{121}$$

Case of Surface Work

If the surface area of a pure substance is A and the interfacial tension or interfacial energy is γ, the variation in the internal energy if the area is changed is

$$dE = TdS - PdV + \mu\,dn + \gamma\,dA \tag{122}$$

The variation in the Gibbs energy is

$$dG = -SdT + VdP + \mu\,dn + \gamma\,dA \tag{123}$$

At constant T and P,

$$dG = \mu\,dn + \gamma\,dA \tag{124}$$

These relationships are utilized in the study of adsorption at interfaces (Chapter 10) and the effects of particle size on solubility (Chapter 5).

2.7 PARTIAL MOLAR QUANTITIES

Partial molar quantities of chemical components or species are defined at constant T, P, and composition by

$$\bar{X}_i = \left(\frac{\partial X}{\partial n_i}\right)_{T,P,n_j} \tag{125}$$

where X is an extensive property and $\bar{X}_i$ is the corresponding partial molar quantity. They represent the change in an extensive property per mole of added component or species and are themselves *intensive* properties of a system. The chemical potential (equation 93) is a partial molar quantity, the partial molar Gibbs free energy

$$\mu_i = \left(\frac{\partial G}{\partial n_i}\right)_{T, P, n_j} = \bar{G}_i \tag{126}$$

μ_i plays a role for a homogeneous mixture analogous to that of the molar free energy for a pure substance. Other important molar quantities are the partial molar entropy $\bar{S}_i$, the partial molar enthalpy $\bar{H}_i$, the partial molar heat capacity $\bar{C}_p$, and the partial molar volume $\bar{V}_i$. These quantities may be visualized as the rate of change of the extensive property of the system, X, with respect to the mole number of one component. If X is plotted versus n_i, the value of $\bar{X}_i$ is the slope of the function for fixed values of P, T, and all other mole numbers n_j, that is, the composition. For pure substances, the $\bar{X}_i$ are simply *molar* quantities. For solutions (gaseous, liquid, or solid) the various partial molar quantities are fixed when the temperature, pressure, and composition of the phase are fixed. For example, in a solution of k components

$$\mu_i \equiv \bar{G}_i = \bar{G}_i(T, P, n_1, n_2, \ldots, n_k)$$
$$\bar{V}_i = \bar{V}_i(T, P, n_1, n_2, \ldots, n_k)$$
$$\bar{H}_i = \bar{H}_i(T, P, n_1, n_2, \ldots, n_k)$$

and so forth. The intensive partial molar quantities are fixed once the state of the phase is defined.

There are important additional relationships between the partial molar quantities μ_i, $\bar{V}_i$, and $\bar{S}_i$ paralleling those between G, V, and S (equations 103 and 104). Among these are

$$\left(\frac{\partial \mu_i}{\partial P}\right)_T = \bar{V}_i \tag{127}$$

$$\left(\frac{\partial \mu_i}{\partial T}\right)_P = -\bar{S}_i \tag{128}$$

and

$$\left(\frac{\partial \mu_i/T}{\partial T}\right)_P = -\frac{\bar{H}_i}{T^2} \tag{129}$$

These provide the basis for evaluating the effects of temperature and pressure on chemical equilibria.

Total Gibbs Energy of a Phase

At constant temperature and pressure the variation in G is due only to variations in the mole numbers of the phase:

$$dG = \sum_i \mu_i dn_i$$

By the definitions of intensive and extensive properties, extensive ones depend directly on the mole numbers:

$$G(T, P, \lambda n_1, \lambda n_2, \ldots, \lambda n_k) = \lambda G(T, P, n_1, n_2, \ldots, n_k) \tag{130}$$

In mathematical terms, G is homogeneous and of degree 1 in mole number.

If $\lambda - 1$, that is, for a phase of a single component,

$$dG = \mu_1 dn_1 = \bar{G}_1 dn_1$$

and

$$G = \bar{G} n_1 = \mu_1 n_1 \tag{131}$$

The total Gibbs energy of the pure phase is the product of the molar free energy and the mole number.

For a system of several pure phases,

$$G = \sum_r^p \mu_r n_r \tag{132}$$

where r refers to each of the p phases.

Solution Phase. For a solution phase at constant T and P the chemical potentials of the components (or species) are fixed when the composition is fixed. The composition is specified by the complete set of mole numbers n_1, $n_2, \ldots, n_k$ for the solution, with the mole fraction of each

$$x_i = \frac{n_i}{n_1 + n_2 + \cdots + n_k} = \frac{n_i}{n} \tag{133}$$

where $n = \sum_i^k n_i$. If the mole fractions are *fixed* and the *total* number of moles n is varied,

$$dn_1 = \frac{n_1}{n} dn \qquad dn_2 = \frac{n_2}{n} dn \qquad \cdots \qquad dn_i = \frac{n_i}{n} dn$$

$$dG = \sum_i^k \mu_i \frac{n_i}{n} dn \tag{134}$$

The μ_i are constant at fixed values of n_i/n, so that a formal integration from zero moles to n moles for the phase gives

$$G = \sum_i^k \mu_i n_i \tag{135}$$

A theorem due to Euler yields this result directly. For a function f which is homogeneous and of degree 1 in a variable u (as G is in n_i), $u(\partial f/\partial u) = f$. Therefor $G = \sum_i n_i(\partial G/\partial n_i)_{T,P,n_j} = \sum_i n_i \mu_i$. (For a detailed consideration of Euler's theorem in relation to extensive and intensive properties, see for example, Prigogine and Defay [5] or Guggenheim [1].) Note that the internal energy E is a function solely of extensive variables (S, V, n_i). Therefore the analog of equation 130 for E is

$$E(\lambda S, \lambda V, \lambda n_1, \lambda n_2, \ldots, \lambda n_k) = \lambda E(S, V, n_1, n_2, \ldots, n_k) \tag{136}$$

and E is therefore

$$E = TS - PV + \sum_i^k \mu_i n_i \tag{137}$$

consistent with the expression for G and the definition for G in terms of E.

The physical content of equations 135 and 137 is the constancy of the μ_i with variation in the total mole numbers, that is, the size of the system, at *constant composition*, n_i/n. For example, in a solution of NaCl in water, consisting of 1 mol of NaCl and 1 kg of water (55.51 mol), the composition is defined by a single mole fraction, the sum of mole fractions being unity. Taking 2 mol of NaCl and 2 kg yields the *same composition*, but *twice as great* a Gibbs energy. Equation 135 simply describes the expected character of an extensive property: it varies with the *size* of the system, that is, with the mole number, at fixed chemical composition, that is at fixed μ_i.

Several Phases.   For a system of several different phases, r, whether pure substances or solutions, the total Gibbs energy is then the sum over all p phases:

$$G = \sum_r^p \sum_i^k \mu_i^r n_i^r \tag{138}$$

A system consisting of pure solid phase, I, an aqueous solution phase, II, and a gaseous phase, III, with components 1 and 2 distributed between all phases, has a Gibbs energy given by

$$G = n_1^I \mu_1^I + n_1^{II} \mu_1^{II} + n_1^{III} \mu_1^{III} + n_2^I \mu_2^I + n_2^{II} \mu_2^{II} + n_2^{III} \mu_2^{III}$$

A system consisting of elemental mercury, Hg, in three phases: I, an aqueous solution; II, a gas phase; and III, an organic liquid, the total free energy of the elemental mercury alone is

$$G_{Hg} = \mu_{Hg}^I n_{Hg}^I + \mu_{Hg}^{II} n_{Hg}^{II} + \mu_{Hg}^{III} n_{Hg}^{III}$$

The various μ_{Hg} depend on the mole fractions of Hg in the respective phases. The equilibrium condition is that for which G is a minimum, and therefore $dG_{Hg} = 0$ for all variations in the composition (n_{Hg} in each phase will then have an equilibrium value). Therefore $\sum \mu dn_{Hg} = 0$ for all transfers, and $\mu_{Hg}^{I} = \mu_{Hg}^{II} = \mu_{Hg}^{III}$.

Chemical Potential Variations for a Single Phase at Constant T and P: The Gibbs–Duhem Relationship

Differentiation of the expression for the internal energy of a phase, (equation 137) yields

$$dE = TdS + SdT - PdV - VdP + \sum_{i}^{k} \mu_i dn_i + \sum_{i}^{k} n_i d\mu_i \qquad (139)$$

But, from the first principle applied to a multicomponent system,

$$dE = TdS - PdV + \sum_{i}^{k} \mu_i dn_i \qquad (140)$$

Therefore

$$0 = SdT - VdP + \sum_{i}^{k} n_i d\mu_i \qquad (141)$$

In a system whose state is fixed by T, P, and k chemical potentials, of $k + 2$ variables only $k + 1$ can be varied independently.

For a constant T and P system,

$$\sum_{i}^{k} n_i d\mu_i = 0 \qquad (142)$$

Dividing n_i by $n = \sum_{i} n_i$ to obtain the mole fractions x_i

$$\sum x_i d\mu_i = 0 \qquad (143)$$

Alternatively, differentiating the expression for G (equation 135),

$$dG = \sum_{i} \mu_i dn_i + \sum n_i d\mu_i \qquad (144)$$

And comparing with the definition of dG for any process,

$$dG = -SdT + VdP + \sum_{i} \mu_i dn_i \qquad (145)$$

The result is

$$0 = SdT - VdP + \sum_{i} n_i d\mu_i$$

which is identical to equation 141. Equations 142 and 143 are different forms of the Gibbs–Duhem equation which must be satisfied for each phase in a system.

For a two-component solution equation 142 takes the simple form

$$n_1 d\mu_1 + n_2 d\mu_2 = 0 \qquad (146)$$

This expression is useful in finding the variation in the chemical potential of one component of a solution with composition when the variation of that of the second component is known. Thus, if the variation of the vapor pressure of the solvent with solution composition is known, the corresponding variation of solute chemical potential $d\mu_2$ can be found via the Gibbs–Duhem equation.

Variations in Partial Molar Quantities: General Relationship

For any partial molar quantity $\overline{X}_i$ the general relationships at constant T and P are found to be

$$X = \sum_i^k \overline{X}_i n_i \tag{147}$$

and

$$\sum_i n_i d\overline{X}_i = 0 \tag{148}$$

Thus

$$V = \overline{V}_1 n_1 + \overline{V}_2 n_2 + \cdots + \overline{V}_k n_k$$

and

$$n_1 d\overline{V}_1 + n_2 d\overline{V}_2 + \cdots + n_k d\overline{V}_2 = 0$$

and similarly for enthalpy and entropy.

2.8   CHEMICAL POTENTIALS IN VARIOUS PHASES

For pure solids and pure liquids the chemical potential is the molar free energy $\overline{G}$ and is a function of T and P alone:

$$\overline{G} = \mu = \mu(T, P)$$

It is conventional to refer to the value of μ for a component at 1 atm as $\mu°$, the standard-state chemical potential. Then

$$\mu = \mu°(T)$$

For a pure solid we write $\mu°(s)$, and for a liquid $\mu°(l)$, omitting the (T) for convenience.

Chemical Potential of an Ideal Gas

For n moles of pure gas at constant temperature,

$$dG = V dP \tag{149}$$

From the ideal gas law,

$$V = nRT \frac{1}{P} \tag{150}$$

Then, from 149

$$dG = nRT \frac{dP}{P} \tag{151}$$

or

$$dG = nRTd \ln P \tag{152}$$

Integrating from a standard pressure $P°$ to a pressure P yields

$$G - G° = nRT \ln \frac{P}{P°} \tag{153}$$

Dividing by n,

$$\frac{G}{n} - \frac{G°}{n} = RT \ln \frac{P}{P°}$$

The chemical potential for a one-component phase is G/n. Therefore

$$\mu(g) - \mu°(g) = RT \ln \frac{P}{P°} \tag{154}$$

when $P = P°, \mu(g) = \mu°(g)$. Or

$$\mu = \mu°(g) + RT \ln \frac{P}{P°}$$

If $P° = 1$,

$$\mu = \mu°(g) + RT \ln P \tag{155}$$

The standard state value $\mu°(g)$ is a function of T but is independent of P.

Chemical Potential in an Ideal Gas Mixture

An ideal gas mixture, like an ideal gas, has molecules with negligible volume and negligible interaction energies. In an ideal mixture, each component behaves as if the other components were not there. Therefore we expect for the chemical potential of a component an expression of the same form as equation 155, with the *partial pressure* P_i of the component taking the place of the total pressure,

$$\mu_i(g) = \mu_i°(g) + RT \ln P_i \tag{156}$$

If x_i is the mole fraction of component i:

$$P_i = x_i P \tag{157}$$

then the expanded expression is

$$\mu_i(g) = \mu_i^\circ(g) + RT \ln P + RT \ln x_i \tag{158}$$

where P° has been chosen as unity (as in equation 155). When $x_i = 1$, the expression becomes identical with equation 155 for a one-component ideal gas. The thermodynamic basis of the result is the expression for the variation of the chemical potential with pressure:

$$\left(\frac{\partial \mu_i}{\partial P}\right)_T = \bar{V}_i \tag{127}$$

and the recognition that $\bar{V}_i = V/n = RT/P$, where $n = \sum_i n_i$, the total mole number. For constant x_i,

$$d \ln P_i = d \ln P$$

and

$$d\mu_i = RT d \ln P_i$$

which is integrated to yield the expression for $\mu_i(g)$.

Chemical Potentials in Ideal Solutions

For an ideal solution the form of the expression for the chemical potential closely parallels that for an ideal gas mixture. The expression is

$$\mu_i = \mu_i^\circ(\text{sol}) + RT \ln x_i \tag{159}$$

where $\mu_i^\circ(\text{sol}) = \mu_i^\circ(T, P)$ and x_i is the mole fraction in solution. For a solution component i in equilibrium with a vapor phase, the general criterion of equilibrium, $dG = \sum \mu_i dn_i = 0$, yields as the condition of two-phase equilibrium

$$\mu_i(\text{sol}) = \mu_i(g)$$

Assuming that the vapor phase is an ideal gas mixture, we obtain

$$\mu_i^\circ(\text{sol}) + RT \ln x_i = \mu_i^\circ(g) + RT \ln P_i \tag{160}$$

Rearranging,

$$\frac{P_i}{x_i} = \exp\left[\frac{\mu_i^\circ(\text{sol}) - \mu_i^\circ(g)}{RT}\right] \tag{161}$$

Equation 161 states that the vapor pressure of component i is *proportional to* its mole fraction in a solution behaving ideally. Two important observational laws of physical chemistry, Raoult's law (1887) and Henry's law (1803), are contained in equation 161.

Raoult's Law.   The ideal gas vapor pressure of an ideal solution component is equal to the vapor pressure of the pure liquid component multiplied by the mole fraction of the component in the solution, or

$$P_i = x_i P_i^\circ \tag{162}$$

where P_i is the partial pressure, x_i is the mole fraction, and P_i° is the partial vapor pressure of pure component i. Comparing equations 161 and 162,

$$P_i^\circ = \exp\left[\frac{\mu_i^\circ(\text{sol}) - \mu_i^\circ(\text{g})}{RT}\right] \tag{163}$$

When $x_i = 1$,

$$\mu_i^\circ(\text{sol}) = \mu_i^\circ(\text{g}) + RT \ln P_i^\circ$$

Thus the standard-state chemical potential for the ideal solution component depends on the ideal gas standard state and the convention that $x_i = 1$ is the standard state for the component in solution. For real solutions, Raoult's law behavior is approached as $x_i \to 1$.

Henry's Law.   For a dilute solution the ideal gas vapor pressure of a volatile solute is proportional to its mole fraction in the solution; that is, escaping tendency of the solute molecules is proportional to their mole fraction. Henry's law can be expressed as

$$P_i = K_i x_i \tag{164}$$

The meaning of K_i is that it is the *slope* of a plot of P_i versus x_i for sufficiently dilute solutions, that is, where solute-solute interaction forces in the solution are negligible. Comparison of equations 164 and 161 shows that

$$K_i = \exp\left[\frac{\mu_i^\circ(\text{sol}) - \mu_i^\circ(\text{g})}{RT}\right] \tag{165}$$

For real solutions, Henry's law behavior is approached as $x_i \to 0$.

It is important to note that $\mu_i^\circ(\text{sol})$ for a Henry's law solute reflects the properties of a *dilute* solution, while $\mu_i^\circ(\text{sol})$ for a Raoult's law component (generally, the "solvent") reflects the properties of a *pure* component. For example, a solution of molecular oxygen, O_2, in water can be described in terms of both Raoult's law and Henry's law. For dilute solutions, say mole fractions of O_2 less than 1×10^{-4}, Henry's law is obeyed by the solute oxygen; that is, there is a proportionality between partial pressure and solution concentration

$$P_{O_2} = K_{O_2} x_{O_2}$$

and Raoult's law is obeyed by the solvent water:

$$P_{H_2O} = P_{H_2O}^\circ x_{H_2O}$$

The relevant standard chemical potential terms are, for the oxygen $\mu_{O_2}^\circ(\text{sol})$ and $\mu_{O_2}^\circ(\text{g})$, and for the water $\mu_{H_2O}^\circ(\text{sol}) = \mu_{H_2O}^\circ(l)$ and $\mu_{H_2O}^\circ(\text{g})$. In practice, the mole fraction scale is seldom used for an aqueous solute such as O_2; the molal scales or molar scales (discussed later) are used, with corresponding numerical values of $\mu_i^\circ(\text{sol})$.

2.9 CHEMICAL POTENTIALS IN REAL MIXTURES

For ideal gases, ideal gas mixtures, and ideal liquid solutions we now have as general forms for the chemical potential of a component

$$\mu_i = \mu_i^\circ + RT \ln P_i \quad \text{(gases)} \tag{156}$$

and

$$\mu_i = \mu_i^\circ + RT \ln x_i \quad \text{(solutions)} \tag{159}$$

For real nonideal gases and gas mixtures the form of the chemical potential expression is retained, and the pressure is replaced by an idealized pressure, called by G. N. Lewis the fugacity f_i. Thus, for a real gas mixture,

$$\mu_i = \mu_i^\circ + RT \ln f_i \tag{166}$$

The fugacity approaches the partial pressure of the component as the mixture becomes infinitely dilute, that is, as the pressure goes to zero

$$\frac{f_i}{P_i} \to 1 \quad \text{as} \quad P \to 0 \tag{167}$$

The relationship of fugacity and partial pressure may be expressed as a *fugacity coefficient* γ_{f_i}, where

$$f_i = \gamma_{f_i} P_i \tag{168}$$

Fugacity coefficients differ significantly from unity only at high total pressures. Near 1 atm pressure, $\gamma_{f_i} \simeq 1$ for real gases. Fugacity has the dimension of pressure.

Correspondingly, for real solutions, Lewis proposed that actual concentrations, for example, mole fractions, be replaced by *idealized* ones, or *activities*. Thus, for a real solution, instead of equation 159 we have

$$\mu_i = \mu_i^\circ + RT \ln a_i \tag{169}$$

where a_i is the activity. (We will also use the symbol { } for activity; e.g., $\{M_i\}$, or just $\{i\}$ is the activity of component i with molecular formula M_i.) The relationship between the actual concentration, say x_i, and the activity a_i is expressed by an *activity coefficient* γ_i:

$$a_i = \gamma_i x_i \tag{170}$$

The chemical potential in a real solution is then

$$\mu_i = \mu_i^\circ + RT \ln \gamma_i x_i \tag{171}$$

Expanding the logarithm shows the significance of γ_i as an energy term:

$$\mu_i = \mu_i^\circ + RT \ln x_i + RT \ln \gamma_i \tag{172}$$

The expression $RT \ln \gamma_i$ can be seen as the partial molar free energy of the interactions that occur in nonideal mixtures.

Volatile Solution Components

The activity coefficient for a volatile solution component can be interpreted physically with regard to ideal behavior of the solvent (Raoult's law) or solute (Henry's law). Consider equilibrium between a real solution and a real vapor. The condition of equilibrium for volatile component i is

$$\mu_i(\text{sol}) = \mu_i(\text{g})$$

or

$$\mu_i^\circ(\text{sol}) + RT \ln \gamma_i x_i = \mu_i^\circ(\text{g}) + RT \ln p_i \gamma_{f_i} \tag{173}$$

For the solvent, say the component $i = 1$, for which x_i can approach unity, the result is (compare equation 161 for ideal gas mixtures and solutions)

$$\frac{P_1 \gamma_{f_1}}{x_1 \gamma_1} = \exp \left[\frac{\mu_1^\circ(\text{sol}) - \mu_1^\circ(\text{g})}{RT} \right] = P_1^\circ \tag{174}$$

If the pressure of the vapor is low enough so that $\gamma_{f_i} = 1$, the solution-phase activity coefficient is

$$\gamma_1 = \frac{P_1 / P_1^\circ}{x_1} \tag{175}$$

and

$$a_1 = \gamma_1 x_1 = \frac{P_1}{P_1^\circ} \tag{176}$$

In other words, the activity of the solvent is described by the ratio of its vapor pressure over a solution to the vapor pressure of pure solvent. For an *ideal* solution, the activity is precisely x_1, as described by Raoult's law, and $\gamma_1 = 1$. For *real* solutions, activity coefficients are not unity and vary with composition. For example, note the following experimental results from Robinson and Stokes [6] for binary solutions of sucrose (a neutral nonvolatile molecule) and water (25°C) in Table 2.4.

TABLE 2.4 PROPERTIES OF AQUEOUS SUCROSE SOLUTIONS

$x_{sucrose}$	x_{H_2O}	$P_{H_2O}/P_{H_2O}^\circ$	γ_{H_2O}
1.80×10^{-3}	0.99820	0.99819	0.9999
5.40×10^{-3}	0.9946	0.9945	0.9999
1.25×10^{-2}	0.9875	0.9867	0.9992
1.77×10^{-2}	0.9823	0.9806	0.9983
3.48×10^{-2}	0.9652	0.9581	0.9926
5.13×10^{-2}	0.9487	0.9328	0.9832

The activity coefficient of the solvent water approaches unity as its mole fraction approaches unity.

For a volatile solute, say component $i = 2$, for which Henry's law can be tested in dilute solution, the result from equation 161 is

$$\frac{P_2 \gamma_{f_2}}{x_2 \gamma_2} = \exp\left[\frac{\mu_2^\circ(sol) - \mu_2^\circ(g)}{RT}\right] = K_2 \tag{177}$$

where K_2 is a constant. If $\gamma_{f_2} = 1$, the activity coefficient is again evaluated by comparing the actual partial pressure with the ideal (Henry's law) value by

$$\gamma_2 = \frac{P_2}{K_2 x_2} \tag{178}$$

The ideal vapor pressure from Henry's law is $K_2 x_2$, and ideal behavior is approached for the solute as the solution is made more and more dilute: $\gamma_2 \to 1$ as $x_2 \to 0$. The value of K_2 can be obtained by measuring equilibrium partial pressures and solute concentrations in dilute solutions and evaluating the slope of P_2 versus x_2. For O_2 dissolved in water at 25°C, K_{O_2} is reported to be 4.3×10^4 atm. (Note that K_2 has the dimension of pressure; it can be thought of as the pressure that pure O_2 would have if it had the same properties as in a dilute aqueous solution.) The value of K_{O_2} given predicts a dissolved oxygen mole fraction of 4.9×10^{-6} for an O_2 partial pressure of 0.21 atm (normal air). A liter of water contains approximately 55.4 mol, so the corresponding O_2 concentration in mol/liter^{-1} would be 2.6×10^{-4}. For O_2 in pure water under this condition, γ_{O_2} is very close to 1.

Reference States and Standard States for Real Solutions

For real solutions the chemical potential of a component will be given by an equation such as equation 171:

$$\mu_i = \mu_i^\circ + RT \ln \gamma_i x_i$$

or by an expression in which a different concentration scale, for example, the *molar* scale, is used. The molal scale is commonly used for solutes in aqueous solutions. The *molar* scale is also in wide use, for analytical reasons. These concentration scales are defined as follows. The *molality m* of a solute in solution is the number of moles of solute per 1000 g of solvent. If a solution contains n_1 moles of solvent of gram molecular (or formula) weight M_1 and n_i ($i > 1$) moles of solute, the molality m_i is

$$m_i = \frac{1000 n_i}{M_1 n_1} \tag{179}$$

The molarity M_i (also commonly represented by C_i or [i]) is moles of solute per liter of solution:

$$M_i = \frac{n_i}{V} \tag{180}$$

where V is volume in liters. The molarity scale is temperature-dependent; the molality scale is not. In very dilute solutions, molality and molarity are both proportional to the mole fraction.

In terms of the *molal* scale, the chemical potential of a solute can be expressed as

$$\mu_i = \mu_i^* + RT \ln \gamma_i' m_i \tag{181}$$

and in terms of the *molar* scale it can be expressed as

$$\mu_i = \mu_i^\bullet + RT \ln \gamma_i''[i] \tag{182}$$

The value of μ_i is well-defined with respect to a datum, say by a phase equilibrium. But the standard potentials and the activity coefficients must in general have different values for the various scales. Specification of a *reference state* and a *standard state* for each scale is required to define the standard-state potential and the activity coefficient. The *reference state* is the dilute solution limit at which concentration equals activity, that is, where the activity coefficient approaches unity. The *standard state* is one of unit concentration *and* unit activity coefficient. The choice of standard state for solutions is equivalent to a choice of concentration units.

With reference to equation 171 for the solvent, component 1, the conventional definition of the reference state is

$$\gamma_1 \to 1 \quad \text{as} \quad x_1 \to 1$$

and therefore $\mu_1 = \mu_1^\circ$ when $a_1 = \gamma_1 x_1 = 1$ (pure solvent).

For a solute, say component 2, whose chemical potential is described on the molal scale, the reference-state limit in which $\gamma_2' \to 1$ is the infinitely dilute solution. This can be expressed as

$$\gamma_2' \to 1 \quad \text{as} \quad m_2 \to 0$$

Alternatively, for binary systems it could be expressed as

$$\gamma_2' \to 1 \quad \text{as} \quad x_1 \to 1$$

The standard state for the solute on the molal scale is one for which γ_2' and m_2 are both 1, and therefore for which $a_2 = \gamma_2' m_2 = 1$. Then $\mu_2 = \mu_2^*$. Similar considerations apply for solute activity expressed on the other scales; for example,

$$\gamma_2 \to 1 \quad \text{as} \quad x_2 \to 0$$

$$\gamma_2'' \to 1 \quad \text{as} \quad [2] \to 0$$

It is important to note that the standard state for the solute is not defined simply as that for which $a_2 = 1$; an actual combination of $\gamma_2 x_2$, $\gamma_2' m_2$, or γ_2'' [2] might give $a_2 = 1$. The standard state of the solute is one of unit concentration on the scale used *and* having the ideal properties of the very dilute solution; that is, $\gamma_2 = 1$. Since actual solutions of unit concentration do not in general exhibit ideal solution behavior, the standard state for solutes is characterized as a hypothetical solution. The key point concerning the choice of an *ideal* solution for the standard state is that all the properties are thus defined to be those of the ideal Henry's law solution, including zero ΔH of dilution and zero ΔV of mixing.

The numerical relationship between standard-state chemical potentials on the mole fraction (X_i) and molal (m_i) scales for a solute can be easily seen by equating equations 171 and 181 and considering the dilute solution limit where the activity coefficients go to unity. Then

$$\mu^\circ - \mu^* = RT \ln \frac{1000}{M_1} \tag{183}$$

where μ° stands for a mole fraction standard-state potential and μ^* for a molal one. For example, the standard chemical potential on the molal scale for CO_2 dissolved in water, $CO_2(aq)$, is -92.31 kcal mol^{-1} or -386.23 kJ mol^{-1} (the standard state is a hypothetical 1 molal solution). For water, M_1 is 18.01 g mol^{-1}, so the standard chemical potential for aqueous CO_2 on the *mole fraction* scale would be -89.93 kcal mol^{-1} or -376.27 kJ mol^{-1}.

Nonvolatile Solutes in Aqueous Solutions

The chemical potential of a solute that does not have an appreciable vapor pressure is still given on the molal scale by an expression like

$$\mu_i = \mu_i^\circ + RT \ln \gamma_i m_i \tag{184}$$

where we now adopt the symbol μ_i° for the standard state and γ_i for the activity coefficient. An important approach to the determination of $\gamma_i = \gamma_2$ for a non-volatile solute in a two-component solution is based on equilibrium between the solvent (component 1) and its vapor and application of the Gibbs–Duhem equation. Briefly, the method consists of determining experimentally the activity of the solvent at various solution compositions from

$$a_1 = \frac{P_1}{P_1^\circ} \qquad (185)$$

The chemical potential of the solvent is

$$\mu_1 = \mu_1^\circ + RT \ln a_1 \qquad (186)$$

The chemical potential of the solute is

$$\mu_2 = \mu_2^\circ + RT \ln \gamma_2 m_2 \qquad (187)$$

The Gibbs–Duhem relationship requires

$$n_1 d\mu_1 + n_2 d\mu_2 = 0 \qquad (188)$$

The variations in the two potentials are

$$d\mu_1 = RT d \ln a_1 \qquad (189)$$

and

$$d\mu_2 = RT d \ln \gamma_2 + RT d \ln m_2 \qquad (190)$$

The key expression is obtained by combining equations 188–190, yielding for 1000 g of solvent

$$m_2 d \ln m_2 + m_2 d \ln \gamma_2 = -\frac{1000}{M_1} d \ln a_1 \qquad (191)$$

Vapor pressure measurements over solutions yield pairs of values for m_2 and a_1. M_1 is the molecular weight of the solvent. Integration of equation 191 from $m_2 = 0$ to $m_2 = m_2$ provides a means of computing γ_2 for each m_2. However, the chemical potential of a solute goes to minus infinity as the molality goes to zero, so practical difficulties are encountered. The *molal* (or "practical") osmotic coefficient,

$$\phi = -\frac{(1000/M_1) \ln a_1}{m_2} \qquad (192)$$

provides a useful change in variable and, together with equation 191, yields

$$d \ln \gamma_2 = (\phi - 1) d \ln m_2 + d\phi \qquad (193)$$

Integration gives

$$\ln \gamma_2 = \int_0^{m_2} \frac{(\phi - 1)}{m_2} dm_2 + \phi - 1 \qquad (194)$$

This equation, and variations thereon, allow determinations of a_2 and γ_2 for a nonvolatile solute from the activity of the solvent. Note that the osmotic coefficient is a sensitive measure of γ_1, the activity coefficient of the solvent.

Dissociation of the Solute

Variations of the vapor pressure, boiling point, and freezing point of the solvent of a solution with variations of solute concentrations are examples of *colligative properties* of solutions. These arise because dissolution of *solute* causes a decrease in chemical potential of the *solvent*. If a solute, A_v, introduced to a solvent dissociates in solution, for example,

$$A_v = vA$$

the mole fraction of solute in the solution is $n_{A_v}/(n_1 + vn_{A_v})$, but the mole fraction of mobile species, A, is $vn_{A_v}/(n_1 + vn_{A_v})$, where v is the number of moles of solute species formed per mole of component A_v. For example, if $v = 2$, and 1 mol of solute A_v is introduced into 1000 g of water to form a 1 m solution, the mole fraction of species A in solution is $2/(55.51 + 2)$, whereas the mole fraction of solute is $1/(55.51 + 2)$. The ideal (Raoult's law) vapor pressure lowering for a solution of fully dissociated A_2 in water will be twice as great as for a solution of undissociated A_2. The ideal vapor pressure lowering is proportional to the mole fraction of actual solute species, or

$$\frac{P° - P}{P°} = vX_{A_v} = X_A \tag{195}$$

Similarly, for a volatile solute which dissociates in solution, Henry's law should apply to the mole fractions or concentrations of *actual* solute species in solution, not to the mole fraction of the component introduced. Thus, for the chemical potentials of the species A_v and A in a dilute aqueous solution (γ approximately unity) we have, on the molal scale,

$$\mu_{A_v} = \mu_{A_v}° + RT \ln m_{A_v} \tag{196}$$

and

$$\mu_A = \mu_A° + RT \ln m_A \tag{197}$$

For the ideal gas phase the chemical potential of A_v is

$$\mu_{A_v}(g) = \mu_{A_v}°(g) + RT \ln P_{A_v} \tag{198}$$

Application of the equilibrium condition at constant T and P, $\sum_i \mu_i dn_i = 0$ for both the dissolution and the dissociation yields

$$\frac{P_{A_v}}{m_{A_v}} = K'_{A_v} \tag{199}$$

and

$$\frac{m_A^v}{m_{A_v}} = K_d \tag{200}$$

where K'_{A_v} is a molal-scale Henry's law constant and K_d is a solution-phase dissociation constant.

Electrolytes

Electrolytes represent an important class of dissociating solutes in solution. It is instructive to start with some simple examples before giving a more general definition and description. Consider gaseous hydrogen chloride, $HCl(g)$, which dissolves in water as hydrochloric acid, $HCl(aq)$. Hydrochloric acid is a *strong* electrolyte in aqueous solution; that is, it is completely dissociated to yield the species $H^+(aq)$ and $Cl^-(aq)$ in solution. If $HCl(g)$ is equilibrated with water to yield various values of total HCl concentration, m_{HCl}, in solution and P_{HCl} in the gas phase, and if P_{HCl} is then plotted versus m_{HCl}, a finite limiting slope is not obtained as $m_{HCl} \rightarrow 0$; that is, Henry's law is not obeyed. The reason is that m_{H^+} and m_{Cl^-} represent the concentrations of species actually formed from HCl in solution. A plot of P_{HCl} versus $m_{H^+} \times m_{Cl^-} = m_{HCl}^2$ *does* yield a limiting slope as m_{HCl} approaches zero. For this system, a plot of the vapor pressure of the water P_{H_2O} versus the mole fraction of water X_{H_2O}, where

$$X_{H_2O} = \frac{1000/M_1}{1000/M_1 + 2m_{HCl}}$$

yields a well-defined limiting slope as $X_{H_2O} \rightarrow 1$; that is, Raoult's law is obeyed in dilute solution.

Solutions of NaCl in water cause vapor pressure lowering and freezing point depression in accord with Raoult's law in dilute solutions, with the effect proportional to $2X_{NaCl}$ or $2m_{NaCl}$. Thus the uni-univalent electrolytes HCl and NaCl behave essentially the same with respect to their colligative properties. The chemical potential of the solvent water in solution for each of these solutes can be described in the limit of dilute solutions by an expression of the form

$$\mu_1 = \mu_1^\circ + RT \ln X_1 \tag{201}$$

The value of X_1 can be replaced by $1 - 2X_2$, on the basis of the observation that colligative properties depend on $2X_2$ for the fully dissociated uni-univalent electrolytes, where X_2 is the mole fraction of total dissolved electrolyte:

$$\mu_1 = \mu_1^\circ + RT \ln (1 - 2X_2) \tag{202}$$

and

$$d\mu_1 = RT d \ln (1 - 2X_2) = \frac{RT d(1 - 2X_2)}{X_1} \tag{203}$$

or

$$\frac{d\mu_1}{dX_2} = -\frac{2RT}{X_1} \tag{204}$$

The Gibbs–Duhem relationship requires that

$$d\mu_2 = -d\mu_1 \frac{X_1}{X_2}$$

Therefore

$$d\mu_2 = \frac{2RT}{X_1} \frac{X_1}{X_2} = \frac{2RT}{X_2} \tag{205}$$

The expression for $d\mu_2$ indicates that μ_2 is of the form

$$\mu_2 = \mu_2^\circ + RT \ln X_2^2 \tag{206}$$

On a molal scale the corresponding expression for a very dilute uni-univalent electrolyte of molal concentration m_2 would be

$$\mu_2 = \mu_2^* + RT \ln m_2^2 \tag{207}$$

The result is a particular instance of a more general one and is a direct consequence of the Gibbs–Duhem relationship: Solutions that obey Raoult's law for the solvent also obey Henry's law for the solute. The factor of 2 in the expression for vapor pressure lowering demands a power of 2 for the solute concentration in the expression for the chemical potential of the solute. This result for electrolytes was anticipated by the discussion of dissociation in general.

The chemical potential of aqueous NaCl must then be the *sum* of the chemical potentials of the $Na^+(aq)$ and $Cl^-(aq)$ ionic species:

$$\mu_2 \equiv \mu_{NaCl} = \mu_{Na^+} + \mu_{Cl^-} \tag{208}$$

In dilute solution, *theoretical* expressions for the individual ions are

$$\mu_{Na^+} = \mu_{Na^+}^\circ + RT \ln m_{Na^+} \tag{209}$$

$$\mu_{Cl^-} = \mu_{Cl^-}^\circ + RT \ln m_{Cl^-} \tag{210}$$

with

$$\mu_{Na^+}^\circ + \mu_{Cl^-}^\circ \equiv \mu_{NaCl(aq)}^\circ \tag{211}$$

Therefore the *total* chemical potential is of the form

$$\mu_{NaCl(aq)} = \mu_{NaCl(aq)}^\circ + RT \ln m_{Na^+} m_{Cl^-} \tag{212}$$

Adopting the convenient brief notation

$$m_+ = m_{Na^+}$$

$$m_- = m_{Cl^-}$$

and recognizing that stochiometry requires

$$m_{NaCl} = m_{Na^+} = m_{Cl^-}$$

the result is

$$\mu_{NaCl} = \mu_{NaCl}^\circ + RT \ln m_+ m_-$$
$$= \mu_{NaCl}^\circ + RT \ln m_{NaCl}^2 \tag{213}$$

which is identical with equation 207, with $\mu_2^* = \mu_{NaCl}^\circ$ and $m_{NaCl} = m_2$.

For a nonideal electrolyte solution, for example, NaCl(aq), the chemical potential on the molal scale must then be expressed as

$$\mu_{NaCl} = \mu_{NaCl}^\circ + RT \ln a_+ a_- \tag{214}$$

or

$$\mu_{NaCl} = \mu_{NaCl}^\circ + RT \ln \gamma_+ m_+ \gamma_- m_- \tag{215}$$

where a_+ and a_-, and γ_+ and γ_-, represent the molal-scale activities and activity coefficients for $Na^+(aq)$ and $Cl^-(aq)$, respectively, and $m_+ = m_- = m$. The product $a_+ a_-$ is denoted by a_{NaCl}, the *activity* of the aqueous solute, or $a_2 = a_+ a_-$. The *mean activity* (a geometric mean) is defined by $a_\pm = (a_+ a_-)^{1/2} = a_2^{1/2}$. The quantity that can be determined experimentally is a_2 or $a_\pm$. The individual activities a_+ and a_- are not independently measurable quantities.

The product $\gamma_+ \gamma_-$ is experimentally measurable. The quantity $(\gamma_+ \gamma_-)^{1/2}$ is referred to as the *mean activity coefficient* $\gamma_\pm$. The mean ionic molality $m_\pm$ is defined as $(m_+ m_-)^{1/2}$ and is simply m for a uni-univalent electrolyte. Summarizing these definitions for a nonideal, *uni-univalent* solution, where the solute is component 2,

$$a_2 = a_+ a_-$$

$$a_\pm = (a_+ a_-)^{1/2} = a_2^{1/2}$$

$$a_+ = \gamma_+ m_+$$

$$a_- = \gamma_- m_-$$

$$\gamma_\pm = (\gamma_+ \gamma_-)^{1/2}$$

$$a_\pm = \gamma_\pm m_\pm = \gamma_\pm m$$

For *multivalent symmetric* electrolytes ($v_+ = v_-$), for example, $CaSO_4$, $MgCO_3$, the molality–activity–activity coefficient relationships parallel those for uni-univalent electrolytes. For *nonsymmetric* electrolytes, for example, $CaCl_2$, Na_2SO_4, $Al_2(SO_4)_3$, the stoichiometric relationships of cations and anions require close attention. For all strong electrolytes, the essential relationships are

$$v = v_+ + v_-$$

$$m_+ = mv_+ \quad \text{and} \quad m_- = mv_-$$

$$m_\pm = (m_+^{v_+} m_-^{v_-})^{1/v} = m(v_+^{v_+} v_-^{v_-})^{1/v}$$

$$a_\pm = (a_+^{v_+} a_-^{v_-})^{1/v} = m_\pm \gamma_\pm$$

$$\gamma_\pm = (\gamma_+^{v_+} \gamma_-^{v_-})^{1/v}$$

$$\mu = v_+ \mu_+^\circ + v_- \mu_-^\circ + RT \ln \gamma_\pm^v m_\pm^v$$

or

$$\mu = \mu_\pm^\circ + RT \ln \gamma_\pm^v m_\pm^v$$

where

$$\mu_\pm^\circ = v_+ \mu^\circ + v_- \mu_-^\circ$$

These definitions are needed in interpreting experimental data on physical chemistry of sample (binary) electrolyte solutions. The physically significant (measurable) quantities are $\mu_\pm^\circ$, $\gamma_\pm$, and $m_\pm$ (or m).

Example 2.1   Activity and Activity Coefficients of Aqueous Sodium Chloride

Water vapor pressures have been measured over NaCl solutions of varying molal concentrations. Results of such measurements allow calculations of relative vapor pressure lowering, a_w, ϕ, and $\gamma_\pm$, for a range of NaCl concentrations in water. Robinson and Stokes [6] provide such data for concentrations ranging from 0.1 to 6.0 molal. Table 2.5 shows results for three different concentrations. The mole fraction of H_2O, X_w, is also included. From such data the activity of aqueous NaCl can be computed.

**TABLE 2.5   WATER   ACTIVITIES,   OSMOTIC   COEF-
FICIENTS, AND RELATIVE MOLAL VAPOR PRESSURE
LOWERINGS   FOR   NaCl   AT   25°C[a]**

m	a_w	X_w	ϕ	$\gamma_\pm$	$\dfrac{P^\circ - P}{mP^\circ}$
0.1	0.996646	0.99641	0.9324	0.778	0.03354
0.3	0.99009	0.98931	0.9215	0.709	0.03303
1.0	0.96686	0.96522	0.9355	0.657	0.03314

[a] From Robinson and Stokes [6].

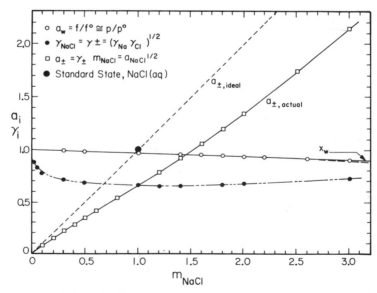

Figure 2.4.   (Example 2.1). Activities of aqueous NaCl (actual and ideal), activity coefficient for NaCl, and the activity of water at various molal concentrations. The *standard state* ($m = 1$, $a = 1$, $\gamma = 1$) is a hypothetical solution.

Figure 2.4 shows the activity of water, a_w, the activity coefficient for NaCl, $\gamma_\pm$, and the mean activity of aqueous NaCl, $a_\pm = (a_+a_-)^{1/2} = a_{NaCl}$, as a function of the molality of NaCl in water at 25°C. Also shown is the "ideal" (i.e., extrapolated from infinity-dilute solution) activity $a_{\pm,\,ideal}$ versus molal concentration. The standard state, for which $a_\pm = 1$, $\gamma_\pm = 1$, and $m = 1$, is clearly not a real state of the system but a hypothetical one. The actual solution of $m = 1$ has an activity of 0.657. A composition of approximately 1.5 molal has an actual activity of 1. This is not the standard state; it does not have the properties of an infinitely-dilute solution. Note that the activity of water, a_w, is very close to the mole fraction of water, X_w, up to about 2.5 molal NaCl. The activity coefficient of the solvent γ_w is actually greater than unity over most of the range shown (compare a_w and X_w, Table 2.5), but the difference from unity is extremely small, for example, 1.00169 at 1 molal. The osmotic coefficient rather than γ_w, is therefore used to describe the nonideal solvent behavior.

Example 2.2 Properties of HCl(aq) from Galvanic Electrochemical Cells and Aqueous–Vapor Phase Equilibrium

Activity coefficients and activities for HCl in dilute and concentrated solutions have been obtained with high precision by measurement of the electric potential under conditions of electrochemical equilibrium in a cell consisting of $H_2(g)$, HCl(aq), AgCl(s), and Ag(s). The hydrogen electrode has the components $H_2(g)$ and HCl at a platinum electrode; the silver–silver chloride electrode has the components of Ag, AgCl, and HCl. Detailed discussion of the cell is available in, for example, Harned and Owen [7] and Robinson and Stokes [6]. The basic equation of electrochemical equilibrium in an electro-chemical system is equation 121:

$$\Delta G = -nFE$$

For the process

$$\tfrac{1}{2}H_2(g) + AgCl(s) = HCl(aq) + Ag(s) \qquad (i)$$

in which 1 mol of electric charge is transferred reversibly, if $P_{H_2} = 1$ and HCl(aq) is present at total molal concentration m_{HCl}, the applicable expression obtained from equation 121, with $\Delta G = \sum_i \mu_i \Delta n_i = \sum_i \mu_i \nu_i$ (see equation 223), and from the definitions for $\mu, \mu_+, \mu_-, a_\pm,$ and $\gamma_\pm$ already given is

$$E = E^\circ - \frac{RT}{F} \ln \gamma_\pm^2 m_{HCl}^2 \qquad (ii)$$

or

$$E + \frac{2RT}{F} \ln m_{HCl} = E^\circ - \frac{2RT}{F} \ln \gamma_\pm \qquad (iii)$$

TABLE 2.6 MEAN MOLAL ACTIVITY COEFFICIENTS FOR HCl(aq) OBTAINED FROM GALVANIC MEASUREMENTS ON THE CELL $H_2(g)$, $HCl(m)$, $AgCl(s)$, Ag $(25°C)^a$

m_{HCl}	$\gamma_{\pm, HCl}$
0.0001	0.9891
0.0005	0.9752
0.002	0.9521
0.01	0.9048
0.05	0.8304
0.1	0.7964
0.5	0.7571
1.0	0.8090
2.0	1.009
4.0	1.762

^{a} Harned and Owen [7].

where $E°$ is the standard electric potential of reaction i, corresponding to $-\Delta G°$. The terms on the left side of equation iii are measured quantities. Extrapolation to infinite dilution, $m_{HCl} \rightarrow 0$, $\gamma_\pm \rightarrow 1$, yields $E°$. Then, $\gamma_\pm$ can be calculated for actual m_{HCl} values from equation iii. The value of $E°$ allows calculation of the chemical potential *difference*

$$\mu°_{H^+} + \mu°_{Cl^-} - \tfrac{1}{2}\mu°_{H_2(g)} - \mu°_{AgCl}$$

If three of the standard-state chemical potentials are known, or defined as a reference, the fourth can be calculated.

At 25°C the value of $E°$ is found to be 0.22240 V [7]. Some values of $\gamma_\pm$ for HCl(aq) calculated from E, $E°$, and m_{HCl} are shown in Table 2.6.

The standard-state chemical potential for aqueous HCl, $\mu°_{HCl(aq)}$, can be obtained from equilibrium measurements for HCl solutions and HCl vapor, together with values of γ_{HCl} from galvanic cells. (See, for example, the treatments by Klotz [8] and Denbigh [2]). The basic relationship to be used is Henry's law, in the form

$$\frac{P_{HCl}}{\gamma_\pm^2 m_{HCl}^2} = \exp\left[\frac{\mu°_{HCl(aq)} - \mu°_{HCl(g)}}{RT}\right] \qquad \text{(iv)}$$

If $\mu°_{HCl(g)}$ is known at 25°C and 1 atm pressure, $\mu°_{HCl(aq)}$ can be computed from measured pairs of P_{HCl} and m_{HCl}, the activity coefficient $\gamma_\pm$ having been obtained independently. For 4 molal HCl, P_{HCl} is 0.2395×10^{-4} atm [3]. The value of $\gamma_\pm$, HCl is 1.762 [6]. For the process

$$\tfrac{1}{2}H_2(g) + \tfrac{1}{2}Cl_2(g) = HCl(g)$$

$\Delta G°$ was found independently to be -22.77 kcal mol^{-1}. By definition $\mu_{Cl_2}° = \mu_{H_2}° = 0$ (elements in their standard states). Thus $\mu_{HCl(g)}° = -22.77$ kcal mol^{-1} or -95.27 kJ mol^{-1}. The calculated value of $\mu_{HCl(aq)}°$ is therefore -31.35 kcal mol^{-1} or -131.17 kJ mol^{-1}.

The individual standard chemical potentials $\mu_{H^+}°$ and $\mu_{Cl^-}°$ are not measurable, but they must sum to $\mu_{HCl(aq)}°$. Therefore

$$\mu_{HCl}° = \mu_{H^+}° + \mu_{Cl^-}° \tag{v}$$

If $\mu_{H^+}° \equiv 0$, as a convention for all temperature and pressures, then $\mu_{HCl}° = \mu_{Cl^-}°$. In this way, individual ion chemical potentials have been assigned. The experimental bases for establishing standard free energy changes $\Delta G°$ and thus the $\mu_i°$ for components and ions, are phase equilibria and reaction equilibria. Important phase equilibria include aqueous-solid, aqueous-gas, gas-solid, and membrane equilibria. Among the methods of great use are electrochemical measurements, colligative property measurements, calorimetry (for $\Delta H°$), determination of distribution and partition coefficients (Henry's law, Raoult's law), and determination of equilibrium constants for chemical reactions. We need to examine the general relationships between $\mu_i°$, $\Delta G°$, and other thermodynamic quantities, particularly the equilibrium constant.

2.10   CHEMICAL REACTIONS AND EQUILIBRIA

We have already stated the general condition for chemical equilibrium in an isothermal, isobaric system:

$$\sum_i \mu_i dn_i = 0$$

which was obtained from the basic condition for equilibrium in terms of the internal entropy production dS_{int}

$$dS_{int} = 0$$

We also have the result that, for a spontaneous or natural process,

$$dS_{int} > 0$$

or

$$\sum_i \mu_i dn_i < 0$$

Now we apply these results to the particular case of chemical reactions.

A single chemical reaction can be represented as

$$v_A A + v_B B + \cdots = v_C C + v_D D + \cdots \tag{216}$$

where the v's are the stoichiometric coefficients (relative molar numbers) of the reactants and products of the reaction and A, B, ..., C, D, ... stand for the chemical symbols (molecular weights) of the reactants and products. More concisely, the reaction can be written as

$$\sum_i v_i M_i = 0 \qquad (217)$$

where M_i is the molecular weight of component i and v_i is the stoichiometric coefficient of i. The v_i are positive for products and negative for reactants. The change in the Gibbs free energy resulting from changes in the number of moles of each component dn_i as a result of chemical reaction is

$$dG = \sum_i \mu_i dn_i \qquad (106)$$

where the summation is taken over all components and all phases.

We introduce a variable called the extent of the reaction ξ which is defined by

$$dn_i = v_i d\xi \qquad (218)$$

and

$$n_i = n_i^\circ + v_i \xi \qquad (219)$$

where n_i is the number of moles of i at any extent of reaction and n_i° is the number of moles of i under initial conditions ($\xi = 0$). The units of ξ are *moles*; the v_i are dimensionless numbers. Note that ξ and v_i are arbitrary; the product of the two is physically meaningful for a reaction in a defined system (specified chemical content in terms of moles).

The total free energy of the system in which the chemical reaction occurs is

$$G = \sum_i n_i \mu_i = \sum_i (n_i^\circ + v_i \xi)\mu_i \qquad (220)$$

From equation 106 the change in the Gibbs energy is, in terms of the extent of reaction,

$$dG = \sum_i v_i \mu_i d\xi \qquad (221)$$

The internal entropy production is then given by

$$dS_{int} = -\frac{dG}{T} = -\frac{1}{T}\sum_i v_i \mu_i d\xi \qquad (222)$$

The quantity $\sum_i v_i \mu_i$ is called the *free energy change of the reaction* ΔG by Lewis and Randall. The quantity $-\sum_i v_i \mu_i$ is called the affinity A by de Donder. Thus

$$\Delta G = \sum_i v_i \mu_i = -A \qquad (223)$$

From equation 221

$$\frac{dG}{d\xi} = \Delta G = -A \qquad (224)$$

The free energy of the reaction is the rate of change of the Gibbs free energy of the system with respect to the extent of the reaction. ΔG and A are *intensive-state* variables of a chemical reaction system.

From equations 222 and 223,

$$dS_{int} = -\frac{\Delta G}{T}\,d\xi = \frac{A}{T}\,d\xi \tag{225}$$

Therefore, at equilibrium for the reaction, $dS_{int} = 0$, $\Delta G = 0$, or $A = 0$; the rate of change of the Gibbs energy is zero, and the system free energy is a minimum at equilibrium. For a spontaneous reaction, $dS_{int} > 0$ and $\Delta G < 0$ or $A > 0$.

Summarizing, for a chemical reaction $\sum_i v_i M_i = 0$

at equilibrium: $\qquad dS_{int} = 0 \qquad \Delta G = \sum_i v_i \mu_i = 0 \tag{226}$

if spontaneous: $\qquad dS_{int} > 0 \qquad \Delta G = \sum_i v_i \mu_i < 0$

It is useful to distinguish immediately between chemical reactions involving pure substances only (solids and liquids) and chemical reactions involving mixtures (gases, liquid solutions, solid solutions). For the former type of reaction $\mu_i = \mu_i^\circ(P, T)$, and at given temperature and pressure application of equation 226 leads to a particularly simple result: $\Delta G = \sum_i v_i \mu_i^\circ$ is constant for any extent of reaction at fixed T and P. The reaction is either at equilibrium ($\Delta G = 0$) or spontaneous in one direction ($\Delta G < 0$). For $\Delta G > 0$, the reaction is impossible; the reverse reaction is then spontaneous.

Example 2.3.

What is the stability relationship for the reaction involving the oxides pyrolusite, magnetite, manganite, and hematite

$$2\,MnO_2(s) + 2\,Fe_3O_4(s) = Mn_2O_3(s) + 3\,Fe_2O_3(s)$$

at 25°C and 1 atm? All reactants and products are pure phases, so $\mu_i = \mu_i^\circ = \bar{G}_{f,i}$. Therefore

$$\Delta G = \sum_i v_i \mu_i = \sum_i v_i \mu_i^\circ = \sum_i v_i \bar{G}_{f,i}^\circ$$

The following free energy data are available:

	$\bar{G}_f^\circ$, kJ mol^{-1}
$MnO_2(s)$	-465.14
$Mn_2O_3(s)$	-881.07
$Fe_3O_4(s)$	-1012.57
$Fe_2O_3(s)$	-742.68

$$\Delta G = -2(-465.14) - 2(-1012.57) + 1(-881.07) + 3(-742.68)$$
$$\Delta G = -153.69 \text{ kJ mol}^{-1}$$

Since $\Delta G < 0$ for the reaction as written pyrolusite and magnetite are unstable with respect to manganite and hematite. There is no equilibrium constant for this process.

The Equilibrium Constant. For mixtures, the relationship between the Gibbs free energy of a reaction and the composition of the system is obtained by substituting the expression for the chemical potential in terms of the activity of a species i:

$$\mu_i = \mu_i^\circ + RT \ln\{i\} \tag{227}$$

into the expression for ΔG:

$$\Delta G = \sum_i v_i \mu_i \tag{223}$$

The direct result is

$$\Delta G = \sum_i v_i \mu_i^\circ + RT \sum_i v_i \ln\{i\} \tag{228}$$

or

$$\Delta G = \Delta G^\circ + RT \ln \prod_i \{i\}^{v_i} \tag{229}$$

where

$$\Delta G^\circ = \sum_i v_i \mu_i^\circ \tag{230}$$

is the *standard* Gibbs free energy of the reaction. Or, in terms of reaction 216, equation 229 becomes

$$\Delta G = \Delta G^\circ + RT \ln \frac{\{C\}^{v_C}\{D\}^{v_D} \cdots}{\{A\}^{v_A}\{B\}^{v_B} \cdots} \tag{230}$$

The reaction quotient Q is defined as

$$Q = \prod_i \{i\}^{v_i} = \frac{\{C\}^{v_C}\{D\}^{v_D} \cdots}{\{A\}^{v_A}\{B\}^{v_B} \cdots} \tag{231}$$

Then

$$\Delta G = \Delta G^\circ + RT \ln Q \tag{232}$$

At equilibrium $\Delta G = 0$ and the numerical value of Q becomes K, the equilibrium constant:

$$K \equiv Q_{eq} = \left(\frac{\{C\}^{v_C}\{D\}^{v_D} \cdots}{\{A\}^{v_A}\{B\}^{v_B} \cdots}\right)_{eq} \tag{233}$$

Then

$$\Delta G^\circ = -RT \ln K \tag{234}$$

This is a central relationship in the chemical thermodynamics of mixtures. Under any conditions

$$\Delta G = RT \ln \frac{Q}{K} \tag{235}$$

comparison of Q (actual composition) with the value of K (equilibrium composition) provides a test for equilibrium ($\Delta G = 0$).

If solutions do not behave ideally, the activity is not equal to the concentration. Then

$$\{i\} = [i]\gamma_i \tag{236}$$

where $[i]$ is concentration and γ_i is an activity coefficient. The chemical potential is then given by

$$\mu_i = \mu_i^\circ + RT \ln [i]\gamma_i \tag{237}$$

The reaction quotient then takes the form

$$Q = \prod_i [i]^{v_i} \prod_i \gamma_i^{v_i} \tag{238}$$

At equilibrium,

$$K = \left(\frac{[C]^{v_C}[D]^{v_D} \cdots}{[A]^{v_A}[B]^{v_B} \cdots} \times \frac{\gamma_C^{v_C}\gamma_D^{v_D} \cdots}{\gamma_A^{v_A}\gamma_B^{v_B} \cdots} \right)_{equil} \tag{239}$$

The activity coefficients are functions of the composition of the solution and become unity in the reference state.

Gibbs Energy of a System. To conclude our discussion of equilibrium and phases of variable composition we treat the variation of the Gibbs free energy of reacting mixtures with the extent of the reaction. Substitution of equation 227 into equation 200 yields

$$G = \sum_i (n_i^\circ + v_i \xi)(\mu_i^\circ + RT \ln \{i\}) \tag{240}$$

The Gibbs free energy versus ξ for reactions involving solutions and gas mixtures is characterized by a minimum value at the equilibrium composition. We have already seen that the minimum value of G corresponds to $\Delta G = 0$, ΔG being the derivative of G with respect to ξ. The $(n_i^\circ + v_i \xi)RT \ln \{i\}$ terms in equation 240 account for the free energy of mixing. For reactions between pure substances these terms are absent and there is no minimum in the G function. Figure 2.5 depicts generalized relationships between the G function, the Gibbs free energy of reaction, and the extent of reaction for a chemical reaction mixture. Point e corresponds to the equilibrium state for the reaction. Point f represents a nonequilibrium state from which the system may proceed to equilibrium spontaneously, the reaction proceeding from left to right as written.

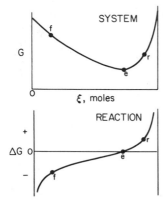

Figure 2.5.   Variation of the Gibbs free energy function G and the Gibbs free energy change of the reaction ΔG for a single reaction in a system of variable composition, for example, reactions in solution.

Point r represents a state in which the reaction is spontaneous in the reverse direction (right to left as written).

Example 2.4   Energetics of CO_2 Dissolution in Water

Describe the variations in the entropy, enthalpy, and Gibbs energy with extent of CO_2 dissolution for a two-phase system comprising a gas phase and an aqueous phase. Find the equilibrium state. Initially, a liter of gas at 1-atm total pressure contains 2×10^{-5} mol of CO_2. It is brought into contact with a liter of pure water. The dissolution process is

$$CO_2(g) = CO_2(aq) \tag{i}$$

The standard-state thermodynamic data at 25°C are:

Species	$\bar{H}_f^\circ$ (kcal mol^{-1})	$\bar{G}_f^\circ$ (kcal mol^{-1})	$\bar{S}^\circ$ (cal deg^{-1} mol^{-1})
$CO_2(g)$	-94.05	-94.26	51.06
$CO_2(aq)$	-98.69	-92.31	29.0

The Gibbs energy of the two-phase system is given by

$$G = n_{CO_2(g)}\mu_{CO_2(g)} + n_{CO_2(aq)}\mu_{CO_2(aq)} \tag{ii}$$

The mole fraction of the solvent water is essentially unity, so a term is not included for the water in ii. The gas phase and solution phase will be approximately ideal (why?), so the chemical potentials can be well-approximated by

$$\mu_{CO_2(g)} = \mu_{CO_2(g)}^\circ + RT \ln p_{CO_2} \tag{iii}$$

and

$$\mu_{CO_2(aq)} = \mu_{CH_2(aq)}^\circ + RT \ln m_{CO_2} \tag{iv}$$

with

$$m_{CO_2} \simeq [CO_2(aq)] = n_{CO_2(aq)}$$

for an aqueous volume of 1 liter.

The extent of reaction ξ is numerically equal to $n_{CO_2(aq)}$, and $n_{CO_2(g)} = 2 \times 10^{-5} - \xi$. The partial pressure of $CO_2(g)$ is calculated from the equation of state (ideal gas law):

$$p_{CO_2} V_g = n_{CO_2(g)} RT \qquad (v)$$

Thus G can be readily calculated as a function of the extent of reaction, ξ.

In similar fashion, the system enthalpy H for the dilute gas and solution is given by

$$H = n_{CO_2(g)} \bar{H}^\circ_{CO_2(g)} + n_{CO_2(aq)} \bar{H}^\circ_{CO_2(aq)} \qquad (vi)$$

and change in the quantity TS for the system can be obtained from the relationship

$$T\Delta S = \Delta H - \Delta G \qquad (vii)$$

The results for $G - G_0$, $H - H_0$, and $T(S - S_0)$ are shown in Figure 2.6. (G_0, H_0, and S_0 are values at $\xi = 0$.) The dissolution of CO_2 under these conditions is favored by a decrease in enthalpy and opposed by a decrease in entropy. The net effect is a decrease in G in proceeding from $\xi = 0$ to the equilibrium value, $\xi = 9.5 \times 10^{-6}$ mol. At equilibrium, $dG/d\xi = 0$. The dissolution of CO_2 in the system is accompanied by the release of 44 mcal of heat to the surroundings. The *decrease* in free energy driving dissolution, 8 mcal, corresponds to an internal entropy *increase* of 0.027 mcal deg^{-1}.

The standard free energy change ΔG° for the dissolution process is $\mu^\circ_{CO_2(aq)} - \mu^\circ_{CO_2(g)}$, or 1.95 kcal mol^{-1}. We note that, even though $\Delta G^\circ > 0$, the actual free energy change ΔG is negative for the dissolution process between $\xi = 0$ and $\xi = 9.5 \times 10^{-6}$ mol. The actual free energy change is, according to equation 232,

$$\Delta G = \Delta G^\circ + RT \ln \frac{[CO_2(aq)]}{p_{CO_2}} \qquad (viii)$$

At equilibrium, $\Delta G = 0$, and the equilibrium constant, obtained from $\Delta G^\circ = -RT \ln K$, is $K = 3.72 \times 10^{-2}$ mol liter^{-1} atm^{-1}. At equilibrium,

$$[CO_2(aq)]/p_{CO_2} = 3.72 \times 10^{-2}$$

for the temperature and total pressure considered. Note that an infinite set of $CO_2(aq)$ concentrations and partial pressures would satisfy the equilibrium constant. The mole constraints, $\sum n_{CO_2} = 2 \times 10^{-5}$, and the volume constraints define a unique equilibrium state.

It is of interest to note that neutral CO_2 distributes itself approximately equally between the gas phase and aqueous phase at ordinary temperatures.

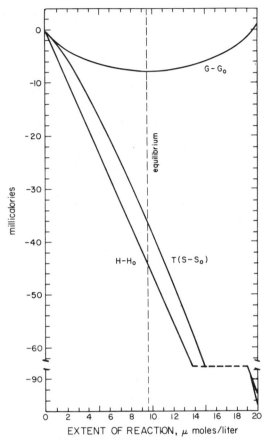

Figure 2.6. Gibbs function, system enthalpy, and system entropy variations with the extent of reaction for the dissolution of gaseous CO_2 in water: $CO_2(g) = CO_2(aq)$ at 25°C in a two-phase system. The total moles of CO_2 is 2×10^{-5} mol, the gas-phase volume is 1 liter, and the water volume is 1 liter. The extent of reaction is given by the moles of CO_2 dissolved. At equilibrium in the system, $[CO_2(aq)] = 9.5 \, \mu M$. The dissolution is favored by $dH/d\xi$ and opposed by $dS/d\xi$.

Here, $[CO_2(aq)]/[CO_2(g)] = 9.5 \times 10^{-6}/1.05 \times 10^{-5} \simeq 1$. ($[CO_2(g)]$ represents mol liter^{-1} of gaseous CO_2.)

Example 2.5.   Reaction of CO_2 to Yield Bicarbonate

Describe $CO_2(aq)$ reacting in solution to yield HCO_3^- in terms of the Gibbs energy function, the Gibbs free energy of the reaction, and the related enthalpy and entropy terms. Assume the concentration of dissolved CO_2 is $1 \times 10^{-5} \, M$.
   The reaction (25°C, 1 atm) is

$$CO_2(aq) + H_2O = HCO_3^- + H^+$$

The initial ($\xi = 0$) composition per liter of solution is (neglecting the ionization of water)

$$n_{CO_2(aq)} = 1 \times 10^{-5} \text{ mol}$$
$$n_{H_2O} = 55.4 \text{ mol}$$
$$n_{HCO_3} = 0 \text{ mol}$$
$$n_{H^+} = 0 \text{ mol}$$

We omit computation of the hydration of $CO_2(aq)$ and we treat CO_2 as nonvolatile for purposes of the example. The stoichiometric coefficients are: $CO_2(aq)$, -1; H_2O, -1; HCO_3^-, $+1$; and H^+, $+1$. The standard partial molar free energies of formation, standard partial molar enthalpies of formation, and standard partial molar entropies for the reactants and products are as follows (from Latimer [9]):

Species	$\bar{G}_f^\circ$ (kcal mol^{-1})	$\bar{H}_j^\circ$ (kcal mol^{-1})	$\bar{S}^\circ$ (cal mol^{-1} deg^{-1})
$CO_2(aq)$	-92.31	-98.69	29.0
$H_2O(l)$	-56.690	-68.317	16.716
$HCO_3^-(aq)$	-140.31	-165.18	22.7
$H^+(aq)$	0.0	0.0	0.0

The extent of reaction ξ for this system is numerically equal to the number of moles of HCO_3^- in solution. When $\xi = 1 \times 10^{-5}$ mol, the reactants are completely converted to products. The Gibbs free energy of the solution may be computed by application of equation 240, which gives

$$G = n_{CO_2}(\mu_{CO_2}^\circ + RT \ln \{CO_2\}) + n_{H_2O}(\mu_{H_2O}^\circ + RT \ln \{H_2O\})$$
$$+ n_{HCO_3^-}(\mu_{HCO_3^-}^\circ + RT \ln \{HCO_3^-\}) + n_{H^+}(\mu_{H^+}^\circ + RT \ln \{H^+\})$$

We have $n_{CO_2} = 1 \times 10^{-5} - \xi$, $n_{H_2O} = 55.4 - \xi$, $n_{HCO_3} = \xi$, and $n_{H^+} = \xi$. Substitution of various values for ξ leads to the computed Gibbs function for the reaction system shown in Figure 2.7, where G is plotted with reference to the initial free energy G_0. Analogous computations have been made for the system enthalpy, $H - H_0$, and the product of the system entropy and the absolute temperature, $T(S - S_0)$. These functions also are shown in Figure 2.7. Recalling that for a system of constant temperature and pressure the equilibrium condition in terms of the system free energy is $dG/d\xi = 0$, we find that the equilibrium extent of reaction is 1.8 μmol liter^{-1}, that is, 1.8 μM HCO_3^-. The $CO_2(aq)$ concentration at equilibrium is $1 \times 10^{-5} - \xi_{\text{equil}}$, or 8.2 μM.

In proceeding from the initial state to the equilibrium state ($dG/d\xi < 0$ for $0 < \xi < \xi_{\text{equil}}$) the closed constant-temperature system absorbs 3.3 mcal of heat per liter of solution from its surroundings (e.g., a heat reservoir). The entropy of the closed system increases by 0.019 mcal deg^{-1}, and the Gibbs free energy of the closed system decreases by 2.4 mcal liter^{-1}. In terms of the *total isolated system* consisting of the *solution plus surroundings* the equilibrium state

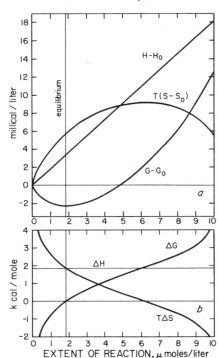

Figure 2.7. (a) Gibbs function, system enthalpy, and system entropy variations for the reaction $CO_2(aq) + H_2O = HCO_3^- + H$ at 25°C in a closed aqueous system at various extents of reaction. The total concentration is 1×10^{-5} M. $CO_2(aq)$ is assumed to be nonvolatile. The extent of reaction is numerically equal to the concentration of HCO_3^- in solution. Equilibrium is at the minimum value of G and corresponds to 1.8 μM HCO_3^-. (b) Gibbs free energy change ΔG, enthalpy change ΔH, and entropy change expressed as $T\Delta S$ for the reaction $CO_2(aq) + H_2O = HCO_3^- + H^+$. At equilibrium $\Delta G = 0$ and $\Delta H = T\Delta S$. ΔH is independent of composition for the dilute solution involved.

represents a state of maximum entropy increase from the initial state to 0.008 mcal deg^{-1}. Note that the maximum increase in the entropy of the closed system itself ($\xi \simeq 6 \mu M$) does *not* represent equilibrium: The entropy of the system itself continues to increase for compositions beyond the equilibrium state. Note also that there is an *increase* in enthalpy in going from the initial state to the equilibrium state. The driving force is an increase in system entropy and a consequent decrease in system free energy. We may summarize the changes between the initial state and the equilibrium state for a liter of solution as follows:

heat absorbed: $\qquad\qquad\qquad\qquad\qquad\qquad\qquad q = H_{eq} - H_0 = 3.3$ mcal

entropy change: $\qquad\qquad\qquad\qquad\qquad\qquad\qquad S_{eq} - S_0 = 0.019$ mcal deg^{-1}

environmental entropy change: $\qquad (S_{eq} - S_0)_e = \dfrac{q}{T} = \dfrac{H_{eq} - H_0}{T} = 0.011$ mcal deg^{-1}

internal entropy change: $\qquad\qquad (S_{eq} - S_0)_i = \dfrac{-(G - G_0)}{T} = 0.008$ mcal deg^{-1}

If we were to start with pure products—HCO_3^- and H^+ in aqueous solution— the approach to equilibrium would be favored both by a decrease in the enthalpy and an increase in the entropy: The decrease in free energy in going from products to the equilibrium state is 15.1 mcal liter^{-1}.

Having looked at the system in terms of the total Gibbs energy, enthalpy, and entropy we turn to the Gibbs free energy of the reaction. Applied to this example, we have

$$\Delta G = \mu^{\circ}_{HCO_3} + \mu^{\circ}_{H^+} - \mu^{\circ}_{CO_2} - \mu^{\circ}_{H_2O} + RT \ln \frac{\{HCO_3^-\}\{H^+\}}{\{CO_2\}\{H_2O\}}$$

or, for a liter of dilute solution, with $\{H_2O\} = 1$ and activity equal to concentrations,

$$\Delta G = \mu^{\circ}_{HCO_3} + \mu^{\circ}_{H^+} - \mu^{\circ}_{CO_2} - \mu^{\circ}_{H_2O} + RT \ln \frac{(\xi)(\xi)}{(1 \times 10^{-5} - \xi)}$$

ΔG is computed as a function of ξ upon substitution of the $\bar{G}^{\circ}_f$ for the reactants and products. The result is plotted in Figure 2.7. Also plotted are ΔH and $T\Delta S$ for the reaction. The equilibrium state is at ΔG equal to zero. As $\Delta G = dG/d\xi$, we note the expected correspondence of the minimum in the $G - G_0$ function and the zero point of ΔG. Recalling the definition of G, we note that $\Delta H = T\Delta S$ at equilibrium (1.8 μM HCO_3^-). The standard free energy change of the reaction ΔG° is

$$\Delta G^{\circ} = \sum_i v_i \mu^{\circ}_i = \sum_i v_i \bar{G}^{\circ}_{f,i}$$

$$= 1(-140.31) + 1(0.0) - 1(-92.31) - 1(-56.690)$$
$$= 8.69 \text{ kcal mol}^{-1}$$

The equilibrium constant can be computed from equation 234:

$$\ln K = \frac{-\Delta G^{\circ}}{RT}$$

$$\log K = \frac{-8.69}{1.364} = -6.36$$

The equilibrium composition ξ is readily computed from

$$K = \frac{\{HCO_3^-\}\{H^+\}}{\{CO_2\}\{H_2O\}} \simeq \frac{[HCO_3^-][H^+]}{[CO_2]}$$

or

$$K \simeq \frac{(\xi)(\xi)}{1 \times 10^{-5} - \xi}$$

The result is of course

$$\xi = 1.8 \times 10^{-6} \ M$$

Driving Force for Chemical Change

The driving force change is the Gibbs free energy ΔG. The free energy change is composed of an enthalpy and an entropy contribution:

$$\Delta G = \Delta H - T\Delta S \tag{68}$$

or, for standard-state conditions,

$$\Delta G° = \Delta H° - T\Delta S° \tag{241}$$

The driving force for a reaction (stability of products with respect to reactants) can be the result of a negative ΔH, a positive ΔS, or both. The condition for equilibrium is that $\Delta G = \Delta H - T\Delta S = 0$ or

$$\Delta H = T\Delta S$$

The magnitude and direction of the driving force for a reaction depend upon the magnitude and sign of the enthalpy change, the magnitude and sign of the entropy change, and the temperature. For $\Delta H < 0$ and $\Delta S < 0$, as well as for $\Delta H > 0$ and $\Delta S > 0$, the possibility of spontaneous reaction depends upon the temperature.

The driving force for the ionization of aqueous CO_2 (Example 2.5) was found to be the result of an increase in entropy: $T\Delta S > \Delta H$ at 25°C. Under standard-state conditions at 25°C, the process

$$CO_2(g) = CO_2(aq)$$

is favored by an enthalpy decrease: $\Delta H° = -4.64$ kcal mol^{-1}. However, the decrease in entropy results in $T\Delta S° = -6.59$, and the standard free energy change is $+1.94$ kcal mol^{-1}.

The small stability difference in favor of calcite with respect to aragonite:

$$CaCO_3(\text{aragonite}) = CaCO_3(\text{calcite}) \qquad \Delta G° = -0.25 \text{ kcal mol}^{-1}$$

is associated with an entropy increase of 1 cal mol^{-1} deg^{-1}.

The precipitation of barium sulfate

$$Ba^{2+} + SO_4^{2-} = BaSO_4(s)$$

is characterized by the following thermodynamic properties (kcal mol^{-1}, 25°C):

$$\Delta G° = -12.1 \qquad \Delta H° = -4.6 \qquad T\Delta S° = +7.5$$

For the precipitation of ferric phosphate

$$Fe^{3+} + PO_4^{3-} \pm FePO_4(s)$$

$$\Delta G° = -24.4 \qquad \Delta H° = +18.7 \qquad T\Delta S° = 43.1$$

The reaction is endothermic. The precipitation of ferric hydroxide is favored both by $T\Delta S° = 30.1$ kcal mol^{-1} and by $\Delta H° = -20.6$ kcal mol^{-1}. with a resulting $\Delta G°$ of -50.7 kcal mol^{-1} (25°C). Large standard entropy changes in the precipitation of a metal ion are associated with an increase in the randomness of water (decreased aquation of metal ions).

Ion association reactions and chelation reactions of aqueous metal ions are generally characterized by significant entropy increases (decreased orientation of solvent molecules and configurational entropy). For example, the ion-pair reaction

$$Co(NH_3)_5H_2O^{3+} + SO_4^{2-} = Co(NH_3)_5(H_2O)SO_4^{1+}$$

has $\Delta H° = 0$ kcal mol^{-1} and $\Delta S° = +16.4$ cal mol^{-Z} deg^{-1} at 25°C. The association reaction

$$Cu^{2+} + P_3O_{10}^{5-} - CuP_3O_{10}^{3-}$$

has $\Delta H = +4.9$ kcal mol^{-1} and $\Delta S = +59$ cal mol^{-1} deg^{-1} with a resulting ΔG of -12.48 kcal mol^{-1} (20°C, 0.1 M constant ionic medium).

The chelation reactions of Ca^{2+} and Mg^{2+} with ethylenediaminetetraacetate (Y^{4-}) have the following thermodynamic properties (0.1 M constant ionic medium, 25°C):

$$Ca^{2+} + Y^{4-} = CaY^{2-}$$

$\Delta H = -8$	$\Delta S = 22$	$\Delta G = -14.2$	$\log K = 10.4$

$$Mg^{2+} + Y^{4-} = MgY^{2-}$$

$\Delta H = 2$	$\Delta S = 47$	$\Delta G = -11.8$	$\log K = 8.6$

The binding of aqueous protons to the anions of weak acids is generally favored by sizable entropy increases. For example, the reaction

$$SO_4^{2-} + H^+ = HSO_4^-$$

has $\Delta S° = 26.3$ cal deg^{-1} mol^{-1} and $\Delta H° = 5.2$ kcal mol^{-1}.

It is clear that the enthalpy and entropy changes of reactions in aqueous systems vary greatly. The enthalpy change does not serve as a criterion of spontaneous reaction. Free energy changes provide the only general description of the driving force of reactions.

Example 2.6.   Solution of NH₃ in Water

From enthalpy and entropy data [9] compute the equilibrium constant for the solution of gaseous ammonia in water at 25°C. The standard molar enthalpies of formation and the standard molar entropies are

Species	$\bar{H}_f°$ (kcal mol^{-1})	$\bar{S}°$ (cal deg^{-1} mol^{-1})
$NH_3(g)$	-11.04	46.01
$NH_3(aq)$	-19.32	26.3

For the process

$$NH_3(g) = NH_3(aq)$$

$$\Delta H^\circ = -19.32 - (-11.04) = -8.28 \text{ kcal mol}^{-1}$$

$$\Delta S^\circ = 26.3 - 46.01 = -19.7 \text{ cal mol}^{-1} \text{ deg}^{-1}$$

$$T\Delta S^\circ = 298.16(-19.7) = -5.88 \text{ kcal mol}^{-1}$$

The dissolution process is favored by a negative enthalpy change but opposed by a decrease in entropy. Decrease in entropy upon dissolution is characteristic of uncharged solutes. The standard free energy change is

$$\Delta G^\circ = \Delta H^\circ - T\Delta S^\circ = -8.28 - (-5.88)$$
$$= -2.40 \text{ kcal mol}^{-1}$$

The equilibrium constant is computed from

$$-RT \ln K = \Delta G^\circ$$

or

$$\log K = \frac{-\Delta G^\circ}{1.364} = \frac{2.40}{1.364} = 1.76$$

From this result we may estimate the equilibrium partial pressure of $NH_3(g)$ over a solution containing 5×10^{-9} M $NH_3(aq)$, as has been found for many acid rains with a high NH_4^+ concentration. Assuming ideal behavior,

$$\frac{[NH_3(aq)]}{p_{NH_3}} = K = 57.5$$

$$p_{NH_3} = \frac{5 \times 10^{-9}}{56} = 8.7 \times 10^{-11} \text{ atm}$$

2.11   INFLUENCE OF TEMPERATURE ON EQUILIBRIA

The basic relationship for the influence of temperature is the variation of the chemical potential with temperature at constant pressure

$$\left(\frac{\partial \mu_i}{\partial T}\right)_P = -\bar{S}_i \tag{178}$$

or

$$\left(\frac{\partial \mu_i/T}{\partial T}\right)_P = -\frac{\bar{H}_i}{T^2} \tag{129}$$

For a chemical reaction $\sum_i M_i \nu_i = 0$, $\Delta H^\circ = \sum_i \nu_i \bar{H}_i^\circ$ is the standard enthalpy change of the reaction, and the variation of the standard Gibbs free energy of the reaction is given by (see equation 82)

$$\left(\frac{\partial(\Delta G^\circ/T)}{\partial T}\right)_P = -\frac{\Delta H^\circ}{T^2} \tag{242}$$

The equilibrium constant for the reaction K is related to the standard free energy by $\Delta G^\circ = -RT \ln K$. Therefore, at constant pressure,

$$\frac{d \ln K}{dT} = \frac{\Delta H^\circ}{RT^2} \tag{243}$$

which is the van't Hoff equation.

The variations of enthalpy and entropy with temperature are dependent on the heat capacity:

$$H_2 - H_1 - \int_{T_1}^{T_2} C_p dT \tag{244}$$

$$S_2 - S_1 - \int_{T_1}^{T_2} C_p d \ln T \tag{245}$$

The Equilibrium Constant. Table 2.7 summarizes the pertinent relationships for describing the influence of temperature on the equilibrium constant of a chemical reaction or phase equilibrium. The thermodynamic information of interest includes ΔH° for the reaction, ΔC_p° for the reaction, and the variation of ΔC_p° with temperature. For a number of reactions of interest in aquatic systems heat capacity data are limited or unavailable. However, sufficient enthalpy data are available for most reactions of interest to provide at least a partial assessment of temperature influences. Enthalpy data are available in Latimer's text [9], the compilation of Robie, Hemingway and Fisher [10], in NBS Circular 500 [11], and to a limited extent in *Stability Constants of Metal–Ion Complexes* [12]. Nancollas [13] has summarized and discussed enthalpy data for ion pairs and complexes. Enthalpy data are obtained by direct calorimetry and by measurements of equilibrium properties of chemical reactions, electrochemical cells, and transfer processes.

Equations (4), (7) and (10) of Table 2.7 suggest that a plot of the logarithm of the equilibrium constant (or a representative equilibrium activity) of a reaction versus the reciprocal of absolute temperature can yield information concerning ΔH°. For many reactions ΔC_p° is close to zero and ΔH° is essentially independent of temperature, and a linear plot of log K versus $1/T$ is obtained over an appreciable temperature range. The equilibrium constant can then be computed readily by the simple relationship of (3) in Table 2.7. When ΔC_p° is constant over a range of temperature, (6) of Table 2.7 can be used to compute the equilibrium constant-temperature coefficient. The most general case, in which ΔC_p° is a function of temperature, is described by (10) of Table 2.7.

TABLE 2.7 INFLUENCE OF TEMPERATURE ON THE EQUILIBRIUM CONSTANT

The basic relationships are

$$\frac{d \ln K}{dT} = \frac{\Delta H^\circ}{RT^2} \tag{1}$$

$$\ln \frac{K_2}{K_1} = \int_{T_1}^{T_2} \frac{\Delta H^\circ}{RT^2} \, dT \tag{2}$$

When ΔH° is independent of temperature,

$$\ln \frac{K_2}{K_1} = \frac{\Delta H^\circ}{R} \left(\frac{1}{T_1} - \frac{1}{T_2} \right) \tag{3}$$

or

$$\ln K = -\frac{\Delta H^\circ}{RT} + \text{constant} \tag{4}$$

When the heat capacity of the reaction, ΔC_p°, is independent of temperature,

$$\Delta H_2^\circ = \Delta H_1^\circ + \Delta C_p^\circ (T_2 - T_1) \tag{5}$$

Integration of (1) then yields

$$\ln \frac{K_2}{K_1} = \frac{\Delta H_1^\circ}{R} \left(\frac{1}{T_1} - \frac{1}{T_2} \right) + \frac{\Delta C_p^\circ}{R} \left(\frac{T_1}{T_2} - 1 - \ln \frac{T_1}{T_2} \right) \tag{6}$$

or

$$\ln K = B - \frac{\Delta H_0}{RT} + \frac{\Delta C_p^\circ}{R} \ln T \tag{7}$$

where ΔH_0 and B are constants
When ΔC_p° is a function of temperature,
If the heat capacity of each reactant and product is given by an expression of the form

$$C_p^\circ = a_i + b_i T + c_i T^2 \tag{8}$$

then the heat capacity of the reaction is given by

$$\frac{d \Delta H^\circ}{dT} = \Delta C_p^\circ = \Delta a + \Delta b T + \Delta c T^2 \tag{9}$$

Integration of (9) and (1) yields

$$\ln K = B - \frac{\Delta H_0}{RT} + \frac{\Delta a}{R} \ln T + \frac{\Delta b}{2R} T + \frac{\Delta c}{6R} T^2 \tag{10}$$

where ΔH_0 and B are constants and $\Delta a = \sum_i v_i a_i$, and so on

Example 2.7.   Effect of Temperature on Aqueous Equilibria

To illustrate the influence of temperature on the equilibrium of reactions and processes in aquatic systems we have selected representative examples. Figure 2.8 contains plots of log K versus the reciprocal of absolute temperature for seven different aqueous equilibria.

1   O_2 solubility in water
2   Vaporization of water
3   Ionization of water
4   Ionization of CO_2(aq) in water
5   Dissolution of solid $CaCO_3$ in water
6   CO_2 solubility in water
7   Ionization of acetic acid in water

The temperature range covered is from 0 to 50°C. An arbitrary constant (specific for each system) has been added to log K in order to obtain a convenient plot.

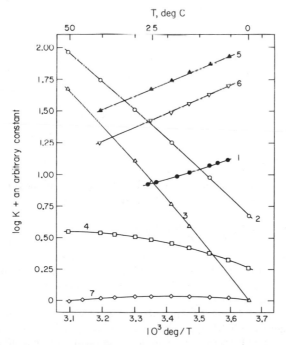

Figure 2.8.   Plots of the log of the equilibrium constant (mol liter^{-1}, atm) versus reciprocal of absolute temperature for seven aqueous physicochemical processes. The processes and the values of the arbitrary constant C are: (1) $O_2(g) = O_2(aq)$, $C = 3.8$; (2) $H_2O(l) = H_2O(g)$, $C = 0$; (3) $H_2O(l) = H^+ + OH^-$, $C = +14.93$; (4) $CO_2(aq) + H_2O = HCO_3^- + H^+$, $C = +6.85$; (5) $CaCO_3(s) = Ca^{+2} + CO_3^-$, $C = 10.01$; (6) $CO_2(g) = CO_2(aq)$, $C = +2.89$; (7) $CH_3COOH(aq) = CH_3COO^- + H^+$, $C = +4.7807$.

The equilibrium constants refer to the conventional standard states and reference states for solids, liquids, gases, and solutions. Solution activities are on the infinite dilution scale. With the exception of the data for O_2 solubility [14], the equilibrium data have been taken from the compilations already referred to. Processes 2, 5, and 6 have essentially constant $\Delta H°$ values over the temperature range considered, while processes 3, 4, and 7 show a significant variation of $\Delta H°$ with temperature. In the case of 7, the ionization of acetic acid, there is a reversal in sign of $\Delta H°$ as the temperature varies. Process 1 shows a slight $\Delta H°$ variation with temperature.

From the slope of the plot for oxygen solubility at 25°C, one obtains $\Delta H° \simeq -3.9$ kcal mol^{-1}. From the enthalpy data in Latimer [9], the computed value is $\Delta H° = \bar{H}°_{f,O_2(aq)} - \bar{H}°_{f,O_2(g)} = -3.8 - 0 = -3.8$ kcal mol^{-1}. The heat capacity of $O_2(g)$ is a known function of T, but there appear to be no $\bar{C}°_P$ data for $O_2(aq)$ available.

The equation describing the ion product of water K_W (process 3) is [17] $\log K_W = -4470.99/T + 6.0875 - 0.01706T$ [compare (10), Table 2.7].

From the slope of the straight line fitted to the data for the solubility constant of calcite (process 5) we may obtain an estimate for $\Delta H°$ for the solution reaction. The slope is $+1050$ deg. From (4) in Table 2.7, the slope is equal to $-\Delta H°/2.303R$. Therefore we have

$$- \frac{\Delta H°}{2.303 \times 1.987 \text{ cal mol}^{-1} \text{ deg}^{-1}} = 1050 \text{ deg}$$

and

$$\Delta H° = -4.8 \text{ kcal mol}^{-1}$$

Activity Coefficients and Temperature. If $\bar{H}_i$ is the partial molal enthalpy in the *actual* solution and $\bar{H}°_i$ is the standard-state value, the variation of the conventional infinite dilution scale activity coefficient γ_i with temperature is described by

$$\frac{d \ln \gamma_i}{dT} = \frac{\bar{H}°_i - \bar{H}_i}{RT^2} \tag{246}$$

For electrolytes, the difference between $\bar{H}°_i$ and $\bar{H}_i$ becomes appreciable at concentrations greater than about 0.1 molal. In seawater and similar solutions, activity coefficients of ions are significantly influenced by temperature.

In summarizing our brief discussion we want to emphasize that aqueous equilibria may be shifted to the right or to the left by an increase in temperature. The direction and magnitude of the shift depend upon the sign and magnitude of the heat of the reaction as a function of temperature, and the magnitude is affected by the salt content of the water, ΔH being a function of the ionic medium. From (1) of Table 2.7 the *approximate* magnitude of the change in $\log K$ per degree is $0.0029 \Delta H$ at 0°C, 0.0025 at 25°C, and 0.0021 at 50°C. In the examples given in Figure 2.8 we have illustrated a range of possibilities for the effect of temperature on equilibrium in the range from 0 to 50°C.

2.12 INFLUENCE OF PRESSURE ON CHEMICAL EQUILIBRIA

The variation of the chemical potential of a solution species with pressure at constant temperature is given by

$$\left(\frac{\partial \mu_i}{\partial P}\right)_T = \bar{V}_i \qquad (127)$$

where $\bar{V}_i$ is the partial molal volume ($cm^3 \ mol^{-1}$) under actual conditions. If a variable-pressure standard state is adopted, then $\mu_i^\circ = \mu_i^\circ(T, P)$ and

$$\left(\frac{\partial \mu_i^\circ}{\partial P}\right)_T = \bar{V}_i^\circ \qquad (247)$$

where $\bar{V}_i^\circ$ is the standard-state partial molal volume. For a chemical reaction $\sum_i \nu_i M_i = 0, \Delta G^\circ = \sum_i \nu_i \mu_i^\circ$ and $\Delta V^\circ = \sum_i \nu_i \bar{V}_i^\circ$. Using $\Delta G^\circ = -RT \ln K$, where K is the conventional thermodynamic equilibrium constant, the variation of K with pressure is obtained:

$$\left(\frac{\partial \ln K}{\partial P}\right)_T = -\frac{\Delta V^\circ}{RT} \qquad (248)$$

If a fixed-pressure standard state is adopted, as discussed by Rock [15], then $\mu_i^\circ = \mu_i^\circ(T)$. According to this convention the thermodynamic constant K is independent of pressure, $(\partial \ln K/\partial P)_T = 0$, and the effect of pressure on μ_i, hence on equilibria, is described by variation of the activity of the reference state Γ_i with pressure. For a solution, the activity of a solute can then be expressed as

$$\{i\}^\bullet = m_i \gamma_i^\bullet \Gamma_i \qquad (249)$$

where m_i is the molal concentration and $\gamma_i^\bullet$ is an activity coefficient describing nonideal behavior at a fixed (standard) pressure. The chemical potential must then be given by

$$\mu_i = \mu_i^\circ(T) + RT \ln \Gamma_i + RT \ln m_i + RT \ln \gamma_i^\bullet \qquad (250)$$

which is to be compared with the corresponding expression for a variable standard state:

$$\mu_i = \mu_i^\circ(T, P) + RT \ln m_i + RT \ln \gamma_i \qquad (251)$$

The activity coefficient γ_i varies with pressure, while $\gamma_i^\bullet$ does not, by definition. On the variable-pressure standard-state basis, the activity is given by

$$\{i\} = m_i \gamma_i$$

The actual variation of μ_i, hence equilibrium composition with pressure, is of course independent of the choice of convention.

Table 2.8 summarizes the essential relationships for pressure effects on chemical equilibrium for the variable-pressure standard-state convention. Note that these relationships can apply to any consistent choice of standard partial

TABLE 2.8 INFLUENCE OF PRESSURE ON CHEMICAL EQUILIBRIUM IN AQUEOUS SYSTEMS

Basic relationships

$$\left(\frac{\partial \mu_i}{\partial P}\right)_T = \bar{V}_i \tag{1a}$$

$$\left(\frac{\partial \mu_i^\circ}{\partial P}\right)_T = \bar{V}_i^\circ \tag{1b}$$

where V_i and V_i° are the partial molar volumes of i under actual conditions and under defined standard-state conditions, respectively,

$$\bar{k}_i^\circ = -\left(\frac{\partial \bar{V}_i^\circ}{\partial P}\right)_T \tag{2}$$

where $\bar{k}_i^\circ$ is the standard partial molar compressibility, the rate of change of molar volume with pressure,

$$\Delta V = \sum v_i \bar{V}_i \tag{3a}$$

$$\Delta V^\circ = \sum v_i \bar{V}_i^* \tag{3b}$$

where $\Delta \bar{V}$ and $\Delta \bar{V}^\circ$ are the volume changes of reaction under actual and under defined standard-state conditions, respectively

$$\left(\frac{\partial \ln K}{\partial P}\right)_T = -\frac{\Delta V^\circ}{RT} \tag{4}$$

where K is the equilibrium constant.

When ΔV° is independent of pressure

$$\ln \frac{K_P}{K_1} = -\frac{\Delta V^\circ (P - 1)}{RT} \tag{5}$$

When Δk° is independent of pressure

$$\ln \frac{K_P}{K_1} = -\frac{1}{RT} [\Delta V^\circ (P - 1) - \tfrac{1}{2}\Delta k^\circ (P - 1)^2] \tag{6}$$

where $\Delta k^\circ = \sum_i v_i \bar{k}_i^\circ$

When Δk° is a function of pressure

$$RT \ln \frac{K_P}{K_1} = -\Delta V^\circ (P - 1) + \Delta k^\circ (B + 1)(P - 1) - (B + 1)^2 \ln \left(\frac{B + P}{B + 1}\right) \tag{6a}$$

where B is independent of pressure and depends upon the temperature [20]

74

For aqueous solutions specifically

$$\mu_i = \mu_i^\circ + RT \ln \gamma_i m_i \tag{7}$$

$$\left(\frac{\partial \ln K}{\partial P}\right)_{T,m} = -\frac{\Delta V^\circ}{RT} \tag{8}$$

$$\left(\frac{\partial \ln \gamma_i}{\partial P}\right)_{T,m} = \frac{\bar{V}_i - \bar{V}_i^\circ}{RT} \tag{9}$$

$$\left(\frac{\partial \ln K'}{\partial P}\right)_{T,m} = \frac{\Delta V}{RT} \tag{10}$$

where $K' = \prod_i m_i^{\gamma_i} = K/\prod_i \gamma_i^{\gamma_i}$, the concentration product at equilibrium

---

molar volumes, for example, one for which an ionic medium such as seawater is adopted as the solute reference state. For detailed discussion of applications to seawater see, for example, Millero [16], Whitfield [17], and Disteche [18]. A comprehensive treatment of the physicochemical effects of pressure has been provided by Hamann [19]. Calculations of the effect of pressure on ionic equilibria in solutions have been detailed by Owen and Brinkley [20]. The reader should consult the references mentioned for details of measurement techniques and for compilations of data.

Example 2.8.   Effect of Pressure on the Chemical Potential of Water

Consider the effect of pressure on the chemical potential of water. What is the approximate chemical potential of liquid water at 25°C and 1000 atm? We will obtain an approximate answer by assuming that the partial molar compressibility is independent of pressure. Applying (1b) and (2) of Table 2.8. we obtain

$$\mu_P - \mu_1^\circ = \int_1^P \bar{V}_{H_2O} dP$$

$$\simeq \bar{V}_{H_2O}^\circ(P - 1) - \tfrac{1}{2}\bar{k}_{H_2O}^\circ(P - 1)^2$$

The molar volume of water at 1 atm is 18.02 cm^3 mol^{-1}, and $\bar{k}_{H_2O}^\circ$ is 8.2 × 10^{-4} cm^3 mol^{-1} atm^{-1} [20]. Therefore

$$\mu_P - \mu_1 \simeq 17,595 \text{ cm}^3 \text{ atm mol}^{-1}$$

$$\simeq 0.44 \text{ kcal mol}^{-1}$$

The chemical potential, or molar Gibbs free energy, is approximately 0.44 kcal mol^{-1} more positive at 1000 atm than at 1 atm. Recalling that for any substance we can write

$$\mu_i = \mu_i^\circ + RT \ln\{i\}$$

it is a matter of convention whether we now increase the value of the standard-state chemical potential (1 atm) by 0.44 kcal mol^{-1} or define the activity of the water at 1000 atm to be

$$\{H_2O\} \simeq \ln^{-1}\left(\frac{0.44}{0.001987 \times 298.15}\right)$$

$$\simeq 2.1$$

compared to unity at 1 atm. A more accurate answer (corresponding to an activity of 2.06) is obtained by graphical integration of

$$\int_1^{1000} \overline{V}_{H_2O}\,dP$$

Note that, for the fixed-pressure standard state, $\Gamma_{H_2O} = 2.1$.

Owen and Brinkley [20] computed the effect of pressure on the equilibrium constant for the ionization of water, both for pure water and for 0.725 M sodium chloride as ionic medium. For pure water they obtained, in part,

P (bars)*	$K_{W,P}/K_{W,1}$		
	5°C	25°C	45°C
1	1	1	1
200	1.24	1.202	1.16
1000	2.8	2.358	2.0

The results obtained for 0.725 M NaCl were not greatly different from these. At 25°C and 1000 atm the ratio K_P/K_1 was 2.1 instead of 2.358.

The standard partial molar volumes of the participants in the solution (ionization) of calcite at 25°C

$$CaCO_3(s) = Ca^{2+} + CO_3^{2-}$$

are, according to Owen and Brinkley [20],

Species	$\overline{V}°$ (cm^3 mol^{-1})
CaCO$_3$(s)	36.9
Ca^{2+}†	−17.7
CO$_3^{2-}$	−3.7

* 1 atm = 1.01325 bar.

† The partial molar volumes of ions tabulated by Owen and Brinkley are based on the convention $\overline{V}°_{H^+} = 0$.

Neglecting the compressibility of calcite we may estimate the change in chemical potential of the solid at a pressure of, say, 1000 atm. We have

$$\mu_P - \mu_1 = \int_1^P \overline{V}_{CaCO_3(s)} \, dP \simeq \overline{V}_{CaCO_3(s)}(P - 1)$$

$$\simeq 36.9 \times 999 \simeq 36{,}860 \text{ cm}^3 \text{ atm mol}^{-1} \simeq 0.89 \text{ kcal mol}^{-1}$$

This increase in chemical potential corresponds to an activity (relative to the 1-atm condition) of about 4.5.

The equilibrium constant for the solution of $CaCO_3(s)$ at elevated pressure may be estimated by applying the appropriate equations in Table 2.8. The volume change of the reaction is

$$\Delta V° = -17.7 - 3.7 - 36.9 = -58.3 \text{ cm}^3 \text{ mol}^{-1}$$

At a pressure of 1000 atm the ratio of the calcite solubility equilibrium constant at that pressure to that at 1 atm is

$$\frac{K_P}{K_1} = 8.1$$

The volume change of reaction is negative for most ionization processes. Therefore the effect of increased pressure is generally to increase the extent of ionization; for example, Owen and Brinkley [20] report the following computed pressure effects in pure water medium for 1000 atm and 25°C.

Reaction	K_P/K_1
$CaSO_4(s) = Ca^{2+} + SO_4^{2-}$	5.8
$H_2O + CO_2(aq) = HCO_3^- + H^+$	3.2
$HCO_3 = CO_3^{2-} + H^+$	2.7
$CH_3COOH = CH_3COO^- + H^+$	1.4

Since the partial molar volumes of dissolved species are a function of the ionic strength of the medium, it is expected that the pressure effects upon equilibrium constants will be different in pure water and in a medium such as seawater (or other high salt solution). Some of the results of Owen and Brinkley [20] will serve to indicate the extent of the differences for calcite solubility:

	$K_{calcite, P}/K_{calcite, 1}$ at 25°C	
P (bars)	Pure Water	0.725 M NaCl
1	1	1
500	3.2	2.8
1000	8.1	6.7

The basis for these differences is seen in (6a), (9), and (10) of Table 2.8. On the basis of these results supersaturation of seawater would be expected to decrease significantly with depth. A complete analysis of supersaturation and equilibrium requires consideration of other reactions or other species in addition to the simple ions (e.g., ion pairs).

From this brief discussion of pressure effects on equilibria we can conclude that only in the deep oceans (and to a lesser degree in aquifers under high pressures) are such effects of great significance.

2.13   THERMODYNAMIC DATA

The basic data needed for equilibrium calculations and for ascertaining the direction of spontaneous reaction are the partial molar free energies of formation of substances under well-defined conditions. The partial molar free energy of formation $\bar{G}_f^\circ$ is the free energy change accompanying the formation of a substance from the elements in their standard states. For ions in aqueous solutions the free energies of formation are based on the arbitrary assignment of zero values to $\bar{G}_f^\circ$, $\bar{H}_f^\circ$, and $\bar{S}^\circ$ for the aqueous hydrogen ion. The partial molar free energy of formation at a given temperature is equal to the standard chemical potential at the same temperature and 1 atm pressure:

$$\mu_i^\circ(T, 1 \text{ atm}) = \bar{G}_{f,i}^\circ(T, 1 \text{ atm})$$

Entropies are referenced to $\bar{S}^\circ = 0$ at absolute zero temperature (see Section 2.4).

A rather complete thermodynamic description of each substance involved in a chemical reaction, allowing its chemical behavior to be described over a wide range of conditions, includes the following: $\bar{G}_f^\circ$, $\bar{H}_f^\circ$, $\bar{S}^\circ$, C_P° (the heat capacity), and $\bar{V}^\circ$ (the partial molar volume). These thermodynamic data permit calculation or estimation of equilibrium constants for reactions under a variety of conditions. Extensive, critical compilations of thermodynamic data for substances are available. Three sources are Latimer's text [9], the National Bureau of Standards Circular 500 [11], and the summary of thermodynamic data by Robie, Hemingway, and Fisher [10].

Thermodynamic data have been obtained in a variety of ways: calorimetry, chemical measurements on reactions at equilibrium, electromotive force measurements of galvanic cells, calculations of the entropy based on the third law of thermodynamics, and estimations of ion entropies. Thus, through application of the basic relationship

$$\Delta G^\circ = \Delta H^\circ - T\Delta S^\circ$$

standard energies, hence equilibrium constants, can be calculated for many chemical reactions which have not been studied directly. On the other hand, many enthalpy and entropy values have been obtained from chemical equilibrium measurements rather than from thermal data.

Stability constants of metal–ion complexes, dissociation constants of acids, solubility products of solids, and equilibrium constants or potentials for oxidation–reduction reactions have been compiled. Extensive collections of equilibrium constant data available in *Stability Constants of Metal–Ion Complexes*, compiled by Sillén and Martell [12], and in Supplement No. 1 to that volume [21]. A critical and unique compilation of metal complex equilibrium constants has been prepared by Martell and Smith [22]. The data comprise equilibrium constants (free energies), enthalpies, and entropies for solution, complexes, and solids. The Appendix to this book contains free energies, enthalpies, and entropies of important species.

Generally speaking, the equilibrium constants found in the chemical literature are defined in terms of either the infinite dilution reference state or the constant ionic medium reference state. In terms of the Gibbs free energies of reaction

$$\Delta G^\circ \text{ (infinite dilution scale)} = \Delta G' \text{ (constant ionic medium scale)} + \text{constant}$$

where the constant reflects the differences in the activity coefficients of the reactants and products as defined on each scale.

2.14 CHEMICAL KINETICS

Chemical thermodynamics, or energetics, provides the means to determine the *direction* and the possible *extent* of chemical change. Spontaneous change flows down a chemical potential gradient, and so for a reaction $\sum_i v_i M_i = 0$, when the free energy, $G = \sum_i n_i \mu_i$, decreases with extent of reaction $\xi : dG/d\xi < 0$, and that reaction is possible. At equilibrium, $dG/d\xi = 0$, and the composition is obtained from the equilibrium constant K. But a spontaneous process need not be a rapid process. Chemical kinetics, or dynamics, is concerned with the *rate* at which chemical change takes place, that is, with $d\xi/dt$. There is no simple relation between the energetics of overall chemical reactions and their rates. The kinetic properties of reactions, the rate laws, are learned from experiment and interpreted in terms of mechanisms consisting of elementary steps. However, there are important theoretical relationships between the thermodynamic properties of certain elementary processes, namely, those leading to the formation of *activated* intermediate species in the passage from reactants to products and the rate of reaction. Enthalpies, entropies, and free energies of elementary processes of activation play an important part in the understanding reaction rates in solution. Furthermore, for reactions which are reversible a kinetic description of the equilibrium state can provide valuable information about the rates of opposing reactions in solution.

Rate-Controlling Steps

Reactions between molecules or ions in solution (homogeneous reactions) and between species at interfaces (heterogeneous reactions) can be thought of as involving three different steps: (1) transport of the reactants to each other or to an interface, generally by diffusion; (2) chemical transformation of the reactants; and (3) transport of products away from one another. The slow step will control the overall reaction rate. Some reactions are encounter- or diffusion-controlled, others are chemically controlled. The chemical transformation process may consist of a single elementary step, but more often it is complex, that is, it consists of a number of simple or elementary steps. These steps together make up the mechanism for the overall reaction.

The mechanism for an overall chemical reaction in solution may include a combination of simple unimolecular or bimolecular (rarely termolecular) steps taking place in series (consecutive reactions) or in parallel (concurrent reactions). Each step may be reversible (appreciable rates in the forward and reverse directions) or essentially irreversible (negligible reverse rate). The slowest step in a sequence of elementary steps is rate-controlling for that sequence, that is, determines the overall rate in that sequence. The overall rate for elementary processes in parallel is the sum of the rates for the individual paths. For *opposing* (reversible) elementary or overall reactions the *net* reaction rate is the difference between the forward and backward rates.

Thus the experimentally determined rate law for a fixed temperature relates the rate of an overall chemical reaction to the solution (or surface) concentrations of reactants, products, and other species in the system. The rate law reflects the rates of the slow (rate-limiting) elementary steps for consecutive reactions and the total flux through parallel paths. The slow elementary steps can be controlled, by rates of diffusion or by rates of chemical change, frequently the latter.

Elementary Reactions

The *molecularity* of a reaction is defined as the number of molecules of reactant participating in a simple reaction constituting an elementary step. Unimolecular reactions involve a single molecule, for example,

$$A \longrightarrow B \tag{252}$$

$$A \longrightarrow B + C \tag{253}$$

In bimolecular reactions two identical or different species combine to give a product or products. Examples are association reactions

$$A + B \longrightarrow AB \tag{254}$$

$$A + A \longrightarrow A_2 \tag{255}$$

and exchange reactions

$$A + B \longrightarrow C + D \tag{256}$$

$$A + A \longrightarrow C + D \tag{257}$$

Termolecular reactions are unusual, involving the encounter of three species to yield products. Examples include

$$A + B + C \longrightarrow X + \cdots \tag{258}$$

and

$$A + A + B \longrightarrow X + \cdots \tag{259}$$

and so on. The species involved in these elementary molecular reactions may in fact be molecules, ions, free radicals, or atoms. It is important to note that the molecularity of elementary reactions identifies the order of the rate expression; for complex reactions, the order does not necessarily reveal molecularity. Order is not synonymous with molecularity.

Rate Expressions for Elementary Reactions

The rate of a unimolecular reaction

$$A \xrightarrow{k} B \tag{260}$$

can be expressed by the differential equation

$$-\frac{d[A]}{dt} = k[A] \tag{261}$$

where $[A]$ is the molar (M) concentration of A and k is called the first-order *rate constant*. The disappearance of A is said to be first-order in $[A]$. The rate has units of $M \sec^{-1}$ if k has units of $\sec^{-1}$. A similar expression describes the rate of radioactive decay.

For the bimolecular reaction

$$A + B \xrightarrow{k} \text{product} \tag{262}$$

the rate of change of $[A]$ is

$$-\frac{d[A]}{dt} = k[A][B] \tag{263}$$

and k is a *second-order rate constant* with units, for examples of $M^{-1} \sec^{-1}$. The disappearance of A is first-order in both $[A]$ and $[B]$. For the bimolecular reaction

$$A + A \xrightarrow{k} \text{product} \tag{264}$$

the corresponding rate equation is

$$-\frac{d[A]}{dt} = k[A]^2 \tag{265}$$

The disappearance of A is described as second-order in [A]. Similar rate expressions can be written for termolecular reactions, for example,

$$-\frac{d[A]}{dt} = k[A][B][C] \tag{266}$$

Complex Reaction Mechanisms

Mechanisms for overall chemical processes are constructs or models. Proposed elementary steps with rate expressions of well-defined forms are combined so as to predict kinetic behavior and thus provide an explanation for experimentally obtained rate laws. A few representative ways in which elementary reactions can combine to give complex reactions are presented.

Opposing Reactions   The opposing reactions

$$A \underset{k_{-1}}{\overset{k_1}{\rightleftharpoons}} B \tag{267}$$

represent a combination of two elementary unimolecular reactions. (This reaction was introduced in Section 2.1 to characterize differences between closed-system and open-system models for the chemical composition of aquatic systems.) The rate expression at constant temperature is

$$-\frac{d[A]}{dt} = k_1[A] - k_{-1}[B] \tag{268}$$

At *equilibrium*, in a closed reaction system,

$$\frac{d[A]}{dt} = \frac{d[B]}{dt} = 0$$

and

$$\frac{[B]}{[A]} = \frac{k_1}{k_{-1}}$$

The equilibrium constant for the opposing reactions, K, is [B]/[A]. Thus $K = k_1/k_{-1}$, and the ratio of the *rate constants* for the opposing elementary reactions is the *equilibrium constant*.

The opposing elementary bimolecular processes

$$A + B \overset{k}{\rightleftharpoons} C + D \tag{269}$$

are described by the rate expression

$$-\frac{d[A]}{dt} = -k_1[A][B] + k_{-1}[C][D] \tag{270}$$

and at equilibrium

$$\frac{d[A]}{dt} = \frac{d[B]}{dt} = \frac{d[C]}{dt} = \frac{d[D]}{dt} = 0$$

and

$$\frac{k_1}{k_{-1}} = \frac{[C][D]}{[A][B]} = K \tag{271}$$

Consecutive Opposing Reactions   As an example consider a sequence of opposing elementary reactions such as

$$A + B \; \underset{k_{-1}}{\overset{k_1}{\rightleftharpoons}} \; C \tag{272}$$

$$C \; \underset{k_{-2}}{\overset{k_2}{\rightleftharpoons}} \; D \tag{273}$$

At complete equilibrium *each* elementary reaction and the reverse reaction must occur at the same rate. This requirement is known as the principle of microscopic reversibility [23]. As a consequence,

$$\frac{[C]}{[A][B]} = \frac{k_1}{k_{-1}} = K_1 \tag{274}$$

and

$$\frac{[D]}{[C]} = \frac{k_2}{k_{-2}} = K_2 \tag{275}$$

so that

$$\frac{[D]}{[A][B]} = K_{12} = \frac{k_1 k_2}{k_{-1} k_{-2}} \tag{276}$$

In general, the principle of microscopic reversibility leads to the result

$$K_{1n} = \prod_{i}^{n} \left(\frac{k_i}{k_{-i}} \right) \tag{277}$$

for n consecutive opposing elementary processes.

The net rate of change of [A] for this mechanism is given by

$$-\frac{d[A]}{dt} = k_1[A][B] - k_{-1}[C] \tag{278}$$

Consecutive Irreversible Reactions   A sequence of elementary unimolecular reactions, for example,

$$A \xrightarrow{\ k_1\ } B \xrightarrow{\ k_2\ } C \tag{279}$$

is described kinetically by the equations

$$\frac{d[A]}{dt} = -k_1[A] \tag{280}$$

$$\frac{d[B]}{dt} = k_1[A] - k_2[B] \tag{281}$$

$$\frac{d[C]}{dt} = k_2[B] \tag{282}$$

A sequence of bimolecular and unimolecular reactions, for example,

$$2A \xrightarrow{\ k_1\ } B \xrightarrow{\ k_2\ } C \tag{283}$$

would be described by the equations

$$\frac{d[A]}{dt} = -k_1[A]^2 \tag{284}$$

$$\frac{d[B]}{dt} = k_1[A]^2 - k_2[B] \tag{285}$$

$$\frac{d[C]}{dt} = k_2[B] \tag{286}$$

Benson [24], Frost and Pearson [25], Laidler [23], and Amdur and Hammes [26] discuss a variety of combinations of consecutive first- and second-order reactions.

Concurrent Reactions   A simple case is a mechanism in which a single reactant gives different products

$$A \xrightarrow{\ k_1\ } B$$

$$A \xrightarrow{\ k_2\ } C$$

$$A \xrightarrow{\ k_3\ } D$$

For which the rate equation is

$$\frac{-d[A]}{dt} = (k_1 + k_2 + k_3)[A] \tag{287}$$

Another simple combination of unimolecular reactions involves different reactants to yield a single product, for example,

$$A \xrightarrow{\ k_1\ } C \tag{288}$$

$$B \xrightarrow{\ k_2\ } C \tag{289}$$

with

$$-\frac{d[A]}{dt} = k_1[A] \tag{290}$$

$$-\frac{d[B]}{dt} = k_2[B] \tag{291}$$

and

$$\frac{d[C]}{dt} = -\frac{d[A]}{dt} - \frac{d[B]}{dt} \tag{292}$$

A combination of concurrent bimolecular and unimolecular steps to form a common product, for example,

$$A + B \xrightarrow{\ k_1\ } C \tag{293}$$

$$A \xrightarrow{\ k_2\ } C \tag{294}$$

yields the rate equation

$$\frac{d[C]}{dt} = k_1[A][B] + k_2[A] \tag{295}$$

Concurrent Consecutive Mechanisms.   There are essentially unlimited combinations of elementary reactions to yield complex mechanisms for overall reactions. Representative patterns for solutions are presented by Frost and Pearson [25], King [27], Benson [28], Sykes [29], and Edwards [30].

A fairly simple mechanism of interest in aqueous systems is the case of concurrent sequences of opposing reactions, such as

$$A + B \underset{k_{-1}}{\overset{k_1}{\rightleftharpoons}} C \underset{k_{-2}}{\overset{k_2}{\rightleftharpoons}} E \tag{296}$$

$$A + D \underset{k_{-3}}{\overset{k_3}{\rightleftharpoons}} E \tag{297}$$

Consecutive Opposing and Irreversible Reactions.   The opposing reactions

$$A \underset{k_{-1}}{\overset{k_1}{\rightleftharpoons}} B \tag{298}$$

followed by the irreversible reaction

$$B \xrightarrow{\ k_2\ } C \tag{299}$$

represent a mechanism encountered frequently in overall reactions. If the opposing reactions are rapid and the second reaction slow, a simple result is obtained by the approximation $d[B]/dt \simeq 0$, that is, that the intermediate B changes concentration very slowly during the reaction. This approximation is known as the stationary-state assumption. Then

$$\frac{d[B]}{dt} = k_1[A] - k_{-1}[B] - k_2[B] = 0 \tag{300}$$

and

$$[B] = \frac{k_1[A]}{k_{-1} + k_2} \tag{301}$$

The rate equation is then

$$\frac{d[C]}{dt} = k_2[B] \tag{302}$$

or

$$\frac{d[C]}{dt} = \frac{k_2 k_1}{k_{-1} + k_2}[A] \tag{303}$$

If $k_{-1} \gg k_2$,

$$\frac{d[C]}{dt} = \frac{k_2 k_1}{k_{-1}}[A] = k_2 K_1[A] \tag{304}$$

where $K_1 = k_1/k_{-1}$ for the equilibrium between A and B (from the principle of microscopic reversibility).

Rate Expressions for Overall Reactions

If a reaction is a single elementary reaction, the form of the rate expression observed experimentally will be precisely that expected from the stoichiometry of the elementary reaction.

For the reversible ionization of acids in water, for example, carbonic acid, with the overall reaction being

$$H_2CO_3 + H_2O = HCO_3^- + H_3O^+ \tag{305}$$

where H_3O^+ denotes the hydrated proton, experimental evidence supports a one-step mechanism

$$H_2CO_3 + H_2O \underset{k_b}{\overset{k_f}{\rightleftharpoons}} HCO_3^- + H_3O^+ \tag{306}$$

with the rate law

$$\frac{d[HCO_3^-]}{dt} = k_f[H_2CO_3][H_2O] - k_b[HCO_3^-][H_3O^+] \tag{307}$$

Experiments at 25°C result in value of $k_f[H_2O] \cong 8 \times 10^6$ sec^{-1}, $k_b = 4.7 \times 10^{10}$ M^{-1} sec^{-1}, and an equilibrium constant

$$K[H_2O] = \frac{[HCO_3^-][H^+]}{[H_2CO_3]} \simeq 1.7 \times 10^{-4} \tag{308}$$

(The concentration of water is essentially constant in dilute solutions, and is generally incorporated into the equilibrium constant.)

Although a number of important aqueous reactions, such as dissociations, hydrations, substitutions, and ion-pair formation reactions, consists of single-step mechanisms, the bulk of solution and heterogeneous reactions take place by complex mechanisms, and the overall stoichiometric equation for the reaction generally will *not* predict the form of the observed rate equation. For example, the rate of decomposition of nitrous acid in water (which appears to be a significant kinetic process in acid rain formation)

$$3\,HNO_2 + H_2O = H_3O^+ + NO_3^- + 2\,NO(g) + H_2O \tag{309}$$

is described by the expression

$$v = \frac{d[NO_3^-]}{dt} = \frac{k_f[HNO_2]^4}{P_{NO}^2} - k_b[H_3O^+][NO_3^-][HNO_2] \tag{310}$$

where k_f and k_b are rate constants, P_{NO} is the nitric oxide partial pressure, and v is the net rate of reaction, or reaction velocity. The form of the rate law shows that the reaction is reversible, the first term describing a forward velocity and the second term a backward velocity [27].

The observed *orders* of the several concentration terms indicate the complexity of the reaction: The forward rate is fourth-order in $[HNO_2]$ and inverse second-order in P_{NO}; the backward rate is first-order in $[HNO_2]$. At equilibrium, $v = v_f - v_b = 0$, where v_f and v_b are the forward rate and backward rates, respectively. Thus

$$k_f\frac{[HNO_2]^4}{P_{AO}^2} - k_b[H_3O^+][NO_3^-][HNO_2] = 0 \tag{311}$$

and

$$\frac{k_f}{k_b} = \frac{[H_3O^+][NO_3^-]P_{NO}^2}{[HNO_2]^3} = K \tag{312}$$

The following mechanism is consistent with the observed rate law [27]:

$$2\,HNO_2 \underset{k_{-1}}{\overset{k_1}{\rightleftharpoons}} NO_2 + NO + H_2O \tag{313}$$

$$2\,NO_2 \underset{k_{-2}}{\overset{k_2}{\rightleftharpoons}} N_2O_4 \tag{314}$$

$$N_2O_4 + 2\,H_2O \underset{k_{-3}}{\overset{k_3}{\rightleftharpoons}} H_3O^+ + NO_3^- + HNO_2 \tag{315}$$

If the third step is slow, hence rate-determining,

$$v_b = k_{-3}[H_3O^+][NO_3^-][NHO_2] \tag{316}$$

and

$$v_f = k_3[N_2O_4] \tag{317}$$

If the first two steps are at equilibrium,

$$[N_2O_4] = \left(\frac{k_1}{k_{-1}}\right)^2 \frac{k_2}{k_{-2}} \frac{[HNO_2]^4}{P_{NO}^2} \tag{318}$$

so that

$$v_f = k_3\left(\frac{k_1}{k_{-1}}\right)^2 \frac{k_2}{k_{-2}} \frac{[HNO_2]^4}{P_{NO}^2} \tag{319}$$

Therefore, if the mechanism is the correct one,

$$k_f = k_3\left(\frac{k_1}{k_{-1}}\right)^2 \frac{k_2}{k_{-2}} = k_3 K_1^2 K_2 \tag{320}$$

and

$$k_b = k_{-3} \tag{321}$$

This example illustrates that the order of a complex rate expression is not revealed by overall stoichiometry but depends on the sequence of rapid and slow elementary reactions in the actual mechanism. The proposed mechanism involves an intermediate species which is not represented in the stoichiometry. It also shows that the forward and backward velocity expressions for a reversible path lead to an equilibrium constant for the overall reaction. This result is an important general property of correct forward and reverse rate laws for reversible reaction paths in an overall reaction, where the reaction mechanism applies at equilibrium and far from it (see e.g., Benson [24] and Laidler [23].)

A reaction of considerable importance in nature is the hydration of aqueous carbon dioxide

$$CO_2(aq) + 2H_2O = HCO_3^- + H_3O^+ \tag{322}$$

an experimentally derived rate expression for which is

$$v = -\frac{d[CO_2]}{dt} = (k_{f1} + k_{f2}[OH^-])[CO_2] - (k_{b1} + k_{b2}[H^+])[HCO_3^-] \tag{323}$$

A mechanism consistent with the rate expression consists of the concurrent reversible reactions

$$CO_2 + H_2O \underset{k_{-1}}{\overset{k_1}{\rightleftarrows}} H_2CO_3 \tag{324}$$

$$H_2CO_3 + H_2O \underset{k_{-2}}{\overset{k_2}{\rightleftarrows}} HCO_3^- + H_3O^+ \tag{325}$$

$$CO_2 + OH^- \underset{k_{-3}}{\overset{k_3}{\rightleftarrows}} HCO_3^- \tag{326}$$

The first and third steps are slow at room temperature, while the second is extremely rapid. The relative importance of the third step to the overall hydration rate increases with increasing pH in the range of natural waters.

A form of rate expression which requires additive terms involving concentrations of reactant, product, and other species in the denominator is encountered for many reactions. For example, the oxidation of iodide by ferric iron [29, 31] in acid solution

$$2I^- + 2Fe^{3+} = I_2 + 2Fe^{2+} \tag{327}$$

is described by a rate expression of the form

$$v = -\frac{d[Fe^{3+}]}{dt} = \frac{a[I^-]^2[Fe^{3+}]}{b[Fe^{2+}]/[Fe^{3+}] + 1} \tag{328}$$

A mechanism involving two reversible steps followed by the rate-limiting step is consistent with the rate law

$$I^- + Fe^{3+} \underset{k_{-1}}{\overset{k_1}{\rightleftharpoons}} FeI^{2+} \tag{329}$$

$$I^- + FeI^{2+} \underset{k_{-2}}{\overset{k_2}{\rightleftharpoons}} Fe^{2+}I_2^- \tag{330}$$

$$I_2^- + Fe^{3+} \xrightarrow{k_3} I_2 + Fe^{2+} \tag{331}$$

The term $[Fe^{2+}]/[Fe^{3+}]$ in the denominator of the rate expression reflects the reverse reactions preceding the rate-limiting step. Equation 328 is obtained from the mechanism by means of steady-state assumptions.

Order of Reaction

For a general stoichiometric reaction

$$v_A A + v_B B + \cdots = v_P P + \cdots \tag{332}$$

or

$$\sum_i v_i M_i = 0$$

the rate v is given by

$$v = \frac{1}{v_A}\frac{d[A]}{dt} = \frac{1}{v_B}\frac{d[B]}{dt} = \frac{1}{v_P}\frac{d[P]}{dt} = \frac{1}{v_i}\frac{d[M_i]}{dt} \tag{333}$$

where the stoichiometric coefficient v_i is negative for reactants and positive for products. Often the observed rate law takes the simple form

$$v = k[A]^{v_A}[B]^{v_B}\cdots[P]^{v_P}\cdots[X]^{v_X}\cdots \tag{334}$$

where the exponents are determined by experiment and may be positive or negative integers or fractions. The species X whose concentration appears in the rate law need not be a reactant or product in the overall stoichiometric reaction. The reaction is γ_A-order in [A], γ_B-order in [B], and γ_X-order in [X], and so on. The total order of the rate law is said to be $\gamma = \sum_i \gamma_i$, where γ_i is the exponent of species i.

For example, the rapid oxidation of iodide by hypochlorite in alkaline solution

$$\text{I}^- + \text{OCl}^- = \text{OI}^- + \text{Cl}^- \tag{335}$$

has the rate law

$$v = \frac{d[\text{OI}]}{dt} = k\,\frac{[\text{I}^-][\text{OCl}^-]}{[\text{OH}^-]} \tag{336}$$

The rate is first-order in $[\text{I}^-]$, first-order in $[\text{OCl}^-]$, and inverse first-order in $[\text{OH}^-]$. The inverse dependence on hydroxide suggests a rapid equilibrium preceding a slow step, namely,

$$\text{OCl}^- + \text{H}_2\text{O} \; \underset{}{\overset{K}{\rightleftharpoons}} \; \text{HOCl} + \text{OH}^- \tag{337}$$

General Character of Rate Laws

In general, rate expressions observed or deduced from proposed mechanisms may consist of additive terms of the kind in equation 295 (for concurrent mechanisms), subtractive terms (opposing or reversible overall reactions), or quotients of additive and subtractive terms (mechanisms which may comprise concurrent paths, each of which may include rapid and slow reversible or irreversible steps). Thus the rate law for a reaction with a complex mechanism should be represented by a general expression of the form

$$v = \frac{\sum_j a_j \prod_{i,j} [X_i]^{\gamma_{ij}}}{\sum_m b_m \prod_{i,m} [X_i]^{\gamma_{im}}} \tag{338}$$

where $[X_i]$ represents the concentration of any species (reactant, product, or intermediate) and the coefficients a_j and b_m are constant terms, individual rate constants, products, or quotients. While this expression is not meant to exhaust the possibilities for rate expressions, it should serve to suggest the variety of expressions to be encountered. Rate laws of the relatively simple form

$$v = k \prod_i [X_i]^{\gamma_i} \tag{339}$$

are commonly dealt with in simple kinetics, frequently giving readily integrated equations useful in extracting rate constants from experimental data and in describing the time course of reactions.

Integrated Rate Laws

A few cases of readily integrated rate laws will be mentioned, for example, the reaction

$$\text{A} \xrightarrow{\;k\;} \text{B} + \cdots \tag{340}$$

If the rate law is found to be

$$v = -\frac{d[A]}{dt} = k[A] \tag{341}$$

($\nu_A = -1, \gamma_A = 1, [X_i] = [A]$), then an integrated expression for this first-order rate expression is

$$[A] = [A]_0 e^{-kt} \tag{342}$$

where $[A]_0$ is the initial concentration. The half-time of reaction, $t_{1/2}$, is $\ln 2/k$.
    For a reaction with a second-order rate expression of the form

$$-\frac{d[A]}{dt} = k[A]^2 \tag{343}$$

an integrated expression is

$$\frac{1}{[A]} - \frac{1}{[A]_0} = kt \tag{344}$$

and a half-time is $1/k[A]_0$.
    For a reversible overall reaction

$$A \underset{k_b}{\overset{k_f}{\rightleftharpoons}} B \tag{345}$$

whose rate law is found to be

$$v = v_f - v_b = -\frac{d[A]}{dt} = k_f[A] - k_b[B] \tag{346}$$

and if only A is present initially,

$$[A]_0 - [A] = [B] \tag{347}$$

the integrated expression is

$$\frac{(k_f + k_b)[A] - k_b[A]_0}{k_f[A]_0} = \exp - (k_f + k_b)t \tag{348}$$

The equilibrium value of $[A]$, $[A]_e$, is readily found from $v_f = v_b$ and the stoichiometry $[A_0] - [A] = [B]$. The result is

$$[A]_e = \frac{k_b}{k_f + k_b} [A]_0 \tag{349}$$

and

$$\frac{[A] - [A]_e}{[A]_0 - [A]_e} = e^{-(k_f + k_b)t} \tag{350}$$

The approach to equilibrium is a first-order process with apparent rate constant $k_f + k_b$.

Reaction Variable and Degree of Advancement   It is sometimes convenient to express reaction rates in terms of a concentration variable χ which describes the concentration *change* in time t.

For example, returning to the first-order reaction

$$A \xrightarrow{\quad k \quad} B$$

$$\text{at time } 0: [A] = [A]_0, [B] = 0$$

$$\text{at time } t: [A] = [A]_0 - \chi, [B] = \chi$$

and the integrated expression becomes

$$[B] = \chi = [A]_0(1 - e^{-kt}) \tag{351}$$

For the reaction

$$A \underset{k_b}{\overset{k_f}{\rightleftharpoons}} B \tag{352}$$

$\chi = [B]$ if $[B] = 0$ at $t = 0$, and the rate expression is

$$v = \frac{d\chi}{dt} = k_f([A]_0 - \chi) - k_b\chi \tag{353}$$

At equilibrium, $\chi = \chi_e$ and the integrated expression obtained is

$$\frac{\chi_e - \chi}{\chi_e} = e^{-(k_f + k_b)t} \tag{354}$$

We may note that the kinetic reaction variable χ is related to the degree of advancement ξ. For constant volume V the rate expression for a reaction, $\sum_i v_i M_i = 0$, may be written
$= 0$, may be written

$$v = \frac{1}{v_i}\frac{d[M_i]}{dt} = \frac{d\chi}{dt} = \frac{1}{V}\frac{1}{v_i}\frac{dn_i}{dt} = \frac{1}{V}\frac{d\xi}{dt} \tag{355}$$

so that the rate of change of the degree of advancement with time is proportional to the reaction velocity.

Zero-Order Reactions   Some reaction rates are zero-order in the concentrations of a reactant; that is, the reaction rate is independent of that reactant. For a process

$$A \xrightarrow{\quad k \quad} P + \cdots$$

where the rate law is

$$\frac{-d[A]}{dt} = k$$

the rate is constant (k may include other reaction species at constant concentrations) and the half-time for reaction is $t_{1/2} = [A]_0/2k$.

Pseudo Order of Reaction

If certain species concentrations $[X_i]$ are kept constant over the course of a reaction, the factors $[X_i]^{y_i}$ remain constant and become incorporated in the rate constant. Thus, if the rate law is

$$v = -\frac{d[A]}{dt} = k[A][B]^2[C]$$

and the concentrations $[B]$ and $[C]$ remain constant (through being in great excess, replenished by exchange, etc.), then the apparent rate law is

$$v = k'[A]$$

and the reaction is *pseudo first order* in A. The use of excess concentrations to reduce the total order of a reaction is known as the method of isolation.

Example 2.9.   Rates of Reaction and Approach to Equilibrium for the Hydration of Carbon Dioxide

The hydration of $CO_2(aq)$ follows two paths, as discussed previously. In pure water the rate data indicate that the dominant path consists of the reaction sequence

$$CO_2(aq) + H_2O \underset{k_{-1}}{\overset{k_1}{\rightleftharpoons}} H_2CO_3 \qquad \text{(i)}$$

$$H_2CO_3 \underset{k_{-2}}{\overset{k_2}{\rightleftharpoons}} HCO_3^- + H^+ \qquad \text{(ii)}$$

The values of the rate constants (25°C) are $k_1 = 3.0 \times 10^{-2}$ sec^{-1}, $k_{-1} = 11.9$ sec^{-1}, $k_2 = 8 \times 10^6$ sec^{-1}, and $k_{-2} = 4.7 \times 10^{10}$ M^{-1} sec^{-1} [32]. Note that k_1 is a pseudo-first-order rate constant, the concentration of water being essentially constant in dilute reaction solutions. The rate constant values for reaction ii are much greater than those for reaction i. Reaction i is the slow step, hence is rate-controlling. Reaction ii can be considered to be at equilibrium. The two reactions can be combined to describe the overall hydration process:

$$CO_2(aq) + H_2O \underset{k_b}{\overset{k_f}{\rightleftharpoons}} HCO_3^- + H^+ \qquad \text{(iii)}$$

with the rate constants $k_f = 3.0 \times 10^{-2}$ (equal to k_1 for the slow step) and $k_b = 7.0 \times 10^4$ M^- sec^{-1} (equal to $k_{-2}k_{-1}/k_2$).

The conditions chosen will be the same as for Example 2.5, in which the hydration was examined thermodynamically. Initially, $[CO_2(aq)] = 1 \times 10^{-5}$ M. The system is closed, so no CO_2 is lost to or gained from the atmosphere. We neglect contributions of H^+ from $H_2O = H^+ + OH^-$.

The reversible reaction is of the type

$$A \underset{k_b}{\overset{k_f}{\rightleftharpoons}} B + C \qquad \text{(iv)}$$

for which an integrated solution has been obtained (see, e.g., Benson [28], Frost and Pearson [25], or Capellos and Bielski [33]). The solution for $[B]_0 = [C]_0 = 0$, with $\chi_e = [B]_e$ the equilibrium advancement, is

$$k_f t = \frac{\chi_e}{(2[A]_0 - \chi_e)} \ln \frac{[A]_0 \chi_e + \chi([A]_0 - \chi_e)}{[A]_0(\chi_e - \chi)} \tag{v}$$

The value of χ_e is found from $v_f = v_b$, where

$$v_f = k_f([A]_0 - \chi) \tag{vi}$$

$$v_b = k_b \chi^2 \tag{vii}$$

so that $\chi_e = [HCO_3^-] = [H^+] = 1.8 \times 10^{-6}\ M$ and $[A] = [A]_0 - \chi_e = 8.2 \times 10^{-6}\ M$.

Figure 2.9 shows concentrations of $CO_2(aq)$ and HCO_3^- (equal to H^+ concentration) as a function of time, and Figure 2.9b shows forward, backward, and

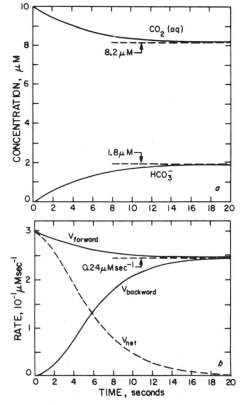

Figure 2.9. (a) Computed concentrations of $CO_2(aq)$ and HCO_3^- as a function of time for the reversible reaction $CO_2(aq) + H_2O = HCO_3^- + H^+$ at 25°C in a closed aqueous system. The total concentration $C_T = [CO_2(aq)] + [HCO_3^-]$, is $1 \times 10^{-5}\ M$. $CO_2(aq)$ is assumed to be nonvolatile. (b) Computed velocities, $v_{forward}$, $v_{backward}$, and v_{net} in the reaction mixture as a function of time for $CO_2(aq) + H_2O = HCO_3^- + H^+$. At equilibrium, the velocity in both directions is 0.24 $\mu M\ sec^{-1}$.

net velocities computed from equations vi and vii. Equilibrium is approached closely (to 1%) in a reaction time of 20 sec. It is interesting to note that v_f at equilibrium is close to 80% of the initial value.

This reaction is quite slow in comparison with many other acid–base reactions, for example, ionization of acetic acid or hydrolysis of ammonia, for which equilibrium is closely approached in microsceonds to milliseconds.

The $[H^+]$ range of this system is from about 10^{-7} to 2×10^{-6} M, so that the $[OH^-]$ concentration is in the range from 10^{-8} to 10^{-7}. The approximate magnitude of the forward reaction velocity for the second hydration path, $CO_2 + OH^- = HCO_3^-$, can be found from $v'_f = k'_f[CO_2][OH^-]$. The value of k'_f is approximately 8.5×10^3 M^{-1} sec^{-1}, so that $v'_f \simeq 8.5 \times 10^3 \times 10^{-5} \times 10^{-7} \simeq 8.5 \times 10^{-9}$ $M\,sec^{-1}$, which is small compared to v_f for the path we have considered. However, at higher pH, hence greater $[OH^-]$, the second path becomes more significant. (The kinetics of CO_2 reaction and transport in solution are discussed in Chapter 4.)

2.15   REACTION RATES OF ELEMENTARY PROCESSES

The key to understanding the rates of overall chemical reactions is an understanding of the factors that affect the elementary steps of transport and chemical transformation. Among the important factors that influence the specific rates (i.e., the rate constants) of these steps in aqueous solution are temperature, size, shape, and electric charge of reaction solute species, ionic strength of the solution, viscosity, and pressure. For heterogeneous reactions, for example, adsorption, surface catalysis, dissolution, and precipitation, additional factors of importance include reactive surface site densities, interfacial free energies, configuration of surface species, and the mixing energy in the bulk solution. In this section we review a few elementary models for specific reaction rates in solution, using the frameworks of *collision theory* and *absolute reaction rate theory* to gain an impression concerning upper limits of rates of reactions, and the expected influences of temperature, ionic strength, and pressure.

The Arrhenius Equation

Arrhenius (1889) proposed the equation

$$\frac{d \ln k}{dT} = \frac{E_a}{RT^2} \tag{356}$$

to describe the influence of temperature on the second-order rate constant k. E_a is the *activation energy*, R the gas constant, and T the absolute temperature. If E_a is temperature-independent (not always the case), equation 326 can be integrated to obtain

$$k = Ae^{-E_a/RT} \tag{357}$$

Arrhenius' concept was that reactant molecules must attain an activated state before they can react to form products. His results derived from analysis of a *reversible equilibrium*

$$A + B \underset{k_b}{\overset{k_f}{\rightleftharpoons}} B + C \tag{358}$$

and drew upon the van't Hoff equation (Section 2.11), which gives the temperature dependence of $K = k_f/k_b$ for the overall equilibrium. The term A in equation 357 is called a "frequency factor," with units, for example, of $M^{-1} \sec^{-1}$ for a second-order rate constant and $\sec^{-1}$ for a first-order rate constant. The frequency factor is in general a function of temperature, but usually not a strong one. Practical use of the Arrhenius equation is made by determining the rate constant k at several temperatures. A linear plot of $\log k$ versus $1/T$ yields the activation energy, which can be used to calculate k at intermediate temperatures. Changes in slope on an Arrhenius plot have been used to diagnose possible changes in reaction mechanism with temperature. For example, activation energies may change as temperature is varied because of a change in rate control from a chemical to a transport step or because of a change from a homogeneous to a heterogeneous reaction. For aqueous solution reactions experimental values of E_a are found to range from a few kcal mol^{-1} (ca. 10 kJ) to above 25 kcal (ca. 100 kJ).

It needs to be pointed out that an experimental rate constant often reflects prior equilibrium steps. These are incorporated into apparent rate constants, that is, $k_a = k_0 \prod_n K_n$, where K_n is an equilibrium constant, with a temperature dependence.

Collisions in Solution

Interpretation of the Arrhenius equation basically consists in understanding the factors that influence the frequency factor A and the activation energy E_a. Calculation of collision rates in solution gives some insight into A; an understanding of the E_a/RT term requires an examination of the energetics of activated complex formation from reactants.

The bimolecular collision rate for uncharged particles in solution can be calculated from the Smoluchowski expression for encounters through diffusion [25, 34]. The bimolecular rate constant for collisions between A and B is

$$k = \frac{4\pi N}{1000} (D_A + D_B) r_{AB} \tag{359}$$

where D_A and D_B are the diffusion coefficients in solution, r_{AB} is the distance of separation upon collision, and N is Avogadro's number. The diffusion coefficient of a spherical molecule (particle) is given by the Stokes–Einstein equation:

$$D = \frac{RT}{6\pi \eta r N} \tag{360}$$

where r is the radius. D depends upon T directly and through the influence of T on viscosity η. Hence there is an "activation energy" for diffusion of about 3–5 kcal mol^{-1}. Choosing $D_A = D_B = 10^{-5}$ cm^2 sec^{-1} and $r_{AB} = 5 \times 10^{-8}$ cm gives, at 25°C, $k \simeq 10^{10}$ M^{-1} sec^{-1} for the *first* encounters between molecules (*repeated* encounters tend to take place because of the "solvent cage" effect).

In the presence of long-range forces, for example, electrostatic forces, the collision rate will be decreased by repulsive forces and increased by attractive forces. For charged species the Debye–Smoluchowski equation [35] for the bimolecular collision is

$$k = \frac{4\pi N Z_A Z_B e^2 (D_A + D_B)}{1000 \varepsilon k T (\exp(Z_A Z_B e^2/\varepsilon k T r_{AB}) - 1)} \tag{361}$$

where Z_A and Z_B are charge numbers, e is the unit of charge, ε is the dielectric coefficient, and k is the Boltzmann constant. Choosing $r_{AB} = 7.5 \times 10^{-8}$ cm, the bimolecular rate constant for oppositely charged univalent ions at 25°C is

$$k \simeq 9 \times 10^{14} (D_A + D_B) \tag{362}$$

For the reaction

$$H_3O^+ + OH^- = 2H_2O$$

with $D_{H_3O^+} = 9.3 \times 10^{-5}$ cm^2 sec^{-1} and $D_{OH^-} = 5.1 \times 10^{-5}$ cm^2 sec^{-1} [23, 35], $k \simeq 1.3 \times 10^{11}$ M^{-1} sec^{-1}. The attractive electrostatic force increases the calculated bimolecular rate by a factor of about 10. The value obtained by experiment is 1.4×10^{11} M^{-1} sec^{-1}. The theoretical estimate and the experimental result give some appreciation of "how fast is fast?" for elementary processes in aqueous solutions. The Debye–Smoluchowski equation also predicts a *slowing down* of collisions for similarly charged species.

A large number of proton-transfer reactions in dilute aqueous solutions are found to have rate constants close to those predicted by the diffusion model. For example, consider the following proton transfers and the observed rate constants. [23]

Reaction	k (M^{-1} sec^{-1})
$H_3O^+ + OH^- = 2H_2O$	1.4×10^{11}
$H_3O^+ + SO_4^{2-} = H_2O + HSO_4^-$	1×10^{11}
$OH^- + NH_4^+ = H_2O + NH_3$	3.3×10^{10}
$HCO_3^- + H_3O^+ = H_2CO_3 + H_2O$	4.7×10^{10}

Equilibrium models for such fast reactions will thus give accurate results if thermodynamic information is accurate.

Slow Reactions. Not all reactions are encounter-controlled and hence fast. Reported rate constants for second-order reactions in solution at 25°C range from the diffusion-controlled value of $\sim 10^{10}$ M^{-1} sec^{-1} to $\sim 10^{-11}$ M^{-1} sec^{-1}, and first-order constants range from 10^{12} to 10^{-11} sec^{-1}.

The slow reactions are controlled by chemical steps and depend on entropic (solvation, desolvation, configurational, etc) and bonding (new bonds, structural rearrangements, etc.) changes in "activating" the reactant species to a new energetic state from which products can be formed. For example, we have seen that hydration of CO_2 in neutral solution

$$CO_2 + H_2O \underset{k_b}{\overset{k_f}{\rightleftharpoons}} H_2CO_3$$

has a pseudo-first-order k_f value of 3.0×10^{-2} sec^{-1} at 25°C (Example 2.9). The equivalent second-order rate constant is then $\sim 3.0 \times 10^{-2}/55.4$ or $\sim 5 \times 10^{-4}$ M^{-1} sec^{-1}, which is $\sim 10^{12}$ times slower than the diffusion-controlled reaction value calculated earlier. For a constant A in the Arrhenius equation, a slowing down of 10^{12} corresponds to $E_a \sim 70$ kJ mol^{-1}. The exchange of water molecules in the hexaaquo Cr(III) complex

$$Cr(H_2O)_6^{3+} + H_2O^* \longrightarrow Cr(H_2O)_5(H_2O^*) + H_2O$$

has a second-order rate constant of $\sim 5 \times 10^{-8}$ M^{-1} sec^{-1} (25°C). This is $\sim 10^{17}$ times slower than a diffusion-controlled bimolecular process (corresponding to $E_a \sim 100$ kJ mol^{-1} for constant A).

Theories for Slow Reactions

In the framework of the classic Arrhenius expression, $k = Ae^{-E_a/RT}$, slow reactions are characterized by small values of A, the frequency factor, large values of E_a, the activation energy, or both. Two theories of elementary reactions have been developed to describe rates of slow reactions: the collision theory and the transition-state activated complex or, as it is sometimes known, absolute reaction rate theory.

Collision Theory. In collision theory the rate constant for a bimolecular reaction is described by

$$k = pZe^{-E/RT} \tag{363}$$

where p is a "steric factor," Z is a specific collision rate or collision frequency (M^{-1} sec^{-1}), and E is the energy associated with the fraction of collisions, $e^{-E/RT}$, having sufficient energy to enter the activated state [23, 36]. The steric factor p is an empirically determined dimensionless factor intended to account for the fraction of collisions having proper orientation, configuration, and so on, to allow reaction to occur. The energy E and the activation energy of the Arrhenius equation are approximately equal ($E \sim E_a + \frac{1}{2}RT$; since $RT \sim 0.6$ kcal mol^{-1} or ~ 2.5 kJ mol^{-1}, $E \sim E_a$). Therefore the bimolecular rate constant can be written

$$k \simeq pZe^{-E_a/RT} \tag{364}$$

and, in terms of the Arrhenius equation, $A \simeq pZ$. What is the magnitude of Z? For aqueous solutions, with the apparent activation energy for collision by diffusion in the range 3–5 kcal mol^{-1} (12.6–21 kJ mol^{-1}), $Z \simeq 10^{12}$ to 10^{13} M^{-1} sec^{-1}, and the bimolecular rate constant for uncharged species, with $p = 1$, is then $k \simeq 10^{10}$ M^{-1} sec^{-1}, as already estimated from the Smoluchowski equation. Clearly, Z depends on the size of the reacting species, viscosity, separation distance, charge, and so on. The significance of equation 364 is that it suggests a way of attributing the specific rate of a rate-limiting step to combined collisional, entropic, and energetic influences. If E_a is found from rate experiments at different temperatures, the factor pZ can be calculated. There is no general way within the collision theory framework to predict p. If Z is calculated from a model, p can then be estimated from pZ. Values are generally much smaller than unity for slow reactions with small activation energies, but p values much greater than unity are sometimes found. Collision theory does not appear to offer an explanation for the variations in p in different systems.

Transition-State Theory. In this approach, which is now widely used to interpret kinetic observations in solution, the reactants are in *equilibrium* with a high-energy species known as the *activated complex*. The energy state of this complex is the *transition state*, thus one name for the theory. The activated complex reacts to form products at a translational frequency v given by

$$v = \frac{\mathbf{k}T}{h} \tag{365}$$

where **k** is the Boltzmann constant and h is Planck's constant. The reaction rate can thus be calculated from two fundamental constants and T, if the concentration of activated complex can be calculated from equilibrium considerations. Thus this rate theory consists of a "prior equilibrium" theory for calculating the concentration of activated complex, coupled with a universal frequency factor for the rate of product formation from the complex.

The equilibrium process between reactants and activated complex for an elementary bimolecular reaction is

$$\text{A} + \text{B} \;\xrightleftharpoons{\;K^{\ddagger}\;}\; \text{X}^{\ddagger} \tag{366}$$

where $\text{X}^{\ddagger}$ stands for the complex and $K^{\ddagger}$ the equilibrium constant. It is assumed that equilibrium is maintained among A, B, and $\text{X}^{\ddagger}$. In an ideal system the relation

$$[\text{X}^{\ddagger}] = K^{\ddagger}[\text{A}][\text{B}] \tag{367}$$

represents this equilibrium. The formation of products

$$\text{X}^{\ddagger} \;\longrightarrow\; \text{C} + \text{D} + \cdots \tag{368}$$

proceeds at an elementary rate

$$v = \frac{d[C]}{dt} = v[X^\ddagger] = \frac{\mathbf{k}T}{h}[X^\ddagger] \tag{369}$$

Substituting from equation 337,

$$v = \frac{\mathbf{k}T}{h} K^\ddagger[A][B] \tag{370}$$

Therefore, the bimolecular rate constant, k, is given by

$$k = \frac{\mathbf{k}T}{h} K^\ddagger \tag{371}$$

The energetics of forming the activated complex are obtained by a result from equilibrium thermodynamics

$$\Delta G^\ddagger = -RT \ln K^\ddagger \tag{372}$$

The *standard* free energy of activation, $\Delta G^\ddagger$, is related to the *standard* entropy and enthalpy of activation, $\Delta S^\ddagger$ and $\Delta H^\ddagger$, respectively, by

$$\Delta G^\ddagger = \Delta H^\ddagger - T\Delta S^\ddagger \tag{373}$$

The rate constant is thus

$$k = \frac{\mathbf{k}T}{h} e^{\Delta S^\ddagger/R} e^{-\Delta H^\ddagger/RT} \tag{374}$$

For liquid phase reactions, the enthalpy of activation and activation energy are related by

$$\Delta H^\ddagger = E_a - RT \tag{375}$$

which is obtained by differentiating the logarithmic form of equation 374, and substituting the Arrhenius and van't Hoff expressions for $d \ln k/dT$ and $d \ln K^\ddagger/dT$. (At $300°K$ $\Delta H^\ddagger$ and E_a differ by ~ 0.6 kcal or 2.5 kJ.) Therefore the rate constant is finally expressed as

$$k = e \frac{\mathbf{k}T}{h} e^{\Delta S^\ddagger/R} e^{-E_a/RT} \tag{376}$$

In terms of the Arrhenius equation, then

$$A = e \frac{\mathbf{k}T}{h} e^{\Delta S^\ddagger/R} \tag{377}$$

At 25°C, the term $\mathbf{k}T/h$ is 1.3805×10^{-23} J deg^{-1} $\times$ 298.15 deg $\div$ 6.6256 $\times$ 10^{-34} J sec^{-1} = 6.21×10^{12} sec^{-1}, so that

$$A \simeq 2 \times 10^{13} e^{\Delta S^\ddagger/R} \; M^{-1} \; sec^{-1}$$

for a mol/liter^{-1} concentration scale (numerical values of entropies depend on choice of standard state). We note that the collision theory and transition-state theory can be related through

$$pZ = e\,\frac{kT}{h}\,e^{\Delta S^{\ddagger}/R}$$

Smaller than normal ($p \sim 1$) values of p correspond to *negative* values of $\Delta S^{\dagger}$. The activated complex represents a state with less disorder and fewer degrees of freedom than that of the reactants. Forming an activated complex from two oppositely charged solution species is expected to lead to a *positive* entropy of activation (net release of water molecules from the reactants and a complex and overall increase in disorder), whereas combination of two species of like charge will produce a more highly charged species and net increased ordering of solvent molecules, leading to a negative entropy of activation. Experimental values of $\Delta S^{\ddagger}$ for ion–ion reactions decrease from ~ -10 eu (entropy units, in cal deg^{-1} mol^{-1}) to ~ -40 eu as the absolute value of the charge on the activated complex increases from 2 to 4 [25]. The electrostatic entropy contibution has been estimated *approximately* by $\Delta S_{cl}^{\ddagger} \simeq -10 Z_A Z_B$ (eu) [28]. Size and shape of aqueous molecules and ions can also influence $\Delta S^{\ddagger}$ considerably.

An activation entropy change of $+10$ eu (42 J deg^{-1} mol^{-1}) represents an *increase* in the rate constant by a factor of 1.5×10^2; an activation entropy change of -10 eu *decreases* the rate constant by the same factor. At 25°C, an increase in E_a of 10.5 kJ mol^{-1} lowers k by the same factor. Thus, for $\Delta S^{\ddagger} = -126$ J mol^{-1} deg^{-1} and $E_a = 63$ kJ mol^{-1}, the rate is lowered to 10^{-14} of the diffusion-controlled value.

By making kinetic experiments at several temperatures in the range of interest it is possible to obtain the value of E_a (or more than one value, if there is a change in mechanism with temperature). Application of equation 346 at a given temperature then yields $\Delta S^{\ddagger}$. Experimental values of E_a, $\Delta H^{\ddagger}$, $\Delta S^{\ddagger}$, and $\Delta G^{\ddagger}$ are thus all interrelated. Reliable activation energy data are valuable for kinetic interpretation of a reaction in terms of the transition-state theory parameters. Where $\Delta S^{\ddagger}$ and $\Delta H^{\ddagger}$ can be estimated from model calculations of entropies and bonding, rate constants can then be estimated. Such estimates are not highly developed for aqueous solutions.

Example 2.10.   Hydration of CO_2 in Terms of Transition-State Parameters

The slow step in CO_2 hydration in neutral solution, $CO_2 + H_2O \rightarrow H_2CO_3$, has a second-order rate constant (25°C) of 5×10^{-4} M^{-1} sec^{-1}. The reported activation energy is 63 kJ mol^{-1}. Calculate $\Delta H^{\ddagger}$, $\Delta S^{\ddagger}$, and $\Delta G^{\ddagger}$ for the reaction. Estimate the rate constant at 10°C.

The enthalpy of activation $\Delta H^{\ddagger}$ is given by $\Delta H^{\ddagger} = E_a - RT$. At 25°C, $T = 298.15$ deg; the gas constant R is 8.31 J mol^{-1} deg^{-1}. Therefore $\Delta H^{\ddagger}$ is 60.5 kJ mol^{-1}.

From equation 376, with $k = 5 \times 10^{-4}\ M^{-1}\ \sec^{-1}$, $T = 298.15$ deg, and $E_a = 63$ kJ mol^{-1}.

$$e^{\Delta S^{\ddagger}/R} = ke^{-1}\left(\frac{\mathbf{k}T}{h}\right)^{-1} e^{E_a/RT} = 2 \times 10^{-6} \tag{i}$$

and $\Delta S^{\ddagger} = -109$ J mol^{-1} deg^{-1} (or -26 eu). A decrease in entropy is in accord with the increase in order with hydration in forming the activated complex. (Note that the p of collision theory would be approximately 2×10^{-6}.)

The standard free energy of activation $\Delta G^{\ddagger}$ is computed from equation 373, and

$$\Delta G^{\ddagger} = \Delta H^{\ddagger} - T\Delta S^{\ddagger} = 28 \text{ kJ mol}^{-1} \tag{ii}$$

The equilibrium constant for forming the activated complex $K^{\ddagger}$ is obtained from Equation 372 in the form

$$K^{\ddagger} = e^{-\Delta G^{\ddagger}/RT} \tag{iii}$$

and $K^{\ddagger}$ is $1.2 \times 10^{-5}\ M^{-1}$. We may note that $\Delta G^{\ddagger}$ represents the "energy hill" that reactants must surmount in forming the activated complex. (Another process in which the free energy of activation figures largely is homogeneous nucleation, which is treated in Chapter 5.) The approximate free energy change in going from $CO_2(aq)$ to $H_2CO_3(aq)$ is about 14 kJ mol^{-1} (based on the equilibrium constant for hydration of CO_2, about 2.6×10^{-3}). Thus $\Delta G^{\ddagger}$ is roughly twice ΔG° for the overall endergonic reaction.

For 10°C (283.15 K), assuming that $\Delta H^{\ddagger}$ and $\Delta S^{\ddagger}$ are approximately constant with temperature, the rate constant calculated from equation 346 is

$$k = e\frac{k}{h}283.15 \cdot 2 \times 10^{-6} \exp -(63/8.31 \times 10^{-3} \times 283.15) \tag{iv}$$

$$k = 7.5 \times 10^{-5}\ M^{-1}\ \sec^{-1}$$

The rate is lowered by a factor of 6.7 for a decrease in temperature from 25 to 10°C.

Influence of Nonideal Behavior on Solution Kinetics

In a solution of electrolytes the activities of species may differ from their concentrations, as discussed in connection with the chemical potentials in nonideal solutions. Two different experimental approaches to studying rates of reactions involving ionic species are found valuable. One is to vary the compositions of fairly dilute reaction solutions (less than $\sim 0.01\ M$) and observe the effect upon the rate; another is to "swamp" the variations in ionic composition caused by progress of the reaction by use of an ionic medium which effectively fixes the electrolyte composition of the system. In the "ionic medium" approach, activity coefficients of reacting species are essentially constant.

The equilibrium between reactants and the activated complex in nonideal solution is described by the expression

$$K^\ddagger = \frac{\{X^\ddagger\}}{\{A\}\{B\}} = \frac{[X^\ddagger]}{[A][B]} \frac{\gamma^\ddagger}{\gamma_A \gamma_B} \tag{378}$$

Then, the rate expression (370) becomes

$$v = \frac{kT}{k}[X^\ddagger] = \frac{kT}{h} K^\ddagger \frac{\gamma_A \gamma_B}{\gamma_\ddagger}[A][B] \tag{379}$$

and the rate constant is

$$k = \frac{kT}{h} K^\ddagger \frac{\gamma_A \gamma_B}{\gamma_\ddagger} \tag{380}$$

Denoting the ideal-solution value of the rate constant k_0 the result is

$$k = k_0 \frac{\gamma_A \gamma_B}{\gamma_\ddagger} \tag{381}$$

A constant ionic medium serves to fix the activity coefficient quotient at a constant value. If the concentration of *indifferent* electrolytes in solution is varied, theoretical or empirical expressions for γ_i may be used to predict changes in k. For ions, the ionic strength, $I = \frac{1}{2}\sum_i C_i Z_i^2$, can be used to calculate γ_i for a *dilute* solution from the Debye–Hückel limiting law:

$$\log \gamma_i = -AZ_i^2 \sqrt{I} \tag{382}$$

in which A is a function of temperature and the dielectric constant of the solution. Combining the limiting law expressions for γ_A, γ_B, and $\gamma_\ddagger$ (and recognizing that $Z_\ddagger = Z_A + Z_B$) with the logarithmic form of equation 351 yields

$$\log \frac{k}{k_0} = 2AZ_A Z_B \sqrt{I} \tag{383}$$

known as the Brønsted–Bjerrum equation ($A \sim \frac{1}{2}$ for water at 25°C). The predicted linear variation of log k with $\sqrt{I}$, the slope being $2AZ_A Z_B$, has been confirmed for a number of ionic reactions for ionic strength less than $\sim 0.05\ M$ [23, 29]. For ions of like charge, the rate constant increases with ionic strength; for oppositely charged reactant ions the rate constant decreases. At $I = 0.01\ M$, $k/k_0 \simeq 1.3$ for $Z_A Z_B = 1$, and 1.6 for $Z_A Z_B = 2$; for $Z_A Z_B = -4$, $k/k_0 \simeq 0.4$, and so on. For a neutral reactant $Z_i = 0$, the rate is not influenced by I (but this holds only in dilute solutions). The effect described by the Brønsted–Bjerrum equation is known as the "primary salt effect" in that it pertains to reactant species and activated complex. In addition, there can be a secondary salt effect upon other solution equilibria linked to the main reaction, for example, catalysts such as H^+, OH^-, and other acids and bases. The activity coefficients of neutral reactants are also affected by ionic strength, although not as strongly as for ionic species. Many neutral species are "salted out" by ionic strength increases, for example, they are made less soluble, corresponding to an increase in activity coefficient with increasing ionic strength.

Influence of Pressure on Rates

Transition-state theory accounts for pressure effects on solution reaction rates through the effect of pressure on the equilibrium constant $K^{\ddagger}$. At constant temperature and electrolyte concentration (constant ionic strength),

$$\left(\frac{d \ln K^{\ddagger}}{dP}\right) = -\frac{\Delta V^{\ddagger}}{RT} \tag{384}$$

where $\Delta V^{\ddagger}$ is the volume of activation, $\bar{V}_{\ddagger} - \bar{V}_A - \bar{V}_B$, for $A + B \rightarrow X_{\ddagger}$. Therefore the rate constant variation with pressure is given by

$$\left(\frac{d \ln k}{dP}\right)_{T,I} = -\frac{\Delta V^{\ddagger}}{RT} \tag{385}$$

and the integration (for $\Delta V^{\ddagger}$ assumed independent of pressure; see Table 2.8 for the basis of the more general case)

$$\frac{k_P}{k_1} = \exp\left(-\frac{\Delta V^{\ddagger}(P-1)}{RT}\right) \tag{386}$$

The volume of activation can be obtained from a plot of $\log k_P$ versus P. If $\Delta V^{\ddagger} < 0$, the rate constant increases with pressure, and vice versa. For a $\Delta V^{\ddagger}$ of, say, -10 cm^3 mol^{-1}, $P = 500$ atm or 507 bars, and $T = 298.15$ K, $k_P/k_1 \simeq 1.25$; a $\Delta V^{\ddagger}$ of -50 cm^3 mol^{-1} for the same conditions gives $k_P/k_1 \simeq 2.25$. Negative activation volumes, like negative activation entropies, are indicative of "slow" reactions (compared to diffusion-limited, or "fast" reactions).

Catalysis

Slow reactions can be accelerated by a catalyst, a substance that alters the velocity of overall reaction and is both a reactant and product of the reaction. In the framework of transition-state theory, a catalyst provides an alternative pathway which lowers $\Delta G^{\ddagger}$; in terms of the Arrhenius equation, a lower activation energy is required for a catalyzed pathway if the frequency factor is unchanged. Lowering the activation energy from, say, 100 to 30 kJ mol^{-1} increases the reaction rate by a factor of $\sim 10^{12}$.

A general mechanism for catalysis has been outlined by Laidler [23] for simple (nonchain) reactions involving a single reactant, S, and a catalyst:

$$C + S \underset{k_{-1}}{\overset{k_1}{\rightleftharpoons}} X + Y \tag{387}$$

$$X + W \xrightarrow{k_2} P + Z + C \tag{388}$$

where C is the catalyst, P and Z are products, W is a second reactant, and Y is an intermediate.

In the case of a simple enzyme-catalyzed reaction, Y, W, and Z are nonexistent; $C \equiv E$, the enzyme, and $X \equiv ES$, the enzyme–substrate complex. Initially, $[S]_0 = [S] + [ES] \approx [S]$; $[E]_0 = [E] + [ES]$ for all conditions. Solving for the steady-state value of $[ES]$ and substituting in $v = k_2[ES]$ gives for the rate of reaction, under the steady-state assumption, $d[ES]/dt = 0$,

$$v = -\frac{d[S]}{dt} = \frac{d[P]}{dt} = \frac{k_2[E]_0[S]_0}{[S]_0 + K_m} \tag{389}$$

the Michaelis–Menten equation, in which $[E]_0$ and $[S]_0$ are *total* concentrations. More generally, $v = k_2[E]_0[S]/([S] + K_m)$, because $[S]_0 - [S] + [ES] + [P]$. K_m, the Michaelis constant, is equal to $(k_{-1} + k_2)/k_1$, which becomes k_{-1}/k_1 for the *equilibrium* case $(k_{-1} \gg k_2)$. An enzyme-catalyzed reaction is of apparent first-order in S for low concentrations ($[S] \ll K_m$) and of apparent zero-order in S at high concentrations ($[S] \gg K_m$). Equations like equation 389 are applicable to a variety of heterogeneous and homogeneous catalysis reactions.

Acid–Base Catalysis.   Catalysis of homogeneous solution reactions by acids and bases is often found. Both specific acid–base catalysis and general acid–base catalysis occur. In specific acid–base catalysis the velocity of the reaction depends upon $[H^+]$ to a higher power than indicated in the overall stoichiometric equation. In terms of the general scheme of reactions 387 and 388,

$$S + H^+(aq) \xrightarrow[k_{-1}]{k_1} SH^+ \qquad K_1 = \frac{k_1}{k_{-1}} \tag{390}$$

$$SH^+ + W \xrightarrow{k_2} P + H^+ \tag{391}$$

and

$$v = k_2 K_1[S][H^+][W] \tag{392}$$

The reaction is *specifically catalyzed* by $[H^+]$. A two-term rate law is possible if both the uncatalyzed and catalyzed reactions have appreciable rates. In *general acid* catalysis,

$$S + HA \xrightarrow[k_{-1}]{k_1} SH^+ + A^- \qquad K_1 = \frac{k_1}{k_{-1}} \tag{393}$$

$$SH^+ + A^- + W \xrightarrow{k_2} P + HA \tag{394}$$

and

$$v = k_2 K_1[HA][S][W] \tag{395}$$

For a reaction subject to both specific acid catalysis and general acid catalysis from acids $HA_1, HA_2, \ldots,$

$$k = k_0 + k'[H^+] + k_1[HA_1] + k_2[HA_2] + \cdots \tag{396}$$

and similarly for specific and general base catalysis.

2.16 HETEROGENEOUS PROCESSES

Reactions at interfaces such as the solution–solid interface and the gas–solution interface can be rate-limited by diffusion or by an elementary chemical step. If the solution is considered well mixed to within a small distance from the interface, δ, the steady diffusion flux of species i across the small layer or film between the phase boundary and the well-mixed solution, F_i, is given by

$$F_i = \frac{D_i}{\delta} \Delta C_i \tag{397}$$

where D_i is the diffusion coefficient and ΔC_i is the concentration difference across the distance δ. Flux has dimensions, for example, of mol cm^{-2} sec^{-1}. This equation is a simple, one-dimensional version of Fick's first law, $F_i = -D_i dC_i/dx$, in which the flux is in the direction of decreasing concentration ($dC_i/dx < 0$) and the gradient is in general a function of time. Fick's second law, $dC_i/dt = D(d^2C_i/dx^2)$, gives the rate of change in concentration with time. Equation 397 takes $d^2C_i/dx^2 = 0$ in the diffusion film and $dC_i/dx = 0$ in the well-mixed solution; in the film, $dC_i/dx = \Delta C_i/\delta$. The film diffusion approach replaces the concentration gradients of the real system with a simple model.

If the concentration at the interface (say $x = 0$, with x increasing toward the solution) is C_i^s and the concentration in the well-mixed bulk solution (say $x = \delta$) is C_i^b, then equation 397 can be written

$$F_i = \frac{D_i}{\delta} (C_i^s - C_i^b) \tag{398}$$

If $C_i^s > C_i^b$, transport is toward the solution; if $C_i^s < C_i^b$, transport is from the solution to the interface.

Rate-Limiting Step

A chemical reaction at the interface ($x = 0$) may have characteristics which make it rate-limiting for the overall sequence: interfacial chemical reaction → diffusion → solution species. Denbigh [37] has shown that, for a first-order chemical reaction at the interface, for which the two-dimensional surface reaction rate is

$$r_i^s = F_{i,ch} = k^s C_i^s \tag{399}$$

and the diffusion flux of reactant to the interface is, from equation 398,

$$F_{i,d} = \frac{D_i}{\delta}(C_i^b - C_i^s)$$

the steady-state rate r is

$$r = \frac{k_i^s(D_i/\delta)}{k_i^s + (D_i/\delta)} C_i^b \qquad (400)$$

When $D_i/\delta \gg k_i^s$, the rate is chemically controlled; when $k_i^s \gg D_i/\delta$, the rate is diffusion-controlled.

Chemically Controlled Processes

For a slow chemical step in general, the overall rate law may be complex and of any order; sequential or parallel pathways may be involved. For example, dissolution of the iron sufide mineral mackinawite [38] is found to have a total flux described by the rate expression

$$F = k_1[H^+] + k_2$$

indicating that two concurrent processes contribute to the total reaction. Rate constants for heterogeneous processes are expressed in terms of the interfacial area, for example, k_1 in units of cm sec^{-1}, k_2 in units of mol cm^{-2} sec^{-1}. The dissolution rate law reported appears as the sum of a first-order process in $[H^+]$ and a zero-order process in $[H^+]$. The variation of rate constants of heterogeneous processes with temperature to obtain E_a values may serve to distinguish between diffusion control and chemical control. Intensity of mixing in solution affects rates of diffusion-controlled processes through changes in the apparent value of δ.

Diffusion-Controlled Processes

The change in solution concentration as a result of a diffusion-controlled heterogeneous process in a closed, well-mixed system can be described by

$$\frac{dC_i^b}{dt} = F_i \frac{A_s}{V_b} = \frac{D_i}{\delta}\frac{A_s}{V_b}(C_i^s - C_i^b) \qquad (401)$$

where A_s is the interfacial area, V_b is the bulk solution volume, and C_i^b is the uniform bulk solution concentration. This expression can be integrated for certain simple boundary conditions, for example, where C_i^s is constant and $C_i^b = C_{i0}^b$, at $t = 0$. Two situations corresponding to these conditions are the diffusion-controlled dissolution of a solid with A_s essentially constant and the diffusion-controlled transfer of a gas between the gas phase and water. Berner [39, 40] has discussed dissolution rates of solids.

Gas Transfer

Transfer of gases between the atmosphere and waters has been described in terms of the diffusion film model by a number of workers [41], [42], [43].

In general, it is necessary to consider *two* diffusion films, one in the liquid phase and one in the gas phase, each phase being regarded as well-mixed by turbulence to within a small distance of the interface. The steady flux for a gas species into solution, $F = F_i^q = F_i^L$, is given by

$$F = k_g(C_g - C_g^s) = k_L(C_L^s - C_L) \tag{402}$$

in which $k_g = D_g/\delta_g$, $k_L = D_L/\delta_L$, C_g and C_L are the bulk phase concentrations in gas and liquid, respectively, and C_g^s and C_L^s are the corresponding interfacial concentrations in each phase. The quantities k_g and k_L are exchange rate constants (cm sec^{-1}). The two-phase equilibrium condition for the species being transported can be expressed by a variation on Henry's law (Section 2.8)

$$\frac{C_g^s}{C_L^s} = H \quad \text{(dimensionless)} \tag{403}$$

Substitution of equation 403 into equation 402 allows elimination of the interface concentrations and, upon rearrangement, gives two equivalent equations for the flux, each in terms of C_g, C_L, k_L, k_G, and H. The result is

$$F = K_L\left(\frac{C_g}{H} - C_L\right) \tag{404}$$

in which

$$\frac{1}{K_L} = \frac{1}{k_L} + \frac{1}{Hk_g} \tag{405}$$

or, in terms of apparent *series* resistances,

$$R_L = r_L + r_g \tag{406}$$

This result [42] shows that, for a large liquid film resistance, $r_L \gg r_g$, the flux is controlled by a small $k_L = D_L/\delta_L$, whereas for $r_g \gg r_L$ the flux is controlled by a small product of H and k_g. For comparable values of $k_g = D_g/\delta_g$, gases with small H values will tend to be controlled by gas-phase resistance to diffusion. For gases of comparable H values, those that react rapidly in water to yield more than one diffusing species (e.g., CO_2, SO_2, NH_3) will tend to have larger k_L values, hence be gas resistance-controlled. For example, SO_2 reacts very rapidly in water to give HSO_3^- and SO_3^{2-} ions. The pseudo-first-order rate constant for hydration of SO_2

$$SO_2 + H_2O \longrightarrow HSO_3^- + H^+$$

is $\sim 10^6$ sec^{-1}; the corresponding rate constant for CO_2 is $\sim 3 \times 10^{-2}$ sec^{-1}.

For several diffusing species derived by very rapid chemical reaction of the dissolved gas,

$$F_{total} = \frac{1}{\delta} \sum_i D_i (C_i^s - C_i) \tag{407}$$

Emerson [44] has developed a general model for the treatment of CO_2 transfer to lake water; CO_2 hydration reactions are not very rapid, and so Fick's second law was used together with the rate of chemical change of each species, v_i:

$$\frac{dC_i}{dt} = D_i \frac{d^2 C_i}{dx^2} + v_i \tag{408}$$

Among the gases for which liquid film diffusion appears to control are O_2, N_2, CO_2, CH_4, and N_2O; gas-phase resistance appears to control transport of SO_2, NH_3, SO_3, and HCl [42]. Gas transport is discussed further in Chapters 4 (carbon dioxide) and 11 (synthetic chemical pollutants).

Dimensionless Distribution Coefficients for Aqueous-Gas Equilibria $A(aq) \rightleftharpoons A(g)$ at 25°C

Compound	H^a	Compound	H^a
N_2	63.9^b	CO_2	1.2^f
H_2	$50.4^{c,d}$	CCl_4	1.1^e
CO	42.8^b	Hg	0.47^g
O_2	32.4^b	H_2S	0.40^h
CH_4	27.2^c	CH_3Cl	0.16^i
NO	21.7^b	SO_2	0.033^j
N_2O	1.6^e	NH_3	0.00071^k

[a] The dimensionless constant $H = C_{A(g)}/C_{A(aq)}$ is related to other equilibrium parameters as follows: $K_H = 1/HRT$, where K_H is a "Henry's law constant", with dimensions M atm^{-1}; $K = HRTM_W$, where K is the conventional equilibrium constant of Henry's law, with dimensions of atm and M_W is the molecular weight of water.
[b] *Handbook of Chemistry and Physics*, 39th ed., Chemical Rubber Publishing Co., Cleveland (1957).
[c] Thibodeaux, L. J. *Chemodynamics*, Wiley-Interscience, N.Y. (1979).
[d] 20°C.
[e] Liss, P. S., and P. G. Slater, *Nature*, **247**, 181 (1974).
[f] See Table 4.7.
[g] MacKay, D. and P. J. Leinonen, *Environ. Sci. Technol.*, **9**, 1178 (1975).
[h] Smith, R. M., and A. E. Martell, *Critical Stability Constants*, Plenum, N.Y. (1976).
[i] Dilling, W. L., *Environ. Sci. Technol.*, **11**, 405 (1977).
[j] Roberts, D., Ph.D. Thesis, Calif. Institute of Technology (1978).
[k] Liljestrand, H. M. and J. J. Morgan, *Environ. Sci. Technol.*, **15**, 333 (1981).

2.17 EQUILIBRIUM MODELS AND COMPUTATION OF COMPLEX EQUILIBRIA

Equilibrium is the time-invariant state of a closed system. Finding the equilibrium composition of an aqueous system, such as that depicted in Figure 2.2, is a mathematical problem which can be expressed simply: Minimize the Gibbs free energy of the system subject to the constraints of mass balance [45]. The general problem of finding the equilibrium composition of complex systems has been reviewed by Zeleznik and Gordon [46] and van Zeggeren and Storey [47]. Minimization of the Gibbs free energy of the system for a system of n independent reactions

$$\frac{dG}{d\xi_n} = 0$$

is mathematically equivalent to finding the composition that makes $\Delta G_n = 0$ for each of the reactions. The Rand programs [48] involve iterative computation (starting with an initial guess for the equilibrium composition which satisfies stoichiometric constraints) for obtaining minimization of the free energy and thus the correct equilibrium composition. Examples 2.4 and 2.5 give extremely simple examples of explicit minimization of the Gibbs free energy.

A number of equilibrium programs for aqueous systems, for example, [45], [49], [50], and [51], employ the so-called equilibrium constant approach in which the set of nonlinear equilibrium constant expressions of the general form

$$K_n = \prod_i \{M_i\}^{\nu_{i,n}} \tag{409}$$

for basic components and derived species are incorporated in the set of linear stoichiometric mass conservation equations for the basic components, which simply state that each component must be found in free or uncombined form and in any species derived from it by reaction. The stoichiometric equations are then solved iteratively for the correct equilibrium composition. The equilibrium constant approach is used throughout this book to find the composition. In general, iteration is also required to find the nonideality corrections for solution species, which may be thought of as the activity coefficient quotient $Q_{\gamma,n}$, where

$$Q_{\gamma,n} = \prod_i \gamma_i^{\nu_{i,n}} \tag{410}$$

In an ionic medium, $Q_{\gamma,n}$ is essentially constant, and the concentration equilibrium constant ^cK_n is

$$^cK_n = K_n Q_{\gamma,n}^{-1} \tag{411}$$

Otherwise, $Q_{\gamma,n}$ is a function of the equilibrium composition resulting from chemical reaction. Note that the stoichiometric equations for components must be expressed in terms of *concentrations* of components and the derived species, whereas the equilibrium constants are fundamentally expressed in terms of *activities*. (For elaboration on activities versus concentrations, see the appendix to Chapter 6).

An example of a program used successfully for finding the equilibrium compositions of complex aqueous systems is REDEQL [52, 53], which was developed along the lines of a coordination-chemical framework: Metals and ligands (M_i and L_j) are chosen as basic constituents or components, and other species, or complexes, such as aqueous solution complexes, solids, gases, and oxidized and reduced forms, are formed from these; for example, if M_i is Ca^{2+} and L_j is

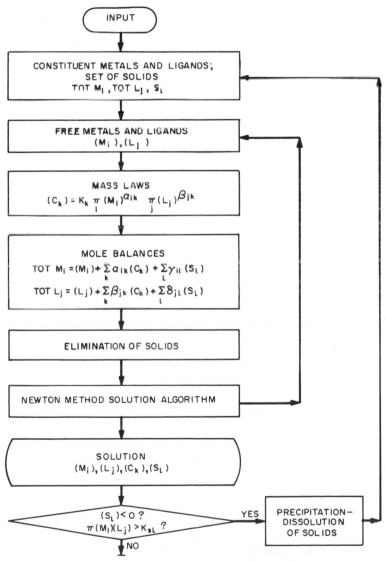

Figure 2.10.   Organization of the equilibrium program REDEQL, which uses a modified Newton method [52] for iterative solution: This program was used, for example, to perform the computations presented in Figure 6.17, Table 6.14, Figure 6.22, and Table 6.15.

CO_3^{2-}, the aqueous complex $CaCO_3(aq)$ is C_{ij} and the solid $CaCO_3(s)$ is S_{ij}, and so forth. In this example, Ca^{2+} and CO_3^{2-} are *components* and $CaCO_3(s)$ and $CaCO_3(aq)$ are derived *species*. Figure 2.10 shows the organization of REDEQL, which uses a modified Newton method [52] for iterative solution. This program was used, for example, to perform the computations presented in Figure 6.17, Table 6.14, Figure 6.22, and Table 6.15.

2.18   NATURAL WATER SYSTEMS AND MODELS: EQUILIBRIA AND RATES

Equilibrium Models

Real systems may or may not be closely approximated by equilibrium models. Discrepancies between actual chemical composition and the composition predicted from an equilibrium model are sometimes encountered. One hopes that an equilibrium model may give a useful first approximation to many features of the real system. There are several reasons for differences between the real system and the equilibrium model. Briefly, the causes of the discrepancies can be categorized as follows.

Limitations of Thermodynamic Information

Pertinent chemical equilibria may have been ignored in formulating the model, or important species in solid or solution phases not considered. Thermodynamic data on assumed species and phases may be incorrect or inadequate. Temperature, pressure, and activity corrections may require refinement. Significant temperature gradients exist at times in nearly all bodies of water. Adequate data on temperature dependence of equilibria may not be available. Pressure correction requires knowing the partial molar volumes of the species involved.

Need for Chemical Characterization of Species in the Real System

Analytical data on some chemical elements may be inadequate to distinguish between various physical and chemical forms (dissolved versus suspended, oxidized versus reduced, monomeric versus polymeric). Analytical deficiencies often relate to highly reactive nonconservative elements: for example, iron, manganese, phosphorus, carbon, nitrogen, aluminum, and other trace metal ions.

Slow Rates of Chemical Reaction

The rates of some chemical reactions are such that equilibrium is only slowly attained in the real system. When a *closed* system represents a reasonable approximation to the physical nature of the system, the slowness of some chemical reactions will cause discrepancies between the real system and an equilibrium model for the species involved in these reactions.

Chemical Reaction Times

For *irreversible* reactions in closed systems there is no equilibrium state; such reactions progress toward complete exhaustion of the reactant that is stoichiometrically limiting. The time scales for irreversible reactions are easily identified from their rate constants. As noted above, observed first-order rate constants for aqueous reactions range from 10^{11} sec^{-1} to less than 10^{-11} sec^{-1}, with corresponding half-times $t_{1/2}$ ranging from picoseconds to greater than 2000 years. An activation energy of 150 kJ mol^{-1} corresponds to a $t_{1/2}$ of $\sim$ 100,000 years; an activation energy of 100 kJ mol^{-1} corresponds to a $t_{1/2}$ of $\sim$ 1 day; an activation energy of 85 kJ mol^{-1} gives a $t_{1/2}$ of $\sim$ 1 min. Similarly, for second-order reactions, where $t_{1/2} \sim 1/[A]k$, half-times can range from nanoseconds to millions of years.

Among irreversible reactions that are slow (half-times of minutes or longer) in particular aquatic environments are certain metal-ion oxidations, oxidation of sulfides, sulfate reduction by organic substances, various metal ion polymerizations (e.g., aluminum ion), aging of hydroxide and oxide precipitates, precipitation of metal-ion silicates and carbonates (e.g., dolomites), conversions among aluminosilicates (e.g., feldspar-kaolinite), solution or precipitation of quartz, and nonbiological hydrolysis of polymers (e.g., polyphosphates). Some of these reactions can be accelerated greatly by biological catalysis (e.g., sulfate reduction, metal-ion oxidations). A few examples for uncatalyzed reactions are:

Reactants	$t_{1/2}$ (sec)
$Fe^{2+} + O_2$ (pH < 4)	$\sim 10^8$
$Fe^{2+} + O_2$ (pH $\sim$ 7.5)	$\sim 10{-}10^3$
$Mn^{2+} + O_2$ (pH $\sim$ 8)	$\sim 10^{10}$
$H_2S + O_2$	$\sim 10^2{-}10^5$
$P_2O_7^{4-} + H_2O \rightarrow 2PO_4^{3-} + 2H^+$	$\sim 10^6$

Reversible Reactions

Example 2.7 showed that the time for 99% approach to equilibrium in a relatively slow hydration reaction, $CO_2 + H_2O = HCO_3^- + H^+$, with $k_f = 3 \times 10^{-2}$ sec^{-1} and $k_b \simeq 7 \times 10^4$ M^{-1} sec^{-1}, was about 20 sec and that the equilibration half-time was about 4 sec. Most simple hydration reactions are more rapid than this (CO_2 hydration is slow because the linear CO_2 molecule must be rearranged into the trigonal H_2CO_3 molecule before transferring a proton). For example, the reaction $NH_4^+ = NH_3 + H^+$ has a $t_{1/2}$ of less than 50 msec. Dissociation of highly stable complexes, that is, those with large equilibrium constants of formation, are relatively slow in comparison to simple hydrations and dissociations. For example, the dissociation of the iron(III) EDTA complex $Fe(EDTA)OH^{2-}$ (EDTA is a complexing agent that binds the iron at six sites)

$$Fe(EDTA)OH^{2-} + H_2O \underset{k_b}{\overset{k_f}{\rightleftharpoons}} Fe(OH)_2^+ + H(EDTA)^{3-}$$

has a k_f of $\sim 10^{-8}$ to 10^{-6} M^{-1} sec^{-1} (estimated from $K \sim 10^{14}$ and $k_b \sim 10^6$ to 10^8 sec^{-1}). The dissociation has a much longer characteristic time than has the association.

Closed-System Reversible Kinetics

The applicability of a complete equilibrium model to a closed system obviously depends upon the time allowed for reaction. Some reactions are sufficiently rapid in both forward and reverse directions that only short elapsed times are necessary for a close approach to equilibrium. A kinetic model approach to equilibrium in a closed system requires, in general, information on forward and reverse rate laws and initial concentrations (see Example 2.9, on CO_2 hydration). Integrated rate expressions are available for simpler reversible reactions such as $A \rightleftharpoons B + C$, $A + B \rightleftharpoons C$, and $A + B \rightleftharpoons C + D$ [23, 24, 33]. Numerical integration is often feasible for more complex cases. The half-times of high-order reactions depend upon initial concentrations. The first-order reaction

$$A \underset{k_b}{\overset{k_f}{\rightleftharpoons}} B$$

has a half-time independent of concentration and serves as a simple illustration of characteristic times for reversible reactions.

On the basis of equation 350,

$$t_{1/2} = \frac{0.69}{k_f(K + 1)/K} \tag{412}$$

where K is the equilibrium constant. The half-time depends upon k_f and K. Bada [54] studied isomerization of amino acids for which $K = 1$ and calculated half-times as $0.69/2k_f$. Table 2.9 gives some values of $t_{1/2}$ for the reaction $A \rightleftharpoons B$ for different values of k_f and K. For $K \gg 1$, $t_{1/2} \sim 0.69/k_f$. For *very* slow, but thermodynamically favorable processes, for example, $k_f \sim 10^{-10}$ sec^{-1}, $t_{1/2} \sim$ 200 years.

Equilibrium Models with Kinetic Constraints

It is sometimes the case that one key reaction in a system of coupled reactions is slow and the rest of the reactions sufficiently rapid that equilibrium is an accurate

TABLE 2.9 HALF-TIMES OF EQUILIBRATION FOR THE
REACTION A ⇌ B FOR DIFFERENT VALUES OF THE FOR-
WARD RATE CONSTANT, k_f AND THE EQUILIBRIUM CON-
STANT K

K	k_f (sec^{-1})	$t_{1/2}$
$\frac{1}{10}$	10^{-3}	~1 min
	10^{-5}	~100 min
	10^{-7}	~7 days
1	10^{-3}	~6 min
	10^{-5}	~10 hr
	10^{-7}	~40 days
10	10^{-3}	~10 min
	10^{-5}	~17 hr
	10^{-7}	~75 days

description of them. Then the slow reaction can be modeled kinetically and the
output used to define a set of mass-balance constraints for the rest of the system.
These constraints change with time, permitting an equilibrium approach in
describing the composition of the parts of the system that have sufficiently short
chemical time scales.

Differences between Open Systems and Closed Systems

Most natural water systems are continuous, open systems. Flows of matter and
energy exist in the real system. The time-invariant state of a continuous system
with flows at the boundaries is the *steady state*. This state might be poorly
approximated by the *equilibrium* state of a closed system. In Figure 2.1 we
indicated the important features of an open-system model with material fluxes
and chemical reactions. The simple reversible reaction

$$A \; \underset{k_b}{\overset{k_f}{\rightleftharpoons}} \; B$$

was introduced to illustrate the elementary differences between closed- and open-
system models. Solution for the steady-state concentration values of A and B
gave the results

$$C_A = \frac{r\bar{C}_{A,0} + k_b(\bar{C}_{A,0} + \bar{C}_{B,0})}{k_f + k_b + r} \tag{11}$$

$$C_B = \frac{r\bar{C}_{B,0} + k_f(\bar{C}_{A,0} + \bar{C}_{B,0})}{k_f + k_b + r} \tag{12}$$

We now examine the *ratio* of C_B and C_A and compare it with the ratio expected for chemical equilibrium. The ratio C_B/C_A for the steady state is, dividing equation 12 by equation 11,

$$\frac{C_B}{C_A} = \frac{r\bar{C}_{B,0} + k_f(\bar{C}_{A,0} + \bar{C}_{B,0})}{r\bar{C}_{A,0} - k_b(\bar{C}_{A,0} + \bar{C}_{B,0})}. \tag{413}$$

Equation (413) shows that C_B/C_A will tend toward k_f/k_b as the material flows to the system becomes small, that is, as $r\bar{C}_{B,0}$ and $r\bar{C}_{A,0}$ vanish. For $r = Q/V = 0$, the system becomes a closed system, and $C_B/C_A = k_f/k_b = K$, the equilibrium constant. The quantity $r = Q/V$ is the reciprocal of the fluid residence time τ_R of the well-mixed system: $r = \tau_R^{-1}$. As τ_R tends to very large values, the steady-state concentrations of the system approach the equilibrium values.

A simple result is obtained for only A entering the system. Then, $\bar{C}_{B,0} = 0$ and equation 413 reduces to

$$\frac{C_B}{C_A} = \frac{k_f\bar{C}_{A,0}}{r\bar{C}_{A,0} + k_b\bar{C}_{A,0}} = \frac{k_f}{r + k_b} \tag{414}$$

The steady-state concentration ratio quotient depends on the chemical rate constants and the flux rate constant. For $r \ll k_b$, $C_B/C_A \simeq k_f/k_b = K$; if $r \gg k_b$, then $C_B/C_A \simeq k_f/r$. In terms of the residence time, τ_R and the half-time of the backward reaction τ_b, the steady-state ratio approximates the equilibrium ratio if $\tau_R \gg \tau_b$.

Example 2.11. Steady-State Composition of an Open Completely Mixed System with the Reaction A $\rightleftharpoons$ B as a Function of Residence Time

Assume that $\bar{C}_{B,0} = 0$, $k_f = 10^{-5}$ sec^{-1}, and $k_b = 10^{-6}$ sec^{-1}, so that $K = 10$. The residence time is varied over the range from 10^5 to 10^8 sec. Table 2.10 shows the calculated ratio of steady-state C_B/C_A to K for a 1000-fold variation in the residence time relative to the characteristic reaction time ($\sim k_b^{-1}$).

The analysis can be readily extended to other simple reversible reactions, for example, A + B $\rightleftharpoons$ C and A + B $\rightleftharpoons$ C + D (where quadratic expressions are obtained for A, B, and so on, in the general case), with the same conclusion: Long residence times τ_R compared to the characteristic chemical times $\sim \tau_{1/2, chem}$ are required for agreement of equilibrium and steady-state models.

The equilibrium assumption is justified for many reactions with short τ_{chem} and long τ_R. Rate data for a large number of first- and second-order aqueous reactions indicate τ_{chem} less than seconds to minutes, and many other aqueous reactions have τ_{chem} less than hours to days [55]. The residence times of a number of freshwater systems are greater than these ranges [56]. Critical attention needs to be directed to slow chemical reactions for which $\tau_{chem} \gtrsim \tau_R$.

For many systems it is known that there exist regions or environments in which the time-invariant condition closely approaches equilibrium. The concept

TABLE 2.10 COMPOSITION[a] OF A COMPLETELY MIXED OPEN SYSTEM AT STEADY STATE FOR

$$A \underset{k_b}{\overset{k_f}{\rightleftharpoons}} B$$

τ_R (sec)	r (sec^{-1})	r/k_b	$\dfrac{C_B/C_A}{K}$
10^5	10^{-5}	10	0.09
2×10^5	5×10^{-6}	5	0.17
5×10^5	2×10^{-6}	2	0.33
10^6	1×10^{-6}	1	0.50
2×10^6	5×10^{-7}	0.5	0.67
5×10^6	2×10^{-7}	0.2	0.83
10^7	1×10^{-7}	0.1	0.91
10^8	1×10^{-8}	0.01	0.99

[a] $(C_B/C_A)/K$ versus τ_R; $k_f = 1 \times 10^{-5}$ sec^{-1}; $k_b = 1 \times 10^{-6}$ sec^{-1}.

of local equilibrium is important in examining complex systems. Local equilibrium conditions are expected to develop, for example, for kinetically rapid species and phases at sediment–water interfaces in fresh, estuarine, and marine environments. In contrast, other local environments, such as the photosynthetically active surface regions of nearly all lakes and ocean waters and the biologically active regions of soil–water systems are clearly far removed from total system equilibrium.

REFERENCES

1 E. A. Guggenheim, *Thermodynamics: An Advanced Treatment for Chemists and Physicists*, 5th ed., North Holland, Amsterdam, 1967.

2 K. G. Denbigh, *The Principles of Chemical Equilibrium: With Applications in Chemistry and Chemical Engineering*, 3rd ed., Cambridge University Press, Cambridge, 1971.

3 G. N. Lewis and M. Randall, *Thermodynamics*, 2nd ed., revised by K. S. Pitzer and L. Brewer, McGraw-Hill, New York, 1961.

4 F. T. Wall, *Chemical Thermodynamics*, 2nd ed., Freeman, San Francisco, 1965.

5 I. Prigogine and R. Defay, *Chemical Thermodynamics*, translated by D. H. Everett, McKay, New York, 1954.

6 R. A. Robinson and R. H. Stokes, *Electrolyte Solutions: The Measurement and Interpretation of Conductance, Chemical Potential and Diffusion in Solutions of Simple Electrolytes*, 2nd ed., Butterworths, London, 1959.

7 H. S. Harned and B. B. Owen, *The Physical Chemistry of Electrolytic Solutions*, 3rd ed., Van Nostrand, Reinhold, New York, 1958.

8 I. M. Klotz, *Chemical Thermodynamics: Basic Theory and Methods*, revised ed., W. A. Benjamin, Menlo Park, Calif., 1964.

9 W. M. Latimer, *The Oxidation States of the Elements and Their Potentials in Aqueous Solutions*, 2nd ed., Prentice-Hall, Englewood Cliffs, N.J., 1952.

10 R. A. Robie, B. S. Hemingway, and J. R. Fisher, *Thermodynamic Properties of Minerals and Related Substances at 298.15K and 1 Bar Pressure and at Higher Temperatures, Geological Survey Bulletin*, No. 1452, US G.P.O., Washington, 1978.

11 F. D. Rossini et al., *Selected Values of Chemical Thermodynamic Properties, National Bureau of Standards Circular*, No. 500, U.S. Department of Commerce, Washington, D.C., 1952.

12 L. G. Sillén and A. E. Martell, *Stability Constants of Metal-Ion Complexes*, Special Publication, No. 17, Chemical Society, London, 1964.

13 G. H. Nancollas, *Interactions in Electrolyte Solutions*, Elsevier, New York, 1966.

14 J. C. Morris, W. Stumm, and H. Galal, *J. Sanit. Eng. Div. Am. Soc. Civ. Eng.*, **87**, 81 (1961).

15 P. A. Rock, *J. Chem. Educ.*, **44**, 104, 1967.

16 F. Millero, *J. Limnol. Oceanogr.*, **14**, 376 (1969).

17 M. Whitfield, in *Chemical Oceanography*, J. P. Riley and Skirrow, Eds., Vol. 1, 2nd ed., Academic, New York, 1975, pp. 44–171.

18 A. Disteche, in *The Sea*, E. D. Goldberg, Ed., Vol. 5, Wiley-Interscience, 1974, pp. 81–121.

19 S. D. Hamann, *Physico-Chemical Effects of Pressure*, Butterworths, London, 1957.

20 B. B. Owen and S. R. Brinkley, Jr., *Chem. Rev.*, **29**, 461, 1941.

21 L. G. Sillén and A. E. Martell, *Stability Constants of Metal-Ion Complexes*, Supplement No. 1, Chemical Society, London, 1971.

22 R. M. Smith and A. E. Martell, *Critical Stability Constants*, Vols. 1–4, Plenum, New York, 1976.

23 K. J. Laidler, *Chemical Kinetics*, 2nd ed., McGraw-Hill, New York, 1965.

24 S. W. Benson, *The Foundations of Chemical Kinetics*, McGraw-Hill, New York, 1960.

25 A. A. Frost and R. G. Pearson, *Kinetics and Mechanism*, 2nd ed., Wiley, New York, 1961.

26 I. Amdur and G. G. Hammes, *Chemical Kinetics*, McGraw-Hill, New York, 1966.

27 E. L. King, *How Chemical Reactions Occur*, W. A. Benjamin, Menlo Park, Calif., 1964.

28 D. Benson, *Mechanisms of Inorganic Reactions in Solution*, McGraw-Hill, New York, 1968.

29 A. G. Sykes, *Kinetics of Inorganic Reactions*, Pergamon, Elmsford, N.Y., 1966.

30 J. O. Edwards, *Inorganic Reaction Mechanisms: An Introduction*, W. A. Benjamin, Menlo Park, Calif., 1964.

31 T. S. Lee, in *Treatise on Analytical Chemistry*, Part I, Vol. 1, I. M. Kolthof and P. J. Elving, Eds., Interscience, New York, 1959, pp. 185–275.

32 J. T. Edsall, in *CO_2: Chemical, Biochemical and Physiological Effects*, R. E. Foster, et al., Eds., NASA SP-188, Washington, D.C. 1969.

33 C. Capellos and B. H. J. Bielski, *Kinetic Systems*, Wiley-Interscience, New York, 1972.

34 D. Eisenberg and D. Crothers, *Physical Chemistry*, W. A. Benjamin, Menlo Park, Calif., 1979.

35   H. Eyring and E. M. Eyring, *Modern Chemical Kinetics*, Van Nostrand Reinhold, New York, 1963.

36   H. E. Avery, *Basic Reaction Kinetics and Mechanisms*, Macmillan, New York, 1974.

37   K. G. Denbigh, *Chemical Reactor Theory*, Cambridge University Press, 1965.

38   J. F. Pankow and J. J. Morgan, *Environ. Sci. Technol.*, **13**, 1248 (1979).

39   R. A. Berner, *Principles of Chemical Sedimentology*, McGraw-Hill, New York, 1971.

40   R. A. Berner, *Amer. J. Sci.*, **278**, 1235 (1978).

41   W. S. Broecker and T. H. Peng, *Tellus*, **26**, 21 (1974).

42   P. S. Liss and P. G. Slater, *Nature*, **247**, 181 (1974).

43   A. Lerman, *Geochemical Processes*, Wiley, New York, 1979.

44   S. Emerson, *Limnol. Oceanogr.*, **20**, 743 (1975).

45   J. C. Westall, J. I. Zachary, and F. M. M. Morel, *MINEQL, A Computer Program for the Calculation of Chemical Equilibrium Composition of Aqueous Systems*, **TN-18**, Parsons Laboratory, M.I.T. Cambridge, Mass. (1976).

46   F. J. Zeleznik and S. Gordon, *Ind. Eng. Chem.*, **60**, 27 (1968).

47   Van Zeggeren, F., and S. H. Storey, *The Computation of Chemical Equilibria*, Cambridge University Press (1970).

49   N. Ingri, W. Kakolowicz, L. G. Sillén, B. Warnquist, *Talanta*, **14**, 1261 (1967).

50   D. D. Perrin and I. G. Sayce, *Talanta*, **14**, 833 (1967).

51   A. H. Truesdell and B. F. Jones, *J. Res. U.S. Geol. Surv.*, **2**, 233 (1974).

52   F. M. M. Morel and J. J. Morgan, *Environ. Sci. Technol.*, **6**, 58 (1972).

53   F. M. M. Morel, R. E. McDuff, and J. J. Morgan, *Mar. Chem.*, **4**, 1 (1976).

54   J. I. Bada, in *Nonequilibrium Systems in Natural Water Chemistry*, Advances in Chemistry Series, No. 106, American Chemical Society, Washington, D.C., 1971, pp. 309–334.

55   M. R. Hoffmann, *Environ. Sci. Tech.*, **15**, 345 (1981).

56   D. M. Imboden and A. Lerman, in *Lakes*, A. Lerman, Ed., Springer-Verlag, New York, (1979), pp. 341 56.

READING SUGGESTIONS

Adamson, A. W., *A Textbook of Physical Chemistry*, 2nd ed., Academic, New York, 1979.

Avery, H. E., *Basic Reaction Kinetics and Mechanisms*, Macmillan, New York, 1974.

Benson, D., *Mechanisms of Inorganic Reactions in Solution*, McGraw-Hill, New York, 1968.

Benson, S. W., *The Foundations of Chemical Kinetics*, McGraw-Hill, New York, 1960.

Berner, R. A., *Principles of Chemical Sedimentology*, McGraw-Hill, New York, 1971.

Broecker, W. S., and V. M. Oversby, *Chemical Equilibria in the Earth*, McGraw-Hill, New York, 1971.

Capellos, C., and B. H. J. Bielski, *Kinetic Systems*, Wiley-Interscience, New York, 1972.

Denbigh, K. G., *The Thermodynamics of the Steady State*, Wiley, New York, 1951.

Denbigh, K. G., *The Principles of Chemical Equilibrium: with Applications in Chemistry and Chemical Engineering*, 3rd ed., Cambridge University Press, Cambridge, 1971.

Edwards, J. O., *Inorganic Reaction Mechanisms: An Introduction*, W. A. Benjamin, Menlo Park, Calif., 1964.

Eisenberg, D., and D. Crothers, *Physical Chemistry*, W. A. Benjamin, Menlo Park, Calif., 1979.

Eyring, H., and E. M. Eyring, *Modern Chemical Kinetics*, Van Nostrand, Reinhold, New York, 1963.

Frost, A. A., and R. G. Pearson, *Kinetics and Mechanism*, 2nd ed., Wiley, New York, 1961.

Garrels, R. M., and C. L. Christ, *Solutions, Minerals, and Equilibria*, Harper and Row, New York, 1965.

Guggenheim, E. A., *Thermodynamics: An Advanced Treatment for Chemists and Physicists*, 5th ed., North-Holland, Amsterdam, 1967.

Harned, H. S., and B. B. Owen, *The Physical Chemistry of Electrolytic Solutions*, 3rd ed., Van Nostrand, Reinhold, New York, 1958.

King, E. L., *How Chemical Reactions Occur*, W. A. Benjamin, Menlo Park, Calif., 1964.

Klotz, I. M., *Chemical Thermodynamics: Basic Theory and Methods*, revised ed., W. A. Benjamin, Menlo Park, Calif., 1964.

Laidler, K. J., *Chemical Kinetics*, 2nd ed., McGraw-Hill, New York, 1965.

Latimer, W. M., *The Oxidation States of the Elements and Their Potentials in Aqueous Solutions*, 2nd ed., Prentice-Hall, Englewood Cliffs, N.J., 1952.

Lee, T. S., "Chemical Equlibrium and the Thermodynamics of Reactions." In *Treatise on Analytical Chemistry*, Part I, Vol. 1, I. M. Kolthof and P. J. Elving, Eds., Interscience, New York, 1959, pp. 185–275.

Lerman, A., *Geochemical Processes*, Wiley, New York, 1979.

Lewis, G. N., and M. Randall, *Thermodynamics*, 2nd ed., revised by K. S. Pitzer and L. Brewer, McGraw-Hill, New York, 1961.

Millero, F. J., "Thermodynamic Models for the State of Metal Ions in Seawater." In *The Sea*, E. D. Goldberg, et al., Eds., Vol. 6, Wiley-Interscience, New York, 1978, pp. 653–692.

Prigogine, I., and R. Defay, *Chemical Thermodynamics*, translated by D. H. Everett, McKay, New York, 1954.

Robinson, R. A., and R. H. Stokes, *Electrolyte Solutions: The Measurement and Interpretation of Conductance, Chemical Potential and Diffusion in Solutions of Simple Electrolytes*, 2nd ed., Butterworths, London, 1959.

Rock, P. A., "Fixed Pressure Standard States in Thermodynamics and Kinetics," *J. Chem. Educ.*, **44**, 104, 1967.

Staples, B. R., "Equilibrium Constants and Thermochemical Data," *Environ. Sci. Technol.*, **12**, 339, 1978.

Sykes, A. G., *Kinetics of Inorganic Reactions*, Pergamon, Elmsford, N.Y., 1966.

Van Zeggeren, F., and S. H. Storey, *The Computation of Chemical Equilibria*, Cambridge University Press, Cambridge, 1970.

Wall, F. T., *Chemical Thermodynamics*, 2nd ed., Freeman, San Francisco, 1965.

Whitfield, M., "Seawater as an Electrolyte Solution." In *Chemical Oceanography*, J. P. Riley and G. Skirrow, Eds., Vol. 1, 2nd ed., Academic, New York, 1975, pp. 44–171.

Wilkins, R. G. *The Study of Kinetics and Mechanisms of Reactions of Transition Metal Complexes*, Allyn and Bacon, Boston, 1974.

3

Acids and Bases

3.1 INTRODUCTION

pH values for most mineral-bearing waters are known to lie generally within the narrow range 6 to 9 and to remain very nearly constant for any given water. The composition of natural waters is influenced by the interactions of acids and bases. According to Sillén [1] one might say that the ocean is the result of a gigantic acid–base titration; acids that have leaked out of the interior of the earth are titrated with bases that have been set free by the weathering of primary rock. $[H^+]$ of natural waters is of great significance in all chemical reactions associated with the formation, alteration, and dissolution of minerals. The pH of the solution will determine the direction of the alteration process.

Biological activities such as photosynthesis and respiration, and physical phenomena such as natural or induced turbulence with concomitant aeration, influence pH regulation through their respective abilities to decrease and increase the concentration of dissolved carbon dioxide. Besides photosynthesis and respiration, other biologically mediated reactions affect the H^+ ion concentrations of natural waters. Oxygenation reactions lead to a decrease in pH, whereas processes such as denitrification and sulfate reduction tend to increase pH.

Because of the ubiquitousness of carbonate rocks and the equilibrium reactions of CO_2, bicarbonate and carbonate are present as bases in most natural waters. In addition small concentrations of the bases borate, phosphate, arsenate, ammonia, and silicate may be present in the solution. Volcanoes and certain hot springs may yield strongly acid water by adding gases like HCl and SO_2. Free acids also enter natural water systems as a result of the disposal of industrial waters. Similarly, acidity is imparted through hydrolysis of multivalent metal ions. Other acidic constituents are boric acid, silicic acid, and ammonium ions. From the point of view of interaction with solid and dissolved bases, the most important acidic constituent is CO_2, which forms H_2CO_3 with water.

Proton transfer reactions are usually very fast (half-lives less than milliseconds). Equilibria characterizing H^+ ion transfer reactions are among the simplest types of equilibrium models. In this chapter we devote considerable space to develop the idea of chemical equilibrium using acid–base reactions as examples. This gives us an opportunity to discuss the various activity conventions in some detail. We also demonstrate numerical and graphical methods to compute the equilibrium composition of simple acid–base transfer systems. The graphical procedures using pH as a master variable permit us to survey conveniently the interrelationships of the equilibrium concentrations of individual solute species.

Hydrogen ion regulation in natural waters is provided by numerous homogeneous and heterogeneous buffer systems. It is important to distinguish in these systems between intensity factors (pH) and capacity factors (e.g., the total acid- or base-neutralizing capacity). The buffer intensity is found to be an implicit function of both these factors. In this chapter we discuss acid–base equilibria primarily from a general and didactic point of view. In Chapter 4 we address ourselves more specifically to the dissolved carbonate system.

3.2   THE NATURE OF ACIDS AND BASES

Arrhenius (1887) proposed that an acid was a substance whose water solution contained an excess of hydrogen ions and that a base contained an excess of hydroxide ions in aqueous medium. Arrhenius assumed that the excess H^+ ions or OH^- ions in a water solution resulted from dissociation of the acid or base as it was introduced into the water. In the light of more recent knowledge the Arrhenius concept is limited. First, it applies to ionic solutions only. Second, the dissociation theory is no longer tenable, since we now know, for example, that bases such as NaOH exist as ions even in the crystalline state. Furthermore, it is well known today that a hydrogen ion, that is, a proton, cannot exist as a bare ion in water solution. Theoretical calculations show that a proton would strongly react with a water molecule to form a hydrated proton, a hydronium or a hydroxonium ion (H_3O^+). Actually the H_3O^+ ion in an aqueous solution is itself associated through hydrogen bonds with a variable number of H_2O molecules: $(H_7O_3)^+$, $(H_9O_4)^+$, and so on. The formula H_3O^+ or H^+ is generally used, however, to denote a hydrated hydrogen ion. Formulas analogous to H_3O^+ for the solvated proton are used for other solvents; for example, NH_4^+ in liquid ammonia or $C_2H_5OH_2^+$ in ethanol. The hydroxide ion is also strongly hydrated in aqueous solutions. Similarly, metal ions do not occur as bare metal ions but as aquo complexes.

The Brønsted Concept.   The fact that hydrogen ions cannot exist unhydrated in water solution is incompatible with the notion of a simple dissociation of an acid. The ionization of an acid in water may more logically be represented as a reaction of the acid with the water

$$HCl + H_2O = H_3O^+ + Cl^- \tag{1}$$

The function of a base in its reaction with water is the opposite of that of an acid

$$NH_3 + H_2O = NH_4^+ + OH^- \tag{2}$$

This general idea has led to the very broad concept of acids and bases proposed by Brønsted. (The same concept was suggested by Lowry.) Accordingly an acid is simply defined as any substance that can donate a proton to any other substance, and a base as any substance that accepts a proton from another substance; that is, an acid is a proton donor and a base is a proton acceptor. Thus a proton transfer can occur only if an acid reacts with a base:

$$\begin{array}{c} \text{Acid}_1 = \text{Base}_1 + \text{proton} \\ \underline{\text{proton} + \text{Base}_2 = \text{Acid}_2} \\ \text{Acid}_1 + \text{Base}_2 = \text{Acid}_2 + \text{Base}_1 \end{array} \tag{3}$$

Further illustrations of such proton transfers are:

$$\text{Acid}_1(A_1) + \text{Base}_2(B_2) \rightleftharpoons \text{Acid}_2(A_2) + \text{Base}_1(B_1)$$
$$\text{(solvent)}$$

Perchloric acid	$HClO_4$	$+ H_2O$	$= H_3O^+$	$+ ClO_4^-$	(4a)
Carbonic acid	H_2CO_3	$+ H_2O$	$= H_3O^+$	$+ HCO_3^-$	(4b)
Bicarbonate	HCO_3^-	$+ H_2O$	$= H_3O^+$	$+ CO_3^{2-}$	(4c)
Ammonium	NH_4^+	$+ H_2O$	$= H_3O^+$	$+ NH_3$	(4d)
Ammonium	NH_4^+	$+ C_2H_5OH$	$= C_2H_5OH_2^+$	$+ NH_3$	(4e)
Acetic acid[a]	HAc	$+ NH_3$	$= NH_4^+$	$+ Ac$	(4f)
Water	H_2O	$+ H_2O$	$= H_3O^+$	$+ OH^-$	(4g)
Water	H_2O	$+ NH_3$	$= NH_4^+$	$+ OH^-$	(4h)

[a] HAc and Ac⁻ stand for acetic acid and the acetate ion, respectively.

In the same way the reaction of bases accepting protons from acids can be illustrated:

$$B_1 + A_2 \rightleftharpoons B_2 + A_1$$
$$\text{(solvent)}$$

Ammonia[a]	NH_3	$+ H_2O$	$= OH$	$+ NH_4^+$	(5a)
Cyanide	CN^-	$+ H_2O$	$= OH^-$	$+ HCN$	(5b)
Bicarbonate	HCO_3^-	$+ H_2O$	$= OH^-$	$+ H_2CO_3$	(5c)
Carbonate	CO_3^{2-}	$+ H_2O$	$= OH^-$	$+ HCO_3^-$	(5d)
Ammonia	NH_3	$+ C_2H_5OH$	$= C_2H_5O^-$	$+ NH_4^+$	(5e)
Amine	RNH_2	$+ HAc$	$= Ac^-$	$+ RNH_3^+$	(5f)
Hydroxide	OH^-	$+ NH_3$	$= NH_2^-$	$+ H_2O$	(5g)

[a] The ammonia molecule is represented by NH_3 rather than NH_4OH. The best evidence available indicates that no NH_4OH molecule exists. NH_3 may be loosely bound (hydrogen bonding) to a number of H_2O molecules.

Reaction 4g illustrates the self-ionization of the solvent water, that is, water is both a proton donor and a proton acceptor, an acid and a base in the Brønsted sense. Similarly, self-ionization in liquid NH_3 is represented by

$$NH_3 + NH_3 = NH_4^+ + NH_2^-$$

and self-ionization in the solvent H_2SO_4 is represented by

$$H_2SO_4 + H_2SO_4 = H_3SO_4^+ + HSO_4^-$$

Reactions of a salt constituent, cation or anion, with water (equations 4d, 5b, 5c, 5d) have been referred to as *hydrolysis* reactions. Within the framework of the Brønsted theory, the term hydrolysis is no longer necessary, since in principle there is no difference involved in the protolysis of a molecule and that of a cation or anion to water.

Metal ions, like hydrogen ions, exist in aqueous medium as hydrates. Many metal ions coordinate four or six molecules of H_2O per ion. H_2O can act as a weak acid. The acidity of the H_2O molecules in the hydration shell of a metal ion is much larger than that of water. This enhancement of the acidity of the coordinated water may, in a primitive model, be visualized as the result of the repulsion of the protons of H_2O molecules by the positive charge of the metal ion or as a result of the immobilization of the lone electron pair of the hydrate–H_2O molecule. Thus hydrated metal ions are acids:

$$[Al(H_2O)_6]^{3+} + H_2O = H_3O^+ + [Al(OH)(H_2O)_5]^{2+} \qquad (6)$$

To a first approximation their acidity increases with the decrease in the radius and an increase in the charge of the central ion. Similarly, the acidity of boric acid H_3BO_3 or $B(OH)_3$ can be represented formally as

$$H_3BO_3(H_2O)_x + H_2O = B(OH)_4^-(H_2O)_{x-1} + H_3O^+ \qquad (7)$$

because the borate ion is $B(OH)_4^-$ and not $H_2BO_3^-$. Acidic properties of some substances in aqueous solutions can be interpreted in terms of proton transfer only by assuming hydration of the substance and loss of protons from the primary hydration shell. This is demonstrated in the acidity of metal ions.

In the illustration given above for proton transfer reactions (equations 4 and 5) $Acid_1$ and $Base_1$ or $Acid_2$ and $Base_2$ form *conjugate acid–base pairs*. Thus chloride is the conjugate base of hydrogen chloride; the latter is the conjugate acid of chloride. The conjugate acid of the base water is the hydronium ion, and the conjugate base of the acid water is the hydroxide ion.

Many acids can donate more than one proton. Examples are H_2CO_3, H_3PO_4, and $[Al(H_2O)_6]^{3+}$. These acids are referred to as *polyprotic acids*. Similarly, bases that can accept more than one proton, for example, OH^-, CO_3^{2-}, and NH_2^-, are polyprotic bases. Many important substances, for example, proteins or polyacrylic acids, so-called polyelectrolytic acids or bases, contain a large number of acidic or basic groups.

The *Lewis concept* of acids and bases (G. N. Lewis, 1923) interprets the combination of acids with bases in terms of the formation of a coordinate covalent bond. A Lewis acid can accept and share a lone pair of electrons donated by a Lewis base. Because protons readily attach themselves to lone electron pairs, Lewis bases are also Brønsted bases. Lewis acids, however, include a large number of substances in addition to proton donors: for example, metal ions, acidic oxides, or atoms. The Lewis concept will be discussed more fully in Chapter 6.

3.3   THE STRENGTH OF AN ACID OR BASE

The strength of an acid or base is measured by its tendency to donate or accept a proton, respectively. Thus a weak acid is one that has a weak proton-donating tendency; a strong base is one that has a strong tendency to accept protons. It is, however, difficult to define the "absolute" strength of an acid or base, since the extent of proton transfer (protolysis) depends not only on the tendency of proton donation by $Acid_1$ but also on the tendency of proton acceptance by

Base$_2$. Under these circumstances the *relative* strengths of acids are measured with respect to a standard Base$_2$—usually the solvent. In aqueous solutions, the acid strength of a conjugate acid–base pair, HA–A$^-$, is measured relative to the conjugate acid–base system of H_2O, that is, H_3O^+–H_2O. In a similar way the relative base strength of a conjugate base–acid pair, B–HB$^+$, is defined in relation to the base–acid system of water, OH$^-$–H_2O.

The rational measure of the strength of the acid HA relative to H_2O as proton acceptor is given by the equilibrium constant for the proton transfer reaction

$$HA + H_2O = H_3O^+ + A^- \qquad K_1 \qquad (8)$$

which may be represented formally by two steps:

$$HA = proton + A^- \qquad K_2 \qquad (9)$$

$$H_2O + proton = H_3O^+ \qquad K_3 \qquad (10)$$

Because the concentration (activity) of water is essentially constant in dilute aqueous solutions ($\sim 55\ M$), the hydration of the proton can be ignored in defining acid–base equilibria. Because the equilibrium activity of the proton and of H_3O^+ are not known separately, the thermodynamic convention sets the standard free energy change $\Delta G°$ for reaction 10 equal to zero, that is, $K_3 = 1$. In dealing with dilute solutions we can, because of this convention, represent the aquo hydrogen ion by H^+(aq), or more conveniently by H^+; that is,

$$[H^+] \equiv [H^+(aq)] = \sum_n [H(H_2O)_n^+(aq)] \qquad (11)$$

and the free energy change ΔG involved in the proton transfer reaction 8 may be expressed in terms of the equilibrium constant of reaction 9, that is, the acidity constant of the acid HA, K_{HA}. Ignoring activity coefficients, we have

$$K_2 = K_1 = K_2 K_3 = K_{HA} = \frac{[H^+][A^-]}{[HA]} \qquad (12)$$

which upon rearrangement gives

$$pH = pK_{HA} + \log \frac{[A^-]}{[HA]} \qquad (13)$$

Strong Acids and Bases.   In aqueous systems some acids are stronger than H_3O^+ and some bases are stronger than OH$^-$. Water exerts a leveling influence because of its very high concentration, and pH values much lower than zero or much higher than 14 cannot be achieved in dilute aqueous solutions. In such solutions acids stronger than H_3O^+ and bases stronger than OH$^-$ are not stable as protonated or deprotonated species, respectively.

Strong Acids in Rainwater.   Rainwaters often contain strong acids. These acids are formed from atmospheric pollutants (HCl, HNO_3, H_2SO_4) (Figure 3.1).

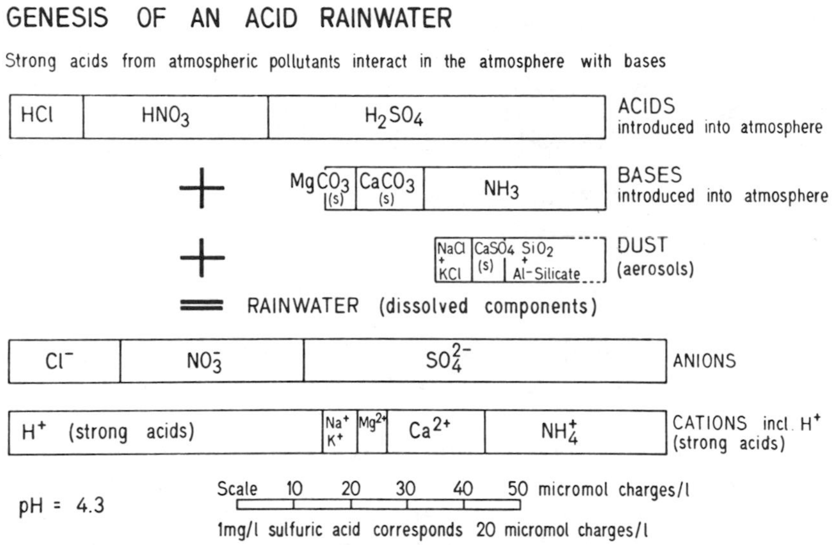

GENESIS OF AN ACID RAINWATER

Strong acids from atmospheric pollutants interact in the atmosphere with bases

Figure 3.1 Strong acids in rainwater. The acid–base reaction involved in the genesis of a typical acid rainwater. Acids formed from atmospheric pollutants react in the atmosphere with bases and dust particles. The resulting rainwater contains an excess of strong acids. H_2SO_4 originates mostly from S in fossil fuels; after combustion the SO_2 formed is oxidized to SO_3 which gives, with H_2O, H_2SO_4; HNO_3 originates from NO and NO_2. These molecules are formed in the combustion of fossil fuels and to a large extent in the combustion of the automobile engine. For each molecule of NO one of HNO_3 is formed, for example, $NO + O_3 \rightarrow NO_2 + O_2$; $3 NO_2 + H_2O \rightarrow 2 HNO_3 + NO$. HCl may largely originate from the combustion of Cl-bearing polymers, for example, polyvinyl chloride, in refuse incinerations. Most bases in the atmosphere are often of natural origin. Atmospheric dust may contain carbonates (calcite and dolomite). NH_3 is released from many soils (together with urea). (From J. Zobrist and W. Stumm, *Forschung und Technik, Neue Zürcher Zeitung*, June 1979.)

Self-ionization of Water.   In all aqueous solutions the autoprotolysis

$$H_2O + H_2O = H_3O^+ + OH^- \tag{14}$$

has to be considered. In dilute aqueous solutions ($\{H_2O\} = 1$) the equilibrium constant for equation 14, usually called the ion product of water, is

$$K_w = \{OH^-\}\{H_3O^+\} \equiv \{OH^-\}\{H^+\} \tag{15}$$

At 25°C, $K_w = 1.008 \times 10^{-14}$ and the pH = 7.00 corresponds to exact neutrality in pure water ($[H^+] = [OH^-]$). Because K_w changes with temperature (Table 3.1), the pH of neutrality also changes with temperature.

TABLE 3.1.   ION PRODUCT OF
WATER[a]

°C	K_w	pK_w
0	0.12×10^{-14}	14.93
5	0.18×10^{-14}	14.73
10	0.29×10^{-14}	14.53
15	0.45×10^{-14}	14.35
20	0.68×10^{-14}	14.17
25	1.01×10^{-14}	14.00
30	1.47×10^{-14}	13.83
50	5.48×10^{-14}	13.26

[a] $\log K_w = -4470.99/T + 6.0875 - 0.01706T$ (T = absolute temperature). From Harned and Owen, *The Physical Chemistry of Electrolytic Solutions*, van Nostrand Reinhold, New York, 1958. Reproduced with permission from Reinhold Publishing Corporation.

For seawater (34.82% salinity, 25°C) C. Culberson and R. M. Pytkowicz [*Mar. Chem.*, **1**, 309 (1973)] determined $\log[H^+][OH^-] = -13.199$.

Pressure dependence of the ion product of water has been measured by M. Whitfield (*J. Chem. Eng. Data* **17**, 1972, pp. 124–128) and calculated by F. J. Millero, E. V. Hoff and L. Kahn (*J. Solution Chem.*, **1**, 309 (1972)] at 15°C and $I = 0.1$; values for $K_{w(p)}/K_{w(l)}$ are, respectively, for 200 atm, 1.20; 400 atm, 1.41; 600 atm, 1.62; 800 atm, 1.91; and 1000 atm, 2.19.

Equation 15 interrelates the acidity constant of an acid with the basicity constant of its conjugate base; for example, for the acid–base pair, HB^+-B, the basicity constant for the reaction

$$B + H_2O = OH^- + BH^+$$

is

$$K_B = \frac{\{HB^+\}\{OH^-\}}{\{B\}} \tag{16}$$

and the acidity constant for

$$HB^+ + H_2O = H_3O^+ + B$$

is

$$K_{HB^+} = \frac{\{H^+\}\{B\}}{\{HB^+\}} \tag{17}$$

Thus

$$K_{HB^+} = \frac{\{H^+\}\{B\}}{\{HB^+\}} = \frac{\{H^+\}\{OH^-\}}{K_B}$$

or

$$K_W = K_{HB^+} \cdot K_B \qquad (18)$$

Thus either the acidity or basicity constant describes fully the protolysis properties of an acid–base pair. The stronger the acidity of an acid, the weaker the basicity of its conjugate base, and vice versa. For illustration purposes Table 3.2 lists a series of acids and bases in the order of their relative strength. Complications of acidity constants valid at different temperatures and ionic strengths are contained in [2] and [3].

Composite Acidity Constants

It is not always possible to specify a protolysis reaction unambiguously in terms of the actual acid or base species. As it is possible to ignore the extent of hydration of the proton in dilute aqueous solutions, the hydration of an acid or base species can be included in a composite acidity constant; for example, it is difficult analytically to distinguish between $CO_2(aq)$ and H_2CO_3. The equilibria are

$$H_2CO_3 = CO_2(aq) + H_2O \qquad K = \frac{\{CO_2(aq)\}}{\{H_2CO_3\}} \qquad (19)$$

$$H_2CO_3 = H^+ + HCO_3^- \qquad K_{H_2CO_3} = \frac{\{H^+\}\{HCO_3^-\}}{\{H_2CO_3\}} \qquad (20)$$

and a combination of equations 19 and 20 gives

$$\frac{\{H^+\}\{HCO_3^-\}}{(\{H_2CO_3\} + \{CO_2(aq)\})} = \frac{K_{H_2CO_3}}{1 + K} = K_{H_2CO_3^*} \qquad (21)$$

where $K_{H_2CO_3^*}$ is the composite acidity constant. Under conditions in which activities can be considered equal to concentration, the sum $\{H_2CO_3\} + \{CO_2\}$ is approximately equal to the sum of the concentrations, $[H_2CO_3] + [CO_2(aq)]$. $[H_2CO_3^*]$ is defined as the analytic sum of $[CO_2(aq)]$ and $[H_2CO_3]$. The true H_2CO_3 is a much stronger acid ($pK_{H_2CO_3} = 3.8$) than the composite $H_2CO_3^*$ ($pK_{H_2CO_3^*} = 6.3$), because less than 0.3 % of the CO_2 is hydrated at 25°C.

An amino acid can lose two protons by two different paths:

$$^+NH_3R\ COOH \underset{NH_2RCOOH}{\overset{NH_3^+RCOO^-}{\rightleftharpoons}} NH_2RCOO^- \qquad (22)$$

Four microscopic constants can be defined, but potentiometrically only two composite (macroscopic) acidity constants can be determined. The acid–base behavior of polymers will be discussed in Section 6.9.

TABLE 3.2   ACIDITY AND BASICITY CONSTANTS OF ACIDS AND BASES IN AQUEOUS SOLUTIONS (25°C)

Acid^a		$-$Log Acidity Constant, pK (approximate)	Base^b	$-$Log Basicity Constant, pK (approximate)
$HClO_4$	Perchloric acid	-7	ClO_4^-	21
HCl	Hydrogen chloride	~ -3	Cl^-	17
H_2SO_4	Sulfuric acid	~ -3	HSO_4^-	17
HNO_3	Nitric acid	-1	NO_3^-	15
H_3O^+	Hydronium ion	0	H_2O	14
HSO_4^-	Bisulfate	1.9	SO_4^{2-}	12.1
H_3PO_4	Phosphoric acid	2.1	$H_2PO_4^-$	11.9
$[Fe(H_2O)_6]^{3+}$	Aquo ferric ion	2.2	$[Fe(H_2O)_5(OH)]^{2+}$	11.8
CH_3COOH	Acetic acid	4.7	CH_3COO^-	9.3
$[Al(H_2O)_6]^{3+}$	Aquo aluminum ion	4.9	$[Al(H_2O)_5(OH)]^{2+}$	9.1
$H_2CO_3^*$	Carbon dioxide^c	6.3	HCO_3^-	7.7
H_2S	Hydrogen sulfide	7.1	HS^-	6.9
$H_2PO_4^-$	Dihydrogen phosphate	7.2	HPO_4^{2-}	6.8
$HOCl$	Hypochlorous acid	7.6	OCl^-	6.4
HCN	Hydrogen cyanide	9.2	CN^-	4.8
H_3BO_3	Boric acid	9.3	$B(OH)_4^-$	4.7
NH_4^+	Ammonium ion	9.3	NH_3	4.7
$Si(OH)_4$	O-Silicic acid	9.5	$SiO(OH)_3^-$	4.5
HCO_3^-	Bicarbonate	10.3	CO_3^{2-}	3.7
H_2O_2	Hydrogen peroxide	—	HO_2^-	2.3
$SiO(OH)_3^-$	Silicate	12.6	$SiO_2(OH)_2^{2-}$	1.4
HS^-	Bisulfide	14	S^{2-}	0
H_2O	Water	14	OH^-	0
NH_3	Ammonia	~ 23	NH_2^-	-9
OH^-	Hydroxide ion	~ 24	O^{2-}	~ -10
CH_4	Methane	~ 34	CH_3^-	~ -20

^{a} In order of decreasing acid strength. ^{b} In order of increasing base strength. ^{c} Total unionized CO_2 in water.

129

3.4 ACTIVITY AND pH SCALES

In dealing with quantitative aspects of chemical equilibrium, we inevitably are faced with the problem either of evaluating or maintaining constant the activities of the ions under consideration. G. N. Lewis (1907) defined the chemical activity of a solute A, {A}, and its relationship to chemical concentration of that solute, [A], by

$$\mu_A = k_A + RT \ln\{A\} = k_A + RT \ln[A] + RT \ln f_A \tag{23}$$

where μ_A is the chemical potential of species A and k_A is a constant that identifies the concentration scale adopted (mol liter^{-1}, mol kg^{-1}, or mole fraction) and corresponds to the value of μ at {A} = 1. It will be recognized that k_A corresponds to the more familiar standard-state chemical potential μ_A° and to the standard partial molar Gibbs energy $\bar{G}_A^\circ$.

A more extensive discussion on the various scales for equilibrium constants, activity coefficients, and pH will be given in the appendix to Chapter 6.

Concentration Scales

Any activity can be written as the product of a concentration and activity coefficient. Here we usually express concentration in terms of mol liter^{-1} of solution (molarity, M). Concentration may also be expressed in terms of mol kg^{-1} of solvent, that is, water (molality, m). The *molal scale* gives concentrations that are independent of temperature and pressure and is used in many precise physicochemical calculations†. The difference between molarity and molality is small in dilute solutions, especially in comparison to the uncertainties involved in determining equilibrium constants or in estimating activity coefficients; for example, a 1.00 molal solution of NaCl is 0.98 M (25°C, 1 atm). Fundamentally the concentration scale most suitable for expressing deviations from ideality is the *mole fraction* scale $\left[x_i = n_i \middle/ \left(n_i + \sum_j n_j \right) \right.$, i.e., moles of solute per mole of solution$\left.\vphantom{\sum}\right]$; but this scale is inconvenient to use experimentally. Molality [weight solute × 1000/(formula weight solute × weight water)] can be converted into molarity [weight solute × 1000/(formula weight solute × volume solution)] by

$$m = M \frac{\text{weight solution}}{\text{weight solution} - \text{weight solutes}} \times \frac{1}{\text{density}} \tag{24}$$

As already summarized in Table 2.3, two activity scales are useful:

---

† In seawater analysis one often uses the units mol kg^{-1} seawater, which is also pressure- and temperature-independent.

The Infinite Dilution Scale.   This activity convention is defined in such a way that the activity coefficient, $f_A = \{A\}/[A]$, approaches unity as the concentration of all solutes approaches zero; that is,

$$f_A \to 1 \quad \text{as} \quad \left(c_A + \sum_i c_i\right) \to 0$$

where c is the concentration in molar or molal units.

The Ionic Medium Scale.   This convention can be applied to solutions that contain a "swamping" concentration of inert electrolyte in order to maintain a constant ionic medium. The activity coefficient, $f' = \{A\}/[A]$, becomes unity as the solution approaches the pure ionic medium, that is, when all concentrations other than the medium ions approach zero:

$$f'_A \to 1 \quad \text{as} \quad c_A \to 0 \quad \text{in a solution}$$

where the total concentration is still $\sum_i c_i$. If the concentration of the medium electrolyte is more than approximately 10 times the concentration of the species under consideration, activity coefficients remain very close to 1. Thus generally no extrapolation is necessary (Figure 3.2).

As equation 23 illustrates, a change in the activity scale convention merely changes k_A. In an ideal constant ionic medium equation 23 becomes

$$\mu_A = k'_A + RT \ln[A]$$

As pointed out by Sillen [4], both activity scales are thermodynamically equally well defined. In constant ionic medium, activity ($\approx$ concentration) can frequently be determined by means of emf methods. In the last few decades an increasing number of researchers in aqueous solution chemistry have been using the ionic medium activity scale.

pH Conventions.   The original definition of pH (Sørensen, 1909) was

$$pH = -\log[H^+] \tag{25}$$

Within the infinite dilution concept pH may be defined in terms of hydrogen ion activity:

$$p^aH = -\log\{H^+\} = -\log[H^+] - \log f_{H^+} \tag{26}$$

Within the ionic medium convention, Sørensen's original definition (equation 25) may be used operationally because one can usually measure $-\log[H^+] = -\log\{H^+\}$ rather accurately (Sillén [4]). An electrode system is calibrated with solutions of known concentrations of strong acid (e.g., $HClO_4$), which are

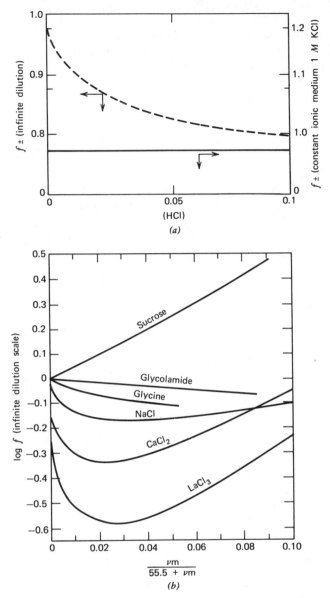

Figure 3.2 Activity coefficients depend on the selection of the reference state and standard state. (a) On the *infinite dilution scale* the reference state is an infinitely dilute aqueous solution; the standard state is a hypothetical solution of concentration unity and with properties of an infinitely dilute solution). For example, the activity coefficient of H^+ in a HCl solution, $f\pm_{HCl}$, varies with [HCl] in accordance with a Debye–Hückel equation dashed line, left ordinate) (see Table 3.3); only at very great dilutions does f become unity. On the *ionic medium scale*, for example, in 1 M KCl, the reference state is the ionic medium (i.e., infinitely diluted with respect to HCl only). In such a medium f_{HCl} (solid line, right ordinate) is very nearly constant, that is, $f_{HCl} = 1$. Both activity coefficients are thermodynamically equally meaningful (Modified from P. Schindler). (b) A comparison of activity coefficients (infinite dilution scale) of electrolytes and nonelectrolytes as a function of concentration (mole fraction of solute) m = moles of solute per kg of solvent (molality); v = number of moles of ions formed from 1 mol of electrolyte; 1 kg solvent contains 55.5 mol of water. [From Robinson and Stokes, *Electrolyte Solutions* (1959). Reproduced with permission from Butterworths, Inc., London.]

adjusted with an electrolyte to the appropriate ionic strength,† the observed potentiometer (pH meter) reading being compared with [H$^+$].

An operational definition endorsed by the International Union of Pure and Applied Chemistry (IUPAC) and based on the work of Bates [5] determines pH relative to that of a standard buffer (where pH has been estimated in terms of p^{a}H) from measurements on cells with liquid junctions; the NBS (National Bureau of Standards) pH scale. This operational pH is not rigorously identical with p^{a}H defined in equation 26 because liquid junction potentials and single ion activities cannot be evaluated without nonthermodynamic assumptions. In dilute solutions of simple electrolytes (ionic strength, $I < 0.1$) the measured pH corresponds to within ± 0.02 to p^{a}H [5]. Measurement of pH by emf methods is discussed in Chapter 7.

Activity of Water.   If the standard state for water in aqueous solutions is taken as the pure solvent, then for all dilute solutions $\{H_2O\} - 1$. In salt solutions (e.g., seawater) the activity of water may then be defined as the ratio of the vapor pressure of the salt solution p_s to that of pure water p_{H_2O}, that is, $\{H_2O\} = p_s/p_{H_2O}$, but even in seawater $\{H_2O\}$ does not fall below 0.98.

Operational Acidity Constants

According to the different activity conventions the following equilibrium expressions may be defined.

1   For the infinite dilution scale,

$$K = \frac{\{H^+\}\{B\}}{\{HB\}} \tag{27}$$

In this and subsequent equations charges are omitted: B can be any base of any charge.

2   In a constant ionic medium, the concentration quotient becomes the equilibrium constant

$$^cK = \frac{[H^+][B]}{[HB]} \tag{28}$$

More exactly, on the ionic medium scale the equilibrium constant may be defined as the limiting value (as the solution composition approaches that of the pure ionic medium) for the equilibrium concentration quotient L. In usual measurements at low reactant concentrations, the deviations of L from

---

† Ionic strength I is a measure of the interionic effect resulting primarily from electrical attraction and repulsions between the various ions; it is defined by the equation $I = \frac{1}{2}\sum_i c_i Z_i^2$. The summation is carried out for all types of ions, cations, and anions in the solution.

^cK are usually smaller than the experimental errors; hence it is preferable to set $L = K$ rather than to extrapolate (Sillén [4]).

3 A so-called mixed acidity constant is frequently used:

$$K' = \frac{\{H^+\}[B]}{[HB]} \tag{29}$$

This convention is most useful when pH is measured according to the IUPAC convention (pH $\approx$ p^{a}H), but the conjugate acid–base pair is expressed in concentrations.

In a way similar to the expression of activity as a product of concentration and activity coefficient, an acidity constant can be expressed in terms of a product of an equilibrium concentration quotient $L = ([H^+][B])/[HB]$ and an activity coefficient factor. Correspondingly we can interrelate the various constants defined above in the following way:

$$K = L \frac{f_{H^+} f_B}{f_{HB}} \tag{30}$$

$$^cK = \lim L \qquad \text{(pure ionic medium)} \tag{31}$$

$$K' = K \frac{f_{HB}}{f_B} \tag{32}$$

Expressions for Activity Coefficients

The theoretical expressions based on the Debye–Hückel limiting law together with more empirical expressions are given in Table 3.3. In defining the mean activity coefficient $f_\pm$ of a solute, z^2 in the equations of Table 3.3 should be replaced by $z_+ z_-$ where the charges are taken without regard to sign.

Single-ion activity coefficients are constructs; they are not measurable individually; only ratios or products of ionic activity coefficients are measureable. The use of single-ion activity coefficients greatly simplifies calculations.

Table 3.4 lists some activity coefficients as calculated from the extended Debye–Hückel limiting law for various values of I. In dilute solutions such calculated values agree well with experimental data for mean activity coefficients of simple electrolytes. At higher concentrations the Davies equation usually represents the experimental data better.

In natural water systems and under many experimental conditions, several electrolytes are present together. The limiting laws for activity coefficients can no longer be applied satisfactorily for electrolyte mixtures of unlike charge types. For estimating unknown activity coefficients, Güntelberg proposed that a in (2) of Table 3.3 be taken as 3.0 Å, resulting in a formula containing no adjustable parameter. For aqueous systems containing several electrolytes, (3) of Table 3.3 is most useful.

TABLE 3.3 INDIVIDUAL ION ACTIVITY COEFFICIENTS

Approximation	Equation[a]		Approximate Applicability [ionic strength (M)]
Debye–Hückel	$\log f = -Az^2\sqrt{I}$	(1)	$<10^{-2.3}$
Extended Debye–Hückel	$= -Az^2 \dfrac{\sqrt{I}}{1 + Ba\sqrt{I}}$	(2)	$<10^{-1}$
Güntelberg	$= -Az^2 \dfrac{\sqrt{I}}{1 + \sqrt{I}}$	(3)	$<10^{-1}$ useful in solutions of several electrolytes
Davies	$= -Az^2 \left(\dfrac{\sqrt{I}}{1 + \sqrt{I}} - 0.2I\right)$	(4)[b]	< 0.5

[a] I (ionic strength) $= \frac{1}{2}\sum C_i z_i^2$; $A = 1.82 \times 10^6 (\varepsilon T)^{-3/2}$ (where $\varepsilon =$ dielectric constant); $A \approx 0.5$ for water at 25°C; $z =$ charge of ion; $B = 50.3(\varepsilon T)^{-1/2}$; $B \approx 0.33$ in water at 25°C; $a =$ adjustable parameter (angstroms) corresponding to the size of the ion. (See Table 3.4.)

[b] Davies has proposed 0.3 (instead of 0.2) as a coefficient for the last term in parentheses.

TABLE 3.4 PARAMETER a AND INDIVIDUAL ION ACTIVITY COEFFICIENTS

Ion Size Parameter, a (Å)[a]	Ion	Activity Coefficients Calculated with (2) of Table 3.3 for Ionic Strength				
		10^{-4}	10^{-3}	10^{-2}	0.05	10^{-1}
9	H^+	0.99	0.97	0.91	0.86	0.83
	Al^{3+}, Fe^{3+}, La^{3+}, Ce^{3+}	0.90	0.74	0.44	0.24	0.18
8	Mg^{2+}, Be^{2+}	0.96	0.87	0.69	0.52	0.45
6	Ca^{2+}, Zn^{2+}, Cu^{2+}, Sn^{2+}, Mn^{2+} Fe^{2+}	0.96	0.87	0.68	0.48	0.40
5	Ba^{2+}, Sr^{2+}, Pb^{2+}, CO_3^{2-}	0.96	0.87	0.67	0.46	0.39
4	Na^+, HCO_3^-, $H_2PO_4^-$, CH_3COO^-	0.99	0.96	0.90	0.81	0.77
	SO_4^{2-}, HPO_4^{2-}	0.96	0.87	0.66	0.44	0.36
	PO_4^{3-}	0.90	0.72	0.40	0.16	0.10
3	K^+, Ag^+, NH_4^+, OH^-, Cl^- ClO_4^-, NO_3^-, I^-, HS^-	0.99	0.96	0.90	0.80	0.76

[a] After J. Kielland, *J. Am. Chem. Soc.*, **59**, 1675 (1937). Reproduced with permission from American Chemical Society.

Activity coefficients defined within the infinite dilution activity scale cannot be formulated theoretically for the ionic medium of seawater. Since the oceans contain an ionic medium of practically constant composition, the ionic medium activity scale might be used advantageously in studying acid–base and other equilibria in seawater (see also the appendix to Chapter 6).

Example 3.1. Individual Activity Coefficients

Estimate p^aH of a 10^{-3} M HCl solution that is 0.05 M in NaCl. The ionic strength is 0.051. Using (3) of Table 3.3, $-\log f_{H^+} = 0.092$. Thus

$$p^aH = -\log[H^+] + 0.092 = 3.09_2.$$

Approximately the same result would be obtained from the extended Debye–Hückel limiting law [(1), Table 3.3] $p^aH = -\log[H^+] + 0.113 = 3.11_3$.

Salt Effects on Acidity Constants

The equilibrium condition of an acid–base reaction is influenced by the ionic strength of the solution. With the help of the formulations of the Debye–Hückel theory or with empirical expressions (Table 3.3) an estimate of the magnitude of the salt effect can be obtained. For example, the mixed acidity constant K' has been related to K (i.e., the acidity constant valid at infinite dilution) by equation 32. Using the Güntelberg approximation for the single-ion activities we can write instead of equation 32

$$pK' = pK + \frac{0.5(z_{HB}^2 - z_B^2)\sqrt{I}}{1 + \sqrt{I}} \tag{33}$$

The numerical values for the correction term (i.e., the second term in equation 33) are given in Table 3.5. Since we are frequently interested in the equilibrium

TABLE 3.5 NONIDEALITY CORRECTIONS FOR MIXED ACIDITY CONSTANTS; NUMERICAL VALUES OF SECOND TERM IN EQUATION 33

Charge Acid, Z_{HB}	Charge Base, Z_B	Ionic Strength, $I(M)$							
		0.0001	0.0005	0.001	0.002	0.005	0.01	0.05	0.1
+1 0	0 −1	+ − 0.005	0.01	0.015	0.02	0.03	0.05	0.09	0.12
+2 −1	+1 −2	+ − 0.015	0.03	0.05	0.06	0.10	0.14	0.27	0.36
+3 −2	+2 −3	+ − 0.02	0.05	0.08	0.11	0.17	0.23	0.46	0.60

concentrations of the various species in a solution of a given pH ($\approx p^a$H), it is convenient to convert K to K'. With the operational K', calculations can be carried out for all species in terms of concentrations with the exception of H^+. In equilibrium calculations, concentration conditions and charge balance or proton conditions are correct only if formulated in terms of concentrations.

Example 3.2. Effect of Ionic Strength on pK'

Estimate the effect of ionic strength on the successive mixed acidity constants (pK' values) of a dilute ($< 10^{-3}$ M), neutralized tribasic acid (e.g., H_3PO_4) in a constant ionic medium of 0.01 M Na_2SO_4 solution. The contribution of the acid and its bases to the ionic strength can be neglected. $I = \frac{1}{2}([Na^+] + 4[SO_4^{2-}]) = 0.03$ M. Using equation 33, we have

$$pK_1' = pK_1 - 0.07$$
$$pK_2' = pK_2 - 0.21$$
$$pK_3' = pK_3 - 0.35$$

3.5 NUMERICAL EQUILIBRIUM CALCULATIONS

The quantitative evaluation of the systematic relations that determine equilibrium concentrations (or activities) of a solution constitutes a purely mathematical problem which is amenable to exact and systematic treatment.

Any acid–base equilibrium can be described by a system of fundamental equations. The appropriate set of equations is comprised of the equilibrium constant (or mass law) relationships (which define the acidity constants and the ion product of water) and any two equations describing the constitution of the solution, for example, equations describing a concentration and an electroneutrality or proton condition. Table 3.6 gives the set of equations and their mathematical combination for pure solutions of acids, bases, or ampholytes in monoprotic or diprotic systems. The problem of the precise calculation in all possible cases has been treated in a truly general way by Ricci [6].

The principle of such equilibrium calculations is best explained by a series of illustrative examples. In this and subsequent examples, a temperature of 25°C is assumed and conditions such that all activity coefficients are equal to 1.

Example 3.3. pH and Equilibrium Composition of a Monoprotic Acid

Calculate the pH ($= -\log [H_3O^+]$) of a 5×10^{-4} M aqueous boric acid solution $[B(OH)_3]$.

TABLE 3.6 [H⁺] OF PURE AQUEOUS ACIDS, BASES OR AMPHOLYTES

I. Monoprotic

Species[a]

$$HA \quad A \quad H^+ \quad OH^-$$

Equilibrium constants[b]

$$[H^+][A]/[HA] = K \tag{1}$$
$$[H^+][OH^-] = K_w \tag{2}$$

Concentration condition

$$[HA] + [A] = C \tag{3}$$

	Acid	Base
Proton condition[c]	$[H^+] = [A] + [OH^-]$ (4)	$[HA] + [H^+] = [OH^-]$ (5)
Numerical solution	$[H^+]^3 + [H^+]^2 K - [H^+](CK + K_w) - KK_w = 0$ (6)	$[H^+]^3 + [H^+]^2(C + K) - [H^+]K_w - KK_w = 0$ (7)

II. Diprotic

Species[a]

$$H_2X \quad HX \quad X \quad H^+ \quad OH^-$$

Equilibrium constants[b]

$$[H^+][HX]/[H_2X] = K_1 \tag{8}$$
$$[H^+][X]/[HX] = K_2 \tag{9}$$
$$[H^+][OH^-] = K_w \tag{2}$$

Concentration condition

$$[H_2X] + [HX] + [X] = C \tag{10}$$

	Acid (H₂X)	Ampholyte (NaHX)	Base (Na₂X)
Proton condition[c]	$[H^+] = [HX] + 2[X] + [OH^-]$ (11)	$[H_2X] + [H^+] = [X] + [OH^-]$ (12)	$2[H_2X] + [HX] + [H^+] = [OH^-]$ (13)
Numerical solution	$[H^+]^4 + [H^+]^3 K_1 + [H^+]^2 \times (K_1 K_2 - CK_1 - K_w) - [H^+]K_1(2CK_2 + K_w) - K_1 K_2 K_w = 0$ (14)	$[H^+]^4 + [H^+]^3(C + K_1) + [H^+]^2 \times (K_1 K_2 - K_w) - [H^+]K_1(CK_2 + K_w) - K_1 K_2 K_w = 0$ (15)	$[H^+]^4 + [H^+]^3(2C + K_1) + [H^+]^2 \times (CK_1 + K_1 K_2 - K_w) - K_1 K_w[H^+] - K_1 K_2 K_w = 0$ (16)

[a] Charges are omitted for acid or base species. Equations given are independent of charge type of the acid.
[b] Equilibrium constants are either ${}^cK$ or are defined in terms of the constant ionic medium activity scale.
[c] Instead of the proton condition, the electroneutrality equation can be used. Independent of charge type, a combination of electroneutrality and concentration condition gives the proton condition. Na in NaHX or Na₂X is used as a symbol of a nonprotolyzable cation (Li⁺, K⁺, ...).

138

In attacking such a problem, it is convenient to proceed systematically through a number of steps.

1   Establish all the *species present* in the solution: H_3O^+, OH^-, $B(OH)_3$, $B(OH)_4^-$. For brevity we call $B(OH)_3 = HB$ and $B(OH)_4^- = B^-$. Since four chemical species in addition to H_2O are involved in solution, four independent mathematical equations are necessary to interrelate the equilibrium concentrations.

2   The *equilibrium constants* that relate the concentrations of the various species must be found. For the boric acid solution the following equilibrium constants can be used:

$$[H^+][OH^-] = K_W = 10^{-14} \tag{i}$$

$$\frac{[H^+][B^-]}{[HB]} = K = 7 \times 10^{-10} \tag{ii}$$

3   A *concentration condition* or a mass balance must be established. Since the analytical concentration C of the HB solution is known, we can write

$$C = 5 \times 10^{-4} \, M = [HB] + [B^-] \tag{iii}$$

The analytical concentration is the total number of moles of a pure substance that has been added to 1 liter of solution. In our example 5×10^{-4} mol of HB has been added per liter of solution. Some of the HB has protolyzed to form B^-, but the sum of $[HB] + [B^-]$ must equal the number of moles (5×10^{-4}) that was added originally.

So far we have established three independent equations. One additional relation has to be found in order to have as many equations as unknowns.

4   This additional equation follows from the fact that the solution must be electrically neutral. This can be expressed in a charge balance or in an *electroneutrality equation*. The total number of positive charges per unit volume must equal the total number of negative charges; that is, the molar concentration of each species is multiplied by its charge:

$$[H^+] = [B^-] + [OH^-] \tag{iv}$$

Instead of writing the electroneutrality equation we can derive a relation called the *proton condition*. If we started making our solution from pure H_2O and HB, we can state that after equilibrium has been reached the number of excess protons must be equal to the number of proton deficiencies. Excess or deficiency of protons is counted with respect to a "zero level" representing the species that were added, that is, H_2O and HB. The number of excess protons is equal to $[H^+]$; the number of proton deficiencies must equal $[B^-] + [OH^-]$. This proton condition gives, as in (iv), $[H^+] = [B^-] + [OH^-]$.

5   Four equations (i to iv) for four unknown concentrations must be solved simultaneously. The exact solution, although straightforward, is tedious. Frequently the problem is readily solved by making suitable approximations. But in order to know what approximations can be made, we have to have a qualitative knowledge of the chemical system within the given concentration range.

We first discuss the exact numerical solution; then we illustrate what we mean by making approximations. In the next section we discuss a method of graphical representation which is expedient in making such calculations.

Exact Numerical Solution.   A numerical approach may start out by eliminating $[OH^-]$ in equations (i and iv). After this substitution we have

$$[H^+] = \frac{K_W}{([H^+] - [B^-])} \tag{v}$$

We now solve equation iii for [HB] and substitute in equation ii, eliminating [HB]:

$$[H^+][B^-] = K(C - [B^-]) \tag{vi}$$

Now we solve equation v for $[B^-]$ and substitute the result in equation vi to obtain a single equation in $[H^+]$:

$$[H^+]^3 + K[H^+]^2 - [H^+](CK + K_W) - KK_W = 0 \tag{vii}$$

A value for $[H^+]$ can be obtained by trial and error from this equation:

$$[H^+]^3 + [H^+]^2(7 \times 10^{-10}) - [H^+](3.6 \times 10^{-13}) - (7 \times 10^{-24}) = 0 \tag{viia}$$

Most numerical methods for solving polynomial equations are based on an initial guess for the answer followed by some iterative procedure for obtaining successively better approximations. A preliminary guess may be obtained by neglecting, one or more terms in equation viia. A convenient way to obtain a root for this equation is to plot various values of $f([H^+])$ versus $[H^+]$ and to locate $[H^+]$ where $f([H^+])$ crosses the zero axis. Programmable scientific calculators can readily be used to make such trial-and-error calculations; the Newton–Raphson method [7] as well as computer methods [8] can be used to achieve a rather systematic convergence of the successive approximations to the required root.)

The value obtained by trial and error ($[H^+] = 6.1 \times 10^{-7}$ M) is substituted into equation i to give $[OH^-]$ and into equation v to obtain $[B^-]$. Then [HB] is calculated from equation iii. The following results obtain:

$$[H^+] = 6.10 \times 10^{-7} \, M \, (pH = 6.21) \qquad [OH^-] = 1.64 \times 10^{-8} \, M$$

$$[B^-] = 5.94 \times 10^{-7} \, M \qquad [HB] = 4.99 \times 10^{-4} \, M$$

Approximate Solution.   The equations to be solved are the same as those given before. We look for terms in additive equations that are negligible. (Multiplicative terms, even if very small cannot of course be neglected.) It is always necessary to check the final result to determine whether the assumptions made were justifiable.

Our solution presumably is acid, that is, $[H^+] > [OH^-]$; we might therefore try to neglect $[OH^-]$ in equation iv as a first approximation. Then we obtain

$$[H^+] = [B^-] \tag{viii}$$

Considering equation viii a combination of equations ii and iii gives .

$$\frac{[H^+]^2}{C - [H^+]} = K \tag{ix}$$

which after conversion to the general quadratic equation, $ax^2 + bx + c = 0$, gives

$$[H^+]^2 + [H^+]K - KC = 0 \tag{x}$$

The solution of this quadratic equation according to

$$x = \frac{[-b \pm (b^2 - 4ac)^{1/2}]}{2a}$$

gives

$$[H^+] = \frac{-K + (K^2 + 4KC)^{1/2}}{2}$$

Only the positive root is physically meaningful. Substitution of $K = 7 \times 10^{-10}$ and $C = 5 \times 10^{-4}$ yields $[H^+] = 5.92 \times 10^{-7}$.

Frequently an approximation can be made in the concentration condition iii. Either the acid is completely protolyzed ($[B^-] \gg [HB]$) or remains predominantly in the acid form ($[HB] \gg [B^-]$). Since HB is a rather weak acid, the latter conditions may be true:

$$C = [HB] = 5 \times 10^{-4} \, M \tag{xi}$$

Substituting this together with the assumption made in equation viii into equation ii gives $[H^+]^2/C = K$ and $[H^+] = 5.92 \times 10^{-7} \, M$. This answer is the same as that obtained using the approximation of equation viii only. With the approximations used, the following are obtained

$[H^+] = 5.92 \times 10^{-7} \, M$ (pH = 6.23) $[OH^-] = 1.69 \times 10^{-8} \, M$

$[B^-] = 5.92 \times 10^{-7} \, M$ $[HB] = 5.0 \times 10^{-4} \, M$

Comparing the approximate results with the "exact" results given before, we see that the values for $[H^+]$ and $[OH^-]$ are off by about 3%. For most calculations of practical interest such an error is tolerable and frequently smaller than the uncertainty of the equilibrium constants used or the activity corrections made, so we may say that the approximations introduced by equations viii and xi were justified.

Example 3.4. pH of a Strong Acid

Compute the equilibrium concentrations of all the species of a 2×10^{-4} M HCl solution (25°C). This problem is essentially analogous to the previous one and can be approached the same way:

1   Mass laws:

$$[H^+][OH] = K_W = 10^{-14} \tag{i}$$

$$\frac{[H^+][Cl^-]}{[HCl]} = K = 10^{+3.0} \tag{ii}$$

2   Concentration condition:

$$[HCl] + [Cl^-] = C = 2 \times 10^{-4} \tag{iii}$$

3   Electroneutrality or proton condition:

$$[H^+] = [Cl^-] + [OH^-] \tag{iv}$$

As in Example 3.3, the exact numerical solution would lead first to

$$[H^+]^3 + K[H^+]^2 - (K_W + CK)[H^+] - K_W K = 0 \tag{v}$$

Solving this by trial and error gives

$$[H^+] = 2.0 \times 10^{-4} \ M \qquad [HCl] = 4 \times 10^{-11} \ M$$

$$[Cl^-] = 2.0 \times 10^{-4} \ M \qquad [OH^-] = 5.0 \times 10^{-11} \ M$$

Since HCl has a large protolysis constant, $[H^+] \gg [OH^-]$ and correspondingly $[OH^-] \ll [Cl^-]$, the electroneutrality condition reduces to $[H^+] = [Cl^-]$. Furthermore, $[HCl] \ll [Cl^-]$; correspondingly the reaction $HCl + H_2O = H_3O^+ + Cl^-$ has gone very far to the right. As shown in this example, the ratio of acid to base, that is, $[HCl]/[Cl^-]$, is extremely small. Strong acids are virtually completely protolyzed.

Example 3.5. Weak Base

Compute the equilibrium concentrations of a $10^{-4.5}$ M sodium acetate (NaAc) solution.

1   Species:

$$HAc, \ Ac^-, \ H^+, \ OH^-, \ Na^+$$

2 Constants:

$$\frac{[H^+][Ac^-]}{[HAc]} = K \quad \text{or} \quad \frac{[HAc][OH^-]}{[Ac^-]} = K_B \qquad (i)$$

$$pK = 4.70 \qquad pK_B = 9.30$$

3 Concentration condition:

$$[HAc] + [Ac^-] = C = [Na^+] \qquad (ii)$$

4 Electroneutrality:

$$[Na^+] + [H^+] = [Ac^-] + [OH^-]; \text{ since } [Na^+] = [HAc] + [Ac^-] \quad (iii)$$

this becomes

$$[HAc] + [H^+] = [OH^-] \qquad (iv)$$

and is identical with the proton condition.

Approximations in the proton condition similar to those used in the previous examples are not permissible in this case. However, an approximation might be possible in the concentration condition. Because Ac^- is a weak base, $[Ac^-] >$ $[HAc]$ and $[Ac^-] \approx C$. If we combine this approximation with the proton condition and the two equilibrium constants, we obtain

$$[H^+]^2(C + K) = KK_W \qquad (v)$$

which results in $[H^+] = 10^{-7.2}$. Subsequent calculations give $[HAc] = 10^{-7.01}$ and $[Ac^-] = 10^{-4.51}$, thus confirming that the result is consistent with the approximation made.

If one is not aware of the possibility of an approximation, one can always attempt to solve the exact equation [(7). Table 3.6].

Example 3.6. Diprotic System; Ampholyte

Calculate the equilibrium pH of a solution prepared by diluting $10^{-3.7}$ mol of sodium hydrogen phthalate to 1 liter with water:

$$C_6H_4 \underset{\diagdown COONa}{\overset{\diagup COOH}{}} = NaHP$$

The acidity constants at the appropriate temperature (25°C) for H_2P and HP^- are $10^{-2.95}$ and $10^{-5.41}$, respectively.

1 Species:

$$H_2P, HP^-, P^{2-}, H^+, OH^-, Na^+$$

2 Equilibrium constants:

$$\frac{[H^+][HP^-]}{[H_2P]} = K_1 \qquad \frac{[H^+][P^{2-}]}{[HP^-]} = K_2 \qquad (i)$$

$$[H^+][OH^-] = K_W$$

3 Concentration condition:

$$P_T = [H_2P] + [HP^-] + [P^{2-}] \qquad (ii)$$

4 Proton condition:

$$[H_2P] + [H^+] = [P^{2-}] + [OH^-] \qquad (iii)$$

Solution of (15) (from Table 3-6) by trial and error gives pH = 4.55, and a closer analysis shows that $[H_2P] < [HP^-] > [P^{2-}]$ and that the proton condition can be approximated by $[H^+] > [H_2P]$ and $[P^{2-}] \gg [OH^-]$.

Example 3.7. Mixture of Acid and Base

Calculate the pH of a solution containing 10^{-3} mol of NH_4Cl and 2×10^{-4} mol of NH_3 per liter of aqueous solution.

1 Species:

$$NH_4^+, NH_3, H^+, OH^-, HCl, Cl^-$$

2 Equilibria, in addition to the ion product of water:

$$\frac{[NH_3][H^+]}{[NH_4^+]} = K = 10^{-9.3} \qquad (i)$$

As shown by Example 3.4, the acidity equilibrium for HCl can be ignored because $[HCl] \ll [Cl^-]$. Similarly HCl can be neglected in the subsequent concentration and proton conditions.

3 Concentration condition:

$$[NH_4^+] + [NH_3] = C_0(NH_4Cl) + C_0(NH_3) = 1.2 \times 10^{-3} \, M \qquad (ii)$$

4 Electroneutrality condition:

$$[NH_4^+] = [Cl^-] + [OH^-] - [H^+] \qquad (iii)$$

Because $[Cl^-] = C_0(NH_4Cl)$ and considering the concentration condition we can also write

$$[NH_3] = C_0(NH_3) - [OH^-] + [H^+] \qquad (iv)$$

Combining these equations with the acidity equilibrium we obtain

$$[H^+] = K \frac{C_0(NH_4Cl) + [OH^-] - [H^+]}{C_0(NH_3) - [OH^-] + [H^+]} \qquad (v)$$

Neglecting as a justified approximation $[H^+]$ and $[OH^-]$ in the numerator and denominator gives

$$[H^+] = 2.5 \times 10^{-9} \qquad pH = 8.6$$

Example 3.8.   Volatile Acid or Base

Estimate the pH of an aqueous electrolyte solution exposed to a partial pressure of NH_3 of 10^{-4} atm. Equilibrium constants valid at this temperature are $p^c K_{NH_4^+}$ = 9.5; $\log K_H$ = 1.75 (Henry's law constant $K_H = [NH_3(aq)]/p_{NH_3}$); $p^c K_w$ = 14.2. The information given by the equilibrium constants can be rearranged as

$$NH_3(aq) + H^+ = NH_4^+ \qquad -\log{}^c K_{NH_4^+} = 9.5 \qquad (i)$$

$$NH_3(g) = NH_3(aq) \qquad \log K_H = 1.75 \qquad (ii)$$

$$H_2O = H^+ + OH^- \qquad \log{}^c K_w = -14.2 \qquad (ii)$$

Summing up the reaction formulas and the $\log K$ values we obtain

$$NH_3(g) + H_2O = NH_4^+ + OH^- \qquad \log K = -2.95 \qquad (iv)$$

that is, $[NH_4^+][OH^-]/p_{NH_3} = 10^{-2.95}$. At equilibrium the proton condition is $[NH_4^+] + [H^+] = [OH^-]$ or $[NH_4^+] \approx [OH^-]$. Thus $[OH^-]^2/10^{-4} = 10^{-2.95}$ which gives $[OH^-] = 10^{-3.5}$ and pH $\simeq$ 10.7.

$[H^+]$ of Pure Aqueous Acids, Bases, or Ampholytes

As derived in the preceding examples, exact algebraic solutions for $[H^+]$ of monoprotic and diprotic acid–base systems are given in Table 3.6 (p. 138).

3.6   pH AS A MASTER VARIABLE; EQUILIBRIUM CALCULATIONS USING A GRAPHICAL APPROACH

Surveys of the influence of master variables, such as pH, and the rapid solution of even complicated equilibria can be accomplished with relative facility by graphic representation of equilibrium data. The concepts of graphical representation of equilibrium relationships were first introduced by Bjerrum [9] in 1914 and have more recently been developed and popularized by Sillén [10].

As we have seen, a direct numerical approach is often quite difficult because rigorous simultaneous solutions of equilibrium relationships lead to equations of third, fourth, or higher order; these equations are obviously not amenable to convenient numerical resolution.

The simplest example of the application of graphical representation of equilibrium data is that for acid–base equilibria involving a monoprotic acid, such as the acid HA, for which the equilibrium expression for solution in water may be written in terms of a concentration acidity constant, that is, an acidity constant valid at the appropriate temperature and corrected for activity by, for example, the Güntelberg approximation:

$$^cK = \frac{[H^+][A^-]}{[HA]} \tag{34}$$

For the purpose of illustration it has been assumed that ^cK has the value 10^{-6} ($p^cK = 6$) and that a quantity of HA sufficient to give an exactly $10^{-3}\ M$ solution has been added to the water; the total concentration C_T of soluble A-containing species in the water at any position of equilibrium is then $10^{-3}\ M$, or

$$C_T = 10^{-3}\ M = [HA] + [A^-] \tag{35}$$

The control variable in any acid–base equilibrium is pH; hence it is desirable to represent graphically the equilibrium relationships of all species as functions of pH. For any value of pH the unknowns in the present example are of course [HA] and [A$^-$], each of which may now be expressed in terms of the known quantities C_T and [H$^+$] by combining equations 34 and 35 as follows:

$$[HA] = \frac{C_T[H^+]}{^cK + [H^+]} \tag{36}$$

and

$$[A^-] = \frac{C_T\ ^cK}{^cK + [H^+]} \tag{37}$$

It is convenient in the construction of the equilibrium diagram to consider first the asymptotes of the individual curves of solute concentration against pH and in this manner to determine the slopes of the separate sections of each curve. After examining one or two examples of this, the method becomes quite obvious and it is not usually necessary to go through any computations; however, for purposes of illustration, the method is discussed one step at a time (see Figure 3.3).

For values of pH less than p^cK, the values of ^cK in the denominators of equations 36 and 37 are much smaller than the asymptotic value of [H$^+$] and may be neglected. Taking logarithms, equation 36 then becomes

$$\log[HA] = \log C_T \tag{38}$$

Thus the slope of that portion of the curve of HA against pH in the region pH $< p^cK$ is zero.

Similarly equation 37, relating [A$^-$] to pH, may be written

$$\log[A^-] = \log C_T - p^cK + pH \tag{39}$$

Differentiation of equation 39 with respect to pH yields $d\log[A^-]/dpH = 1$; thus the slope of the part of the curve representing the variation of [A$^-$] with pH in the region pH $< p^cK$ is unity.

Consideration of the asymptotes of the sections of the two curves in the region pH $> p^cK$ permits us to ignore the quantity [H$^+$] in the denominators of

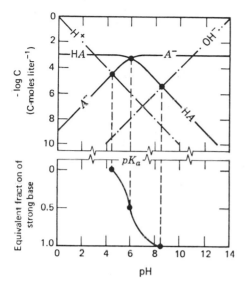

Figure 3.3 Construction of logarithmic diagram and titration curve for a monoprotic
acid. The method of constructing logarithmic
equilibrium diagrams and titration curves by
graphical representation of mass-law and
electroneutrality relations is illustrated. For
this example a monoprotic acid (HA) with a
p^cK value of 6 and a total concentration of
10^{-3} mol liter^{-1} was used. The three points on
the curve for titration of the acid with strong
base are given by (1) $f = 0$: $[H^+] = [A^-] +$
$[OH^-]$, pH = 4.5 neglect $[OH^-]$); (2) $f =$
0.5: pH = pK = 6, (3) $f = 1.0$: $[HA] +$
$[H^+] = [OH^-]$, pH = 8.5 (neglect $[H^+]$).

equations 36 and 37. Equation 36 then becomes

$$\log[A^-] = \log C_T \tag{41}$$

hence $d \log[A^-]/d\mathrm{pH} = 0$. In a similar way it can be shown that, for pH > p^cK,
$d \log[HA]/d\mathrm{pH} = -1$.

The slopes of the straight-line plots for each solute species against pH have
now been calculated from the asymptotic values for the curves in the regions
of pH < p^cK and pH > p^cK. None of these curve asymptotes is rigorous in
the immediate region of pH = p^cK. At this point $\log[HA] = \log[A^-] =$
$\log(C_T/2)$. Therefore, at pH = p^cK, the curves must intersect at an ordinate
value of $\log C_T - \log 2$, or 0.3 unit below the ordinate value of $\log C_T$.

Computing the equilibrium composition of a 10^{-3} M HA solution, we
simply have to find where on the graph the appropriate proton condition is
fulfilled. The condition to be satisfied is

$$[H^+] = [A^-] + [OH^-] \tag{42}$$

Equation 42 is fulfilled at the intersection of the $[H^+]$ line with the $[A^-]$ line
because obviously at this point $[OH^-] \ll [A^-]$. Equilibrium $[H^+]$ and the
concentrations of all other species at this $[H^+]$ can be read directly from the
logarithmic concentration diagram:

$$-\log[H^+] = -\log[A^-] = 4.5 \qquad -\log[HA] = 3.0$$

If the diagram is drawn on graph paper (where one logarithmic unit corresponds to about 2 cm) the result can be read within an accuracy of better
than ± 0.05 logarithmic units and the relative error $(d\Delta[X]/d[X] = 2.3 \times$
$\log \Delta[X])$ is smaller than 10%. The slight loss of accuracy involved in substituting graphical for numerical procedures is usually not significant. If a

very exact answer is necessary, the graphical procedure will immediately show which concentrations can be neglected in the numerical calculations.

The same graph can be used to compute the equilibrium concentrations of a 10^{-3} M solution of NaA. In this case the proton condition is

$$[HA] + [H^+] = [OH^-] \tag{43}$$

This condition is fulfilled at the intersection $[HA] = [OH^-]$; since $[H^+]$ is 1000 times smaller than $[HA]$, it can be neglected in equation 43. This point gives $-\log[H^+] = 8.5$, $-\log[HA] = 5.5$, and $-\log[A^-] = 3.0$.

The proton conditions of equations 42 and 43 correspond to the two equivalence points in acid–base titration systems. The half-titration point is usually (not always) given by $pH = pK$. Thus the qualitative shape of the titration curve can be sketched readily along these three points (Figure 3.3).

Diprotic Acid–Base Systems

For more complicated equilibria the merit of the graphical method is obvious. Figure 3.4 illustrates a logarithmic pH–concentration diagram for a *diprotic acid*

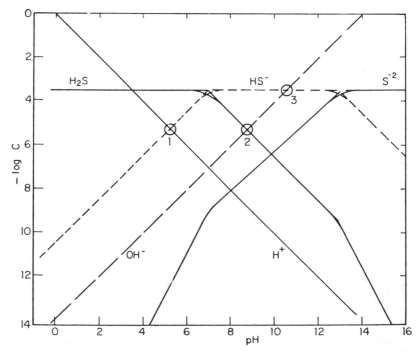

Figure 3.4   Equilibrium diagram for a diprotic acid system. Proton conditions: 1. Solution of H_2S: $[H^+] = [HS^-] + 2[S^{2-}] + [OH^-]$. 2. Solution of NaHS: $[H_2S] + [H^+] = [S^-] + [OH^-]$. 3. Solution of Na_2S: $2[H_2S] + [HS^-] + [H^+] = [OH^-]$. Equilibrium composition: 1. pH = pHS⁻ = 5.3; $pH_2S = 3.5$; $pS^{2-} = 12.3$. 2. pH = 8.7; $pH_2S = 5.3$; $pS^{2-} = 7.5$; pHS⁻ = 3.5. 3. pH = 10.5; $pS^{2-} = 5.8$; $pH_2S = 7$; pHS⁻ = 3.5.

with acidity constants ($pK_1 = 7.0$ and $pK_2 = 13.0$) representative of hydrogen sulfide, $H_2S(aq)$. A combination of the equilibrium expressions for the two acidity constants with the concentration condition ($S_T = [H_2S] + [HS^-] + [S^{2-}]$) gives the equations that define the log concentration–pH dependence of $[H_2S]$, $[HS^-]$ and $[S^{2-}]$:

$$[H_2S] = \frac{S_T}{1 + K_1/[H^+] + K_1K_2/[H^+]^2} \tag{44}$$

$$[HS^-] = \frac{S_T}{[H^+]/K_1 + 1 + K_2/[H^+]} \tag{45}$$

$$[S^{2-}] = \frac{S_T}{[H^+]^2/K_1K_2 + H^+/K_2 + 1} \tag{46}$$

Considering equation 44, it is apparent that the log H_2S–pH line can be constructed as a sequence of three linear asymptotes that prevail in the three pH regions:

I: $\qquad pH < pK_1 < pK_2 \qquad \log[H_2S] = \log S_T$

$$\frac{d\log[H_2S]}{dpH} = 0 \tag{47}$$

II: $\qquad pK_1 < pH < pK_2 \qquad \log[H_2S] = pK_1 + \log S_T - pH$

$$\frac{d\log[H_2S]}{dpH} = -1 \tag{48}$$

III: $\quad pK_1 < pK_2 < pH \quad \log[H_2S] = pK_1 + pK_2 + \log S_T - 2pH$

$$\frac{d\log[H_2S]}{dpH} = -2 \tag{49}$$

These linear portions can be readily constructed; they change their slopes from 0 to -1 and from -1 to -2 at $pH = pK_1$ and $pH = pK_2$, respectively.

Similar considerations apply to the plotting of equations 45 and 46. The sections having slopes of -2 or $+2$ are usually unimportant because they occur only at extremely small concentrations. Diagrams of the types given in Figures 3.3 and 3.4 are not only useful in evaluating specific positions of equilibrium, but they permit us to survey the entire spectrum of equilibrium conditions as a function of pH as a master variable.

Example 3.9. Strong Acid

Estimate the equilibrium composition of a 10^{-2} M HCl and 10^{-2} M NaCl solution, respectively. (Assume an acidity constant of $K \approx 10^{+3}$; see Figure 3.5.)

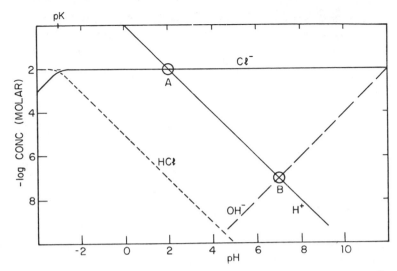

Figure 3.5. Equilibrium diagram for a strong acid (25°C) (Example 3.9). A: 10^{-2} M HCl; $[H^+] = [Cl^-] = 10^{-2}$ M, $[HCl] = 10^{-7}$ M. B: 10^{-2} M NaCl; $[H^+] = [OH^-] = 10^{-7}$ M, $[Cl^-] =$ 10^{-2} M, $[HCl] = 10^{-12}$ M.

For the 10^{-2} M acid solution the proton condition (A) is $[H^+] = [Cl^-] +$ $[OH^-]$, which is given by the intersection $[H^+] = [Cl^-]$. For the corresponding salt solution, the proton condition (B) is $[HCl] + [H^+] = [OH^-]$ which becomes, since HCl is negligible, $[H^+] = [OH^-]$.

Example 3.10. Weak Base

Compute the equilibrium concentrations in a $10^{-4.5}$ M sodium acetate (NaAc) solution; pK (25°C) = 4.70. Figure 3.6 plots the expressions for the acidity constant and the ion product of water as well as the concentration condition. For a pure solution of NaAc the following proton condition is valid: $[HAc] +$ $[H^+] = [OH^-]$.

It is obvious from the graph that no approximation in the proton condition is possible. In order to find the point where the proton condition is fulfilled, we move slightly to the right of the intersection of log[HAc] with log[OH$^-$] and find by trial and error where the proton condition is fulfilled, that is, at pH = 7.2: $10^{-7.0} + 10^{-7.2} \approx 10^{-6.8}$.

Example 3.11. Mixture of Two Acids

Find the pH of a 10^{-3} M NH$_4$Cl solution to which 4×10^{-5} M methyl orange indicator in the acid form (HIn) has been added: p$K_{NH_4^+} = 9.2$; p$K_{HIn} = 4.0$.

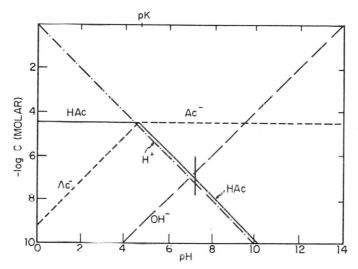

Figure 3.6 Equilibrium composition of $10^{-4.5}$ M NaAc (Example 3.10). Proton condition; $[HAc] + [H^+] = [OH^-]$. $[H^+] = 10^{-7.2}$ M; $[HAc] = 10^{-7.0}$ M; $[HAc] = 10^{-4.5}$ M.

The logarithmic diagram is constructed by superimposing the equilibrium diagrams for the two acid–base systems (Figure 3.7). The proton condition $[H^+] = [NH_3] + [In^-] + [OH^-]$ defines the composition of the solution. The first intersection of the $[H^+]$ line is with the $[In^-]$ line. The other species on the right-hand side of the proton condition can be neglected. At equilibrium we have: $pH = 4.5$; $pIn^- = 4.5$; $pHIn = 4.9$; $pNH_4^+ = 3$; $pNH_3 = 7.7$.

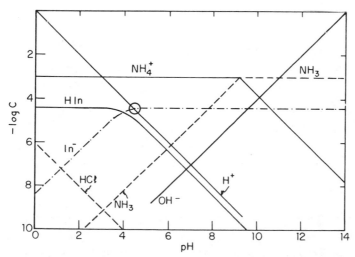

Figure 3.7 Equilibrium composition of solution containing 10^{-3} M NH_4Cl and $4 \times 10^{-4} M$ methyl orange (HIn) (Example 3.11).

It is interesting to note, and it should be kept in mind that, if one attempts to measure pH with the indicator, the addition of an indicator acid to a poorly buffered solution may markedly affect the pH of the solution. The 10^{-3} M NH$_4$Cl solution in the absence of an indicator (proton condition: $[H^+]$ = $[NH_3] + [OH^-]$) has a pH of 6.1.

Example 3.12. Diprotic Base

Compute the equilibrium composition of a $10^{-1.3}$ M sodium phthalate (Na$_2$P) solution (25°C). pK_1 = 2.95; pK_2 = 5.41. Using the Güntelberg approximation, we first convert the pK values into p^cK values. In order to calculate the ionic strength, we assume that $[P^{2-}] > [HP^-]$. Thus $I = \frac{1}{2}([Na^+] + 4[P^{2-}])$ = 0.15.

$$p^cK_1 = pK_1 - \frac{\sqrt{I}}{1 + \sqrt{I}} = 2.95 - 0.28 = 2.67$$

$$p^cK_2 = pK_2 - \frac{2\sqrt{I}}{1 + \sqrt{I}} = 5.41 - 0.56 = 4.85$$

$$p^cK_w = pK_w - \frac{\sqrt{I}}{1 + \sqrt{I}} = 14.00 - 0.28 = 13.72$$

The logarithmic pH concentration diagram (Figure 3.8) is now constructed using p^cK values. Note that the log[OH$^-$] line intersects the log $C = 0$ at

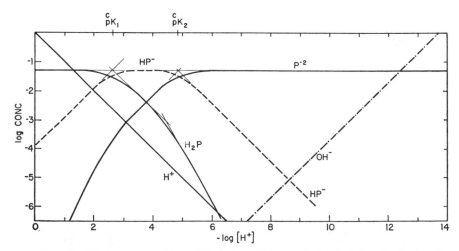

Figure 3.8 Equilibrium composition of diprotic acid (Example 3.12). A 5×10^{-2} M sodium phthalate solution (Na$_2$P) has a pH of 8.6. Proton condition: $2[H_2P] + [HP^-] + [H^+] = [OH^-]$; $[HP^-] \approx [OH^-]$.

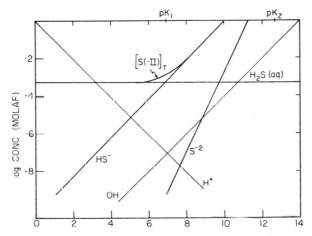

Figure 3.9 Distribution of S($-$II) species of a solution in equilibrium with $p_{H_2S} = 5 \times 10^{-1}$ atm (Example 3.13).

pH $= 13.72$. The proton condition for a solution of Na_2P is: $2[H_2P] + [HP^-] + [H'] = [OH^-]$. This proton condition is fulfilled at $-\log[H^+] = 8.60$. The other species are present at the following concentrations:

$$[P^{2-}] = 10^{-1.3}\ M \qquad [HP^-] = 10^{-5.1}\ M \qquad [H_2P] < 10^{-6}\ M\ (10^{-8.85}\ M)$$

$$-\log\{H'\} = -\log[H'] + \frac{0.5\sqrt{I}}{1 + \sqrt{I}} - 8.76$$

Example 3.13. Volatile Diprotic Acid

Estimate the distribution of H_2S, HS^-, and S^{2-} as a function of pH for an aqueous solution (20°C) in equilibrium with an atmosphere containing 0.5% (by volume) of H_2S. Equilibrium constants: $p^cK_{H_2S} = 6.9$; $p^cK_{HS^-} = 12.7$; $\log K'_H = \log([H_2S]/p_{H_2S}) = -1.0$. Since Henry's law is fulfilled over the entire pH range, $[H_2S] = 10^{-3.3}$; $p_{H_2S} = 10^{-2.3}$ (see Figure 3.9). The line for $[HS^-]$ is defined by $[HS^-] = K_{H_2S}[H_2S]/[H^+]$; ($d\log[HS^-]/d$pH $= +1$). Similarly the pH dependence for S^{2-} is given by $[S^{2-}] = K_{H_2S}K_{HS^-}[H_2S]/[H^+]^2$; ($d\log[S^{2-}]/d$pH $= +2$).

The sum of all the S(-II) species ($[S(\text{-II})_T] = [H_2S] + [HS^-] + [S^{2-}]$) is also given in the diagram as a function of pH.

3.7 IONIZATION FRACTIONS OF ACIDS, BASES, AND AMPHOLYTES

The relation between the concentration condition, $C = [HB] + [B]$, and the acidity constant, $K = [H^+][B]/[HB]$, permits us to calculate the relative distribution of acid and conjugate base as a function of pH:

$$\alpha_B = \alpha_1 = \frac{[B]}{C} = \frac{K}{K + [H^+]} \tag{50}$$

$$\alpha_{HB} = \alpha_0 = \frac{[HB]}{[C]} = \frac{[H^+]}{K + [H^+]} \tag{51}$$

where $\alpha_1 + \alpha_0 = 1$. Historically, α_1 has been called the degree of dissociation; it might be better to call it the ionization fraction or the degree of protolysis. α_0 has been called the degree of formation of the acid, and a plot of α_0 versus pH has been called the formation function, or better the *distribution diagram*. Figure 3.10a gives a schematic distribution diagram. Sometimes it is more convenient to plot the logarithms of the ionization fractions as a function of pH.

Since the ionization fractions are independent of total concentration, their tabulation or graphical representation is very convenient when calculations with the same equilibrium system have to be carried out repeatedly or with more complicated systems. The computation of α values can also be readily programmed on programmable scientific calculators. The concentration of the species [HB] and [B] can then always be represented by $[HB] = C\alpha_0$ and $[B] = C\alpha_1$, respectively. The logarithmic equilibrium diagram, discussed in the preceding section, is essentially the additive combination of the line, $\log C =$ const., and the logarithmic distribution diagram (Figure 3.10b). In a diprotic acid–base system we define similarly

$$[H_2A] = C\alpha_0 \tag{52}$$

$$[HA^-] = C\alpha_1 \tag{53}$$

$$[A^{2-}] = C\alpha_2 \tag{54}$$

the subscript α refers to the number of protons lost from the most protonated species. The α values are implicit functions of $[H^+]$:

$$\alpha_0 = \frac{1}{1 + K_1/[H^+] + K_1K_2/[H^+]^2} \tag{55}$$

$$\alpha_1 = \frac{1}{[H^+]/K_1 + 1 + K_2/[H^+]} \tag{56}$$

$$\alpha_2 = \frac{1}{[H^+]^2/K_1K_2 + [H^+]/K_2 + 1} \tag{57}$$

The α values are interrelated by

$$\alpha_0 + \alpha_1 + \alpha_2 = 1 \tag{58}$$

$$\alpha_0 = \frac{[H^+]}{K_1}\alpha_1 \tag{59}$$

$$\alpha_1 = \frac{[H^+]}{K_2}\alpha_2 \tag{60}$$

α values for polyprotic acid–base systems can be readily derived.

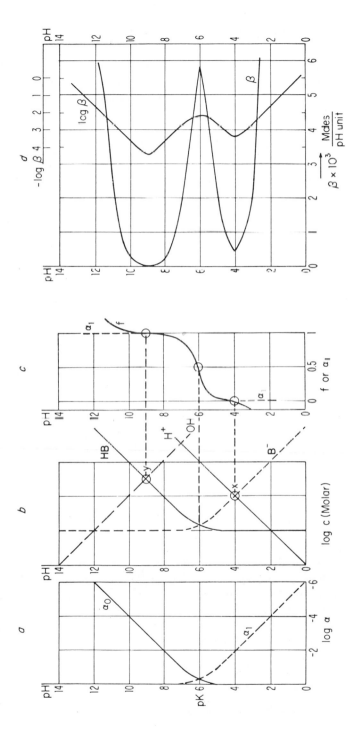

Figure 3.10 The titration curve and the buffer intensity are related to the equilibrium species distribution. For a monoprotic acid–base system (HB-B) (a) gives a semilogarithmic distribution diagram. α_1 and α_0 are the ionization fractions of B and HB respectively. (b) The log concentration-pH diagram is the superposition of the semilogarithmic distribution diagram (α) and the concentration condition (log C = constant). Points X and Y correspond to the equivalence points in alkalimetric or acidimetric titration curve. (c) Plot of the titration curve. The equivalence points (X and Y) and the half-titration point pH = pK are as given in (b). The equivalence fraction of the titrant added, f, shows, over a significant portion of the titration curve (0.1 < f < 0.9), the same dependence upon pH as α_1. (d) The buffer intensity β, corresponding to the inverse slope of the titration curve (dc_B/dpH), can be computed from a log concentration-pH diagram (b) by multiplying by 2.3 the sum of all concentrations represented by a line of slope +1 or −1 at that particular pH in the diagram. (See Section 3.9.)

155

3.8   TITRATION OF ACIDS AND BASES

In the titration of an aqueous solution containing C moles per liter of an acid HA with a quantity of strong base (C_B), such as NaOH, the titration curve can be readily deduced because at any point in the titration the following condition of electroneutrality must be fulfilled:

$$[Na^+] + [H^+] = [A^-] + [OH^-]$$

or   (61)

$$C_B = [A^-] + [OH^-] - [H^+]$$

With this equation and the logarithmic concentration–pH diagram, the titration curve relating pH to the quantity of base added can be constructed. In Figure 3.10c, pH is plotted as a function of the equivalent fraction of the titrant (strong base) added:

$$f = \frac{C_B}{C} = \frac{[Na^+]}{C} \tag{62}$$

Equation 61 can be rearranged into

$$C_B = C\alpha_1 + [OH^-] - [H^+] \tag{63}$$

$$f = \alpha_1 + \frac{[OH^-] - [H^+]}{C} \tag{64}$$

where α_1 is the degree of protolysis (ionization fraction) (see Section 3.7); $\alpha_1 = K/(K + [H^+]) = [A^-]/C$.

 If we want to consider any dilution resulting from the addition of V milliliters of strong base to V_0 milliliters of solution containing a concentration C_0 of acid HA before dilution, we simply have to introduce a dilution factor and substitute for C in the above equations:

$$C = C_0 \frac{V_0}{V + V_0} \tag{65}$$

 For the titration of a C molar solution of the conjugate base (i.e., the salt of a strong base with the weak acid HA, such as KA) with a strong acid (i.e., HCl), the curve for variation of pH with quantity of acid (C_A) added can be derived similarly from the electroneutrality condition:

$$[K^+] + [H^+] = [A^-] + [OH^-] + [Cl^-] \tag{66}$$

$$C + [H^+] = [A^-] + [OH^-] + C_A \tag{67}$$

$$C_A = [HA] + [H^+] - [OH^-] \tag{68}$$

$$C_A = C\alpha_0 + [H^+] - [OH^-] \tag{69}$$

where $\alpha_0 = [H^+]/(K + [H^+])$ (see Section 3.7).

The equivalent fraction of the titrant (strong acid) added, $g = C_A/C$, can be given as an implicit function of H^+ by

$$g = \frac{C_A}{C} = \alpha_0 + \frac{[H^+] - [OH^-]}{C} \tag{70}$$

Comparison of equation 70 with equation 64 shows that $g = 1 - f$. Equations 63 and 69 can be generalized into

$$C_B - C_A = C\alpha_1 + [OH^-] - [H^+]$$

or $\tag{71}$

$$C_A - C_B = C\alpha_0 + [H^+] - [OH^-]$$

Either of equations 71 can be used to evaluate pH changes that result from the addition of strong acid or strong base to a monoprotic weak acid–base system, or to characterize the titration curve of a mixture of a strong acid or base with a weak acid or base. The equivalence points marked X and Y in Figure 3.10b correspond to equilibrium conditions prevailing in pure equimolar solutions of HA ($g = 1$ and $f = 0$) and NaA ($f = 1$ and $g = 0$). In other words, at the equivalence point ($f = 1$) of an alkalimetric titration of HA with NaOH, the solution cannot be distinguished from an equimolar solution of the salt NaA (proton condition: $[HA] + [H^+] = [OH^-]$). Correspondingly, in the acidimetric titration of the salt, NaA, with HCl, the proton condition at the equivalence point ($f = 0$) is identical with that of an equimolar solution of HA ($[H^+] = [A^-] + [OH^-]$).

Example 3.14.   Acidimetric Titration of Strong and Weak Base

Describe the acidimetric titration curve for a solution of the following composition: $[Na^+] = 3.0 \times 10^{-3} M$; $[HOCl] + [OCl^-] = 2.0 \times 10^{-3} M$; $[OH^-] = 1.0 \times 10^{-3} M$.

The titration curve can be drawn with the help of equation 70. $C_B = 1.0 \times 10^{-3} M$. The equivalence points at $C_A = 1.0 \times 10^{-3} M$ and $C_A = 3.0 \times 10^{-3} M$ are at pH values 9.5 and 5.2, respectively, the second equivalence point being somewhat sharper than the first ones (Figure 3.11).

Because proton conditions corresponding to equivalent points $f = 0$ and $f = 1$ can be readily identified, titration curves can be sketched expediently with the help of the log concentration diagram.

The principles outlined above can be readily extended to *multiprotic acids*. The alkalimetric titration of an acid H_2L^+ added as the salt $H_2L^+X^-$ (e.g., an amino acid, $RNH_2COOH = HL$) is given by the electroneutrality condition

$$C_B + [H_2L^+] + [H^+] = [L^-] + [OH^-] + [X^-] \tag{72}$$

which can be rearranged with the concentration condition

$$C = [H_2L^+] + [HL] + [L^-] = [X^-] \tag{73}$$

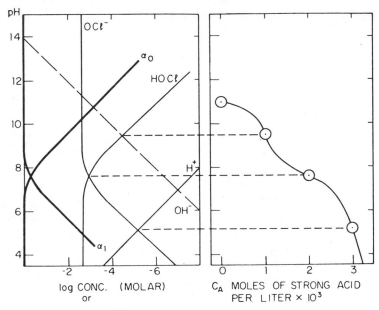

Figure 3.11 Acidimetric titration of a strong base and a weak base (Example 3.14). The solution to be titrated corresponds to "Eau de Javel" made by dissolving Cl_2 in NaOH. $[Na^+] = 3 \times 10^{-3}\ M$; $[HOCl] + [OCl^-] = 2 \times 10^{-3}\ M$, $[OH^-] = 1 \times 10^{-3}\ M$. The first equivalence point (equivalent to the strong base) is at pH $\simeq$ 9.5.

to give

$$C_B = [HL] + 2[L^-] + [OH^-] - [H^+] \tag{74}$$

The relation is expressed more generally by

$$C_B = C(\alpha_1 + 2\alpha_2) + [OH^-] - [H^+] \tag{75}$$

$$f = \frac{C_B}{C} = \alpha_1 + 2\alpha_2 + \frac{[OH^-] - [H^+]}{C} \tag{76}$$

where the α values are defined as $\alpha_1 = [HL]/C$ and $\alpha_2 = [L^-]/C$ and $\alpha_0 = [H_2L^+]/C$. In such a diprotic system three equivalence points may be defined. In Figure 3.12a the points x, y, and z correspond to the equivalence points $f = 0$ (proton condition: $[H^+] = [HL] + 2[L^-] + [OH^-]$); $f = 1$ $([H_2L^+] + [H^+] = [L^-] + [OH^-])$; and $f = 2$; $(2[H_2L^+] + [HL] + [H^+] = [OH^-])$, respectively.

Similarly, the equation describing the acidimetric titration curve of the base ML (where M^+ does not protolyze) can be derived from the concentration condition 73 and the electroneutrality condition:

$$[M^+] + [H_2L^+] + [H^+] = [L^-] + [OH^-] + C_A \tag{77}$$

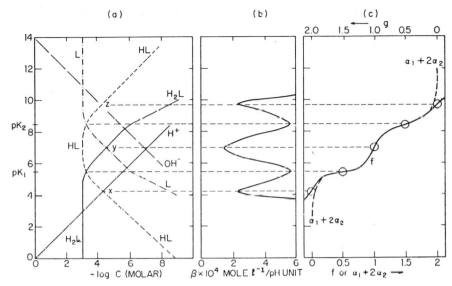

Figure 3.12   Equilibrium composition, buffer intensity and titration curve of diprotic acid–base system. (*a*): Species distribution. (*b*) Buffer intensity. (*c*) Titration curve. The equivalence points, *x*, *y*, and *z* (*a*) are representative of the composition of pure solutions of H_2L, NaHL and Na_2L, respectively, and correspond to minima in the buffer intensity. The smaller the buffer intensity, the steeper is the titration curve.

The morphology of the titration curve is given by

$$g = 2 - f = \frac{C_A}{C} = 2\alpha_0 + \alpha_1 + \frac{[H^+] - [OH^-]}{C} \tag{78}$$

The alkalimetric or acidimetric titration of the ampholyte HL is described simply by the appropriate portions of equations 76 and 78, respectively.

For a mixture of protolysis systems the variation in $[H^+]$ as a result of the addition of strong acid and/or strong base can be defined by

$$C_B - C_A = {}^IC({}^I\alpha_1 + 2{}^I\alpha_2 + \cdots)$$
$$+ {}^{II}C({}^{II}\alpha_1 + 2{}^{II}\alpha_2 + \cdots) + \cdots + [OH^-] - [H^+] \tag{79}$$

3.9   BUFFER INTENSITY AND NEUTRALIZING CAPACITY

The slope of a titration curve (pH versus C_B) is related to the tendency of the solution at any point in the titration curve to change pH upon addition of base. The buffer intensity at any point of the titration is inversely proportional to the slope of the titration curve at that point and may be defined as

$$\beta = \frac{dC_B}{dpH} = -\frac{dC_A}{dpH} \tag{80}$$

where dC_B and dC_A are the numbers of mole liter^{-1} of strong acid or strong base required to produce a change in pH of dpH. β has also been called the buffer capacity or the buffer index.

Buffer Intensity

Obviously the buffer intensity can be expressed numerically by differentiating the equation defining the titration curve with respect to pH. For a monoprotic acid–base system (see equations 61 and 63),

$$\beta = \frac{dC_B}{dpH} = \frac{d[A^-]}{dpH} + \frac{d[OH^-]}{dpH} - \frac{d[H^+]}{dpH}$$

$$= C\frac{d\alpha_1}{dpH} + \frac{d[OH^-]}{dpH} - \frac{d[H^+]}{dpH} \tag{81}$$

The terms on the right-hand side of equation 81 can be differentiated as follows:

$$-\frac{d[H^+]}{dpH} = \frac{-d[H^+]}{-(1/2.3)\, d \ln [H^+]} = 2.3[H^+] \tag{82}$$

$$\frac{d[OH^-]}{dpH} = \frac{d[OH^-]}{(1/2.3)\, d \ln [OH^-]} = 2.3[OH^-] \tag{83}$$

$$C\frac{d\alpha_1}{dpH} = C\frac{d[H^+]}{dpH}\frac{d\alpha_1}{d[H^+]} = 2.3C\frac{K[H^+]}{(K + [H^+])^2} \tag{84}$$

Because $\alpha_0 = [H^+]/(K + [H^+])$ and $\alpha_1 = K/(K + [H^+])$, the right-hand side of equation 84 can be expressed in terms of ionization fractions or concentrations of [HA] and [A$^-$]:

$$C\frac{d\alpha_1}{dpH} = 2.3\alpha_1\alpha_0 \qquad C = 2.3\frac{[HA][A^-]}{C} \tag{85}$$

Summing up the individual terms of equation 81 results in

$$\beta = \frac{dC_B}{dpH} = 2.3([H^+] + [OH^-] + C\alpha_1\alpha_0)$$

$$= 2.3\left([H^+] + [OH^-] + \frac{[HA][A^-]}{[HA] + [A^-]}\right) \tag{86}$$

The terms on the right-hand side of equation 86 are in the logarithmic concentration–pH diagram, and β can readily be computed (Figure 3.10d). Maximum buffer intensity occurs (inflection point of titration curve) where $d^2\alpha_1/d(\mathrm{pH})^2 = 0$. This occurs when $\alpha_1 = \alpha_0$ or where [HA] = [A$^-$] and pH = pK. Accordingly, *buffers* usually made by mixing an acid and its conjugate base have their maximum buffer intensity at a pH where [HA] = [A$^-$]. The pH of a solution containing $C_{HA} M$ HA and $C_{NaA} M$ NaA corresponds to a point in the titration curve and can be readily computed by equation 87 (which can be derived with the help of the electroneutrality condition; see Example 3.7):

$$[H^+] = K \frac{C_{HA} - [H^+] + [OH^-]}{C_{NaA} + [H^+] - [OH^-]} \qquad (87)$$

Aqueous solutions are well buffered at either extreme of the pH scale. If in an alkalimetric or acidimetric titration curve the pH at the equivalence point falls into a pH range where the buffer intensity caused by $[H^+]$ or $[OH^-]$ exceeds that of the other protolytes, obliteration of a pH jump at the equivalence point results (see Figure 3.13). The concept of pH buffers can be extended to ions other than H^+. Metal-ion buffers will be discussed in Chapter 6.

If various acid–base pairs, HA, A; HB, B; and so on, are present in the solution the buffer intensity is given by

$$\beta = 2.3\{[H^+] + [OH^-] + C_A \alpha_{HA} \alpha_A + C_D \alpha_{HD} \alpha_D + \cdots\}$$

$$= 2.3\left\{[H^+] + [OH^-] + \frac{[HA][A]}{[HA] + [A]} + \frac{[HB][B]}{[HB] + [B]} + \cdots\right\} \qquad (88)$$

Polyprotic Systems. In the same fashion as equation 86 has been derived, expressions for the buffer intensity of polyprotic acid–base systems can be

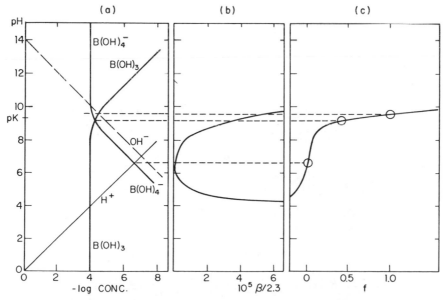

Figure 3.13 Alkalimetric titration of a weak acid ($10^{-4} M$ boric acid). (*a*) Equilibrium distribution. (*b*) Buffer intensity. (*c*) Alkalimetric titration. No pH jump occurs at the equivalence point ($f = 1$) because of buffering by OH^- ions.

developed. In Table 3.7 the buffer intensity of a diprotic acid–base system is derived. A polyprotic acid can be treated the same way as a mixture of individual monoprotic acids; for example, for the dibasic acid H_2C,

$$\beta \simeq 2.3 \left\{ [H^+] + [OH^-] + \frac{[H_2C][HC^-]}{[H_2C] + [HC^-]} + \frac{[HC^-][C^{2-}]}{[HC^-] + [C^{2-}]} \right\} \quad (89)$$

This approximate equation holds with an error of less than 5% if $K_1/K_2 > 100$ (Ricci, *loc. cit.*). Note that in the last term of equation 86 and in the two last terms of equations 88 and 89, one of the concentrations in the denominator can usually be neglected: for example, at $pH < pK$, equation 86 becomes $\beta \approx 2.3\,([H^+] + [OH^-] + [A^-])$; at $pH > pK$, it reduces to $\beta \approx 2.3\,([H^+] + [OH^-] + [HA])$. In accord with this well-justified approximation, the buffer intensity of a solution can be computed from a logarithmic equilibrium diagram by multiplying by

TABLE 3.7   TITRATION CURVE AND BUFFER INTENSITY OF A TWO-PROTIC ACID (H_2C)^a

I. *Definitions*

$$C = [H_2C] + [HC^-] + [C^{2-}] \qquad \alpha_0 = 1/[1 + K_1/[H^+] + K_1K_2/[H^+]^2]$$
$$[H_2C] = C\alpha_0 \qquad\qquad\qquad \alpha_1 = 1/[1 + [H^+]/K_1 + K_2/[H^+]]$$
$$[HC^-] = C\alpha_1 \qquad\qquad\qquad \alpha_2 = 1/[1 + [H^+]/K_2 + [H^+]^2/K_1K_2]$$
$$[C^{2-}] = C\alpha_2$$

II.

$$\frac{d\alpha_1}{dpH} = 2.3\alpha_1(\alpha_0 - \alpha_2); \qquad \frac{d\alpha_2}{dpH} = 2.3\alpha_2(\alpha_1 + 2\alpha_0); \qquad \frac{d\alpha_0}{dpH} = -2.3\alpha_0(\alpha_1 + 2\alpha_2)$$

III. *Titration Curve*

$$C_B - C_A = [ANC] = [HC^-] + 2[C^{2-}] + [OH^-] - [H^+]$$
$$C_B - C_A = 2C - [BNC] = C(\alpha_1 + 2\alpha_2) + [OH^-] - [H^+]$$

IV. *Buffer Intensity*

$$\beta = \frac{dC_B}{dpH} = \frac{-dC_A}{dpH} = \frac{d[ANC]}{dpH} = \frac{-d[BNC]}{dpH}$$

$$= \frac{d[OH^-]}{dpH} - \frac{d[H^+]}{dpH} + \frac{Cd(\alpha_1 + 2\alpha_2)}{dpH}$$

$$= 2.3\{[H^+] + [OH^-] + C[\alpha_1(\alpha_0 + \alpha_2) + 4\alpha_2\alpha_0]\}$$

$$\beta = 2.3\left\{ CK_1[H^+] \frac{[H^+]^2 + 4K_2[H^+] + K_1K_2}{([H^+]^2 + K_1[H^+] + K_1K_2)^2} + [H^+] + [OH^-] \right\}$$

^{a} The equations given are rigorous. In very good approximation, a polyprotic acid can be treated the same way as a mixture of individual monoprotic acids (see equation 89).

2.3 the sum of all concentrations represented by a line of slope $+1$ or -1 at that particular pH in the diagram.

The concept of buffer intensity considered above may be extended and defined in a generalized way for the incremental addition of a constituent to a closed system at equilibrium. Thus in addition to the buffer intensity with respect to strong acids or bases, buffer intensities with respect to weak acids and bases and for heterogeneous systems may be defined. In general

$$\beta_{C_j}^{C_i} = \frac{dC_i}{dpH} \tag{90}$$

where $\beta_{C_j}^{C_i}$ is the buffer intensity for adding C_i incrementally to a system of constant C_j; for example, $\beta_{CaCO_{3}(s)}^{C_{CO_2}}$ measures the tendency of a solution in contact and equilibrium with solid $CaCO_3$ to resist a pH change resulting from the addition or withdrawal of CO_2. In principle, the concept can be further extended to the buffering of metal ions, that is, to the stability of water with respect to the concentration of other ions and parameters such as

$$\beta = \frac{dC_i}{dpCa}$$

can be elucidated. The buffer intensity can always be found analytically by differentiating the appropriate function of C_i for the system with respect to the pH or pMe. The buffer intensity is an intrinsic function of the pH or pMe.

Acid- and Base-Neutralizing Capacity

Operationally we might define as a base-neutralizing capacity [BNC] the equivalent sum of all the acids that can be titrated with a strong base to an equivalence point. Similarly, the acid-neutralizing capacity [ANC] can be determined from the titration with strong acid to a preselected equivalence point. At every equivalence point a particular proton condition defines a reference level of protons. Conceptually [BNC] measures the concentration of all the species containing protons in excess minus the concentration of the species containing protons in deficiency of the proton reference level; that is, it measures the net excess of protons over a reference level of protons. Similarly [ANC] measures the net deficiency of protons.

In an aqueous monoprotic acid–base system [ANC] is defined by the right-hand side of equation 61 or 63 in Section 3.8.

$$[ANC] = [A^-] + [OH^-] - [H^+]$$

or
$$\tag{91}$$

$$= C\alpha_1 + [OH^-] - [H^+]$$

The reference level is defined by the composition of a pure solution of HA in H_2O ($f = 0$; $[ANC] = 0$), which is defined by the proton condition, $[H^+] = [A^-] + [OH^-]$. (In this and subsequent equations, the charge type of the acid is unimportant; the equation defining the net proton excess or deficiency can always be derived from a combination of the concentration condition and the condition of electroneutrality.) Thus in a solution containing a mixture of HA and NaA, $[ANC]$ is a conservative capacity parameter. It must be expressed in concentrations (and not activities). Addition of HA (a species defining the reference level) does not change the proton deficiency and thus does not affect $[ANC]$.

In the same monoprotic acid–base system the base–neutralizing capacity with respect to the reference level ($f = 1$) of a NaA solution (proton condition: $[HA] + [H^+] = [OH^-]$) is defined by

$$[BNC] = [HA] + [H^+] - [OH^-]$$
$$= C\alpha_0 + [H^+] - [OH^-] \tag{92}$$

Example 3.15. [ANC] of Buffer Solutions

Compare $[ANC]$ of the following NH_4^+–NH_3 buffer solutions

$$\text{a. } [NH_4^+] + [NH_3] = 5 \times 10^{-3} \, M, \text{pH} = 9.3$$
$$\text{b. } [NH_4^+] + [NH_3] = 10^{-2} \, M, \text{pH} = 9.0$$

Despite the lower pH, solution b has a slightly larger acid-neutralizing capacity than solution a. With a pK value of 9.3, the α values are 0.5 and 0.33, and the corresponding $[ANC]$ capacities are 2.5×10^{-3} and 3.3×10^{-3} equivalents per liter, for a and b, respectively.

In a *multiprotic acid–base system* various reference levels ($f = 0, 1, 2, \ldots$) may be defined; for example, in a sulfide-containing solution the acid-neutralizing capacity with reference to the equivalence point defined by the pH of a pure H_2S solution ($f = 0, g = 2$) is

$$[ANC]_{f=0} = [HS^-] + 2[S^{2-}] + [OH^-] - [H^+]$$
$$= S_T(\alpha_1 + 2\alpha_2) + [OH^-] - [H^+] \tag{93}$$

where S_T is the sum of the S(-II) species ($[H_2S] + [HS^-] + [S^{2-}]$).

The base-neutralizing capacity of the phosphoric acid system with reference to the equivalence point, $f = 2$ (solution of Na_2HPO_4 with the proton condition: $2[H_3PO_4] + [H_2PO_4^-] + [H^+] = [PO_4^{3-}] + [OH^-]$), is given by

$$[BNC]_{f=2} = 2[H_3PO_4] + [H_2PO_4^-] + [H^+] - [PO_4^{3-}] - [OH^-]$$
$$= P_T(2\alpha_0 + \alpha_1 - \alpha_3) + [H^+] - [OH^-] \tag{94}$$

These relations can be generalized into

$$[BNC]_{f=n} = C[n\alpha_0 + (n-1)\alpha_1 + (n-2)\alpha_2 + (n-3)\alpha_3 + \cdots]$$
$$+ [H^+] - [OH^-] \tag{95}$$

and

$$[ANC]_{f-n} = C[-n\alpha_0 + (1-n)\alpha_1 + (2-n)\alpha_2$$
$$+ (3-n)\alpha_3 + \cdots] - [H^+] + [OH^-] \tag{96}$$

(As mentioned before, α_x refers to the ionization fraction of the species that has lost x protons from the most protonated acid species; f defines the equivalence points, the point at the lowest pH being $f = 0$.)

The ANC and BNC concept can be extended readily to mixed–base systems. For example, a natural carbonate-bearing water containing some NH_4^+ and borate has an $[ANC]_{f=0}$ (reference, pure CO_2 solution) of the equivalent sum of all the bases that have proton-unpopulated energy levels of HCO_3^- or less, minus the equivalent sum of all the acids of energy levels higher than $H_2CO_3^*$; that is,

$$[ANC]_{f=0} = [HCO_3^-] + 2[CO_3^{2-}] + [NH_3] + [B(OH)_4^-]$$
$$+ [OH^-] - [H^+] \tag{97}$$
$$= C_T(\alpha_1 + 2\alpha_2) + [NH_3] + [B(OH)_4^-] + [OH^-] - [H^+]$$

[ANC] and [BNC] are very useful in defining and characterizing an acid–base system. The proton-free energy level, that is, the pH of the system, is independent of the quantity of the solution (*intensity factors*). On the other hand, the number of protons (added coulometrically or with strong acids) required to attain a certain pH represents a *capacity factor* because it is proportional to the quantity of the solution. [ANC] is an integration of the buffer intensity over a pH range:

$$[ANC] = \int_{f=n}^{f=x} \beta d\mathrm{pH} \tag{98}$$

and gives us a conservative parameter that is not affected by temperature and pressure.

Any acid–base system of unknown distribution can be characterized fully with the help of two parameters. For example, in a solution of phosphates (Na salts) the equilibrium composition with regard to the six species (H_3PO_4, $H_2PO_4^-$, HPO_4^{2-}, PO_4^{3-}, H^+, and OH^-) can be resolved completely if the concentration of at least two of the species or two of certain combinations thereof are evaluated analytically; capacity factors such as [ANC] and [BNC] are especially valuable for defining acid–base systems in terms of conservative parameters. They can be determined frequently with ease and relatively good accuracy; thus the discrepancy between conceptual and operational definition is very small.

In carbonate systems and in natural waters [ANC] is referred to as *alkalinity*, while [BNC] is called *acidity*. In the context of natural waters these terms will be discussed in the next chapter.

REFERENCES

1   L. G. Sillén, "The Physical Chemistry of Seawater," *Oceanography*, American Association for the Advancement of Science, Publ. 67, Washington, D.C., 1961, p. 549.

2   L. G. Sillén and A. E. Martell, *Stability Constants of Metal Ion Complexes*, Chemical Society, London, 1964 and 1971.

3   R. M. Smith and A. E. Martell, *Critical Stability Constants*, Vol. 4, Plenum, New York, 1976.

4   L. G. Sillén, in *Equilibrium Concepts in Natural Water Systems*, Advances in Chemistry Series, No. 67, American Chemical Society, Washington, D.C., 1967, p. 45.

5   R. Bates, in *Treatise on Analytical Chemistry*, I. M. Kolthoff and P. J. Elving, Eds., Part I, Vol. 1, Interscience, New York, 1959, p. 361.

6   J. E. Ricci, *Hydrogen Ion Concentration*, Princeton University Press, Princeton, N.J., 1952.

7   J. G. Eberhart and T. R. Sweet, *J. Chem. Educ.*, **37**, 422 (1966).

8   A. J. Bard and D. M. King, *J. Chem. Educ.*, **42**, 127 (1965).

9   N. Bjerrum, *Sammlung Chem. Chem.-techn. Vorträge*, **21**, 575 (1914).

10 L. G. Sillén, in *Treatise in Analytical Chemistry*, I. M. Kolthoff and P. J. Elving, Eds., Interscience, New York, 1959.

READING SUGGESTIONS

Breneman, G. L., "A General Acid–Base Titration Curve Computer Program," *J. Chem. Educ.*, **812** (1974).

Butler, J. N., *Ionic Equilibrium, a Mathematical Approach*, Addison-Wesley, Reading, Mass., 1964. (This book presents a unified rigorous treatment of equilibria with a large number of realistic examples and problems used throughout the text.)

Covington, A. K., and J. Candle, "Acid Dissociation Constant of the Ammonium Ion Using the Glass Electrode," *J. Chem. Educ.*, **49**, 552 (1977). (An example for determining an acidity constant potentiometrically.)

Drago, R. S., "A Modern Approach to Acid–Base Chemistry," *J. Chem. Educ.*, **51**, 300 (1974). (This article discusses the scope of Lewis acid–base interactions in aqueous and nonaqueous solutions.)

Garrels, R. M., and C. L. Christ, *Solutions, Minerals and Equilibria*, Harper and Row, New York, 1965. (This very important book includes a detailed interpretation of activity coefficients in mixed electrolyte solutions.)

King, E. J., *Acid-Base Equilibria*, Macmillan, New York, 1965. (Comprehensive survey of current experimental investigations and interpretations of acid–base equilibria; pays special attention to polyelectrolytic acids.)

Laitinen, H. A., and W. E. Harris, *Chemical Analysis: An Advanced Text and Reference*, 2nd ed., McGraw-Hill, New York, 1975. (The principles of acid–base equilibria are covered in a rigorous fashion.)

Morris, J. C., "The Acid Ionization Constant of HOCl from 5 to 35°C," *J. Phys. Chem.*, **70**, 3798 (1966). (An example for measuring acidity constant from spectrophotometric data.)

Perrin, D. D., and B. Dempsey, *Buffers for pH and Metal Ion Control*, Chapman and Hall, London, 1974. (This is a practical book giving numerous buffer tables and recommendations for buffers in biology.)

Ramette, R. W., "Equilibrium Constants from Spectrophotometric Data," *J. Chem. Educ.*, **44**, 647 (1967).

Ricci, J. E., *Hydrogen Ion Concentration*, Princeton University Press, Princeton, N. J., 1952. (A most systematic, uniform, and rigorous presentation of the purely mathematical problems involved in the quantitative relations determining the H^+ ion concentration in aqueous solutions.)

Robinson, R. A., and R. H. Stokes, *Electrolyte Solutions*, Butterworths, London, 1959. (Detailed review on the nature of electrolyte solutions and theoretical treatment based on the theory of Debye–Hückel and its later developments by Onsager and Fuoss.)

Sillén, L. G., "Graphic Presentation of Equilibrium Data" In *Treatise on Analytical Chemistry*, Part I, Vol. 2, I. M. Kolthoff and P. J. Elving, Eds., Interscience, New York, 1959, Chapter 8. (Logarithmic and distribution diagrams with a master variable as well as predominance area diagrams with two master variables and their use are fully discussed.)

Sillén, L. G., "Master Variables and Activity Scales." In *Equilibrium Concepts in Natural Water Systems*, Advances in Chemistry Series, No. 67, American Chemical Society, Washington, D.C., 1967. (Compares merits of infinite dilution activity scale and ionic medium activity scale.)

Smith, R. M., and D. E. Martell, *Critical Stability Constants*, Vol. 4, *Inorganic Complexes*, Plenum, New York, 1976. (Critical compilation of equilibrium (including acidity) constants valid at various ionic strengths; the information includes $\Delta H°$ values or K values determined at different temperatures.)

Waser, J., "Acid–Base Titration and Distribution Curves," *J. Chem. Educ.*, **44**, 275 (1967).

PROBLEMS

3.1 Two acids, of approximately $10^{-2} M$ concentration, are titrated separately with a strong base and show the following pH at the end point (equivalence point, $f = 1$):

$$HA: pH = 9.5$$

$$HB: pH = 8.5$$

(a) Which one (HA or HB) is the stronger acid?
(b) Which one of the conjugate bases (A^- or B^-) is the stronger base?
(c) Estimate the pK values for the acids HA and HB.

3.2 Hypochlorous acid (HOCl) has an acidity constant $K = 3 \times 10^{-8}$ (25°C). The strength of a sodium hypochlorite solution can be determined by titration with a strong acid. Sketch a titration curve for a $10^{-3} M$ solution indicating pH at the beginning and at the end point of the titration.

3.3 A $4 \times 10^{-3} M$ solution of an acid HX has a pH of 2.4. What is the pH of an equimolar solution of the Na^+ salt of its conjugate base?

3.4 Calculate $-\log[H^+]$ and $-\log[OH^-]$ of a solution of $0.14\ M$ ammonia:
(a) at 25°C
(b) at 100°C
(c) (^cK_B at 25°C and at 100°C $= 1.8 \times 10^{-5}$).

3.5 What is the acidity produced by the addition of 1 g of pyrite agglomerate (FeS_2) to 1 liter of distilled water? Assume that $Fe(II)$ and S_2^{2-} are oxidized to $Fe(III)$ and SO_4^{2-}, respectively.

3.6 Report qualitatively the following titration curves:
(a) pH versus f, strong acid titrated with strong base
(b) $[H^+]$ versus f, strong acid titrated with strong base
(c) pH versus f, weak base titrated with strong acid
(d) pH versus f, weak acid titrated with strong base

3.7 Arrange the following solutions in order of increasing buffer intensity:
(a) $10^{-3}\ M\ NH_3-NH_4^+$, pH $= 7$
(b) $10^{-3}\ M\ NH_3-NH_4^+$, pH $= 9.2$
(c) $10^{-3}\ M\ H_2CO_3^*-HCO_3^--CO_3^{2-}$, pH $= 8.2$
(d) $10^{-3}\ M\ H_2CO_3^*-HCO_3^-$, pH $= 6.3$

3.8 SO_2 and CO_2 are important atmospheric species. Their dissolution in water is a key step in atmospheric-aquatic cycles. On the basis of the thermodynamic data provided in the appendix at the end of the book, evaluate the following:
(a) The respective equilibrium constants ($M\ atm^{-1}$) for

$$CO_2(g) + H_2O = H_2CO_3^*(aq)$$
$$SO_2(g) + H_2O = H_2SO_3^*(aq)$$

where $H_2SO_3^*$ is defined (similarly to $H_2CO_3^*$) as the analytical sum of $SO_2(aq)$ and H_2SO_3 (true).
(b) The relative strengths of the aqueous acids $H_2SO_3^*$, H_2SO_3, $H_2CO_3^*$, and H_2CO_3.

3.9 Benzoic acid, pK_a 4.2, can be used to inhibit biological growth. The following data show the dependence of the minimum required toxic concentration on the pH of the solution.

pH	Minimum benzoic acid required (mM)
3.5	1.2
4.0	1.6
4.5	3.0
5.0	7.3
5.5	21.0
6.0	64.0

Interpret the concentration required for growth inhibition in terms of the acid–base chemistry of benzoic acid. Which is the effective species, benzoic acid or benzoate anion?

3.10 The following data describe rainfall composition averaged over a year at Ithaca, New York, 1973–1974.

Ion	mg liter^{-1}
Na^+	0.15
K^+	0.09
Ca^{2+}	0.83
Mg^{2+}	0.08
NH_4^+	0.32
SO_4^{2-}	4.96
NO_3^-	2.88
Cl^-	0.47

The average (geometric) p^{a}H for the rainfall, measured with a glass electrode calibrated with an NBS (activity) buffer, was 4.05.

(a) On the basis of the composition data, estimate the hydrogen-ion concentration of the rainfall, assuming that all important ions have been accounted for.

(b) What is the ionic strength of the rainfall?

(c) Compare the measured p^{a}H with that which you estimate from the composition data.

(d) Discuss the relative importance of these atmospheric components in establishing the acidity of the rainfall: (i) HNO_3, (ii) H_2SO_4, and (iii) CO_2.

ANSWERS TO PROBLEMS

3.1 (a) HB; (b) A^-; (c) $pK_{HA} = 7$, $pK_{HB} = 5$.

3.2 $g = 0$, pH $= 9.3$; $g = 0.5$, pH $= 7.5$; $g = 1.0$, pH $= 5.3$.

3.3 pH $= 7$.

3.4 (a) $-\log[OH^-]\,(25°C) = -\log[OH^-]\,(100°C) = 2.8$.
(b) $-\log[H^+]\,(25°C) = 11.2$; $-\log[H^+]\,(100°C) = 9.2$.
For the calculation $^cK_W = 10^{-14}$ (25°C) and $^cK_W = 10^{-12}$ (100°C) have been used.

3.5 3.3×10^{-2} eq liter^{-1}. (For stoichiometry of the reaction see equations 11 to 13 in Section 10.)

3.6

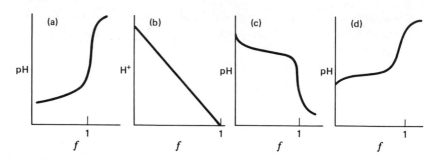

3.7 $a < c < d < b$.

3.8 (a) 3.4×10^{-2}, 1.23.

(b) 1.7×10^{-2}, 6.8×10^{-2}, 4.2×10^{-7}, 3×10^{-4}.

4

Dissolved Carbon Dioxide

4.1 INTRODUCTION

Inorganic constituents dissolved in fresh water and sea water have their origin in minerals and the atmosphere. Carbon dioxide from the atmosphere provides an acid that reacts with the bases of the rocks. The water may also lose dissolved carbon to the sediments by precipitation reactions. Representative dissolution and precipitation reactions with $CaCO_3(s)$ and a feldspar are

$$CaCO_3(s) + CO_2 + H_2O \rightleftharpoons Ca^{2+} + 2HCO_3^- \tag{1}$$
$$\text{calcite}$$

$$NaAlSi_3O_8(s) + CO_2 + \tfrac{11}{2}H_2O \rightleftharpoons$$
$$\text{albite}$$

$$Na^+ + HCO_3^- + 2H_4SiO_4 + \tfrac{1}{2}Al_2Si_2O_5(OH)_4(s) \tag{2}$$
$$\text{kaolinite}$$

In these and similar reactions HCO_3^- and CO_3^{2-} (alkalinity) are imparted to or withdrawn from the water. CO_2 is also added to the atmosphere by volcanic activity and through the combustion of fossil fuels. Carbon dioxide is reduced in the course of photosynthesis and set free during respiration and oxidation of organic matter; it thus occupies a unique position in the biochemical exchange between water and biomass. Dissolved carbonate species participate in homogeneous and heterogeneous acid–base and exchange reactions with the lithosphere and the atmosphere. Such reactions are significant in regulating the pH and the composition of natural waters. Table 4.1 gives a survey of the distribution of carbon in its various forms in the atmosphere and biosphere. A discussion of the geochemical significance of carbon and its cycle is postponed until Chapter 8.

In this chapter we describe the distribution of CO_2, H_2CO_3, HCO_3^-, and CO_3^{2-} in natural waters, examine the exchange of CO_2 between atmosphere and waters, evaluate the buffering mechanisms of fresh waters and seawater, and define their capacities for acid and base neutralization.

Two idealized equilibrium models—a system closed and a system open to the atmosphere—will be emphasized in order to account for the distribution of the carbonate species. The results will be applied in discussing the major acid–base system of fresh water, seawater, and rainwater.

Of serious concern is human alteration of the natural CO_2 cycle (combustion of fossil fuel and deforestation) and the resulting progressive increase in the CO_2

TABLE 4.1   CARBON IN SEDIMENTARY ROCKS, HYDROSPHERE, ATMO-SPHERE, AND BIOSPHERE[a]

	Total on Earth (10^{18} mol C)	Total on Earth (units of atmospheric CO_2, A_0)
Sediments		
Carbonate	5,100	94,000
Organic carbon	1,000	18,500
Land		
Organic carbon	0.1	1.8
Ocean		
$CO_2 + H_2CO_3$	0.019	0.35
HCO_3^-	2.9	54
CO_3^{2-}	0.36	7
Dead organic	0.4	7
Living organic	0.0007	0.01
Atmospheric		
$CO_2(A_0)$	0.0540	1.0

Representative Concentrations	$C_T(M)$[b]	Alkalinity (eq liter^{-1})[c]
Seawater	2.3×10^{-3}	2.5×10^{-3}
River waters, average	$\sim 10^{-3}$	$\sim 10^{-3}$
River waters, typical range	10^{-4}–5×10^{-3}	$10^{-4} - 5 \times 10^{-3}$
Groundwaters, typical range, United States[d]	5×10^{-4}–8×10^{-3}	$10^{-4} - 5 \times 10^{-3}$
Rain water, typical range	$10^{-5} - 5 \times 10^{-5}$	$0 - 4 \times 10^{-5}$[e]

Atmospheric $CO_2 = 0.033\%$ by volume in dry air; preindustrial $p_{CO_2} = 2.9 \times 10^{-4}$ atm; $p_{CO_2} = 3.3 \times 10^{-4}$ atm

[a] Compare Figure 8.1.
[b] $C_T = [CO_2(aq)] + [H_2CO_3] + [HCO_3^-] + [CO_3^{2-}]$ (Oceanographers often use the symbol $\sum CO_2$.)
[c] Alkalinity, the acid neutralizing capacity, is expressed as mol protons per liter, or equivalents per liter.
[d] Compare Figure 9.1.
[e] Some rainwaters contain mineral acidity of up to 10^{-3} eq liter^{-1}.

concentration in the atmosphere. We will evaluate the acidification of surface waters and of the surface layers of the oceans resulting from an increase in atmospheric CO_2. We will consider some rate factors of the carbonate systems, especially the kinetics of CO_2 hydration and dehydration reactions and of the CO_2 gas–water transfer. Finally, some aspects of carbon isotopes and their fractionation will be discussed.

4.2   DISSOLVED CARBONATE EQUILIBRIA (CLOSED SYSTEM)

We first consider a system that is closed to the atmosphere; that is, we treat $H_2CO_3^*$ as a nonvolatile acid. For simple aqueous carbonate solutions the interdependent nature of the equilibrium concentrations of the six solute components—CO_2, H_2CO_3, HCO_3^-, CO_3^{2-}, H^+, and OH^-—can be described completely by a system of six equations. The appropriate set of equations is comprised of four equilibrium relationships [which define the hydration equilibrium of CO_2 ($H_2CO_3 = CO_2(aq) + H_2O$), the first and second acidity constants of H_2CO_3 and the ion product of water] and any two equations describing a concentration and an electroneutrality or proton condition. As shown earlier (Section 3.3), it is convenient to define for the aqueous carbonate system a composite constant for all dissolved CO_2, hydrated or not. For the total analytical concentration of dissolved CO_2 we write $[H_2CO_3^*] = [CO_2(aq)]$ + $[H_2CO_3]$; the number of equations necessary to describe the distribution of solutes reduces to five. These equations together with the relationships that describe the distribution of solutes are given in Table 4.2 which is patterned after Table 3.6.

Equilibrium constants valid for different temperatures and activity conventions are tabulated in Tables 4.7 to 4.9.

As we have seen (Section 3.3) the equilibrium for the reaction

$$H_2O + CO_2(aq) = H_2CO_3(aq)$$

lies rather far to the left, and by far the greater fraction of unionized CO_2 is present in the form of $CO_2(aq)$. The commonly used "first acidity constant" K_1 is a composite constant for the protolysis of $H_2CO_3^*$ reflecting both the hydration reaction and the protolysis of true H_2CO_3. The acidity constants of true H_2CO_3, $K_{H_2CO_3}$, and the composite acidity constant of $H_2CO_3^*$, K_1, are interrelated [see (1), (2), and (2a), Table 4.2] by

$$K_1 = \frac{K_{H_2CO_3}}{1 + K} \tag{3}$$

where K is the constant describing the hydration equilibrium [(1) in Table 4.2]. At 25°C K is on the order of 650.† Thus equation 3 can often be simplified to

$$K_1 \simeq \frac{K_{H_2CO_3}}{K} \tag{4}$$

Correspondingly, the concentration of $CO_2(aq)$ is nearly identical to the analytical concentration of $H_2CO_3^*$. K_1 is the equilibrium constant known, from direct experimental determination, with a high degree of accuracy.

Representative Equilibrium Diagrams for Fresh Water and Seawater

Figure 4.1 illustrates the equilibrium distribution of the carbonate solutes as a function of pH (cf. Sections 3.6 to 3.9). The pH values of the pure solutions of the

TABLE 4.2 THE EQUILIBRIUM DISTRIBUTION OF SOLUTES IN AQUEOUS CARBONATE SOLUTION (SYSTEM CLOSED TO THE ATMOSPHERE)

Species:

$$CO_2(aq), H_2CO_3, HCO_3^-, CO_3^{2-}, H^+, OH^-$$

$$[H_2CO_3^*] = [CO_2 \cdot aq] + [H_2CO_3]$$

Equilibrium constants:[a]

$$[CO_2(aq)]/[H_2CO_3] = K \tag{1}$$

$$[H^+][HCO_3^-]/[H_2CO_3^*] = K_1 \tag{2}$$

$$[H^+][HCO_3^-]/[H_2CO_3] = K_{H_2CO_3} \tag{2a}$$

$$[H^+][CO_3^{2-}]/[HCO_3^-] = K_2 \tag{3}$$

$$[H^+][OH^-] = K_w \tag{4}$$

Concentration condition:

$$C_T = [H_2CO_3^*] + [HCO_3^-] + [CO_3^{2-}] \tag{5}$$

Ionization fractions:[b]

$$[H_2CO_3^*] = C_T\alpha_0 \qquad [HCO_3^-] = C_T\alpha_1 \qquad [CO_3^{2-}] = C_T\alpha_2$$

$$\alpha_0 = \left(1 + \frac{K_1}{[H^+]} + \frac{K_1 K_2}{[H^+]^2}\right)^{-1} \tag{6}$$

$$\alpha_1 = \left(\frac{[H^+]}{K_1} + 1 + \frac{K_2}{[H^+]}\right)^{-1} \tag{7}$$

$$\alpha_2 = \left(\frac{[H^+]^2}{K_1 K_2} + \frac{[H^+]}{K_2} + 1\right)^{-1} \tag{8}$$

Proton conditions of pure solutions (equivalence points): (e.g., of $H_2CO_3^*$, $NaHCO_3$, Na_2CO_3, respectively):[c]

$$[H^+] = [HCO_3^-] + 2[CO_3^{2-}] + [OH^-] \tag{9}$$

$$[H_2CO_3^*] + [H^+] = [CO_3^{2-}] + [OH^-] \tag{10}$$

$$2[H_2CO_3^*] + [HCO_3^-] + [H^+] = [OH^-] \tag{11}$$

with respective values of $[H^+]$ at equivalence points:[d]

$$[H^+] = (C_T K_1 + K_w)^{0.5} \tag{12}$$

$$[H^+] \simeq [K_1(K_2 + K_w/C_T)]^{0.5} \tag{13}$$

$$[H^+] \simeq K_w/2C_T + [K_w^2/4C_T^2 + K_2 K_w/C_T]^{1/2} \tag{14}$$

As corresponding approximate values:

$$[H^+] \simeq (C_T K_1)^{0.5} \tag{15}$$

$$[H^+] \simeq (K_1 K_2)^{0.5} \tag{16}$$

$$[H^+] \simeq (K_2 K_w/C_T)^{0.5} \tag{17}$$

174

Titration:
Alkalimetric

$$f - C_B/C_T - \alpha_1 + 2\alpha_2 + ([OH^-] - [H^+])/C_T \tag{18}$$

Acidimetric

$$g = C_A/C_T = 2 - f = 2\alpha_0 + \alpha_1 - ([OH^-] - [H^+])/C_T \tag{19}$$

Acid- and base-neutralizing capacity·[e]
Alkalinity

$$[Alk] = C_T(\alpha_1 + 2\alpha_2) + [OH^-] - [H^+] \tag{20}$$

Acidity

$$[Acy] - C_T(\alpha_1 + 2\alpha_0) + [H^+] - [OH^-] \tag{21}$$

Buffer intensity[f]

$$\beta_{C_T}^{C_B} = 2.3\{[H^+] + [OH^-] + C_T[\alpha_1(\alpha_0 + \alpha_2) + 4\alpha_2\alpha_0]\} \tag{22}$$

---

[a] Equilibrium constants are defined for a constant ionic medium activity scale (Section 3.4).
[b] Equations (6) to (8) are derived as in Section 3.7.
[c] Na' (in $NaHCO_3$ or Na_2CO_3) is used as a symbol of a nonprotolyzable cation (Na^+, Li^+, K^+, ...). In Figure 4.1 the equivalence points corresponding to (9), (10), and (11) are marked x, y and z, respectively.
[d] The exact numerical solution is of the fourth degree in $[H^+]$ (see Table 3.6). For many practical purposes ($C_T > 10^{-6} M$), (12) to (14) are sufficiently exact. If $C_T > 10^{-5} M$, (15), and if $C_T \geq 10^{-3} M$, (16) and (17), may be used.
[e] For derivation and further discussion see Sections 3.9 and 4.4.
[f] See Sections 3.9 and 4.5.

acid, the ampholyte, and the base, that is, of pure solutions of $H_2CO_3^*$, $NaHCO_3$, and Na_2CO_3, correspond to the *equivalence points, x, y, and z,* in alkalimetric and acidimetric titrations of natural waters; they are defined by the appropriate proton conditions [(9) to (11) of Table 4.2, Example 4.1] of $C_T = 10^{-3} M$ ($I = 10^{-3}$, 25°C, $p = 1$ atm) or $C_T = 2.3 \times 10^{-3} M$ ($I =$ seawater conditions, 10°C, $p = 1$ atm) solutions, respectively, of $H_2CO_3^*$, $NaHCO_3$, and Na_2CO_3.

Obviously the diagrams for "fresh water" and for "seawater" are very similar. Because of the ionic strength effects the operational acidity constants are larger for seawater than for fresh waters; that is, the pK' values and thus the pH at the equivalence point—especially at the equivalence point y—are lower for seawater than for fresh water. Seawater contains, in addition to dissolved CO_2, boric acid, H_3BO_3 (representative concentration of total boron $= 4.1 \times 10^{-4} M$). Its presence does not contribute markedly to the buffering of seawater. At the pH of seawater (pH $= 8.1$), its buffer intensity is near the minimum.

† At 25°C reported values of K range from 350 to 990. The corresponding range in $pK_{H_2CO_3}$ for true carbonic acid is from 3.8 to 3.4.

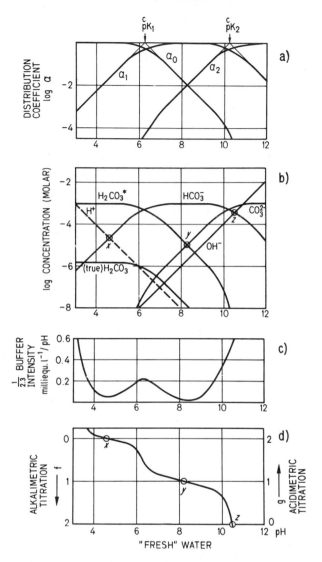

Figure 4.1 Distribution of solute species and buffering in aqueous carbonate systems. (*a–d*) Fresh water, 25°C; (*e* and *f*) seawater, 10°C. This figure has been constructed under the assumption that $C_T = [H_2CO_3^*] + [HCO_3^-] + [CO_3^{2-}] = \text{constant}$ (10^{-3} M in fresh water and 2.3×10^{-3} M in seawater). The following equilibrium constants corrected for salt effects have been used. Fresh water: $I = 10^{-3}$, $pK_1' = 6.3$, $pK_2' = 10.25$ (25°C, $p = 1$ atm); seawater $p^cK_1 = 6.1$, $p^cK_2 = 9.3$ (10°C, $p = 1$ atm), $pK_{H_3BO_3}'$ (10°C) $= 8.8$. The hydration constant K [(1) in Table 4.2] was taken to be 630. (*a*) Ionization fractions as a function of pH. (*b*) Logarithmic equilibrium diagram for fresh water. Because $[CO_2(aq)] \gg [H_2CO_3]$, $[CO_2 \text{ aq}] \approx [H_2CO_3^*]$. Note that H_2CO_3 is a much stronger acid than $CO_2(aq)$ or $H_2CO_3^*$. Pure H_2CO_3 has a pK value (where $[H_2CO_3] = [HCO_3^-]$) of $pK_{H_2CO_3} \simeq 3.5$. The equivalence points corresponding to pure solutions (C_T molar) of $H_2CO_3^*$, $NaHCO_3$, and Na_2CO_3 are marked *x*, *y* and *z*, respectively. (*c*) Buffer intensity is plotted as a function of pH. (*d*) Alkalimetric or acidimetric titration curve. Note that no pH jump occurs at the equivalence point *z*, because at this point the buffer intensity caused by high $[OH^-]$ is too large.

176

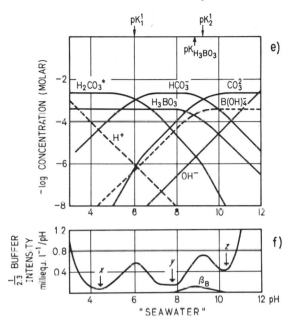

Figure 4.1 (*Continued.*) (*e*) Logarithmic equilibrium diagram for seawater. Because seawater contains 4.1×10^{-4} M boric acid and borate [H_3BO_3 or $B(OH)_3$ and $B(OH)_4^-$], the distribution of these species is also given, (*f*) Buffer intensity of seawater. Note that seawater has its minimum buffer intensity in the slightly alkaline pH range (endpoint *y*) approximately half a pH unit lower than fresh water. The H_3BO_3- $B(OH)_4^-$ couple does not contribute significantly to the total buffer intensity (β_B for the contribution of aqueous B to the buffer intensity).

Obviously, natural waters are not closed systems. The idealized model discussed so far is still useful because a natural water sample in the laboratory, for example, during acid–base titration, or waters in groundwater systems or in water supply distribution systems often behave, in first approximation, as in a closed system.

Example 4.1. Equivalence Points in Alkalimetric and Acidimetric Titrations: pH of Pure CO_2, $NaHCO_3$, and Na_2CO_3 Solutions

Estimate the pH as a function of concentration for pure solutions of CO_2, $NaHCO_3$, and Na_2CO_3, respectively. (Assume a closed system; i.e., treat $H_2CO_3^*$ as a nonvolatile acid.) The answer is given in Figure 4.2, where pH is plotted for the different solutions as a function of log C_T. The numerical calculation of these curves is based on (12) to (14) of Table 4.2, which characterize the appropriate proton conditions. The pH values at the equivalence points can also be readily estimated from logarithmic equilibrium sketches (Figure 4.2*b* to *d*). For example the proton condition of a pure $NaHCO_3$ solution is

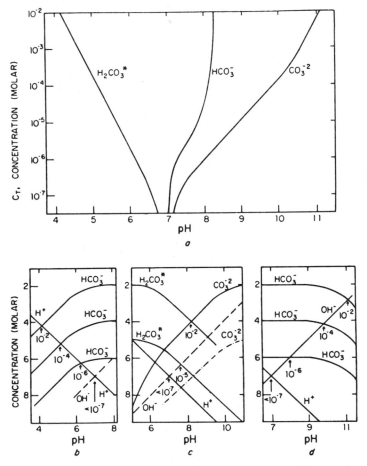

Figure 4.2 pH of pure solutions of $CO_2(H_2CO_3^*)$, $NaHCO_3$, and Na_2CO_3 at various dilutions (see Example 4.1). The curves in (a) can be computed by equations 12–14, Table 4–2, or with the help of logarithmic equilibrium diagrams. (b–d) Sketches for pure CO_2, $NaHCO_3$, and Na_2CO_3 solutions, respectively. For a few concentrations the intersections that characterize the appropriate proton conditions are indicated by arrows.

given by $[H_2CO_3^*] + [H^+] = [CO_3^{2-}] + [OH^-]$. As Figure 4.2c illustrates, at high concentrations ($C_T > 10^{-3}$ M) the equilibrium is characterized by the simplified condition $[H_2CO_3^*] \approx [CO_3^{2-}]$. In this concentration range, $NaHCO_3$ solutions maintain a constant pH (ignoring small changes that may result from activity variations). In a more dilute concentration range (10^{-4} $M > C_T > 10^{-7}$ M) the pH of pure $NaHCO_3$ solutions is characterized by the appropriate proton condition $[H_2CO_3^*] \approx [OH^-]$. As $C_T \to 0$, the electroneutrality condition becomes $[H^+] \approx [OH^-]$ and neutrality (pH $\simeq$ 7) exists. Pure solutions of CO_2 ($C_T > 10^{-6}$ M) and of Na_2CO_3 ($C_T > 10^{-3}$ M) have $d\text{pH}/d \log C_T$ values of -0.5 and $+0.5$, respectively.

4.3 DISSOLUTION OF CO_2 (OPEN SYSTEM)

If we now open the aqueous carbonate system to the atmosphere, we have to consider the equilibrium between the gas and the solution phase. As we have seen in Section 2.8, such an equilibrium can be formulated in terms of a mass law equilibrium relationship. Because of different conventions used in expressing the concentration (activity) in the gas and the solution phases, equilibrium constants with different dimensions are commonly used to characterize the dissolution equilibrium. Table 4.3 gives the various equilibrium expressions and shows how the equilibrium constants are interrelated.

Ideally these expressions should be written in terms of activities and fugacities; then the gas solution equilibrium is independent of the salinity of the solution.

TABLE 4.3 SOLUBILITY OF CO_2

Example:[a]

$$CO_2(g) \rightleftharpoons CO_2(aq)$$

Assumptions: Gas behaves ideally; $[CO_2(aq)] - [H_2CO_3^*]$

I. *Expressions for Solubility Equilibrium:*[b]
   (1) Distribution (mass law) constant, K_D:

$$K_D - [CO_2(aq)]/[CO_2(g)] \quad \text{(dimensionless)} \tag{1}$$

   (2) Henry's law constant, K_H:
       In (1), $[CO_2(g)]$ can be expressed by Dalton's law of partial pressure:

$$[CO_2(g)] = p_{CO_2}/RT \tag{2}$$

   Combination of (1) and (2) gives

$$[CO_2(aq)] = (K_D/RT)p_{CO_2} = K_H p_{CO_2} \tag{3}$$

   where $K_H = K_D/RT$ (mol liter^{-1} atm^{-1})
   (3) Bunsen absorption coefficient, α_B:

$$[CO_2(aq)] = (\alpha_B/22.414)p_{CO_2} \tag{4}$$

   where $22.414 = RT/p$ (liter mol^{-1}) and

$$\alpha_B = K_H \times 22.414 \text{ (atm}^{-1}) \tag{5}$$

II. *Partial Pressure and Gas Composition*

$$p_{CO_2} = x_{CO_2}(P_T - w) \tag{6}$$

where x_{CO_2} = mole fraction or volume fraction in dry gas, P_T = total pressure, and w = water vapor pressure

---

[a] Same types of expressions apply to other gases.
[b] The equilibrium constants defined by (1) to (4) are actually constants only if the equilibrium expressions are formulated in terms of activities and fugacities.

For example, the activity of $CO_2(aq)$ ($= H_2CO_3^*$) is identical for fresh water and for seawater that has been equilibrated with an atmosphere containing the same p_{CO_2}.† However, since activity coefficients for uncharged species become larger than 1 (salting-out effect), the concentration of $CO_2(aq)$ is smaller in the salt solution than in the dilute aqueous medium.

The concentration (activity) of CO_2 (or any other gas or volatile substance) in solution can always be expressed either in terms of concentration units or in terms of the "partial pressure of the gas in solution," that is, that pressure of CO_2 in the gas phase with which the sample would be in equilibrium. $CO_2(aq)$ and p_{CO_2} are interrelated by Henry's law [(3) in Table 4.3]. If the aqueous system is in equilibrium with the gas phase, the "partial pressure of the gas in solution" is equal to the partial pressure of the gas in the gas phase.

Aqueous Carbonate System Open to the Atmosphere with Constant p_{CO_2}

A very simple model showing some of the characteristics of the carbonate system of natural waters is provided by equilibrating pure water with a gas phase (e.g., the atmosphere) containing CO_2 at a constant partial pressure. One may then vary the pH by the addition of strong base or strong acid, thereby keeping the solution in equilibrium with p_{CO_2}. This simple model has its counterpart in nature when CO_2 reacts with bases of rocks (i.e., with silicates, clays).

Figure 4.3 shows the distribution of the solute species of such a model. A partial pressure of CO_2 ($p_{CO_2} = 10^{-3.5}$ atm) representative of the atmosphere and equilibrium constants valid at 25°C has been assumed. The equilibrium concentration of the individual carbonate species can be expressed as a function of p_{CO_2} and pH. By combining (6), (7), or (8) (from Table 4.2) with Henry's law,

$$[H_2CO_3^*] = K_H p_{CO_2} \tag{5}$$

one obtains

$$C_T = \frac{1}{\alpha_0} K_H p_{CO_2} \tag{6}$$

$$[HCO_3^-] = \frac{\alpha_1}{\alpha_0} K_H p_{CO_2} = \frac{K_1}{[H^+]} K_H p_{CO_2} \tag{7}$$

and

$$[CO_3^{2-}] = \frac{\alpha_2}{\alpha_0} K_H p_{CO_2} = \frac{K_1 K_2}{[H^+]^2} K_H p_{CO_2} \tag{8}$$

It follows from these equations that in a logarithmic concentration–pH diagram (Figure 4.3) the lines of $H_2CO_3^*$, HCO_3^-, and CO_3^{2-} have slopes of 0, +1, and +2, respectively (compare Figure 3.9).

---

† This has been confirmed experimentally, for example, for oxygen by Mancy [1] Amperometric membrane electrodes measure oxygen activity because the response is proportional to the activity gradient across the membrane.

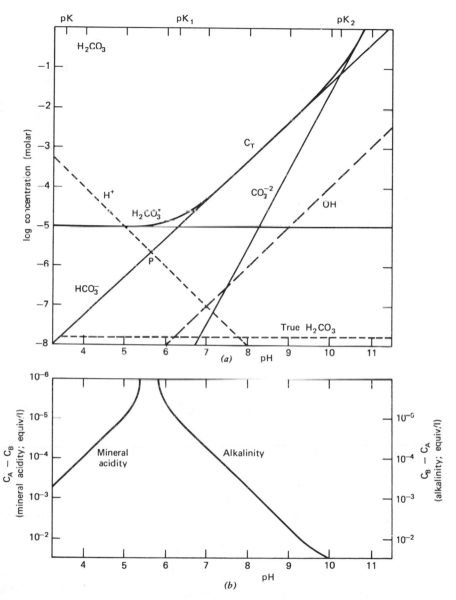

Figure 4.3 Aqueous carbonate equilibrium; Constant p_{CO_2}. (*a*) Water is equilibrated with the atmosphere ($p_{CO_2} = 10^{-3.5}$ atm), and the pH is adjusted with strong base or strong acid. Equations 5 to 8 with the constants (25°C) $pK_H = 1.5$, $pK_1 = 6.3$, $pK_2 = 10.25$, pK (hydration of CO_2) = -2.8 have been used. The pure CO_2 solution is characterized by the proton condition $[H^+] = [HCO_3^-] + 2[CO_3^{2-}] + [OH^-]$ (see point *P*) and the equilibrium concentrations $-\log[H^+] = -\log[HCO_3^-] = 5.65$; $\log[CO_2aq] = -\log[H_2CO_3^*] = 5.0$; $-\log[H_2CO_3] \approx 7.8$; $-\log[CO_3^{-2}] = 10.4$. (*b*) At pH values different from that of a pure CO_2 solution, the solution contains either alkalinity or mineral acidity, depending on the concentration of strong acid C_A or strong base C_n that had to be added to reach these pH values. In accordance with equation 11, alkalinity and acidity (logarithmic units) are plotted as a function of pH.

181

If we equilibrate pure water with CO_2, the system is defined by two independent variables (in addition to total pressure),† for example, temperature and p_{CO_2}. In other words, the equilibrium concentrations of all the solute components can be calculated with the help of Henry's law, the acidity constants, and the proton condition or charge balance if, in addition to temperature, one variable, such as p_{CO_2}, $[H_2CO_3^*]$, or $[H^+]$, is given.

The pH of a pure aqueous CO_2 solution is defined by the charge balance (see Figure 4.3) $[H^+] = [HCO_3^-] + 2[CO_3^{2-}] + [OH^-]$. If we keep the system in equilibrium with a constant p_{CO_2}, we can vary the pH only by adding a base C_B or an acid C_A. The electroneutrality condition must be adjusted for such an addition.

Considering that C_B is equivalent to the concentration of a monovalent cation (e.g., $[Na^+]$ from NaOH) and that C_A is equivalent to the concentration of a monovalent anion (e.g., $[Cl^-]$ from HCl),

$$C_B + [H^+] = [HCO_3^-] + 2[CO_3^{2-}] + [OH^-] + C_A \qquad (9)$$

which can be rearranged to

$$C_B - C_A = C_T(\alpha_1 + 2\alpha_2) + [OH^-] - [H^+] \qquad (10)$$

$$[Alk] = C_B - C_A = \frac{K_H p_{CO_2}}{\alpha_0}(\alpha_1 + 2\alpha_2) + [OH^-] - [H^+] \qquad (11)$$

$C_B - C_A$ is the acid-neutralizing capacity of the solution with respect to the pure solution of CO_2 and is thus the *alkalinity*, [Alk]. If $C_A > C_B$, then $C_A - C_B$ represents the base-neutralizing capacity (with respect to the same reference point) or the *mineral acidity*. For the conditions selected, alkalinity and mineral acidity are plotted as a function of pH in Figure 4.3*b*.

The model has three independent variables (in addition to total pressure). Therefore it is sufficient to give T, p_{CO_2}, and alkalinity to define the system. As equation 10 shows, for example, the pH (implicity contained in α values) is given by [Alk] and p_{CO_2}.

If bases other than CO_3^{2-}, HCO_3^-, and OH^- are introduced into the system, they have to be considered too in the electroneutrality or proton condition (see equation 29 in chapter 3). For example, for a water containing borate, for example, *seawater*, equations 9 to 11 can be rewritten:

$$C_B + [H^+] = [HCO_3^-] + 2[CO_3^{2-}] + [B(OH)_4^-] + [OH^-] + C_A \qquad (12)$$

$$C_B - C_A = C_T(\alpha_1 + 2\alpha_2) + B_T \alpha_{B^-} + [OH^-] - [H^+] \qquad (13)$$

$$[Alk] = C_B - C_A = \frac{K_H p_{CO_2}}{\alpha_0}(\alpha_1 + 2\alpha_2) + B_T \alpha_{B^-} + [OH^-] - [H^+] \ddagger \qquad (14)$$

where $B_T = [H_3BO_3] + [B(OH)_4^-]$ and $\alpha_B = [B(OH)_4^-]/B_T$.

---

† This is in accord with the Gibbs phase rule. The system consists of two components, $C = 2$ (H_2O and CO_2) and two phases ($P = 2$). The number of independent variables, F, is given by $F = C + 2 - P = 2$. (For further application of the phase rule see Section 5.7.)

‡ We retain the term *alkalinity*, [Alk], (in the literature sometimes referred to as titration alkalinity) to refer to the total ANC of any system with respect to a pure CO_2 solution. The part given by $[HCO_3^-] + 2[CO_3^{2-}]$ is referred to as *carbonate alkalinity*.

Waters in equilibrium with the atmosphere ($p_{CO_2} = 3 \times 10^{-4}$ atm) and having pH values similar to those of natural waters (6.5 to 9) have alkalinities between 10^{-5} and 10^{-2} eq liter^{-1}. *Seawater* having an alkalinity of 2.3 meq liter^{-1} brought into equilibrium with the atmosphere should have a $-\log[H^+]$ of approximately 8.2 (see Example 4.3). The gas composition in soils is quite different from that in the atmosphere; because of respiration by organisms, the CO$_2$ composition in soils is up to a few hundred times higher than that in the atmosphere. Thus, when brought to the surface, *groundwaters* are usually over-saturated with CO$_2$ and tend to lose CO$_2$.

Example 4.2. pH of Solutions at Given p_{CO_2}

Estimate the pH of the following solutions that have been equilibrated with the atmosphere ($p_{CO_2} = 10^{-3.5}$ atm):

  (i) 10^{-3} M KOH

  (ii) 10^{-3} M NaHCO$_3$

  (iii) 5×10^{-4} M Na$_2$CO$_3$

  (iv) 5×10^{-4} M MgO

All the solutions have the same alkalinity; $[Alk] = 10^{-3}$ eq liter^{-1}. They must all have the same pH at equilibrium with the same p_{CO_2} (apart from essentially negligible differences because of ionic strength). Using the constants given in Figure 4.3, we calculate by equation 11 a pH of 8.3. This answer can also be found from Figure 4.3b.

Example 4.3. Equilibration with the Gas Phase

A liter of 2×10^{-3} M NaHCO$_3$ solution is brought into contact with 10 ml nitrogen. How much CO$_2$ will be in the gas phase after equilibration (tem. = 25°C, pK$_H$ = 1.5, pK$_1$ = 6.3, pK$_2$ = 10.2)?

The distribution equilibrium (equation i) and a mass balance (equation ii) need to be considered.

$$\frac{[CO_2(aq)]}{[CO_2(g)]} = K_H RT = 0.75 \tag{i}$$

$$\frac{[CO_2(aq)]}{\alpha_0} + 0.01[CO_2(g)] = 3 \times 10^{-3} \tag{ii}$$

The volume fraction V_{gas}/V_{aq} is 0.01. If this fraction is small, the second term in equation ii is negligible; that is, the solution composition is not changed appreciably and α_0 at equilibrium with the gas phase is equal to α_0 before equilibration with N$_2$ ($\alpha_0 \simeq 1 \times 10^{-2}$). Thus $[CO_2(g)] = 2.67 \times 10^{-5}$ M; that is, the total gas phase contains 2.82×10^{-7} mol CO$_2$.

If the composition of the solution changes appreciably as a result of the transfer of CO_2 to the gas phase, a third equation is necessary. This, most conveniently, is an expression of alkalinity which does not change as a result of CO_2 loss.

Example 4.4a. pH of Seawater

Calculate the pH of seawater at 15°C in equilibrium with the atmosphere ($p_{CO_2} = 3.3 \times 10^{-4}$ atm). The seawater has an alkalinity of 2.47×10^{-3} equiv liter^{-1} and contains a total boron concentration B_T of 4.1×10^{-4} M. Constants corrected for ionic strength effects of seawater at 15°C (cf. Tables 4.7 to 4.10) are $^cK_H = 4.8 \times 10^{-2}$ [M atm^{-1}]; $K_1' = 8.8 \times 10^{-7}$, $K_2 = 5.6 \times 10^{-10}$; the acidity constant for H_3BO_3 is $K' = 1.6 \times 10^{-9}$; $K_w' = 2.0 \times 10^{-14}$.

pH for a given alkalinity is defined by equation 14. One approach would be to solve equation 14 by trial and error. For every assumed [H^+] we can obtain the corresponding α values (either from graphs that plot α values versus pH or with the help of a programmable calculator or minicomputer. As a first approximation we will consider the first term in equation 14 only, neglecting $B_T\alpha_{B^-}$, [OH^-], and [H^+]. We can determine later whether the other terms are indeed negligible.

An alkalinity of 2.47 meq obtains at pH = 8.216. At this pH $\alpha_{B^-} = 0.21$ and the contribution of [$B(OH)_4^-$] = 8.5×10^{-5} M, or about 3% of the total alkalinity. For a more exact result, however, we can include [$B(OH)_4^-$], [OH^-], and [H^+] in the iterative calculation. A pH of 8.203 gives an alkalinity of 2.469×10^{-3} eq liter^{-1}. Further iteration is not necessary. At this pH, the carbonate alkalinity is 2.39×10^{-3} eq liter^{-1}, [$B(OH)_4^-$] is 8.3×10^{-5} M, and [OH^-] is 3.2×10^{-6} M.†

Example 4.4b. pH Change Resulting from p_{CO_2} Increase

The partial pressure of CO_2 in the atmosphere is increasing, mostly as a result of the release of CO_2 from fossil fuel burning, and will double within the next 40 to 70 years (see Section 11.7). What is the pH of surface seawater (as specified in Example 4.4a) in equilibrium with a p_{CO_2} of 6.6×10^{-4} atm? Because the sea is oversaturated with regard to solid $CaCO_3$, it appears justified to disregard any heterogeneous reaction of the seawater with $CaCO_3$; thus we can assume that the alkalinity will remain constant. As before, we can apply equation 14 and, as a first approximation, treat the contribution by [$B(OH)_4^-$] to [Alk] as negligible. A pH of 7.932 obtains; after including the noncarbonate contributions to the alkalinity, a pH of 7.924 results. (At this pH, [$B(OH)_4^-$] amounts to

---

† A 10-line program can compute α_0, α_1, α_2, α_{B^-} and [OH^-] for a series of pH values, then compute [Alk] = [$K_{HPCO_2}(\alpha_1 + 2\alpha_2)/\alpha_0$] + $B_T\alpha_{B^-}$ + K_w'/[H^+] for these pH values, and stop when computed [Alk] = given [Alk], yielding all aqueous species for the given [Alk]. B_T, and p_{CO_2}.

4.8×10^{-5} M, or 2% of the total alkalinity.) Thus the doubling of p_{CO_2} results under equilibrium conditions in a lowering of pH 0.279 pH units or an increase in [H$^+$] by a factor of 1.90.

4.4 CONSERVATIVE QUANTITIES: ALKALINITY, ACIDITY, AND C_T

Acidimetric or alkalimetric titrations of a carbonate-bearing water to the appropriate end points represent operational procedures for determining alkalinity and acidity, that is, the equivalent sum of the bases that are titratable with strong acid and the equivalent sum of the acids that are titratable with strong base. Alkalinity and acidity are then capacity factors that represent, respectively, the acid- and base-neutralizing capacities (ANC and BNC; see Section 3.9) of an aqueous system. For solutions that contain no protolysis system other than that of aqueous carbonate, alkalinity is a measure of the quantity of strong acid per liter required to attain a pH equal to that of a C_T molar solution of $H_2CO_3^*$. Alternatively, acidity is a measure of the quantity per liter of strong base required to attain a pH equal to that of a C_T molar solution of Na_2CO_3.

The pH values at the respective equivalence points (around 4.5 and 10.3) for titrations of alkalinity and acidity represent approximate thresholds beyond which most life processes in natural waters are seriously impaired. Thus alkalinity and acidity are convenient measures for estimating the maximum capacity of a natural water to neutralize acidic and caustic wastes without permitting extreme disturbance of biological activities in the water.

If mineral acid is added to a natural water beyond equivalence point x ($f = 0$) (compare Figures 4.1 and 4.4), the H$^+$ added will remain as such in solution. Such a water is said to contain *mineral acidity*. Correspondingly a water with a pH higher than point z ($f = 2$) contains *caustic alkalinity*.

Titration to the intermediate equivalence point $y(f = 1)$ (the phenolphthalein end point) is a measure of the CO$_2$ acidity in an alkalimetric titration and of p alkalinity in an acidimetric titration.

The equations given in Figure 4.4 are of analytical value because they represent rigorous conceptual definitions of the acid-neutralizing and the base-neutralizing capacities of carbonate systems. As discussed in Section 3.9 the definitions of alkalinity and acidity algebraically express the proton excess or proton deficiency of the system with respect to a reference proton level (equivalence point).† Alkalimetric and acidimetric titrations to the equivalence points $f = 0$ and $f = 1$ give inherently accurate values of base and acid-neutralizing

---

† The formulas for the species may include medium ions. In other words [HCO$_3^-$] includes the concentration of free HCO$_3^-$ as well as the concentration of complex-bound HCO$_3^-$, that is, [HCO$_3^-$] = [true HCO$_3^-$] + [NaHCO$_3$] + [MgHCO$_3^+$] $\cdots$. Similarly [CO$_3^{2-}$] = [true CO$_3^{2-}$] + [MgCO$_3$] + [NaCO$_3^-$].

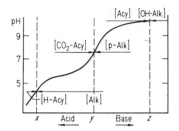

Figure 4.4   Conservative quantities: alkalinity and acidity as acid-neutralizing and base-neutralizing capacity. These parameters can be determined by acidimetric and alkalimetric titration to the appropriate end points. The equations given below define the various capacity factors of an aqueous carbonate system rigorously. If the solution contains protolytic systems other than that of aqueous carbonate, these equations have to be corrected; for example, in the presence of borate one has to add $[B(OH)_4^-]$ to the right-hand side of (3) and $[H_3BO_3]$ to the right-hand side of (6).

Term	End point	Definition	
		I Acid-Neutralizing Capacity (ANC)	
Caustic alkalinity	$f = 2$	$[OH^- \text{-Alk}] = [OH^-] - [HCO_3^-] - 2[H_2CO_3^*] - [H^+]$	(1)
p-Alkalinity	$f = 1$	$[p\text{-Alk}] = [OH^-] + [CO_3^{2-}] - [H_2CO_3^*] - [H^+]$	(2)
Alkalinity	$f = 0$	$[\text{Alk}] = [HCO_3^-] + 2[CO_3^{2-}] + [OH^-] - [H^+]$	(3)
		II Base-Neutralizing Capacity (BNC)	
Mineral acidity	$f = 0$	$[\text{H-Acy}] = [H^+] - [HCO_3^-] - 2[CO_3^{2-}] - [OH^-]$	(4)
CO_2-acidity	$f = 1$	$[CO_2\text{-Acy}] = [H_2CO_3^*] + [H^+] - [CO_3^{2-}] - [OH^-]$	(5)
Acidity	$f = 2$	$[\text{Acy}] = 2[H_2CO_3^*] + [HCO_3^-] + [H^+] - [OH^-]$	(6)

III   Combinations

$f = 2 - g$	(7)	$[\text{Alk}] + [CO_2\text{-Acy}] = C_T$	(11)
$[\text{Alk}] + [\text{H-Acy}] = 0$	(8)	$[\text{Alk}] + [\text{Acy}] = 2C_T$	(12)
$[\text{Acy}] + [\text{OH-Alk}] = 0$	(9)	$[\text{Alk}] - [p\text{-Alk}] = C_T$	(13)
$[p\text{-Alk}] + [CO_2\text{-Acy}] = 0$	(10)	$[CO_2\text{-Acy}] - [\text{H-Acy}] = C_T$	(14)

capacity. As Figure 4.4 and (11) to (14) of this figure suggest, $C_T\,(=[H_2CO_3^*] + [HCO_3^-] + [CO_3^{2-}])$ can also be obtained from an acidimetric or alkalimetric titration (difference between the equivalence points $f = 0$ and $f = 1$.)

The Effect of Other Bases.   Many other dissolved buffer components occur in natural waters. Dissolved silicates, borates, ammonia, organic bases, sulfides, and phosphates are among the bases that may be titrated in an acidimetric titration of alkalinity. Similarly, noncarbonic acids such as H_2S, polyvalent metal ions, and organic acids may influence the base-neutralizing capacity of natural waters. Usually, however, the concentrations of each of these noncarbonate components are very small in comparison to the carbonate species.

Perhaps the most relevant species in seawater are the borates ($10^{-3.4}\ M$) and in many fresh waters the silicates (10^{-4} to $10^{-3}\ M$).

In the pH range of most natural waters, usually only a small fraction of the silica Si_T or boron species B_T become part of alkalinity or acidity. Because boric acid $[B(OH)_3]$ and orthosilicic acid $[Si(OH)_4]$ are rather weak acids $[pK_{B(OH)_3} = 8.9; pK_{Si(OH)_4} = 9.5]$, they have very little effect on the acidimetric titration curve (see Figure 4.1). Applying equation 12, we have

$$[\text{Alk}] = C_T(\alpha_1 + 2\alpha_2) + [\text{OH}^-] - [\text{H}^+] + B_T\alpha_{B^-} + Si_T\alpha_{Si^-} \qquad (15)$$

where

$$\alpha_{B^-} = \frac{[B(OH)_4^-]}{B_T} \qquad \alpha_{Si^-} = \frac{[SiO(OH)_3^-]}{Si_T}$$

Because in waters of pH < 9, α_{B^-} and α_{Si^-} are much smaller than one, these constituents usually become only a minor portion of alkalinity. For precise calculations of carbonate equilibria a correction must be applied:

$$[\text{Alk}] - X = C_T(\alpha_1 + 2\alpha_2) + [\text{OH}^-] - [\text{H}^+] \qquad (16)$$

where X now includes the equivalent sum of the bases that can be titrated with strong acid, that is, $B(OH)_4^-$ and $SiO(OH)_3^-$. In seawater about 20% of the boron species are present as borate. Because silica exists predominantly as o-silicic acid in most natural waters, it usually does not become a part of the alkalinity.

Conservative Properties

Although individual concentrations or activities, such as $[H_2CO_3^*]$, and pH are dependent on pressure and temperature, ANC, BNC, and C_T are conservative properties that are pressure- and temperature-independent. (Alkalinity, acidity, and C_T must be expressed as a concentration, e.g., molarity or molality.)†

Furthermore, these conservative quantities remain constant for selected changes in the chemical composition. The case of the addition or removal of dissolved carbon dioxide is of particular interest in natural waters. Any increase in carbon dioxide or, more rigorously, any increase in $[H_2CO_3^*]$, $dC_{H_2CO_3^*}$, increases both the acidity of the system and C_T, the total concentration of dissolved carbonic species. Unlike the case for the addition of strong acid, however, alkalinity remains unaffected by increases or decreases in $[H_2CO_3^*]$. The fact that alkalinity is unaffected by CO_2 can be understood if we consider that alkalinity measures the proton deficiency with respect to the reference proton level $CO_2 - H_2O$. An analogous argument considers that the addition of CO_2 does not affect the net charge balance (which inherently defines alkalinity) of

---

† In seawater these capacity factors are rigorously independent of pressure and temperature if expressed in mol kg^{-1} seawater.

TABLE 4.4   CHANGE IN CAPACITY PARAMETERS AS A RESULT OF CHEMICAL CHANGES[a]

Capacity	C_B	C_A	$H_2CO_3^*$ (CO_2)	$NaHCO_3$	$CaCO_3$ or Na_2CO_3
			Addition of Molar Increments of		
C_T, M	0	0	+1	+1	+1
[Alk] (eq liter^{-1})	+1	−1	0	+1	+2
[Acy] (eq liter^{-1})	−1	+1	+2	+1	0
[CO$_2$–Acy]	−1	+1	+1	0	−1
[p–Alk]	+1	−1	−1	0	+1
[H–Acy]	−1	+1	0	−1	−2
[OH–Alk]	+1	−1	−2	−1	0

[a] Examples: $\dfrac{dC_T}{dC_B} = 0$; $d[\text{Acy}]/d[H_2CO_3^*) = +2$; $d[\text{Alk}]/d[\text{NaHCO}_3] = +1$.

the solution.† It can be shown in a similar way that acidity remains unaffected by the addition or removal of $CaCO_3(s)$ or $Na_2CO_3(s)$. Acidity is thus a valuable capacity parameter for solutions in equilibrium with calcite. C_T, on the other hand, remains unchanged in a closed system upon addition of strong acid or strong base. Table 4.4 summarizes the effect of various chemical changes upon the capacity parameters. For each chemical change one capacity parameter that remains independent of this change can be found.

Capacity Diagrams

Graphs using variables with conservative properties (C_T, [Alk], or [Acy]) as coordinates can be used expediently to show contours of pH, $[H_2CO_3^*]$, $[HCO_3^-]$, and so on. On such diagrams, as shown by Deffeyes [2], the addition or removal of base, acid, CO_2, HCO_3^-, or CO_3^{2-} is a vector property. These graphs can be used to facilitate equilibrium calculations.

Figures 4.5 and 4.6 show two such diagrams. In Figure 4.5 [Alk] is plotted as a function of C_T. The construction of the diagram can be derived readily from the definition of [Alk]:

$$[\text{Alk}] = [HCO_3^-] + 2[CO_3^{2-}] + [OH^-] - [H^+] \qquad (17)‡$$

$$[\text{Alk}] = C_T(\alpha_1 + 2\alpha_2) + [OH^-] - [H^+] \qquad (18)$$

---

† Alkalinity may also be defined as the excess of positive charges over the anions of strong acids:
$[\text{Alk}] = [Na^+] + [K^+] + 2[Ca^{2+}] + 2[Mg^{2+}] + \cdots - [Cl^-] - 2[SO_4^{2-}] - [NO_3^-] - \cdots$
‡ [Alk], the alkalinity, is the acid neutralizing capacity with respect to the equivalence point $f = 0$, and is expressed as mole of protons per liter, or eq liter^{-1}.

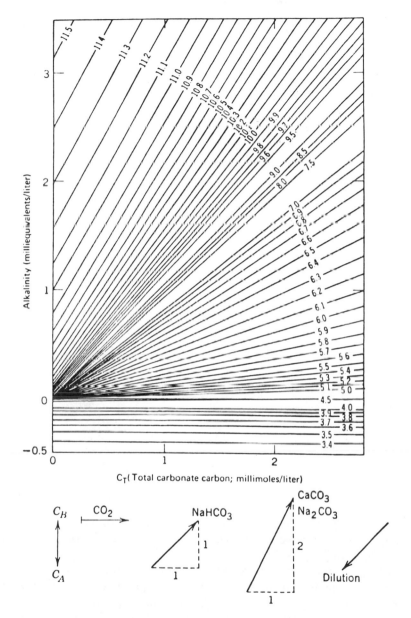

Figure 4.5 pH contours in alkalinity-versus C_T diagram. The point defining the solution composition moves as a vector in the diagram as a result of the addition (or removal) of CO_2, $NaHCO_3$, and $CaCO_3$ (Na_2CO_3) or C_B and C_A. (After K. S. Deffeyes [2].)

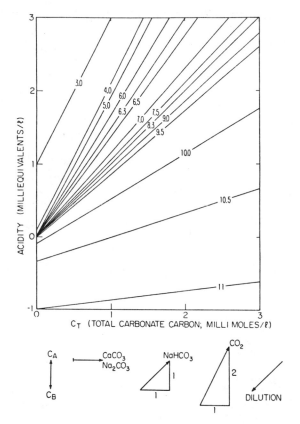

Figure 4.6 pH contours in an acidity-versus-C_T diagram. [Acy] remains unchanged upon addition or removal of Na_2CO_3 or $CaCO_3$. The direction and relative extent with which the solution composition changes as a result of the addition of chemicals is indicated.

It is apparent from equation 18 that, for any $[H^+]$, [Alk] is a linear function of C_T; the system is completely defined by specifying C_T and [Alk]. For each value of C_T and [Alk], pH is fixed.

As Deffeyes points out, most of the usefulness of these diagrams stems from the fact that changes in the solution move the point representing the solution composition in definite directions on the diagram. Adding strong acid decreases [Alk] without changing C_T. Vertical lines on Figure 4.5 therefore give alkalimetric or acidimetric titration curves for each C_T value. On the other hand addition or removal of CO_2 increases or decreases C_T without changing [Alk] and the point moves to the right by the amount of CO_2 added. The change in solution composition (pH) by changes in CO_2, caused for example by respiration or photosynthesis, can be readily elucidated because the horizontal line defines a curve of titration with CO_2 for any alkalinity. Figure 4.5 gives the vector nature of the changes in C_T and [Alk] with the addition or removal of various substances. The diagram is less sensitive in the pH regions of low buffer intensity—pH = 8 and pH = 4.5. Note that the line at pH = 4.5 has a small positive slope and does not intercept exactly at [Alk] = 0 (in fact, the intercept is -3×10^{-5} eq liter^{-1}).

Figure 4.6 gives a plot of [Acy] versus C_T:

$$[Acy] = 2[H_2CO_3^*] + [HCO_3^-] + [H^+] - [OH^-] \qquad (19)$$

$$[Acy] = C_T(2\alpha_0 + \alpha_1) + [H^+] - [OH^-] \qquad (20)$$

In this diagram the ordinate value ([Acy]) is unaffected by the addition or removal of crystalline carbonates such as $CaCO_3$ or $NaCO_3$. This diagram is of value in evaluating the effect of precipitation or dissolution of $CaCO_3$. Examples 4.5 and 4.6 illustrate the application of these diagrams for equilibrium calculations.

In solutions that contain bases and acids other than H^+, OH^-, and the carbonate species, the contours in the diagrams will become displaced vertically. For example, if [Alk] contains other bases (see equation 12), the intercept of the pH lines with the $C_T = 0$ line will be displaced vertically proportional to the concentration of the noncarbonate base.

Algebraic Approach

Graphs cannot always be read with sufficient precision. The uncertainty of the answer is great, especially for points in a weakly buffered pH region (pH 7.5 to 9). The problems can of course be dealt with numerically, and graphs may provide useful guidelines on how to attack a problem numerically.

Specifying any two independent variables at a given temperature and pressure (i.e., equilibrium constants at a given temperature and pressure) defines the composition of a carbonate solution. Recognizing how capacity parameters are affected by chemical changes (Table 4.4), a suitable pair of variables can generally be selected. Table 4.6 gives readily derivable equations that describe the titration of the carbonate system with base (acid), CO_2, or Na_2CO_3 ($CaCO_3$) while either C_T, [Alk], [Acy], or p_{CO_2} is kept constant. Corresponding titration curves are plotted in Figure 4.7. (See Section 4.5.)

Example 4.5. Mixing of Waters

Two waters (A: pH = 6.1, [Alk] = 1.0 meq liter^{-1}; B: pH = 9, [Alk] = 2 meq liter^{-1}) are mixed in equal proportions. What is the pH of the mixture if no CO_2 was lost to the atmosphere? Using Figure 4.5 we find $C_T = 2.8$ mM and $C_T = 1.9$ mM for A and B, respectively. The mixture has [Alk] = 1.5 meq liter^{-1} and $C_T = 2.35$ mM; its pH from Figure 4.5 is 6.6.

Example 4.6. Increase in pH by Addition of Base or Removal of CO_2

The pH of a surface water having an original alkalinity of 1 meq liter^{-1} and a pH of 6.5 (25°C) is to be raised to pH 8.3. The following methods are considered:

(i) increase in pH with NaOH
(ii) increase in pH with Na_2CO_3
(iii) increase in pH by removal of CO_2 (i.e., in a cascade aerator).

For each case the chemical change and the final composition of the solution are given by

			Final Composition	
Change per Liter of Solution	pH	C_T (mM)	Alk (meq liter^{-1})	Acy (meq liter^{-1})
Original solution	6.5	1.7	1.0	2.4
(i) +0.7 mmol NaOH	8.3	1.7	1.7	1.7
(ii) +0.7 mmol Na_2CO_3	8.3	2.4	2.4	2.4
(iii) −0.7 mmol CO_2	8.3	1.0	1.0	1.0

By first using Figure 4.5 we find the intersection of the pH = 6.5 line with the [Alk] = 1 meq liter^{-1} line and see that $C_T = 1.7 \times 10^{-3}$ M. If no CO_2 is exchanged with the atmosphere, the quantity of NaOH (C_B) necessary to reach pH = 8.3 can be found directly from the graph (vertical displacement) as ~0.7 meq liter^{-1}.

For the pH increase with Na_2CO_3 we have to draw a line with slope 2 from the intersection (pH = 6.5; [Alk] = 1 meq liter^{-1}). About 0.7 mM liter^{-1} of Na_2CO_3 is necessary to attain a water of pH = 8.3.

This part of the problem could have been solved more conveniently with the help of Figure 4.6. According to this figure a water of pH = 6.5 and $C_T = 1.7 \times 10^{-3}$ has an [Acy] = 2.35 meq liter^{-1}. By drawing a horizontal line through this point ([Acy] = 2.35 meq liter^{-1}; pH = 6.5), one sees that ~0.7 mM liter^{-1} of Na_2CO_3 is needed to reach pH = 8.3.

Finally, a pH of 8.3 can also be attained by decreasing $C_T(CO_2)$ and maintaining a constant alkalinity. At the horizontal 1.0 meq liter^{-1} line (Figure 4.5) the difference between the points pH 6.5 and 8.3 corresponds to 0.7 mM liter^{-1} of CO_2.

The table summarizing the results of this example illustrates that a given pH change may be attained by different pathways and that the final composition depends on the pathway.

Alkalinity Changes

Obviously alkalinity is also affected by all the processes that yield or consume H^+ or OH^- in stoichiometric equations, for example, the oxygenation of soluble ferrous iron to ferric oxide, $4\,Fe^{2+} + O_2 + 4\,H_2O \rightleftarrows 2\,Fe_2O_3(s) + 8\,H^+$, decreases alkalinity, while the reduction of $MnO_2(s)$ by CH_2O, $2\,MnO_2(s) + CH_2O + 4\,H^+ = CO_2 + 3\,H_2O + 2\,Mn^{2+}$, increases alkalinity. A few other examples are given in Table 4.5. Of particular interest are the alkalinity changes caused by photosynthesis and respiration.

Photosynthesis and Respiration.   As we have seen, the addition or removal of CO_2 has no effect on alkalinity. This would be true for the photosynthesis process only if it were not accompanied by the assimilation of ions such as NO_3^-, NH_4^+, and HPO_4^{2-}. Since alkalinity is associated with charge balance, such assimilation processes must be accompanied by the uptake of H^+ or OH^- (or release of OH^- or H^+), that is, by alkalinity changes. Thus the photosynthetic assimilation of NH_4^+ causes the uptake of OH^- or the release of H^+ ions (Table 4.5). Similarly, alkalinity increases as a result of photosynthetic NO_3^- assimilation; conversely the aerobic bacterial decomposition of biota to NO_3^- is accompanied by a decrease in alkalinity. Such processes occurring in land ecosystems are often not without influence on the pH and alkalinity of the adjoining aquatic ecosystems.

Decrease in alkalinity (acidification and cation depletion in soils) occurs whenever the production of organic matter (assimilation of NH_4^+) is larger than the decomposition. This takes place, for example, when peat bogs or forest peats are formed; these systems are very acidic. The harvest of crops on agricultural and forest land often causes discrepancies between production and decomposition.†

Example 4.7. Decrease in pH from Aerobic Respiration

What is the pH change resulting from the aerobic decomposition of organic matter—6 μg organic carbon—in 1 cm^3 of interstitial lake water (10°C) which initially has an alkalinity of 1.2×10^{-3} eq liter^{-1}, a pH value of 6.90, and an ionic strength of 3×10^{-3}? The respiration releases NH_4^+.

Lacking other information, we may assume that the reaction occurs with the stoichiometry given in (1c) of Table 4.5. The pH change may be estimated in two steps: (i) from the CO_2 increase at constant alkalinity, and (ii) from the alkalinity change (from the uptake of H^+) at constant C_T. We apply equation 18, $[Alk] = C_T(\alpha_1 + 2\alpha_2) + [OH^-] - [H^+]$, and use the equilibrium constants (10°C) corrected for ionic strength effects, $pK_1' = 6.43$, $pK_2' = 10.39$. At the initial

† For a discussion of alkalinity changes resulting from respiration *and* $CaCO_3$ dissolution in deep oceans see C. T. A. Chen, *Science*, **201**, 735 (1978).

TABLE 4.5 PROCESSES AFFECTING ALKALINITY

Process	Alkalinity Change for Forward Reaction
Photosynthesis and Respiration:	
(1a) $nCO_2 + nH_2O \xrightleftharpoons[\text{respir.}]{\text{photos.}} (CH_2O)_n + nO_2$	No change
(1b) $106CO_2 + 16NO_3^- + HPO_4^{2-} + 122H_2O + 18H^+ \xrightleftharpoons[\text{respir.}]{\text{photos.}} \{C_{106}H_{263}O_{110}N_{16}P_1\}_{\text{"algae"}} + 138O_2$	Increase
(1c) $106CO_2 + 16NH_4^+ + HPO_4^{2-} + 108H_2O \xrightleftharpoons[\text{respir.}]{\text{photos.}} \{C_{106}H_{263}O_{110}N_{16}P_1\} + 107O_2 + 14H^+$	Decrease
Nitrification:	
(2) $NH_4^+ + 2O_2 \longrightarrow NO_3^- + H_2O + 2H^+$	Decrease
Denitrification:	
(3) $5CH_2O + 4NO_3^- + 4H^+ \longrightarrow 5CO_2 + 2N_2 + 7H_2O$	Increase
Sulfide Oxidation:	
(4a) $HS^- + 2O_2 \longrightarrow SO_4^{2-} + H^+$	Decrease
(4b) $FeS_2(s) + \frac{15}{4}O_2 + 3\frac{1}{2}H_2O \longrightarrow Fe(OH)_3(s) + 4H^+ + 2SO_4^{2-}$ pyrite	Decrease
Sulfate Reduction:	
(5) $SO_4^{2-} + 2CH_2O + H^+ \longrightarrow 2CO_2 + HS^- + H_2O$	Increase
$CaCO_3$ Dissolution:	
(6) $CaCO_3 + CO_2 + H_2O \rightleftharpoons Ca^{2+} + 2HCO_3^-$	Increase

pH, $\alpha_1 = 0.747$ and α_2 is negligible; thus $C_T = 1.61 \times 10^{-3}$ M. After addition of the CO_2, C_T is 2.11×10^{-3} M, $\alpha_1 = 0.569$, and the new pH = 6.55. Then, at this C_T, alkalinity is changed by $\Delta[Alk] = 5 \times 10^{-4}(\frac{14}{106}) = 6.6 \times 10^{-5}$ eq liter^{-1}. Using equation 18 again, with $[Alk] = 1.266 \times 10^{-3}$ eq liter^{-1}, $\alpha_1 \cong [Alk]/C_T = 0.600$ which corresponds to pH = 6.6.

Example 4.8. pH Change Resulting from Photosynthetic CO_2 Assimilation

1. As a result of photosynthesis with NO_3^- assimilation, a surface water with an alkalinity of 8.5×10^{-4} eq liter^{-1} showed within a 3-hr period a pH variation from 9.0 to 9.5. What is the rate of net CO_2 fixation? (Assume a closed aqueous system, that is, no exchange of CO_2 with the atmosphere and no deposition of $CaCO_3$ (25°C), $pK_1 = 6.3$, $pK_2 - 10.2$.)

The pH change results (i) from CO_2 uptake and (ii) from an uptake of 18 mol of H^+ per 106 mol of CO_2 assimilated (1b of Table 4.5). Initially, $[Alk]_{pH=9} = C_I(\alpha_1 + 2\alpha_2)_{pH=9} + [OH^-]$, and we obtain $C_T = 7.94 \times 10^{-4}$ M.

For the conditions at pH = 9.5 we have two independent equations:

$$[Alk]_{pH=9.5} = [Alk]_{pH=9} + \tfrac{18}{106}\Delta C_T \tag{i}$$

and

$$[Alk]_{pH=9.5} - [OH^-] = (C_T - \Delta C_T)(\alpha_1 + 2\alpha_2)_{pH=9.5} \tag{ii}$$

Solving, $\Delta C_T = 7.9 \times 10^{-5}$ M and the alkalinity change is $+1.3 \times 10^{-5}$. Its effect is negligible; nearly the same result would have been obtained by considering the CO_2 change only.

2. In a more careful analysis it was detected that the alkalinity actually decreased by 3×10^{-5} eq liter^{-1} as a result of $CaCO_3$ precipitation. Considering the equation

$$[Alk]_{final} = [Alk]_{pH=9.5} - 3 \times 10^{-5} \tag{iii}$$

we obtain $[Alk]_{final} = 8.35 \times 10^{-4}$ eq liter^{-1}. At pH = 9.5, this alkalinity corresponds to a $C_{T,final} = 6.9 \times 10^{-4}$ M. Because

$$C_{T,final} = C_{T,pH=9} - \Delta C_{T,photosynthesis} - 1.5 \times 10^{-5} \tag{iv}$$

we calculate $\Delta C_{T,photosynthesis} = 8.9 \times 10^{-5}$ M or an hourly loss of ca. 3×10^{-5} mol CO_2 liter^{-1}.

4.5   BUFFERING

Buffer intensity controls the magnitude of shifts in the pH of solutions. The role of the dissolved or homogeneous carbonate system in effecting buffer action in natural waters is unquestionably significant, especially for waters that have become isolated from their environment.

In Table 4.6 equations for the buffer intensity of the aqueous carbonate system are listed. As explained in Section 3.9, the buffer intensity is found analytically by differentiating the appropriate function of the system (equations for the titration curves, Table 4.6) with respect to pH. Figure 4.7 shows a number of computed buffer intensity versus pH relationships for millimolar (C_T) or millinormal ([Alk] or [Acy]) solutions at 25°C. Besides the familiar buffer intensity with respect to strong acid or base, $\beta^{C_B}_{C_T}$, buffer intensities with respect to

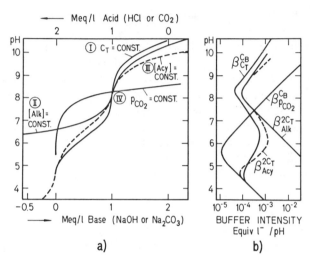

Figure 4.7 Buffering of aqueous carbonate systems of constat C_T, [Alk], [Acy], and p_{CO_2}, respectively. (*a*) Titration curves. I: Titration of 10^{-3} M C_T solution with strong acid or strong base [(1) of Table 4.5]. II: Addition or removal of CO_2 to solution of [Alk] = 10^{-3} eq liter^{-1} [(4) of Table 4.5]. III: Addition or removal of $Na_2CO_3(CaCO_3)$ to solution of [Acy] = 10^{-3} eq liter^{-1} [(7) of Table 4.5]. IV: Titration of solution in equilibrium with p_{CO_2} = $10^{-3.5}$ atm with strong acid or base [(10) of Table 4.5]. Curves constructed with the following constants: pK_H = 1.5; pK_1 = 6.3; pK_2 = 10.2. (*b*) Buffer intensities versus pH.

CO_2 as an acid, $\beta^{2C_T}_{Alk}$ [(5) in Table 4.5] and with respect to Na_2CO_3 as a base, $\beta^{2C_T}_{Acy}$ (8, Table 4.5) are given in this figure. In the lower pH range (pH < 7.5) the buffer intensity of a water with respect to CO_2 is much larger than the buffer intensity with respect to a strong acid. Equations (10) and (11) in Table 4.5 and Figures 4.3 and 4.7 illustrate that for a water which remains in equilibrium with CO_2 of the atmosphere, the huge reservoir of CO_2 imparts a significant buffering action upon waters of high pH (pH > 7.5).

In the pH range of most natural waters, noncarbonate buffer components exert little influence upon pH regulation. The total buffer intensity β^{C_B} is simply the sum of the contributions of the buffer intensities of the individual buffer components.

The Increase in Atmospheric CO_2; the Buffer Factor

The burning of fossil fuel and wood produces an increase in the CO_2 content of the atmosphere (Fig. 4.8, *inset*) [3–6]. The redistribution of the excess CO_2 depends on the *uptake by*

TABLE 4.6 BUFFERING OF AQUEOUS CARBONATE SYSTEM

Titration Curve		Buffer Intensity	
I. C_T = constant Addition of strong acid or base:			
$C_B - C_A = C_T(\alpha_1 + 2\alpha_2) + [OH^-] - [H^+]$ $dC_B = d[Alk]$	(1)	$\dfrac{d[Alk]}{dpH} = \beta_{C_T}^{C_E} = 2.3\{C_T[\alpha_1(\alpha_0 + \alpha_2) + 4\alpha_2\alpha_0] + [H^+] + [OH^-]\}$	(2)
		$\beta_{C_T}^{C_E} \approx 2.3\alpha_1(\alpha_0 + \alpha_2)C_T$	(3)
II. [Alk] = constant Addition or removal of CO_2:			
$C_T = ([Alk] - [OH^-] + [H^+])/(\alpha_1 + 2\alpha_2)$ $dC_T = \tfrac{1}{2}d[Acy]$	(4)	$\dfrac{d[Acy]}{dpH} = \beta_{Alk}^{2C_T} \simeq -4.6\dfrac{\alpha_1\alpha_0 + \alpha_1\alpha_2 + 4\alpha_2\alpha_0}{(\alpha_1 + 2\alpha_2)^2}[Alk]$	(5)
		$\beta_{Alk}^{2C_T} \approx -4.6[(\alpha_0 + \alpha_2)/\alpha_1][Alk]$	(6)
III. [Acy] = constant Addition or removal of $NaCO_3(CaCO_3)$:			
$C_T = ([Acy] - [H^+] + [OH^-])/(\alpha_1 + 2\alpha_0)$ $dC_T = \tfrac{1}{2}d[Alk]$	(7)	$\dfrac{d[Alk]}{dpH} = \beta_{Acy}^{2C_T} \simeq 4.6\dfrac{\alpha_1\alpha_0 + \alpha_1\alpha_2 + 4\alpha_2\alpha_0}{(\alpha_1 + 2\alpha_0)^2}[Acy]$	(8)
		$\beta_{Acy}^{2C_T} \approx 4.6[(\alpha_0 + \alpha_2)/\alpha_1][Acy]$	(9)
IV. p_{CO_2} = constant Addition of strong base or strong acid:			
$C_B - C_A = (K_H(\alpha_1 + 2\alpha_2)/\alpha_0)p_{CO_2} + [OH^-] - [H^+]$ $dC_B = d[Alk]$	(10)	$\dfrac{d[Alk]}{dpH} = \beta_{pCO_2}^{C_B} = 2.3\Big\{K_Hp_{CO_2}\dfrac{\alpha_1(\alpha_0 + \alpha_2) + 4\alpha_2\alpha_0 + (\alpha_1 + 2\alpha_2)^2}{\alpha_0} + [H^+] + [OH^-]\Big\}$	(11)
		$\beta_{pCO_2}^{C_B} \approx 2.3K_Hp_{CO_2}/\alpha_C \approx 2.3C_T$	(12)

the ocean. As pointed out by Broecker *et al.* [3], the following three steps must be considered: First, the relatively thin layer of well-mixed water lying above the seasonal thermocline at the ocean surface will quickly establish equilibrium with the CO_2 in the atmosphere. Next, downward mixing and the sinking of the particles generated by organisms will slowly carry this carbon to the deep sea. Finally, as the CO_2 content of the ocean rises, water masses currently supersaturated with respect to $CaCO_3$ (calcite and aragonite and their solid solutions) will become undersaturated, and the $CaCO_3$ in sediments bathed in these waters will begin to dissolve.

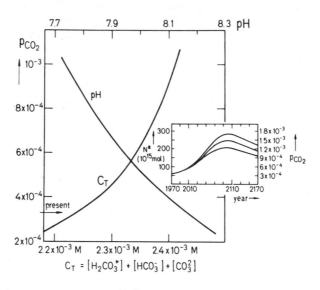

$$C_T = [H_2CO_3^*] + [HCO_3^-] + [CO_3^{2-}]$$

p_{CO_2} ($\times 10^4$ atm)	Carbonate alkalinity (eq liter^{-1}) $\times 10^3$	C_T ($M \times 10^3$)	pH
2.7	2.380	2.199	8.242
3.0	2.386	2.217	8.203
3.3	2.392	2.237	8.168
3.6	2.396	2.256	8.136
4.4	2.406	2.292	8.060
5.0	2.418	2.320	8.012
6.6	2.423	2.356	7.902
10	2.436	2.412	7.735

Amount of CO_2 in the atmosphere for three alternatives for the ultimate amounts of fossil fuel burned (300×10^{21} J; 250×10^{21} J; 200×10^{21} J) assuming a growth coefficient in a logistic function of 4.3% per year. [From K. E. Zimen, P. Offerman, and G. Hartmann, *Z. Naturforsch.*, **32a**, 1544 (1977)].

Figure 4.8 Effect of increasing p_{CO_2} upon C_T and pH of surface ocean water. The calculations have been made for the following conditions: seawater at 15°C, $p_{total} = 1$ atm, [Alk] = constant = 2.47×10^{-3} eq liter^{-1}, $[B(OH)_4^-] + [H_3BO_3] = 4.1 \times 10^{-4}$ M, $^cK_H = 4.8 \times 10^{-2}$ M atm^{-1}, $K_1' = 8.8 \times 10^{-7}$, $K_2' = 5.6 \times 10^{-10}$, $K_{H_1BO_3}' = 1.6 \times 10^{-9}$.

We will primarily consider here the first step: equilibration of surface water will exceed atmospheric CO_2. The increase in C_T resulting from an increase in p_{CO_2} can be formulated by the relationship

$$\frac{\Delta p_{CO_2}}{^0 p_{CO_2}} = \eta \frac{\Delta C_T}{^0 C_T} \qquad (21)$$

where η is the so-called buffer factor, and Δp_{CO_2} and ΔC_T are, respectively, $p_{CO_2} - {}^0 p_{CO_2}$ and $C_T - {}^0 C_T$; ${}^0 p_{CO_2}$ and ${}^0 C_T$ correspond to any previous CO_2 partial pressure and total inorganic carbon in seawater, respectively. For example, they might represent the pre-industrial partial pressure of CO_2 and the preindustrial total inorganic carbon in seawater. If the atmospheric p_{CO_2} increases by X percent, the C_T increases by X/η percent.

We are already acquainted with the mass-law equations that connect atmospheric CO_2 and surface water. Because the CO_2 exchange does not cause any $CaCO_3$ dissolution (at least as long as the waters remain supersaturated with respect to $CaCO_3$), it does not affect alkalinity. Hence the buffer factor in equation 21 can be evaluated under conditions of constant alkalinity.

For a given alkalinity and an appropriate set of equilibrium constants the relationship between p_{CO_2} and C_T be established. With the help of equation 14

$$[Alk] = \frac{K_H p_{CO_2}}{\alpha_0} (\alpha_1 + 2\alpha_2) + [OH^-] - [H^+] + B_T \alpha_{B^-}$$

the H^+ activity compatible with a given set of p_{CO_2} and $[Alk]$ can be computed (see Example 4.4); then $C_T = K_H p_{CO_2}/\alpha_0$ is found.

With the results displayed in Figure 4.8, a buffer factor, $\eta = 9.7$ (15°C), is obtained for a change in p_{CO_2}, from 2.7×10^{-4} to 3.0×10^{-4} atm; in other words, for an increase of 10% in p_{CO_2}, C_T increases by ca. 1%. The buffer factor increases with increasing p_{CO_2}; for example, if p_{CO_2} is increased from 3.3×10^{-4} atm (approximate present global value) to 3.6×10^{-4} atm, $\eta = 15.5$; if p_{CO_2} is increased from 3.3×10^{-4} to 6.6×10^{-4} atm, $\eta = 17.4$. Thus, a doubling of the present atmospheric CO_2 would increase C_T by about 5 to 6%.[†] Similar effects will be observed in surface fresh waters. A doubling of the CO_2 content in the atmosphere will approximately double the $[H^+]$ concentration in these waters. The effect will be smaller in the deeper waters of the ocean because of mixing.

4.6 ANALYTIC CONSIDERATIONS

Complete resolution of simple carbonate systems in terms of the equilibrium concentrations of each of the five individual components (H^+, OH^-, H_2CO_3, HCO_3^-, and CO_3^{2-}) is possible if the concentrations of at least two of these, or two of certain combinations thereof, are evaluated analytically, C_T (total alkalinity, CO_2 acidity, p-alkalinity, and pH are commonly determined in routine water analyses). The measurements of caustic alkalinity and mineral

---

[†] Calculations of the buffer factor of seawater have been described in the literature, especially by B. Bolin and E. Eriksson in *The Atmosphere and the Sea in Motion*, Rossby Memorial Volume 130, Rockefeller Institute, New York, 1959, by Broecker et al. [3], and by Keeling, e.g., R. Bacastow and C. D. Keeling [5].

acidity, in addition to the parameters listed, is often useful in the definition of specific systems.

From values for any two of the quantities listed, the other parameters and the concentrations of the single ions HCO_3^-, CO_3^{2-}, H^+, and OH^- can be computed. It is evident from the conceptual definitions given in Figure 4.4 that no direct rigorous determination for the concentration of any one of the carbonic species is feasible. For example, acidimetric titration to the CO_2 end point (the methyl orange end point), $\sim pH = 4.5$, allows evaluation of alkalinity rather than of the concentration of HCO_3^-. In waters with a pH of less than 9, the alkalinity and concentration of HCO_3^- are of course practically identical. Similarly, alkalimetric titration to the bicarbonate end point gives CO_2 acidity but not the concentration of $H_2CO_3^*$, although these are for all practical purposes the same in waters within the approximate pH range of 5.7 to 7.6.

Precision of Analytical Determinations

Although any two of the quantities may be used in the evaluation of the others, the question as to the choice of the two parameters whose experimental determination permits the most precise characterization of a carbonate system is significant. No general answer applicable to all natural waters can be given, and experimental expediency must be considered.

Because the sharpness of the end point in an alkalimetric or acidimetric titration is related to the slope, dpH/dC_B or $1/\beta$, of the titration curve at the equivalence points, the relative error involved is directly proportional to the buffer intensity at the end point. No inflection point is observed at the carbonate end point ($f = 2$); thus the parameters of acidity and caustic alkalinity, the titrations for which are characterized by this end point, cannot be evaluated by direct titration [7].

Alkalinity

The operational end point of an acidimetric titration for alkalinity generally is not in exact accord with the calculated equivalence point pH, as CO_2 is lost to the atmosphere during titration [8]. The quantity of CO_2 lost in the course of a titration naturally depends on the amount of agitation and on the degree of oversaturation attained at low pH values. It has been the authors' experience, however, that relative CO_2 losses are negligible if smooth stirring such as is obtainable with a magnetic stirrer, is maintained throughout the course of the titration. An independent check on the proper end point pH can be obtained by conductimetric titration of alkalinity with a mineral acid or by a Gran plot as discussed below.

As the end point and the buffer capacity at the end point depend on the concentration of CO_2 at this point in a titration, the sharpness of the end point can be improved if CO_2 is continuously removed (e.g., by scrubbing with CO_2-free gas) from the solution during the titration. With decreasing CO_2 the

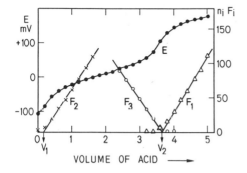

Figure 4.9 Emf titration curve for 154 g of seawater with V milliliters (0.1000 M HCl plus 0.4483 M NaCl); characterization of equivalence points by the Gran method. F_1, F_2, and F_3 are Gran functions, and n_i is an arbitrary scale factor (see appendix at the end of this chapter), for finding the equivalence points V_1 and V_2 corresponding to $f = 1$ and $f = 0$, respectively (see [14]).

end point pH is progressively shifted toward pH 7 where the sharpness of the end point is the same as in the titration of a strong base with a strong acid. Alternatively alkalinity is estimated by treating the sample with a measured excess of strong acid, driving off the liberated CO_2 and backtitrating to a pH of 6 to 7 at which any boric acid is negligibly ionized.

Alkalinity can be determined fairly precisely even without these precautions [10]. For example, an alkalinity of 10^{-3} eq liter^{-1} (50 mg liter^{-1} as $CaCO_3$) can be determined with a relative error of approximately $\pm 3.5\%$ if the end point is recognizable within 0.3 pH unit of the true equivalence point. For the same precision of end point detection the relative error is 10 times as great for a 100-fold dilution. In waters in which low alkalinities (or mineral acidities) have to be determined (e.g., rain water, demineralized water, and steam condensate), the end point recognition must be more precise in order to obtain results with small relative error. A procedure developed by Gran [9] and Larson and Henley [10], discussed by Thomas [11], and applied for seawater by Dyrssen and Hanson [12] and Edmond [13] is based on the principle that added increments of mineral acid linearly increase hydrogen ion concentration. In seawater a precision of $\pm 0.17\%$ can be attained.

A detailed derivation of Gran's graphical method of end point location is given in an appendix to this chapter. A significant advantage of the Gran plot is that it does not matter whether the "pH" values are defined in terms of the IUPAC-NBS convention or in terms of $-\log[H^+]$ units. Figure 4.9 gives an example (from Dyrssen and Sillén [14]) of an emf titration of seawater. This figure illustrates that a single acidimetric titration can give quite accurate data on [p-Alk], [Alk], and C_T.

Carbonate alkalinity, that is, the portion of [Alk] due to $[HCO_3^-]$ + $2[CO_3^{2-}]$ can only be derived from the measured [Alk] by deducting the contribution of $B(OH)_4^-$ and of $[OH^-]$ and other bases (cf. equation 12):

$$[\text{Carbonate Alk}] = [\text{Alk}] - B_T \alpha_{B^-} - ([OH^-] - [H^+])$$

Mineral Acidity. Mineral acidity, exhibited by waters containing acids stronger than $H_2CO_3^*$, and alkalinity are in a sense complementary parameters since $[H^+\text{-acidity}] + [\text{alkalinity}] = 0$. Many acid waters contain metal ions

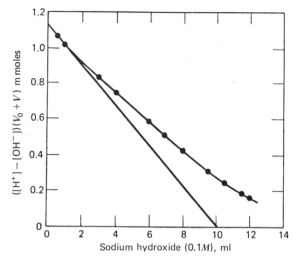

Figure 4.10 Extrapolation method for determining the equivalence point for H^+ acidity in the presence of metal ions or other weak acids. The curve broken by solid circles is for 1×10^{-2} M $Fe(ClO_4)_3$. The equivalence point for the HCl alone can be evaluated by extrapolation to $([H^+] - [OH^-])[V_0 + V] = 0$, because H^+ is titrated preferentially in the initial portion of the titration of the mixed solution. The value of 10 ml of NaOH determined by extrapolation of the data for the combined solution is in good agreement with the value of 9.9 ml obtained from titration of a pure solution of HCl.

capable of neutralizing titrant bases. This particular interference frequently can be avoided by adding complex formers such as fluoride.

On occasion it is also possible to remove interfering metal ions by using ion-exchange resins in the sodium form; however, cation-exchange resins do not always yield quantitative exchanges with multivalent ions, and it is frequently preferable to obtain the value of mineral acidity by a Gran procedure. The procedure, similar to that described for alkalinity, is based on the fact that, in the presence of low concentrations of metal ions or other weak acids, H^+ acidity is titrated preferentially in low-pH regions by initial increments of titrant. Thus the change in $[H^+]$ during the initial stages of an alkalimetric titration of a solution containing metal ions and H^+ acidity is closely proportional to the volume of titrant added.

Experimental data for titrations of 100-ml aliquots of 10^{-2} M HCl and 10^{-2} M HCl + 10^{-3} M $Fe(ClO_4)_3$ with 10^{-1} M NaOH are presented in Figure 4.10.

CO_2 Acidity and p-Alkalinity

There is no direct way to determine rigorously by alkalimetric or acidimetric titration the individual value of $[H_2CO_3^*]$ or $[CO_3^{2-}]$. For many waters without mineral acidity and with pH values below 7.6, CO_2 acidity is practically identical to $[H_2CO_3^*]$. Similarly, for waters not containing caustic alkalinity and with pH values above 9, the p-alkalinity is nearly equal to $[CO_3^{2-}]$. Within the pH range 7.6 to 9, neither CO_2 acidity nor p-alkalinity can be interpreted in terms of $[H_2CO_3^*]$ or $[CO_3^{2-}]$, respectively. For example, a water at pH 8 with a CO_2 acidity of 10^{-3} eq liter^{-1} determined by titration with strong base actually contains 1.28×10^{-3} M $H_2CO_3^*$. Identification of CO_2 acidity with $[H_2CO_3^*]$ in this water would lead to an underestimation of $[H_2CO_3^*]$ of nearly 30%.

In most natural waters the determination of CO_2 acidity is only slightly less precise than an alkalinity determination. For example, CO_2 acidity can be determined in a water containing 10^{-3} eq liter^{-1} of alkalinity and 5×10^{-4} eq liter^{-1} of CO_2 acidity with a relative error of approximately 4% provided the end point can be defined within 0.3 pH unit of the true equivalence point pH; however, the error increases quite rapidly as the ratio of CO_2 acidity to alkalinity decreases.

The relative error in an alkalimetric titration becomes larger if Na_2CO_3 instead of NaOH is used as a titrant. The use of soda leads to a reduced slope at the equivalence point because of increased buffer intensity (Figure 4.7).

In titrations of waters of high hardness and alkalinity, $CaCO_3$ saturation is frequently exceeded prior to the end point (pH about 8.3, $f = 1$). If $CaCO_3$ precipitates before the end point is reached, positive errors occur because of the reaction $Ca^{2+} + HCO_3^- = CaCO_3(s) + H^+$. The possibility of this is more pronounced if Na_2CO_3 is used as the titrant. There are, however, various means for circumventing this interference, such as the addition of a cation-exchange resin in the sodium form.

Other frequently significant errors in CO_2 determinations occur because of CO_2 loss to the atmosphere during sampling and transfer. With many natural waters it is more expedient to determine pH and water temperature in the field and, as alkalinity is not changed by loss of CO_2, this quantity can be determined after samples have been transported to the laboratory. (Also, if $CaCO_3$ precipitates when the samples are stirred, CO_2 can be added prior to the alkalinity determination to redissolve the precipitate.) The free CO_2 (or CO_2 acidity) and CO_3^{2-} (or CO_3^{2-} alkalinity) can then be computed from the measured values of pH and alkalinity with the appropriate constants — that is, those valid at the temperature measured in the field and corrected for activity. If a pH meter with a precision of 0.05 pH unit is used, the relative error in $[H^+]$ amounts to about 10%. As $[CO_2]$ is a function of $[H^+]$, values computed from measurements of the latter include relative errors of at least a similar order.

Total Carbon Dioxide, C_T.   The characterization of a carbonate system by alkalinity and pH has the advantage of expediency and convenience. In most waters, however, precision can be improved if, in addition to [Alk], C_T is determined. C_T is obtainable from a full titration curve (see Figures 4.4 and 4.9) from the difference between the equivalent points $f = 1$ and $f = 0$. Many routine methods are based on determining the total CO_2, stripped from the sample after acidification, by gas chromatography or infrared spectrophotometry [15].

4.7 THE EQUILIBRIUM CONSTANTS

Tables 4.7 to 4.10 give the equilibrium constants for various temperatures ($P = 1$ atm). These constants are needed to evaluate the distribution of carbonate species in a quantitative analytical way. The reader is referred to the original

TABLE 4.7 EQUILIBRIUM CONSTANT FOR CO_2 SOLUBILITY. EQUILIBRIUM: $CO_2(g) + aq = H_2CO_3^*$; HENRY'S LAW CONSTANT: $K_H = \{H_2CO_3^*\}/p_{CO_2}$ (M atm^{-1}).

Temp. (°C)	$-\text{Log } K_H$ $\to 0^a$	$-\text{Log } {}^cK_H$ 1 M NaClO$_4$	$-\text{Log } {}^cK_H$ Seawater, 19‰ Cl$^-$
0	1.11[b]	—	1.19[b]
5	1.19[b]	—	1.27[b]
10	1.27[b]	—	1.34[b]
15	1.32[b]	—	1.41[b]
20	1.41[b]	—	1.47[b]
25	1.47[b]	1.51[d]	1.53[b]
30	1.53[c]	—	1.58[b]
35	—	1.59[d]	—
40	1.64[c]	—	—
50	1.72[c]	—	—

Edmond and Gieskes[e] give the following summarizing equations for the temperature and salinity dependence of $-\log {}^cK_H$ *in* seawater:

$$-\log {}^cK_H = -2385.73/T + 14.0184 - 0.0152642T \tag{i}$$
$$+ I(0.28596 - 6.167 \times 10^{-4} \times T)$$

where $I =$ ionic strength and $T =$ absolute temperature. I is related to chlorinity, Cl(‰), by

$$I = 0.00147 + 0.03592 \times Cl + 0.000068 \times Cl^2 \tag{ii}$$

[a] In this and subsequent tables, →0 represents a constant extrapolated to zero ionic strength.
[b] Values based on data taken from Bohr and evaluated by K. Buch, *Meeresforschung*, **1.15** (1951).
[c] A. J. Ellis, *Amer J. Sci.*, **257**, 217 (1959).
[d] G. Nilsson, T. Rengemo and L. G. Sillén, *Acta Chem. Scand.*, **12**, 878 (1958).
[e] J. M. Edmond, and J. M. Gieskes, *Geochim. Cosmochim. Acta*, **34**, 1261 (1970).

TABLE 4.8 FIRST ACIDITY CONSTANT: $H_2CO_3^* = HCO_3^- + H^+$

$$K_1 = \frac{\{H^+\}\{HCO_3^-\}}{\{H_2CO_3^*\}} \qquad K_1' = \frac{\{H^+\}[HCO_{3_T}^-]}{[H_2CO_3^*]} \qquad {}^cK_1 = \frac{[H^+][HCO_{3_T}^-]}{[H_2CO_3^*]}$$

Temp. (°C)	$-\log K_1$	$-\log K_1'$		$-\log {}^cK_1$	
	→0	Seawater, 19‰ Cl⁻	Seawater, 35‰ Salinity	Seawater	1 M NaClO₄
0	6.579[a]	6.15[b]	6.20[c]	—	—
5	6.517[a]	6.11[b]	6.14[c]	6.01[d]	—
10	6.464[a]	6.08[b]	6.10[c]	—	
14	—	—	6.07[c]	6.02[e]	—
15	6.419[a]	6.05[b]	6.06[c]	—	—
20	6.381[a]	6.02[b]	6.03[c]	—	—
22		6.00[f]	6.02[c]	5.89[f]	—
25	6.352[a]	6.00[b], 6.09[g]	6.00[c]	—	6.04[h]
30	6.327[a]	5.98[b]	5.98[c]	—	—
35	6.309[a]	5.97[b]	5.96[c]	—	—
40	6.298[a]	—	—	—	—
50	6.285[a]	—	—	—	—

Gieskes[i] gives for seawater the following summarizing equations for the dependence of $-\log K_1'$ on pressure, temperature, and chlorinity:

$$-\log K_1' = \frac{3404.71}{T} + 0.032786T - 14.712 - 0.19178 \times Cl^{1/3} \qquad (i)$$

$$\log \frac{(K_1')_P}{(K_1')_{P-1}} = -\frac{\Delta V_1'(P-1)}{2.303RT} \qquad (ii)$$

$$\Delta V_1' = -(24.2 - 0.085 \times t) cm^3\ mol^{-1} \qquad (iii)$$

where P = pressure in atm, T = K, t = °C, Cl = chlorinity (‰), and $2.303RT - 5.634 \times 10^4\ cm^3\ atm\ mol^{-1}$ at 25°C.

[a] H. S. Harned and R. Davies, Jr., *J. Amer. Chem. Soc.*, **65**, 2030 (1943).

[b] After Lyman (1956), quoted in G. Skirrow, *Chemical Oceanography*, Vol. 2, 2nd ed., J. P. Riley and G. Skirrow, Eds., Academic Press, New York, 1975, p. 176.

[c] C. Mehrbach, G. H. Culberson, T. E. Hawley and R. M. Pytkowicz, Limnol. Oceanogr. **18**, 897 (1973).

[d] D. Dyrssen and L. G. Sillén, *Tellus*, **19**, 810 (1967).

[e] D. Dyrssen, *Acta Chem. Scand.*, **19**, 1265 (1965).

[f] A. Distèche and S. Distèche, *J. Electrochem. Soc.*, **114**, 330 (1967).

[g] Calculated as log $(K_1/f_{HCO_3^-})$ from $f_{HCO_3^-}$ as determination by A. Berner, *Geochim. Cosmochim. Acta*, **29**, 947 (1965).

[h] M. Frydman, G. N. Nilsson, T. Rengemo and L. G. Sillén, *Acta Chem. Scand.*, **12**, 878 (1958).

[i] J. M. Gieskes, in *The Sea*, Vol. 5, E. D. Goldberg, Ed., Wiley-Interscience, New York, 1974, p. 125.

TABLE 4.9   SECOND ACIDITY CONSTANT: $HCO_3^- = H^+ + CO_3^{2-}$

$K_2 = \dfrac{\{H^+\}\{CO_3^{2-}\}}{\{HCO_3^-\}}$		$K_2' = \dfrac{\{H^+\}[CO_{3T}^{2-}]}{[HCO_{3T}^-]}$		$^c K_2 = \dfrac{[H^+][CO_{3T}^{2-}]}{[HCO_{3T}^-]}$	
	$-\log K_2$	$-\log K_2'$		$-\log K_2'$	$-\log {}^c K_2$
Temp. (°C)	→0	Seawater, 19‰ Cl⁻	35‰ salinity	0.75 M NaCl	1 M KClO₄
0	10.625[a]	9.40[b]	9.45[c]	—	—
5	10.557[a]	9.34[b]	9.39[c]	—	—
10	10.490[a]	9.28[b]	9.33[c]	—	—
15	10.430[a]	9.23[b]	9.25[c]	—	—
20	10.377[a]	9.17[b]	9.18[c]	—	—
22	—	9.12[d]	9.15[c]	9.49[d]	—
25	10.329[a]	9.10[b]	9.11[c]	—	9.57[e]
30	10.290[a]	9.02[b]	9.06[c]	—	—
35	10.250[a]	8.95[b]	9.01[c]	—	—
40	10.220[a]	—	—	—	—
50	10.172	—	—	—	—

Gieskes[f] gives *for seawater* the following summarizing equations for the dependence of $-\log K_2'$ or temperature, pressure, and chlorinity:

$$-\log K_2' = \frac{2902.39}{T} + 0.02379T - 6.471 - 0.4693 \times Cl^{1/3} \qquad (i$$

$$\log \frac{(K_2')_P}{(K_2')_{P=1}} = \frac{\Delta V_2'(P-1)}{2.303RT} \qquad (ii$$

$$\Delta V_2' = -(16.4 - 0.04t)cm^3\ mol^{-1} \qquad (iii$$

where variables are as in Table 4.8

[a] H. S. Harned and S. R. Scholes, *J. Amer. Chem. Soc.*, **63**, 1706 (1941).
[b] After Lyman (1956), quoted in G. Skirrow, *Chemical Oceanography*, J. P. Riley and G. Skrirow, Eds. Vol. 2, 2nd ed., Academic, New York 1975, p. 176.
[c] C. Mehrbach, C. H. Culberson, J. E. Hawley, and R. M. Pytkowicz, *Limnol. Oceanogr.* **18**, 897 (1973)
[d] A. Distèche and S. Distèche, *J. Electrochem. Soc.*, **114**, 330 (1967).
[e] M. Frydman, G. N. Nilsson, T. Rengemo and L. G. Sillén, *Acta Chem. Scand.*, **12**, 878 (1958).
[f] J. M. Gieskes, in *The Sea*, Vol. 5, E. D. Goldberg, Ed., Wiley-Interscience, New York, 1974, p. 125.

literature or to Skirrow's appendix [16] to find the equilibrium constants as given originally by the authors (also for temperatures other than those listed here). Data for equilibrium constants are given for different activity scales (see Section 3.4). For dealing with freshwater systems the infinite dilution activity scale might be most useful, while in seawater the seawater medium scale is more appropriate.

TABLE 4.10   ACIDITY CONSTANT OF H_3BO_3[a]

$$K_1 = \frac{\{H^+\}\{B(OH)_4^-\}}{\{H_3BO_3\}} \qquad K_1' = \frac{\{H^+\}[B(OH)_{4T}^-]}{[H_3BO_{3T}]}$$

Temp. (°C)	$-\log K_1$, →0	$-\log K_1'$, Seawater, 19‰ Cl^-	$-\log {}^cK$ $I = 0.1$	$-\log {}^cK$ $I = 0.7$
0	9.50	8.95	—	—
5	9.44	8.90	—	—
10	9.38	8.85	—	—
15	9.33	8.80	—	—
20	9.28	8.75	—	—
25	9.24	8.71	8.97[b]	8.85[b]

For seawater, Gieskes[c] gives the following temperature, pressure, and chlorinity dependence:

$$-\log K_1' = \frac{2291.9}{T} + 0.01756T - 3.385 - 0.32051 \times Cl^{1/3} \qquad \text{(i)}$$

$$\log \frac{(K_1')_P}{(K_1')_{P=1}} = -\frac{\Delta V_B'(P-1)}{2.303RT} \qquad \text{(ii)}$$

$$\Delta V_B' = -(27.5 - 0.095 \times t)\text{cm}^3 \text{ mol}^{-1} \qquad \text{(iii)}$$

[a] After Lyman, quoted in G. Skirrow, *Chemical Oceanography*, vol. 2, 2nd ed., J. P. Riley and G. Skirrow, Eds., Academic, New York, 1975, p. 177.
[b] R. M. Smith and A. E. Martell, *Critical Stability Constants*, Vol. 4, Plenum, New York, 1976.
[c] J. M. Gieskes, in *The Sea*, Vol. 5, E. D. Goldberg, Ed., Wiley-Interscience, 1974, p. 125.

Seawater. In seawater precise equilibrium calculations should be made preferably on the constant-medium activity scale. Since the oceans contain an ionic medium of essentially constant composition, and since carbonate and borate species are present as minor constituents (in comparison to other ions), seawater may be defined as the ionic medium.

Even if we adopt seawater as a constant medium, different equilibrium constants have to be defined for different definitions of pH [17]. If pH is defined operationally in terms of a galvanic cell and if the measurement is made by comparing the solution with a standard buffer (NBS), we assume pH $\approx p^aH \approx -\log\{H^+\}$ (IUPAC, NBS) and equilibria defined by mixed constants (K_1', K_2', in Tables 4.8 and 4.9) may be employed.† On the other hand, if one attempts to

† The same constants are obtained independent of whether molar or molal concentration units are used, because the K values contain the ratios of species concentrations while $\{H^+\}$ is based on the NBS pH scale.

measure pH in terms of $-\log[H^+]$, that is, by calibrating the glass electrode with a dilute solution of a strong acid of known $[H^+]$ in the same ionic medium, equilibria defined by $^cK_1$ or $^cK_2$ should be employed. The Hanson pH scale for seawater and equilibrium constants based on this scale will be discussed in the Appendix to Chapter 6.

Complex or Ion-Pair Formation with Seawater Medium Ions. HCO_3^- or CO_3^- ions may form complexes with the medium ions (e.g., $MgCO_3$, $NaCO_3^-$, $CaCO_3$, $MgHCO_3^+$). In the constant ionic medium method, K' and ^cK are defined in terms of species that include an unknown number of medium ions, for example,

$$[CO_{3T}^{2-}] = [CO_3^{2-}] + [MgCO_3] + [CaCO_3] + [NaCO_3^-] \tag{22}$$

This is operationally advantageous because in an equilibrium computation, let us say the estimation of $[H_2CO_3^*]$ from pH and [Alk], one does not have to consider the complex formation equilibria of HCO_3^- and CO_3^{2-} with Ca^{2+}, Mg^{2+}, and Na^+. It may be necessary to limit the validity of a K' or ^cK value to a certain pH range.

That the activity coefficient for CO_3^{2-} as computed from $^cK_2$ (seawater) is lower than the activity coefficient as computed from Debye–Hückel formulas suggests that such complex or ion-pair formation really exists in seawater [18, 19]. This will be discussed further in Section 6.6. A comparison of the K_2' values (Table 4.9) for seawater and a $0.75\ M$ NaCl medium (which is approximately equivalent to the ionic strength of seawater) indicate that Mg^{2+} and Ca^{2+} (in addition to Na^+) must be involved primarily in complex formation.

Complex and ion-pair formation is of less importance in most fresh waters. Considering the stability constants given by Garrels and Thompson [18], we calculate that waters with $[Ca^{2+}] \leq 2 \times 10^{-3}\ M$ and $[Alk] \leq 4 \times 10^{-3}$ eq liter^{-1} do not form bicarbonato or carbonato complexes of Ca^{2+} exceeding 5% of the total.

Fresh Water. In fresh water with $I < 10^{-2}\ M$, equilibrium constants valid for infinite dilution may be used. These constants must be corrected with the help of the Debye–Hückel limiting law; the Güntelberg approximation is especially useful. Larson and Buswell [20] have recommended the following salinity corrections

$$pK_1' = pK_1 - \frac{0.5\sqrt{I}}{(1 + 1.4\sqrt{I})} \tag{23}$$

$$pK_2' = pK_2 - \frac{2\sqrt{I}}{(1 + 1.4\sqrt{I})} \tag{24}$$

where pK and pK' are defined as in Table 4.8. In fresh water whose detailed composition is not known an approximate ionic strength can be estimated from the total hardness $H(H \simeq [Ca^{2+}] + [Mg^{2+}])$ (M) and the alkalinity (eq liter^{-1})

$$I \simeq 4H - \tfrac{1}{2}[\text{Alk}] \tag{25}$$

or from total dissolved solids S (mg liter^{-1})

$$I \simeq S \times 2.5 \times 10^{-5} \tag{26}$$

Pressure Dependence

The effect of pressure on the equilibrium constant can be found with the thermodynamic relation

$$\left(\frac{\partial \ln K}{\partial P}\right)_{T, m} = -\frac{\Delta V}{RT} \tag{27}$$

where ΔV (cm^3 mol^{-1}) is the algebraic difference between the total partial molar volumes of products and reactants. For small changes in pressure it is sufficient to regard ΔV as constant and to use equation 27 in its integrated form:

$$\log \frac{K_{P_1}}{K_{P_2}} = -\frac{\Delta V(P_2 - P_1)}{2.3RT} \tag{28}$$

In freshwater systems pressures rarely exceed 30 atm. For most acids $-\Delta V$ is between 10 and 30 cm^3 mol^{-1}, hence the difference in log K does not exceed one or two hundredths of a unit. For seawater we encounter pressures of a few hundred atmospheres. At high pressure in seawater $H_2CO_3^*$ and HCO_3^- protolyze to a different extent than at zero ionic strength. Because complex formation and ion-pair equilibrium are also pressure-dependent, the change in p^cK or pK' values might be quite different from changes in pK values. In seawater much reliance has been placed on experimental determinations of pressure dependencies [19, 21–23] (see Tables 4.8 to 4.10). Edmond and Gieskes [22] interpreted the available results in the following summary:

$$V(K_1') = -(24.2 - 0.085t) \text{ cm}^3 \text{ mol}^{-1};$$
$$V(K_2') = -(16.4 - 0.040t) \text{ cm}^3 \text{ mol}^{-1}$$
$$V(K_{H_3BO_3}') = -(27.5 - 0.095t) \text{ cm}^3 \text{ mol}^{-1}$$

where t is the temperature ($^\circ$C). Using the experimentally determined values, at the average depth of the ocean ($P = 200$ atm)log K_1, log K_2, and log $K_{H_3BO_3}'$ values will increase by approximately 0.09, 0.06, and 0.1 units, respectively.

Since in many investigations the pH of the sample is measured after decompression to $p = 1$ atm, values for the pressure coefficient of pH are needed. Values of pH$_{1 \text{ atm}}$ − pH$_p$ are on the order of 0.1 and 0.2 pH unit for pressures of

250 and 500 atm, respectively [21]. For reviews on pressure dependence see [16] and [23].

Although equilibrium constants and, as a consequence, activities or concentrations of individual species are dependent upon pressure and temperature, it is appropriate to emphasize once more that, despite these changes, capacity parameters such as C_T, [Alk], and [Acy] are conservative, that is, independent of pressure, temperature, and ionic strength.

4.8   KINETIC CONSIDERATIONS

Many natural waters are not in equilibrium with the atmosphere, partially because of unfavorable mixing conditions but primarily because of the slowness of the gas transfer reaction; this transfer is frequently slower than reactions that produce or consume CO_2 in the aqueous phase [respiration, photosynthesis, precipitation, and dissolution reactions—e.g., $Ca^{2+} + 2HCO_3^- = CaCO_3 + CO_2 + H_2O$—mineral alterations—e.g., $NaAlSi_3O_8(s) + H_2CO_3^* + \frac{9}{2}H_2O = Na^+ + HCO_3^- + 2H_4SiO_4 + \frac{1}{2}Al_2Si_2O_5(OH)_4(s)$].

On the other hand, ionization equilibria in the dissolved carbonate system are established very rapidly. Somewhat slower, however (seconds), is the attainment of equilibrium in the hydration or dehydration reaction of CO_2 [24]

$$CO_2(aq) + H_2O \; \rightleftharpoons \; H_2CO_3 \tag{29}$$

Kinetics of Hydration of CO_2

The hydration of CO_2 leads to the formation of H_2CO_3, but it may also yield H^+ and HCO_3^-. The reaction scheme may be written [25, 26]

$$\text{(1)} \quad H^+ + HCO_3^- \; \underset{k_{21}}{\overset{k_{12}}{\rightleftharpoons}} \; H_2CO_3 \quad \text{(2)}$$

$$k_{13} \Big\backslash\Big\backslash k_{31} \qquad k_{23}\Big/\!\!\Big/ k_{32} \tag{30}$$

$$CO_2 + H_2O$$
$$\text{(3)}$$

The same reaction scheme applies to the hydration of SO_2. Formally we may write the rate law for the disappearance of CO_2:

$$-d[CO_2]/dt = (k_{31} + k_{32})[CO_2] - k_{13}[H^+][HCO_3^-] - k_{23}[H_2CO_3] \tag{31}$$

Considering that k_{21}/k_{12} equals the first acidity constant of "true" H_2CO_3, $K_{H_2CO_3}$ and that both k_{21} and k_{12} are far larger† than any other of the four other rate constants in

---

† M. Eigen and G. Hammes (*Advances in Enzymology*, F. F. Nord, Ed., Wiley, New York, 1963) give for 25°C the value $k_{12} = 4.7 \times 10^{10}\ M^{-1}\ \text{sec}^{-1}$; for $K_{H_2CO_3} \simeq 2 \times 10^{-4}$, $k_{21} \simeq 9 \times 10^6\ \text{sec}^{-1}$.

equation 30, we can replace $[H^+][HCO_3^-]$ with $K_{H_2CO_3}[H_2CO_3]$ in equation 31 and write

$$-\frac{d[CO_2]}{dt} = (k_{31} + k_{32})[CO_2] - (k_{13}K_{H_2CO_3} + k_{23})[H_2CO_3] \tag{32}$$

$$= (k_{31} + k_{32})[CO_2] - [k_{13} + k_{23}(K_{H_2CO_3})^{-1}][H^+][HCO_3^-] \tag{33}$$

Setting

$$k_{CO_2} = k_{31} + k_{32} \tag{34}$$

and

$$k_{H_2CO_3} = k_{13} K_{H_2CO_3} + k_{23} \tag{35}$$

equation 32 can be written as

$$-\frac{d[CO_2]}{dt} = k_{CO_2}[CO_2] - k_{H_2CO_3}[H_2CO_3] \tag{36a}$$

$$= k_{CO_2}[CO_2] - \frac{k_{H_2CO_3}}{K_{H_2CO_3}}[H^+][HCO_3^-] \tag{36b}$$

The rate law (equation 36) corresponds to the simplified scheme

$$CO_2 + H_2O \underset{k_{H_2CO_3}}{\overset{k_{CO_2}}{\rightleftharpoons}} H_2CO_3 \overset{\text{very fast}}{\rightleftharpoons} H^+ + HCO_3^- \tag{37}$$

That is, the hydration reaction is first-order with respect to dissolved CO_2 and has a rate constant of $k_{CO_2} = 0.025$ to 0.04 sec^{-1} (25°C). The activation energy is approximately 15 kcal mol^{-1}. Similarly the rate of dehydration has a first-order rate constant $k_{H_2CO_3}$ of 10 to 20 sec^{-1} (20–25°C); its activation energy is ca. 16 kcal mol^{-1}. Considering the order of magnitude of the reaction rate constants, it is obvious that not more than a few minutes are necessary to establish the hydration equilibrium.

The ratio of these velocity constants permits estimation of the equilibrium constant of equation 37 [cf. (1) and (2) of Table 4.2]:

$$K' = K + 1 = \frac{[CO_2] + [H_2CO_3]}{[H_2CO_3]} = \frac{[H_2CO_3^*]}{[H_2CO_3]} = \frac{K_{H_2CO_3}}{K_1} = \frac{k_{H_2CO_3}}{k_{CO_2}} + 1 \tag{38}$$

K' has reported values of about 350 to 990 at 25°C.

Superimposed on these first-order reactions are the processes

$$CO_2 + OH^- \underset{k_{41}}{\overset{k_{14}}{\rightleftharpoons}} HCO_3^- \tag{39}$$

with approximate constants $k_{14} = 8.5 \times 10^3$ M^{-1} sec^{-1} and $k_{41} = 2 \times 10^{-4}$ sec^{-1} (25°C). Process 39 is kinetically insignificant at pH values below about 8. Above pH = 10, it dominates the hydration reaction. A numerical example on CO_2 hydration kinetics is given in Section 2.14.

Catalysis. Hydration and dehydration reactions are catalyzed by various bases, by certain metal chelates, and by the enzyme carbonic anhydrase. Natural waters may contain some natural catalysts [27]. Reaction 39 is essentially a catalysis of reaction 37. The conversion of CO_2 into H_2CO_3 or HCO_3^- by both mechanisms is

$$\frac{d[CO_2]}{dt} = -(k_{CO_2} + k_{14}[OH^-])[CO_2] + (k_{H_2CO_3}(K_{H_2CO_3})^{-1}[H^+] + k_{41})[HCO_3^-]$$

(40)

CO_2 Transfer at the Gas–Liquid Interface

As discussed in 2.16, the flux of gases from the gas phase into an aqueous solution may be adequately described by a two-layer model at the gas–liquid interface. Accordingly, in the transfer of CO_2 molecules from the atmosphere to the water, the molecules are transported through the gas phase and a gas film and then through a nonmixed liquid layer adjacent to the surface before they are distributed and hydrated. Of these different steps, the transfer through the liquid layer adjacent to the surface is usually slowest and rate-determining. This is also valid for the desorption of CO_2 from the solution into the gas phase. The flux of CO_2 across the interface is given by Fick's first law:

$$F = \frac{D}{\delta} \Delta C = K_L \Delta C = \frac{1}{R_L} \Delta C$$

(41)

where F is the flux in moles (n) per unit area A (cm^2) per unit of time t (sec); D is the diffusion coefficient (cm^2 sec^{-1}) for the region of thickness δ (cm), ΔC is the concentration difference (mol cm^{-3}) across this region, K_L(cm sec^{-1}) is the transfer coefficient, that is, the velocity (cm sec^{-1}) with which the gas molecules traverse the unstirred layer, and $R_L (= 1/K_L = \delta/D)$ (sec cm^{-1}) may be understood as the resistance to transfer in this layer. The thickness of the film, δ, hence the rate of gas exchange, depends on the degree of agitation of the water (wind, waves). The higher the agitation, the thinner the film.

What is ΔC in equation 41? To answer this it is necessary to consider the species that can form when CO_2 dissolves in water and the rates of their formation with respect to the rate of diffusion of CO_2 itself. The species that can form from CO_2 (aq) are HCO_3^- and CO_3^{2-}. Consider the case in which diffusion of CO_2 molecules through the liquid layer is *fast* compared to the rates of formation of HCO_3^- and CO_3^{2-}. For this condition, diffusion of CO_2 alone accounts for transport, and the concentration gradient which corresponds to F_{CO_2} is given by

$$\Delta C = [CO_2(aq)]_{eq.atm} - [CO_2(aq)]_{bulk}$$

(42)

Now consider the case where diffusion of CO_2 is *slow* compared to the hydration and acid–base reactions:

$$CO_2 + H_2O \rightleftharpoons H_2CO_3 \rightleftharpoons H^+ + HCO_3^-$$

$$CO_2 + OH^- \rightleftharpoons HCO_3^-$$

$$HCO_3^- \rightleftharpoons H^+ + CO_3^{2-}$$

Then, HCO_3^- and CO_3^{2-} are formed (or consumed) at rapid rates compared to the rate at which CO_2 moves through the interface, and diffusion of the ions must also be considered in order to determine the flux. That is,

$$F_C = F_{CO_2} + F_{HCO_3^-} + F_{CO_3^{2-}}$$

(43)

There are concentration gradients for each carbonic species, corresponding to $\Delta[CO_2]$, $\Delta[HCO_3^-]$, and $\Delta[CO_3^{2-}]$ (gradients in H_2CO_3 can be neglected because the concentrations are so small). Thus the pertinent concentration difference across the diffusion layer is that for total carbonic species, C_T:

$$\Delta C = C_{T_0} - C_{T, \text{bulk}} \qquad (44)$$

C_T varies with distance in the liquid layer and alkalinity is a conservative property (electroneutrality). Equation 44 is the general expression that applies in the limiting case in which CO_2 itself accounts for all of the flux. Where chemical reactions and diffusion have comparable rates, some of the CO_2 entering the diffusion layer is transformed to HCO_3^-, and total carbon flux is enhanced.

An upper limit to enhancement of the CO_2 flux is established by the chemical equilibrium case, that is, where HCO_3^- and CO_3^{2-} are at equilibrium with CO_2. The residence time of CO_2 in the boundary layer by diffusion is then much longer than by chemical reaction, and

$$F_C > F_{CO_2}$$

Whether physical or chemical (enhanced) transport dominates, the flux of carbon depends on pH (equation 40), on the presence of catalysts, and on film thickness, hence the rate of gas exchange. At low pH and small film thicknesses (high agitation levels), the transport is dominated by CO_2 diffusion. For higher pH, lower agitation levels, catalysts, and higher temperature, transport is enhanced by diffusion of HCO_3^- and ΔC_T governs the rate of gas transfer.

The Exchange Rate of CO_2 between the Sea and the Atmosphere

For seawater (20°C), a $K_L(=D/\delta)$ of about 1000 m year^{-1} (~ 11 cm hr^{-1}) (e.g., corresponding to a diffusivity of CO_2 of about 5×10^{-2} m^2 year^{-1} and a δ of ca. 50 μm) is representative for the world oceans [28]. $[CO_2(aq)]_{eq.atm}$ is, in accordance with Henry's law for $p_{CO_2} = 3 \times 10^{-3}$ atm, ca. 10^{-2} mol m^{-3}. By making $[CO_2(aq)]_{bulk} = 0$, we obtain the exchange rate F_0, i.e., the flux of CO_2 in either direction:

$$F_0 = K_L(CO_2(aq))_{eq.atm} \qquad (45)$$

An exchange rate of ~ 10 mol m^{-2} year^{-1} results. The residence time of CO_2 in the atmosphere, τ_{CO_2}, is obtained by dividing the total mass of CO_2 in the atmosphere per unit area by the exchange rate:

$$\tau_{CO_2} = \frac{m_{CO_2(atm)}}{F_0} \qquad (46)$$

Since there is about 100 mol of CO_2 above each square meter (1 atm $\approx$ 1000 g air cm^{-2}; and the average molecular weight of air is ≈ 29 g; thus there are 1000 g/29 g ≈ 34 mol air molecules above 1 cm^2 of which 0.03 % are CO_2), a residence time of $\tau_{CO_2} = 10$ years is obtained. Similar residence times have been established on the basis of carbon-14 data [28]. Thus, under these circumstances, the chemical enhancement of the exchange rate does not appear to be very significant. Thus the diffusion of CO_2 is more important than the chemical cycle of HCO_3^- and CO_3^{2-} for the ocean–atmosphere CO_2 exchange.

As shown by Broecker and Peng [28], the rate-limiting step for removal of anthropogenic CO_2 from the air is vertical mixing within the sea rather than transfer across the air–sea interface.

Chemical Enhancement of CO_2 Transfer into Lakes

Lakes often become greatly undersaturated in CO_2 during photosynthesis because CO_2 assimilation is faster than the resupply by gas exchange from the atmosphere. Under such circumstances—high pH, higher film thickness—the invasion rates of CO_2 into the lake are more likely to be enhanced by reactions of CO_2 to form HCO_3^-.

The theory of this enhancement has been discussed by various authors [28–33]. Emerson [32, 33] developed a general model to explain the kinetics of reaction and diffusion in the boundary layer of a productive lake. He measured (with the isotope pair radium-226 and randon-222) the gas-exchange mass-transfer coefficient in small lakes and found k values of 0.2 to 0.4 m day^{-1} (0.8 to 1.6 cm hr^{-1}). These coefficients are more than 10 times lower than oceanic values; they correspond to a stagnant boundary layer thickness of 300 to 600 μm. Equilibrium enhancement factors for CO_2 invasion were calculated to be on the order of 20. The CO_2 invasion rates of the lake were found to be four to six times greater than if there were no chemical reaction, in good agreement with the kinetic-diffusion model predictions.

4.9   CARBON ISOTOPES AND ISOTOPE FRACTIONATION

Carbon has two stable isotopes, ^{12}C and ^{13}C, and a radioactive isotope, ^{14}C, with a half-life of 5720 years. These isotopes are present in the following abundances: $^{12}C = 98.89\%$, $^{13}C = 1.11\%$, and $^{14}C \approx 10^{-10}\%$ of the total.

Carbon-14

The main source of natural ^{14}C is the upper atmosphere where it is formed by the interaction of cosmic ray neutrons with ^{14}N; as a β emitter it returns to the original form, ^{14}N, after ejection of an electron from its nucleus. About 100 ^{14}C atoms are being generated over each square centimeter of the earth's surface each minute.†

Despite its low abundance, ^{14}C is an interesting tracer because of its radioactive decay:

$$-\frac{dN}{dt} = \lambda N \tag{47}$$

where N is the number of ^{14}C atoms (or generally, of radioactive elements) and λ is the decay constant ($\lambda = 1.2 \times 10^{-4}$ year^{-1} for ^{14}C). This decay constant corresponds to a half-life $t_{1/2}$ (the time necessary to reduce radioactive atoms present at time zero, N_0, to one-half, $\frac{1}{2}N_0$):

$$t_{1/2} = \frac{1}{\lambda} \ln \frac{N_0}{\frac{1}{2}N_0} = \frac{\ln 2}{\lambda} \tag{48}$$

---

† The atmospheric concentration of ^{14}C is not as was previously supposed. In addition, the output of CO_2 from fossil fuels—which contain no ^{14}C—has tended to dilute the natural level (Suess effect [H. E. Suess, Science **122**, 415 (1955)]), and recent injections of nuclear bomb-produced—artificial— ^{14}C have enhanced the ^{14}C level in the opposite direction [M. S. Baxter and A. Walton, *Proc. Roy. Soc.*, A**321**, 105 (1971)].

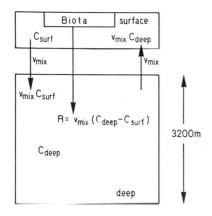

Figure 4.11 Two-box steady-state model for the cycle of Water, Carbon, and ^{14}C between the surface and deep Sea. (See Example 4.9.) (After Broecker [34].)

$t_{1/2}$ is 5720 years for ^{14}C. The mean lifetime, τ, that is, the mean life expectancy of a radioactive atom, is given by the sum of the life expectancies of all the atoms divided by the initial number:

$$\tau = -\frac{1}{N_0} \int_{t-0}^{t=\infty} t \, dN = \frac{1}{N_0} \int_0^\infty t\lambda N \, dt$$

$$= \lambda \int_0^\infty t e^{-\lambda t} \, dt = \frac{1}{\lambda} \tag{49}$$

The average ^{14}C atom has a life expectancy of 8300 years; that is, each year 1 atom out of 8300 undergoes radioactive transformation. In *radiocarbon dating* of sea shells one assumes that the ^{14}C/C ratio in seawater has always been the same as it is today. In accordance with equation 47, the age t is given by

$$t = \frac{1}{\lambda} \ln \frac{(^{14}C/C)_{\text{formation}}}{(^{14}C/C)_{\text{today}}} \tag{50}$$

Example 4.9. Carbon-14 as a Tracer for Oceanic Mixing. In a simplified two-box model of the ocean [34], the warm waters and the cold waters may be subdivided into two well-mixed reservoirs—an upper one a few hundred meters in depth and a lower one of 3200 m depth. The C_T content of the upper and lower reservoirs (corresponding to the Pacific) are, respectively, 1.98×10^{-3} mol liter^{-1} and 2.44×10^{-3} mol liter^{-1} whereas the ^{14}C/C ratios for upper and lower reservoirs are, respectively, 0.92 and 0.77×10^{-12} mol/mol. Estimate from this information the rate of vertical mixing and the residence time of the water in the deep sea [34, 35].

Figure 4.11 sketches the situation. Water conservation requires that the flux of water downwelling equal the flux of water upwelling ($= v_{\text{mix}}$). The input of carbon to the deep reservoir must be balanced by the output of carbon:

$$v_{\text{mix}} C_{\text{surf}} + B = v_{\text{mix}} C_{\text{deep}} \tag{i}$$

B is the carbon added to the deep reservoir each year by the destruction of biological debris, formed in the surface layer, falling from the surface. From the mass balance,

$$B = v_{\text{mix}}(C_{\text{deep}} - C_{\text{surf}}) \tag{ii}$$

We can now write a mass balance for ^{14}C. The radioactive decay of ^{14}C has to be considered an "output" of the deep reservoir; its rate is given by equation 47:

$$\frac{dN_{^{14}C}}{dt} = V_{\text{deep}}[^{14}C_{\text{deep}}]\lambda \tag{iii}$$

where V_{deep} is the volume of the deep reservoir and $[^{14}C]$ is the concentration of ^{14}C. Because laboratory measurements are typically given as ^{14}C/C ratios, we may express the mass balance in these ratios:

$$v_{\text{mix}} C_{\text{surf}} \frac{^{14}C_{\text{surf}}}{C_{\text{surf}}} + B \frac{^{14}C_{\text{surf}}}{C_{\text{surf}}} = v_{\text{mix}} C_{\text{deep}} \frac{^{14}C_{\text{deep}}}{C_{\text{deep}}}$$

$$+ V_{\text{deep}} \frac{^{14}C_{\text{deep}}}{C_{\text{deep}}}$$

Substituting equation ii into equation iv and solving for v_{mix}, one obtains

$$v_{\text{mix}} = \lambda V_{\text{deep}} \frac{^{14}C_{\text{deep}}/C_{\text{deep}}}{^{14}C_{\text{surf}}/C_{\text{surf}} - {}^{14}C_{\text{deep}}/C_{\text{deep}}} \tag{v}$$

V_{deep} can be expressed as area times mean depth. Using 3200 m as a mean and using the data given, we obtain

$$v_{\text{mix}} = 1.2 \times 10^{-4} \times 3.2 \times 10^3 \times A \frac{9.77}{0.92 - 0.77} \tag{vi}$$

$$v_{\text{mix}} = 2 \text{ m year}^{-1} \left(2 \frac{m^3}{m^2} \text{ year}^{-1} \right)$$

that is, the yearly volume of water exchanged between the surface and the deep ocean is equal in volume to a layer 2 m thick with an area equal to that of the ocean. Since the mean thickness of the deep reservoir is 3200 m, the mean residence time of the water in the deep sea is ca. $3200/2 = 1600$ years.

Isotope Effects

Differences exhibited by isotopic species in coexisting phases are primarily caused by

1   Kinetic differences between isotopic molecules resulting in differences in such properties as diffusional velocity and rate of vaporization. For example, for $^{12}CO_2$ and $^{13}CO_2$ the ratio of diffusional flux would be predicted—since the diffusion coefficient, D, is inversely proportional to the square root of the molecular weight—to be

$$\frac{D_{44}}{D_{45}} = \left(\frac{45}{44} \right)^{1/2} = 1.011$$

The vapor pressure of isotopic compounds decreases with increasing isotopic mass; that is, the tendency to evaporate is larger for the compound with the light isotope.

2   Purely chemical differences among isotopic compounds are primarily caused by differences in the vibrational frequencies of isotopic molecules. The electronic structure of a

given element is not changed by isotopic substitution. Thus isotopes of a given element in compounds form the same bonds. Because the energy levels for vibration depend on mass as well as on bond strength, isotopic substitution causes a change in vibration frequency. These differencies lead to small fractionations. The bonds formed by light isotopes are more readily broken than bonds involving heavy isotopes. Isotope fractionation measurements taken during irreversible chemical reactions always show a preferential enrichment of the lighter isotope in the products of the reaction (For a comprehensive discussion of isotope effects see [36].)

Isotopic data are typically recorded as δ values:

$$\delta(\%_0) = \frac{(^{13}C/^{12}C)_{sample} - (^{13}C/^{12}C)_{standard}}{(^{13}C/^{12}C)_{standard}} \times 1000 = \left(\frac{(^{13}C/^{12}C)_{sample}}{(^{13}C/^{12}C)_{standard}} - 1\right)1000$$

(51)

For example, a sample with a $\delta\,^{13}C = $ [or $\delta(^{13}C/^{12}C)$] $= -15\%_0$ indicates that the sample is depleted in ^{13}C relative to the standard such that the $^{13}C/^{12}C$ ratio differs by $15\%_0$. There are some worldwide standards in use: For C it is CO_2 prepared from a Cretaceous belemnite from the Peedee formation of South Carolina; for O the usual standard is standard mean ocean water (SMOW).

Isotopic Fractionation During Evaporation and Condensation

Fractionation within the hydrosphere occurs almost exclusively during vapor-to-liquid or vapor-to-solid phase changes. For example, it is evident from the vapor pressure data for water (21.0, 20.82, and 19.51 mm Hg for $H_2^{16}O$, $H_2^{18}O$, and $HD^{16}O$, respectively) that the vapor phase is preferentially enriched in the lighter molecular species, the extent depending on the temperature (Raleigh distillation). The progressive formation and removal of raindrops from a cloud and the formation of crystals from a solution too cool to allow diffusive equilibrium between the crystal interior and the liquid, that is, isotopic reactions carried out in such a way that the products are isolated immediately after formation from the reactants, show a characteristic trend in isotopic composition [37, 38].

Ocean water of $\delta^{18}O = 0$ stays in equilibrium with a vapor of $\delta^{18}O = -10\%_0$ and $\delta D = -9\%_0$ (at about 10°C). If this vapor condenses during cloud formation, the first droplets formed have $\delta^{18}O = 0$. If the liquid phase is always removed from the vapor phase during condensation, as is approximately the case in the atmosphere, the remaining vapor phase will become more and more depleted in ^{18}O as condensation proceeds, so that the δ values of both the vapor and the liquid will decrease [39].

In the atmosphere–ocean system the subtropical oceanic areas are the main supply regions for atmospheric water vapor. As this vapor spreads north and over the continents and decreases as a result of precipitation, the δ values decrease until they reach values of -6/mil in fresh water in midlatitudes and about -30/mil in polar precipitations [38]. Since the δ values reflect the fraction of moisture removed and since this fraction is controlled by the atmospheric temperature, the δ values of precipitation are strongly correlated with the atmospheric temperature [37–40]. $^{18}O/^{16}O$ ratios serve as oceanic water mass tracers. There is a linear relationship between ^{18}O and salinity in water samples of different geographic locations in the North Atlantic Ocean. This suggests that all these samples are mixtures of fresh water (salinity ≈ 0 and $\delta^{18}O \approx -20$/mil) and water typical of the open ocean (salinity 36/mil and $\delta^{18}O = 1$/mil [34].

Carbon-13

We first consider a typical exchange reaction expressing the distribution of ^{12}C and ^{13}C between two phases:

$$^{13}CO_2(g) + H^{12}CO_3^-(aq) \rightleftharpoons \ ^{12}CO_2(g) + H^{13}CO_3^-(aq) \tag{52}$$

which is characterized by the equilibrium constant

$$K = \frac{[^{12}CO_2(g)][H^{13}CO_3^-(aq)]}{[^{13}CO_2(g)][H^{12}CO_3^-(aq)]}$$

One often defines a fractionation factor $\bar{\alpha}$:

$$\bar{\alpha} = \frac{(^{13}C/^{12}C)_{HCO_3^-}}{(^{13}C/^{12}C)_{CO_{2(g)}}} \tag{53}$$

For equilibrium 52, $K = \bar{\alpha}$. In general, for an isotopic equilibrium reaction,

$$aA_1 + bB_2 \rightleftharpoons aA_2 + bB_1 \tag{54}$$

where A and B are different molecules, both of which contain a common element existing in the light isotopic form (subscript 1) and the heavy isotopic form (subscript 2), the fractionation factor is given by

$$\bar{\alpha} = K^{1/ab} \tag{55}$$

The equilibrium constant K for an isotopic exchange reaction depends upon the vibrational frequency shifts produced upon isotopic substitution and can be calculated in principle from knowledge of the fundamental frequencies of the isotopic species [36]. At high temperatures differences in chemical properties cease.

For reaction 52 the distribution factor has been determined experimentally [41]: $\bar{\alpha} = K = 1.0092$ (0°C), and $= 1.0068$ (30°C). Thus $\Delta_{CO_2-HCO_3^-} = \delta_{HCO_3^-} - \delta_{CO_{2(g)}} = 9.2‰$ (0°C) or 6.8‰ (30°C) [40]. This fractionation occurs predominantly in the hydration stage and not during the passage of the atmospheric carbon dioxide through the air–water interface, that is, the reaction

$$^{13}CO_2(g) + \ ^{12}CO_2(aq) \rightleftharpoons \ ^{12}CO_2(g) + \ ^{13}CO_2(aq) \tag{56}$$

has an equilibrium constant of $K = \alpha = 1.00$.

Example 4.10. $\delta^{13}C$ of Aqueous Carbonate System

Estimate the $\delta^{13}C$ of an aqueous carbonate system of pH $= 7.2$ in equilibrium with a reservoir of a given $p_{CO_2} = 10^{-2}$ atm and a given $\delta^{13}C_{CO_2} = -25‰$ (10°C). (This p_{CO_2} pressure and $\delta^{13}C_{CO_2}$ value may be representative of the gas phase under soil conditions.) At 10°C the following equilibrium constants, using gaseous CO_2 as a reference, interrelating the various carbon species are given [P. Deines, D. Langmuir, and R. S. Harmon, *Geochim. Cosmochim. Acta*, **38**, 1147 (1974)].

$$K_0 = \frac{(^{13}C/^{12}C)_{H_2CO_3^*}}{(^{13}C/^{12}C)_{CO_{2(g)}}} = 0.999 \tag{i}$$

$$K_1 = \frac{(^{13}C/^{12}C)_{HCO_3^-}}{(^{13}C/^{12}C)_{CO_{2(g)}}} = 1.0092 \tag{ii}$$

$$K_2 = \frac{(^{13}C/^{12}C)_{CO_3^{2-}}}{(^{13}C/^{12}C)_{CO_{2(g)}}} = 1.0075 \tag{iii}$$

The composition of the solution can be computed (cf. equations 5 to 8) using the following equilibrium constants valid at 10°C and corrected for a ionic strength of 4×10^{-3} M ($\log K_H = 1.27$, $-\log K_1' = 6.43$, $-\log K_2' = 10.38$) as $C_T = 3.7 \times 10^{-3}$ M, $[H_2CO_3^*]$ $= 5.4 \times 10^{-4}$ M, $[HCO_3^-] = 3.2 \times 10^{-3}$ M, and $[CO_3^{2-}] = 2 \times 10^{-6}$ M.

Combining equation i with the definition of $\delta^{13}C$, we obtain

$$K_0 = \frac{\delta^{13}C_{H_2CO_3^*} + 1000}{\delta^{13}C_{CO_{2(g)}} + 1000} \tag{iv}$$

or

$$\delta^{13}C_{H_2CO_3^*} = K_0\delta^{13}C_{CO_{2(g)}} + (K_0 - 1) \times 1000 = -26\%_{00} \tag{v}$$

and

$$\delta^{13}C_{HCO_3^-} = K_1\delta^{13}C_{CO_2} + (K_1 - 1) \times 1000 = -16\%_{00} \tag{vi}$$

and

$$\delta^{13}C_{CO_3^{2-}} = K_2\delta^{13}C_{CO_2} + (K_2 - 1) \times 1000 = -17.7\%_{00} \tag{vii}$$

The isotopic composition of the carbon in the solution is

$$\delta^{13}C_{sol} = ([H_2CO_3^*]\delta^{13}C_{H_2CO_3^*} + [HCO_3^-]\delta^{13}C_{HCO_3^-}$$
$$+ [CO_3^{2-}]\delta^{13}C_{CO_3^{2-}})/C_T \; \delta^{13}C_{sol} = -17.6\%_{00} \tag{viii}$$

In this example it was assumed that the carbon gas reservoir of a given $\delta^{13}C$ was large in comparison to the C reservoir of the water and the solid phases and was in continuous isotopic exchange with these phases. If such a water becomes isolated from the gas reservoir and then interacts with calcite or dolomite, then the ^{13}C content of the solution will also depend on that of the dissolving carbonate rock.

Biogenic Organic Matter

During photosynthesis plants discriminate against ^{13}C in favor of ^{12}C, and as a result the $^{13}C/^{12}C$ ratios of biogenic materials are lower than those of atmospheric CO_2. The photosynthetic fractionation was shown [42] to be comprised of two steps: (1) preferential uptake of ^{12}C from the atmosphere, and (2) preferential conversion of ^{12}C-enriched dissolved CO_2 to phosphoglyceric acid, the first product of photosynthesis. The subsequent metabolism of photosynthetic products may also be accompanied by isotopic fractionation, but these fractionations are relatively small.

TABLE 4.11 TYPICAL $\delta^{13}C$ VALUES OF SOME MAJOR CARBON RESERVOIRS[a]

Carbon Reservoir	Approximate $\delta(^{13}C/^{12}C)(\%_{oo})$
Marine limestones	0
Bicarbonate in seawater	0
Atmospheric CO_2	−7
Marine organic carbon	−20
Recent marine sediments	−20
Land organic carbon	−25
Recent freshwater deposits	−25

[a] From Degens [43].

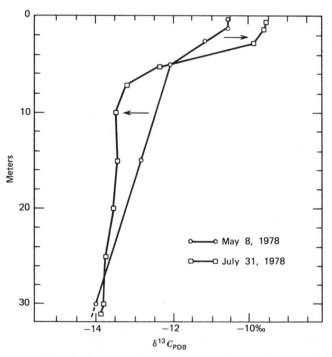

Figure 4.12 Spring–summer depth profiles of the ^{13}C content of total dissolved CO_2 from Greifensee, Switzerland. The shifts toward more positive $\delta^{13}C$ values in the surface waters, which result from biological activity, and toward more negative $\delta^{13}C$ values in the lower waters, which result from the oxidation of sinking organic material, are shown. The lower shift corresponds to depletion of dissolved oxygen in Greifensee during the summer months. (From J. McKenzie, Swiss Federal Institute of Technology, Zürich, personal communication, 1979.)

Degens [43] presents a comprehensive review of the carbon isotopic variations in nature (Table 4.11). The organic carbon of recent fresh water has an isotopic composition that reflects the source material.

In general, the isotopic composition in *precipitated carbonates* is representative of the HCO_3^- ions from which they were precipitated. Thus the $^{13}C/^{12}C$ ratio in shells is close to their ratio in the dissolved C of the water in which the organisms produced the shells. Typically, there is a difference in the $^{13}C/^{12}C$ ratio of organisms living in surface waters (e.g., plankton) and that of organisms living at the sediment–water interface (benthic organisms), which is related to the productivity of the water. This can be accounted for by considering that the biomass settling from the surface waters into the deeper waters is depleted in ^{13}C ($\delta = -20\%_0$) with respect to the C in the water itself (^{12}C is fixed ca. 1.02 times faster during photosynthesis than ^{13}C; Figure 4.12). This in turn leads to a relatively small enrichment of the surface water with ^{13}C. Mineralization of the biological debris in the deeper waters leads to a depletion of ^{13}C in these waters. Such effects are also observed in lakes (Figure 4.12).

Stuiver [44] and Wagener [45] have recently estimated, from ^{13}C and ^{14}C records in the cellulose of trees, changes that have occurred in the major carbon reservoirs, biosphere, atmosphere, ocean, and fossil fuels over the last 100 years.

The appearance of $\delta^{13}C$ values indicative of photosynthesis in Precambrian sedimentary materials 3–3.8×10^9 years old implies that photosynthetic processes must have occurred and that photosynthetic oxygen was produced 3 to 3.8 billions of years ago [46].

REFERENCES

1   K. H. Mancy, W. C. Westgarth, and D. A. Okun, 17th Industrial Waste Conference, Purdue University, Lafayette, Ind., May 1962.

2   K. S. Deffeyes, *Limnol. Oceanogr.*, **10**, 412 (1965).

3   W. S. Broecker, Y. -H. Li, and T. -H. Peng, in *Impingement of Man on the Oceans*, D. W. Hood, Ed., Wiley-Interscience, New York, 1971.

4   C. F. Baes et al., *Amer. Sci.*, **65**, 310 (1977).

5   G. M. Woodwell and E. V. Pecan, Eds., "Carbon and the Biosphere," U.S. Atomic Energy Commission, Washington, D.C., 1973.

6   W. Stumm, Ed., *Global Chemical Cycles and their Alterations by Man*, Dahlem Konferenzen, Berlin, 1977.

7   W. Weber and W. Stumm, *J. Amer. Water Works Assoc.*, **55**, 1560 (1963).

8   *Standard Methods for the Examination of Water and Waste Water*, American Public Health Association, New York, 1960.

9   G. Gran, *Analyst*, **77**, 661 (1952).

10   T. E. Larson and L. Henley, *Anal. Chem.*, **27**, 851 (1965).

11   J. F. Thomas and J. J. Lynch, *J. Amer. Water Works Assoc.*, **52**, 255 (1960).

12   D. Dyrssen and I. Hanson, *Mar. Chem.*, **1**, 137 (1972).

13   J. M. Edmond, *Deep-Sea Res.*, **17**, 737 (1970).

14   D. Dyrssen and L. G. Sillén, *Tellus*, **19**, 113 (1967).

15   T. Takahashi et al., *J. Geophys. Res.*, **75**, 7648 (1970).

16   G. Skirrow, in *Chemical Oceanography*, J. P. Riley and G. Skirrow, Eds., Vol. 2, 2nd ed., Academic, New York, 1975.

17 R. G. Bates, in *The Nature of Seawater*, E. D. Goldberg, Ed., Dahlem Konferenzen, Berlin, 1975.

18 R. M. Garrels and M. E. Thompson, *Amer. J. Sci.*, **260**, 57 (1962).

19 A. Distèche and S. Distèche, *J. Electrochem. Soc.*, **114**, 330 (1967).

20 T. E. Larson and A. M. Buswell, *J. Amer. Water Works Assoc.*, **34**, 1667 (1942).

21 C. Culberson and R. M. Pytkowicz, *Limnol. Oceanogr.*, **13**, 403 (1970).

22 J. M. Edmond, J. M. T. M. Gieskes, *Geochim. Cosmochim. Acta*, **34**, 1261 (1970).

23 A. Distèche, in *The Sea*, Vol. 5, E. D. Goldberg, Ed., Wiley-Interscience, 1974, p. 92.

24 D. M. Kern, *J. Chem. Educ.*, **37**, 14 (1960).

25 M. Eigen, K. Kustin, and G. Maass, *Z. Phys. Chem. N.F.*, **30**, 130 (1961).

26 J. T. Edsall, in CO_2: *Chemical, Biochemical and Physiological Aspects*, R. E. Forster, J. T. Edsall, A. B. Otis, and F. J. W. Roughton, Eds., NASA SP-188, Washington, D.C., 1969, p. 15. (This publication reviews the experimental findings on the kinetics of CO_2 hydration.)

27 R. Berger and W. F. Libby, *Science*, **164**, 1395 (1969).

28 W. S. Broecker and T. -H. Peng, *Tellus*, **26**, 21 (1974).

29 B. Bolin, *Tellus*, **12**, 274 (1960).

30 T. E. Hoover and D. C. Berkshire, *J. Geophys. Res.*, **74**, 456 (1969).

31 J. A. Quinn and N. C. Otto, *J. Geophys. Res.*, **76**, 1539 (1971); N. C. Otto and J. A. Quinn, *Chem. Eng. Sci.*, **26**, 949 (1971).

32 S. Emerson, W. S. Broecker, and D. W. Schindler, *J. Fish. Res. Board Can.*, **30**, 1475 (1973).

33 S. Emerson, *Limnol. Oceanogr.*, **20**, 743 (1975).

34 W. S. Broecker, *Chemical Oceanography*, Harcourt Brace Jovanovich, New York, 1974.

35 W. S. Broecker and Y. -H. Li, *J. Geophys. Res.*, **75**, 3545 (1970).

36 W. S. Broecker and V. M. Oversby, *Chemical Equilibria in the Earth*, McGraw-Hill, New York, 1971, pp. 150–170.

37 W. Dansgard et al., *Science*, **166**, 377 (1969).

38 U. Siegenthaler, in *Isotope Geology*, E. Jäger and J. C. Hunziker, Eds., Springer, Berlin, 1979, p. 264.

39 C. Junge, in *Global Chemical Cycles and Their Alterations by Man*, W. Stumm, Ed., Dahlem Konferenzen, Berlin, 1977, p. 33.

40 S. Epstein, P. Thompson, and C. J. Yapp, *Science*, **198**, 1209 (1977).

41 W. G. Deuser and E. T. Degens, *Nature*, **215**, 1033 (1967).

42 R. Park and S. Epstein, *Geochim. Cosmochim. Acta*, **21**, 110 (1960).

43 E. T. Degens, in *Organic Geochemistry*, G. Eglington and M. T. J. Murphy, Eds., Springer, New York, 1969, p. 304.

44 M. Stuiver, *Science*, **199**, 253 (1978).

45 K. Wagener, *Rad. Environ. Biophys.*, **15**, 101 (1978).

46 C. Junge, M. Schidlowski, R. Eichmann, and H. Pietreck, *J. Geophys. Res.*, **80**, 4542 (1975); personal communication, 1979.

READING SUGGESTIONS

Carbonate Equilibria

Deffeyes, K. S., "Carbonate Equilibria: A Graphic and Algebraic Approach," *Limnol. Oceanogr.*, **10**, 412 (1965).

Distèche, A., "The Effect of Pressure on Dissociation Constants and Its Temperature Dependency." In *The Sea*, Vol. 5, E. D. Goldberg, Ed., Wiley-Interscience, New York, 1974. p. 81–121.

Garrels, R. M., and C. Christ, *Minerals, Solutions and Equilibria*, Harper and Row, New York, 1965. (Treatment of carbonate equilibria in Chapter 3.)

Gieskes, J. M., "The Alkalinity—Total Carbon Dioxide System in Seawater." In *The Sea*, Vol. 5, E. D. Goldberg, Ed., Wiley-Interscience, New York, 1974, p. 123–151.

Carbon Dioxide in Natural Waters and in the Atmosphere: Geochemical Considerations

Bolin, E., E. T. Degens, S. Kempe, and P. Ketner, Eds., *The Global Carbon Cycle*, Wiley and Sons, New York, 1979. Published on behalf of the Scientific Committee on Problems of the Environment (SCOPE 13).

Broecker, W. S., *Chemical Oceanography*, Harcourt Brace Jovanovich, New York, 1974., (The reader is urged to read this book as an independent source for self-education. Wide in scope yet simple in presentation, it tries to summarize how the distribution of chemical species in the water and in the sediment is largely generated by an interaction between mixing cycles and biological cycles. It is shown how radioisotope measurements can be used to establish the time scale of these processes.)

Broecker, W. S., and T.-H. Peng, "Gas Exchange Rates between Air and the Sea," *Tellus* **26**, 21–35 (1974). (A discussion of the measurement of the gas-exchange rate between ocean and atmosphere.)

Dyrssen, D., and L. G. Sillén, "Alkalinity and Total Carbonate in Seawater. A Plea for p T independent Data," *Tellus*, **19**, 110 (1967). (It is shown by practical example how Gran's graphical method can be applied for determining [Alk] and C_T.)

Hoefs, J., *Stable Isotope Geochemistry*, Springer, New York, 1973. (This book surveys the most important results on isotope fractionation from a geological point of view.)

Hutchinson, G. E., "The Biochemistry of the Terrestrial Atmosphere." In *The Earth as a Planet*, G. Kuiper, Ed., University of Chicago Press, Chicago, 1954, pp. 371–427.

Mehrbach, C., C. H. Culberson, J. E. Hawley, and R. M. Pytkowicz, "Measurement of the Apparent Dissociation Constant of Carbonic Acid in Seawater at Atmospheric Pressure," *Limnol. Oceanogr.*, **18**, 897–907 (1973).

Rubey, W. W., "Geological History of Seawater: An Attempt to State the Problem," *Bull. Geol. Soc.*, **62**, 1111 (1951). (Considers the role of volatile materials from the earth's interior.)

Stuiver, M., "Atmospheric Carbon Dioxide and Carbon Reservoir Changes," *Science*, **199**, 253–258 (1978).

Woodwell, G. M., "The Carbon Dioxide Question," *Sci. Amer.*, **238**, 34 (1978).

PROBLEMS

4.1 Does the alkalinity of a natural water (isolated from its surroundings) increase, decrease, or stay constant upon addition of small quantities to the following:

(a) HCl

(b) NaOH

(c) Na_2CO_3

(d) $NaHCO_3$

(e) CO_2

(f) $AlCl_3$

(g) Na_2SO_4

4.2 If deep ocean water were stored in the laboratory

(a) At 5°C, 1 atm pressure

(b) At 20°C, 1 atm pressure

how would its pH, [Alk], and $[CO_3^{2-}]$ change?

4.3 The following short method for the determination of alkalinity has been proposed: Add a known quantity of standard mineral acid and measure the final pH. (Mineral acid quantity should preferably be such that final pH is between 3 and 4.3.) Present the theory and show how you compute the alkalinity.

4.4 A water containing 1.0×10^{-4} mol CO_2 liter^{-1} and having an alkalinity of 2.5×10^{-4} eq liter^{-1} has a pH of 6.7. The pH is to be raised to pH 8.3 with NaOH.

(a) How many moles of NaOH per liter of water are needed for this pH adjustment? ($pK_1 = 6.3$, $pK_2 = 10.3$)

(b) How many moles of lime ($Ca(OH)_2$) would be required?

4.5 An industrial waste from a metals industry contains approximately 5×10^{-3} M H_2SO_4. Before being discharged into the stream, the water is diluted with tap water in order to raise the pH. The tap water has the following composition: pH = 6.5, alkalinity = 2×10^{-3} eq liter^{-1}. What dilution is necessary to raise the pH to approximately 4.3?

4.6 S. L. Rettig and B. F. Jones [*U. S. Geol. Surv. Paper* 501D, D134 (1964)] designed a manometric method for determining alkalinity. Increments of strong acid (H_2SO_4) are added to the solution in a closed system. After each addition the pressure (predominantly p_{CO_2}) is measured.

(a) Develop an equation that describes the essential features of the partial pressure of CO_2 as a function of g (equivalent fraction of titrant added) and give a semiquantitative plot.

(b) Discuss the advantages and disadvantages of the method.

4.7 A surface water has an alkalinity of 2 meq liter^{-1} and a measured pH of 7.8.

(a) What is the direction of the flux of CO_2 across the air–water interface?

(b) Calculate the instantaneous flux.

(c) What is the time scale for reaching 99% of air–water equilibrium with respect to CO_2? The film thickness is estimated at 300 μm. The molecular diffusion coefficient of CO_2 in water is 1.9×10^{-5} cm^2 sec^{-1}. Assume that the partial pressure of CO_2 in the air is 0.00033 atm.

4.8 A cascade aerator operating on a well-water will reduce the dissolved CO_2 content from 45 to 18 mg liter^{-1} in one pass. The atmospheric saturation value of CO_2 may be assumed to be 0.5 mg liter^{-1}. Laboratory tests indicate that the gas transfer coefficient for H_2S is about 80% that for CO_2. Assume that enough H_2S will be present in the air around the aerator to give a saturation value of 0.1 mg liter^{-1}. Estimate the effluent concentration of H_2S if the well water contains 12 mg liter^{-1} of H_2S.

4.9 How is the relative distribution of $H_2CO_3^*$, HCO_3^-, and CO_3^{2-} in a solution with [Alk] $= 2.5 \times 10^{-3}$ eq liter^{-1} affected by a variation in partial pressure of CO_2? Plot distribution versus p_{CO_2}.

4.10 Compute the pH variation resulting from isothermal evaporation (25°C) of an incipiently 10^{-5} M NaHCO$_3$ solution that remains in equilibrium with the partial pressure of CO_2 of 3×10^{-4} atm.

4.11 Estimate the flux of CO_2 carried into a surface water by rainfall. Compare this flux with the exchange rate at the air–water interface. An annual rainfall of 100 cm may be assumed.

ANSWERS TO PROBLEMS

4.1 Decrease for (a) and (f); increase for (b), (c), and (d); no change for (e) and (g).

4.3 $\bar{C}_A[V/(V_0 + V)] - [H^+] \simeq$ [Alk] ($V_0 =$ original volume of sample, $V =$ volume of strong acid added, $\bar{C}_A =$ molarity of strong acid). The strong acid that has been added is equivalent to the alkalinity originally contained in the sample plus the mineral acidity that remains after the acid was added. The mineral acidity can be estimated by measuring pH, because [H-Acy] $\approx [H^+]$.

4.4 (a) 1.0×10^{-4}; (b) 0.5×10^{-4}. Note that pH $= 8.3$ corresponds to the equivalence point ($f = 1$). Cf. Figure 4-1 (point y).

4.5 Fivefold dilution. Note that pH $= 4.3$ corresponds to the equivalence point ($f = 0$, x in Figure 4.1) for the titration of alkalinity. The problem is equivalent to the titration of [Alk] of tap water with the acid of the waste water.

4.8 6 mg liter^{-1} H_2S. The rate of decrease in the concentration of the gas will be proportional to the oversaturation, that is, $-(C_0 - C_T)/(C_0 - C_3) = \exp(-kt)$.

4.9   Use equation 11 and solve by trial and error. Perhaps more conveniently diagrams such as Figure 4.3 may be constructed in order to compute the equilibrium concentrations.

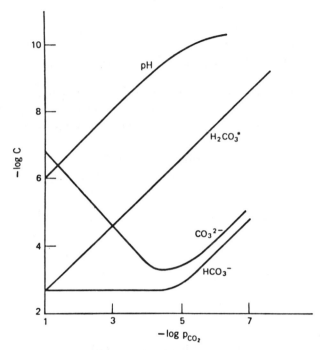

Figure 4.14

4.10   Use equation 11 or read result from Figure 4.3. For $[Na^+] = 10^{-5}\ M$, pH = 6.3; $[Na^+] = 10^{-3}\ M$, pH = 8.3; $[Na^+] = 10^{-2}\ M$, pH = 9.2; $[Na^+] = 10^{-1}\ M$, pH = 9.9.

APPENDIX   GRAN TITRATION

Consider Figure 4.13 and the following definitions.

Symbols: v_0, original volume of sample

  v, volume of strong acid added

  $\bar{c}_A$, molarity of strong acid

  H_2C, HC^-, and C^{2-} = $H_2CO_3^*$, HCO_3^-, and CO_3^{2-}, respectively

Capacities: $$C_T = [H_2C] + [HC^-] + [C^{2-}] \tag{1}$$

$$[Alk] = [HC^-] + 2[C^{2-}] + [OH^-] - [H^+] \tag{2}$$

$$[H^+\text{-Acy}] = [H^+] - [HC^-] - 2[C^{2-}] - [OH^-] \tag{3}$$

$$[CO_2\text{-Acy}] = [H_2C] + [H^+] - [C^{2-}] - [OH^-] \tag{4}$$

$$[\text{Acy}] = 2[H_2C] + [HC^-] + [H^+] - [OH^-] \tag{5}$$

$$[\text{p-Alk}] = [C^{2-}] + [OH^-] - [H^+] - [H_2C] \tag{6}$$

$$[OH^-\text{-Alk}] = [OH^-] - [HC^-] - 2[H_2C] - [H^+] \tag{7}$$

At the various equivalence points the following equalities exist:

$$v_0[\text{Alk}] = v_2 \bar{c}_A \tag{8}$$

$$v_0[\text{p-Alk}] = v_1 \bar{c}_A \tag{9}$$

$$v_0[OH^-\text{-Alk}] = x_0 \bar{c}_A \tag{10}$$

$$v_0 C_T = (v_2 - v_1)\bar{c}_A \tag{11}$$

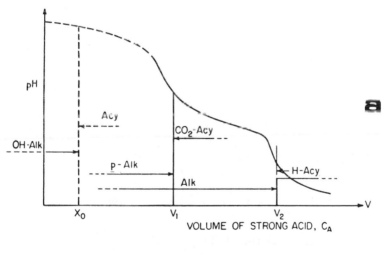

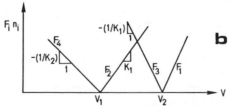

Figure 4.13 Sketch of an acidimetric titration curve (a). In (b) the results of (a) are plotted in terms of Gran functions; F_i is multiplied by scale factors n_i. The F_i values are defined by (18) and (21) to (24) of the appendix to this chapter. x_0, v_1, and v_2 are the volumes of strong acid corresponding to the equivalence points $f = 2$, $f = 1$, and $f = 0$, respectively.

At any point in the titration curve the following equalities exist:

$$(v_0 + v)[H^+\text{-Acy}] = (v - v_2)\bar{c}_A \tag{12}$$

$$(v_0 + v)[CO_2\text{-Acy}] = (v - v_1)\bar{c}_A \tag{13}$$

$$(v_0 + v)[\text{Alk}] = (v_2 - v)\bar{c}_A \tag{14}$$

$$(v_0 + v)[\text{Acy}] = (v - x_0)\bar{c}_A \tag{15}$$

$$(v_0 + v)[\text{p-Alk}] = (v_1 - v)\bar{c}_A \tag{16}$$

I. Beyond v_2, (12) can be simplified to:

$$(v_0 + v)[H^+] \simeq (v - v_2)\bar{c}_A \tag{17}$$

because for $v > v_2$

$$[H^+] \gg [HC^-] + 2[C^{2-}] + [OH^-]$$

The equivalence point v_2 can be obtained by plotting the left-hand side of (17) versus v.

Instead of $[H^+]$ we may write $10^{-\text{"pH"}}$. Independent of any pH or activity conventions, operationally one can define "pH" = $pH^\circ - \log[H^+]$ where pH° is a constant. If the pH meter reading is made in volts one can also use: $[H^+] = 10^{-\text{"pH"}} = 10^{(E-e)/k}$ because $E = e + k \log [H^+]$ where $k = 2.3$ RTF^{-1} (where F = faraday). Rewriting (17),

$$F_1 = (v_0 + v)10^{-\text{"pH"}} \simeq (v - v_2)\bar{c}_A \tag{18}$$

where $F_1 = 0$ for $v = v_2$.

II. Between v_1 and v_2, (13) and (14) can be simplified because in this range $[H_2C] \gg [H^+] - [C^{2-}] - [OH^-]$ and $[HC^-] \gg [C^{2-}] + [OH^-] - [H^+]$. Accordingly, instead of (13) and (14), respectively,

$$(v_0 + v)[H_2C] \simeq (v - v_1)\bar{c}_A \tag{19}$$

and

$$(v_0 + v)[HC^-] \simeq (v_2 - v)\bar{c}_A \tag{20}$$

Substituting $[H_2C] = (1/K_1)[H^+][HC^-]$ in (19) and combining with (20) gives:

$$F_2 = (v_2 - v)10^{-\text{"pH"}} = (v - v_1)K_1 \tag{21}$$

$F_2 = 0$ for $v = v_1$, the slope of F_2 versus v being K_1.

III. Equation 21 can be rearranged:

$$F_3 = (v - v_1)10^{\text{"pH"}} = (v_2 - v)K_1^{-1} \tag{22}$$

F_3 gives an additional check on v_2, the slope being $-1/K_1$.

IV. Another linear function can be derived, for $x_0 < v < v_1$, where $[C^{2-}] > [OH^-] - [H^+] - [H_2C]$, so that (16) can be written approximately as

$$(v_0 + v)[CO_3^{2-}] \simeq (v_1 - v)\bar{c}_A \tag{23}$$

An expression for $[HCO_3^-]$ can be derived from already estimated values of v_1 and v_2, giving

$$(v_0 + v)[HCO_3^-] \simeq (v_2 - 2v_1 + v)\bar{c}_A \tag{24}$$

combining (23) and (24) yields

$$\frac{[CO_3^{2-}]}{[HCO_3^-]} = \frac{K_2}{[H^+]} \simeq \frac{v_1 - v}{v_2 - 2v_1 + v} \tag{25}$$

Define

$$F_4 = (v_2 - 2v_1 + v)10^{\text{"pH"}} = \frac{v_1 - v}{K_2} \tag{26}$$

For $F_4 = 0$, $v = v_1$, the slope of F_4 versus v being $-1/K_2$.

5

Precipitation and Dissolution

5.1 INTRODUCTION

The hydrological cycle interacts with the cycle of rocks. Minerals dissolve in or react with the water. Under different physicochemical conditions minerals are precipitated and accumulate on the ocean floor and in the sediments of rivers and lakes. Dissolution and precipitation reactions impart to the water constituents that modify its chemical properties. Natural waters vary in chemical composition; consideration of solubility relations aids in the understanding of these variations. This chapter sets forth principles concerning reactions between solids and water. Here again the most common basis is a consideration of the equilibrium relations.

Dissolution or precipitation reactions are generally slower than reactions among dissolved species, but it is quite difficult to generalize about rates of precipitation and dissolution. There is a lack of data concerning most geochemically important solid–solution reactions, so that kinetic factors cannot be assessed easily. Frequently the solid phase formed incipiently is metastable with respect to a thermodynamically more stable solid phase. Examples are provided by the occurrence under certain conditions of aragonite instead of stable calcite or by the quartz oversaturation of most natural waters. This oversaturation occurs because the rate of attainment of equilibrium between silicic acid and quartz is extremely slow.

The solubilities of many inorganic salts increase with temperature, but a number of compounds of interest in natural waters ($CaCO_3$, $CaSO_4$) decrease in solubility with an increase in temperature. Pressure dependence of solubility is slight but must be considered for the extreme pressures encountered at ocean depths. For example, the solubility product of $CaCO_3$ will increase with increased pressure (by approximately 0.2 logarithmic units for a pressure of 200 atm†).

Heterogeneous Equilibria. The extent of the dissolution or precipitation reaction for systems that attain equilibrium can be estimated by considering the equilibrium constants. A simple example is the solubility of silica:

$$SiO_2(s) + 2H_2O \underset{\text{precipitation}}{\overset{\text{dissolution}}{\rightleftarrows}} H_4SiO_4 \tag{1}$$

† $(\partial \log K_{s0}/\partial P)_T = -\Delta V^0/RT \ln 10$, where ΔV^0, the change in molar volume for the reaction $CaCO_3(s)$ (calcite) $= Ca^{2+} + CO_3^{2-}$ is ca. -45 ml.

230

The solubility equilibrium for pure $SiO_2(s)$ is defined by

$$K_{s0} = \{H_4SiO_4\} \tag{2}$$

Equation 2 is obeyed regardless of whether H_4SiO_4 has been added to the solution or comes from the dissolution of $SiO_2(s)$.†

If a solution contains an activity of orthosilicic acid larger than K_{s0}, it is oversaturated; from a thermodynamic point of view $SiO_2(s)$ will precipitate. On the other hand, if $\{H_4SiO_4\}$ is smaller than K_{s0}, the solid phase will dissolve until $\{H_4SiO_4\} = K_{s0}$. It is obvious that the concentration of dissolved SiO_2 does not depend on the quantity of $SiO_2(s)$ in contact with the solution. The equilibrium condition of equation 2 is meaningful only in the context of a heterogeneous aqueous solution–solid equilibrium in which the activity of the pure solid phase is constant and can thus by convention be set equal to unity.

Reaction 1 may be compared with the application of the mass law to the following simple heterogeneous equilibria:

$$H_2O(1) = H_2O(g) \qquad\qquad K = p_{H_2O} \tag{3}$$

$$CaCO_3(s) = CaO(s) + CO_2(g) \qquad K = p_{CO_2} \tag{4}$$

$$MgCO_3(s) + 3H_2O(g) = MgCO_3 \cdot 3H_2O(s) \qquad K = p_{H_2O}^{-3} \tag{5}$$

in which the activity of a phase involved in the reaction is constant. The vapor pressure of pure water (reaction 3) ($\{H_2O(1)\} = 1$), however, is higher than the vapor pressure of a solution ($\{H_2O(1)\} < 1$). Thus, in a more general way, the equilibrium of equation 3 can be given by

$$K = \frac{p_{H_2O}}{\{H_2O(1)\}} \tag{6}$$

where it is understood that, in the limiting case of infinite dilution (pure solvent), $\{H_2O(1)\} = 1$. In an analogous way, a solid solution, for example, a solid in which certain lattice sites are occupied by foreign atoms or ions, has a smaller activity than the pure solid. Thus, in a most general way, we can write for the equilibrium of equation 1:

$$K = \frac{\{H_4SiO_4\}}{\{SiO_2(s)\}} \tag{7}$$

The reference state again is taken as an infinitely dilute solid solution; that is, for the pure solid $SiO_2(s)$, $\{SiO_2(s)\} = 1$. As long as the dissolution reaction is considered a physical process, the solubility may be characterized by the quantity of material that is soluble in a given volume of solvent. Frequently, however, the dissolution reaction is a heterogeneous chemical reaction; then a simple solubility datum is not sufficient to define the solution–solid equilibrium. It is necessary to characterize the solubility by a solubility product;

† Orthosilicic acid can be formulated as H_4SiO_4 or $Si(OH)_4$.

with an equilibrium constant one can characterize the solubility as well as predict how solution variables change the solubility.

In a general way for an electrolyte that dissolves in water according to the reaction

$$A_m B_n(s) \rightleftharpoons mA^{+n}(aq) + nB^{-m}(aq) \tag{8}$$

the equilibrium condition is

$$\{A_m B_n(s)\} = \{A^{+n}(aq)\}^m \{B^{-m}(aq)\}^n$$

The conventional solubility expression

$$K_{s0} = \{A^{+n}(aq)\}^m \{B^{-m}(aq)\}^n$$

results if the activity of the pure solid phase is set equal to unity and if the common standard-state convention for aqueous solutions is adopted; that is, the solubility product is constant for varying compositions of the liquid phase under constant temperature and pressure and for a chemically pure and compositionally invariant solid phase. Furthermore, the solid should be uniform and of large grain size (cf. Section 5.6). Only in some cases can the solubility of a salt be calculated from its solubility product alone. Generally one deals with the solubility of a salt in solutions that contain a common ion, that is, an ion that also exists in the ionic lattice of the solid salt.

Example 5.1a. Solubility of Sulfates, Chlorides, Fluorides, and Chromates

Characterize the solubility of the following salts as a function of the concentration of the common anion from the respective solubility products (K_{s0}):

$$CaSO_4(pK_{s0} = 4.6), \quad SrSO_4(6.2), \quad BaSO_4(9.7), \quad AgCl(10.0),$$

$$PbCl_2(4.8), \quad Ag_2CrO_4(12.0), \quad MgF_2(8.1), \quad CaF_2(10.3)$$

Figure 5.1 is a graphical representation of the solubility product, where the log of the metal-ion concentration is plotted as a function of the $-\log$ of the common anion, for example,

$$\log[Ca^{2+}] = -pK_{s0(CaSO_4)} + pSO_4^{2-} \tag{i}$$

$$\log[Mg^{2+}] = -pK_{s0(MgF_2)} + 2pF^- \tag{ii}$$

$$\log[Ag^+] = -\tfrac{1}{2}pK_{s0(Ag_2CrO_4)} + \tfrac{1}{2}pCrO_4^{2-} \tag{iii}$$

Correspondingly, in Figure 5.1 the $\log[Me^{z+}]$ lines have slopes of 1, 2, and $\tfrac{1}{2}$ and intercepts of $-pK_{s0}$, $-pK_{s0}$, and $-\tfrac{1}{2}pK_{s0}$ for salts of the $1:1$, $1:2$, and $2:1$ type, respectively.

The cations and anions of these salts do not undergo protolysis reactions to any appreciable extent in solutions that are near neutrality. Furthermore, complex formation (or ion-pair binding) between cation and anion may be

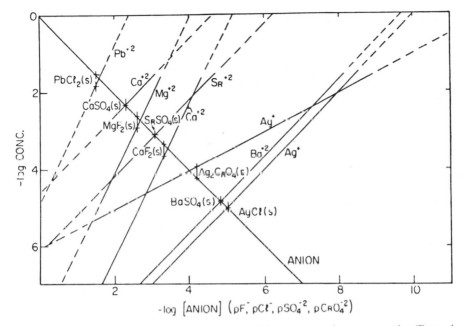

Figure 5.1 Solubility of "simple" salts as a function of the common anion concentration (Example 5.1). The cations and anions of these salts do not protolyze in the neutral pH range. The equilibrium solubility is given by the metal–ion concentration. At high anion or cation concentration, complex formation or ion-pair binding becomes possible (dashed lines). If the salt is dissolved in pure water (or in an inert electrolyte), the solubility is defined by the electroneutrality $z[Me^{n+}] = n[anion^{n-}]$. If $z = n$ (e.g., $BaSO_4$), the solubility is given by the intercept (+). If $z \neq n$, the electroneutrality condition is fulfilled at a point slightly displaced from the intersection ($\updownarrow$).

assumed to be negligible as long as free metal-ion and free anion concentration is small (ca. $< 10^{-1.5}$ M). Hence the solubility is characterized by the metal ion concentration. (Dashed lines in Figure 5.1 indicate where the foregoing assumptions are no longer valid.)

In the absence of a common anion, for example, if the salt is dissolved in pure water, the solubility is given by the electroneutrality requirement. Considering that $[H^+] \simeq [OH^-]$ we have, for example,

for $CaSO_4(s)$: $\qquad [Ca^{2+}] = [SO_4^{2-}]$ $\qquad\qquad$ (iv)

for $MgF_2(s)$: $\qquad 2[Mg^{2+}] = [F^-]$ $\qquad\qquad$ (v)

for $Ag_2CrO_4(s)$: $\qquad [Ag^+] = 2[CrO_4^{2-}]$ $\qquad\qquad$ (vi)

Hence the solubility of a 1:1 salt in pure water is defined by the intersection of the lines of $\log[Me^{2+}]$ and $\log[anion]$. For a salt of the 1:2 and 2:1 types the solubility is slightly displaced from the intersection in such a way that $\log[Me^{z+}] = +0.3 + \log[anion]$ and $\log[Me^{z+}] = -0.3 + \log[anion]$, respectively.

However, the case in which the solubility of a solid can be calculated from the known analytical concentration of added components and from the solubility product alone is very seldom encountered. Ions that have dissolved from a crystalline lattice frequently undergo chemical reactions in solution, and therefore other equilibria in addition to the solubility product have to be considered. The reaction of the salt cation or anion with water to undergo acid-base reactions is very common. Furthermore, complex formation of salt cation and salt anion with each other and with one of the constituents of the solution has to be considered. For example, the solubility of FeS(s) in a sulfide-containing aqueous solution depends, in addition to the solubility equilibrium, on acid–base equilibria of the cation (e.g., $Fe^{2+} + H_2O = FeOH^+ + H^+$) and of the anion (e.g., $S^{2-} + H_2O = HS^- + OH^-$ and $HS^- + H_2O = H_2S + OH^-$), as well as on equilibria describing complex formation (e.g., formation of $FeHS^+$ or FeS_2^{2-}).

Data for Solubility Constants. Available compilations of solubility products [1–6] illustrate that values given by different authors for the same solubility products often differ markedly. Differences of a few orders of magnitude are not uncommon. For example, data for the solubility product of $FePO_4(s)$ vary over a range of 10^{13}. There are various reasons for such discrepancies: (a) The formation of a sparingly soluble phase and its equilibrium with the solution is a more complicated process than equilibration reactions in a homogeneous solution phase. (b) The composition and properties, that is, reactivity, of the solids vary for different modifications of the same compound or for different active forms of the same modification. (c) Species influencing the solubility equilibrium (e.g., species formed by complex formation) are overlooked.

Activity Corrections. The solubility equilibrium is influenced by the ionic strength of the solution. Often it is most convenient to use operational equilibrium constants, that is, constants expressed as concentration quotients and valid for a medium of given ionic strength (Section 3.4). For freshwater conditions, the Debye–Hückel theory [or the Güntelberg or Davies equation (Table 3.3)] may be used to convert the solubility equilibrium constant given at infinite dilution or at a specified I to an operational constant, ^cK, valid for the ionic strength of interest. In seawater solubility equilibrium constants, experimentally determined in seawater, may be used. For example, the $CaCO_3$ solubility in seawater of specified salinity may be defined by $^cK'_{s0} = [Ca_T^{2+}][CO_{3_T}^{2-}]$, where $[Ca_T^{2+}]$ and $[CO_{3_T}^{2-}]$ are the total concentrations of calcium and carbonate ions, e.g., in mole liter^{-1}, in seawater (cf. Section 5.7).

Precipitation. An appreciation of the various types of precipitates that may be formed and an understanding of the changes the precipitates undergo in

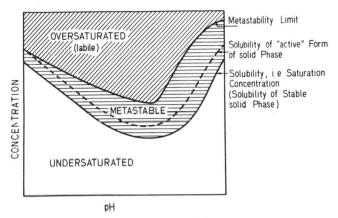

Figure 5.2 Solubility and saturation. A schematic solubility diagram showing concentration ranges versus pH for supersaturated, metastable, saturated, and undersaturated solutions. A super-saturated solution in the labile concentration range forms a precipitate spontaneously; a meta-stable solution may form no precipitate over a relatively long period. Often an "active" form of the precipitate, usually a very fine crystalline solid phase with a disordered lattice is formed from over-saturated solutions. Such an active precipitate may persist in metastable equilibrium with the solution; it is more soluble than the stable solid phase and may slowly convert into the stable phase. Compare also Figure 5.25.

aging are prerequisites for understanding and interpreting solubility equilibrium constants. A lucid treatment of these considerations, especially emphasizing precipitation and dissolution of metal hydroxides, has been given by Feitknecht and Schindler [3]. Figure 5.2 illustrates in a schematic way some of the domains of precipitation and solubility.

An "active" form of the compound, that is, a very fine crystalline precipitate with a disordered lattice, is generally formed incipiently from strongly over-saturated solutions. Such an active precipitate may persist in metastable equilibrium with the solution and may convert ("age") only slowly into a more stable "inactive" form. Measurements of the solubility of active forms give solubility products that are higher than those of the inactive forms. Inactive solid phases with ordered crystals are also formed from solutions that are only slightly oversaturated.

Hydroxides and sulfides often occur in amorphous and several crystalline modifications. Amorphous solids may be active or inactive. As Feitknecht and Schindler [3] point out, initially formed amorphous precipitates or active forms of unstable crystalline modifications may undergo two kinds of changes during aging. Either the active form of the unstable modification becomes inactive or a more stable modification is formed. With amorphous compounds deactivation may be accompanied by condensation [e.g., $Fe(OH)_3$]. If a metal oxide is more stable than the primarily precipitated hydroxide, dehydration may occur. When several of the processes mentioned take place together, nonhomogeneous solids are formed upon aging. In dissolution experiments with such nonhomogeneous solids, the more active components are dissolved more readily; that is, the solubility may depend upon the quantity of the solid present. Feitknecht and Schindler give the following example:

$$
\text{(amorph)Fe(OH)}_3\text{(active)} \quad
\begin{cases}
\nearrow & \text{(amorph)FeO}_{n/2}\text{(OH)}_{3-n}\text{(inactive)} \\
\longleftrightarrow & \alpha\text{-FeOOH(active)} \\
\searrow & \alpha\text{-Fe}_2\text{O}_3
\end{cases}
$$

In determining solubility equilibrium constants, many investigators have been motivated by a need to gain information that is pertinent primarily for the relatively short-term conditions (minutes to hours) typically encountered in the laboratory. In operations of analytical chemistry, for example, precipitates are frequently formed from strongly oversaturated solutions; the conditions of precipitation of the incipient active compound rather than the dissolution of the aged inactive solid are often of primary interest. Most solubility products measured in such cases refer to the most active component.[†] On the other hand, in dealing with heterogeneous equilibria of natural water systems, the more stable and inactive solids are frequently more pertinent.[‡] Aging often continues for geological time spans. Furthermore, the solid phase has frequently been formed in nature under conditions of slight supersaturation. Solubility constants determined under conditions where the solid has been identified by X-ray diffraction are especially valuable.

Solubility Product and Saturation

In order to test whether a solution, or a natural water, is over- or undersaturated, we inquire whether the free energy of dissolution of the solid phase is positive, negative, or zero; that is, for the solubility equilibrium

$$
\text{CaCO}_3(s) = \text{Ca}^{2+} + \text{CO}_3^{2-} \qquad K_{s0} \tag{9}
$$

the free energy of dissolution is given by (Section 2.10)

$$
\begin{aligned}
\Delta G &= RT \ln \frac{Q}{K} \\
&= RT \ln \frac{\{\text{Ca}^{2+}\}_{\text{act}}\{\text{CO}_3^{2-}\}_{\text{act}}}{\{\text{Ca}^{2+}\}_{\text{eq}}\{\text{CO}_3^{2-}\}_{\text{eq}}} = \frac{\{\text{Ca}^{2+}\}\{\text{CO}_3^{2-}\}}{K_{s0}}
\end{aligned} \tag{10}
$$

The actual ion activity product, IAP, may be compared with K_{s0}. The state of saturation of a solution with respect to a solid is defined:

$$
\begin{aligned}
\text{IAP} > K & \quad \text{(oversaturated)} \\
\text{IAP} = K & \quad \text{(equilibrium, saturated)} \\
\text{IAP} < K & \quad \text{(undersaturated)}
\end{aligned}
$$

[†] Strictly speaking, solubility products for active solid compounds are, because of their time dependence, not equilibrium constants; they are of operational value to estimate the conditions (pMe, pH, etc.) under which precipitation occurs.
[‡] A critical review [5] on thermodynamic properties of minerals is very useful.

By comparing Q with K (equation 10), we can define the state of saturation for all reactions that involve a solid phase. For example, the solubility equilibrium could also be written as

$$CaCO_3(s) + H^+ = Ca^{2+} + HCO_3^- \qquad *K_s \qquad (11)$$

Then

$$\Delta G = RT \ln \frac{\{Ca^{2+}\}_{act}\{HCO_3^-\}_{act}\{H^+\}_{eq}}{\{H^+\}_{act}\{Ca^{2+}\}_{eq}\{HCO_3^-\}_{eq}} = \frac{\{Ca^{2+}\}_{act}\{HCO_3^-\}_{act}}{\{H^+\}_{act} *K_s} \qquad (12)$$

If in equation 11 $Q/*K_s < 1$ (ΔG is negative), $CaCO_3(s)$ will dissolve; if $Q/*K_s > 1$, $CaCO_3$ will precipitate. The saturation test may often be made by simple comparing the activity (or concentration) of an individual reaction component, for example, H^+ in equation 11, with the activity (or concentration) this component would have if it were in hypothetical solubility equilibrium. Thus the state of saturation of $CaCO_3(s)$ may be interpreted in the following way:

$$pH_{act} > pH_{eq} \qquad \text{(oversaturated)}$$

$$pH_{act} = pH_{eq} \qquad \text{(saturated)}$$

$$pH_{act} < pH_{eq} \qquad \text{(undersaturated†)}$$

In a similar way the $CaCO_3(s)$ saturation could be assessed by comparing $[H_2CO_3^*]_{act}$ with $[H_2CO_3^*]_{eq}$ and considering the reaction $CaCO_3(s) + H_2CO_3^* = Ca^{2+} + 2HCO_3^-$

Example 5.1b. Solubility of Anhydrite and Gypsum

Compare the solubility ($25°C$, 1 atm) of anhydrite ($CaSO_4(s)$, $K_{s0} = 4.2 \times 10^{-5}$) with that of gypsum ($CaSO_4 \cdot 2H_2O(s)$, $K_{s0} = 2.5 \times 10^{-5}$) and determine the state of saturation of seawater of normal salinity ($\{Ca^{2+}\} = 2.4 \times 10^{-3}$ M; $\{SO_4^{2-}\} = 1.9 \times 10^{-3}$ M) with respect to these solid phases.

The IAP of seawater (4.7×10^{-6}) is smaller than either solubility product. Hence seawater is undersaturated ca. 5 times and ca. 9 times, respectively, with regard to gypsum and anhydrite. Since gypsum is less soluble ($25°C$, 1 atm), it should be more stable in seawater. The equilibrium constant for the conversion of gypsum into anhydrite can be obtained from

$$CaSO_4 \cdot 2H_2O(s) = Ca^{2+} + SO_4^{2-} + 2H_2O \qquad \log K_{s0gyps} = -4.60 \qquad (i)$$

$$Ca^{2+} + SO_4^{2-} = CaSO_4(s) \qquad \log K_{s0anhyd} = +4.38 \qquad (ii)$$

$$\overline{CaSO_4 \cdot 2H_2O(s) = CaSO_4(s) + 2H_2O \qquad \log K = -0.22 \qquad (iii)}$$

† The extent of saturation could then be expressed by a saturation index, $pH_{act} - pH_{eq}$. This saturation index can also be determined with relative ease experimentally by measuring the change in pH upon addition of $CaCO_3(s)$.

The equilibrium of equation iii (assuming activities of solid phases = 1) is defined by

$$\{H_2O\}^2 = 0.6 \quad \text{or} \quad \{H_2O\} = 0.78 \tag{iv}$$

As long as the activity of water is larger than 0.78, gypsum at this T and P is more stable than anhydrite. The activity of normal seawater is $\{H_2O\} \approx 0.98$. Upon evaporation of seawater to the point where gypsum becomes saturated, the activity of H_2O ($\{H_2O\} \approx 0.93$) is still much larger than the equilibrium H_2O activity for the conversion. Hence gypsum is more stable than anhydrite even in partially evaporated seawater. Gypsum may, however, become less stable than anhydrite upon burial in sediments (increase in temperature, pressure, and salinity).

5.2   THE SOLUBILITY OF OXIDES AND HYDROXIDES

Many minerals with which water comes into contact are oxides, hydroxides, carbonates, and hydroxide-carbonates. The same ligands (hydroxides and carbonates) are dissolved constituents of all natural waters. The solid and solute chemical species under consideration belong to the ternary system $Me^{z+}-H_2O-CO_2$. In this section a discussion of solubility equilibria is given. Complications arising in treating such solubility equilibria from nonpure solid phases (solid solutions) and from the possible occurrence of carbonate, mixed carbonato-hydroxo, and polynuclear complexes are neglected, but will be taken up in Sections 5.5 and 6.5. In this section we restrict ourselves to applying operational equilibrium constants (K' or ^cK).

If a pure solid oxide or hydroxide is in equilibrium with free ions in solution, for example,

$$Me(OH)_2(s) = Me^{2+} + 2OH^-$$
$$MeO(s) + H_2O = Me^{2+} + 2OH^- \tag{13}$$

the conventional solubility product is given by

$$^cK_{s0} = [Me^{2+}][OH^-]^2 \; \text{mol}^3 \; \text{liter}^{-3} \tag{14}$$

The subscript zero indicates that the equilibrium of the solid with the simple (uncomplexed) species Me^{2+} and OH^- is considered.† Sometimes it is more appropriate to express the solubility in terms of reaction with protons, since the equilibrium concentrations of OH^- ions may be extremely small, for example,

$$Me(OH)_2(s) + 2H^+ = Me^{2+} + 2H_2O \tag{15}$$
$$MeO(s) + 2H^+ = Me^{2+} + H_2O \tag{16}$$

---

† The complete dissolution reaction considers the water that participates in the dissolution reaction: $Me(OH)_z(s) + (x + zy)H_2O(l) = Me(H_2O)_x^{z+} + z(OH^-)(H_2O)_y$.

Then the solubility equilibrium can be characterized by

$$^{c}*K_{s0} = \frac{[Me^{2+}]}{[H^+]^2} \; mol^{-1} \; liter \qquad (17)$$

or, for a solubility equilibrium with a trivalent metal ion,

$$FeOOH(s) + 3H^+ = Fe^{3+} + 2H_2O$$

by

$$^{c}*K_{s0} = \frac{[Fe^{3+}]}{[H^+]^3} \; mol^{-2} \; liter^2 \qquad (18)$$

The definitions for the solubility equilibrium contained in equations 14 and 17 are interrelated. For $MeO_{z/2}$ or $Me(OH)_z$ the following general equation obtains:

$$^{c}*K_{s0} = \frac{^{c}K_{s0}}{K_w^z} \qquad (19)$$

where K_w is the ion product of water. A few representative solubility products are given in Table 5.1. From these equilibrium constants, $[Me^{z+}]$ in equilibrium with the pure solid phase can be computed readily as a function of pOH or pH. Especially convenient is the graphical representation in a $\log[Me^{z+}]$—pH diagram for which the relationship of equation 20 can be derived:

$$\log[Me^{z+}] = \log {}^{c}*K_{s0} - zpH$$
$$\log[Me^{z+}] = \log {}^{c}K_{s0} + zpK_w - zpH \qquad (20)$$

Equation 20 is plotted for a few oxides or hydroxides in Figure 5.3. Obviously, $\log[Me^{z+}]$ plots linearly as a function of pH with a slope of $-z$ ($d \log[Me^{z+}]/dpH = -z$) and an intercept at $\log[Me^{z+}] = 0$, with the value $pH = -(1/z)p^{c}*K_{s0}$.

The relations depicted in Figure 5.3 or characterized by equation 14 do not fully describe the solubility of oxides or hydroxides. We have to consider that the solid can be in equilibrium with hydroxo metal–ion complexes[†] $[Me(OH)_n]^{z-n}$.

We already have encountered such hydroxo complexes as conjugate (Brønsted) bases of aquo metal ions, for example,

$$Zn^{2+} + H_2O = ZnOH^+ + H^+ \qquad *K_1 \qquad (21)$$

and

$$FeOH^{2+} + H_2O = Fe(OH)_2^+ + H^+ \qquad *K_2 \qquad (22)$$

---

† More generally and more exactly, polynuclear complexes $[Me_m(OH)_n]^{zm-n}$, and complexes with other ligands, L^{y-}, in the solution, $[Me_pL_q]^{(zp-yq)}$, and mixed complexes with OH^- and L^{y-} have to be considered too. In the examples discussed here, the possible occurrence of these more complicated species has little effect upon the solubility characteristics (see Section 6.5).

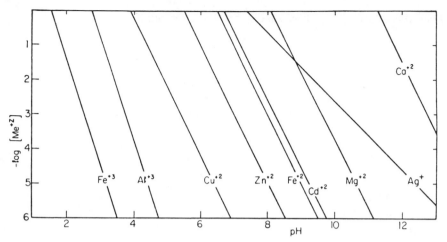

Figure 5.3 Solubility of oxides and hydroxides. Free metal–ion concentration in equilibrium with solid oxides or hydroxides. The occurrence of hydroxo metal complexes must be considered for evaluation of complete solubility.

We can characterize the solubility of the metal oxide or hydroxide, Me_T, by

$$Me_T = [Me^{z+}] + \sum_1^n [Me(OH)_n^{z-n}] \tag{23}$$

Figure 5.4 gives examples for the solubility of amorphous $Fe(OH)_3$,[†] ZnO, and CuO.

Example 5.2. Graphical Representation of Solubility of ZnO(s)

The construction of a logarithmic diagram will be illustrated for the solubility equilibrium of ZnO. It is convenient to write all the possible reactions for solid phase with solute species, for example, at 25°C

$$ZnO(s) + 2H^+ = Zn^{2+} + H_2O \qquad \log {}^*K_{s0} = 11.2 \qquad \text{(i)}$$

$$ZnO(s) + H^+ = ZnOH^+ \qquad \log {}^*K_{s1} = 2.2 \qquad \text{(ii)}$$

$$ZnO(s) + 2H_2O = Zn(OH)_3^- + H^+ \qquad \log {}^*K_{s3} = -16.9 \qquad \text{(iii)}$$

$$ZnO(s) + 3H_2O = Zn(OH)_4^{2-} + 2H^+ \qquad \log {}^*K_{s4} = -29.7 \qquad \text{(iv)}$$

[†] Oxide hydroxides of Fe(III) precipitated in natural waters are often mixtures of amorphous material and goethite (α-FeOOH). The thermodynamic stability of these precipitates may be characterized by their solubility products ($[Fe^{3+}][OH^-]^3 = K_{s0}$). According to Langmuir and Whittemore [*Adv. Chem. Ser.*, **106** 209 (1971)] fresh precipitates had $-\log K_{s0}$ values ranging from 37.3 to 43.3.

TABLE 5.1 CONSTANTS FOR SOLUBILITY EQUILIBRIA[a]

I. Oxides and Hydroxides

	Symbol for Equilibrium Constants	$\log K$ at 25°C	I
$H_2O(l) = H^+ + OH^-$	K_W	-14.00	0
$(am)Fe(OH)_3(s) = Fe^{3+} + 3OH^-$	K_{s0}	-38.7	$1\,M$ NaClO$_4$
$(am)Fe(OH)_3(s) = FeOH^{2+} + 2OH^-$	K_{s1}	-27.5	$3\,M$ NaClO$_4$
$(am)Fe(OH)_3(s) = Fe(OH)_2^+ + OH^-$	K_{s2}	-16.6	$3\,M$ NaClO$_4$
$(am)Fe(OH)_3(s) + OH^- = Fe(OH)_4^-$	K_{s4}	-4.5	$3\,M$ NaClO$_4$
$2(am)Fe(OH)_3(s) = Fe_2(OH)_2^{4+} + 4OH^-$	K_{s22}	-51.9	$3\,M$ NaClO$_4$
$(am)FeOOH(s) + 3H^+ = Fe^{3+} + 2H_2O$	$*K_{s0}$	3.55	$3\,M$ NaClO$_4$
$\alpha\text{-}FeOOH(s) + 3H^+ = Fe^{3+} + 2H_2O$	$*K_{s0}$	1.6	$3\,M$ NaClO$_4$
$Fe^{3+} + H_2O = FeOH^{2+} + H^+$	$*K_1$	-2.19	0
$Fe^{3+} + 2H_2O = Fe(OH)_2^+ + 2H^+$	$*\beta_2$	-5.67	0
$Fe^{3+} + 3H_2O = Fe(OH)_3(aq) + 3H^+$	$*\beta_3$	<-12	0
$Fe^{3+} + 4H_2O = Fe(OH)_4^- + 4H^+$	$*\beta_4$	-21.6	0
$2Fe^{3+} + 2H_2O = Fe_2(OH)_2^{4+} + 2H^+$	$*\beta_{22}$	-2.95	0
$\alpha\text{-}FeOOH(s) + 3H^+ = Fe^{3+} + 2H_2O$	$*K_{s0}$	0.5	0
$(am)FeOOH(s) + 3H^+ = Fe^{3+} + 2H_2O$	$*K_{s0}$	2.5	0
$Fe^{2+} + H_2O = FeOH^+ + H^+$	$*K_1$	-9.5	0
$Fe(OH)_2(s) \text{ (active)} + 2H^+ = Fe^{2+} + 2H_2O$	$*K_{s0}$	12.85	0
$Al^{3+} + H_2O = AlOH^{2+} + H^+$	$*K_1$	-4.97	0
$Al^{3+} + 2H_2O = Al(OH)_2^+ + 2H^+$	$*\beta_2$	-9.3	0
$Al^{3+} + 3H_2O = Al(OH)_3(aq) + 3H^+$	$*\beta_3$	-15.0	0
$Al^{3+} + 4H_2O = Al(OH)_4^- + 4H^+$	$*\beta_4$	-23.0	0
$2Al^{3+} + 2H_2O = Al_2(OH)_2^{4+} + 2H^+$	$*\beta_{22}$	-7.7	0
$3Al^{3+} + 4H_2O = Al_3(OH)_4^{5+} + 4H^+$	$*\beta_{43}$	-13.9	0
$13Al^{3+} + 28H_2O = Al_{13}O_4(OH)_{24}^{7+} + 32H^+$	$\beta_{32,13}$	-98.7	0

(continued)

241

TABLE 5.1 (continued)

	Symbol for Equilibrium Constants	log K at 25°C	I
$\alpha\text{-Al(OH)}_3(s) + 3H^+ = Al^{3+} + 3H_2O$	$*K_{s0}$	8.5	0
$Al(OH)_3(\text{amorph}) + 3H^+ = Al^{3+} + 3H_2O$	$*K_{s0}$	10.8	0
$CuO(s) + 2H^+ = Cu^{2+} + H_2O$	$*K_{s0}$	7.65	0
$Cu^{2+} + OH^- = CuOH^+$	K_1	6.0 (18°C)	0
$Cu^{2+} + 2OH^- = Cu(OH)_2$	K_2	12.8	1
$2Cu^{2+} + 2OH^- = Cu_2(OH)_2^{2+}$	K_{22}	17.0 (18°C)	0
$Cu^{2+} + 3OH^- = Cu(OH)_3^-$	K_3	15.2	0
$Cu^{2+} + 4OH^- = Cu(OH)_4^{2-}$	K_4	16.1	0
$Zn^{2+} + H_2O = ZnOH^+ + H^+$	$*K_1$	-8.96	0
$Zn^{2+} + 2H_2O = Zn(OH)_2 + 2H^+$	$*\beta_2$	-16.9	0
$Zn^{2+} + 3H_2O = Zn(OH)_3^- + 3H^+$	$*\beta_3$	-28.4	0
$Zn^{2+} + 4H_2O = Zn(OH)_4^- + 4H^+$	$*\beta_4$	-41.2	0
$ZnO + 2H^+ = Zn^{2+} + H_2O$	$*K_{s0}$	11.14	0
$Zn(OH)_2(\text{amorph}) + 2H^+ = Zn^{2+} + 2H_2O$	$*K_{s0}$	12.45	0
$Cd^{2+} + H_2O = CdOH^+ + H^+$	$*K_1$	-10.1	0
$Cd^{2+} + 2H_2O = Cd(OH)_2(aq) + 2H^+$	$*\beta_2$	-20.4	0
$Cd^{2+} + 3H_2O = Cd(OH)_3^- + 3H^+$	$*\beta_3$	< -33.3	0
$Cd^{2+} + 4H_2O = Cd(OH)_4^{2-} + 4H^+$	$*\beta_4$	-47.4	0
$\beta\text{-Cd(OH)}_2(s) + 2H^+ = Cd^{2+} + 2H_2O$	$*K_{s0}$	13.65	0
$Mn^{2+} + H_2O = MnOH^+ + H^+$	$*K_1$	-10.6	0
$Mn^{2+} + 3H_2O = Mn(OH)_3^- + 3H^+$	$*\beta_3$	~ -35	0
$Mn^{2+} + 4H_2O = Mn(OH)_4^{2-} + 4H^+$	$*\beta_4$	-48.3	0
$Mn(OH)_2(s) \text{ (active)} + 2H^+ = Mn^{2+} + H_2O$	$*K_{s0}$	15.2	0
$Hg^{2+} + H_2O = HgOH^+ + H^+$	$*K_1$	-3.4	0

Reaction		Constant	log value	Medium
$Hg^{2+} + 2H_2O = Hg(OH)_2(aq) + 2H^+$		$*\beta_2$	-6.2	0
$Hg^{2+} + 3H_2O = Hg(OH)_3^- + 3H^+$		$*\beta_3$	-21.1	0
$HgO + 2H^+ = Hg^{2+} + H_2O$		$*K_{s0}$	2.56	0
$Mg^{2+} + H_2O = MgOH^+ + H^+$		$*K_1$	-11.44	0
$Mg(OH)_2(s)$ (brucite) $+ 2H^+ = Mg^{2+} + 2H_2O$		$*K_{s0}$	16.84	0
$Ca^{2+} + H_2O = CaOH^+ + H^+$		$*K_1$	-12.85	0
$Ca(OH)_2(s) + 2H^+ = Ca^{2+} + H_2O$		$*K_{s0}$	22.8	

II. Carbonates and Hydroxide Carbonates

Reaction		Constant	log value	Medium
$CaCO_3(calcite) = Ca^{2+} + CO_3^{2-}$		K_{s0}	-8.42	0
			-6.2	Seawater
$CaCO_3(aragonite) = Ca^{2+} + CO_3^{2-}$		K_{s0}	-8.22	0
			-6.05	Seawater
$SrCO_3(s) = Sr^{2+} + CO_3^{2-}$		K_{s0}	-9.03	0
			-6.8	Seawater
$BaCO_3(s) = Ba^{2+} + CO_3^{2-}$		K_{s0}	-8.30	0
$ZnCO_3(s) + 2H^+ = Zn^{2+} + H_2O + CO_2(g)$		$*K_{ps0}$	7.95	0
$Zn(OH)_{1.2}(CO_3)_{0.4}(s) + 2H^+ = Zn^{2+} + {}_{1.6}H_2O + {}_{0.4}CO_2(g)$		$*\acute{K}_{ps0}$	9.8	0
$Cu(OH)(CO_3)_{0.5}(s) + 2H^+ = Cu^{2+} + \frac{3}{2}H_2O + \frac{1}{2}CO_2(g)$		$*K_{ps0}$	6.49	0
$Cu(OH)_{0.67}(CO_3)_{0.67}(s) + 2H^+ = Cu^{2+} + \frac{4}{3}H_2O - \frac{2}{3}CO_2(g)$		$*K_{ps0}$	6.47	0
$MgCO_3(magnesite) = Mg^{2+} + CO_3^{2-}$		K_{s0}	7.46^b	0
$MgCO_3 \cdot 3H_2O = Mg^{2+} + CO_3^{2-} + 3H_2O$		K_{s0}	-5.19	0
$Mg_4(CO_3)_3(OH)_2 \cdot 3H_2O(hydromagnesite) = 4Mg^{2+} + 3CO_3^{2-} + 2OH^{-\,c}$		K_{s0}	-29.5	0
$CaMg(CO_3)_2(dolomite) = Ca^{2+} + Mg^{2+} + 2CO_3^{2-}$		K_{s0}	-16.7	0
$FeCO_3(siderite) = Fe^{2+} + CO_3^{2-}$		K_{s0}	-10.7	0
$MnCO_3(s) = Mn^{2+} + CO_3^{2-}$		K_{s0}	-10.4	0
$CdCO_3(s) + 2H^+ = Cd^{2+} + H_2O + CO_2(g)$		$*K_{ps0}$	6.44	$1\,M\,NaClO_4$
$PbCO_3(s) = Pb^{2+} + CO_3^{2-}$		K_{s0}	-13.1	0

(continued)

243

TABLE 5.1 (continued)

III. CaCO₃ Solubility

Reaction	−log K', infinite dilution[d]						−log $^cK'_{s0}$, seawater[e]	
	5°C	10°C	15°C	20°C	25°C	40°C	5°C	25°C
$CaCO_3(s) = Ca^{2+} + CO_3^{2-}$ (calcite)	8.35, 8.37[f]	8.36	8.37	8.39	8.42, 8.45[f]	8.53	6.10[f]	6.23[f], 6.34[g]
$CaCO_3(s) + H^+ = Ca^{2+} + HCO_3^-$ (calcite)	−2.22	−2.13	−2.06	−1.99	−1.91	−1.69	—	—
$CaCO_3(s) = Ca^{2+} + CO_3^{2-}$ (aragonite)	—	—	—	—	8.22	—	5.94[f]	6.05[f]

Summarizing Equations:

Infinite dilution (Jacobson and Langmuir[d]):

$$-\log K'_{s0} = 13.870 - 3059/T - 0.04035T \qquad \text{(i)}$$
$$-\log K_{s0} = 13.543 - 3000/T - 0.04001T \qquad \text{(ii)}$$

Seawater (pressure, salinity, and temperature dependence (J. M. Gieskes, in *The Sea*, Vol. 5, E. D. Goldberg, Ed., Wiley-Interscience, New York, 1974)

Calcite	$-\log {}^cK' = (0.1614 + 0.02892\ Cl - 0.0063t) \times 10^{-6}$	(iii)
Aragonite	$-\log {}^cK' = (0.5115 + 0.02892\ Cl - 0.0063t) \times 10^{-6}$	(iv)
	$\log ({}^cK'_i)_P/({}^cK'_i)_{P=1} = -\Delta V'(P-1)/2.303\,RT$	(v)

where

$$\Delta V' \text{ (calcite)} = -(47.5 - 0.23t)\ cm^3\ mol^{-1} \qquad \text{(vi)}$$
$$\Delta V' \text{ (aragonite)} = -(45.0 - 0.23t)\ cm^3\ mol^{-1} \qquad \text{(vii)}$$

where P = pressure in atm, T = K, t = °C, Cl = chlorinity (‰), and $2.303\,RT$ at 25°C = $5.636 \times 10^4\ cm^3\ atm\ mol^{-1}$

IV. Sulfides

$-\log K$ 25°C, $I = 0$

	$-\log K$
MnS(s) (green) = $Mn^{2+} + S^{2-}$	13.5
FeS(s) = $Fe^{2+} + S^{2-}$	18.1
ZnS(s)(α) = $Zn^{2+} + S^{2-}$	24.7
CdS(s) = $Cd^{2+} + S^{2-}$	27.0
CuS(s) = $Cu^{2+} + S^{2-}$	36.1
Cu_2S(s) = $2Cu^+ + S^{2-}$	48.5
Ag_2S(s) = $2Ag^+ + S^{2-}$	50.1
HgS(s) = $Hg^{2+} + S^{2-}$	52.7
PbS(s) = $Pb^{2+} + S^{2-}$	27.5
H_2S(aq) = $H^+ + HS^-$	7.02
H_2S(g) = H_2S(aq)	0.99
HS^- = $H^+ + S^{2-}$	13.9

V. Phosphates

$\log K$ at 25°C, $I = 0$

	$\log K$ at 25°C, $I = 0$
$FePO_4 \cdot 2H_2O$(s) (Strengite) = $Fe^{+3} + PO_4^{-3} + 2H_2O$	-26
$AlPO_4 \cdot 2H_2O$(s) (Variscite) = $Al^{+3} + PO_4^{-3} + 2H_2O$	-21
$CaHPO_4$(s) = $Ca^{+2} + HPO_4^{-2}$	-6.5
$Ca_4H(PO_4)_3$(s) = $4Ca^{+2} + 3PO_4^{-3} + H^+$	-46.9
$Ca_{10}(PO_4)_6(OH)_2$(s) = $10Ca^{+2} + 6PO_4^{-3} + 2OH^-$	-114
$Ca_{10}(PO_4)_6(F)_2$(s) = $10Ca^{+2} + 6PO_4^{-3} + 2F^-$	-118
$Ca_{10}(PO_4)_6(OH)_2$(s) + $6H_2O$ = $4[Ca_2(HPO_4)(OH)_2] + 2Ca^{+2} + 2HPO_4^{-2}$	-17
$CaHAl(PO_4)_2$(s) = $Ca^{+2} + Al^{+3} + H^+ + 2PO_4^{-3}$	-39
CaF_2(s) = $Ca^{+2} + 2F^-$	-10.4
$MgNH_4PO_4$(s) = $Mg^{+2} + NH_4^+ + PO_4^{-3}$	-12.6
$FeNH_4PO_4$(s) = $Fe^{+2} + NH_4^+ + PO_4^{-3}$	~ -13
$Fe_3(PO_4)_2$(s) = $3Fe^{+2} + 2PO_4^{-3}$	~ -32

(continued)

245

TABLE 5.1 (continued)

	log K at 25°C, $I = 0$
Acid–Base and Complex Formation	
$H_2CO_3 = H^+ + HCO_3^-$	-6.3
$HCO_3^- = H^+ + CO_3^{-2}$	-10.3
$H_3PO_4 = H^+ + H_2PO_4^-$	-2.2
$H_2PO_4^- = H^+ + HPO_4^{-2}$	-7.2
$HPO_4^{-2} = H^+ + PO_4^{-3}$	-12.3
$CaPO_4^- = Ca^{+2} + PO_4^{-3}$	-6.5
$CaHPO_4^0 = Ca^{+2} + HPO_4^{-2}$	-2.7
$CaH_2PO_4^+ = Ca^{+2} + H_2PO_4^-$	-1.4
$MgHPO_4^0 = Mg^{+2} + HPO_4^{-2}$	-2.5
$FeHPO_4^+ = Fe^{+3} + HPO_4^{-2}$	-8.3
$FeH_2PO_4^{+2} = Fe^{+3} + H_2PO_4^-$	-1.8
$AlHPO_4^+ = Al^{+3} + HPO_4^{-2}$	-8
$AlH_2PO_4^{+2} = Al^{+3} + H_2PO_4^-$	-3
$CaP_2O_7^{-2} = Ca^{+2} + P_2O_7^{-4}$	-5.6
$CaHP_2O_7^{-1} = Ca^{+2} + HP_2O_7^{-3}$	-3.6
$MgP_2O_7^{-2} = Mg^{+2} + P_2O_7^{-4}$	-5.7
$NaP_2O_7^{-3} = Na^+ + P_2O_7^{-4}$	-2.2
$Fe(HP_2O_7)_2^{-3} = Fe^{+3} + 2HP_2O_7^{-3}$	-22.0
$CaP_3O_{10}^{-3} = Ca^{+2} + P_3O_{10}^{-5}$	-6.9
$CaP_4O_{12}^{-2} = Ca^{+2} + P_4O_{12}^{-4}$	-4.9

[a] Most of the constants given here are taken from quotations or selections in [1–6]. The reader should bear in mind that the quality of these data is highly variable; for a critical assessment and for a more complete compilation [1–6] should be consulted. Symbols used are those given in [1]. For an additional survey of solubility products of oxides and hydroxides and first-hydrolysis constants, see Figures 6.4b and 6.6. Additional equilibrium constants can be assessed from the thermodynamic data in the Appendix at the end of the book.

[b] Various solubility products have been reported; [5] gives log $K_{s0} = -8.20$.

[c] The solubility equilibrium for hydromagnesite has also been formulated [5] as $Mg_5(CO_3)_4(OH)_2 \cdot 4H_2O = 5Mg^{2-} + 4CO_3^{2-} + 2OH^- + 4H_2O$.

[d] K' is defined so as to ignore the existence of $CaCO_3$ or $CaHCO_3^+$ ion pairs. Unless otherwise noted data for the solubility product are from R. L. Jacobson and D. Langmuir (*Geochim. Cosmochim. Acta*, **38**, 301 (1974). The equilibrium constant for the reaction $CaCO_3(s) + H^+ = Ca^{2+} + HCO_3^-$ is obtained by dividing the solubility product by the second acidity constant K_2' of $H_2CO_3^*$ (Table 4.10).

[e] $K_{s0}' = [Ca_T^{2+}][CO_{3T}^{2-}]$ (concentrations are expressed in volumetric units (molar) with exception of the value cited in note g, which is molal units).

[f] From R. Berner, *Amer. J. Sci.*, **276**, 713 (1976).

[g] S. E. Ingle, C. H. Culberson, J. E. Hawley, and R. M. Pytkowicz, *Mar. Chem.*, **1**, 295 (1973).

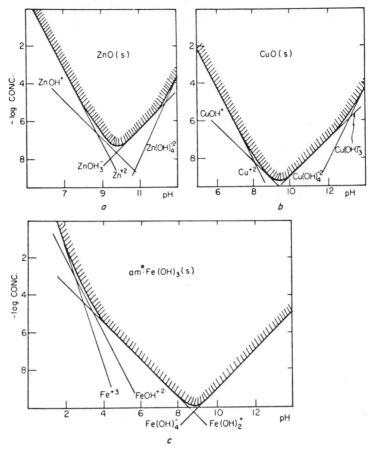

Figure 5.4 Solubility of amorphous $Fe(OH)_3$, ZnO, and CuO. The construction of the diagram is explained in Example 5.2. The possible occurrence of polynuclear complexes, for example, $Fe_2(OH)_2^{4+}$, $Cu_2(OH)_2^{2+}$, has been ignored. Such complexes do not change the solubility characteristics markedly for the solids considered here.

The constants given here are equivalent to those given in Table 5.1. The equilibrium constant of equation ii obtains from the following combination:

$$ZnO(s) + 2H^+ = Zn^{2+} + H_2O \qquad \log {}^*K_{s0} = 11.2 \qquad \text{(v)}$$

$$Zn^{2+} + OH^- = ZnOH^+ \qquad \log K_1 = 5.0 \qquad \text{(vi)}$$

$$H_2O = H^+ + OH^- \qquad \log K_w = -14 \qquad \text{(vii)}$$

$$\overline{ZnO(s) + H^+ = ZnOH^+ \qquad \log {}^*K_{s1} = 2.2 \qquad \text{(ii)}}$$

In accordance with equations i to iv the lines characterizing the logarithmic concentrations of Zn^{2+}, $Zn(OH)^+$, $Zn(OH)_3^-$, and $Zn(OH)_4^{2-}$ as a function of pH have slopes of -2, -1, $+1$, and $+2$, respectively. The intercepts are also defined readily; for example, for $Zn(OH)^+$, $\log[ZnOH^+] = 0$ when pH = $-p*K_{s1}$ (equation ii); or $\log[ZnOH^+] = \log[Zn^{2+}]$ when pH = $p*K_1$ (equation 22). Summing up all the soluble zinc species gives the line that surrounds the shaded area. Numerically, $[Zn(II)_T]$ is given by

$$Zn(II)_T - *K_{s0}[H^+]^2 + *K_{s1}[H^+] + *K_{s3}[H^+]^{-1} + *K_{s4}[H^+]^{-2}. \quad (24)$$

As Figure 5.4 illustrates in a general way, solid oxides and hydroxides have amphoteric characteristics; they can react with protons or hydroxide ions. There is a pH value for which the solubility is at a minimum.† In more alkaline or more acidic pH regions, the solubility becomes greater.

At this point it is necessary to remind ourselves that the solubility predictions based on solubility products given in Table 5.1 or depicted in Figures 5.1 and 5.3 are valid only for solutions in which, under the appropriate condition, these metal oxides or hydroxides are thermodynamically stable or at least metastable. The affinity of ligands other than OH^- (i.e., S^{2-}, CO_3^{2-}, PO_4^{3-}) present in natural waters may preclude the existence of stable oxides or hydroxides (Section 5.4).

5.3 CARBONATES

In the $Me^{2+}-H_2O-CO_2$ system in the presence of the earth's atmosphere carbonates are frequently more stable than oxides or hydroxides as solid phases. Thus in natural water systems the concentration of some metal ions is controlled by the solubility of metal carbonates. The equilibrium constants used to characterize solubility equilibria of carbonates are summarized in Table 5.2. The various solubility expressions [(6) to (11), Table 5.2] are interrelated and can all be expressed in terms of the conventional solubility product K_{s0}. A listing of the different formulations should indicate merely that the solubility can be characterized by different experimental variables. For example, we can fully define a solubility equilibrium with a solid carbonate by p_{CO_2}, $[Me^{2+}]$, and $[H^+]$ [(10), Table 5.2]; by p_{CO_2}, $[Me^{2+}]$, and $[HCO_3^-]$ [(9), Table 5.2] or by $[H^+]$, $[Me^{2+}]$, and $[HCO_3^-]$ [(7), Table 5.2]. Parameters such as these are more accessible to direct analytical determination than $[CO_3^{2-}]$. Table 5.1 lists solubility products for a few metal carbonates.

In treating heterogeneous equilibria of the system $Me^{2+}-CO_2-H_2O$ it is important to distinguish two cases: (a) systems that are closed to the atmosphere (we consider only the solid phase and the solution phase; that is, we treat $H_2CO_3^*$ as a nonvolatile acid) and (b) systems that include a (CO_2-containing) gas phase in addition to the solid and solution phase. For each of the two cases

---

† Solubilities near this minimum usually cannot be determined experimentally. They are obtained by extrapolating from measurements in pH regions where solubilities are higher.

TABLE 5.2   EQUILIBRIUM CONSTANTS DEFINING THE SOLUBILITY OF CARBONATES

I. Equilibrium Among Solutes and $CO_2(g)$

$H_2O = H^+ + OH^-$	K_W	(1)
$CO_2(g) + H_2O = H_2CO_3^*(aq)$	K_H	(2)
$CO_2(g) + H_2O = HCO_3^- + H^+$	$K_{p1} = K_H K_1$	(3)
$H_2CO_3^* = HCO_3^- + H^+$	K_1	(4)
$HCO_3^- = CO_3^{2-} + H^+$	K_2	(5)

II. Solid-Solutions Equilibria

$MeCO_3(s) = Me^{2+} + CO_3^{2-}$	K_{s0}	(6)
$MeCO_3(s) + H^+ = Me^{2+} + HCO_3^-$	$*K_s = K_{s0}K_2^{-1}$	(7)
$MeCO_3(s) + H_2CO_3^* = Me^{2+} + 2HCO_3^-$	$^+K_{s0} = K_{s0}K_1K_2^{-1}$	(8)
$MeCO_3(s) + H_2O + CO_2(g) = Me^{2+} + 2HCO_3^-$	$^+K_{ps0} = {}^+K_{s0}K_H$	(9)
$MeCO_3(s) + 2H^+ = Me^{2+} + CO_2(g) + H_2O$	$*K_{ps0} = K_{s0}K_2^{-1}K_{p1}^{-1}$	(10)
$MeCO_3(s) + 2H^+ = Me^{2+} + H_2CO_3^*$	$*K_{s0} = K_{s0}K_2^{-1}K_1^{-1}$	(11)

a few representative models will be discussed (Table 5.3). Because of its significance in natural water systems, calcite will be used as an example of the solid phase.

Model I: System Closed to Atmosphere

(a) *Solubility of $CaCO_3$ for $C_T = Constant$.* What is the maximum soluble metal-ion concentration as a function of C_T and pH? This example is analytically quite important. We have a water of a given analytical composition, and we inquire whether the water is oversaturated or undersaturated with respect to a solid metal carbonate. In other words, we compute the equilibrium solubility. Numerically, for any pH and C_T, the equilibrium solubility must be maintained. In the case of calcite,

$$[Ca^{2+}] = \frac{K_{s0}}{[CO_3^{2-}]} = \frac{K_{s0}}{C_T \alpha_2} \tag{25}$$

Because α_2 is known for any pH, equation 25 gives the equilibrium saturation value of Ca^{2+} as a function of C_T and pH. An analogous type of equation can be written for any $[Me^{2+}]$ in equilibrium with $MeCO_3(s)$. Equation (25) is amenable to simple graphical representation in a $\log[Me^{2+}]$ versus pH diagram (Figure 5.5).

The graphical representation consists essentially of a superimposition of the equation for the solubility product upon the carbonate equilibria (cf. Figure 4.1); the product of $[Ca^{2+}]$ and $[CO_3^{2-}]$ must be constant (K_{s0}) (Figure 5.5, inset). Thus at high pH (pH > pK_2), where the $\log[CO_3^{2-}]$ line has a slope of

TABLE 5.3 SOLUBILITY EQUILIBRIA IN THE SYSTEM CaO–CO_2–H_2O

Components in Addition to CaO, CO_2 and H_2O	Selected Variables Necessary to Define System in Addition to Pressure and Temperature	Equation Defining Composition	Composition, $P = 1$ atm, $25°C$
Model 1 System Closed to Atmosphere (Phases: Calcite and Aqueous Solution)			
(a) Acid or base	$C_T = $ constant	(25)	Figure 5.4
(b) 0	$C_T = [Ca^{2+}]$	(30)	$pH = 9.9, pCa = 3.9, pHCO_3^- = 4.05,$ $pCO_3^{2-} = 4.4, pAlk = 3.62$
(c) Acid or base	$C_T = [Ca^{2+}]$ $C_A - C_B$	(31)	Figure 5.5
Model 2 System in Equilibrium with Gas Phase (Phases: Calcite, Aqueous Solution and $CO_2(g)$)			
(a) 0	$p_{CO_2}{}^a$	(39)	$pH = 8.4, pCa = 3.3, pHCO_3^- = 3.0,$ $pCO_3^{2-} = 5.0, pH_2CO_3^* = 5.0, pAlk = 3.0$
(b) Acid or base	$C_A - C_B$	(40)	Figure 5.6

^{a} p_{CO_2} is selected as a variable, in addition to total pressure.

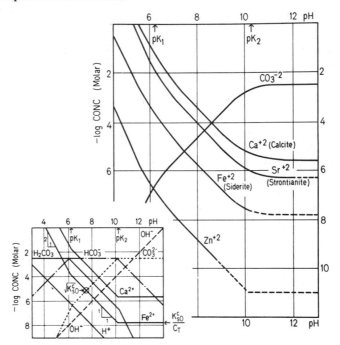

Figure 5.5 Model Ia. Solubility of $MeCO_3(s)$ for a closed system with C_T = constant $(= 3 \times 10^{-3} M)$. The diagram gives the maximum soluble $[Me^{+2}]$ as a function of pH for a given C_T. Dashed portions of the curves indicate conditions under which $MeCO_3(s)$ is not thermodynamically stable. The inset gives the essential features for the construction of the diagram.

zero, the $\log[Ca^{2+}]$ line must also have a slope of zero. Here the saturation concentration $[Ca^{2+}]$ must be equal to $K_{s0}/[CO_3^{2-}]$. In the region where $pK_1 < pH < pK_2$, $\log[CO_3^{2-}]$ has a slope of $+1$; correspondingly, $\log[Ca^{2+}]$ must have a slope of -1. At $pH < pK_1$, $\log[CO_3^{2-}]$ has a slope of $+2$; in order to maintain constancy of the product, $[Ca^{2+}][CO_3^{2-}]$, $\log[Ca^{2+}]$ must have a slope of -2. Figure 5.5 shows the equilibrium saturation values of a few metal ions with respect to their carbonates for $C_T = 10^{-2.5} M$. Such a diagram is well suited for comparing the solubility of various metal carbonates and their pH dependence. The relations characterized by equation 25 are valid only for conditions (pH, C_T, temp.) under which the corresponding solid metal carbonates are stable or metastable. Criteria for determining the most stable solid phase will be discussed in Section 5.4.

(b) Dissolution of $CaCO_3(s)$ in Pure Water. The following solute species will be encountered: Ca^{2+}, $H_2CO_3^*$, HCO_3^-, CO_3^{2-}, H^+, and OH^-. Since we have six unknowns, at a given pressure and temperature, we need six equations to define the solution composition. Four mass laws interrelate the equilibrium concentrations of the solutes [first and second acidity constant of $H_2CO_3^*$, ion product of water, and solubility product of $CaCO_3(s)$]. An additional equation obtains if one considers that all Ca^{2+} that becomes dissolved must equal in concentration the sum of the dissolved carbonic species

$$[Ca^{2+}] = C_T \tag{26}$$

Furthermore, the solution must fulfill the condition of electroneutrality:

$$2[Ca^{2+}] + [H^+] = [HCO_3^-] + 2[CO_3^{2-}] + [OH^-] \tag{27}$$

The set of six equations has to be solved simultaneously. It may be convenient to start with the solubility product

$$[Ca^{2+}] = \frac{K_{s0}}{[CO_3^{2-}]} \tag{28}$$

Using ionization fractions (Table 4.2),

$$[CO_3^{2-}] = C_T \alpha_2 \qquad [HCO_3^-] = C_T \alpha_1 \qquad [H_2CO_3^*] = C_T \alpha_0$$

Equation 28, rewritten as $[Ca^{2+}] = K_{s0}/C_T \alpha_2$, can be combined with equation 26 to give

$$[Ca^{2+}] = C_T = \left(\frac{K_{s0}}{\alpha_2}\right)^{0.5} \tag{29}$$

This can be substituted into the charge condition equation 27:

$$\left(\frac{K_{s0}}{\alpha_2}\right)^{0.5} (2 - \alpha_1 - 2\alpha_2) + [H^+] - \frac{K_w}{[H^+]} = 0 \tag{30}$$

Equation 30 can be solved by trial and error. For calcite, the result is given in Table 5.3.

(c) *CaCO$_3$ plus Acid or Base.* The addition of acid C_A or base C_B will change the pH and the solubility relations. The problem is analogous to the alkalimetric or acidimetric titration of a $CaCO_3$ suspension. The addition of acid (i.e., $C_A = [HCl]$) or base ($C_B = [NaOH]$) will shift the charge balance (equation 30) to the following:

$$C_A - C_B = \left(\frac{K_{s0}}{\alpha_2}\right)^{0.5} (2 - \alpha_1 - 2\alpha_2) + [H^+] - \frac{K_w}{[H^+]} \tag{31}$$

With the help of this equation it is possible to compute the quantity of acid or base, C_A or C_B, needed per liter of solution [which remains in equilibrium with $CaCO_3(s)$] to reach a given pH value (Figure 5.6). For each pH value, the solubility of $CaCO_3(s)$, for example, $[Ca^{2+}]$ or C_T, can be calculated with

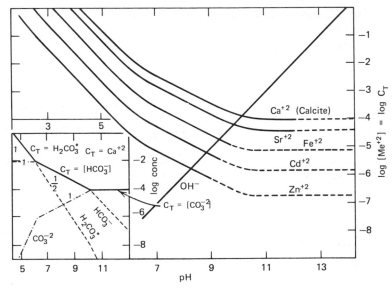

Figure 5.6 Models Ib and Ic. Solubility of metal carbonates in a closed system: $[Me^{2+}] = C_T$. The inset gives the essential features for the construction of the diagram for $CaCO_3(s)$ and equilibrium concentrations of all the carbonate species. A suspension of pure $MeCO_3(s)(C_B - C_A = 0)$ is characterized by the intersection of $[OH^-]$ and $[Me^{2+}] = C_T$. Dashed portions of the curves indicate conditions under which $MeCO_3(s)$ is not thermodynamically stable.

equation 29, which is still valid under our assumptions. Then the equilibrium concentrations of the other solutes can be readily obtained.

It is convenient to plot equation 29 graphically in a logarithmic concentration–pH diagram. The construction of the plot is facilitated by considering that the following conditions exist in the various pH regions:

for $pH > pK_2$:

$$d \log \frac{\alpha_2}{d\mathrm{pH}} = 0 \qquad d \log \frac{C_T}{d\mathrm{pH}} = 0 \qquad -\log C_T = \tfrac{1}{2}pK_{s0} \qquad (32)$$

for $pK_1 < pH < pK_2$:

$$d \log \frac{\alpha_2}{d\mathrm{pH}} = +1 \qquad d \log \frac{C_T}{d\mathrm{pH}} = -\tfrac{1}{2} \qquad (33)$$

for $pH < pK_1$:

$$d \log \frac{\alpha_2}{d\mathrm{pH}} = +2 \qquad d \log \frac{C_T}{d\mathrm{pH}} = -1 \qquad (34)$$

Figure 5.6 gives solubility diagrams for some other metal carbonates.

Model II: System in Equilibrium with CO_2 (g)

(a) $CaCO_3(s)$–CO_2–H_2O. We open the system previously discussed to the atmosphere. Specifically, we prepare a solution by adding $CaCO_3(s)$ (calcite) to pure H_2O and expose this solution to a gas phase containing CO_2. This model is representative of conditions encountered typically in fresh water. For our example we select a partial pressure of CO_2 corresponding to that of the atmosphere ($-\log p_{CO_2} = 3.5$). We can write, because of the additional phase, in addition to the four mass laws, an independent relationship for the solubility of CO_2 in the aqueous solution (e.g., Henry's law). If we specify the temperature, total pressure and a partial pressure of CO_2, the system is completely defined. Furthermore, an electroneutrality condition can be formulated.

Because of the equilibrium with CO_2 in the gas phase, $[Ca^{2+}]$ is no longer equal to C_T, but the same electroneutrality condition (equation 27) pertains:

$$2[Ca^{2+}] + [H^+] = C_T(\alpha_1 + 2\alpha_2) + [OH^-] \tag{35}$$

Furthermore, $[Ca^{2+}]$ can be expressed as a function of C_T and $[H^+]$, and C_T is defined by p_{CO_2};

$$[Ca^{2+}] = \frac{K_{s0}}{C_T \alpha_2} \tag{36}$$

$$C_T = \frac{K_H p_{CO_2}}{\alpha_0} \tag{37}$$

$$[CO_3^{2-}] = \frac{K_H p_{CO_2} \alpha_2}{\alpha_0} \tag{38}$$

$$[Ca^{2+}] = \frac{(K_{s0}/K_H p_{CO_2})\alpha_0}{\alpha_2} \tag{39}$$

With the substitution of equations 37 and 39, equation 35 can now be solved for $[H^+]$. The result is given in Table 5.3. Comparing this result with that of Model 1b, we see that the influence of atmospheric CO_2 has depressed the pH markedly and that $[Ca^{2+}]$ and $[Alk]$ have been raised to values very representative of those in natural waters.

(b) $CaCO_3$–H_2O–CO_2 plus Acid or Base. pH changes can occur upon addition of acid or bases (e.g., addition of wastes, dissolution of volcanic volatile compounds like HCl, biological reactions) to the model discussed before. The electroneutrality condition can be formulated most generally by considering that the charge balance of equation 35 has been shifted because of the addition of C_A (acid) or C_B (base):

$$C_B + 2[Ca^{2+}] + [H^+] = C_T(\alpha_1 + 2\alpha_2) + [OH^-] + C_A \tag{40}$$

Since $[Ca^{2+}]$ and C_T in this equation can again be expressed as a function of p_{CO_2} and $[H^+]$, the "titration curve" of a $CaCO_3$ suspension that remains in equilibrium with $CO_2(g)$ can be computed.

Equation (39) can be plotted graphically (Figure 5.7). The distribution of the dissolved carbonate species at a given temperature is defined entirely by p_{CO_2}. Hence, Figure 5.7 is essentially the superposition of equation 40 upon Figure

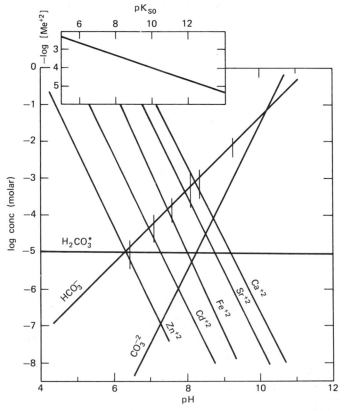

Figure 5.7 Models IIa and IIb. Solubility of $MeCO_3(s)$ as a function of pH at constant p_{CO_2}. Here $-\log p_{CO_2} = 3.5$ (corresponding to partial pressure of CO_2 in atmosphere). If no excess acid or base is added ($C_B - C_A = 0$), the equilibrium composition of the solution is given by the electroneutrality condition $2[Me^{2+}] \simeq [HCO_3^-]$. This condition is indicated by a vertical dash slightly displaced from the intersection of $[Me^{2+}]$ with $[HCO_3^-]$. The inset gives $-\log[Me^{2+}]$ for pure $MeCO_3(s)$ suspensions in equilibrium with $p_{CO_2} = 10^{-3.5}$ atm as a function of pK_{s0}.

4.3. The $\log[Ca^{2+}]$ line has a slope of -2 with respect to pH. Differentiating equation 39 with respect to pH and considering that $d\log(\alpha_0/\alpha_2)/pH = -2$ leads to a slope of -2 with respect to pH for the $\log[Ca^{2+}]$ line in the diagram.

Example 5.3a. Calcite in Seawater

Compare the composition of a $CaCO_3(s)$ (calcite)–CO_2–H_2O "seawater" model system, made by adding calcite to pure H_2O containing the seawater electrolytes (but incipiently no Ca^{2+} and no carbonates and, for simplicity, no borate) and by equilibrating this solution at $25°C$ and 1 atm total pressure with the atmosphere ($p_{CO_2} = 3.3 \times 10^{-4}$ atm), with the composition of a real surface seawater whose carbonate alkalinity, Ca(II) concentration, and pH have been determined as 2.4×10^{-3} eq liter^{-1}, 1.06×10^{-2} M, and 8.2, respectively. Estimate the extent of oversaturation of this seawater with respect to calcite. The solubility of calcite at $25°C$ is given by $^cK'_{s0} = [Ca_T][CO_{3_T}^{2-}] = 5.94 \times 10^{-7}$, where $[Ca_T]$ and $[CO_{3_T}^{2-}]$ are the concentration of total soluble Ca(II) ($[Ca^{2+}]$ plus concentration of Ca complexes with medium ions) and of total soluble carbonate ($[CO_3^{2-}]$ and concentration of carbonate complexes with medium ions), respectively (Table 5.1, III). The other constants needed, Henry's law constant and the acidity constant of $H_2CO_3^*$, are taken from Tables 4.7 to 4.9: $p^cK_H = 1.53$; $pK'_1 = 6.00$; $pK'_2 = 9.11$. The model system is characterized by the electroneutrality (equation 35)

$$2[Ca^{2+}] + [H^+] = C_T(\alpha_1 + 2\alpha_2) + [OH^-] \qquad (i)$$

where in accordance with equations 36 to 39 the following substitutions can be made:

$$[Ca_T] = \frac{K'_{s0}}{C_T\alpha_2} \qquad (ii)$$

$$C_T = \frac{^cK_H\, p_{CO_2}}{\alpha_0} \qquad (iii)$$

$$[CO_{3_T}^{2-}] = \frac{^cK_H\, p_{CO_2}\alpha_2}{\alpha_0} \qquad (iv)$$

$$[Ca_T] = \frac{(^cK'_{s0}/K_H\, p_{CO_2})\alpha_0}{\alpha_2} \qquad (v)$$

Equation i is then solved by trial and error. It is convenient to plot the various α values for a set of $\{H^+\}$ values (or to calculate these α values with the help of a programmable calculator) (cf. equations iv and v). We obtain pH = 8.36. For this pH, $[Ca_T] = 1.50 \times 10^{-3}$ M, $[CO_{3_T}^{2-}] = 3.95 \times 10^{-4}$ M, $[HCO_{3_T}^-] = 3.13 \times 10^{-3}$ M, and $[Alk] = 3.92 \times 10^{-3}$.

For surface seawater with pH $= 8.2$ and $[\text{Carb-Alk}] = 2.4 \times 10^{-3}$ eq liter^{-1}, we obtain for $[CO_{3_T}^{2-}] = C_T\alpha_2 = [\text{Carb-Alk}]\alpha_2/(\alpha_1 + 2\alpha_2) = 3.87 \times 10^{-4}$ M. Since $[Ca_T] = 1.06 \times 10^{-2}$ M, we obtain for the ion concentration product

$$[Ca_T]_{\text{act}}[CO_{3_T}^{2-}]_{\text{act}} = 4.1 \times 10^{-6}$$

This may be compared with the equilibrium solubility product K'_{s0}

$$\frac{[Ca_T]_{\text{act}}[CO_{3_T}^{2-}]_{\text{act}}}{[Ca_T]_{\text{eq}}[CO_{3_T}^{2-}]_{\text{eq}}} = \frac{[Ca_T][CO_{3_T}^{2-}]}{{}^cK'_{s0}} = 6.9 \tag{vi}$$

Thus the surface seawater, at $25°C$ ($p = 1$ atm), is oversaturated by a factor of ca. 7 with respect to calcite. The model equilibrium system has a slightly higher pH value and a $[Ca_T]$ approximately seven times smaller than that of the surface seawater.

Example 5.3b. Effect of Pressure and Temperature on Calcite Solubility in Seawater

An enclosed sample of the surface seawater, as discussed in Example 5.3a ($25°C$; pH $= 8.2$; $[Ca^{2+}] = 1.06 \times 10^{-2}$ M; $[\text{Carb-Alk}] = 2.4 \times 10^{-3}$ eq liter^{-1}), is cooled to $5°C$ and then subjected to increases in total pressure of up to 1000 atm (equivalent to exposing the sample to increased water depths of approximately 10,000 m). How does the composition, pH, $[CO_3^{2-}]$, $[Ca^{2+}]$, and extent of oversaturation change as a result of the temperature change at 1 atm and as a result of the pressure change at $5°C$? The water is incipiently oversaturated with respect to calcite. Assume that $CaCO_3$ does not precipitate or dissolve and that the presence of borate does not affect the calculation significantly.

$[Ca_T^{2+}]$, C_T, and $[\text{Carb-Alk}]$ remain constant and independent of pressure and temperature.† At any pressure and temperature the following relationship must hold:

$$[\text{Carb-Alk}] = C_T(\alpha_1 + 2\alpha_2) \tag{i}$$

α_1 and α_2 are calculated with K'_1 and K'_2 values valid for seawater and corrected for pressure; Tables 4.9 and 4.10 also give the equations for pressure dependence of these constants:

$$\log \frac{(K'_i)_P}{(K'_i)_{P=1}} = \frac{-\Delta V'_i(P-1)}{2.303RT} \tag{ii}$$

---

† This is rigorous only if concentrations are expressed in mol kg^{-1} (i.e., in molal units). The mixed acidity constants K'_1 and K'_2 (Tables 4.9 and 4.10) are the same whether molar or molal concentration units are used. ${}^cK_{s0}$ for calcite is given in molar units. We could correct by considering the density of the seawater at any t and p, but the correction is much smaller ($<3\%$) than the uncertainty in the ${}^cK'_{s0}$ value.

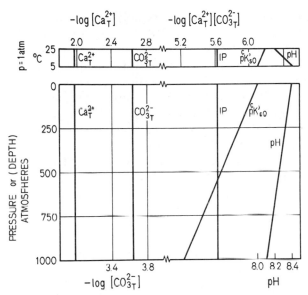

Figure 5.8 Effect of temperature and pressure upon composition and extent of calcite over-saturation of an enclosed seawater sample. IP = Ion product ($[Ca_T^{2+}][CO_3^{2-}]$). Initially the seawater, corresponding to a "typical" surface seawater, has the following composition at 25°C: pH = 8.2, $C_T = 2.18 \times 10^{-3}$ M, [Carb-Alk] = 2.4 × 10⁻³ eq liter⁻¹, $[Ca_T^{2+}] = 1.06 \times 10^{-2}$ M. See Example 5.3b. In real seawater the undersaturation with respect to calcite is due to both a decrease in pK_{s0} caused by increased pressure and a decrease in IP because of biologically mediated CO_2 production. The calculation also illustrates that the *in situ* pH of deep seawater can be estimated from the pH measured in samples brought to the surface: pH ≈ $pH_S - 3 \times 10^{-4}$ P, where pH_S = pH measured on a sample brought to the surface from a depth of pressure P. C. Culberson and R. M. Pytkowicz [*Limnol. Oceanogr.*, **13**, 403 (1968)] found in laboratory pressure measurements a pressure coefficient of 4 × 10⁻⁴ pH units atm⁻¹.

where $\Delta V_i'(pK_i') = -(24.2 - 0.085t)$ cm³ mol⁻¹ and $\Delta V_2'(pK_2) = -(16.4 - 0.040t)$ cm³ mol⁻¹ and where $t = $ °C, $T = $ K, and $R = 82.05$ cm³ atm mol⁻¹ deg⁻¹.

pH values compatible with a given C_T and [Carb-Alk] can be computed from equation i; then $[CO_3^{2-}] = C_T \alpha_2$ is calculated. The ion product, $[Ca_T^{2+}]$ $[CO_3^{2-}]$, can then be compared with $^cK_{s0}'$ of calcite at 25°C and at 5°C; its pressure dependence may be obtained from Gieskes' summarizing equation (Table 5.1, III). $\Delta V_{calcite}'$ in equation ii is given as $\Delta V_{calcite}'(p^cK_{s0}') = -(47.5 - 0.23t)$ cm³ mol⁻¹. With Berner's values for $^cK_{s0}'$, the results are given in Figure 5.8.

Temperature and pressure have a pronounced effect on pH but cause little variation in $[CO_3^{2-}]$.† Since $[Ca_T^{2+}]$ is conservative, the ion product $[Ca_T^{2+}]$

---

† $[CO_3^{2-}]$ is relatively insensitive to pressure and temperature, because at the pH of seawater [Carb-Alk]/$C_T \approx \alpha_1 \cong$ constant. Thus $[CO_3^{2-}]/[HCO_3^{-}] = \alpha_2/\alpha_1 = K_2'/\{H^+\} \approx$ constant; therefore the effect of P and t upon K' becomes largely reflected in changes in $\{H^+\}$.

$[CO_{3_T}^{2-}]$ also does not change appreciably. The extent of oversaturation, however, charges markedly because $p^c K'_{s0}$ (calcite) decreases strongly with both temperature and with pressure.

This calculation illustrates that a decrease in temperature and an increase in pressure increase the calcite solubility. In the real ocean the C_T and [Alk] change with depth. Photosynthesis in the upper layers leads to a consumption of CO_2. The organic matter generated by photosynthesis leads, after settling through the water column, to a consumption of O_2 and to an increase in C_T. Some of the $CaCO_3$ formed in the surface waters becomes dissolved, thus increasing C_T and [Alk], in the deeper waters. The extent of over- and under-saturation is influenced by kinetic factors (inhibition of nucleation and crystal growth of calcite in surface waters and retardation of $CaCO_3$ dissolution in deep waters) (Section 5.9).

Example 5.4a. Reaction Paths for Calcite Dissolution in Groundwaters

The chemical composition of a newly formed groundwater is initially determined by rainwater (sometimes also by river water) that becomes exposed to increased partial pressure of CO_2 (from the microbially mediated oxidation of organic matter in the soil horizon) after infiltration into the soil. The CO_2-enriched water dissolves minerals such as aluminum silicates, $CaCO_3$, and $CaMg(CO_3)_2$.

How does the groundwater composition change during the dissolution of $CaCO_3$ (calcite)? We assume that $CaCO_3$ is the only mineral being dissolved. The temperature is 10°C.

Two idealized cases may be distinguished (R. M. Garrels and C. L. Christ, *Solutions, Minerals and Equilibria*, Harper and Row, New York, 1965): (i) During the process of $CaCO_3$ dissolution, the water remains in contact and equilibrium with a relatively large reservoir of CO_2 of fixed partial pressure. (ii) An initially CO_2-rich water becomes isolated from the $CO_2(g)$ reservoir. We calculate for both cases the reaction progress—the change in composition as a function of the extent of the reaction ($CaCO_3$ dissolution)—for a few selected initial conditions. Equilibrium constants valid at 10°C and corrected for an ionic strength of $I = 4 \times 10^{-3}$ M† are: $-\log K_H = 1.27$; $-\log K'_1 = 6.43$; $-\log K'_2 = 10.38$, $-\log K'_{s0(\text{calcite})} = 7.95$.

(i) Reservoir with Constant p_{CO_2}. During dissolution of $CaCO_3$, conditions characterizing an aqueous carbonate system open to the atmosphere (equations 5 to 8 in Chapter 4) prevail:

$$[HCO_3^-] = \frac{\alpha_1}{\alpha_2} K_H p_{CO_2} \tag{i}$$

For every p_{CO_2}, the linear relationship between $[HCO_3^-]$ and pH can be plotted (Figure 5.9; cf. Figure 4.3). The extent of $CaCO_3$ dissolution is equivalent to the increase in

$$\Delta[Ca^{2+}] = \tfrac{1}{2}\Delta[\text{Alk}] \cong \tfrac{1}{2}\Delta[HCO_3^-] \tag{ii}$$

† For more exact calculations, the ionic strength needs to be calculated iteratively for every reaction step.

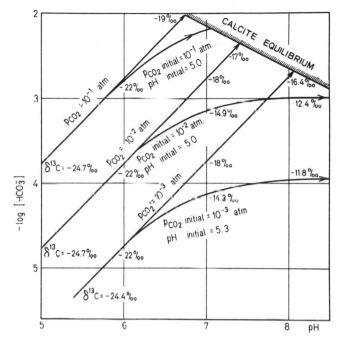

Figure 5.9 Dissolution paths of $CaCO_3$ (calcite) (Example 5.4*a* and *b*). Two idealized cases are distinguished: (i) dissolution of $CaCO_3$ in systems with a reservoir of constant p_{CO_2} (the diagonal straight lines represent dissolution paths); (ii) dissolution of $CaCO_3$ in a system (of different initial conditions) that becomes enclosed (dissolution paths are given by the curved lines). Calculated $\delta^{13}C$ values of the dissolved carbon (with reference to the BDD standard) are given, assuming $\delta^{13}C = -25\%_0$ for the CO_2 reservoir and $\delta^{13}C = 1\%_0$ for the $CaCO_3(s)$. [A similar figure has been given by Deines et al., *Geochim. Cosmochim. Acta*, **38**, 1147 (1974).]

That is, the solution paths are along the diagonal lines in Figure 5.9. Solubility equilibrium is attained where condition iv is attained:

$$CaCO_3(s)(\text{calcite}) + H^+ = Ca^{2+} + HCO_3^- \qquad (iii)$$

$$*K_s' = \frac{K_{s0}'}{K_2'} = \frac{[Ca^{2+}][HCO_3^-]}{\{H^+\}} \qquad (iv)$$

If [Alk] results only from $CaCO_3$ dissolution,

$$2[Ca^{2+}] = [HCO_3^-] \qquad (v)$$

Equation iv can be rewritten as

$$[HCO_3^-] = (*K_s' \times 2\{H^+\})^{1/2} \qquad (vi)$$

(ii) Enclosed System. When the infiltrated CO_2-enriched water becomes separated from the CO_2 reservoir, the dissolved carbon dioxide, $H_2CO_3^*$, reacts with $CaCO_3(s)$ in a closed system:

$$H_2CO_3^* + CaCO_3(s) = Ca^{2+} + 2HCO_3^- \qquad (vii)$$

The extent of $CaCO_3$ dissolution is given by equation ii. The acidity of the system no longer changes with $CaCO_3(s)$ dissolution (Section 4.4); thus the acidity, [Acy], acquired at the time of separation from the CO_2 reservoir, remains constant:

$$[Acy] = 2[H_2CO_3^*] + [HCO_3^-] + [H^+] - [OH^-] = \text{constant} \tag{viii}$$

$$[Acy] \cong C_T(2\alpha_0 + \alpha_1) \cong \text{constant} \tag{ix}$$

We compute [Acy] for the selected initial condition with

$$C_T = \frac{K_H p_{CO_2}}{\alpha_0} \tag{x}$$

(cf. equation 6 in Chapter 4) and then compute, with the help of the constraint of equation ix for selected pH values, values for C_T and $[HCO_3^-] = C_T\alpha_1$. Results are plotted in Figure 5.9. Reaction progress toward solubility equilibrium follows upward along the curved lines in the $-\log[HCO_3^-]$ versus pH plots. The boundary for $CaCO_3(s)$ (calcite) saturation is given by equation iv.

Example 5.4b. Carbon-13 Isotopes as Indicators of the Existence of a Gas Phase in the Evolution of Groundwaters

P. Deines, D. Langmuir and R. Harmon [*Geochim. Cosmochim. Acta*, **38**, 1147 (1974)] have illustrated that the ^{13}C content of dissolved carbonate species can aid our understanding of the evolution of a carbonate groundwater (see Section 4.9). For the two cases discussed above and illustrated in Figure 5.9, the ^{13}C content of the waters can be computed considering appropriate values of $\delta^{13}C$ in the CO_2 reservoir and in calcite. We adopt for the CO_2 reservoir $\delta^{13}C_{CO_2(g)} = -25\%_0$, and for $CaCO_3(s)$, $\delta^{13}C_{CaCO_3} = +1\%_0$. The calculation of $\delta^{13}C$ for the open system (reservoir with constant p_{CO_2}) is the same as that described in Example 4.10:

$$\delta^{13}C_{sol} = \frac{([H_2CO_3^*]\delta^{13}C_{H_2CO_3^*} + [HCO_3^-]\delta^{13}C_{HCO_3^-} + [CO_3^{2-}]\delta^{13}C_{CO_3^{2-}})}{C_T} \tag{xi}$$

where $\delta^{13}C$ values for $H_2CO_3^*$, HCO_3^-, and CO_3^{2-} are computed from the equilibrium constants given in Example 4.10.

For the enclosed system the total ^{13}C content of the solution is composed of the ^{13}C content of the solution at the time of separation from the CO_2 reservoir and the ^{13}C content of the carbon resulting from the $CaCO_3$ dissolution:

$$\delta^{13}C_{sol} = C_{T, \text{initial}} \delta^{13}C_{\text{initial}} + C_{T, \text{from CaCO}_3} \delta^{13}C_{CaCO_3} \tag{xii}$$

where

$$C_{T, \text{from CaCO}_3} = \Delta[Ca^{2+}] = \tfrac{1}{2}\Delta[Alk] \cong \tfrac{1}{2}\Delta[HCO_3^-] \tag{xiii}$$

and where $\delta^{13}C_{\text{initial}}$ is calculated using equation xi.

For both cases of $CaCO_3$ dissolution representative values of $\delta^{13}C_{sol}$ are given in Figure 5.9. A comparison of these figures illustrates that ^{13}C information may aid in deciding whether carbonate rock dissolution occurs mainly under open- or closed-system conditions.

Fresh surface waters have typically $\delta^{13}C$ values between $-5\%_0$ and $-11\%_0$, and measurements of ^{13}C in groundwaters can also sometimes be used to evaluate the extent of river water infiltration.

Example 5.5. Solubility of Pure $MeCO_3(s)$ Suspensions

Derive an equation that shows how the solubility of various bivalent metal carbonates varies with their solubility product.

Consider the reaction

$$MeCO_3(s) + CO_2(g) + H_2O = Me^{2+} + 2HCO_3^- \qquad {}^+K_{ps0} \qquad (i)$$

We can represent its equilibrium constant by

$$ {}^+K_{ps0} = K_{s0}K_1K_HK_2{}^{-1} \qquad (ii)$$

This becomes evident from the addition of the following equilibrium reactions

$$MeCO_3(s) = Me^{2+} + CO_3^{2-} \qquad K_{s0} \qquad (iii)$$

$$CO_2(g) + H_2O = H_2CO_3^* \qquad K_H \qquad (iv)$$

$$H_2CO_3^* = H^+ + HCO_3^- \qquad K_1 \qquad (v)$$

$$CO_3^{2-} + H^+ = HCO_3 \qquad K_2^{-1} \qquad (vi)$$

The electroneutrality equation can be approximated by

$$2[Me^{2+}] \simeq [HCO_3^-] \qquad (vii)$$

which can now be substituted into the equilibrium expression of equation i:

$$\frac{[Me^{2+}][HCO_3^-]^2}{P_{CO_2}} \simeq \frac{4[Me^{2+}]^3}{P_{CO_2}} \simeq {}^+K_{ps0} \qquad (viii)$$

$$[Me^{2+}] \simeq 0.63\,{}^+K_{ps0}^{1/3} P_{CO_2}^{1/3} \qquad (ix)$$

Hence in a plot of $\log[Me^{2+}]$ versus $\log K_{s0}$ a slope of $\frac{1}{3}$ obtains (see Figure 5.7, inset).

Hydroxide Carbonates. Some metal ions form solid hydroxide carbonates. Examples are hydromagnesite $Mg_4(CO_3)_3(OH)_2 \cdot 3H_2O$, azurite $Cu_3(OH)_2(CO_3)_2$, malachite $Cu_2(OH)_2CO_3$, and hydrozincite $Zn_5(OH)_6(CO_3)_2$. In evaluating solubility characteristics of such hydroxide carbonates the same principles apply as were discussed before. For example, the solubility of hydrozincite can be characterized by the reaction

$$Zn_5(OH)_6(CO_3)_2(s) = 5Zn^{2+} + 2CO_3^{2-} + 6OH^- \qquad (41)$$

where

$$K_{s0} = \{Zn^{2+}\}^5\{OH^-\}^6\{CO_3^{2-}\}^2 \qquad \log K_{s0} = -68.2 \qquad (I = 0.1, 25°C, p = 1 \text{ atm})$$
$$(42)$$

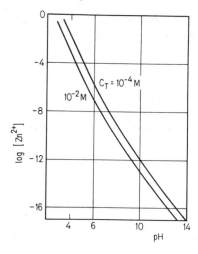

Figure 5.10 $[Zn^{2+}]$ in equilibrium with hydrozincite (25°C, $I = 4 \times 10^{-3}$, $p^cK_{s0} = 67$). Graphical representation of equation 43. (Hydrozincite is not stable over the entire pH range shown; its solubility is also determined by hydroxo Zn complexes [cf. Figure 5.15]).

$[Zn^{2+}]$ in equilibrium with hydrozincite is given by

$$[Zn^{2+}] = \frac{{}^cK_{s0}^{1/5}[H^+]^{6/5}}{C_T^{2/5}\alpha^{2/5}K_w^{6/5}} \tag{43}$$

Figure 5.10 plots equation 43. The slopes of the $\log[Zn^{2+}]$ versus pH line are -2, $-\frac{8}{5}$, and $-\frac{6}{5}$ for pH $<$ pK_1, p$K_1 <$ pH $<$ pK_2, and pH $>$ pK_2, respectively.

5.4 THE STABILITY OF HYDROXIDES, CARBONATES, AND HYDROXIDE CARBONATES

In the previous sections we have applied equilibrium constants for heterogeneous equilibria in a rather formal way. Thermodynamically meaningful conclusions are justified only if, under the specified conditions (concentration, pH, temperature, pressure), the solutes are in equilibrium with the solid phase for which the mass-law relationship has been formulated or if under the specified conditions the assumed solid phase is really stable or at least metastable. It remains to be illustrated how we can establish which phase predominates under a given set of conditions.

Which Solid Phase Controls the Solubility of Fe(II)?

In order to exemplify how to find out which solid predominates as a stable phase for selected conditions, we may consider the solubility of Fe(II) in a carbonate-bearing water of low redox potential ($P = 1$ atm, 25°C, $I = 6 \times$

10^{-3} M). We may first consider the solubility products of $Fe(OH)_2(s)$ and of $FeCO_3(s)$:

$$K_{s0}(Fe(OH)_2) \simeq 10^{-14.7} \text{ mol}^3 \text{ liter}^{-3}$$

$$K_{s0}(FeCO_3) \simeq 10^{-10.7} \text{ mol}^2 \text{ liter}^{-2}$$

The numerical values of these equilibrium constants cannot be compared directly. Note that the constants have different units. It would be incorrect and misleading to infer from the numerical values of the solubility products that $Fe(OH)_2(s)$ is less soluble than $FeCO_3(s)$. It is appropriate rather to inquire which solid controls the solubility for a given set of conditions (P, T, pH, C_T, [Alk], or p_{CO_2}), that is, gives the smallest concentration of soluble Fe(II).

Example 5.6.   Control of Solubility

Is it $FeCO_3(s)$ or $Fe(OH)_2(s)$ that controls the solubility of Fe(II) in anoxic water of $[Alk] = 10^{-4}$ eq liter^{-1} and pH $= 6.8$? We estimate maximum soluble [Fe(II)] by considering the solubility equilibrium (25°C) with (i) $FeCO_3(s)$ as well as that with (ii) $Fe(OH)_2(s)$.
(i) Assuming solubility equilibrium with $FeCO_3(s)$

$$FeCO_3(s) = Fe^{2+} + CO_3^{2-} \qquad \log {}^cK_{s0} \simeq -10.4 \qquad (i)$$

$$H^+ + CO_3^{2-} = HCO_3^- \qquad -\log {}^cK_2 \simeq +10.1 \qquad (ii)$$

$$FeCO_3(s) + H^+ = Fe^{2+} + HCO_3^- \qquad \log {}^{c*}K_s \simeq -0.3 \qquad (iii)$$

Correspondingly, $\log[Fe^{2+}] = \log {}^{c*}K_s - pH - \log[HCO_3^-]$; and since at this pH, $[Fe^{2+}] \simeq [Fe(II)]$ and $[HCO_3^-] \simeq [Alk]$, we obtain $\log[Fe(II)] = -3.1$.
(ii) Assuming solubility equilibrium with $Fe(OH)_2(s)$

$$Fe(OH)_2(s) = Fe^{2+} + 2OH^- \qquad \log {}^cK_{s0} = -14.5 \qquad (iv)$$

$$2H^+ + 2OH^- = 2H_2O \qquad -2\log {}^cK_w = +27.8 \qquad (v)$$

$$Fe(OH)_2(s) + 2H^+ = Fe^{2+} + 2H_2O \qquad \log {}^{c*}K_{s0} = +13.3 \qquad (vi)$$

Thus $\log[Fe^{2+}] = \log *K_{s0} - 2pH$; and $\log[Fe(II)] \simeq -0.3$. Because $[Fe^{2+}]$ (or Fe(II)]) is smaller for hypothetical equilibrium with $FeCO_3(s)$ than with $Fe(OH)_2$, siderite $[FeCO_3(s)]$ is more stable than $Fe(OH)_2(s)$.

Solubility, Predominance, and Activity Ratio Diagrams.   From thermodynamic information, diagrams can be constructed that circumscribe the stability boundaries of the solid phases. Depending on the variables used, different kinds of predominance diagrams can be constructed.

For systems closed to the atmosphere, a *solubility diagram* (e.g., $\log[Fe^{2+}]$, $\log[H^+]$, at fixed $\log C_T$) can conveniently illustrate the conditions under which a particular solid phase predominates. Figure 5.11 gives a solubility diagram for Fe(II) considering $FeCO_3(s)$ and $Fe(OH)_2(s)$ as possible solid phases. Construction of the diagram consists essentially of the superposition of a pH-dependent $Fe(OH)_2(s)$ solubility diagram (Figure 5.2) and a pH- and C_T-dependent $FeCO_3(s)$ solubility diagram (Figure 5.5). For a given $[H^+]$ and C_T the solid compound giving the smaller Fe^{2+} is more stable. Thus, for the conditions in Figure 5.11a, $FeCO_3(s)$ dictates the maximum concentration of Fe(II) below pH values of approximately 10, and $Fe(OH)_2(s)$ limits soluble iron above pH 10.

The same information can be gained from an *activity ratio diagram*. The construction is very simple and is illustrated in Figure 5-11b. We again choose pH as a master variable and make our calculation for a given C_T. In this figure we plot the ratios between the activities of the various soluble and solid species as a function of pH. In our case one of the species $\{Fe^{2+}\}$ (or $[Fe^{2+}]$) is chosen as a reference state. Thus the ordinate values are $\log\{A_i\}/\{Fe^{2+}\}$. Because

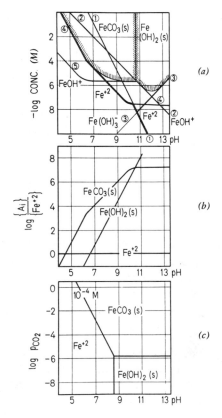

Figure 5.11   Stability of $Fe(OH)_2(s)$ and $FeCO_3(s)$ (siderite) ($25°C$; $I = 0$). With the help of such diagrams it is possible to evaluate the conditions (pH, C_T, $[Fe^{2+}]$, p_{CO_2}) under which a solid phase [$Fe(OH)_2(s)$ or $FeCO_3(s)$] is stable. (*a*) Solubility diagram of Fe(II) in a $C_T = 10^{-3}\,M$ carbonate system. The numbers on the curves refer to equations describing the respective equilibria in this system as follows: (1) $Fe(OH)_2(s) = Fe^{2+} + 2OH^-$, $K_{s0} = 2 \times 10^{-15}$; (2) $Fe(OH)_2(s) = [Fe(OH)]^+ + OH^-$, $K_{s1} = 4 \times 10^{-10}$; (3) $Fe(OH)_2(s) = [Fe(OH)_3]^-$, $K_{s3} = 8.3 \times 10^{-6}$; (4) $FeCO_3(s) = Fe^{2+} + CO_3^{2-}$, $K_{s0} = 2.1 \times 10^{-11}$; (5) $FeCO_3(s) + OH^- = [Fe(OH)]^+ + CO_3^{2-}$, $K_{s1} = 0.1 \times 10^{-5}$. (*b*) Activity ratio diagram for $C_T = 10^{-3}\,M$. The same conclusions as under (*a*) apply: Above pH $\simeq$ 10, $Fe(OH)_2(s)$ has the highest relative activity; it can precipitate as a pure phase. Below pH $\simeq$ 10, $FeCO_3$ becomes more stable than $Fe(OH)_2(s)$ and controls the solubility of Fe(II). (*c*) $\log p_{CO_2}$–pH predominance diagram. $FeCO_3(s)$ is the stable phase at p_{CO_2} larger than 10^{-6} atm, that is, in normal atmosphere.

the diagram gives activities on a relative scale (relative to $[Fe^{2+}]$) we treat the activities of solid phases formally in the same way as the activities (or concentrations) of the solutes. Two equations determine the ratios $\{Fe(OH)_2(s)/\{Fe^{2+}\}$ and $\{FeCO_3(s)\}/\{Fe^{2+}\}$. For the solubility of $Fe(OH)_2(s)$ we have

$$\frac{\{Fe^{2+}\}}{\{H^+\}^2\{Fe(OH)_2(s)\}} = {}^*K_{s0Fe(OH)_2} \tag{44}$$

Thus $\log(\{Fe(OH)_2(s)\}/\{Fe^{2+}\})$ plots as a function of pH as a straight line with a slope of $+2$:

$$\log \frac{\{Fe(OH)_2(s)\}}{\{Fe^{2+}\}} = p^*K_{s0Fe(OH)_2} + 2pH \tag{45}$$

and an intercept pH $- -\frac{1}{2}p^*K_{s0Fe(OH)_2}$.
  For the solubility of $FeCO_3$,

$$\frac{\{Fe^{2+}\}\{CO_3^{2-}\}}{\{FeCO_3(s)\}} = K_{s0FeCO_3} \tag{46}$$

Because $\{CO_3^{2-}\} = C_T\alpha_2$, equation 46 can be rearranged to

$$\log \frac{\{FeCO_3(s)\}}{\{Fe^{2+}\}} = pK_{s0FeCO_3} + \log C_T + \log \alpha_2 \tag{47}$$

Equation 47 can be plotted considering that $\log \alpha_2 = 0$ at $pH > pK_2$ and that $d(\log \alpha_2)/dpH$ is $+1$ or $+2$ in the pH regions $pK_1 < pH < pK_2$ and $pH < pK_1$, respectively.

At any pH the ordinate values on the activity ratio diagram give the activities for the various species on a relative scale. Thus in Figure 5.1 at pH $= 12$, $Fe(OH)_2(s)$ has the highest relative activity. This solid phase will precipitate at this pH; as a pure solid its activity will be unity, and $FeCO_3(s)$ must have an activity of much less than unity and cannot exist as a pure solid phase. Figure 5.11 shows that, for $C_T = 10^{-3} M$, $FeCO_3(s)$ is stable below ca. pH $= 10$.

For *open systems* it is convenient to select $\log p_{CO_2}$ and $-\log[H^+]$ as variables. An assumption must then be made about the concentrations of the solute. The computation of the straight lines in Figure 5.11 is based on the following equations. For the coexistence of Fe^{2+} and $Fe(OH)_2(s)$,

$$\log {}^*K_{s0Fe(OH)_2} - 2pH + pFe^{2+} = 0 \tag{48}$$

The thermodynamic coexistence of Fe^{2+} and $FeCO_3(s)$ as a function of pH and p_{CO_2} is expressed, perhaps most conveniently, by considering the solubility equilibrium in the form of the reaction

$$FeCO_3(s) + 2H^+ = Fe^{2+} + CO_2(g) + H_2O \qquad {}^*K_{ps0FeCO_3}$$

$$\log p_{CO_2} = \log {}^*K_{ps0FeCO_3} - 2pH + pFe^{2+} \tag{49}$$

The coexistence of $FeCO_3(s)$ and $Fe(OH)_2(s)$ is given by equilibrium

$$FeCO_3(s) + H_2O = Fe(OH)_2(s) + CO_2(g) \tag{50}$$

where (compare equations 48 and 49)

$$K = \frac{*K_{ps0\,FeCO_3}}{*K_{s0\,Fe(OH)_2}} = p_{CO_2} \tag{51}$$

The equilibrium partial pressure of CO_2 for equation 51 is approximately 10^{-6} atm. At p_{CO_2} higher than 10^{-6} atm, for example, in systems exposed to the atmosphere, $Fe(OH)_2$ is not stable and will be converted to $FeCO_3(s)$. In Figure 5.11 the enhancement of the solubility caused by the formation of $FeOH^+$ $(= 10^{-4}\ M)$ also has been considered.

Example 5.7. Mixing of Groundwaters

Two groundwaters and both exposed to and in equilibrium with an atmosphere exhibiting a p_{CO_2} of 10^{-2} atm. Groundwater I enters a stratum containing siderite and no calcite. Groundwater II enters a stratum containing only calcite. Solutions I and II, each having been in equilibrium with their respective solid phase, are intercepted by a well and thereby mixed in equal proportions. A problem of this type has been discussed by J. D. Hem (in *Principles and Applications of Water Chemistry*, S. D. Faust and J. V. Hunter, Eds., Wiley, New York, 1967).

1 Compute $[Me^{2+}]$, $[HCO_3^-]$, and $[H^+]$ for each type of groundwater.
2 Compute the composition of the mixture.
3 Is the mixture stable with respect to precipitation of $FeCO_3$? Assume $25°C$.

The composition of solutions I and II may be computed with the help of equations 35 and 39 or with equation ix of Example 5.5. Using the latter equation as a first approximation,

$$[Me^{2+}] \simeq 0.63^+ K_{ps0}^{1/3}\, p_{CO_2}^{1/3} \tag{i}$$

with $^+K_{ps0} = K_{s0} K_1 K_H K_2^{-1}$. Using constants from Table 5.1 for the carbonate and $CaCO_3$ system and a value of 10.24 for pK_{s0} of $FeCO_3$, the following values obtain $(25°C, I = 0)$:

$$\log {}^+K_{ps0}(CaCO_3) = -5.83$$
$$\log {}^+K_{ps0}(FeCO_3) = -7.73$$

Considering furthermore that

$$[HCO_3^-] = 2[Me^{2+}] \tag{ii}$$
$$[H^+] = K_1 K_H p_{CO_2}[HCO_3^-]^{-1} \tag{iii}$$

the following results can be tabulated

	−log Concentration (M) Calculated for 25°C			
	HCO_3^-	Ca^{2+}	Fe^{2+}	H^+
Groundwater I	(3.14) 3.10	—	(3.44) 3.40	6.7
Groundwater II	(2.50) 2.44	(2.80) 2.74	—	7.3
Mixed water	2.66	3.04	3.70	7.1
After precipitation	(2.81)	(3.11)	(4.11)	(7.0)

Figures in parentheses give results obtained without correcting the constants for activities. With these results the ionic strength can be computed and equilibrium constants corrected for ionic strength. The following corrections have been applied:

$$p^{c+}K_{ps0} = p^+K_{ps0} - \frac{3\sqrt{I}}{1 + \sqrt{I}} \tag{iv}$$

$$p^cK_1 = pK_1 - \frac{\sqrt{I}}{1 + \sqrt{I}} \tag{v}$$

$$p^{c*}K_s = p^*K_s - \frac{2\sqrt{I}}{1 + \sqrt{I}} \tag{vi}$$

where $*K_s = K_{s0}K_2^{-1}$.

The mixed water is unstable because it is oversaturated with siderite. If it were in equilibrium with $FeCO_3(s)$, its equilibrium concentration would then be $-\log[Fe^{2+}] = 4.1$.

Free Energy of Formation of a Metal Carbonate

The Gibbs free energy of formation of a metal carbonate can be computed from the free energy of formation of MeO(s) and $CO_2(g)$ and the free energy of the reaction $MeO(s) + CO_2(g) = MeCO_3(s)$.

For example, the following information is available that permits calculation of the free energy of formation of $CdCO_3(s)$ [7]:

$$C(s) + O_2(g) = CO_2(g) \qquad \Delta G_1^\circ = -94,260 \text{ cal} \tag{52a}$$

$$Cd(s) + \tfrac{1}{2}O_2(g) = CdO(s) \qquad \Delta H_2^\circ = -62,360 \text{ cal} \tag{52b}$$

$$\Delta S_2^\circ = -23.7 \text{ cal deg}^{-1}$$

$$CdO(s) + CO_2(g) = CdCO_3(s) \qquad \Delta G_3^\circ = -12,145 \text{ cal} \tag{52c}$$

For equation 52b, ΔG_2° can be computed from ΔH_2° and $T\Delta S_2^\circ$. The former has been determined calorimetrically from the heat of combustion; the latter has been estimated from specific heat data (C_p). Using $\Delta G^\circ = \Delta H^\circ - T\Delta S^\circ$, we obtain $\Delta G_2^\circ = -62,360 + 7,080 = -55,280$ (cal). Summing up equations 52a to 52c, one obtains

$$Cd(s) + C(s) + \tfrac{3}{2}O_2(g) = CdCO_3(s) \qquad \Delta G^\circ = -161,690 \text{ cal} \quad (52d)$$

Gamsjäger, Stuber, and Schindler determined the solubility of $CdCO_3(s)$ and evaluated the free energy of formation from the following cycle

$$C(s) + O_2(g) = CO_2(g) \qquad\qquad \Delta G_1^\circ = -94,260 \text{ cal}$$
$$(53a)$$

$$H_2(g) + \tfrac{1}{2}O_2(g) = H_2O(g) \qquad\qquad \Delta G_5^\circ = -54,635 \text{ cal}$$
$$(53b)$$

$$H_2O(g) = H_2O(l)_I \qquad\qquad \Delta G_6^\circ = -2,130 \text{ cal}$$
$$(53c)$$

$$Cd_I^{2+} + CO_2(g) + H_2O(l)_I = CdCO_3(s) + 2H_I^+ \qquad \Delta G_7^{\circ\prime} = RT \ln {}^{c*}K_{ps0}$$
$$= 8,830 \text{ cal}$$
$$(53d)$$

$$Cd(s) + 2H_I^+ = Cd_I^{2+} + H_2(g) \qquad \Delta G_8^{\circ\prime} = 2FE_{\text{cell}(I)}^{\circ\prime}$$
$$= -18,980 \text{ cal}$$
$$(53e)$$

$$Cd(s) + C(s) + \tfrac{3}{2}O_2(g) = CdCO_3(s) \qquad\qquad \Delta G_4^\circ = -161,170 \text{ cal}$$
$$(53f)$$

$$(\Delta G_4^\circ = \Delta G_1^\circ + \Delta G_5^\circ + \Delta G_6^\circ + \Delta G_7^{\circ\prime} + \Delta G_8^{\circ\prime})$$

(Solute species present in the ionic medium of ionic strength I are indicated by subscript I.) The equilibria of equations 53d and 53e have been determined by solubility measurements (${}^{c*}K_{ps0} = [Cd^{2+}]p_{CO_2}/[H^+]^2$) and from emf measurements (electrochemical cell), respectively, in solutions of constant ionic medium (3 M $NaClO_4$); the free energy of evaporation of water [reaction 53c] is also given for the same ionic medium. Note that ΔG_4°, the free energy of formation, is unaffected by the ionic medium used in the experimental determination of the solubility constant (equation 53d) and the standard potential (equation 53e). The same principle can be applied for determination of the free energy of formation of sulfides.

Example 5.8. The Stability of Carbonates in the System Mg^{2+}–CO_2–H_2O

Below are listed standard free energies of formation, $\bar{G}_f^\circ$, of Mg^{2+}-bearing minerals:

$$\bar{G}_f^\circ \text{ at } 25°C \text{ (kcal mol}^{-1})$$

$Mg(OH)_2(s)$ (brucite)	-200.0
$MgCO_3(s)$ (magnesite)	-245.3
$MgCO_3 \cdot 3H_2O(s)$ (nesquehonite)	-411.7
$Mg_4(CO_3)_3(OH)_2 \cdot 3H_2O(s)$ (hydromagnesite)†	-1100.1
$Mg^{2+}(aq)$	-109.0
$CO_3^{2-}(aq)$	-126.2
$OH^-(aq)$	-37.6
$H_2O(l)$	-56.69

On the basis of these data,† establish stability domains as a function of variables C_T, pH, and p_{CO_2} for the minerals listed.

We first compute equilibrium constants using the relationships $\Delta G^\circ = RT \ln K$ and obtain (at 25°C)

	ΔG° (kcal mol^{-1})	$-\log K_{s0}$
$Mg(OH)_2(s) = Mg^{2+} + 2OH^-$ (brucite)	15.8	11.6
$MgCO_3(s) = Mg^{2+} + CO_3^{2-}$ (magnesite)	10.2	7.5
$MgCO_3(s) \cdot 3H_2O(s) = Mg^{2+} + CO_3^{2-} + 3H_2O$ (nesquehonite)	6.4	4.7
$Mg_4(CO_3)_3(OH)_2 \cdot 3H_2O = 4Mg^{2+} + 3CO_3^{2-} + 2OH^- + 3H_2O$ (hydromagnesite)	40.2	29.5

In order to obtain a first survey on the stability relationships, we construct an activity ratio diagram using Mg^{2+} as a reference state. The equations are

For brucite:

$$\log \frac{\{Mg(OH)_2(s)\}}{\{Mg^{2+}\}} = pK_{s0} - 2pK_W + 2pH$$

$$= -16.4 + 2pH \tag{i}$$

For magnesite:

$$\log \frac{\{magnesite(s)\}}{\{Mg^{2+}\}} = pK_{s0} + \log C_T + \log \alpha_2$$

$$= 7.5 + \log C_T + \log \alpha_2 \tag{ii}$$

---

† The values given are at variance with data by others. Robie et al. [5] list the $\bar{G}_f^\circ$ values for $MgCO_3$ (magnesite), -1029.48 kJ mol^{-1} (-245.9 kcal mol^{-1}), and for hydromagnesite which is given as $Mg_5(CO_3)_4(OH)_2 \cdot 4H_2O$ [or $(MgO)_5(CO_2)_4(H_2O)_5$], -5864.166 kJ mol^{-1} (-1400 kcal mol^{-1}). According to these data the solubility of magnesite is lower by a factor of 5 than that calculated from the $\bar{G}_f^\circ$ value given in the table; that is, magnesite has at 25°C a solubility ($-\log K_{s0} = 8.2$) only slightly larger than that of calcite ($-\log K_{s0} = 8.42$). The solubility of hydromagnesite, however, is about the same for both stoichiometric formulas.

For nesquehonite:

$$\log \frac{\{\text{nesquehonite(s)}\}}{\{Mg^{2+}\}} = pK_{s0} + \log C_T + \log \alpha_2$$

$$= 4.7 + \log C_T + \log \alpha_2 \quad \text{(iii)}$$

For hydromagnesite:

$$\log \frac{\{\text{hydromagnesite(s)}\}^{1/4}}{\{Mg^{2+}\}} = \tfrac{1}{4}pK_{s0} - \tfrac{1}{2}pK_w + \tfrac{3}{4}\log C_T + \tfrac{3}{4}\log \alpha_2 + \tfrac{1}{2}pH$$

$$= 0.4 + \tfrac{3}{4}\log C_T + \tfrac{3}{4}\log \alpha_2 + \tfrac{1}{2}pH \quad \text{(iv)}$$

These equations are plotted in Figure 5.12a for an assumed value of $\log C_T = -2.5$. It is convenient to start to plot these equations at high pH values where $\log \alpha_2 = 0$. In the pH region $pK_1 < pH < pK_2$, $d \log \alpha_2/dpH$ has a slope of $+1$. Our activity ratio diagram postulates that, for $\log C_T = -2.5$, magnesite

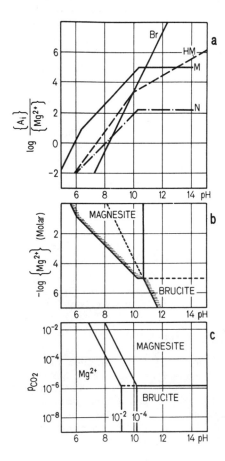

Figure 5.12 Stability in the system $Mg^{2+}-CO_2-H_2O$ ($I = 0$, 25°C) (see Example 5.8). Br, Brucite, $Mg(OH)_2(s)$; HM, hydromagnesite, $Mg_4(CO_3)_3(OH)_2 3H_2O(s)$; N, nesquehonite, $MgCO_3 3H_2O(s)$; M, magnesite, $MgCO_3(s)$. (a) Activity ratio diagram for $-\log C_T = 2.5$. Equations i to iv define relative activity. Stable phases are M(s): pH < 10.7; Br(s): pH > 10.7. (b) Solubility ($-\log\{Mg^{2+}\}$ versus pH) diagram for $-\log C_T = 2.5$. (c) Predominance diagram for $\log\{Mg^{2+}\} = -2$ and -4. Brucite can exist only at low p_{CO_2}.

is stable below pH $\simeq$ 10.7; nesquehonite is less stable than magnesite; brucite becomes stable above pH $\simeq$ 10.7.

In Figure 5.12b a solubility diagram is plotted for $-\log C_T = 2.5$.

For an open system, activity ratio or solubility diagrams for given p_{CO_2} values can be constructed. Figure 5.12c gives $\log p_{CO_2}$–pH predominance diagrams for $pMg^{2+} = 2$ and 4.0. The dissolution reaction may be rearranged to give mass-law expressions with p_{CO_2} and $[H^+]$, for example:

For magnesite:

$$MgCO_3(s) = Mg^{2+} + CO_3^{2-} \qquad\qquad \log K_{s0} \quad = -7.5$$

$$CO_3^{2-} + 2H^+ = H_2CO_3^* \qquad\qquad -\log (K_1 K_2) = \quad 16.6$$

$$H_2CO_3^* = CO_2(g) + H_2O \qquad\qquad -\log K_H \quad = \quad 1.5$$

$$MgCO_3(s) + 2H^+ = Mg^{2+} + CO_2(g) + H_2O \qquad \log {}^*K_{ps0} \quad = \quad 10.6$$
$$\text{(v)}$$

Similarly one derives for hydromagnesite:

$$Mg_4(CO_3)_3(OH)_2 3 H_2O(s) + 8H^+ = 4Mg^{2+} + 3CO_2(g) + 5H_2O$$

$$\log {}^*K_{ps0} = 52.8 \quad \text{(vi)}$$

And for the conversion of brucite into magnesite:

$$Mg(OH)_2(s) + CO_2(g) = MgCO_3(s) + H_2O \qquad \log K = 5.7 \quad \text{(vii)}$$

The equilibrium constant for equation vii is $p_{CO_2} = 10^{-5.7}$ atm. If p_{CO_2} (actual) is larger than $10^{-5.7}$ atm, brucite is converted to magnesite. On the other hand, hydromagnesite at this temperature is unstable (or metastable) with respect to magnesite:

$$Mg_4(CO_3)_3(OH)_2 \cdot 3 H_2O(s) + CO_2(g) = 4MgCO_3(s) + 4H_2O$$

$$K = 10^{10.4} \quad \text{(viii)}$$

As Figure 5.12c suggests, brucite is unstable if brought into contact with humid air.

Two mixed carbonates of Ca^{2+} and Mg^{2+} occur in nature: dolomite $CaMg(CO_3)_2(s)$ and huntite $CaMg_3(CO_3)_4(s)$. Dolomite constitutes a large fraction of the total quantity of carbonate rocks. The conditions under which dolomite precipitates in nature are not well understood. Attempts to precipitate a dolomite phase from oversaturated solutions under atmospheric conditions have been unsuccessful. Dolomite may be formed by the reaction of Mg^{2+}-rich waters with carbonate sediments.

Although precipitation of dolomite from most natural waters appears to be unimportant as a controlling factor in carbonate equilibria, the extensive areas underlain by dolomite bed rock suggest that the solubility characteristics are important [8]. Thus water in equilibrium with dolomite rocks may leave a composition quite different from that of waters in equilibrium with $CaCO_3$.

The Solubility of Dolomite.   The lack of understanding of the dolomite precipitation process is reflected in the discrepancy of solubility products reported by different investigators. Published figures range from $10^{-16.5}$ to $10^{-19.5}$. As mentioned before, solubility equilibrium can be reached (under atmospheric conditions) only from undersaturation. The time of approaching equilibrium is unknown. Thus it is very difficult to ascertain equilibrium in laboratory experiments.

Nature, however, has provided us with a long-term solubility experiment. As shown by Hsu [9], the well waters of Florida show a constant ratio of magnesium to calcium ($[Mg^{2+}]/[Ca^{2+}] = 0.8 \pm 0.1$). Hsu points out that the tendency for subsurface waters to have such a nearly constant magnesium–calcium ratio suggests that waters in porous dolomitic limestones might have equilibrated with both the calcite and dolomite phases.

The solubility product of dolomite is given by

$$K_{s0} = \{Ca^{2+}\}\{Mg^{2+}\}\{CO_3^{2-}\}^2 \tag{54}$$

For the reaction

$$2\,CaCO_3(s)(calcite) + Mg^{2+} = CaMg(CO_3)_2(s) + Ca^{2+} \tag{55}$$

the equilibrium constant K is defined in terms of the activity ratio of Ca^{2+} and Mg^{2+} and the K_{s0} values of dolomite and calcite:

$$K = \frac{K^2_{s0_{CaCO_3}}}{K_{s0_{dolomite}}} = \frac{\{Ca^{2+}\}}{\{Mg^{2+}\}} \tag{56}$$

For solution in contact and equilibrium with dolomite and calcite, the activity ratio is a constant at any temperature and pressure. In the Florida aquifer waters the concentration ratio remains nearly constant ($f_{Ca^{2+}}/f_{Mg^{2+}} \simeq 1$) even though $[Ca^{2+}]$ and $[Mg^{2+}]$ vary from less than 10^{-3} to 10^{-2} M. With an average value of $[Mg^{2+}]/[Ca^{2+}]$ of 0.78 and $K_{s0_{calcite}} = 5 \times 10^{-9}$ (25°C), a $K_{s0_{dolomite}} = 2.0 \times 10^{-17}$ can be calculated [9]. The same value (25°C, $P = 1$ atm) has been obtained from laboratory investigations [10]. With this constant and the constants given earlier, a predominance diagram for the system Mg^{2+}, Ca^{2+}, H_2O, CO_2 can be constructed. The diagram is shown in Figure 5.13. Obviously dolomite is thermodynamically the stable phase in seawater of average salinity and average calcium to magnesium ratio. But despite an abundance of dolomite in ancient sedimentary rocks, no unequivocal proof of recent dolomite in seawater of typical composition has been given.

Dolomites found in nature seldom have exact stoichiometric composition and are frequently structurally rich in calcium (protodolomite). Dolomite, as well as calcite, has a tendency to form solid solutions with many metal ions. Calcite has a tendency to accommodate Mg^{2+} in its structure to form *magnesian calcite*. Kinetically, the deposition of magnesian calcite may be more favorable than the deposition of dolomite.

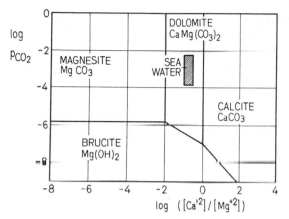

Figure 5.13   Stability relations in the system $Ca^{2+}-Mg^{2+}-CO_2-H_2O$, based on free energy values given in Example 5.6 and a solubility product of dolomite $K_{s0} = 2 \times 10^{-17}$ (25°C, 1 atm). Seawater is within the thermodynamic predominance of dolomite as a solid phase, but no convincing evidence for the formation of recent dolomite in seawater of normal salinity has been demonstrated.

Example 5.9.   Solubility of Dolomite

Consider fresh water containing 5×10^{-4} M Mg^{2+} and total carbonate carbon, $C_T = 2 \times 10^{-3}$ M. Assuming a closed system, define the pH range in which the solubility of Ca^{2+} should be dominated by dolomite solubility.

If the solubilities of dolomite and calcite (25°C, $P = 1$ atm) are $K_{s0} = 2 \times 10^{-17}$ and 5×10^{-9}, respectively, the pH at the dolomite–calcite predominance boundary is defined by $[Ca^{2+}]/[Mg^{2+}] = 1.25$. This corresponds to a pH where $[Ca^{2+}] = 6.25 \times 10^{-4}$ M $= K_{s0}/C_T\alpha_2$, that is, where $-\log \alpha_2 = 2.4$. Thus at pH values above pH $= 7.8$ dolomite should be more stable than calcite. A graphical representation is given in Figure 5.14.

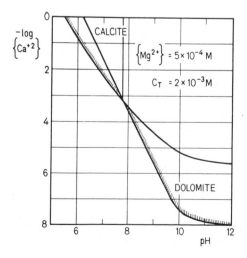

Figure 5.14   Comparison of the stability of calcite and dolomite (Example 5.9). Equilibrium solubility of calcite and dolomite for $Mg^{2+} = 5 \times 10^{-4}$ and $C_T = 2 \times 10^{-3}$ M (25°C, $I = 0$). Dolomite becomes more stable than calcite at pH values above 7.8.

5.5 THE SOLUBILITY OF SULFIDES AND PHOSPHATES; THE CONDITIONAL SOLUBILITY PRODUCT

We have illustrated that the solubility of many salts is not governed by the solubility product alone because other equilibria besides the solubility equilibrium occur in solution. The cation of the salt can react with water to produce hydroxo complexes, and the anion may also undergo acid–base reactions. A related phenomenon is that the cation and anion of the salt can form complexes with each other or with other species present in solution. All these side reactions tend to remove the ions of the slightly soluble salt from the solution and thus to increase the solubility.

Conditional Constants

In order to make solubility predictions, all these side reactions have to be considered. The required equilibrium data may be found frequently in *Stability Constants of Metal–Ion Complexes* [1] and other compilations [2, 5]. However, the involvement of so many species may lead to rather complicated calculations. In order to simplify the calculations, the concept of "conditional constants," that is, equilibrium constants that hold only under given experimental conditions (e.g., at a given pH) has been introduced by Schwarzenbach [11]. Other terms such as "apparent constant" and "effective constant" have been used instead of "conditional constant." The concept of conditional constants has been discussed extensively by Ringbom [12].

Effects of complex formation involving ligands and metal ions other than OH^- and H^+ upon solubility will be discussed in Chapter 6. In order to illustrate the usefulness of the conditional constant in treating side reactions involving H^+ and OH^- ions and to clarify its nature, the application to solubility equilibria will be discussed here.

Solubility Product Valid at a Given pH. Consider the solubility equilibrium

$$M_n A_m(s) = mA^{-n} + nM^{+m}$$

with the solubility product

$$^c K_{s0} = [A^{-n}]^m [M^{+m}]^n \tag{57}$$

The conditional solubility product is defined by the equation

$$P_s = A_T^m M_T^n = \frac{^c K_{s0}}{\alpha_M^n \alpha_A^m} \tag{58}$$

where M_T is the total concentration of the metal and A_T the total concentration of the anion of slightly soluble salt, irrespective of the form in which each is present, for example (omitting charges),

$$M_T = [M] + [MOH] + [M(OH)_2] + \cdots + [M(OH)_n] \tag{59}$$

$$A_T = [A] + [HA] + [H_2A] + \cdots + [H_nA] \tag{60}$$

The α values are similar to those defined previously for acid-base equilibria:

$$\alpha_M = \frac{[M]}{M_T} \qquad \alpha_A = \frac{[A]}{A_T} \tag{61}$$

They are functions of equilibrium constants, for example,

$$
\begin{aligned}
\alpha_M &= \frac{[M]}{[M] + [MOH] + [M(OH)_2] + \cdots + [M(OH)_n]} \\
&= \left(1 + \frac{{}^*K_1}{[H^+]} + \frac{{}^*K_1\,{}^*K_2}{[H^+]^2} + \cdots + \frac{{}^*K_1\,{}^*K_2\cdots{}^*K_n}{[H^+]^n}\right)^{-1}
\end{aligned} \tag{62}
$$

$$
\begin{aligned}
\alpha_A &= \frac{[A]}{[A] + [HA] + [H_2A] + \cdots + [H_nA]} \\
&= \left(1 + \frac{[H^+]}{K_n} + \frac{[H^+]^2}{K_n K_{n-1}} + \cdots + \frac{[H^+]^n}{K_n K_{n-1}\cdots K_1}\right)^{-1}
\end{aligned} \tag{63}
$$

where ${}^*K_1, {}^*K_2, \ldots$ and $K_1, K_2, \ldots$ are acidity constants of the metal ion (hydrolysis constants) and acidity constants of the base, respectively. It is convenient to represent α values as a function of pH in graphs or tables. For the α values known for any pH of interest the conditional solubility product (equation 57) is readily obtained. The conditional constant gives the relationship between the quantities that are of direct interest; in other words, a complicated solubility equilibrium may be reduced to a form analogous to that encountered when the cation and anion do not undergo any chemical side reactions (Example 5.1). The following simple examples illustrate the principle. In these examples, a conventional approach may not be any more troublesome numerically than the use of conditional constants, but with the concept of conditional constants even very complicated systems may be attacked more readily.

Example 5.10. The Solubility of MgNH$_4$PO$_4$(s)

In which pH range is the precipitation of $MgNH_4PO_4$ possible from a water containing $Mg_T(=[Mg^{2+}] + [MgOH^+]) = 10^{-2}$ M, $N_T(=[NH_4^+] + [NH_3])$ $= 10^{-3}$ M, and $P_T(=[H_3PO_4] + [H_2PO_4^-] + [HPO_4^{2-}] + [PO_4^{3-}]) = 10^{-4}$ M? The conditional solubility product of $MgNH_4PO_4$(s) is defined by

$$P_s = \frac{K_{s0\,MgNH4PO4}}{\alpha_{Mg}\,\alpha_N\,\alpha_P} = Mg_T \times N_T \times P_T$$

The following constants are available ($25°C$, $I = 0$):

$$\log K_{s0_{MgNH_4PO_4}} = -12.6$$

$$\log *K_{1(\text{hydrolysis of } Mg^{+2})} = -11.4$$

$$\log K_{1(NH_4^+)} = -9.24$$

$$\log K_{1(H_3PO_4)} = -2.1$$

$$\log K_{2(H_2PO_4^-)} = -7.2$$

$$\log K_{3(HPO_4^{2-})} = -12.3$$

We first use these K values without correcting for salinity. (But for a more rigorous answer the K valucs must be converted into ^cK values.)

$$\alpha_{Mg} = \left(1 + \frac{*K_1}{[H^+]}\right)^{-1}$$

$$\alpha_N = \left(1 + \frac{K_{1(NH_4^+)}}{[H^+]}\right)^{-1}$$

$$\alpha_P = \left(1 + \frac{[H^+]}{K_3} + \frac{[H^+]^2}{K_2 K_3} + \frac{[H^+]^3}{K_1 K_2 K_3}\right)^{-1}$$

The α values are plotted as a function of pH in Figure 5.15. The conditional solubility product reaches its minimum at pH $= 10.7$ $[=\frac{1}{2}(pK_{3(HPO_4^{2-})} + pK_{(NH_4^+)})]$. Thus precipitation of $MgNH_4PO_4(s)$ is favored in alkaline solutions. The conditional solubility product can now be compared with the product of the actual concentrations, $Q_{sT} = N_T \times P_T \times Mg_T = 10^{-9}$. As Figure 5.15 shows, only within the pH range 9 to 12 is $Q_{sT} > P_s$; in principle, a precipitation is possible. However, because the difference between pQ_{sT} and pP_s is quite small (<0.6) no efficient precipitation appears possible. Furthermore, the effect of ionic strength has been ignored. When a Güntelberg approximation and an ionic strength of $I = 0.1$ are used, pK_{s0} (hence p^cP_{sT}) becomes smaller by ca. 1.6 units; the solution actually does not become oversaturated with respect to $MgNH_4PO_4(s)$.

Example 5.11. Solubility in the System Zn^{2+}-CO_2-H_2O

Calculate conditional solubility constants for the solubility of $ZnCO_3(s)$ and hydrozincite, $Zn_5(OH)_6(CO_3)_2(s)$. The following equilibrium constants are available ($I = 0.2$, $25°C$):

$Zn(OH)_2(\text{amorph}) + 2H^+ = Zn^{2+} + 2H_2O$	$\log {}^{c*}K_{s0} = 12.7$	(i)
$ZnO(s) + 2H^+ = Zn^{2+} + H_2O$	$\log {}^{c*}K_{s0} = 11.4$	(ii)
$ZnCO_3(s) + 2H^+ = Zn^{2+} + H_2CO_3^*$	$\log {}^{c*}K_{s0} = 6.7$	(iii)

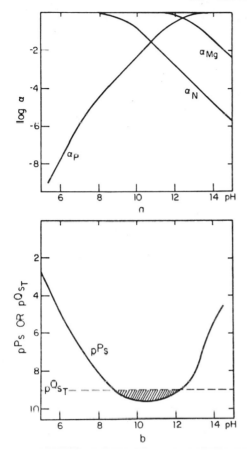

Figure 5.15 Solubility of $MgNH_4PO_4$ (Example 5.10). The conditional solubility product $P_S = P_T \times Mg_T$ (b) is readily calculated from the α values (a). Minimum solubility at pH $\simeq 10.7$.

$$Zn(OH)_{1.2}(CO_3)_{0.4}(s) + 2H^+$$

$$= Zn^{2+} + 0.4H_2CO_3^* + 1.2H_2O \qquad \log {}^{c*}K_{s0} = 9.4 \qquad \text{(iv)}$$

$$Zn^{2+} + H_2O = ZnOH^+ + H^+ \qquad \log {}^{c*}K_1 = -9.4 \qquad \text{(v)}$$

$$Zn^{2+} + 3H_2O = Zn(OH)_3^- + 3H^+ \qquad \log {}^{c*}\beta_3 = -28.2 \qquad \text{(vi)}$$

$$Zn^{2+} + 4H_2O = Zn(OH)_4^{2-} + 4H^+ \qquad \log {}^{c*}\beta_4 = -40.5 \qquad \text{(vii)}$$

We may define the following α values:

$$\alpha_{Zn} = \frac{[Zn^{2+}]}{Zn_T} = \left(1 + \frac{{}^{c*}K_1}{[H^+]} + \frac{{}^{c*}\beta_3}{[H^+]^3} + \frac{{}^{c*}\beta_4}{[H^+]^4}\right)^{-1} \qquad \text{(viii)}$$

$$\alpha_0 = \frac{[H_2CO_3^*]}{C_T} = \left(1 + \frac{K_1}{[H^+]} + \frac{K_1K_2}{[H^+]^2}\right)^{-1} \qquad \text{(ix)}$$

Thus the conditional solubility products for equation iii, $ZnCO_3(ZC)$, and equation iv, hydrozincite (HZ), are

$$P_{s(ZC)} = Zn_T \times C_T = \frac{{}^{c*}K_{s0zc}[H^+]^2}{\alpha_{Zn}\alpha_0} \tag{x}$$

or

$$pP_{s(ZC)} = -6.7 + 2pH + \log \alpha_{Zn} + \log \alpha_0 \tag{xi}$$

$$P_{s(HZ)} = Zn_T \times C_T^{0.4} = \frac{{}^{c*}K_{s0HZ}[H^+]^2}{\alpha_{Zn}\alpha_0^{0.4}} \tag{xii}$$

or

$$pP_{s(HZ)} = -9.4 + 2pH + \log \alpha_{Zn} + 0.4 \log \alpha_0 \tag{xiii}$$

$\log \alpha$ and pP values are plotted in Figure 5.16.

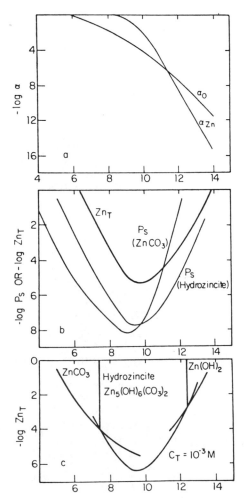

Figure 5.16 Conditional solubility products of $ZnCO_3(s)$ and hydrozincite $[Zn_5(OH)_6 \cdot (CO_3)_2(s)]$. (*a*) α_{Zn} and α_0 versus pH. (*b*) The conditional solubility products and largest possible Zn_T. (*c*) Maximum soluble Zn_T in a closed system ($C_T = 10^{-3}$ M).

There is an additional constraint: Because the solubility of $Zn(OH)_2(s)$ [or the more stable $ZnO(s)$] must not be exceeded, Zn_T in equations viii, x, and xii cannot assume any arbitrary value:

$$Zn_T \leq \frac{*K_{sOZn(OH)_2}[H^+]^2}{\alpha_{Zn}} \tag{xiv}$$

Equation xiv derives from equations i and viii. Figure 5.16b gives a plot of the largest possible concentration of Zn_T as a function of pH.

The use of the conditional constants may be illustrated by computing with the help of equations x, xii, and xiv the maximum total zinc concentration that can be maintained in a carbonate-bearing water ($C_T = 10^{-3}$ M) without becoming oversaturated with respect to zinc carbonate, hydrozincite, or amorphous $Zn(OH)_2$. The result is given in Figure 5.16c.

The stability boundary between $Zn(OH)_2$ and hydrozincite depends on which polymorphous form of $Zn(OH)_2(s)$ is assumed. Considering a more stable (and less soluble) form such as $ZnO(s)$ will shift the boundary to a less alkaline value.

Sulfides

The principles considered so far for hydroxides and carbonates can be applied to salts containing other anions. Of special importance are the sulfides [13]. Solubility of sulfides, however, is frequently complicated by the formation of thio complexes (see Section 6.7).

Example 5.12.   Solubility of α-CdS

1   Estimate the solubility of α-Cds at pH $= 4.5$ and $p_{H_2S} = 1$ atm (25°C).
2   Construct a predominance diagram for the system Cd^{2+}–H_2S–CO_2–H_2O with $p_{CO_2} = 10^{-3.5}$ atm; the following information [W. Kraft, H. Gamsjäger and E. Schwarz-Bergkampf, *Monatsh. Chem.*, 97, 1134 (1966)] is available for $I = 1$ M $NaClO_4$:

$$\log {}^{c*}K_{psO(CdS)} = -5.8 \qquad \log {}^{c*}K_{psO(CdCO_3)} = 6.44$$

$$\log {}^{c}K_{p12(H_2S)} = -1.05\dagger \qquad \log {}^{c}K_{1(H_2S)} = -6.90$$

$$\log {}^{c}K_{2(HS^-)} = -14.0 \qquad \log {}^{c}K_{p12(CO_2)} = -1.51\dagger$$

$$\log {}^{c}K_{1(H_2CO_3)} = -6.04 \qquad \log {}^{c}K_{(HCO_3^-)} = -9.57$$

† K_{p12} is Henry's law constant for the dissolution of the gas.

It is assumed (but not confirmed experimentally) that under the conditions specified, no thio or hydroxo complexes are formed.

1 The solubility is given by $[Cd^{2+}]$ in equilibrium in α-CdS(s) and $p_{H_2S} = 1$.

$$^{c*}K_{ps0} = \frac{[Cd^{2+}]p_{H_2S}}{[H^+]^2}$$

Hence at pH $= 4.5$, $[Cd^{2+}] = 10^{-14.8} M$.

2 The predominance diagram is characterized mainly by the reaction

$$CdCO_3(s) + H_2S(g) = CdS(s) + H_2O(l) + CO_2(g) \qquad (i)$$

The equilibrium constant of this reaction is given by

$$\log \frac{^{c*}K_{ps0}(CdCO_3)}{^{c*}K_{ps0}(CdS)} = 12.24 = \log \frac{p_{CO_2}}{p_{H_2S}} \qquad (ii)$$

The coexistence of Cd^{2+} with CdS(s) and with $CdCO_3$(s), respectively, is given by the equilibria defined by the respective $^{c*}K_{ps0}$ values:

$$6.44 = \log[Cd^{2+}] + 2pH + \log p_{CO_2} \qquad (iii)$$

$$-5.8 = \log[Cd^{2+}] + 2pH + \log p_{H_2S} \qquad (iv)$$

For a fixed $\log p_{CO_2} = -3.5$, these equations become

$$\log[Cd^{2+}] = 9.94 - 2pH$$

$$\log[Cd^{2+}] = -2.3 - 2pH + \log \frac{p_{CO_2}}{p_{H_2S}} \qquad (v)$$

These relations are plotted in Figure 5.17 for the two conditions $\log[Cd^{2+}] = -4$ and 0. Extremely small partial pressures of H_2S are sufficient to convert $CdCO_3$ into CdS. As pointed out by Kraft, Gamsjäger, and Schwarz-Bergkampf, natural greenocite (α-CdS) is quite resistant to weathering and carbonation.

Phosphates

The distribution of the several acid and base species of orthophosphates and condensed phosphates in solution is governed by pH. Information on the pH-dependent distribution of the several species is required in interpreting the solubility behavior, complex formation, and sorption processes of phosphorus in water. Table 5.1, V contains some pertinent equilibrium constants. The predominant dissolved orthophosphate species over the pH range 5 to 9 are $H_2PO_4^-$ and HPO_4^{2-}. By employing the solubility equilibrium constants and the acidity constants it is possible to compute total phosphate solubility (P_T) under specified conditions (pH, calcium concentrations, etc.). For example,

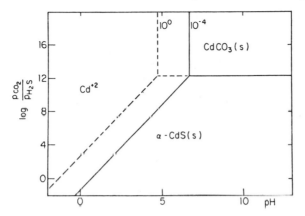

Figure 5.17 Predominance diagram in the system $Cd^{2+}-H_2S(g)-CO_2(g)-H_2O$ (see Example 5.11). $CdCO_3(s)$ becomes unstable with respect to $CdS(s)$ at extremely small partial pressures of H_2S. Based on data by Kraft, Gamsjäger, and Schwarz-Bergkampf for $I = 1\ M$ (NaClO$_4$) and 25°C. A fixed p_{CO_2} of $10^{-3.5}$ atm and $[Cd^{2+}] = 10^{-4}$ and $= 10^{0}$ are assumed.

we can compute soluble P_T for pure $AlPO_4(s)$ in contact with pure water whose pH is adjusted by addition of acid or base. When $Al(OH)_3(s)$ or $Al_2O_3(s)$ forms, the soluble P_T is then governed by an additional equilibrium condition. Similar considerations apply to $FePO_4(s)$ solubility. Figure 5.18 gives solubility diagrams for a few solid phosphate phases. These diagrams show that $FePO_4(s)$ (strengite) and $AlPO_4(s)$ (variscite) are the stable solid phases if phosphate is precipitated in the low-pH range; the pH of minimum $AlPO_4(s)$ solubility occurs at about 1 pH unit higher than that of $FePO_4(s)$. In the neutral pH range (on the right side of the solubility lines of $FePO_4$ and $AlPO_4$ in Figure 5.18), metastable hydroxophosphate Al(III) or Fe(III) precipitates can be formed.

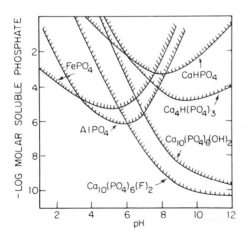

Figure 5.18 Solubility of the metal phosphates. The solubility of $AlPO_4$ and $FePO_4$ has been calculated on the basis of the equilibria (Table 5.1,V), assuming that $FePO_4(s)$ or $AlPO_4(s)$ can be converted incongruently into $Fe(OH)_3(s)$ [or α-FeOOH(s)] or $Al(OH)_3(s)$. The solubility of the calcium phosphate phases has been calculated under the assumption that $[Ca^{2+}] = 10^{-3}\ M$ and that F^- is regulated by the solubility of $CaF_2(s)$.

Calcium forms several insoluble solid phases with phosphate. Thermodynamically stable over most of the range of interest (Figure 5.18) is apatite, $Ca_{10}(PO_4)_6X_2(s)$, where X is usually OH^- or F^-; in carbonate-bearing waters a carbonato apatite may also be formed. Some investigators have postulated that $CaHPO_4(s)$ is precipitated, at least initially as a metastable phase. Usually variable Ca/P ratios are found in most naturally occurring, often poorly crystallized, apatites.

Example 5.13. Conversion of Calcite into Apatite

Under what condition (pH, concentration of inorganic P) can calcite be converted into hydroxylapatite? The simplifying assumptions $P_T \ll [Ca^{2+}]$ and $[Ca^{2+}] = C_T$ may be used.

The reaction may be written as

$$10CaCO_3(s) + 2H^+ + 6HPO_4^{2-} + 2H_2O$$
$$= Ca_{10}(PO_4)_6(OH)_2(s) + 10HCO_3^- \quad (i)$$

The equilibrium constant K for this reaction may be computed conveniently from

$$\log K = -\log K_{s0(apatite)} + 10 \log K_{s0(calcite)}$$
$$- 10 \log K_{HCO_3^-} + 6 \log K_{HPO_4^{2-}} + 2 \log K_w \quad (ii)$$

The equilibrium constant for equation i is of course at least as uncertain as the solubility product of apatite. With a value of K_{s0} for $Ca_{10}(PO_4)_6(OH)_2(s)$ of 10^{-114}, an equilibrium constant of $K = 10^{32}$ is obtained. Hence the free energy of this conversion, $\Delta G = RT \ln(Q/K)$, where Q is the quotient of the reactants, may be used to predict under what conditions reaction i is possible. For example, at pH = 8 and $[HCO_3^-] = 10^{-3} \, M$, a $10^{-4} \, M$ solution of HPO_4^{2-} would, thermodynamically speaking, convert calcite into apatite because ΔG is approximately -30 kcal mol^{-1} of apatite formed. Figure 5.19 plots concentrations of soluble phosphate, P_T necessary to convert $CaCO_3$ into apatite as a function of pH.

Although equation i represents an oversimplification (apatite is not necessarily formed as a pure solid phase), the tentative result obtained suggests that the phosphorus concentration at the sediment–water interface, especially at higher pH values, is buffered by the presence of hydroxyapatite.

J. O. Leckie and W. Stumm (in *The Changing Chemistry of the Oceans*, D. Dyrssen and D. Jagner, Eds., Almqvist and Wiksell, Stockholm, 1972) have shown that apatite can be formed on the surface of calcite.

The mineral or chemical composition of phosphorus compounds tends to be different in different environments. Variscite ($AlPO_4 \cdot 2H_2O$) and strengite ($FePO_4 \cdot 2H_2O$), appear to be more common in soils and freshwater sediments, whereas apatite prevails in certain marine sediments. However, the solubility

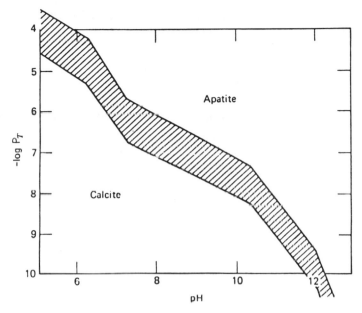

Figure 5.19 Phosphate necessary to convert $CaCO_3(s)$ into $Ca_{10}(PO_4)_6(OH)_2(s)$ (Example 5.13). Calculated by assuming that the solution remains in saturation equilibrium with $CaCO_3(s)$; furthermore, $P_T \ll C_T$ and $[Ca^{2+}] = C_T$. The equilibrium constant for equation i has been taken as 10^{28} (upper line) and 10^{30} (lower line), respectively.

product of apatite does not appear to be exceeded on the deep-ocean floor. In areas of high organic productivity in sediments of eutrophic lakes and of the shallow areas of the ocean, especially the tropical ocean, calcium phosphate minerals (substituted apatites, such as francolite, $Ca_{10}[PO_4CO_3]_6F_2$) are deposited. The main sink of phosphates in the ocean and in many lakes consists of iron(III) oxides on the surface of which phosphates become chemisorbed.

5.6 THE ACTIVITY OF THE SOLID PHASE

If a solid and a solution are in equilibrium, all components occurring in both phases have, according to the Gibbs equilibrium condition, the same chemical potential in both phases. The solubility of a salt depends on the activity of the solid phase. There are various factors that affect the activity of the solid phase: (a) the lattice energy, (b) the degree of hydration, (c) solid solution formation, and (d) the free energy of the surface.

In defining a solid–solution equilibrium, the pure solid phase is taken as a reference state and its activity is, because of its constancy, set equal to unity. This means that the activity of the solid phase is implicitly contained in the solubility equilibrium constant, and this can thus be used to deduce information about the constitution of the solid phase.

Polymorphism

A chemical substance can exist in more than one crystalline form. These polymorphic forms are also called allotropic modifications. Except at a transition temperature only one form of crystal is thermodynamically stable for a given chemical composition of a system. The metastable modifications are more soluble than the stable one; for example, the solubilities of quartz and of amorphous silica in H_2O (25°C) are approximately 2×10^{-4} M and 2×10^{-3} M, respectively, as orthosilicic acid. Obviously quartz is the stable solid phase; its free energy of formation is ca. 1.36 kcal smaller (more negative) than that of amorphous SiO_2

$$SiO_2(amorph) = SiO_2(quartz) \qquad \Delta G° = -1.36 \, kcal \qquad (64)$$

Which of these solid phases shall we consider in making "equilibrium" calculations? At low temperature and pressure amorphous silica precipitates from oversaturated silica solutions; the rate of quartz formation is extremely slow and cannot be observed under these conditions. Obviously it would be wrong to assume that only the thermodynamically stable modification prevails as the solid phase in the system under consideration. It is necessary to adduce some information on the kinetics of the solid–solution interactions and to have, if possible, some information on the type of solid phases that are in contact with the actual solution or with the natural water in order to postulate which solid phase may persist in metastable existence for the periods of time under consideration. The existence of a metastable modification for geological time spans is not uncommon.

There are at least five polymorphic forms of $CaCO_3$ that occur in nature. Calcite and aragonite are especially important. At $P = 1$ atm and 25°C, calcite is less soluble than aragonite, hence is the stable form thermodynamically. Nevertheless, recent carbonate sediments are composed to a large extent of aragonite (Bricker and Garrels [8]). Aragonite is being deposited in vast amounts under conditions of temperature and pressure where calcite is the thermodynamically stable phase. Furthermore, aragonite is the dominant form of $CaCO_3$ deposited by organisms. Organisms can also form calcite in which part of the Ca^{2+} is substituted by Mg^{2+} (magnesian calcite).

The solid phase that precipitates first will be governed by kinetic factors. Trace elements can influence the kinetics significantly. Aragonite has usually a higher Sr^{2+} content than calcite; the inference has been made that the presence of Sr^{2+} in the water favors the precipitation of aragonite and retards or "prevents" the conversion into calcite.

With respect to $CaCO_3$, seawater appears to behave quite differently from fresh water, and Mg^{2+} has been suggested as one of the factors responsible for the anomalous behavior of $CaCO_3$ in seawater.

Organisms have the possibility of carrying out endergonic reactions either coupled with exergonic ones or driven by the energy of photosynthesis. Thus they can excrete or produce solid phases from solutions undersaturated with respect to these solid phases.

Various Hydrates of a Solid. In a similar way as different allotropic modifications have different activities, the various hydrates of a solid have different solubilities, and the one with the lowest solubility at a given temperature, pressure, and water activity (e.g., vapor pressure) is the most stable one. The degree of hydration can have a pronounced effect on the rate of precipitation or on the rate of dissolution. Thus it is not necessarily the most stable hydrate of a solid that is kinetically most persistent within a certain observation period.

Solid Solution Formation

The solids occurring in nature are seldom pure solid phases. Isomorphous replacement by a foreign constituent in the crystalline lattice is an important factor by which the activity of the solid phase may be decreased. If the solids are homogeneous, that is, contain no concentration gradient, one speaks of homogeneous solid solutions. The thermodynamics of solid solution formation has been discussed by Vaslow and Boyd [14] for solid solutions formed by AgCl(s) and AgBr(s).

To express theoretically the relationship involved we consider a two-phase system where AgBr(s) as solute becomes dissolved in solid AgCl as solvent. This corresponds to the reaction that takes place if AgCl(s) is shaken with a solution containing Br^-. The reaction might formally be characterized by the equilibrium

$$AgCl(s) + Br^- = AgBr(s) + Cl \tag{65}$$

The equilibrium constant for equation 65, that is, the distribution constant D,

$$D = \frac{\dfrac{\{AgBr(s)\}}{\{AgCl(s)\}}}{\dfrac{\{Br^-\}}{\{Cl^-\}}} \approx \frac{\left(\dfrac{[Br^-]}{[Cl^-]}\right)_{solid}}{\left(\dfrac{[Br^-]}{[Cl^-]}\right)_{liquid}} \tag{66}$$

corresponds to the quotient of the solubility product constants of AgCl(s) and AgBr(s), where { } denotes activity and [ ] the concentration in a phase.

$$\frac{\{Ag^+\}\{Cl^-\}}{\{AgCl(s)\}} = K_{s0_{AgCl}} \qquad \frac{\{Ag^+\}\{Br^-\}}{\{AgBr(s)\}} = K_{s0_{AgBr}} \tag{67}$$

$$\frac{\{Cl^-\}\,\{AgBr(s)\}}{\{Br^-\}\,\{AgCl(s)\}} = \frac{K_{s0_{AgCl}}}{K_{s0_{AgBr}}} = D \tag{68}$$

The activity ratio of the solids may be replaced by the ratio of the mole fractions $(X_{AgCl} = n_{AgCl}/n_{AgCl} + n_{AgBr})$ multiplied by activity coefficients:

$$\frac{\{Cl^-\}}{\{Br^-\}} \frac{X_{AgBr}}{X_{AgCl}} \frac{f_{AgBr}}{f_{AgCl}} = \frac{K_{s0_{AgCl}}}{K_{s0_{AgBr}}}$$

or   (69)

$$\frac{X_{AgBr}}{X_{AgCl}} = \frac{K_{s0_{AgCl}}}{K_{s0_{AgBr}}} \frac{\{Br^-\}}{\{Cl^-\}} \frac{f_{AgCl}}{f_{AgBr}}$$

According to this equation the extent of dissolution of Br^- in solid AgCl (X_{AgBr}/X_{AgCl}) is a function of (a) the solubility product ratio of AgCl to AgBr; (b) the solution composition, that is, the activity ratio of Br^- to Cl^-; and (c) a solid solution factor, given by the ratio of the activity coefficients of the solid solution components (f_{AgCl}/f_{AgBr}).

As a first approximation, we may assume that f_{AgCl}/f_{AgBr} is equal to unity (ideal solid solution) and that the activity ratio of the species in the fluid may be replaced by the concentration ratio

$$\frac{[Cl^-]}{[Br^-]} \frac{X_{AgBr}}{X_{AgCl}} \approx \frac{K_{s0_{AgCl}}}{K_{s0_{AgBr}}} = D$$ (70)

The qualitative significance of solid solution formation can be demonstrated with the help of this simplified equation, using the following numerical example.

Consider a solid solution of 10% AgBr in AgCl (90%) which is in equilibrium with Cl^- and Br^-; the composition of the suspension is:

Aqueous Phase	Solid Phase
$[Cl^-] = 10^{-4.9}\ M$	$X_{AgCl} = 0.9$
$[Br^-] = 10^{-8.4}\ M$	$X_{AgBr} = 0.1$
$[Ag^+] = 10^{-4.9}\ M$	

In accordance with equation 70, $D = 400$ ($\simeq$ quotient of the K_{s0} values at 25°C; $pK_{s0_{AgCl}} = 9.7$ and $pK_{s0_{AgBr}} = 12.3$).

The composition of the equilibrium mixture shows that Br^- has been enriched significantly in the solid phase in comparison to the liquid phase $(D > 1)$. If one considered the concentrations of aqueous $[Br^-]$ and $[Ag^+]$, one would infer, by neglecting to consider the presence of a solid solution phase, that the solution is undersaturated with respect to AgBr ($[Ag^+][Br^-]/K_{s0_{AgBr}} = 0.1$). Because the aqueous solution is in equilibrium with a solid solution, however, the aqueous solution is saturated with Br^-. Although the solubility of the salt that represents the major component of the solid phase is only slightly affected by the formation of solid solutions, the solubility of the minor component is appreciably reduced. The observed occurrence of certain metal ions in sediments formed from solutions that appear to be formally

TABLE 5.4   SOLID SOLUTION FORMATION

Solid Solution	t, °C	$D_{obs}{}^a$	D^b	Ref.^{c}
AgBr in AgCl	30	211.4	315.7	1
Ra(IO$_3$)$_2$ in Ba(IO$_3$)$_2$	25	1.42	1.32	2
RaSO$_4$ in BaSO$_4$	20	1.8	5.9	3
MnCO$_3$ in calcite	25	17.4	525	4
SrCO$_3$ in calcite	25	0.14	10	5^d
SrCO$_3$ in aragonite	25	1.1	11	5
MgCO$_3$ in calcite	20	0.02	0.16	6

^{a} For solid solution of B in A, [A(s) + B(aq) = B(s) + A(aq)]:

$$D_{obs} = \frac{[A(aq)] \, X_B}{[B(aq)] \, X_A}$$

^{b} D = quotient of solubility products, $D = K_{s0(A)}/K_{s0(B)}$.
^{c} References: (1) F. Vaslow and G. E. Boyd, *J. Amer. Chem. Soc.*, **75**, 4691 (1952); (2) A. Polessitsky and A. Karataewa, *Acta Physicochim. (U.R.S.S.)*, **8**, 259 (1938); (quoted as cited in ref. 1); (3) O. Hahn, *Applied Radiochemistry*, Cornell University Press, Ithaca, N.Y., 1933, p. 88; (4) D. J. J. Kinsman and H. D. Holland, *Geochim. Cosmochim. Acta*, **33**, 1 (1969); (5) H. D. Winland, *J. Sediment. Petrol.*, **39**, 1579 (1969).
^{d} Obtained by direct CaCO$_3$ precipitation from solutions. In later experiments, Katz et al. [17] obtained much lower values for D under conditions of replacement of aragonite by calcite in the presence of aqueous solutions.

(in the absence of any consideration of solid solution formation) unsaturated with respect to the impurity can, in many cases, be explained by solid solution formation.

Usually, however, the distribution coefficients determined experimentally are not equal to the ratios of the solubility product because the ratio of the activity coefficients of the constituents in the solid phase cannot be assumed to be equal to 1. A comparison of solubility product quotients and actually observed D values, given in Table 5.4, illustrates that activity coefficients in the solid phase may differ markedly from 1. Let us consider, for example, the coprecipitation of MnCO$_3$ in calcite. Assuming that the ratio of the activity coefficients in the aqueous solution is close to unity, the equilibrium distribution may be formulated as (cf. equation 68)

$$\frac{K_{s0\,CaCO_3}}{K_{s0\,MnCO_3}} = D = \frac{[Ca^{2+}]}{[Mn^{2+}]} \frac{X_{MnCO_3}}{X_{CaCO_3}} \frac{f_{MnCO_3}}{f_{CaCO_3}} \tag{71}$$

$$D = D_{obs} \frac{f_{MnCO_3}}{f_{CaCO_3}} \tag{72}$$

The solubility product quotient at 25°C ($pK_{s0_{MnCO_3}} = 11.09$, $pK_{s0_{CaCO_3}} = 8.37$) can now be compared with an experimental value of D. The data of Bodine, Holland, and Borcsik [15] (see also Figure 5.20) give $D_{obs} = 17.4$ (25°C). Because D is smaller than the ratio of the K_{s0} values, the solid solution factor acts to lower the solution of $MnCO_3$ in $CaCO_3$ significantly from that expected if an ideal mixture had been formed. If it is assumed that in dilute solid solutions (X_{MnCO_3} very small) the activity coefficient of the solvent is close to unity ($X_{CaCO_3} f_{CaCO_3} \simeq 1$), an activity coefficient for the solute is calculated to be $f_{MnCO_3} = 31$. Qualitatively, such a high activity coefficient reflects a condition similar to that of a gas dissolved in a concentrated electrolyte solution where the gas, also characterized by an activity coefficient larger than unity, is "salted out" from the solution. Thermodynamically, the activity coefficient of a solid solute is related to the decrease in partial molar free energy resulting from the transfer of 1 mol of solute from a large quantity of an ideal solid solution to the real solid solution of the same mole fraction (Vaslow and Boyd [14]). This excess free energy of solid solution will be small if an ion of similar size and

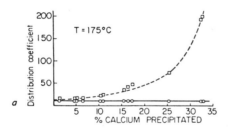

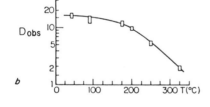

Figure 5.20 Solid solution formation; dissolution of $MnCO_3(s)$ in $CaCO_3(s)$ (calcite). From Bodine, Holland, and Borcsik [15]. (*a*) The distribution coefficient plotted against percent calcium precipitated. D has been calculated for homogeneous solid solution formation (squares)

$$D = \left(\frac{m_{Mn^{2+}}}{m_{Ca^{2+}}}\right)_{solid} \times \frac{[Ca^{2+}]}{[Mn^{2+}]}$$

and for heterogeneous solid solution formation (circles)

$$D = \frac{\log \dfrac{[Mn^{2+}]_0}{[Mn^{2+}]_f}}{\log \dfrac{[Ca^{2+}]_0}{[Ca^{2+}]_f}}$$

(*b*) The distribution coefficient plotted as a function of temperature. (*c*) The enrichment of the calcite in manganese relative to the composition in the aqueous solution at 25°C.

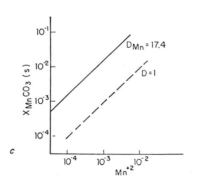

bond type, forming crystals with similar lattice parameters, becomes iso-morphously substituted.

The experimentally determined value of D may depend on the conditions, especially on the kinetics of precipitate formation [16, 17].

Example 5.14.   Solid Solution of Sr^{2+} in Calcite

It has been proposed (e.g., Schindler [18]) that a solid solution of Sr^{2+} in calcite might control the solubility of Sr^{2+} in the ocean. Estimate the com position of the solid solution phase (X_{SrCO_3}). The following information is available: The solubilities of $CaCO_3(s)$ calcite and $SrCO_3(s)$ in seawater at 25°C are characterized by $p^rK_{s0} = 6.1$ and 6.8, respectively. The equilibrium concentration of CO_3^{2-} is $[CO_3^{2-}] = 10^{-3.6}$ M. The actual concentration of Sr^{2+} in seawater is $[Sr^{2+}] \simeq 10^{-4}$ M. According to the value quoted in Table 5.4.

$$\left(\frac{[Ca^{2+}]}{[Sr^{2+}]}\right)\left(\frac{X_{SrCO_3}}{X_{CaCO_3}}\right) = 0.14 \qquad (25°C)$$

Assuming a saturation equilibrium of seawater with Sr^{2+}–calcite, the equilibrium concentrations would be $[Ca^{2+}] \simeq 10^{-2.5}$ ($= {}^cK_{s0}/[CO_3^{2-}]$) and $[Sr^{2+}] = 10^{-4}$ M ($=$ actual concentration). Thus $X_{SrCO_3}/X_{CaCO_3} = 0.004$ and $X_{CaCO_3} = 0.996$.

It may be noted that, since the distribution coefficient is smaller than unity, the solid phase becomes depleted in strontium relative to the concentration in the aqueous solution. The small value of D may be interpreted in terms of a high activity coefficient of strontium in the solid phase, $f_{SrCO_3} \approx 38$. If the strontium were in equilibrium with strontianite, $[Sr^{2+}] \simeq 10^{-3.2}$ M, that is, its concentration would be more than six times larger than at saturation with $Ca_{0.996}Sr_{0.004}CO_3(s)$. This is an illustration of the consequence of solid solution formation where with $X_{CaCO_3} f_{CaCO_3} \simeq 1$:

$$[Sr^{2+}] = \frac{X_{SrCO_3} f_{SrCO_3} {}^cK_{s0SrCO_3}}{[CO_3^{2-}]}$$

that is, the solubility of a constituent is greatly reduced when it becomes a minor constituent of a solid solution phase.

Heterogeneous Solid Solutions

Besides homogeneous solid solutions, *heterogeneous arrangement* of foreign ions within the lattice is possible. While homogeneous solid solutions represent a state of true thermodynamic equilibrium, heterogeneous solid solutions can persist in metastable equilibrium with the aqueous solution. Hetero-geneous solid solutions may form in such a way that each crystal layer as it forms is in distribution equilibrium with the particular concentration of the aqueous solution existing at that time (Doerner and Hoskins [19]; Gordon, Salutsky, and Willard [20]). Correspondingly, there will be a concentration gradient in the solid phase from the center to the periphery. Such a gradient results from very slow diffusion within the solid phase. Following the treatment given by Doerner and Hoskins and by Gordon, Salutsky, and Willard, the distribution equilibrium for the reaction

$$CaCO_3(s) + Mn^{2+} = MnCO_3(s) + Ca^{2+} \tag{73}$$

is written as in equation 68, but we consider that the crystal surface is in equilibrium with the solution:

$$\left(\frac{[MnCO_3]}{[CaCO_3]}\right)_{\text{crystal surface}} \times \frac{[Ca^{2+}]}{[Mn^{2+}]} = D' \tag{74}$$

If $d[MnCO_3]$ and $d[CaCO_3]$, the increments of $MnCO_3$ and $CaCO_3$ deposited in the crystal surface layer, are proportional to their respective solution concentrations, equation 75 is obtained

$$\frac{d[MnCO_3]}{d[CaCO_3]} = D' \frac{[Mn^{2+}]_0 - [Mn^{2+}]}{[Ca^{2+}]_0 - [Ca^{2+}]} \tag{75}$$

or, after rearrangement,

$$\frac{d[MnCO_3]}{[Mn^{2+}]_0 - [Mn^{2+}]} = D' \frac{d[CaCO_3]}{[Ca^{2+}]_0 - [Ca^{2+}]} \tag{76}$$

where $[Ca^{2+}]_0$ and $[Mn^{2+}]_0$ represent initial concentrations in the aqueous solution. Integration of equation 76 leads to

$$\log \frac{[Mn^{2+}]_0}{[Mn^{2+}]_f} = D' \log \frac{[Ca^{2+}]_0}{[Ca^{2+}]_f} \tag{77}$$

where $[Mn^{2+}]_f$ and $[Ca^{2+}]_f$ represent final concentrations in the aqueous solutions. Figure 5.20 illustrates that experimentally determined distribution coefficients for the coprecipitation of Mn(II) in calcite give constant values for heterogeneous solid solution formation only.

Most of the distribution coefficients measured to date for a variety of relatively insoluble solids are characterized by the Doerner–Hoskins relation. This relationship is usually obeyed for crystals that have been precipitated from homogeneous solution or under conditions similar to those encountered in precipitation from homogeneous solution (Gordon, Salutsky, and Willard [20]). If the precipitation occurs in such a way that the aqueous phase remains as homogeneous as possible and the precipitant ion is generated gradually throughout the solution, large, well-formed crystals likened to the structure of an onion [21] are obtained. Each infinitesimal crystal layer is equivalent to a shell of an onion. As each layer is deposited, there is insufficient time for reaction between solution and crystal surface before the solid becomes coated with succeeding layers. Kinetic factors make the metastable persistence of such compounds possible for relatively long—often for geological—time spans.

Stoichiometric Saturation.   The term stoichiometric saturation refers to (metastable) equilibrium between an aqueous phase and a multicomponent solid in situations where, because of kinetic restrictions, the composition of the solid remains invariant, even though the solid phase may be part of a continuous compositional series [21, 22]. Consider the stoichiometric, congruent dissolution of such a multicomponent solid, for example, of a magnesium calcite:

$$Ca_{(1-x)}Mg_xCO_3(s) = (1 - x)Ca^{2+} + xMg^{2+} + CO_3^{2-} \tag{78}$$

where x is the mole fraction of Mg in the solid. If this solid does not change composition during the dissolution (or precipitation), it can be treated thermodynamically as a one-component phase; that is, its activity is unity†; the equilibrium constant for the solubility equilibrium (equation 78) is given by

$$K_{eq(x)} = \{Ca^{2+}\}^{(1-x)}\{Mg^{2+}\}^x\{CO_3^{2-}\} \tag{79}$$

which may be rearranged [21] [multiply and divide equation 79 by $\{Ca^{2+}\}^x$] to

$$K_{eq(x)} = \{Ca^{2+}\}\{CO_3^{2-}\} \frac{\{Mg^{2+}\}^x}{\{Ca^{2+}\}^x} \tag{80}$$

Since $\{Mg^{2+}\}^x/\{Ca^{2+}\}^x$ is usually smaller than one, the solubility of Mg calcite is usually smaller than that of calcite.

Equilibrium constants for the solubility equilibrium of Mg calcites at stoichiometric saturation based on data by Plummer and Mackenzie [23] are given in Figure 5.21.

Thorstenson and Plummer [21] have pointed out that changes in the composition of binary solid solution compounds are very slow at low temperature (25°C); therefore the assumption of thermodynamic equilibrium in experimental work is almost certainly invalid, and it is questionable in many natural environments. Under these conditions, the behavior of the solid phases may often operationally be interpreted on the basis of stoichiometric saturation.

Magnesian Calcite

Analyses of natural calcites known to have been formed at low temperatures show $MgCO_3$ contents as high as 20 mol % [9, 21]. Many marine organisms secrete hard parts of magnesian calcite.

---

† Magnesian calcite may not be eligible as an invariant thermodynamic phase; although the composition of the solid may not so much depend on the ratio of the cations in solution but it may depend on other factors such as pH and $[CO_3^{2-}]$ which influence the kinetics of the system.

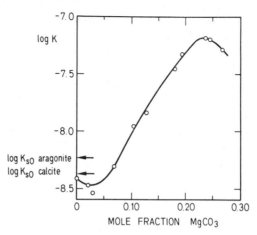

Figure 5.21 Plot of log $K_{eq(x)}$ against mole fraction $MgCO_3$, x, in magnesian calcites. $K_{eq(x)} = \{Ca\}^{(1-x)}\{Mg^{2+}\}^x\{CO_3^{2-}\}$ at stoichiometric saturation (25°C) based on data by Plummer and Mackenzie [23]. Arrows give log K_{so} for aragonite and calcite.

As pointed out by Bricker and Garrels and others, magnesian calcites are metastable under earth surface conditions and lose Mg^{2+} within geologically short time spans. Calcites from Mesozoic and older rocks are very low in Mg^{2+} except under unusual circumstances (Goldsmith et al. [24]). Partial substitution of Mg^{2+} in calcite changes the solubility. Although no true metastable equilibrium solubility can be established, Chave et al. [25] and Plummer and Mackenzie [23] have measured the ultimate pH of suspensions of magnesian calcites with various Mg^{2+} content exposed to a CO_2 pressure of 1 atm. The stoichiometric saturation apparent equilibrium constant (equation 79) is compared with the solubility product of calcite and aragonite.

We may compare the solubility of Mg-calcite with that of $CaCO_3(s)$ (calcite or aragonite):

$$Ca_{(1-x)}Mg_xCO_3(s) = (1-x)Ca^{2+} + xMg^{2+} + CO_3^{2-}$$
$$K_{eq(x)} \qquad (81)$$

$$Ca^{2+} + CO_3^{2-} = CaCO_3(s) \qquad K_{s0_{CaCO_3}}^{-1} \qquad (82)$$

$$Ca_{(1-x)}Mg_xCO_3(s) + xCa^{2+} = CaCO_3(s) + xMg^{2+} \qquad K_{eq(x)} \cdot K_{CaCO_3}^{-1} \quad (83a)$$

or

$$\left(\frac{\{Mg^{2+}\}}{\{Ca^{2+}\}}\right)^x = K_{eq(x)}K_{CaCO_3}^{-1} \qquad (83b)$$

If a Mg calcite is in contact with a solution of which the $(\{Mg^{2+}\}/\{Ca^{2+}\})^x$ ratio is smaller than $K_{eq(x)}K_{CaCO_3}^{-1}$, the Mg-calcite is less stable than $CaCO_3(s)$. For example, a 10% Mg-Calcite [log $K_{eq}(10) \approx 8.0$] is less stable than calcite ($-\log K_{s0} = 8.35$) or aragonite ($-\log K_{s0} = 8.22$) if the activity ratio

$\{Mg^{2+}\}/\{Ca^{2+}\}$ is smaller than 10^2 or $10^{3.5}$, respectively. Thus high Mg-calcite should be converted in marine sediments into calcite or aragonite. The conversion to aragonite has been observed [26]. As the Mg-calcite is dissolved, Mg^{2+} becomes enriched in the solution (=*incongruent dissolution*), and a purer $CaCo_3(s)$ (calcite or aragonite) is precipitated. Low-Mg-calcites ($x < 3$ to 4 mol %) are probably stable in comparison to calcite. Higher-Mg-calcites, although thermodynamically unstable, may persist (in stoichiometric saturation) for considerable time periods. Surface seawater is oversaturated with regard to calcite and aragonite. An extensive controversy among various authors [G. M. Lafon, R. Garrels and R. Wollast, R. A. Berner, D. C. Thorstenson and L. N. Plummer, *Amer. J. Sci.*, **278**, 1455–1488 (1978)] has developed concerning the question whether the chemistry of magnesian calcites in natural waters is governed by thermodynamic equilibrium or whether it reflects kinetic phenomena. More recently J. W. Morse et al. [*Science*, **205**, 904 (1979)] has reported that magnesian calcite overgrowth containing 4 (± 2) mol % $MgCO_3$ forms on calcite exposed to natural seawater near the ocean surface. This magnesian calcite is ca. 30 % less soluble in seawater than pure calcite. The formation of the magnesian calcite of reduced solubility may have an influence on calcite accumulation in deep-sea sediments.

Accessible Mg-calcite may be an important component of the (heterogeneous) CO_2 buffer system of the sea.

Occlusion and Surface Adsorption. In occlusion and surface adsorption the impurities become trapped in the solids or adsorbed on the surface under *nonequilibrium conditions*. The extent of occlusion depends upon the rate of precipitation. In adsorption the microcomponent is adsorbed or undergoes chemical reaction (surface complex formation) on the solid. Occasionally it is possible to distinguish between solid solution formation and incorporation of foreign ions by occlusion or adsorption by the use of X-ray diffraction patterns. Although the method is not very sensitive, solid compound formation will cause a change in the lattice dimension; the latter is proportional to the concentration of the solute. No observable changes in the X-ray diffraction pattern result from surface adsorption.

Solubility of Fine Particles

Finely divided solids have a greater solubility than large crystals. As a consequence small crystals are thermodynamically less stable and should recrystallize into large ones. For particles smaller than about 1 μm or of specific surface area greater than a few square meters per gram, surface energy may become sufficiently large to influence surface properties. Similarly the free energy of a solid may be influenced by lattice defects such as dislocations and other surface heterogeneities.

The change in the free energy ΔG involved in subdividing a coarse solid suspended in aqueous solution into a finely divided one of molar surface S is given by

$$\Delta G = \tfrac{2}{3}\bar{\gamma}S \tag{84}$$

where $\bar{\gamma}$ is the mean free surface energy (interfacial tension) of the solid–liquid interface.

Equation 84 can be derived [27] from the thermodynamic statement that at constant temperature and pressure and assuming only one "mean" type of surface

$$dG = \mu_0\,dn + \bar{\gamma}ds \tag{85}$$

or

$$\mu = \mu_0 + \frac{\bar{\gamma}ds}{dn} \tag{86}$$

This can be rewritten as

$$\mu = \mu_0 + \frac{M}{\rho}\bar{\gamma}\frac{ds}{dv} \tag{87}$$

where M = formula weight, ρ = density, n = number of moles, and s = surface area of a single particle.

Because the surface and the volume of a single particle of given shape are $s = kd^2$ and $v = ld^3$,

$$\frac{ds}{dv} = \frac{2s}{3v} \tag{88}$$

Since the molar surface is $S = Ns$ and the molar volume $V = Nv = M/\rho$, where N is particles per mole, equation 87 can be rewritten as

$$\mu = \mu_0 + \tfrac{2}{3}\bar{\gamma}S \tag{89}$$

or, in terms of equation 84,

$$\frac{d\ln K_{s0}}{dS} = \frac{2}{3}\frac{\bar{\gamma}}{RT} \tag{90}$$

or

$$\log K_{s0(S)} = \log K_{s0(S=0)} + \frac{\tfrac{2}{3}\bar{\gamma}}{2.3RT}S \tag{91}$$

The specific surface effect can also be expressed by substituting $S = M\alpha/\rho d$ where α is a geometric factor which depends on the shape of the crystals ($\alpha = k/l$).

Schindler [28] and Schindler et al. [29] have investigated the effect of particle size and molar surface on the solubilities of ZnO, $Cu(OH)_2$, and CuO. An

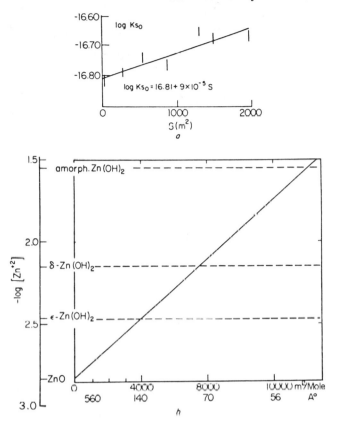

Figure 5.22 Effect of particle subdivision upon of solubility of ZnO. (*a*) Effect of molar surface upon solubility product of ZnO (25°C, $I = 0$). (From Schindler [28]) (*b*) Comparison of the solubility (pH = 7) of ZnO, as it is influenced by molar surface or particle diameter, with the solubility of polymorphous modifications of Zn(OH)$_2$ (see scale on the left). (From data by Schindler et al., [29]) ZnO is the most stable of the modifications, but if its molar surface area is larger than 4000 m^2 ($d < 140$ Å) it becomes less stable than (coarse) ε-Zn(OH)$_2$. (Reproduced with permission from American Chemical Society.)

example of their data is given in Figure 5.22 where the solubility constant of ZnO is plotted, in accordance with equation 91, semilogarithmically as a function of the molar surface. The mean suface tension $\bar{\gamma}$ can be computed from the slope ($\bar{\gamma}$ for interface ZnO–0.2 M NaClO$_4$ = $770 \pm 300 \times 10^{-7}$ J cm^{-2}).

Metastability and Particle Size. The particle size may play a significant role in the inversion from one polymorphous form to another. Following arguments presented by Schindler [28] and Schindler et al. [29] we may consider the reaction

$$Cu(OH)_2(s) = CuO(s) + H_2O(l) \tag{92}$$

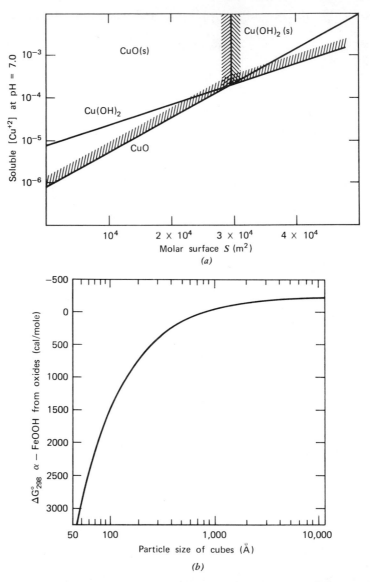

Figure 5.23 (a) Influence of molar surface upon solubility of CuO and of Cu(OH)$_2$ at pH $= 7.0$. (From data on solubility constants and surface tensions by Schindler et al., [29]) (The relations depicted have been validated experimentally only for $S < 10^4$ m^2.) The figure suggests that Cu(OH)$_2$(s) becomes more stable than CuO(s) for very finely divided CuO crystals ($S > 3 \times 10^4$ m^2, $d < 40$ Å). Plausibly, in precipitating Cu(II), Cu(OH)$_2$(s) may be precipitated ($d =$ very small), but CuO(s) becomes more stable than Cu(OH)$_2$ upon growth of the crystals, and an inversion of Cu(OH)$_2$ into the more stable phase becomes possible. (b) $\Delta G°$ for the reaction $\frac{1}{2}\alpha$-Fe$_2$O$_3 + \frac{1}{2}$H$_2$O $= \alpha$-FeOOH is plotted as a function of particle size assuming equal particle size for geoethite and hematite. For equal-sized hematite and goethite crystals, goethite is more stable than hematite when the particle size exceeds 760 Å but less stable than hematite at smaller particle sizes. (From D. Langmuir and D. O. Whittemore, in *Nonequilibrium Systems in Natural Water Chemistry*, Advances in Chemistry Series, No. 106, American Chemical Society, Washington, D.C., 1971, p. 209.)

Schindler et al. determined the solubility constants and the influence of molar surface upon solubility:

$$Cu(OH)_2: \log {}^*K_{s0} = 8.92 + 4.8 \times 10^{-5}S \qquad \bar{\gamma} = 410 \pm 130 \times 10^{-7}\,\text{J cm}^{-2}$$
$$CuO: \log {}^*K_{s0} = 7.89 + 8 \times 10^{-5}S \qquad \bar{\gamma} = 690 \pm 150 \times 10^{-7}\,\text{J cm}^{-2}$$

Thus the solubility of $Cu(OH)_2$ is approximately 10 times greater than that of CuO. The inversion of $Cu(OH)_2$ into CuO (reaction 92) should occur exergonically. However, if CuO is very finely divided ($S_{CuO} > 12{,}000\ \text{m}^2$), it becomes less stable than coarse $Cu(OH)_2$ ($S_{Cu(OH)_2} = 0$):

$$\Delta G\,(\text{cal}) = -1400 + 0.109 S_{CuO} - 0.065 S_{Cu(OH)_2} \tag{93}$$

Figure 5.23a plots the enhancement of the solubility of CuO and $Cu(OH)_2$ (pH $= 7.0$) with an increase in molar surface. We may recall that the solid phase which gives the smaller solubility is the thermodynamically more stable phase. As Figure 5.23 illustrates, at sufficiently large subdivision of the solids ($S_{CuO} = S_{Cu(OH)_2} > 31{,}000\ \text{m}^2; d \simeq 40\ \text{Å}$) $Cu(OH)_2(s)$ may become more stable (less soluble) than CuO(s); this results from the larger interfacial energy of the oxide as compared to the hydroxide. As pointed out by Schindler et al. [29], this dependence of stability (solubility) upon particle size may be one of the reasons why crystal nuclei of $Cu(OH)_2$ are formed incipiently when Cu(II) is precipitated; upon subsequent crystal growth, and a corresponding decrease in S, CuO becomes more stable and $Cu(OH)_2$ is inverted into CuO(s). These arguments are schematic and simplified. The assumption that $\bar{\gamma}$ remains independent of surface area probably breaks down when we deal with nuclei consisting of a very small number of ions or molecules. Furthermore, the relations given are valid only for particles of uniform size.

Figure 5.23b shows variation of the reaction free energy with particle size for the conversion of hematite to goethite. The stability relation depends upon particle size.

5.7 THE EFFECT OF INERT ELECTROLYTE ON SOLUBILITY; NONIDEALITY CORRECTIONS

If we consider a solid–solution equilibrium, any decrease in the activity of the solute species is accompanied by an increase in solubility. Hence, in dilute solutions ($I < 1$), the solubility of a salt is enhanced in the presence of inert electrolytes. In more concentrated solutions activity coefficients may become larger than 1, and in very concentrated electrolytes ($I > 1$) the solubility of a salt may become smaller again.

In this section we discuss briefly how to apply nonideality corrections using the Debye–Hückel equation and how to convert activities from the infinite dilution activity scale to the constant ionic medium activity scale without making nonthermodynamic assumptions. But we should keep in mind that the nonideality effects are frequently small in comparison to the other un-certainties in the calculations of solid–solution equilibria. Frequently the solubility constants are not known with sufficient accuracy and side reactions are disregarded or unknown. If values for the same constant reported by different authors vary by a few orders of magnitude and we have no basis to select critically the "best" value, it makes little sense to apply salinity corrections. In dealing with solid–solution equilibria there are many causes in addition to effects by inert electrolytes which may limit the accuracy and the applicability of equi-librium calculations. Despite such limitations, and even if we ignore corrections for nonideality, equilibrium calculations may be useful as an aid in appreciating the effect of pertinent variables and may help in the interpretation of observed facts.

Solubility and Debye–Hückel Theory

The solubility product of equilibrium $M_n N_m(s) = nM + mN$ can be formulated:

$$K_{s0} = [M]^n [N]^m f_M^n f_N^m \tag{94}$$

By using the extended Debye–Hückel law, equation 94 can be rearranged to

$$\log K_{s0} = \log {}^c K_{s0} - (nz_M^2 + mz_N^2)\left(\frac{A\sqrt{I}}{1 + Ba\sqrt{I}}\right) \tag{95}$$

In dealing with media containing mixed electrolytes it may be appropriate to set $A = 0.5$ and $Ba = 1$ in equation 95; then

$$pK_{s0} = p^c K_{s0} + (nz_M^2 + mz_N^2)\left(\frac{0.5\sqrt{I}}{1 + \sqrt{I}}\right) \tag{96}$$

Instead of equation 90, we may use the Davies equation:

$$pK_{s0} = p^c K_{s0} + (nz_M^2 + mz_N^2)\left(\frac{0.5\sqrt{I}}{1 + \sqrt{I}} - 0.2I\right) \tag{97}$$

Both equations 96 and 97 illustrate that the solubility increases with increasing I and that the effect is especially pronounced for salts containing multivalent ions [e.g., the correction factor is nine times larger for $FePO_4(s)$ than for $AgCl(s)$].

Elucidation of Standard Free Energy of Formation from Measurements in an Ionic Medium.   The experimental evaluation of a solubility constant obviously has to be carried out in solutions of finite concentration. Most solubility (and other equilibrium) constants are reported as so-called thermodynamic constants; the experimental constants are extrapolated to infinite dilution (zero ionic strength) or the constants are corrected to $I = 0$ by the application of a theoretical or empirical formula (usually the Debye–Hückel theory).

Within the last 35 years many solution chemists have preferred to determine so-called stoichiometric constants; these are measured in the presence of a large excess of inert electrolyte (e.g., KNO_3 or $NaClO_4$). In this way a constant ionic environment can be maintained even if one component of the solution is varied; hence ionic strength and all activity coefficients remain constant even if equilibria are shifted with respect to other more dilute solution conditions. As pointed out in Chapter 3, concentration and activity become identical, that is, analytically determined concentration equals potentiometrically determined concentration (or activity) if the same ionic medium is used as a reference. These stoichiometric solubility constants can be determined quite accurately; thermodynamically they are equally well defined as the so-called thermodynamic equilibrium constants valid at infinite dilution. One serious drawback is that the stoichiometric constants are strictly valid only for the specified ionic medium. The standard free energy associated with a solubility equilibrium valid at a given ionic strength, $\Delta G^{\circ\prime}$, is shifted by a constant from the standard free energy defined by the conventional thermodynamic scale, ΔG°.

As Schindler [30] has shown, the free energy of formation of a compound can be obtained from characteristic stoichiometric equilibrium constants without making nonthermodynamic assumptions. Let us consider, for example (see Table 5.5), the determination of the free energy of formation of zinc oxide or zinc hydroxide.

The standard free energy of formation of ZnO or $Zn(OH)_2$ [ΔG° of (5) or (6) of Table 5.5] can be evaluated from measurements carried out with solutions of a particular ionic medium. Note that ΔG° is unaffected by the choice of the ionic medium.

TABLE 5.5   EVALUATION OF STANDARD FREE ENERGY OF FORMATION FROM HETEROGENEOUS EQUILIBRIA IN CONSTANT IONIC MEDIUM[a]

Reaction	Free Energies	
$Zn^{2+}(aq)_{(I)} + H_2O_{(I)} = ZnO(s) + 2H_{(I)}^+$	$\Delta G_{1a}^{\circ\prime} = RT \ln {}^{c*}K_{s0(ZnO)}$	$(1a)$
$Zn^{2+}(aq)_{(I)} + 2H_2O_{(I)} = Zn(OH)_2(s) + 2H_{(I)}^+$	$\Delta G_{1b}^{\circ\prime} = RT \ln {}^{c*}K_{s0(Zn(OH)_2)}$	$(1b)$
$Zn(s) + 2H^+(aq)_{(I)} = Zn_{(I)}^{2+} + H_2(g)$	$\Delta G_2^{\circ\prime} = 2FE_{cell(I)}^{\circ}$	$(2)^b$
$H_2(g) + \frac{1}{2}O_2(g) = H_2O(g)$	$\Delta G_3^{\circ} = -54,635$ cal	(3)
$H_2O(g) = H_2O(l)_{(I)}$	$\Delta G_4^{\circ\prime} = RT \ln p_{H_2O(I)}$	(4)
$Zn(s) + \frac{1}{2}O_2(g) = ZnO(s)$	$\Delta G^{\circ} = \Delta G_{1a}^{\circ} + \Delta G_2^{\circ} + \Delta G_3^{\circ} + \Delta G_4^{\circ}$	$(5)^c$
$Zn(s) + H_2(g) + O_2(s) = Zn(OH)_2(s)$	$\Delta G^{\circ} = \Delta G_{1b}^{\circ} + \Delta G_2^{\circ}$ $+ 2(\Delta G_3^{\circ} + \Delta G_4^{\circ})$	$(6)^c$

[a] Cf. P. Schindler, H. Althaus, and W. Feitknecht, *Helv. Chim. Acta*, **47**, 982 (1964).
[b] $E_{cell(I)}^{\circ}$ corresponds to the standard emf value at a given ionic strength; F = Faraday.
[c] ΔG° for ZnO(s) or $Zn(OH)_2(s)$ is obtained from measurements in solutions of constant ionic medium (I).

Values for $\Delta G_{1a}^{\circ\prime}$ or $\Delta G_{1b}^{\circ\prime}$ are obtained from the corresponding solubility constants ($^{c*}K_{s0}$). $\Delta G_2^{\circ\prime}$ results from emf measurements of a cell,† $Pt(H_2)/$ solution of ionic strength I, Zn^{2+}/Zn, where the solution contains the same ionic medium used for the determination of the solubility constant. Excellent literature data are available for ΔG_3° and ΔG_4°. In Section 5.4 it was illustrated how the free energy of formation of a metal carbonate could be evaluated.

The Conversion of $^{c}K_s$ into K_s

Such thermodynamic cycles can be used to convert stoichiometric constants into thermodynamic ones, and vice versa (Schindler [28]). The procedure is illustrated in Example 5.15 in which the following equation, generally valid for oxides and carbonates of bivalent metals, is derived:

$$RT \ln \frac{^{*}K_{s0}}{^{c*}K_{s0}} = RT \ln \frac{p_{H_2O(I)}}{p_{H_2O(I=0)}} + 2F(E_{cell(I)}^{\circ} - E_{cell(I=0)}^{\circ}) \qquad (98)$$

Equation 98 interrelates the four basic quantities $^{c*}K_{s0}$, $^{*}K_{s0}$, $E_{(I=0)}^{\circ}$, and $E_{(I)}^{\circ}$.

Example 5.15. The Conversion of $^{c}K_s$ into K_s

Compute $^{*}K_{ps0}$ (25°C) for $ZnCO_3(s)$ from $\log ^{c*}K_{ps0} = 8.17 \pm 0.05$ ($I = 0.2\ M\ NaClO_4$, 25°C) (cf. Schindler [28]). We first calculate the free energy of formation of $ZnCO_3$ by building up $ZnCO_3(s)$ from elements through reactions where equilibrium constants (K or E_{cell}) have been determined for an ionic medium of $0.2\ M\ NaClO_4$:

$$Zn(s) + 2H_{(I)}^{+} = Zn_{(I)}^{2+} + H_2(g) \qquad \Delta G_1^{\circ\prime} = 2FE_{cell(I)}^{\circ} \qquad (i)$$

$$C(s) + O_2(g) = CO_2(g) \qquad \Delta G_2^{\circ} \qquad (ii)$$

$$H_2(g) + \tfrac{1}{2}O_2(g) = H_2O(g) \qquad \Delta G_3^{\circ} \qquad (iii)$$

$$H_2O(g) = H_2O(l)_{(I)} \qquad \Delta G_4^{\circ\prime} = RT \ln p_{H_2O(I)} \qquad (iv)$$

$$Zn_{(I)}^{+2} + CO_2(g) + H_2O_{(I)} = ZnCO_3(s) + 2H_{(I)}^{+} \qquad \Delta G_5^{\circ\prime} = RT \ln {}^{c*}K_{ps0} \qquad (v)$$

$$Zn(s) + C(s) + \tfrac{3}{2}O_2(g) = ZnCO_3(s)$$

$$\Delta G_6^{\circ} = \Delta G_1^{\circ\prime} + \Delta G_2^{\circ} + \Delta G_3^{\circ} + \Delta G_4^{\circ\prime} + \Delta G_5^{\circ\prime} \qquad (vi)$$

With ΔG_6°, the free energy of formation of $ZnCO_3$, $^{*}K_{ps0}$ (valid at $I = 0$) can be computed if reliable values for the standard free energy of formation of $Zn^{2+}(aq)_{(I=0)}$, $CO_2(g)$, and $H_2O(l)_{(I=0)}$ are available. We can also close the

---

† Schindler et al. [29] used zinc amalgam instead of Zn. This difference (less than $10^{-5}\ V$) can be disregarded for the calculation of $\Delta G_2^{\circ\prime}$.

cycle and "break down" $ZnCO_3(s)$ into the elements through reactions where equilibria are considered at infinite dilution:

$$ZnCO_3(s) + 2H^+_{(I=0)} = Zn^{2+}_{(I=0)} + CO_2(g)$$
$$+ H_2O_{(I=0)} \qquad \Delta G^\circ_7 = -RT \ln {}^*K_{ps0} \qquad \text{(vii)}$$

$$Zn^{2+}_{(I=0)} + H_2(g) = Zn(s) + 2H^+_{(I=0)} \qquad \Delta G^\circ_8 = -2FE_{cell(I=0)} \qquad \text{(viii)}$$

$$CO_2(g) = C(s) + O_2(g) \qquad -\Delta G^\circ_2 \qquad \text{(ix)}$$

$$H_2O_{(I=0)} = H_2O(g); \qquad \Delta G^\circ_{11} = -RT \ln p_{H_2O(I=0)} \qquad \text{(x)}$$

$$H_2O(g) = H_2(g) + \tfrac{1}{2}O_2(g) \qquad -\Delta G^\circ_3 \qquad \text{(xi)}$$

$$ZnCO_3(s) - Zn(s) + C(s) \qquad \Delta G^\circ_{13} = \Delta G^\circ_7 + \Delta G^\circ_8 - \Delta G^\circ_2$$
$$+ \tfrac{3}{2}O_2(g) \qquad\qquad + \Delta G^\circ_{11} - \Delta G^\circ_3 \qquad \text{(xii)}$$

Because the sum of all the free energy contributions of equation i to equation v and equation vii to equation xi must equal zero,

$$RT \ln \frac{{}^*K_{ps0}}{{}^{c*}K_{ps0}} = RT \ln \frac{p_{H_2O(I)}}{p_{H_2O(I=0)}} + 2F(E^\circ_{cell(I)} - E^\circ_{cell(I-0)}) \qquad \text{(xiii)}$$

Schindler found $E^\circ_{cell(I=0.2)} = -769.3 \pm 0.3$ mV. The standard potential, however, is $E^\circ_{cell(I=0)} = -762.8$ mV. Because the first term on the right-hand side of equation xiii is essentially zero, we obtain

$$\log \frac{{}^*K_{ps0}}{{}^{c*}K_{ps0}} \simeq \frac{-0.0065}{0.02958} = -0.22 \quad \text{or} \quad \log {}^*K_{ps0} = 7.95 \, (\pm 0.05)$$

Schindler also shows that this value is in agreement with that obtained from converting the thermodynamic constant into the stoichiometric one using the Davies equation:

$$\log {}^{c*}K_{ps0} = \log {}^*K_{ps0} + \frac{\sqrt{I}}{1 + \sqrt{I}} - 0.3(I) \qquad \text{(xiv)}$$

By using equation xiv, the constant valid at infinite dilution becomes $\log {}^*K_{ps0} = 7.92 \pm 0.05$.

5.8 THE COEXISTENCE OF PHASES IN EQUILIBRIUM

We illustrated earlier that in equilibrium calculations the elementary algebraic rule, according to which a set of n unknowns can be resolved if they are connected by n independent equations, can be used to calculate the number of independent variables needed to define an equilibrium system under specified conditions; for example, we considered that in a homogeneous phase carbonate solution, in addition to temperature and pressure, two independent variables, such as [Alk] and pH or C_T and $[CO_3^{2-}]$, were necessary in order to define completely the composition of the solution. Similarly, for a ternary $CaO-CO_2-H_2O$ system it was shown that p_{CO_2}, P and T might be freely chosen in order to fix the composition of an equilibrium mixture containing $CaCO_3(s)$, aqueous solution, and $CO_2(g)$. Considerations such as these are statements similar to those given more formally by the Gibbs phase rule. One object of the phase rule is to determine the conditions of reproducibility of the equilibrium system so that it can be copied at will except as to the size and shape.

The Phase Rule

The phase rule is often stated in the form

$$F = C + 2 - P \tag{99}$$

where F is the number of degrees of freedom, that is, the independent internal variables, C is the number of components, and P is the number of phases present in the system and in equilibrium with each other. A *phase* is a domain with uniform composition and properties. Such specific properties as density, specific heat, and compressibility of a phase are the same throughout its extent. Phases are, for example, a gas, a gaseous mixture, a homogeneous liquid solution, a uniform solid substance, a solid solution. The number of *components* C is defined as the minimum number of substances that we must bring together in order to duplicate the system under consideration. Frequently a variety of choices of components may be made, but it is only important to find the least number of substances capable of expressing the composition of the mixture. For example, a system containing solid CaO and $CaCO_3$ in equilibrium with an aqueous solution and a partial pressure of CO_2 may be copied by mixing either set of the following components: CaO, CO_2, H_2O; $Ca(OH)_2$, H_2CO_3, H_2O; or $CaCO_3$, CaO, H_2O. Thus $CaCO_3$ is equivalent to CaO + CO_2, or the carbon dioxide gas phase is equivalent to $CaCO_3$ − CaO.

The number of *degrees of freedom* or the *variance* of a system is the number of variable factors (independent internal variables) such as temperature, pressure, partial pressure, concentration, mole fraction for which one can choose arbitrary values. For a system in equilibrium, not more arbitrary values than F internal variables can be selected.

Derivation. The phase rule can be derived from exact definitions of thermodynamic functions. Only an outline of such a derivation is given here; rigorous derivations will be found in thermodynamic textbooks.

In general the number of independent variables F must be equal to the number of variables U minus the number of independent equations I:

$$F = U - I \tag{100}$$

In any phase, the composition is defined by $C - 1$ concentration terms. For example, $C - 1$ mole fractions (X) define the composition because one of the mole fractions is given by the condition that the sum of the mole fractions is unity. Hence, in a system of P phases, the internal variables U are, besides pressure and temperature, $C - 1$ concentration terms in each phase:

$$U = P(C - 1) + 2 \tag{101}$$

For a system in equilibrium at constant temperature and pressure, the chemical potential for any given component has the same value in every phase. For each unique phase equilibrium we may write one independent equation

$$I = C(P - 1) \tag{102}$$

Thus the number of variables remaining undetermined according to equation 100 for the system at equilibrium is

$$F = P(C - 1) + 2 - C(P - 1) = C - P + 2 \tag{103}$$

Simple Applications.   The phase rule is a useful tool for organizing equilibrium models. In considering such a model it is expedient to examine the variables and the relations connecting them. A knowledge of the variance (F) is of importance in assessing whether or not sufficient information is available for a definite numerical solution of an equilibrium problem; by assigning appropriate values to independent variables the variance can be reduced. Furthermore, the rule illustrates in a lucid way that systems that are quite different in character may behave in a similar manner.

Some of the most simple applications of the phase rule can perhaps be explained by reference to the examples listed in Table 5.6.

All systems of one component (example 1) can be perfectly defined by giving values to, at most, two variable factors, usually P and T. If any two phases can coexist in equilibrium in a system of $C = 1$, only one independent variable, T or P, can be selected. At a given temperature the pressure has a definite value: $P = f(T)$. An invariant system $(C = 1)$ is characterized by a triple point in a $P-T$ diagram; both P and T are fixed if three phases coexist. The phase rule predicts that in a one-component system four phases (e.g., two crystalline solid phases such as ice I and ice III together with liquid water and water vapor) cannot be in equilibrium with each other.

As a further illustration, consider example 3 of Table 5.6. A gas dissolving in a liquid is a bivariant system. If T is specified, the equilibrium solubility of the gas in the liquid still varies with the partial pressure of the gas (Henry's law).

Example 4 illustrates a two-component hydrate where, for $P = 3$, at a given temperature the vapor pressure of the system must be constant.

In Examples 5 and 6 the reactions

$$CaCO_3(s) = CaO(s) + CO_2(g) \tag{104}$$

$$3\,Fe_2O_3(s) = 2\,Fe_3O_4(s) + \tfrac{1}{2}O_2(g) \tag{105}$$

are compared. Because the solids in equation 105 interact to form a solid solution, the system is bivariant at equilibrium. The system of equation 104, however, has only one degree of freedom because the two solids do not interact and form two solid phases.

Example 7 of Table 5.6 illustrates that one degree of freedom is lost for every new phase that is in thermodynamic coexistence. An equilibrium system containing calcite, aqueous solution, and $CO_2(g)$ is bivariant; if we fix the partial pressure of CO_2 at a given temperature, the composition of the solution (pH, C_T, [Alk], etc.) is constant and remains unchanged with isothermal evaporation or dilution. Even the addition of the base $Ca(OH)_2$ does not change the composition of the system as long as equilibrium between the three phases is taken to exist. Hence in such regard the system behaves as a pH stat or a pCa stat; the buffer intensity,

$$\beta_{CaCO_3(S),\ aq,\ CO_2(g)}^{Ca(OH)_2}$$

is infinite. If we "make" four phases [$Ca(OH)_2(s)$, calcite, aqueous solution, $CO_2(g)$] with the same three components, only one degree of freedom remains. As long as all four phases coexist in equilibrium at a given temperature, the p_{CO_2} remains constant and is independent of the quantities of the two solids.

TABLE 5.6 PHASE RULE RELATIONSHIPS

Components		Phases			Variables		
No. C	Type	No. P	Type, Examples		No. F	Examples of Internal Variables	Examples of Independent Variables
(1) 1	H_2O	1	Liquid		2	P, T	P, T
		2	Ice, liquid		1	P, T	P or T
		2	Liquid, vapor		1	P, T, p_{H_2O}	T or p^a
		3	Ice, liquid vapor		0	—	—
(2) 2	SiO_2, H_2O	1	Aqueous solution		3	$[H_4SiO_4], P, T$	$[H_4SiO_4], P, T$
		2	Quartz, aqueous solution		2	$[H_4SiO_4], P, T$	P, T
		3	Quartz, cristobalite, aqueous solution		1	$[H_4SiO_4], P, T$	T or P
(3) 2	H_2O, CO_2	2	Aqueous solution, $CO_2(g)$		2	$T, P, p_{H_2O}, p_{CO_2}, [CO_2], [H_2CO_3]$	T and $p_{CO_2}{}^a$
		3	Ice, solution, $CO_2(g)$		1	$T, P, p_{H_2O}, p_{CO_2}, [CO_2], [H_2CO_3]$	P or p_{CO_2}
(4) 2	$CuSO_4$, H_2O	2	$CuSO_4(s)$, $H_2O(g)$		2	P, T, p_{H_2O}	T, p_{H_2O}
		2	$CuSO_4 \cdot H_2O(s)$, $H_2O(g)$		2	P, T, p_{H_2O}	T, p_{H_2O}
		3	$CuSO_4(s)$, $CuSO_4 \cdot H_2O(s)$, $H_2O(g)$		1	P, T, p_{H_2O}	T

306

(5)	2	CaO, CO₂; or CaCO₃, CO₂; or CaCO₃, CaO	3	CaCO₃(s), CaO(s), CO₂(g)	1	T, p_{CO_2}, P	T
(6)	2	Fe₃O₄, O₂; or Fe₂O₃, O₂; or Fe₃O₄, Fe₂O₃	2	Solid solution of Fe₂O₃ in Fe₃O₄, O₂(g)	2	$T, p_{O_2}, P, x_{Fe_3O_4}, x_{Fe_2O_3}$	T, p_{O_2}
(7)	3	CaO, CO₂, H₂O; or Ca(OH)₂, CaCO₃, H₂O; or CaCO₃, CaO, H₂O	1	Aqueous solution	4	$P, T, [Ca^{2+}], [Alk], [H^+], [CO_3^{2-}], C_T$	$T, P, [Alk], pH$
			2	Calcite, aqueous solution	3	$P, T, [Ca^{2+}], [Alk], [H^-], [CO_3^{2-}], C_T$	T, P, C_T; or T, P, Alk
			3	Calcite, aqueous solution, CO₂(g)	2	$P, T, [Ca^{2+}], [Alk], [H^+], [CO_3^{2-}], C_T, p_{CO_2}$	T, p_{CO_2}; or T, P
			4	Calcite, Ca(OH)₂(s), aqueous solution, CO₂(g)	1	$P, T, [Ca^{2+}], [Alk], [H^+], [CO_3^{2-}], C_T$	T
			5	Calcite, Ca(OH)₂(s), ice, aqueous solution, CO₂(g)	0	$P, T, [Ca^{2+}], [Alk], [H^+], [CO_3^{2-}], C_T$	—

ᵃNote in examples (1) and (3) that $P = p_{H_2O}$ and $P = p_{CO_2} + p_{H_2O}$, respectively.

Example 5.16. Salt Deposition in Oceans

A large volume of an aqueous solution containing K^+, Mg^{2+}, Na^+, Cl^-, and SO_4^{2-} is concentrated by isothermal evaporation. Find the maximum number of solid phases that can be in equilibrium with the solution. It can be assumed that the temperature of evaporation does not happen to be an exact transition point. (Cf. Van't Hoff, *Zur Bildung der Ozeanischen Salzablagerungen*, Braunschweig, 1905–1909; this problem is taken from L. G. Sillén, P. W. Lange and C. O. Gabrielson, *Problems in Physical Chemistry*, Prentice-Hall, Englewood Cliffs, N.J. 1952.) An aqueous solution that contains a different cations and b different anions contains $(a + b)$ components (H_2O and $a + b - 1$ salts). In our example, $C = 5$ (e.g., H_2O and Na_2SO_4, K_2SO_4, $MgCl_2$, KCl). Because $F = 2(P, T)$ there must be $P = 5$ phases. Accordingly, a maximum number of four solid phases can be in equilibrium with solution.

In constructing a thermodynamic model for a natural water system. we usually incorporate the components and specify the phases to be included. In a stepwise construction the addition of each new component ($\Delta C = 1$) must be accompanied by either a new phase ($\Delta P = 1$) or a new internal variable ($\Delta F = 1$). The creation of a new phase in a system of a given number of components reduces the degree of freedom by 1; more phases demand fewer assumptions for solving an equilibrium model. The maintenance of constant concentration conditions in an aqueous solution can be accomplished in principle by the coexistence in equilibrium of a sufficient number of phases. Hence the solid phases comprised in the lithosphere are of particular significance in natural water models. Although there might not be true equilibrium in any actual aquatic system, the composition of the lithosphere represents a significant regulatory factor for the composition of the hydrosphere.

It is appropriate to mention here the work of Sillén [31], who epitomized the application of equilibrium models for portraying the prominent features of the seawater system. His analysis, which will be discussed in some detail in Chapter 9, indicates that heterogeneous equilibria of solid minerals comprise the principal buffer systems in oceanic waters; according to such a model the CO_2 content of the oceans and of the atmosphere is controlled by equilibria with solids at the sediment–water interface.

Local Equilibrium. If we consider a large volume of sediments or of rock assemblage, we find phases that are thermodynamically incompatible. But, as pointed out by Thompson [32], it is generally possible to regard any part of such a system as substantially in internal equilibrium if that part is made sufficiently small. With respect to phase equilibrium the requirement appears to be met if no mutually incompatible phases are in actual contact and if phases of variable composition show only continuous variation from point to point.

Congruent and Incongruent Solubility. If we add $MgCO_3(s)$ (magnesite) to distilled water (25°C), this solid will dissolve incipiently; but upon further addition, $Mg(OH)_2(s)$ will precipitate because under these conditions $Mg(OH)_2$ will be less soluble than $MgCO_3(s)$ ($pK_{s0\,(Mg(OH)_2)} = 10.9$; $pK_{s0(MgCO_3)} = 4.9$). Such a dissolution behavior is referred to as incongruent solubility. The term incongruent is generally used if a mineral upon dissolution reacts to form a new solid; or if a saturated solution, in the presence of solid phases with which it is in equilibrium, reacts upon isothermal evaporation with a change in the solid phases. In natural environments incongruent solubility is probably more prevalent than congruent solution.

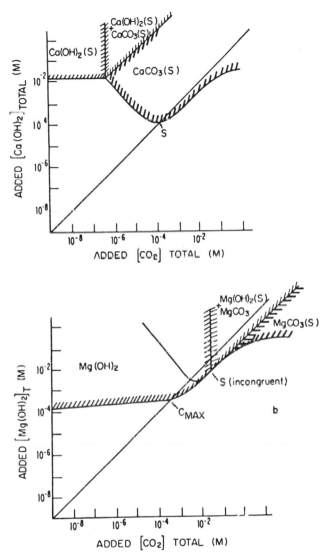

Figure 5.24   Congruent and incongruent solubility of $CaCO_3(s)$ and $MgCO_3(s)$.

From the point of view of the phase rule, dissolution of a simple salt can be interpreted as a system of three components containing not more than two phases (aqueous solution and solid phase; absence of the vapor phase). Hence at a given pressure and temperature, one degree of freedom remains, that is, the concentration of the solution in equilibrium with the solid can undergo change. Ricci [33] has given a lucid, quantitative treatment of incongruent solubility of normal salts. In Figure 5.24 the congruent dissolution behavior of $CaCO_3$ is contrasted with the incongruent dissolution characteristics of $MgCO_3$. A solution of the salt in pure H_2O lies on the diagonal line. The maximum solution concentration of the salt that can be achieved is given by C_{max}. This concentration cannot be exceeded (under conditions of equilibrium). In the case of the $Mg(OH)_2$–CO_2–H_2O system, however, $Mg(OH)_2$ is precipitated. Such a "hydrolytic precipitation" occurs if the solubility of the base C_{max} is smaller than the solubility of the salt S. Thus congruence and incongruence is characterized by $C_{max} > S$ and $C_{max} < S_{inc}$, respectively. Many solid solutions, for example, Mg calcites, being converted into calcite or aragonite, and most aluminum silicates, dissolve incongruently.

5.9   CRYSTAL FORMATION; THE INITIATION AND PRODUCTION OF THE SOLID PHASE

The birth of a crystal and its growth provide an impressive example of nature's selectivity. In qualitative analytical chemistry inorganic solutes are distinguished from each other by a separation scheme based on the selectivity of precipitation reactions. In natural waters certain minerals are being dissolved, while other are being formed. Under suitable conditions a cluster of ions or molecules selects from a great variety of species the appropriate constituents required to form particular crystals.

Minerals formed in natural waters and in sediments provide a record of the physical chemical processes operating during the period of their formation; they also give us information on the environmental factors that regulate the composition of natural waters and on the processes by which elements are removed from the water. The memory record of the sediments allows us to reconstruct the environmental history of the processes that led to the deposition of more ancient minerals and in turn the history of evolution of the ocean and of lakes.

Pronounced discrepancies between observed composition and the calculated equilibrium composition illustrate that the formation of the solid phase, for example, the nucleation of dolomite and calcite in seawater, is often kinetically inhibited.

The Role of Organisms in the Precipitation of Inorganic Constituents

Organisms produce significant chemical differentiation in the formation of solid phases. The precipitation of carbonates, of opal, and of some phosphatic minerals by aquatic organisms has long been acknowledged. Within the last two decades, many different kinds of additional biological precipitates have been found. In a review Lowenstam [34] suggests that life has succeeded in largely substituting for, or displacing to a varying extent, inorganic precipitation processes in the sea in the course of the last 6×10^8 years. It appears [34, 35] that metastable mineral phases and, more commonly, amorphous hydrous phases are the initial nucleation products of crystalline compounds in biological mineral precipitates. Amorphous hydrous substances have been shown to persist in the mature, mineralized hard parts of many aquatic organisms.

The Processes Involved in Nucleation and Crystal Growth

Various processes are involved in the formation of a solid phase from an oversaturated solution (Figure 5.25). Usually three steps can be distinguished:

1   The interaction between ions or molecules leads to the formation of a cluster.

$$X + X \rightleftharpoons X_2$$

$$X_2 + X \rightleftharpoons X_3$$

$$X_{j-1} + X \rightleftharpoons X_j \quad \text{(cluster)}$$

$$\text{Nucleation:} \quad X_j + X \longrightarrow X_{j+1} \quad \text{(nucleus)} \quad (106)$$

Nucleation corresponds to the formation of the new centers from which spontaneous growth can occur. The nucleation process determines the size and the size distribution of crystals produced.

2   Subsequently, material is deposited on these nuclei

$$X_{j+1} + X \rightleftharpoons \text{crystal growth} \quad (107)$$

and crystallites are being formed (crystal growth).

3   Large crystals may eventually be formed from fine crystallites by a process called ripening.

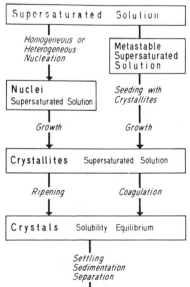

Figure 5.25   Simplified scheme of processes involved in nucleation and crystal growth. A more detailed description of the various processes is given in a similar scheme by Nancollas and Reddy [41].

A better insight into the mechanisms of the individual steps in the formation of crystals would be of great help in explaining the creation and transformation of sedimentary deposits. Valuable reviews are available on the principles of nucleation of crystals and the kinetics of precipitation and crystal growth [22, 36–39]. Only a few important considerations are summarized here to illustrate the wide scope of questions to be answered in order to predict rates and mechanisms of precipitation in natural systems.

Nucleation

If one gradually increases the concentration of a solution, exceeding the solubility product with respect to a solid phase, the new phase will not be formed until a certain degree of supersaturation has been achieved. Stable nuclei can only be formed after an activation energy barrier has been surmounted.

The free energy of the formation of a nucleus, ΔG_j, consists essentially of energy gained (volume free energy) from making bonds and of work required to create a surface:

$$\Delta G_j = \Delta G_{\text{bulk}} + \Delta G_{\text{surf}} \tag{108}$$

For a spherical nucleus, the first quantity (always negative) can be expressed as

$$\Delta G_{\text{bulk}} = -jkT \ln \frac{\text{IAP}_0}{K_{s0}} \tag{109}$$

where j is the number of molecules or ions in the nucleus or, expressed in terms of volume for a spherical nucleus, $j = 4\pi r^3/3V$, where V is the "molecular" volume; IAP_0/K_{s0} is the oversaturation ratio (equation 10) (ion activity product of the supersaturated solution divided by the theoretical solubility for large crystals); the second quantity is given by

$$\Delta G_{\text{surf}} = 4\pi r^2 \bar{\gamma} \tag{110}$$

where $\bar{\gamma}$ is the interfacial energy. Hence the free energy of nucleus formation may be written

$$\Delta G_j = \frac{-4\pi r^3}{3V} kT \ln \frac{\text{IAP}_0}{K_{s0}} + 4\pi r^2 \bar{\gamma} \tag{111}$$

In Figure 5.26a, ΔG_j is plotted as a function of r for assumed constant values† of V, and $\bar{\gamma}$ for a few values of oversaturation. Obviously the activation energy ΔG_a decreases with increasing supersaturation, as does the size of the nucleus, r_j. In Section 5.6 we have seen that small crystallites are more soluble than large

† $\bar{\gamma}$ is not necessarily independent of cluster size.

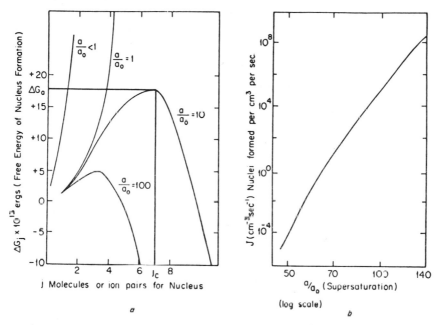

Figure 5.26   Nucleation. The energy barrier and the nucleation rate depend critically on the supersaturation. (a) Free energy of formation of a spherical nucleus as a function of its size, calculated for different supersaturations (a/a_0 or IAP_0/K_{s0}) [cf. (111)]. (b) Double logarithmic plot of nucleation rate versus supersaturation (a/a_0) calculated with equations 114 and 115. The curves have been calculated for the following assumptions: $\bar{\gamma} = 100$ ergs cm^{-2}; "molecular" volume $V = 3 \times 10^{-23}$ cm^3; collision frequency efficiency equation (115) $A = 10^{30}$ cm^{-3} sec^{-1}.

crystals; hence the energy barrier is related to the additional free energy needed to form the more soluble nuclei and is due to the surface energy of these small particles. Equation 91 can be rewritten for an aggregate of spherical shape:

$$kT \ln \frac{IAP_0}{K_{s0}} = \frac{2V\bar{\gamma}}{r} \tag{112}$$

where IAP_0/K_{s0} is the concentration (or activity) ratio (supersaturation); it can also be expressed as a/a_0, where a and a_0 are the actual and equilibrium activities, respectively, of the solutes that characterize the solubility. V is the molecular volume, $\bar{\gamma}$ the interfacial energy, and r the radius. Once nuclei of critical size X_{j+1} (in equation 106) have been formed, crystallization is spontaneous.

The activation energy ΔG_a can be calculated by inserting into equation 111 r_j as obtained from equation 112:

$$r_j = \frac{2\bar{\gamma}V}{kT \ln(a/a_0)} \tag{113}$$

for which

$$\Delta G_a = \frac{16\pi\bar{\gamma}^3 V^2}{3[kT \ln(a/a_0)]^2} \tag{114}$$

The rate at which nuclei form J may be represented according to conventional rate theory as

$$J = A \exp\left(\frac{-\Delta G_a}{kT}\right) \tag{115}$$

where A is a factor related to the efficiency of collisions of ions or molecules. Accordingly, the rate of nucleation is controlled by the interfacial energy, the supersaturation, the collision frequency efficiency, and the temperature. For given values of T, A, and $\bar{\gamma}$, the nucleation rate J (nuclei formed $cm^{-3}\ sec^{-1}$) can be calculated as a function of IAP_0/K_{s0} (Figure 5.26b). J is critically dependent upon the supersaturation. For example, using the conditions specified for Figure 5.26a, one can calculate that nucleation is almost instantaneous at a supersaturation of 100 (10^5 nuclei $cm^{-3}\ sec^{-1}$), while (homogeneous) nucleation should not occur even within geological time spans for a 10-fold supersaturation, 1 nucleus cm^{-3} within 10^{70} sec). It is appropriate to speak of a critical supersaturation that must be exceeded before nucleation will occur.

At a high degree of supersaturation, the nucleation rate is so high that the precipitate formed consists mostly of extremely small crystallites. As discussed in Section 5.6, incipiently formed crystallites might be of a different polymorphous form than the final crystals. If the nucleus is smaller than a one-unit cell, the growing crystallite produced initially is most likely to be amorphous; substances with a large unit cell tend to be precipitated initially as an amorphous phase ("gels").

Heterogeneous nucleation, however, is the predominant formation process for crystals in natural waters. In a similar way as catalysts reduce the activation energy, foreign solids may catalyze the nucleation process by reducing the energy barrier. If the surface of the heteronucleus matches well with the crystal, the interfacial energy between the two solids, $\bar{\gamma}_2$, is smaller than the interfacial energy between the crystal and the solution, $\bar{\gamma}$, and nucleation may take place at a lower concentration on a heteronucleus than in a homogeneous solution. The matching between surfaces is more a matter of agreement in lattice type and atomic distances than chemical similarity. Favorable precipitation sites in the solid substrate will be those where strong adsorption occurs, particularly if bonding with the substrate is possible [39]. The critical supersaturation will decrease with increasing degree of lattice matching between nucleus and substrate [40]. In the extreme case substrate and nucleus are identical and $\bar{\gamma}_2$ (and thus in turn the second term in equation 111) becomes zero.

Phase changes in natural waters are almost invariably initiated by heterogeneous solid substrates. Inorganic crystals, skeletal particles, clays, sand,

and biocolloids can serve as heteronuclei. At the same time trace quantities of surfactants and other material adsorbed at the active sites on potential substrate particles might inhibit or prevent the onset of nucleation.

Crystal Growth

The growth of crystals occurs in successive reaction steps: (a) the transport of solute to the crystal solution interface, (b) the adsorption of solute at the surface, and (c) the incorporation of the crystal constituents into the lattice. Furthermore, as a back-reaction the crystal dissolution must be considered. Growth kinetics depends on the rate-limiting step; essentially two limiting theories on growth kinetics have been established: (a) diffusion-controlled growth and (b) interface-controlled growth.

Diffusion-controlled growth (or dissolution) can be characterized by

$$\frac{dc}{dt} \simeq -ks(c - c_0) \tag{116}$$

where c_0 is the equilibrium concentration, that is, the concentration in the immediate proximity of the crystal surface, and s, the surface area. The rate constant k depends on the diffusion coefficient D and the extent of turbulence in the solution. The crystal grows according to

$$\frac{dr}{dt} = DV(c - c_0) \tag{117}$$

Interface-controlled growth does not depend on the turbulence. The results can be fitted generally to rate laws of the type

$$\frac{dc}{dt} = -k's(c - c_0)^n \tag{118}$$

where n is the order of reaction. n has been determined, for example, for magnesium oxalate ($n = 2$), potassium chloride ($n = 1$), $CaCO_3$ ($n = 2$), and $Mg(OH)_2$ ($n = 1$). Growth of crystals appears to be frequently controlled by interfacial processes such as adsorption or dislocation steps.

Growth of Concretions. In sedimentary rocks we commonly find concretions, that is, material formed by deposition of a precipitate, such as calcite or siderite, around a nucleus of some particular mineral grain or fossil. The origin of most concretions is not known but, as mentioned by Berner [41], their often-spherical shape with concentric internal structure suggests diffusion as an important factor affecting growth. The rate of growth, if diffusion-controlled, is readily amenable to mathematical treatment. Berner [41] has provided some idea of the time scale involved in the growth of postdepositional concretions. His calculations illustrate, for example, that for a typical slowly flowing groundwater, with a supersaturation in $CaCO_3$ of 10^{-4} M (assumed to be constant), the time of growth of calcite concretions ranges from 2500 years for concretions of 1-cm radius to 212,000 years for those of 5-cm radius. Hence concretion growth, if diffusion and convection is the rate-controlling step, is relatively rapid when considered on the scale of geological time.

Some Factors in the Formation and Dissolution of $CaCO_3$

In carbonate diagenesis we deal usually with a combination of low supersaturation and absence of mechanical agitation. Homogeneous nucleation will certainly not occur. The important factors to be investigated are heterogeneous nucleation and rates of growth and dissolution of crystals [42].

Growth Inhibitors.   Because the rate-determining step is frequently controlled at the interface, small amounts of soluble foreign constituents may alter markedly the growth rate of crystals and their morphology. The retarding effect of substances that become adsorbed may be explained as being due primarily to the obstruction by adsorbed molecules to the deposition of lattice ions. In some cases it has been shown that the rate constant for crystal growth is reduced by an amount reflecting the extent of adsorption.

The effects of trace concentrations of dissolved organic matter [43] and of orthophosphates [44] and polyphosphates [41] as "crystal poisons" (e.g., inhibiting the spread of monomolecular steps on the crystal surface by becoming adsorbed on active growth sites such as kinks) on the nucleation and growth of calcite have been investigated in some detail.

Another category of inhibition may be exerted by cations or anions that become adsorbed on active growth sites. Well-hydrated Mg^{2+} interferes with the formation of calcite, apatite, and many other minerals [22, 38, 45–47].

Finally, ions that become adsorbed on the surface of nuclei or crystallites may become incorporated into the growing crystals, that is, solid solutions are formed. These solid solutions may be more soluble than the pure solid phases; the inhibition may then be due to a reduction in the degree of supersaturation. Berner [22] has shown that Mg^{2+} is not adsorbed on the surface of aragonite or incorporated into its crystal lattice. As a result, aragonite crystal growth in seawater is relatively unaffected by the presence of dissolved Mg^{2+}. In contrast, Mg^{2+} is readily adsorbed on the surface of calcite and incorporated into its lattice. The crystal growth of calcite is strongly inhibited by Mg^{2+}†. As we have

---

† R. A. Berner [*Amer. J. Sci.*, **276**, 713 (1976)] has shown that direct measurement of the solubility of calcite in seawater is difficult and irreproducible because of kinetic factors that inhibit the attainment of a true reversible equilibrium. To avoid this problem, he has determined the solubility product of calcite in seawater indirectly by calculation from measurements of the aragonite solubility product in seawater and the differential solubility of calcite and aragonite in carbonated distilled water.

seen (Figure 5.21), Mg-calcites are considerably more soluble than pure calcite. Plausibly, the inhibition of calcite growth is due to this lowering of the extent of supersaturation (Mg^{2+} may, however, interact in a different way in the inhibition of calcite nucleation and in the diagenetic conversion of aragonite to calcite [45]).

It is interesting to note that some crystal poisons may not only interfere with the growth of crystals but may also retard their *dissolution*. Insidious trace quantities of organic matter and phosphate inhibit the dissolution of calcite in undersaturated waters [43, 44]. Apparently precipitation and dissolution of ionic solids proceed by the attachment or detachment of ions at kinks in monomolecular steps on the crystalline surface. The detachment of kinks— points of excess surface energy and preferred sites of chemisorption of crystal poisons—is rendered more difficult by the adsorption of organic solutes, or phosphates, thus giving rise to kink immobilization and retardation of dissolution [48]. Berner and Morse [44] have shown that the critical under saturation of calcite in seawater increases with increasing orthophosphate concentration in solution. Their results provide an explanation for the observed variations of $CaCO_3$ dissolution in the deeper portion of the ocean and for the *lysocline*. The lysocline is a region where the rate of dissolution with depth radically increases.

REFERENCES

1 L. G. Sillén and A. E. Martell, *Stability Constants of Metal-Ion Complexes*, Special Publications, Nos. 17 and 25, Chemical Society, London, 1964 and 1971.

2 W. M. Latimer, *Oxidation Potentials*, Prentice-Hall, Englewood Cliffs, N.J., 1952.

3 W. Feitknecht and P. Schindler, *Solubility Constants of Metal Oxides, Metal Hydroxides and Metal Hydroxide Salts in Aqueous Solution*, Butterworths, London, 1963.

4 R. M. Smith and P. E. Martell, *Critical Stability Constants*, Vol. 4, *Inorganic Complexes*, Plenum, New York, 1976.

5 R. A. Robie, B. S. Hemingway, and J. R. Fisher, *Thermodynamic Properties of Minerals and Related Substances at 298.15 K and 1 Bar (10^5 Pascals) Pressure and at Higher Temperatures*. Geological Survey Bulletin No. 1452, Washington, D.C., 1978.

6 C. F. Baes and R. E. Mesmer, *The Hydrolysis of Cations*, Wiley-Interscience, New York, 1976.

7 H. Gamsjäger, H. U. Stuber, and P. Schindler, *Helv. Chim. Acta*, **48**, 723 (1965).

8 O. P. Bricker and R. M. Garrels, in *Principles and Applications of Water Chemistry*, S. D. Faust and J. V. Hunter, Eds., Wiley, New York, 1967, p. 449.

9 K. J. Hsu, *J. Hydrol.*, **1**, 288 (1963).

10 D. Langmuir, Ph.D. thesis, Harvard University, Cambridge, Mass., 1964.

11 G. Schwarzenbach, *Complexometric Titrations*, Interscience, New York, 1957.

12 A. Ringbom, *Complexation in Analytical Chemistry*, Wiley-Interscience, New York, 1963.

13   P. E. Framson and J. O. Leckie, *Environ. Sci. Technol.*, **12**, p. 465. (1978).

14   F. Vaslow, and G. E. Boyd, *J. Amer. Chem. Soc.*, **74**, 4691 (1952).

15   M. W. Bodine, H. D. Holland and M. Borcsik, *Symposium on Problems of Post-magnative Ore Deposition*, Vol. II, Prague, 1965, p. 407.

16   D. J. J. Kinsman and H. D. Holland, *Geochim. Cosmochim. Acta*, **33**, 1 (1969).

17   A. Katz et al., *Geochim. Cosmochim. Acta*, **36**, 481 (1972).

18   P. Schindler, in *Equilibrium Concepts in Natural Waters*, Advances in Chemistry Series, No. 67, American Chemical Society, Washington, D.C., 1967, p. 196.

19   H. A. Doerner and W. M. Hoskins, *J. Amer. Chem. Soc.*, **47**, 662 (1925).

20   L. Gordon, M. L. Salutsky and H. H. Willard, *Precipitation from Homogeneous Solution*, Wiley, New York, 1959.

21   D. C. Thorstenson and L. N. Plummer, *Amer. J. Sci.*, **277**, 1203 (1977).

22   R. Berner, *Geochim. Cosmochim. Acta*, **39**, 489 (1975).

23   L. N. Plummer and F. T. Mackenzie, *Amer. J. Sci.*, **274**, 61 (1974).

24   J. R. Goldsmith et al., *Geochim. Cosmochim. Acta*, **7**, 212 (1955).

25   K. E. Chave et al., *Science* **137**, 33 (1962).

26   H. D. Winland, *J. Sediment. Petrol.*, **39**, 1579 (1969).

27   B. V. Enüstün and J. Turkevich, *J. Amer. Chem. Soc.*, **82**, 4502 (1960).

28   P. Schindler, in *Equilibrium Concepts in Natural Water Systems*, Advances in Chemistry Series, No. 67, American Chemical Society, Washington, D.C., 1967, p. 196.

29   P. Schindler, H. Althaus, F. Hofer and W. Minder, *Helv. Chim. Acta*, **48**, 1204 (1965).

30   P. Schindler, *Helv. Chim. Acta*, **42**, 577 (1959).

31   L. G. Sillén, in *Oceanography*, M. Sears, Ed., American Association for the Advancement of Science, Washington, D.C., 1961, p. 549.

32   J. B. Thompson, "Local Equilibrium in Metasomatic Processes," in *Researches in Geochemistry*, P. H. Abelson, Ed., Wiley, New York, 1959.

33   J. E. Ricci, *Hydrogen Ion Concentration*, Princeton University Press, Princeton, N.J., 1952.

34   H. A. Lowenstam in *The Sea*, Vol. 5, E. D. Goldberg, Ed., Wiley-Interscience, New York, 1974.

35   H. A. Lowenstam and G. R. Rossman, *Chem. Geol.*, **15**, 15 (1975).

36   A. E. Nielsen, *Kinetics of Precipitation*, Macmillan, New York, 1964.

37   G. H. Nancollas and N. Purdie, *Quart. Rev. (London)*, **18**, 1 (1964).

38   G. H. Nancollas and M. M. Reddy in *Aqueous Environmental Chemistry of Metals*, A. J. Rubin, Ed., Ann Arbor Science Publications, Ann Arbor, Mich., 1974.

39   A. G. Walton, *The Formation and Properties of Precipitates*, Wiley-Interscience, New York, 1979.

40   D. Turnbull and B. Vonnegut, *Ind. Eng. Chem.*, **44**, 1292 (1952); J. B. Newkirk and D. Turnbull, *J. Appl. Phys.*, **26**, 579 (1955).

41   R. A. Berner, *Geochim. Cosmochim. Acta*, **32**, 477 (1968).

42   R. B. de Boer, *Amer. J. Sci.*, **277**, 38 (1977).

43   K. E. Chave and E. Suess, *Limnol. Oceanogr.*, **15**, 633 (1970); E. Suess, *Geochim. Cosmochim. Acta*, **34**, 157 (1970).

44   R. A. Berner and J. W. Morse, *Amer. J. Sci.*, **274**, 108 (1974).

45 J. L. Bishoff, *J. Geophys. Res.*, **73**, 3315 (1968); J. L. Bishoff and W. S. Fyfe, *Amer. J. Sci.*, **266**, 65 (1968).

46 R. M. Pytkowicz, *Amer. J. Sci.*, **273**, 515 (1973).

47 A. Katz, *Geochim. Cosmochim. Acta*, **37**, 1563 (1973).

48 M. B. Ives, *Ind. Eng. Chem.*, **57**, 34 (1965).

READING SUGGESTIONS

Berner, R. A., *Principles of Chemical Sedimentology*, McGraw-Hill, New York, 1971. (This book illustrates how sediments can be studied from a physicochemical view point, that is, by the application of chemical thermodynamics and kinetics to the elucidation of sedimentary problems).

Berner, R. A., "The Role of Magnesium in the Crystal Growth of Calcite and Aragonite in Seawater", *Geochim. Cosmochim. Acta*, **39**, 489–504 (1975).

Bolt, G. H., and M. G. M. Bruggenwert, Eds., *Soil Chemistry, Basic Elements*, Elsevier, Amsterdam, 1976, p. 281. (This book provides a cohesive study of soil chemistry and discusses the common solubility equilibria in soil–water systems.)

Butler, J. N., *Ionic Equilibrium, A Mathematical Approach*, Addison-Wesley, Reading, Mass., 1964.

Case, L. O., "The Phase Rule in Analytical Chemistry." In *Treatise on Analytical Chemistry*, Part I, Vol. 2, I. M. Kolthoff and P. J. Elving, Eds., Wiley-Interscience, New York, 1963, p. 957.

Deines, P., D. Langmuir, and R. S. Harmon, "Stable Carbon Isotope Ratios and the Existence of a Gas Phase in the Evolution of Carbonate Ground Waters," *Geochim. Cosmochim. Acta*, **38**, 1147–1164 (1974).

Drever, J. I., "The Magnesium Problem," in *The Sea*, Vol. 5, E. D. Goldberg, Ed., Wiley-Interscience, New York, 1974, pp. 337–358.

Gieskes, J. M., "The Alkalinity—Total Carbon Dioxide System in Seawater." In *The Sea*, Vol. 5, E. D. Goldberg, Ed., Wiley-Interscience, New York, 1974, pp. 123–152.

Holland, H. D., *The Chemistry of the Atmosphere and Oceans*, Wiley-Interscience New York, 1978. (Describes the processes that control the composition of the atmosphere and oceans; it includes lucid discussions on the weathering of carbonate and silicate rocks and on $CaCO_3$ deposition in the oceans.)

Hömig, H. E., *Physikochemische Grundlagen der Speisewasserchemie*, Classen, Essen, 1963. (Includes excellent discussion on solubility relations at higher temperatures (boilers).)

Leussing, D. L., "Solubility," In *Treatise on Analytical Chemistry*, Part I, Vol. 1, I. M. Kolthoff and P. J. Elving, Eds., Wiley-Interscience, New York, 1963, p. 675.

Lowenstam, H. A., "Impact of Life on Chemical and Physical Processes." In *The Sea*, Vol. 5, E. D. Goldberg, Ed., Wiley-Interscience, New York, 1974, pp. 715–796.

Millero, F. J., "The Thermodynamics of the Carbonate System in Seawater," *Geochim. Cosmochim. Acta*, **43**, 1651–1661 (1979).

Morse, J. W., A. Mucci, and F. J. Millero, "The Solubility of Calcite and Aragonite in Seawater," *Geochim. Cosmochim. Acta.*, **44**, 85 (1980).

Nancollas, G. H. and M. M. Reddy, "Crystal Growth Kinetics of Minerals Encountered in Water Treatment Processes." In *Aqueous Environmental Chemistry of Metals*, A. J. Rubin, Ed., Ann Arbor Science Publications, Ann Arbor, Mich., 1974.

Robie, R. A., B. S. Hemingway, and J. R. Fisher, "*Thermodynamic Properties of Minerals and Related Substances at 298.15°K and 1 Bar (10^5 Pascals) and at Higher Temperatures*," Geological

Survey Bulletin, No. 1452, Washington, 1978. (Contains a critical summary of the available thermo-dynamic data for minerals and related substances in a convenient form for the use of earth scientists.)

Schindler, P. W., "Heterogeneous Equilibria Involving Oxides, Hydroxides Carbonates and Hydroxide Carbonates." In *Equilibria Concepts in Natural Water Systems*, W. Stumm, Ed., Advances in Chemistry Series, No. 67, American Chemical Society, Washington, D.C., 1967, p. 196, (Lucid discussion of evaluation of solubility data emphasizing metastable equilibria with poly-morphous modifications.)

Weber, W. J., Jr., *Physicochemical Processes for Water Quality Control*, Wiley-Interscience, New York, 1972. (This book develops rational bases for the design, interpretation and control of physical chemical processes for effecting or mediating quality transformations in water.)

PROBLEMS

5.1 (a) How much Fe^{2+} (expressed in mol liter^{-1}) could be present in a 10^{-2} M $NaHCO_3$ solution without causing precipitation of $FeCO_3$? The solubility product of $FeCO_3$ is $K_{s0} = 10^{-10.7}$.

(b) How would this maximum soluble Fe^{2+} concentration change upon lowering the pH of that solution by 1 unit?

5.2 C. V. Cole [*Soil Sci.*, **83**, 141 (1956)] established the following equation for soil water:

$$log[Ca^{2+}] + 2pH = const - log\ p_{CO_2}$$

Under what conditions does this equation hold? Express the constant in terms of known equilibrium constants.

5.3 If a sample of deep seawater were returned to the laboratory and stored at 20°C and 1 atm pressure, how would its Ca^{2+} and CO_3^{2-} concentration and pH change?

5.4 How much acid or base has to be added to a saturated solution of $CaCO_3$ to adjust the soluble Ca^{2+} level to

(a) 10^{-3} M

(b) 5×10^{-4} M

if the solution is shielded from the atmosphere?

5.5 Tillmans (1907) made investigations on the solubility of marble ($CaCO_3$) in calcium bicarbonate solutions containing different amounts of $H_2CO_3^*$. He shook his solutions for 10 days in closed flasks with marble chips. After this period equilibrium was attained. He determined experimentally the equilibrium concentrations of the $H_2CO_3^*$ and the alkalinity. For all waters that showed pH values below 8.5, after the saturation with $CaCO_3$, he found the following empirical relation: $[H_2CO_3^*] = K[Alk]^3$.

(a) Show that this relationship can be derived by mass low considerations.

(b) Express K of the above equation in terms of $K_1 =$ first acidity constant of $H_2CO_3^*$, $K_2 =$ second acidity constant of $H_2CO_3^*$, $K_{s0} =$ solubility product of $CaCO_3$.

5.6 10^{-3} mol of $Mg(OH)_2(s)$ is added to 1 liter of water. The system is then exposed to a partial pressure of CO_2 of 10^{-1} atm.

(a) Which solid phase precipitates?

(b) What is the pH of the solution?

(Free energy data or solubility constants are quoted in Example 5.8).

5.7 This problem deals with surveys in anoxic basins [Richards et al., *Limnol. Oceanog.*, **10**, R197 (1965)]. Compute the solubility product of FeS from the figure given below. The computed slope of the line is $[Fe^{2+}][S^{2-}] = 5.3 \times 10^{-18}$. 18°C. $I = 0.06$.

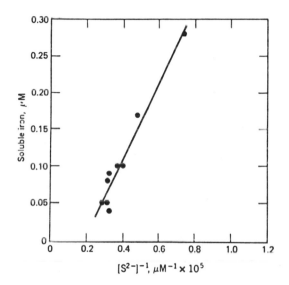

5.8 Some solid $CaCO_3$ is in equilibrium with its saturated solution. How will the amount of Ca^{2+} in the solution be affected (increase, decrease, or no effect) by adding small amounts of the following:

(a) KOH

(b) $CaCl_2$

(c) $(NH_4)_2SO_4$

(d) Sodium metaphosphate $(NaPO_3)_n$

(e) $FeCl_3$

(f) Na_2CO_3

(g) H_2O

(h) by increasing pressure.

5.9 Consider the system containing phases $Ca_{10}(PO_4)_6(OH)_2(s)$, $CaCO_3(s)$, $CaHPO_4(s)$, aqueous solution, and gas $(CO_2(g))$ in equilibrium at a temperature of 10°C.

(a) Give suitable components and degrees of freedom for this system.

(b) Under conditions of equilibrium does $[Ca^{2+}]$ increase, decrease, or stay constant upon addition of small quantities of the following: (i) H_3PO_4; (ii) CO_2, (iii) NaOH, (iv) $Ca(OH)_2$, (v) HCl, (vi) H_2O?

ANSWERS TO PROBLEMS

5.1 (a) $10^{-6.8}$; (b) $10^{-5.8}$.

5.2 Reaction: $CaCO_3(s) + 2H^+ = Ca^{2+} + CO_2(g) + H_2O$. Const $= {}^*K_{ps0}$ $= K_{s0}K_2^{-1}K_H^{-1}K_1^{-1}$ [compare (10) of Table 5.2].

5.3 $[Ca^{2+}]$ and $[CO_3^{2-}]$ do not change appreciably; the pH increases.

5.4 (a) 1×10^{-3} eq liter^{-1}; (b) 5×10^{-4} eq liter^{-1} acid (by use of equation 30).

5.5 Reaction: $CaCO_3(s) + H_2CO_3^* = Ca^{2+} + 2HCO_3^-$; charge balance: $2[Ca^{2+}] \approx [HCO_3^-]$; $[Alk] \approx [HCO_3^-]$. $[HCO_3^-]^3/[H_2CO_3^*] = 2K_{s0}K_1K_2^{-1} = K$.

5.6 No precipitation; pH $= 6.1$.

5.7 ${}^cK_{s0} = 5.3 \times 10^{-18}$; $pK_{s0} \approx p^cK_{s0} + 4\sqrt{I}/(1 + \sqrt{I}) \approx 18.1$. (The value given by the *National Bureau of Standards Circular* 500 for 25°C and $I = 0$ is $pK_{s0} = 17.3$.)

5.8 Increase for (h), (c), (e); decrease for (a), (d), (f); no change for (g).

5.9 (a) CO_2, H_2O, P_2O_5, CaO; five phases, hence one degree of freedom, but specification of temperature makes the system nonvariant.

(b) No change for (i), (ii), (iv), and (vi) (because these represent additions of components); decrease for (iii); increase for (v).

6

Metal Ions in Aqueous Solution:
Aspects of Coordination Chemistry

6.1 INTRODUCTION

All chemical reactions have one common denominator: The atoms, molecules, or ions involved tend to improve the stability of the electrons in their outer shell. In a broad classification of chemical reactions we distinguish between two general groups of reactions by which atoms achieve such stabilization. (a) Redox processes, in which the oxidation states of the participating atoms change, and (b) reactions in which the coordinative relationships are changed. What do we mean by a change in coordinative relations? The coordinative relations are changed if the coordinative partner is changed or if the coordination number† of the participating atoms is changed. This may be illustrated by the following examples.

1. If an acid is introduced into water

$$HClO + H_2O = H_3O^+ + ClO^-$$

   the coordinative partner of the hydrogen ion (which has a coordination number of one) is changed from ClO^- to H_2O.

2. The precipitation that frequently occurs in the reaction of a metal ion with a base

$$Mg \cdot aq^{2+} + 2OH^- = Mg(OH)_2(s) + aq$$

   can be interpreted in terms of a reaction in which the coordinative relations are changed, in the sense that a three-dimensional lattice is formed in which each metal ion is surrounded by and coordinatively "saturated" by the appropriate number of bases.

3. Metal ions can also react with bases without formation of precipitates in reactions such as

$$Cu \cdot aq^{2+} + 4NH_3 = [Cu(NH_3)_4]^{2+} + aq$$

In this simple classification of reactions no distinction needs to be made between acid–base, precipitation, and complex formation reactions; they are all coordinative reactions, hence phenomenologically and conceptually similar.

---

† The coordination number is indicative of the structure and specifies the number of nearest neighbors (ligand atoms) of a particular atom.

Definitions. In the following, any combination of cations with molecules or anions containing free pairs of electrons (bases) is called coordination (or complex formation) and can be electrostatic, covalent, or a mixture of both. The metal cation will be called the *central atom*, and the anions or molecules with which it forms a coordination compound will be referred to as *ligands*. If the ligand is composed of several atoms, the one responsible for the basic or nucleophilic nature of the ligand is called the ligand atom. If a base contains more than one ligand atom, and thus can occupy more than one coordination position in the complex, it is referred to as a *multidentate* complex former. Ligands occupying one, two, three, and so on, positions are referred to as unidentate, bidentate, tridentate, and so on. Typical examples are oxalate and ethylendiamine as bidentate ligands, citrate as a tridentate ligand, ethylene-diamine tetraacetate (EDTA) as hexadentate ligand. Complex formation with multidentate ligands is called *chelation*, and the complexes are called chelates. The most obvious feature of a chelate is the formation of a ring. For example, in the reaction between glycine and $Cu \cdot aq^{2+}$, a chelate with two rings, each of

$$O=C \overset{O}{\underset{H_2C-N}{\diagdown}} \overset{H_2}{\underset{Cu}{\diagup}} \overset{N}{\underset{O-C=O}{\diagdown}} CH_2$$

five members, is formed. Glycine is a bidentate ligand; O— and N— are the donor atoms. If there is more than one metal atom (central atom) in a complex, we speak about *multi-* or *polynuclear complexes.*

One essential distinction between a proton complex and a metal complex is that the *coordination number* of protons is different from that of metal ions. The coordination number of the proton is 1 (in hydrogen bonding, H^+ can also exhibit a coordination number of 2). Most metal cations exhibit an even co-ordination number of 2, 4, 6, or occasionally 8. In complexes of coordination number 2, the ligands and the central ion are linearly arranged. If the coordina-tion number is 4, the ligand atoms surround the central ion either in a square planar or in a tetrahedral configuration. If the coordination number is 6, the ligands occupy the corners of an octahedron, in the center of which stands the central atom. An example is given in Figure 6.18a where the O ligand atoms of trihydroxamic acid surround Fe^{3+} in an octahedral arrangement, thus satisfying the coordinative requirements of Fe^{3+}.

Chemical Speciation. The term *species* refers to the actual form in which a molecule or ion is present in solution. For example, iodine in aqueous solution may conceivably exist as one or more of the species I_2, I^-, I_3^-, HIO, IO^-, IO_3^-, or as an ion pair or complex, or in the form of organic iodo compounds. Figure

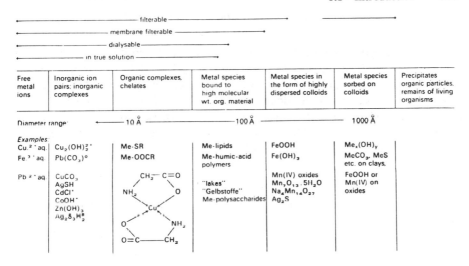

Figure 6.1 Forms of occurrence of metal species.

6.1 shows the various forms in which metals are thought to occur in seawater. It is operationally difficult to distinguish between dissolved and colloidally dispersed substances. Colloidal metal-ion precipitates, such as $Fe(OH)_3(s)$ or $FeOOH(s)$ may occasionally have particle sizes smaller than 100 Å—sufficiently small to pass through a membrane filter. Organic substances can assist markedly in the formation of stable colloidal dispersions. Information on the types of species encountered under different chemical conditions (types of complexes, their stabilities, and rates of formation) is a prerequisite to a better understanding of the distribution and functions of trace elements in natural waters.

Although the importance of species distribution has long been recognized, especially for acids [e.g. H_2CO_3, $B(OH)_3$, $Si(OH)_4$, H_2O, NH_4^+, H_2S] and their conjugate bases, only during the last two decades has a general awareness of the relevance of chemical speciation in natural water been aroused. Goldberg [1] and Krauskopf [2] were among the first to consider the forms of the reacting metal species. Sillén [3] further developed the application of equilibrium models in the portrayal of many aspects of the species composition of seawater, and Garrels [4] demonstrated how chemical relations could be interpreted from mineral equilibria (Garrels and Christ [5]). Identification of species was considered to be one of the most urgent oceanographic analytical problems at the 1971 24th Summer Symposium on Analytical Chemistry.

Baes and Mesmer [6] provide a critical evaluation of the extensive information on the identity of metal-ion species and their hydrolysis products in solutions, as well as the solid oxides and hydroxides they produce. They also provide a critical compilation of hydrolysis equilibrium and oxide and hydroxide solubility constants.

6.2 PROTONS AND METAL IONS

In all solution environments the bare metal ions are in continuous search of a partner. All metal cations in water are hydrated, that is, they form aquo complexes. The coordination reactions in which metal cations participate in aqueous solutions are exchange reactions with the coordinated water molecules exchanged for some preferred ligands. The barest of the metal cations is the free hydrogen ion, the proton. Hence in some regards there is little difference in principle between a free metal ion and a proton.

Brønsted Acidity and Lewis Acidity.   In Figure 6.2 alkalimetric titration curves for the reaction of phosphoric acid and $Fe(H_2O)_6^{3+}$, respectively, with a base (OH$^-$ ion) are compared. Millimolar solutions of H_3PO_4 and ferric perchlorate have a similar pH value. Both acids ($Fe \cdot aq^{3+}$ and H_3PO_4) are multiprotic acids; that is, they can transfer more than one proton.

In Figure 6.3 the titration of H_3O^+ with ammonia is compared with the titration of $Cu \cdot aq^{2+}$ with ammonia. pH and pCu ($= -\log[Cu \cdot aq^{2+}]$) are plotted as a function of the base added. In both cases "neutralization curves" are observed. In the case of the H_3O^+—NH_3 reaction a pronounced pH jump occurs at the equivalence point. The pCu jump is less pronounced in the $Cu \cdot aq^{2+}$—NH_3 reaction because NH_3 is bound to the Cu^{2+} ion in a stepwise consecutive way: ($CuNH_3^{2+}$, $Cu(NH_3)_2^{2+}$, $Cu(NH_3)_3^{2+}$, $Cu(NH_3)_4^{2+}$, $Cu(NH_3)_5^{2+}$) (Figure 6.3d). If, however, four NH_3 molecules are packaged together in one single molecule such as trien (triethylenetetramine, H_2N—

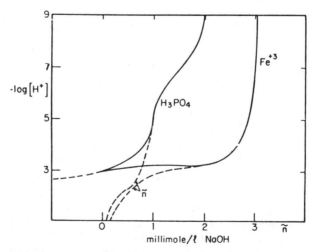

Figure 6.2   Alkalimetric tritration of 10^{-3} M H_3PO_4 and 10^{-3} M $Fe \cdot aq^{3+}$. Both H_3PO_4 and $Fe \cdot aq^{3+}$ are multiprotic Brønsted acids. Millimolar solutions of H_3PO_4 and $Fe(ClO_4)_3$ have similar pH values.

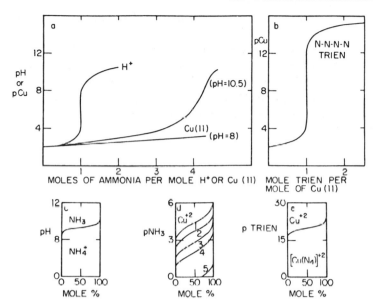

Figure 6.3   Titration of H_3O^+ and Cu aq^{2+} with ammonia (*a*) and with Tetramine (trien) (*b*). Equilibrium diagrams for the distribution of NH_3–NH_4^+ (*c*) of the amino copper(II) complexes (*d*) and of Cu^{2+}, Cu-trien (*e*). The similarity of titrating H$^+$ with a base and titrating a metal ion with a base (Lewis acid–base interaction) is obvious. Both neutralization reactions are used analytically for the determination of acids and metal ions. A pH or pMe indicator electrodes (glass electrode for H$^+$ and copper electrode for Cu^{2+}) can be used for the end point indication.

CH_2—CH_2—NH—CH$_2$ CH$_2$—NH—CH_2—CH_2—NH_2), a 1 : 1 Cu-trien complex is formed and a simple titration curve with a very pronounced pCu jump is observed at the equivalence point (Figure 6.3*b*). In this case the Cu-trien equilibrium (Figure 6.3*e*) is as simple as the H$^+$—NH_3 equilibrium (Figure 6.3*c*). Such neutralization reactions are exploited analytically for the determination of acids or metal ions; a hydrogen ion electrode (glass electrode) and a metal-ion-sensitive electrode (e.g., a copper electrode for Cu^{2+}), or a pH or pMe indicator, are used as sensors for H$^+$ and Me^{+n}, respectively. The examples given illustrate the phenomenological similarity between the "neutralization" of H$^+$ with bases and that of metal ions with complex formers. The bases and that of metal ions with complex formers. The bases, molecules, or ions that can neutralize H$^+$ or metal ions possess free pairs of electrons. Acids are proton donors according to Brønsted. Lewis, on the other hand, has proposed a much more generalized definition of an acid in the sense that he does not attribute acidity to a particular element but to a unique electronic arrangement: the availability of an empty orbital for the acceptance of a pair of electrons. Such acidic or acid analog properties are possessed by H$^+$, metal ions, and other Lewis acids such as $SOCl_2$, $AlCl_3$, SO_2, and BF_3. In aqueous solutions protons and metal ions compete with each other for the available bases.

The Acidity of the Metal Ions

It is frequently difficult to determine the number of H_2O molecules in the hydration shell, but many metal ions coordinate four or six H_2O molecules per ion. Water is a weak acid. The acidity of the H_2O molecules in the hydration shell of a metal ion is much larger than that of water. As pointed out before (Section 3.2), this enhancement of the acidity of the coordinated water may be

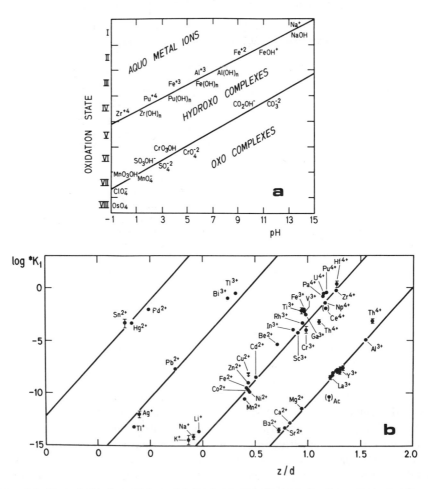

Figure 6.4 Hydrolysis of metal ions. (*a*) Predominant pH range for the occurrence of aquo, hydroxo, hydroxo-oxo and oxo complexes for various oxidation states. The scheme attempts to show a useful generalization, but many elements cannot be properly placed in this simplified diagram because other factors, such as radius and those related to electron distribution, have to be considered in interpreting the acidity of metal ions. A similar diagram has been given by Jørgensen (1963). (*b*) The linear dependence of the first hydrolysis constant $*K_1 = \{MOH^{(z-1)+}\}\{H^+\}/\{M^{z+}\}$ on the ratio of the charge to the M-O distance (z/d) for four groups of cations (25°C). (From Baes and Mesmer [6].) (Note change of zero abcissa for different groups.)

interpreted qualitatively as the result of repulsion of the protons of H_2O molecules by the positive charge of the metal ion.

Hence the acidity of aquo metal ions is expected to increase with a decrease in the radius and an increase in the charge of the central ion. Figure 6.4a attempts to illustrate how the oxidation state of the central atom determines the predominant species (aquo, hydroxo, hydroxo-oxo, and oxo complexes) in the pH range of aqueous solutions. Metal ions with $z = +1$ are generally coordinated with H_2O atoms. Most bivalent metal ions are also coordinated with water up to pH values of 6 to 12. Most trivalent metal ions are already coordinated with OH^- ions within the pH range of natural waters. For $z = +4$ the aquo ions have become too acidic and are out of the accessible pH range of aqueous solutions with few exceptions, for example, Th(IV). There O^{2-} already begins to appear as a ligand, for example, for C(IV) where we have oxo-hydroxo complexes, $H_2CO_3 = CO(OH)_2$ or $HCO_3^- = CO_2(OH)^-$, in the pH range 4.5 to 10; above pH $= 10$, O^{2-} becomes the exclusive ligand (CO_3^{2-}). With even higher oxidation states of the central atom, hydroxo complexes can only occur at very low pH values. The scheme given in Figure 6.4a represents an oversimplification. For every oxidation state, a distribution of acidity according to the ionic radius exists; thus the acidity, as indicated by the pK values given in parentheses, increases in the following series of aquo ions of $z = +2$:

$$Ba^{2+}(14.0), \quad Ca^{2+}(13.3), \quad Mg^{2+}(12.2), \quad Be^{2+}(5.7)\dagger$$

The electrostatic rules given are quite useful, but other factors related to the electron distribution are involved. As Figure 6.4b indicates, different cation groups can be distinguished. The cation group furthest to the right contains cations that are most likely to form ionic M—O bonds, whereas the cation groups on the left, which have a stronger tendency to hydrolyze for their size and charge, tend to form bonds of a more covalent character [6].

Hydrolysis of Metal Ions

More than 30 years ago Brønsted postulated that multivalent metal ions participate in a series of consecutive proton transfers:

$$Fe(H_2O)_6^{3+} = Fe(H_2O)_5OH^{2+} + H^+ = Fe(H_2O)_4(OH)_2^+ + 2H^+$$
$$= Fe(OH)_3(H_2O)_3(s) + 3H^+ = Fe(OH)_4(H_2O)_2^- + 4H^+$$

In the case of Fe(III), hydrolysis can go beyond the uncharged species $Fe(OH)_3(H_2O)_3(s)$ to form anions such as the ferrate(III) ion, probably $[Fe(OH)_4 \cdot 2H_2O]^-$. All hydrated ions can in principle donate a larger number of protons than that corresponding to their charge and can form anionic hydroxo metal complexes but, because of the limited pH range of aqueous solutions, not all elements can exist as anionic hydroxo or oxo complexes.

---

† Be(aq)$^{2+}$ tends to form polynuclear hydrolysis products [e.g., $Be_3(OH)_3^{3+}$] but on a comparable basis it is more acidic than Mg(aq)$^{2+}$.

Polynuclear Hydroxo Complexes

The scheme of a consecutive stepwise hydroxide binding is too simple. Although the hydrolysis products listed for hydrolysis of $Fe \cdot aq^{3+}$ are all known and identified, the intermediate steps are frequently complicated. In a

TABLE 6.1   MULTINUCLEAR, HYDROXO, AND OXO COMPLEXES[a]

Type[b]	Metals Believed to Form Such Complexes[c]
A. Cationic Complexes	
Me—OH—Me	Be(II), Mn^{2+}, Zn(II), Cd(II),
Me⟨OH/OH⟩Me	Cu(II), Fe(III), Hg(II), Sc(II), UO_2^+
Me⟨OH/OH⟩Me⟨OH/OH⟩Me	Hg(II), Sn(II), Pb(II), Sc(III)
Me⟨OH/OH⟩Me⟨OH/OH⟩Me … $(Me(OH)_2)_n$	Sc(III), In(III)
(cube structure) $(Me_4(OH)_4)\parallel$	Pb(II)
Varied	$Be_3(OH)_3^{3+}$, $Bi_6(OH)_{12}^{6+}$, $Pb_6(OH)_8^{4+}$ $Al_7(OH)_{17}^{4+}$, $Al_{13}(OH)_{34}^{5+}$, $Mo_7O_{24}^{6-}$, $V_{10}O_{28}^{6-}$
B. Oxo Complexes of Metals, Metalloids and Nonmetals	

Type	Examples
O_3XOXO_3	$Cr_2O_7^{2-}$, $S_2O_7^{2-}$, $P_2O_7^{4-}$
$O_3X(O_4X)XO_3$	$P_3O_{10}^{5-}$
$(XO_3)_n$	$(PO_3)_n^{n-}$, $(SiO_3)_n^{2n-}$, $CrO_3(s)$
$(X_2O_5)_n$	$P_2O_5(s)$, $(Si_2O_5)_n^{n-}$
$(XO_2)_n$	$SiO_2(s)$

[a] Modified from P. Schindler, personal communication, 1968.
[b] Charges are omitted; the structural arrangement given, although plausible, is hypothetical.
[c] The list is not complete; more detailed information is available from C. F. Baes and R. E. Mesmer [6].

few cases, the main products are monomeric. Polymeric hydrolysis species (isopolycations) have been reported for most metal ions. Thus the existence of multinuclear hydrolysis products is a rather general phenomenon. The hydrolyzed species such as $Fe(H_2O)_5OH^{2+}$ can be considered to dimerize by a

$$2 Fe(H_2O)_5OH^{2+} = [(H_2O)_4Fe \underset{\underset{OH}{\diagdown \diagup}}{\overset{\overset{OH}{\diagup \diagdown}}{}} Fe(H_2O)_4]^{4+} + 2H_2O$$

condensation process. The existence of the dimer has been corroborated experimentally by potentiometric, spectrophotometric, and magnetochemical methods. The dimer may undergo additional hydrolytic reactions which could provide additional hydroxo groups which then could form more bridges. The terms "ol" and "oxo" are often used in referring to the —OH— and —O— bridges. A sequence of such hydrolytic and condensation reactions, sometimes called olation and oxolation,† leads, under conditions of oversaturation with respect to the (usually very insoluble) metal hydroxide, to the formation of colloidal hydroxo polymers and ultimately to the formation of precipitates. In the pH range lower than the zero point of charge of the metal hydroxide precipitate, positively charged metal hydroxo polymers prevail. In solutions more alkaline than the zero point of charge, anionic hydroxo complexes (isopolyanions) and negatively charged colloids exist. Although multinuclear complexes have been recognized for many years for a few hydrolysis systems such as Cr(III) and Be(II) and for anions of Cr(VI), Si(IV), Mo(VI), and V(V), more recent studies have shown that multinuclear hydrolysis products of metallic cations are of almost universal occurrence in the water solvent system. Table 6.1 gives an illustration for some of the polynuclear hydrolysis species reported for various metal ions.

6.3 THE STABILITY OF HYDROLYSIS SPECIES

The establishment of hydrolysis equilibria is usually very fast, as long as the hydrolysis species are simple. The kinetics of typical hydrolysis reactions has been reviewed by Wendt [7]. Polynuclear complexes are often formed rather slowly. Many of these polynuclear hydroxo complexes are kinetic intermediates in the slow transition from free metal ions to solid precipitates and are thus thermodynamically unstable. Some metal-ion solutions "age," that is, they change their composition over periods of weeks because of slow structural transformations of the isopoly ions. Such nonequilibrium conditions can frequently be recognized if the properties of metal-ion solutions (electrode potentials, spectra, conductivity, light scattering, coagulation effects, sedimentation rates, etc.) depend on the history of the solution preparation.

---

† Olation may be followed by oxolation, a process in which the bridging OH group is converted to a bridging O group.

Hydrolysis equilibria can be interpreted in a meaningful way if the solutions are not oversaturated with respect to the solid hydroxide or oxide. Occasionally it is desirable to extend equilibrium calculations into the region of over-saturation; but quantitative interpretations for the species distribution must not be made unless metastable supersaturation can be demonstrated to exist. Most hydrolysis equilibrium constants have been determined in the presence of

TABLE 6.2   FORMULATION OF STABILITY CONSTANTS[a]

I. Mononuclear Complexes

(a) Addition of ligand

$$M \xrightarrow[K_1]{L} ML \xrightarrow[K_2]{L} ML_2 \cdots \xrightarrow[K_i]{L} ML_i \cdots \xrightarrow[K_n]{L} ML_n$$

$$\xrightarrow{\beta_2}$$
$$\xrightarrow{\beta_i}$$
$$\xrightarrow{\beta_n}$$

$$K_i = \frac{[ML_i]}{[ML_{(i-1)}][L]} \tag{1}$$

$$\beta_i = \frac{[ML_i]}{[M][L]^i} \tag{2}$$

(b) Addition of protonated ligands

$$M \xrightarrow[*K_1]{HL} ML \xrightarrow[*K_2]{HL} ML_2 \cdots \xrightarrow[*K_i]{HL} ML_i \cdots \xrightarrow{HL} ML_n$$

$$\xrightarrow{*\beta_2}$$
$$\xrightarrow{*\beta_i}$$
$$\xrightarrow{*\beta_n}$$

$$*K_i = \frac{[ML_i][H^+]}{[ML_{(i-1)}][HL]} \tag{3}$$

$$*\beta_i = \frac{[ML_i][H^+]^i}{[M][HL]^i} \tag{4}$$

II. Polynuclear Complexes

In β_{nm} and $*\beta_{nm}$ the subscripts n and m denote the composition of the complex $M_m L_n$ formed. [If $m = 1$, the second subscript ($=1$) is omitted.]

$$\beta_{nm} = \frac{[M_m L_n]}{[M]^m [L]^n} \tag{5}$$

$$*\beta_{nm} = \frac{[M_m L_n][H^+]^n}{[M]^m [HL]^n} \tag{6}$$

[a] The same notation as that used in L. G. Sillén and A. E. Martell, *Stability Constants of Metal-Ion Complexes*, Special Publications, Nos. 17 and 25, Chemical Society, London, 1964 and 1971, is used.

a swamping "inert" electrolyte of constant ionic strength ($I = 0.1, 1,$ or 3). As we have seen before, the formation of hydroxo species can be formulated in terms of acid–base equilibria. The formulation of equilibria of hydrolysis reactions is in agreement with that generally used for complex formation equilibria (see Table 6.2).

The following rules can be established:

1   The tendency of metal-ion solutions to protolyze (hydrolyze) increases with dilution and with decreasing $[H^+]$.

2   The fraction of polynuclear complexes in a solution decreases on dilution.

The first rule can be illustrated by comparing the equilibria ($I = 0, 25°C$)

$$Mg^{2+} + H_2O = MgOH^+ + H^+; \qquad \log {^*K_1} = -11.4 \qquad (1)$$

$$Cu^{2+} + H_2O = CuOH^+ + H^+; \qquad \log {^*K_1} = -8.0 \qquad (2)$$

At great dilution (pH $\rightarrow$ 7), a substantial fraction of the Cu(II) of a pure Cu-salt solution [e.g., $Cu(ClO_4)_2$] will occur as a hydroxo complex

$$\alpha_{CuOH^+} = \frac{[CuOH^+]}{Cu_T} = \left(1 + \frac{[H^+]}{{^*K_1}}\right)^{-1} = 0.091 \qquad (3)$$

On the other hand, because of the low acidity of Mg^{2+}, even at infinite dilution, the fraction of hydrolyzed Mg^{2+} ions of a solution of an Mg^{2+} salt is very small:

$$\alpha_{MgOH^+} = \frac{[MgOH^+]}{Mg_T} = \left(1 + \frac{[H^+]}{{^*K_1}}\right)^{-1} = 0.00004 \qquad (4)$$

Accordingly, only the salt solutions of sufficiently acid metal ions which fulfill the condition [8]

$$p^*K_1 < \tfrac{1}{2}pK_w \quad \text{or} \quad p^*\beta_n < \frac{n}{2}pK_w \qquad (5)$$

where $^*\beta_n$ is the cumulative acidity constant (see Table 6.2), undergo substantial hydrolysis upon dilution. The progressive hydrolysis upon dilution is the reason that some metal salt solutions tend to precipitate upon dilution. (See Example 6.2.)

Mononuclear Wall

If hydrolysis leads to mononuclear and polynuclear hydroxo complexes, it can be shown that mononuclear species prevail beyond a certain dilution. If we consider, for example, the dimerization of $CuOH^+$:

$$2CuOH^+ = Cu_2(OH)_2^{2+} \qquad \log {^*K_{22}} = 1.5 \qquad (6)$$

it is apparent from the dimensions of the equilibrium constant (conc^{-1}) that the dimerization is concentration-dependent. Thus for a $Cu(II)$ system where $Cu_T = [Cu^{2+}] + [Cu(OH)^+] + 2[Cu_2(OH)_2^{2+}]$, equilibrium 6 can be formulated as

$$\frac{[Cu_2(OH)_2^{2+}]}{[CuOH^+]^2} = \frac{[Cu_2(OH)_2^{2+}]}{(Cu_T - [Cu^{2+}] - 2[Cu_2(OH)_2^{2+}])^2} = {}^*K_{22} \qquad (7)$$

and it becomes obvious that $[Cu_2(OH)_2^{2+}]$ is dependent upon Cu_T. With the help of equations 7 and 2 for each pH, the mononuclear wall (e.g., Cu_T for $[Cu_{dimer}] = 1/100[Cu_{monomer}]$ can be calculated (compare Example 6.1). As pointed out before, for many metals the polynuclear species are formed only under conditions of oversaturation with respect to the metal hydroxide or metal oxide and are thus not stable thermodynamically, for example,

$$Cu_2(OH)_2^{2+} = Cu^{2+}(aq) + Cu(OH)_2(s) \qquad \Delta G^\circ = -2.6 \text{ kcal} \qquad (8)$$

As shown by Schindler [8], multinuclear hydrolysis species usually are not observed during dissolution of the most stable modification of the solid hydroxide or oxide; they are formed, however, by oversaturating a solution with respect to the solid phase. Such polynuclear species, even if thermodynamically unstable, are of significance in natural water systems. Many multinuclear hydroxo complexes may persist as metastable species for years.

Quantitative application of known hydrolysis equilibria is illustrated in the next two examples.

Example 6.1a. The Hydrolysis of Iron(III)

The addition of $Fe(ClO_4)_3$ to H_2O may lead to the following soluble species: Fe^{3+}, $Fe(OH)^{2+}$, $Fe(OH)_2^+$, $Fe(OH)_4^-$, and $Fe_2(OH)_2^{4+}$.
Compute the equilibrium composition of:

1   A homogeneous solution to which 10^{-4} M $(10^{-2}$ $M)$ of iron(III) has been added and the pH adjusted within the range 1 to 4.5 with acid or base;
2   An iron(III) solution in equilibrium with amorphous ferric hydroxide. The following equilibrium constants are available $I = 3(NaClO_4)(25°C)$.

$$Fe^{3+} + H_2O = FeOH^{2+} + H^+ \qquad \log {}^*K_1 = -3.05 \qquad (i)$$

$$Fe^{3+} + 2H_2O = Fe(OH)_2^+ + 2H^+ \qquad \log {}^*\beta_2 = -6.31 \qquad (ii)$$

$$2Fe^{3+} + 2H_2O = Fe_2(OH)_2^{4+} + 2H^+ \qquad \log {}^*\beta_{22} = -2.91 \qquad (iii)$$

$$Fe(OH)_3(s) + 3H^+ = Fe^{3+} + 3H_2O \qquad \log {}^*K_{s0} = 3.96 \qquad (iv)$$

$$Fe(OH)_3(s) + H_2O = Fe(OH)_4^- + H^+ \qquad \log {}^*K_{s4} = -18.7 \qquad (v)$$

1   In the *homogeneous system*, the concentration condition (equations vi or vii) must be fulfilled:

$$Fe_T = [Fe^{3+}] + [FeOH^{2+}] + [Fe(OH)_2^+] + 2[Fe_2(OH)_2^{4+}] \qquad \text{(vi)}$$

$$Fe_T = [Fe^{3+}]\left(1 + \frac{{}^*K_1}{[H^+]} + \frac{{}^*\beta_2}{[H^+]^2} + \frac{2[Fe^{3+}]{}^*\beta_{22}}{[H^+]^2}\right) \qquad \text{(vii)}$$

As with other polyprotic acids we may define successive distribution coefficients: $\alpha_0 = [Fe^{3+}]/Fe_T$, $\alpha_1 = [FeOH^{2+}]/Fe_T$, $\alpha_2 = [Fe(OH)_2^+]/Fe_T$, and $\alpha_{22} = 2[Fe_2(OH)_2^{4+}]/Fe_T$. [$\alpha_{22}$ gives the fraction of iron(III) present in the form of the dimer.]

Inspecting equation vii we note that the last term, proportional to the polymer concentration, is an implicit function of the concentration of iron(III). α_0 may be defined with the help of equations i to iii and vii:

$$\alpha_0 = \left(1 + \frac{{}^*K_1}{[H^+]} + \frac{{}^*\beta_2}{[H^+]^2} + \frac{2Fe_T\alpha_0{}^*\beta_{22}}{[H^+]^2}\right)^{-1} \qquad \text{(viii)}$$

or

$$\frac{\alpha_0^2\, 2Fe_T{}^*\beta_{22}}{[H^+]^2} + \alpha_0\left(1 + \frac{{}^*K_1}{[H^+]} + \frac{{}^*\beta_2}{[H^+]^2}\right) - 1 = 0 \qquad \text{(ix)}$$

Equation ix is written in the form corresponding to the sum $\alpha_{22} + \alpha_0 + \alpha_1 + \alpha_2 - 1 - 0$. Accordingly, the remaining distribution coefficients are defined by

$$\alpha_{22} = \frac{\alpha_0^2\, 2Fe_T\beta_{22}}{[H^+]^2} \qquad \text{(x)}$$

$$\alpha_1 = \frac{\alpha_0{}^*K_1}{[H^+]} \qquad \text{(xi)}$$

$$\alpha_2 = \frac{\alpha_0{}^*\beta_2}{[H^+]^2} \qquad \text{(xii)}$$

Now the computation can be carried out readily, starting with the quadratic equation ix, where we compute α_0 for a given Fe_T and for varying $[H^+]$. The results are plotted in Figure 6.5. It is obvious that the extent of hydrolysis depends on pH and Fe_T. Comparing the distribution diagrams for $Fe_T = 10^{-4}$ M and 10^{-2} M at a given pH, we note that the fraction of the dimer is concentration-dependent. It may also be noted that α_0 is a measure of the relative extent of complex formation by OH^- ions:

$$\log \alpha_0 = pFe_T - pFe^{3+} = \Delta pFe \qquad \text{(xiii)}$$

The higher ΔpM, the better the metal ion is complexed.

2 The species distribution in the *heterogeneous system* can be calculated by considering equation iv in addition to the hydrolysis equilibria. By combining equations i to iii with equation iv, we obtain

$$\log[\text{FeOH}^{2+}] = \log {}^*K_{s0} + \log {}^*K_1 + 2\log[\text{H}^+] \qquad \text{(xiv)}$$

$$\log[\text{Fe(OH)}_2^+] = \log {}^*K_{s0} + \log {}^*\beta_2 + \log[\text{H}^+] \qquad \text{(xv)}$$

$$\log[\text{Fe}_2(\text{OH})_2^{4+}] = 2\log {}^*K_{s0} + \log {}^*\beta_{22} + 4\log[\text{H}^+] \qquad \text{(xvi)}$$

together with equations xvii and xviii which follow from equations iv and v:

$$\log[\text{Fe}^{3+}] = \log {}^*K_{s0} + 3\log[\text{H}^+] \qquad \text{(xvii)}$$

$$\log[\text{Fe(OH)}_4^-] = \log {}^*K_{s4} - \log[\text{H}^+] \qquad \text{(xviii)}$$

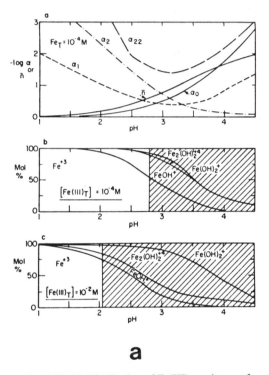

a

Figure 6.5 Hydrolysis of Fe(III). **(a)** Distribution of Fe(III) species as a function of pH (Example 6.1a). The extent of hydrolysis depends on pH and on the total Fe(III) concentration. (*a*) Plots of the average ligand number $\bar{n}$ and the distribution coefficients α, respectively, as a function of pH for a $10^{-4}\,M$ Fe(III) solution. $\bar{n}$ is experimentally accessible from $[\text{H}^+]$ measurements in alkalimetric titrations; α_1 can be determined independently from potentiometric measurements (ferroferri cell) or from spectrophotometric or magnetochemical measurements. Either one or both parameters can be used to arrive at equilibrium constants. Distribution diagrams for the various Fe(III) species are depicted in (*b*) and (*c*) using representative equilibrium constants. In the shaded areas the solution becomes oversaturated with respect to Fe(OH)$_3$(s) ($K_s = 10^{-38}$). Additional polynuclear hydrolysis species occur as kinetic intermediates in the usually slow transition to Fe(OH)$_3$(s) in this pH range.

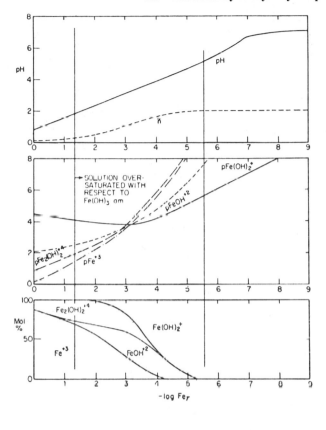

b

Figure 6.5 Hydrolysis of Fe(III). **(b)** Dilution of $Fe(ClO_4)_3$. (Example 6.1b) 1 M $Fe(ClO_4)_3$ is diluted. The extent of hydrolysis is, for example, illustrated by $\bar{n}$ [OH^- bound per Fe(III)]. Fe(III) solutions are not stable with respect to solid amorphous ferric hydroxide over most of the pH range.

In a double logarithmic diagram, equations xiv to xviii can be plotted as straight lines with well-defined slopes of $+4$, $+3$, $+2$, $+1$, and -1, respectively (See Figure 5.4).

Example 6.1b. Dilution of a Ferric Salt Solution

Estimate the species distribution of a pure ferric perchlorate solution (1 M) as a function of dilution. (Perchlorate is chosen as an anion, since it does not appear to form complexes with Fe^{3+}.) Variation in activity coefficients caused by dilution may be ignored; this corresponds to diluting with an inert salt of constant I medium.

The problem consists essentially in simultaneously solving equations i to iii and equation vi of Example 6.1a together with the proton condition

$$[H^+] = [FeOH^{2+}] + 2[Fe(OH)_2^+] + 2[Fe_2(OH)_2^{4+}] + [OH^-] \qquad (i)$$

Equation i also follows from the electroneutrality condition of the solution, considering that $[ClO_4^-] = 3Fe_T$. The existence of $Fe(OH)_4^-$ may be ignored, since it does not occur above concentrations of $10^{-9} M$ in solutions of pH < 7.

There are various operational approaches that can be used to solve the requisite five equations simultaneously. One convenient approach starts by making a guess of $[H^+]$ for a given Fe_T. [This guess may be based on a tentative calculation assuming that only one hydrolysis reaction predominates at a given concentration; for example, one may assume that for concentrated Fe(III) solutions hydrolysis to the dimer (equation iii of Example 6.1a) determines $[H^+]$; while for very dilute solutions $[H^+]$ is controlled by hydrolysis to $Fe(OH)_2^+$ (equation ii of Example 6.1a).] With this guess and with the aid of equation ix of Example 6.1a we calculate a tentative value of α_0. Then tentative values of α_1, α_2 and α_{22} and thus tentative concentrations for the individual species are obtained. The adequacy of the guess is then checked with the help of the proton condition (equation i); with a new and improved guess and subsequent iteration, convergence can be obtained. The results, plotted in Figure 6.5b, illustrate the dilution rules given earlier. The extent of hydrolysis increases upon dilution. In a $10^{-3} M$ solution, assuming metastable oversaturation, approximately 30% of the iron is present as free ferric ion; this fraction is reduced to 4% in a 10^{-4} Fe_T solution. The data given show that, because of a pH increase, the proportion of the dimer increases upon dilution up to about $10^{-2} M$ Fe_T but on further dilution the fraction of Fe(III) present as a dimer decreases.

Figure 6.5b also shows the dilution at which precipitation of amorphous $Fe(OH)_3$ occurs if equilibrium is attained. Accordingly, only concentrated $(10^{-1} M$ and stronger) Fe(III) solutions with pH values below 2 are thermodynamically stable with respect to solid amorphous ferric hydroxide. [Even such solutions, however, are not stable with respect to more stable solid modifications, that is, α-FeOOH (goethite).] Diluting an unacidified Fe(III) salt solution leads to precipitation of $Fe(OH)_3(s)$. For example, a $10^{-3} M$ $Fe(ClO_4)_3$ solution changes its color upon standing; continuously changing spectra (ultraviolet and visible) are observed. Within days such solutions become turbid, and within weeks precipitates of $Fe(OH)_3$ can be observed. Extremely large dilutions of a Fe(III) salt ($Fe_T < 10^{-10} M$) are necessary to maintain iron(III) in solution. As Figure 6.5b suggests, the species that predominates at such dilutions (pH ≈ 7) is $Fe(OH)_2^+$.†

---

† Distribution diagrams similar to those given in Figure 6.5 have been presented for iron(III) complexes in marine systems by D. R. Kester, R. H. Byrne, Jr., and Y. J. Liang, in *Marine Chemistry in the Coastal Environment*, T. M. Church, Ed., American Chemical Society Washington, D.C., 1975.

The Formation Curve

Figure 6.2 illustrates that aquo metal ions can be titrated alkalimetrically. The morphology of the titration curve frequently can give valuable insight into the type of species encountered during the titration. In Section 3.8 it was shown that the alkalimetric titration curve of a polyprotic acid, for instance, H_3PO_4, is given by

$$f = \frac{C_B}{P_T} = \frac{[H_2PO_4^-] + 2[HPO_4^{2-}] + 3[PO_4^{3-}]}{P_T} + \frac{[OH^-] - [H^+]}{P_T} \tag{9}$$

$$= \alpha_1 + 2\alpha_2 + 3\alpha_3 + \frac{[OH^-] - [H^+]}{P_T} \tag{10}$$

Similarly, the alkalimetric titration of aquo ferric ion can be characterized by

$$f = \frac{C_B}{Fe_T} = \frac{[FeOH^{2+}] + 2[Fe(OH)_2^+] + 2[Fe_2(OH)_2^{4+}] + 4[Fe(OH)_4^-]}{Fe_T}$$

$$+ \frac{[OH^-] - [H^+]}{Fe_T} \tag{11a}$$

$$= \alpha_1 + 2\alpha_2 + \alpha_{22} + 4\alpha_4 + \frac{[OH^-] - [H^+]}{Fe_T} \tag{11b}$$

The α values are now defined as in Example 6.1a (note that $\alpha_{22} = 2[Fe_2(OH)_2^{4+}]/Fe_T$). The first term on the right-hand side of equation 11a is a measure of the average number of hydroxide ions bound per iron(III) atom; it is called the *ligand number* or the *formation function* $\bar{n}$.

$$\bar{n} = \frac{[FeOH^{2+}] + 2[Fe(OH)_2^+] + 2[Fe_2(OH)_2^{4+}] + 4[Fe(OH)_4^-]}{Fe_T}$$

$$= \alpha_1 + 2\alpha_2 + \alpha_{22} + 4\alpha_4 \tag{12}$$

The ligand number, generally applicable to any ligand, is usually plotted as a function of the logarithm of the free ligand concentration; hence as a function of $\log[OH^-]$ or $-\log[H^+]$ in the case of hydrolysis equilibria. In the case of H_3PO_4 (equations 9 and 10) $\bar{n}$ may be interpreted similarly as the average number of protons deficient per P atom with respect to H_3PO_4. (In interpreting the solution composition, proton deficiency is mathematically equivalent to hydroxide excess.) In Figure 6.2 the ligand number is plotted as a function of $-\log[H^+]$, and in Figure 6.5 as a function of Fe_T. It is apparent from equation 11a that the ligand number can be obtained directly from potentiometric measurements. For the evaluation of equilibrium constants, attempts are made to interpret data of $\bar{n}$ obtained from measurements over a wide range of Me_T and $[H^+]$ in terms of probable hydrolysis species. A convenient feature of the formation curve of mononuclear systems is that it depends solely on the ligand concentration and not on the total concentration of the metal ion. The existence of polynuclear complexes is detected most readily if the variation in Me_T causes shifts in the formation curve.

The ligand number is also a convenient parameter under nonequilibrium conditions. In a $Me^{z+}-H_2O$ system, $\bar{n}$ also gives the average charge q of the hydrolysis species

$$q = z - \bar{n} \tag{13}$$

Buffer Intensity.   The buffer intensity, $\beta = dC_B/d\text{pH}$, is related to the slope of the formation curve of a hydrolysis system. For the iron(III) system we can formulate (cf. equation 11)

$$\beta = \frac{dC_B}{d\text{pH}}$$

$$= \frac{d[\text{FeOH}^{2+}] + 2d[\text{Fe(OH)}_2^+] + 2d[\text{Fe}_2(\text{OH})_2^{4+}] + 4d[\text{Fe[OH]}_4^+]}{d\text{pH}}$$

$$+ \frac{d[\text{OH}^-] - d[\text{H}^+]}{d\text{pH}} \tag{14}$$

Substituting equation 12 into equation 14 gives

$$\beta = \frac{dC_B}{d\text{pH}} = \text{Fe}_T \frac{d\bar{n}}{d\text{pH}} + \frac{d[\text{OH}^-] - d[\text{H}^+]}{d\text{pH}} \tag{15}$$

where the last term is equal to $2.3([\text{H}^+] + [\text{OH}^-])$. If $[\text{H}^+]$ and $[\text{OH}^-]$ are small in comparison to the total concentration (Fe_T), the slope of the formation curve ($\bar{n}$ versus $-\log[\text{H}^+]$) of a hydrolysis system is a measure of the buffer intensity. Each hydrolysis species contributes to the buffer intensity. Frequently no steps appear in the formation curve, because successive acidity (hydrolysis) constants are very close together; therefore a rather large buffer intensity which remains relatively uniform over a wide pH range may be imparted to the solution by metal-ion hydrolysis (see Figure 6.5a).

Solid Hydroxides and Metal Oxides

It has been pointed out that the formation of a precipitate can often be considered the final stage in the formation of polynuclear complexes. Aggregates of ions that form the building stones in the lattice are produced in the solution, and these aggregates combine with other ions to form neutral compounds.

Plausibly there is a correlation of the solubility of the stable oxide, hydroxide, or oxyhydroxide of a cation with the stability of the first hydrolysis product $\text{MOH}^{(z-1)+}$ (Figure 6.6 [6]). Many multivalent hydrous oxides are amphoteric because of the acid–base equilibria involved in the hydrolysis reactions of aquo metal ions. Alkalimetric or acidimetric titration curves for hydrous metal oxides, that is, formation curves for heterogeneous systems, provide a quantitative explanation for the manner in which the charge of the hydrous oxide depends on the pH of the medium. The amphoteric behavior of solid metal hydroxides becomes evident from such titration curves, From an operational point of view such hydrous oxides can be compared with amphoteric polyelectrolytes and can be considered hydrated solid electrolytes, frequently possessing a variable space lattice in which the proportion of different ions, cations as well as anions, is variable within the limits of electrical neutrality of the solid. These hydroxides show a strong tendency to interact specifically with anions as well as with cations.

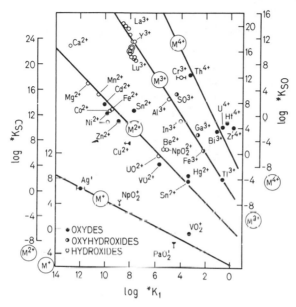

Figure 6.6 Correlation of the solubility product $*K_{s0}$ with the first hydrolysis constant $*K_1$ for M^+, M^{2+}, M^{3+}, and M^{4+} cations. The lines have slopes of -1, -2, -3, and -4 (25°C). This relationship results because the equilibrium constant for the reaction $M(OH)_z(s) + (z-1)M^{z+} = zMOH^{(z-1)+}$, $K = (*K_1)^z K_{s0}$, $K = \{MOH^{(z-1)+}\}^z/\{M^{z+}\}^{(z-1)}$, is often close to $10^{-5.6}$. This relatively low value reflects the general tendency of cations to precipitate shortly after hydrolysis begins unless $[M^z]$ is quite low. The strong tendency for solutions to supersaturate often allows hydrolysis to proceed much further in solution than would be expected from the value of K. (From Baes and Mesmer [6].)

6.4 METAL IONS AND LIGANDS

Considerable emphasis has been placed thus far on hydroxo complexes; this is amply justified by the ubiquitousness of OH^- in water and by the strong affinity of many metal ions for OH^-. But other proton acceptors can serve as electron pair donors and thus coordinate with metal ions. There is no single generally valid correlation between the stability of proton complexes (reciprocal of protolysis constant) and the stability of metal complexes, but some weak bases such as ClO_4^- and NO_3^- have very little tendency to form metal complexes.

Classification of Metal Ions and Ligands

Inorganic and organic ligands contain the following possible donor atoms in the fourth, fifth, sixth, and seventh vertical column of the periodic table:

C	N	O	F
	P	S	Cl
	As	Se	Br
		Te	I

In water, halogens are effective complexing agents only as anions, but not if bonded to carbon. For special reasons cyanide ion is a particularly strong complex former. The more important donor atoms include nitrogen, oxygen, and sulfur [9].

In aqueous solution, preference of a cation for one type of ligand as opposed to another depends on the cation. Ahrland *et al.* [10] divided metal ions into two categories, depending on whether the metal ions formed their most stable complexes with the first ligand atom of each periodic group (i.e. F, O, N) or with a later member of the group (e.g. I, S, P). As Table 6.3 shows, this classification into A- and B-type metal cations is governed by the number of electrons in the outer shell. Class A metal cations have the inert gas type (d^0) electron configuration. These ions may be visualized as being of spherical symmetry; their electron sheaths are not readily deformed under the influence of electric fields such as those produced by adjacent charged ions. They are, as it were, *hard spheres*, whereas class-B metal cations have an electron sheath more readily deformable (higher polarizability) than that of A-type metals and may be visualized as *soft spheres* (nd^{10} and $nd^{10}(n + 1)s^2$ configurations).

Metal cations in class A form complexes preferentially with the fluoride ion and ligands having oxygen as donor atom. Water is more strongly attracted to these metals than ammonia or cyanide. No sulfides (precipitates or complexes) are formed by these ions in aqueous solution, since OH^- ions readily displace HS^- or S^{2-}. Chloro or iodo complexes are weak and occur most readily in acid solutions under which conditions competition with OH^- is minimal. The univalent alkali ions form only relatively unstable ion pairs with some anions; some weak complexes of Li^+ and Na^+ with chelating agents, macrocyclic ligands, and polyphosphates are known. Chelating agents containing only nitrogen or sulfur as ligand atoms do not coordinate with A-type cations to form complexes of appreciable stability. Class-A metal cations tend to form difficultly soluble precipitates with OH^-, CO_3^{2-}, and PO_4^{3-}; no reaction occurs with sulfur and nitrogen donors (addition of NH_3, alkali sulfides, or alkali cyanides produces solid hydroxides). With class-A metals a simple electrostatic picture of the binding of cation and ligand gives a satisfactory first-approximation explanation of complex stability. For example, the stability increases rapidly with an increase in charge on the metal ion, and the ions with the smallest radii form the most stable complexes. Some stability sequences are indicated in Table 6.3.

In contrast, class-B metal ions coordinate preferentially with bases containing I, S, or N as donor atoms. Thus metal ions in this class may bind ammonia more strongly than water and CN^- in preference to OH^- and form more stable I^- or Cl^- complexes than F^- complexes. These metal cations, as well as transition metal cations (Table 6.3), form insoluble sulfides and soluble complexes with S^{2-} and HS^-. In this group, electrostatic forces do not appear to be of primary importance, because neither the charge nor the size of the interacting ions is entirely decisive for the stability sequence; there is superposition of covalent *and* coulombic interactions. Noncolored components often yield a colored

TABLE 6.3 CLASSIFICATION OF METAL IONS

A-Type Metal Cations	Transition-Metal Cations	B-Type Metal Cations
Electron configuration of inert gas; low polarizability; "hard spheres"; (H^+), Li^+, Na^+, K^+, Be^{2+}, Mg^{2+}, Ca^{2+}, Sr^{2+}, Al^{3+}, Sc^{3+}, La^{3+}, Si^{4+}, Ti^{4+}, Zr^{4+}, Th^{4+}	One to nine outer shell electrons, not spherically symmetric; V^{2+}, Cr^{2+}, Mn^{2+}, Fe^{2+}, Co^{2+}, Ni^{2+}, Cu^{2+}, Ti^{3+}, V^{3+}, Cr^{3+}, Mn^{3+}, Fe^{3+}, Co^{3+}	Electron number corresponds to Ni^0, Pd^0 and Pt^0 (10 or 12 outer shell electrons); low electronegativity; high polarizability. "soft spheres"; Cu^+, Ag^+, Au^+, Tl^+, Ga^+, Zn^{2+}, Cd^{2+}, Hg^{2+}, Pb^{2+} Sn^{2+}, Tl^{3+}, Au^{3+}, In^{3+}, Bi^{3+}

According to Pearson's (1963) Hard and Soft Acids[a]

Hard Acids	Borderline	Soft Acids
All A-type metal cations plus Cr^{3+}, Mn^{3+}, Fe^{3+}, Co^{3+}, UO_2^{2+}, VO^{2+} Also species such as BF_3, BCl_3, SO_3, RSO_2^+, RPO_2^+, CO_2, RCO^+, R_3C^+	All bivalent transition metal cations plus Zn^{2+}, Pb^{2+}, Bi^{3+}, SO_2, NO^+, $B(CH_3)_3$	All B-type metal cations minus Zn^{2+}, Pb^{2+}, Bi^{3+} All metal atoms, bulk metals I_2, Br_2, ICN, I^+, Br^+

Preference for ligand atom:

$N \gg P$		$P \gg N$
$O \gg S$		$S \gg O$
$F \gg Cl$		$I \gg F$

Qualitative generalizations on stability sequence:

Cations: Cations:

Stability $\propto \dfrac{\text{charge}}{\text{radius}}$

Irving–Williams order:
$$Mn^{2+} < Fe^{2+} < Co^{2+} < Ni^{2+} < Cu^{2+} > Zn^{2+}$$

Ligands:

$F > O > N$
$= Cl > Br > I > S$
$OH^- > RO^- > RCO_2^-$
$CO_3^{2-} \gg NO_3^-$
$PO_4^{3-} \gg SO_4^{2-} \gg ClO_4^-$

Ligands:
$S > I > Br > Cl$
$= N > O > F$

[a] R. G. Pearson, *J. Amer. Chem. Soc.*, **85**, 3533 (1963).

compound (charge transfer bands), thus indicating a significant deformation of the electron orbital overlap. Hence, in addition to coulombic forces, types of interactions other than simple electrostatic forces must be considered. These other types of interactions can be interpreted in terms of quantum mechanics, and in a somewhat oversimplified picture the bond is regarded as resulting from the sharing of an electron pair by the central atom and the ligand (*covalent bond*). The tendency toward complex formation increases with the capability of the cation to take up electrons (increasing ionization potential of the metal) and with decreasing electronegativity of the ligand (increasing tendency of the ligand to donate electrons). In the series F, O, N, Cl, Br, I, S, the electronegativity decreases from left to right, whereas the stability of complexes with B-type cations increases. However, other factors such as steric hindrance and entropy effects distort the picture, and for this reason stability sequences with cations of the B group are often irregular.

Transition-metal cations have between 0 and 10 d electrons (nd^q configuration, where $0 < q < 10$) (Table 6.3). For these cations, a reasonably well-established rule for the sequence of complex stability, the Irving–Williams [11] order, is valid. According to this rule the stability of complexes increases in the series

$$Mn^{2+} < Fe^{2+} < Co^{2+} < Ni^{2+} < Cu^{2+} > Zn^{2+}.$$

An example is given in Figure 6.7. As a first approximation it might be argued qualitatively that the electrovalent behavior of the bivalent transition-metal cations remains almost constant (A-type character), but that the nonelectrovalent behavior (B-type character) changes markedly, in going from $3d^5$ to the $3d^{10}$ configuration.

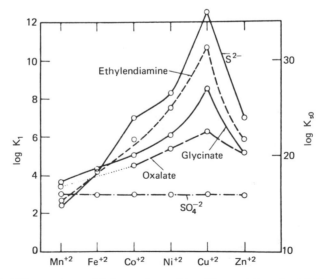

Figure 6.7 Stability constants of 1:1 complexes of transition metals and solubility products of their sulfides (Irving–Williams series).

The Irving–Williams order is usually explained in terms of ligand field or simple-crystal field theory,[†] but other explanations can also account for this order [12]. For other transition elements, the generalization may be made that the B-type character increases slowly with q and markedly with n.

Comparison of a few stability sequences for organic complexes of different metals with given ligands is of interest.

Salicylate (donor groups: O, O): $Ca < Cd < Mn(II) < Fe(II) < Zn < Cu(II)$

Glycine (O, N): $Ca < Mg < Mn(II) < Cd \sim Pb(II) < Hg(II)$

Cysteine (S, N): $Mn(II) < Zn < Fe(II) < Cd < Pb(II) < Hg(II)$

Hydroxide (O): $Ca < Mg < Cd \sim Mn(II) < Zn < Co(II) < Fe(II) < Cu(II) < Pb(II) < Hg(II)$

Soft and Hard Acids and Bases

It is possible to recognize the two following categories of reactions:

1 Reactions of A-type cations with ligands which attach preferentially to A-type ions ($F^- \gg Cl^-$; $OH^- > NH_3$).
2 Reaction of B-type cations with ligands that have a strong tendency to share electrons with B-type ions ($I^- > Cl^- > F^-$; $NH_3 > OH^- > H_2O$).

In Pearson's [13] concept of soft and hard acids and bases (SHAB concept), reactions listed under (1) are classified as hard acid–hard base interactions, whereas reactions under (2) involving B cations fall into the category of soft acid–soft base interactions. Some useful trends, usable for the qualitative prediction of chemical reactions and relative sequences of compound stability are indicated in Table 6.3. Typically, reactions listed under (1) are characterized by a small ΔH and a positive, usually large, $T\Delta S$, whereas those given under (2) show an appreciable negative ΔH and a small positive or negative $T\Delta S$ term. That is, the entropy increase is the primary "drive" behind hard acid–hard base reactions, whereas the free energy change of soft acid–soft base reactions is dominated by a negative enthalpy change.

[†] The crystal field is the electric field, acting at the central metal ion, due to the attached groups or ligands. This field affects the energy of electrons, particularly those in the d orbitals of the central ion. If d electrons can preferentially occupy lower-energy d orbitals, the complex becomes more stable than otherwise by an amount called the crystal field stabilization energy (CFSE). The ions Mn^{2+} (5 d electrons) and Zn^{2+} (10 d electrons) cannot show crystal field stabilization, but CFSE increases progressively from Fe^{2+} (6 electrons) to Cu^{2+} (10 electrons). See for example, F. A. Cotton and G. Wilkinson, *Basic Inorganic Chemistry*, Wiley, New York, 1976, or A. Earnshaw and T. J. Harrington, *The Chemistry of the Transition Elements*, Clarendon, Oxford, 1973.

The principle of hard and soft acids and bases may also be applied to the rates of nucleophilic and electrophilic substitution reactions. Water is a very hard solvent, with respect to both its acidic and basic functions. It is the ideal solvent for hard acids, hard bases, and hard complexes [13]. Modern theories on coordination chemistry are introduced [14,15] and reviewed in [16–18].

Ion Pairs and Complexes

Two types of complex species can be distinguished:†

1 *Ion pairs*: Ions of opposite charge that approach within a critical distance, effectively form an ion pair, and are no longer electrostatically effective. The metal ion or the ligand or both retain the coordinated water when the complex compound is formed; that is, the metal ion and the base are separated by one or more water molecules.
2 *Complexes*: Most stable entities that result from the formation of largely covalent bonds between a metal ion and an electron-donating ligand—the interacting ligand is immediately adjacent to the metal cation—are called complexes (inner-sphere complexes).

Ion pairs are temporary partnerships which are in general formed between "hard" cations and hard anions (Table 6.3); they are also called outer-sphere complexes. In some cases a distinction between the two types of associations is possible through kinetic or spectrophotometric investigation. Kinetically, when true complexes are being formed, a dehydration step must precede the association reaction. Association accompanied by changes in the absorbance of visible light is indicative of complex formation reactions as such, whereas the formation of ion pairs may be accompanied by changes in the ultraviolet region.

Estimates of stability constants of ion pairs can be made on the basis of simple electrostatic models that consider *coulombic* interactions between the ions. Calculations made in this way indicate the following ranges of stability constants (25°C):

For ion pairs with opposite charge of 1, $\log K \simeq 0$ to 1 ($I = 0$); $\log K = -0.5$ to 0.5 (SW)

For ion pairs with opposite charge of 2, $\log K \simeq 1.5$ to 2.4 ($I = 0$); $\log K = 0.1$ to 1.2 (SW)

For ion pairs with opposite charge of 3, $\log K \simeq 2.8$ to 4.0 ($I = 0$)

† Our discussion here is based on Bjerrum's *ion-association model*. An alternative treatment of short-range interaction, the *specific interaction model*, will be discussed in the appendix to this chapter.

The range for seawater (SW) was estimated on the basis of assumed single-ion activity coefficients for a medium of ionic strength 0.7.

The $MgSO_4$ system. An interesting example is given by the interpretation obtained from ultrasonic absorption measurements on 2–2 electrolyte systems. Eigen and Tamm [19] proposed a three-step process, for example,

$$Mg^{2+} \cdot aq. + SO_4^{2-} \cdot aq \underset{k_{2,1}}{\overset{k_{1,2}}{\rightleftharpoons}} \left[Mg^{2+}O \begin{matrix} H\,H \\ H\,H \end{matrix} OSO_4^{2-} \right]aq$$

State 1   State 2

$$Mg^{2+}O\begin{matrix}H\\H\end{matrix}SO_4^{2-} \; aq \underset{k_{4,3}}{\overset{k_{3,4}}{\rightleftharpoons}} [MgSO_4]aq$$

State 3   State 4   (16)

On the basis of this kinetic evidence, three species are supposed to coexist at equilibrium, although equilibrium measurements do not allow these various species to be distinguished. Conductance and emf measurements lead to estimates of association constants that are composites for the formation of several species, namely,

$$K^c = \frac{[MgSO_4] + [Mg(H_2O)SO_4] + [Mg(H_2O)_2SO_4]}{[Mg^{2+}][SO_4^{2-}]} \tag{17}$$

Equilibrium constants computed from the kinetic data are consistent with those derived from conductivity measurements. Fischer [20] has compared sound absorption data of seawater with those of $MgSO_4$ solutions and calculated that 9% of the total Mg ions are associated as $MgSO_4$ ion pairs in seawater (20°C, 1 atm). These results agree well with those obtained by others [21,22].

Example 6.2. Hydrogen Carbonato and Carbonato Complexes of Mg(II)

To what extent are bicarbonato and carbonato Mg complexes significant species in fresh waters? W. Riesen, H. Gamsjäger, and P. W. Schindler [*Geochim. Cosmochim Acta*, **41**, 1193 (1977)] give the following equilibrium constants:

$$Mg^{2+} + HCO_3^- = MgHCO_3^+ \qquad \log \beta_1 = 0.69 \; (25°C) \qquad \text{(i)}$$
$$Mg^{2+} + 2HCO_3^- = Mg(HCO_3)_2^0 \qquad \log \beta_2 = 1.06 \; (25°C) \qquad \text{(ii)}$$
$$Mg^{2+} + CO_3^{2-} = MgCO_3^0 \qquad \log K_{1,CO_3^{2-}} = 2.85 \; (25°C) \qquad \text{(iii)}$$

For our calculation we assume a composition $C_T = 4 \times 10^{-3} \, M$, $Mg(II)_T = 10^{-3} \, M$, $I = 4 \times 10^{-3} \, M$, and 25°C, and vary the pH between 6 and 9.5.

Using the Davis equation (Table 3.3) to correct for ionic strength, we obtain for the carbonate protolysis system $pK'_1 = 6.33$, $pK'_2 = 10.27$; for the complexing with HCO_3^- and CO_3^{2-},

$$\frac{[MgHCO_3^+]}{[Mg^{2+}][HCO_3^-]} = \beta'_1 = 4.1 \tag{iv}$$

$$\frac{[Mg(HCO_3^-)_2^0]}{[Mg^{2+}][HCO_3^-]^2} = \beta'_2 = 8.5 \tag{v}$$

$$\frac{[MgCO_3^0]}{[Mg^{2+}][CO_3^{2-}]} = K'_{1,CO_3^{2-}} = 500 \tag{vi}$$

We might also have to consider the hydrolysis of Mg(II)

$$Mg^{2+} + H_2O = MgOH^+ + H^+ \qquad *K_{1,Mg(II)}$$

$$\frac{[MgOH]\{H^+\}}{[Mg^{2+}]} = *K'_{1,Mg(II)} = 10^{-11.52} \tag{vii}$$

Equations for total concentrations are:

$$[Mg(II)_T] = [Mg^{2+}] + ([MgOH^+] + [MgHCO_3^-]$$
$$+ [Mg(HCO_3^-)_2^0] + [MgCO_3^0]) \tag{viii}$$

and

$$C_T = [H_2CO_3^*] + [HCO_3^-] + [CO_3^{2-}] + ([MgHCO_3^+]$$
$$+ 2[Mg(HCO_3)_2^0] + [MgCO_3^0]) \tag{ix}$$

In principle, equations iv to ix and the two protolysis equilibria of $H_2CO_3^*$) have to be solved simultaneously to establish the equilibrium concentration of the eight species [$H_2CO_3^*$, HCO_3^-, CO_3^{2-}, Mg^{2+}, $MgHCO_3^-$, $Mg(HCO_3)_2^0$, $MgCO_3^0$, $MgOH^+$]. Obviously, the complex Mg species are present at low concentrations relative to Mg^{2+} and C_T; we can assume that in equations viii and ix the terms in parentheses on the right side are negligible. We then can compute [HCO_3^-] and [CO_3^{2-}] and estimate the concentration of the complex Mg species from equations iv to vii by setting [Mg^{2+}] = [$Mg(II)_T$]. We can then test whether our assumptions were correct and, if necessary, recalculate [Mg^{2+}] and $C'_T = [H_2CO_3] + [HCO_3^-] + [CO_3^{2-}]$; with the recalculated concentrations equations iv to vii can be solved again, and occasionally more than one iteration is necessary. The result is given in Figure 6.8 (no iterations were necessary). $MgHCO_3^+$ is the predominant complex species, but it is present at a concentration of less than 1 % of [Mg_T^{2+}]. $MgCO_3^0$ becomes important only at high pH where in our example the solution is already oversaturated with respect to $MgCO_3(s)$. The extent of HCO_3^- and CO_3^{2-} binding of Mg^{2+} is much greater in seawater (Section 6.7).

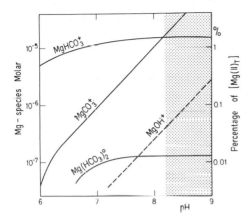

Figure 6.8 Mg(II) species as a function of pH for a system of $Mg(II)_T = 10^{-3} M$, $C_T = 4 \times 10^{-3} M$, 25°C. Under pH conditions of fresh natural waters the $MgHCO_3^+$ complex is more predominant than the species $MgCO_3^0$ and $Mg(HCO_3)_2^0$; the concentration of bicarbonato and carbonato complexes (for the typical conditions specified) are less than a few percent of the total Mg(II) concentration. Above pH 8.2, the solution is supersaturated with magnesite (shaded area).

6.5 COMPLEX FORMATION AND THE SOLUBILITY OF SOLIDS

We have seen that precipitation can be interpreted in terms of formation of polynuclear complexes. As illustrated for OH^- as a ligand (see Figure 5.4), the presence of metal complexing species will affect the solubility of slightly soluble metal salts. It is well known that the presence of foreign ligands increases the solubility. Thus ammonia increases the solubility of silver halides, and citrate or polyphosphates may dissolve $CaCO_3$, but even without foreign ligands complex formation in solution among constituent of the lattice will affect the solubility. Computation of the solubility solely from the solubility product and acid–base equilibria is possible only if the solid phase consists of a salt, that is, if the building stones of the lattice ionize completely in solution. Few of the solid binary compounds actually consist of truly ionic lattices; only then is it possible to calculate the solubility from the free metal-ion concentration $[Me^{+z}]$ in equilibrium with the solid phase:

$$Me_T = [Me^{+z}] \tag{18}$$

In Section 5.2, it was shown that the solubility of oxides and hydroxides could be enhanced by the formation of mononuclear hydroxo complexes. The total solubility was given as

$$Me_T = [Me^{+q}] + \sum_1^n [Me(OH)_n^{q-n}] \tag{19}$$

This equation can now be generalized if we consider the possibility of complex formation in solution with any ligand L or its protonated form $H_j L$. Then the total solubility is given by

$$Me_T = [Me]_{free} + \sum [Me_m H_k L_n(OH)_i] \tag{20}$$

where L represents the different ligand types and where all values of m, n, i, or $k \geq 0$ have to be considered in the summation. For example, the solubilities in marine waters of such "insoluble" substances as Ag_2S ($K_{s0} \approx 10^{-50}$), HgS (10^{-52}), FeOOH (10^{-38}), CuO (10^{-20}), Al_2O_3 (10^{-34}) are most likely to be determined by the presence of, respectively, AgSH, HgS_2^{2-} or HgS_2H^-, $Fe(OH)_2^+$, $CuCO_3$, and $Al(OH)_4^-$.

If a compound Me_mL_n has a very small solubility product K_{s0}, it must be expected that molecular associations between the metal ion and the ligand exist in solution as stable complexes [23]. This is understandable if we consider that the same forces that result in slightly soluble lattices are operative in the formation of soluble complexes. Metal ions of class-B cations (soft acids) are especially capable of forming covalent bonds (σ and π bonds). The presence of covalent bonds is frequently evident when the color of the solids (green copper hydroxide carbonate, yellow AgI, black Ag_2S) is not a composite of the colors of the ions. The following examples will illustrate the significance of complex formation in enhancing the solubility of solids.

The first example is based on a well-documented case history of the solubility of Ag_2S [23]. Ag^+, as a typical class-B cation (soft acid), has a strong tendency to coordinate with sulfur donor atoms. In this instance the concentration of free Ag^+ and the solubility are particularly illustrative. A second example examines the effect of carbonate complexing upon the solubility of Cu(II).

Example 6.3. The Solubility of Ag_2S

1. Compute from the information given in Table 6.4 the concentration of free Ag^+ in a sulfide solution of 0.1 M total sulfur concentration ($S_T = [H_2S] + [HS^-] + [S^{2-}] = 0.1$ M) and pH = 13.

By considering the mass laws of equations 1 to 3, the free Ag^+ concentration can be computed:

$$\log[Ag^+] = \tfrac{1}{2}(\log K_{s0} - \log S_T - \log \alpha_2) \tag{i}$$

where α_2 is defined as usual:

$$\alpha_2 = \frac{[S^{2-}]}{S_T} = \left(1 + \frac{[H^+]}{K_{HS}} + \frac{[H^+]^2}{K_{H_2S}K_{HS}}\right)^{-1}$$

Ignoring the fact that the solution has a slightly higher I than that for which the constants have been determined, one obtains for pH = 13 a $\log \alpha_2 = -1$ and, correspondingly, *pAg = 23.85.*

This result corresponds to *one silver ion per liter.* Somewhat naively one might be inclined to ask the "statistical" question: Where do we find the single Ag^+ if we subdivide a liter of saturated solution into ten 100-ml portions. Interestingly enough, a silver electrode that can sense silver ions potentiometrically will detect pAg = 23.85 in each one of the 100-ml portions. Although a silver electrode is extremely sensitive to free Ag^+, it cannot specifically respond to

**TABLE 6.4   EQUILIBRIA PERTINENT IN EFFECTING THE SOLU-
BILITY OF Ag_2S AND $HgS^{a,b}$**

$H_2S(aq) = H^+ + HS^-$	$\log K_{H_2S} = -6.68$	(1)
$HS^- = H^+ + S^{2-}$	$\log K_{HS} = -14.0$	(2)
$Ag_2S(s) = 2Ag^+ + S^{2-}$	$\log K_{so} = -49.7$	(3)
$Ag^+ + SH^- = AgSH$	$\log K_1 = 13.3$	(4)
$AgSH + SH^- = Ag(SH)_2^-$	$\log K_2 = 3.87$	(5)
$2Ag(SH)_2^- = HSAgSAgSH^{2-} + H_2S(aq)$	$\log K = 3.2$	(6)
$Ag_2S(s) + 2HS^- = HSAgSAgSH^{2-}$	$\log K_{s3} = -4.82$	(7)
$HgS(s) = Hg^{2+} + S^{2-}$	$\log K_{so} = -50.96$	(8)
$Hg^{2+} + 2HS^- = Hg(SH)_2$	$\log \beta_2 = 37.7$	(9)
$Hg(SH)_2 = HgS_2H^- + H^+$	$\log K_{a1} = -6.2$	(10)
$HgS_2H^- = HgS_2^{2-} + H^+$	$\log K_{a2} = 8.3$	(11)
$Hg^{2+} + 2S^{2-} = HgS_2^{2-}$	$\log \beta_2 = 51.5$	(12)

[a] G. Schwarzenbach and M. Widmer [23].
[b] $I = 0.1$ (NaClO$_4$), 20°C, 1 atm.

solitary silver ions. The electrode can of course function only because larger concentrations of Ag(I) are present in the solution of thio silver(I) complexes. We would reach the same conclusions by making a solubility determination using a radiochemical method and Ag_2S tagged with radioactive Ag^1. Indeed, radiochemically we would detect a solubility of approximately $10^{-8} M$. Accordingly the real solubility of Ag_2S under the conditions specified is approximately 14 orders of magnitude larger than that of free $[Ag^+]$.

2. Consider all the equilibria given in Table 6.4 and estimate as a function of pH the concentration of all the species in equilibrium with a saturated solution of $Ag_2S(s)$ in 0.02 M S_T.

Total dissolved silver is given by

$$Ag_T = [Ag^+] + [AgSH] + [Ag(SH)_2^-] + 2[Ag_2S_3H_2^{2-}] \qquad \text{(ii)}$$

where $[Ag^+]$ is defined by

$$[Ag^+] = \left(\frac{K_{so}}{S_T \alpha_2}\right)^{1/2} \qquad \text{(i)}$$

and the concentrations of AgSH and $Ag(SH)_2^-$ can be compared from equilibrium 2, 4 and 5 of Table 6.4:

$$[AgSH] = [Ag^+]K_1 S_T \alpha_1 \qquad \text{(iii)}$$

and

$$[Ag(SH)_2^-] = [Ag^+]K_1 K_2 S_T^2 \alpha_1^2 \qquad \text{(iv)}$$

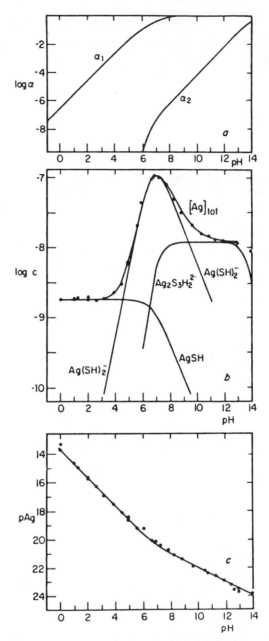

Figure 6.9 The solubility of Ag_2S (see Example 6.3). The solubility is not given by the concentration of free Ag^+; it is determined by the presence of the complexes AgSH, $Ag(SH)_2^-$, and $HSAgSAgSH^{2-}$ (cf. Schwarzenbach and Widmer, *Helv. Chim. Acta,* **49**, 111 (1966). Total concentration Ag_T over precipitated Ag_2S has been measured radiochemically. Free concentration, $[Ag^+]$, has been determined from emf measurements with a silver electrode. The data are valid for a solution containing a total concentration of $S_T (= [H_2S] + [HS^-] + [S^{2-}]) = 0.02\ M$ and an ionic strength of $I = 0.1$ ($NaClO_4$) at 20°C. (*a*) Distribution coefficients of HS^- and S^{2-}, respectively. (*b*) Soluble silver complexes and total Ag(I) as a function of pH. (*c*) Free $[Ag^+]$ versus pH. (Reproduced with permission from Schweiz Chemikerverband, Zürich, Switzerland.)

where

$$\alpha_1 = \frac{[\text{HS}^-]}{S_T} = \left(\frac{[\text{H}^+]}{K_{\text{H}_2\text{S}}} + 1 + \frac{K_{\text{HS}}}{[\text{H}^+]}\right)^{-1}$$

The concentration of HS–AgS–AgSH^{2-} is computed most readily from equilibrium 7 of Table 6.4:

$$[\text{Ag}_2\text{S}_3\text{H}_2^{2-}] = K_{s_3}S_T^2\alpha_1^2 \tag{v}$$

Summing up the terms in equation ii for total soluble silver as a function of S_T and an implicit function of $[\text{H}^+]$, we obtain

$$\text{Ag}_T = \left(\frac{K_{s_0}}{S_T\alpha_2}\right)^{1/2}(1 + K_1 S_T \alpha_1 + K_1 K_2 S_T^2 \alpha_1^2) + 2K_{s3}S_T^2\alpha_1^2 \tag{vi}$$

(see Figure 6.9). For every pH the individual species can be computed readily by the equations given above. In the pH range below 4, Ag_T consists primarily of AgSH. An inspection of equations i and iii shows that $d \log \lfloor\text{AgSH}\rfloor/d\text{pH} = d \log(\alpha_1/\sqrt{\alpha_2})/d\text{pH} - 0$. In the pH range 5 to 9 the solubility of silver is given by Ag(SH)_2^-; its logarithmic concentration plots with a slope of $+1$ and $-\tfrac{1}{2}$ in the pH range up to and beyond $\text{pH} = 6.7(\text{p}K_{\text{H}_2\text{S}})$, respectively; these slopes arise because $d \log[\text{Ag(SH)}_2^-]/d\text{pH} = d \log(\alpha_1^2/\sqrt{\alpha_2})/d\text{pH}$ equals $+1$ and $-\tfrac{1}{2}$ for the regions $\text{pH} < \text{p}K_{\text{H}_2\text{S}}$ and $\text{pH} > \text{p}K_{\text{H}_2\text{S}}$, respectively. Only at high $\text{pH}(>9)$ does the dimer $\text{Ag}_2\text{S}_3\text{H}_2^{2-}$ become significant in controlling the solubility; in this pH range $\alpha_1 = 1$ up to pH 14 ($\text{p}K_{\text{HS}}-$), hence its concentration does not vary with pH.

Schwarzenbach and Widmer obtained Ag_T experimentally by a radiochemical method. They determined free $[\text{Ag}^+]$ from emf measurements with a silver electrode. The existence of the thio complexes was postulated and their stability calculated from a combination of these measurements. In this context it should be mentioned that it is not possible from such solubility experiments to distinguish mathematically between a species Ag(SH)_2^- and polynuclear species of the type $(\text{Ag}_2\text{S})_x \text{Ag(SH)}_2^-$. But, for chemical reasons, the formation of the polynuclear species is improbable, so that x is taken equal to zero.

The solubility of Hg^{2+} and Ag^+ in sulfide-bearing seawater is given in Figure 6.14a and b.

Example 6.4. Solubility of Cu(II) in Natural Water; Effect of Complexing by Carbonate

Estimate the solubility of Cu(II) in carbonate-bearing water of constant $C_T(C_T = 10^{-2}\ M)$ (closed system). The pertinent Cu(II) equilibria are given in Table 6.5.

In order to gain insight into the predominant solid phases and soluble species, it appears expedient to construct first an activity ratio diagram. Taking $\{\text{Cu}^{2+}\}$

TABLE 6.5 Cu(II) EQUILIBRIA[a]

Item	Reaction	Log K^a
1	$CuO(s) + 2H^+ = Cu^{2+} + H_2O$ (tenorite)	7.65
2	$Cu_2(OH)_2CO_3(s) + 4H^+ = 2Cu^{2+} + 3H_2O + CO_2(g)$ (malachite)	14.16
3	$Cu_3(OH)_2(CO_3)_2(s) + 6H^+ = 3Cu^{2+} + 4H_2O + 2CO_2(g)$ (azurite)	21.24
4	$Cu^{2+} + H_2O = CuOH^+ + H^+$	-8
5	$2Cu^{2+} + 2H_2O = Cu_2(OH)_2^{2+} + 2H^+$	-10.95
6	$Cu^{2+} + CO_3^{2-} = CuCO_3(aq)$	6.77
7	$Cu^{2+} + 2CO_3^{2-} = Cu(CO_3)_2^{2-}(aq)$	10.01
8	$CO_2(g) + H_2O = HCO_3^- + H^+$	-7.82
9	$Cu^{2+} + 3H_2O = Cu(OH)_3^- + 3H^+$	-26.3
10	$Cu^{2+} + 4H_2O = Cu(OH)_4^{2-} + 4H^+$	-39.4

[a] $I = 0$; 25°C. Given or quoted by P. Schindler, " Heterogeneous Equilibria Involving Oxides, Hydroxides, Carbonates and Hydroxide Carbonates," in *Equilibrium Concepts in Natural Water Systems*, Advances in Chemistry Series, No. 67, American Chemical Society, Washington, D.C., 1967.

as a reference we obtain the following activity ratios for the equilibria given in Table 6.5:

$$\log \frac{\{CuO(s)\}}{\{Cu^{2+}\}} = -7.65 + 2pH \tag{i}$$

$$\log \frac{\{Cu(OH)(CO_3)_{0.5}(s)\}}{\{Cu^{2+}\}} = -3.17 + 1.5\,pH + 0.5 \log C_T + 0.5 \log \alpha_1 \tag{ii}$$

Equation ii results from a combination of (2) and (8) of Table 6.5.
Similarly combining (3) and (8) gives

$$\log \frac{\{Cu(OH)_{0.67}(CO_3)_{0.67}(s)\}}{\{Cu^{2+}\}} = -1.85 + pH + 0.67 \log C_T + 0.67 \log \alpha_1 \tag{iii}$$

$$\log \frac{\{CuOH^+\}}{\{Cu^{2+}\}} = -8 + pH \tag{iv}$$

$$\log \frac{\{CuCO_3(aq)\}}{\{Cu^{2+}\}} = 6.77 + \log C_T + \log \alpha_2 \tag{v}$$

$$\log \frac{\{Cu(CO_3)_2^{2-}(aq)\}}{\{Cu^{2+}\}} = 10.01 + 2 \log C_T + 2 \log \alpha_2 \tag{vi}$$

$$\log \frac{\{Cu(OH)_3^-\}}{\{Cu^{2+}\}} = -26.3 + 3\,pH \tag{vii}$$

$$\log \frac{\{Cu(OH)_4^{2-}\}}{\{Cu^{2+}\}} = -39.4 + 4\,pH \tag{viii}$$

In a double logarithmic diagram, log activity ratio versus pH (equations i to viii) can be plotted in straight-line portions having readily defined slopes. Figure 6.10*a* shows that under the specified conditions malachite and tenorite qualify as stable solid phases; malachite is stable below pH = 7, while tenorite is more stable in the alkaline region. With respect to dissolved species the activity ratio diagram reveals that under the specified conditions the following species predominate: Cu^{2+} up to pH = 6; $CuCO_3$(aq) in the pH range 6 to 9.3;

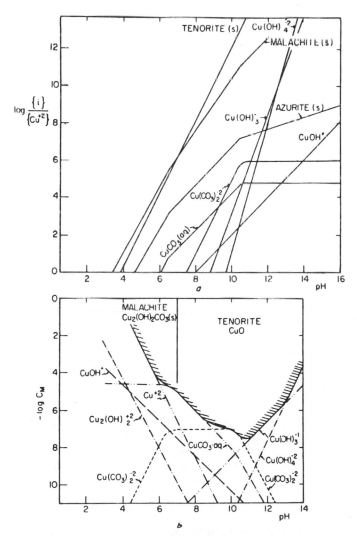

Figure 6.10 Solubility of Cu(II). (*a*) Activity ratio diagram. (*b*) Solubility diagram. The solid line surrounding the shaded area gives the total solubility of Cu(II) which up to pH value of 6.96 is governed by the solubility of malachite [$Cu_2(OH)_2CO_3$(s)]. In the low pH region azurite [$Cu_3(OH)_2(CO_3)_2$(s)] is metastable but may become stable at higher C_T. Above pH 7 the solubility is controlled by the solubility of CuO (tenorite). The predominant species with increasing pH are Cu^{2+}, $CuCo_3$(aq), $Cu(CO_3)_2^{2-}$, and hydroxo copper(II) anions. $C_T = 10^{-2}$ M.

$Cu(CO_3)_2^{2-}$ (aq) in the pH range 9.3 to 10.7; $Cu(OH)_3^-$ and $Cu(OH)_4^{2-}$ above pH 10.7 and 12.9, respectively. With this information a logarithmic solubility diagram can be sketched (Figure 6.10b).

Some Geochemical Implications

The two examples given above underline the need to identify all pertinent species and to consider all reactions and equilibria related to the solubility of a mineral. Calculations from the solubility product alone can often be very misleading; although the solubility products of HgS and PbS may differ by more than 20 orders of magnitude, their solubility does not differ by more than a few orders of magnitude. Complex formation among constituents of the lattice or with constituents of the solution must be taken into account in exploring mechanisms of ore transport by ore-forming fluids. The observation that metal sulfides having the smallest solubility products behave as if they were the most soluble is compatible with the coordination chemical interpretation that, the smaller the solubility product, the stronger the tendency to form soluble associates.

Complex formers present in solution in low concentrations may often have little or no effect on the solubility of solids; they may, however, affect the kinetics of nucleation and of growth and dissolution of crystals.

6.6 CHELATES

One of the reasons why metal ions in solution in natural waters are usually not appreciably complexed by ligands other than H_2O or OH^- is that most bases indigenous to natural waters are unidentate ligands. As we shall see, unidentate ligands form less stable complexes than multidentate ligands, especially in dilute solutions. Carbonate, sulfate, and phosphate can, but seldom do, serve as bidentate ligands. The solid carbonates of Ca^{2+}, Mn^{2+}, Fe^{2+}, Co^{2+}, and Zn^{2+} have six oxygen atoms around the metal ion, but besides $CaCO_3$(aq) and $MgCO_3$(aq) soluble carbonato complexes may exist only at high pH for Be(II), the heavier lanthanides, and Cu(II). Perhaps for steric reasons these anions usually act as unidentate ligands.

As with hydroxide, complexes with other ligands are formed stepwise. In a series such as

$$Cu^{2+}, Cu(NH_3)^{2+}, Cu(NH_3)_2^{2+}, Cu(NH_3)_3^{2+}, Cu(NH_3)_4^{2+}, Cu(NH_3)_5^{2+}$$

the successive stability constants generally decrease; that is, the Cu^{2+} ion takes up one NH_3 molecule after the other. (There are exceptions to this behavior.) As can be seen from Figure 6.2d, relatively high concentrations of NH_3 are necessary to complex copper(II) effectively and to form a tetramine complex. In natural waters the concentrations of the ligands and the affinity of the ligands for the metal ion, with the exception of H_2O and OH^-, are usually sufficiently small so that at best a one-ligand complex may be formed.

The Chelate Effect

Complexes with monodentate ligands are usually less stable than those with multidentate ligands. More important is the fact that the degree of complexation decreases more strongly with dilution for monodentate complexes than for multidentate complexes (chelates). This is illustrated in Figure 6.11, where the degree of complexation is compared as a function of concentration for uni, bi, and tetradentate copper(II) amine complexes. Free $[Cu \cdot aq^{2+}]$ is plotted as a function of dilution in the left-hand graph, while the quantitative degree of complexation, as measured by $\Delta pCu = \log(Cu_T/[Cu^{2+}])$ is given in the right-hand graph. It is obvious from this figure that the complexing effect of NH_3 on Cu^{2+} becomes negligible at concentrations that might be encountered in natural water systems. Chelates, however, remain remarkably stable even at very dilute concentrations.

The curves drawn in Figure 6.11 have been calculated on the basis of constants taken from *Stability Constants of Metal–Ion Complexes*. The calculations are

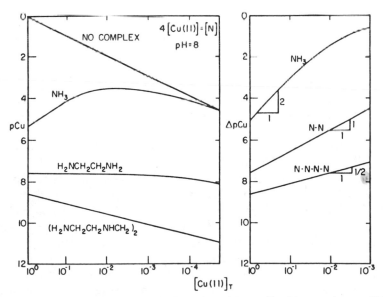

Figure 6.11 The chelate effect on complex formation of $Cu \cdot aq^{2+}$ with monodentate, bidentate, and tetradentate amines. pCu is plotted as a function of concentration in the left-hand diagram. On the right the relative degree of complexation as measured by ΔpCu as a function of concentration is depicted. The extent of complexing is larger with chelate complex formers than with unidentate ligands. Unidentate complexes are dissociated in dilute solutions while chelates remains essentially undissociated at great dilutions.

essentially the same as those outlined for the hydroxo complex formation; although algebraically simple, they are tedious and time-consuming. In the case of the $Cu-NH_3$ system, the following species have to be considered

$$Cu^{2+}, Cu(NH_3)^{2+}, Cu(NH_3)_2^{2+}, Cu(NH_3)_3^{2+}, Cu(NH_3)_4^{2+},$$
$$Cu(NH_3)_5^{2+}, NH_4^+, NH_3$$

Thus, for every $[H^+]$, eight equations have to be solved simultaneously in order to compute the relative concentrations of each species present. Six mass laws (five stability expressions for the five different amine complexes and the acid–base equilibrium of $NH_4^+-NH_3$) and two concentration conditions make up the eight equations. As concentration conditions one might formulate equations defining Cu_T and NH_{3_T}:

$$Cu_T = [Cu^{2+}] + [Cu(NH_3)^{2+}] + [Cu(NH_3)_2^{2+}] + \cdots \tag{21}$$
$$NH_{3_T} = [NH_4^+] + [NH_3] + [Cu(NH_3)^{2+}] + 2[Cu(NH_3)_2^{2+}] + \cdots \tag{22}$$

Guidelines for coping with these and more involved types of calculations have been provided by Ringbom [24], Schwarzenbach [25], and others. Computers are of course also very useful.

In Figure 6.11 a concentration of complex former equivalent to that of metal ion was considered. ΔpM of course increases with increasing concentration of the complex former over the metal. Figure 6.12a shows the effect of various ligands on complex formation with ferric iron. Here the concentration of the complexing agent is kept at a constant value and in excess of $[Fe(III)_T]$. If the ligand is in large excess over the metal, the quantitative degree of complexation ΔpM is independent of the total metal-ion concentration. In this figure we again observe the increase in stability in going from monodentate (F^-, SO_4^{2-}, HPO_4^{2-}) to bidentate (oxalate), to tridentate (citrate), and to hexadentate (EDTA, DCTA) ligands. Figure 6.12 also illustrates that in all aqueous solutions $[H^+]$ and $[OH^-]$ influence markedly the degree of complexation. At low pH, H^+ competes successfully with the metal ions for the ligand. At high pH, OH^- competes successfully with the ligand for the coordinative positions on the metal ion. Furthermore, at low and high pH mixed hydrogen–metal and hydroxide–ligand complexes can be formed. [In the case of EDTA ($=L$), in addition to FeL^- the complexes $FeHL$, $FeOHL^{2-}$, and $Fe(OH)_2L^{3-}$ have to be considered. Because of the competing influence of H^+ or OH^-, the complexing effect cannot be estimated solely from the stability constants.

The calculations predict that a 10^{-2} M solution of sulfate, fluoride, phosphate, oxalate, or citrate can keep 10^{-3} M ferric iron in solution [i.e., prevent precipitation of $Fe(OH)_3(s)$] only up to pH values of 3.3, 4.7, 4.8, 6.9, and 7.6, respectively. From another point of view, the pH of ferric iron precipitation will be strongly affected by the presence of coordinating anions; for example, Figure 6.12a illustrates that in the presence of 10^{-2} M ligand a 10^{-3} M solution of ferric iron will form precipitates of about pH 4.5 if the ligand is phosphate

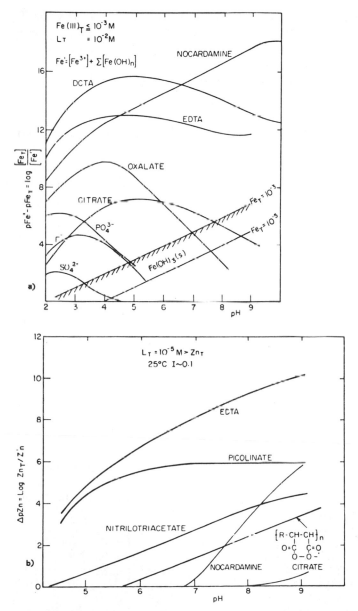

Figure 6.12 (*a*) Complexing of Fe(III). The degree of complexation is expressed in terms of ΔpFe for various ligands (10^{-2} *M*). The competing effect of H^+ at low pH values and of OH^- at higher pH values explains that effective complexation is strongly dependent on pH. Mono-, di-, and tridentate ligands are not able to keep a 10^{-3} *M* Fe(II) in solution at higher pH values. EDTA is Ethylenediaminetetraacetate. DCTA is 1,2-diaminocyclohexane tetraacetate. Nocordamine is a trihydroxamate (see Figure 6.14). (*b*) Chelation of Zn(II). The extent of complex formation (ΔpZn) as a function of pH is strongly influenced by the acid–base behavior of the metal ion and of the complex former. In the case of Zn(II), the tendency to form hydroxo species is less in the pH range of interest than in the case of Fe(III).

(or $H_2PO_4^-$), whereas precipitation will not occur until pH values of about 7.5 if the ligand is citrate. The precipitates formed initially in the presence of such ligands are usually nonstoichiometric mixed precipitates [e.g., phosphato-hydroxo iron(III) precipitates].

Figure 6.12b illustrates complexation of Zn(II) by a few chelate formers.

Example 6.5. The Stability of Ethylenediamine Complexes of Nickel

Estimate the equilibrium composition of a $10^{-4}\ M$ solution of $Ni(en)_3Cl_2$ (en = enthylenediamine) whose pH has been adjusted to (a) pH = 7.2 and (b) pH = 3.0. The following constants are available: (I = 0.1, 25°C)log β_1 = 7.5, log β_2 = 13.8, log β_3 = 18.3, log $K_{H_2en^{2+}}$ = -7.4, log K_{Hen^+} = -10.1.

In addition to the five equations defined by these equilibrium constants, three additional equations can be established (for simplicity, charges are omitted):

$$Ni_T = [Ni] + [Ni(en)] + [Ni(en)_2] + [Ni(en)_3] \tag{i}$$

$$en_T = [en] + [Hen] + [H_2en] + [Ni(en)] + 2[Ni(en)_2] + 3[Ni(en)_3] \tag{ii}$$

Furthermore, for a pure solution of the salt $Ni(en)_3Cl_2$, we obtain the concentration condition

$$3\,Ni_T = en_T \tag{iii}$$

Substituting the formulations for the equilibrium constants into equations i and ii and calculating by trial and error, we obtain the following approximate equilibrium compositions (percent contribution in parentheses):

(a) pH = 7.2	(b) pH = 3.0
$[Ni^{2+}] = 2.12 \times 10^{-5}\ (21\%)$	$[Ni^{2+}] = 9.99 \times 10^{-5}\ (\sim 100\%)$
$[Ni(en)^{2+}] = 6.35 \times 10^{-5}\ (64\%)$	$[Ni(en)^{2+}] = 10^{-11.5}\ (0\%)$
$[Ni(en)_2^{2+}] = 1.20 \times 10^{-5}\ (12\%)$	$[Ni(en)_2^{2+}] = 10^{-20.2}\ (0\%)$
$[Ni(en)_3^{2+}] = 0.003 \times 10^{-5}\ (0\%)$	$[Ni(en)_3^{2+}] = 10^{-30.7}\ (0\%)$
$[en] = 9.5 \times 10^{-8}$	$[en] = 10^{-15}\ (0\%)$
$[Hen^+] = 7.6 \times 10^{-5}$	$[H_2en^{2+}] = 2.99 \times 10^{-4}\ (100\%)$
$[H_2en^{2+}] = 1.2 \times 10^{-4}$	$[Hen^+] = 10^{-7.8}\ (0\%)$

Note the following:

1   The predominant species in the solution is H_2en^{2+}.

2   At about pH = 3, where there is very little complex formation (less than $10^{-11}\ M$), the predominant species is H_2en^{2+}. Even at the higher pH value (7.2), a significant portion of the metal ion remains uncomplexed.

Increase in Entropy Primarily Responsible for the Chelate Effect. A chelate ring is more stable than the corresponding complex with unidentate ligands. The enthalpy change resulting from either complex formation, however, is frequently about the same; for example, ΔH for the formation of a diamine complex is approximately equal to ΔH for the formation of an ethylenediamine complex [26]. Hence essentially the same type of bond occurs in these complexes. That $-\Delta G°$ for the chelate is larger than for the corresponding complex with an unidentate ligand must be accounted for primarily by the fact that the formation of a ring is accompanied by a larger increase in entropy than that encountered in the formation of a nonchelate complex.

Metal-Ion Buffers

The analogy between Me^{z+} and H^+ can be extended to the concept of buffers. pH buffers are made by mixing acids and conjugate bases in proper proportions:

$$[H^+] = \frac{K[HA]}{[A^-]} \qquad (23)$$

Metal ions can be similarly buffered by adding appropriate ligands to the metal-ion solution:

$$[Me] = \frac{K[MeL]}{[L]} \qquad (24)$$

Such pMe buffers resist a change in $[Me^{z+}]$. It is well known that the living cell controls not only pH but also pCu, pMn, pMg, and so on, and that complex formers are used as the buffering component; pMe buffers are convenient tools for investigating phenomena pertaining to metal ions. It is unwise to prepare a pH 6 solution by diluting a concentrated HCl solution, but this mistake is frequently made with metal-ion solutions. If, for example, we want to study the toxic effect of Cu^{2+} on algae, it might be more appropriate to prepare a suitable pCu buffer. If a copper salt solution is simply diluted, the concentration (or activity) of the free Cu^{2+} may, because of hydrolysis and adsorption and other side reactions, be entirely different from that calculated by considering the dilution only.

Example 6.6. pCa Buffer

Calculate pCa of a solution of the following composition: EDTA (ethylenediamine tetraacetate) $= Y_T = 1.95 \times 10^{-2}\,M, Ca_T = 9.82 \times 10^{-3}, pH = 5.13, I = 0.1\ (20°C)$. EDTA is a tetraprotic acid. For the conditions given the four acidity constants are characterized by $pK_1 = 2.0$, $pK_2 = 2.67$, $pK_3 = 6.16$,

$pK_4 = 10.26$. Two Ca complexes are formed with EDTA, $CaHY^-$ and CaY^{2-}. Hence we need the stability constants

$$\frac{[CaY^{2-}]}{[Ca^{2+}][Y^{4-}]} = K_{CaY} = 10^{10.6}$$

$$\frac{[CaHY^-]}{[Ca^{2+}][HY^{3-}]} = K_{CaHY} = 10^{3.5}$$

The computation may start by setting up equations for the concentration conditions (for simplicity charges are omitted):

$$\begin{aligned} Ca_T &= [Ca] + [CaY] + [CaHY] \\ &= [Ca](1 + [Y]K_{CaY} + [Y][H]K_4^{-1}K_{CaHY}) \\ &= [Ca]\alpha_{Ca}^{-1} \end{aligned} \tag{i}$$

and, as before,

$$\alpha_{Ca} = \frac{[Ca]}{Ca_T}$$

$$\begin{aligned} Y_T &= [H_4Y] + [H_3Y] + [H_2Y] + [HY] + [Y] + [CaY] + [CaHY] \\ &= [Y](\alpha_4')^{-1} + [Ca][Y]K_{CaY} + [Ca][H][Y]K_4^{-1}K_{CaHY} \end{aligned} \tag{ii}$$

where

$$\alpha_4' = \frac{[Y]}{\sum_{i=0}^{i=4}[H_iY]} = \left(1 + \frac{[H]}{K_4} + \frac{[H]^2}{K_4K_3} + \frac{[H]^3}{K_4K_3K_2} + \frac{[H]^4}{K_4K_3K_2K_1}\right)^{-1} \tag{iii}$$

α_4' can be plotted most conveniently in a double logarithmic diagram ($\log \alpha_4'$ versus $-\log[H^+]$). Equations i and ii, containing the two unknowns $[Y]$ and $[H]$, are best solved by trial and error. Most conveniently we may start by calculating α_4' from equation iii or taking its value from a graphical plot. Then $[Y]$ may be estimated from equation iii by assuming that $\sum[H_iY] \simeq (Y_T - Ca_T)$.

With this tentative $[Y]$, first values of α_{Ca}^{-1} and $[Ca]$ are estimated, respectively, with the help of equation i. We then may check whether the assumption that $\sum[H_iY] = [Y_T - Ca_T]$ was appropriate. Subsequent reiteration gives the result $[Ca] = 4.12 \times 10^{-5}$ (pCa = 4.39) and $[Y] = 6.05 \times 10^{-9}$. For illustration the concentrations of all the other species are also given

$$[CaY] = 9.66 \times 10^{-3} \qquad [CaHY] = 1.09 \times 10^{-4} \qquad [H_4Y] = 2.26 \times 10^{-8}$$
$$[H_3Y] = 3.07 \times 10^{-5} \qquad [H_2Y] = 8.8 \times 10^{-3} \qquad [HY] = 8.21 \times 10^{-4}$$

Note: If $[Ca^{2+}]$ is lost from the solution, that is, by adsorption at the glass wall, the Ca^{2+} ion buffer described will have a tendency to maintain constant pCa. For example, removal from the solution of $2 \times 10^{-5} M$ Ca^{2+} per liter ($dCa_T = 2 \times 10^{-5} M$, corresponding to approximately 50% of the free $[Ca^{2+}]$)

will change the free $[Ca^{2+}]$ by approximately $4 \times 10^{-7} M$ (dpCa $= 0.005$), that is,

$$\beta_{pH, Y_T}^{Ca_T} = \frac{dCa_T}{d\text{pCa}} \simeq 4 \times 10^{-3} \text{ mol per pCa unit}$$

6.7 INORGANIC COMPLEXES IN NATURAL WATERS

Some stability constants for the association of representative cations and ligands are summarized in Table 6.6. Table 6.6 illustrates some of the features of the classification of cations described in Table 6.3, where the first nine cations listed are those of A-type metals or are classified as hard acids and tend to form stronger complexes with ligands containing O and F donor groups. At the bottom of the list are B-type cations which tend to form chloro and ammonia complexes. These cations form very insoluble sulfides and stable thio complexes. Transition-metal cations are intermediate in their association behavior. The formation constants of OH complexes, especially of class-A cations, are large compared with those of most other ligands. This explains why with most tri- and tetravalent metal ions, the hydroxides or hydrous oxides (or oxide hydroxides) are the only stable precipitates in the pH range of natural waters. However, for many bivalent cations CO_3^{2-}, S^2, S_2^{2-}, and PO_4^{3-} may successfully compete with OH ions to satisfy their coordinative requirements. The distribution of ion pairs and complexes can readily be calculated if all pertinent equilibria can be identified and equilibrium constants valid for solutions of appropriate ionic strength are available.

If two components are present in widely different concentrations, the less abundant component can affect the activity of the other only negligibly through complex formation. For most natural waters, including seawater, the major components are Na^+, Ca^{2+}, Mg^{2+}, K^+, and the ligands Cl^-, HCO_3^-, CO_3^{2-}, SO_4^{2-}, and H_2O (OH^-).

In the freshwater system the complexes are relatively unimportant in the speciation of the major components. In the seawater system, on the contrary, a sizable fraction of the components is found in complex form.

Major Inorganic Ion Association in Fresh Waters. Table 6.6, I gives stability constants for the formation of HCO_3^-, CO_3^{2-}, and SO_4^{2-} complexes of the major cations. For a typical freshwater composition the concentrations of major inorganic complexes as calculated from the stability constants—(Table 6.6, I) corrected with the Davis equation (Table 3.3) for an ionic strength of $I = 3 \times 10^{-3} M$—are given in Table 6.7. (For the calculation, the procedure given in Example 6.2 was used.) Typically the concentrations of HCO_3^-, CO_3^{2-}, and SO_4^{2-} complexes amount to a few percent of the concentrations of the total metals.

TABLE 6.6 STABILITY CONSTANTS ($\log K_1$) FOR ION ASSOCIATION BETWEEN THE MAJOR INORGANIC IONS IN NATURAL WATERS AT 25°C, 1 atm

I. Ionic strength $= 0$ $\log K_1 = \log(\{MX\}/\{M\}\{X\}$

Ion	HCO_3^-	CO_3^{2-}	SO_4^{2-}
Na^+	-0.3^a	0.77^a	0.70
K^+	—	—	0.85
Mg^{2+}	$0.69(1.06)^{b,c}$	2.88^c	2.23
Ca^{2+}	1.0^d	3.15^e	2.31
H^+	6.35	10.33	2.0

II. Seawater $\log K_1^s = \log([MX]/[M][X])$

Ion	HCO_3^-	CO_3^{2-}	SO_4^{2-}	F
Na^+	-0.55^f	0.97^f	0.34^g	—
K^+	—	—	0.13^g	—
Mg^{2+}	0.21^f	2.20^g	1.01^h	1.3
Ca^{2+}	0.29^f	1.89^g	1.03^g	0.62^i
H^+	$6.035^{j,k}$	9.09^k	1.38^l	2.9^m

[a] J. N. Butler and R. Huston, *J. Phys. Chem.*, **24**, 2976 (1970).

[b] $\log[\beta_2 = \{Mg(HCO_3^-)_2^0\}/(\{Mg^{2+}\}\{HCO_3^-\}^2)]$.

[c] W. Riesen et al., *Geochim. Cosmochim Acta*, **41**, 1193 (1977).

[d] R. L. Jacobson and D. Langmuir, *Geochim. Cosmochim. Acta*, **38**, 301 (1974).

[e] E. J. Reardon and D. Langmuir, *Geochim. Cosmochim. Acta*, **40**, 549 (1976).

[f] R. M. Pytkowicz and J. E. Hawley, *Limnol. Oceanogr.*, **19**, 223 (1974).

[g] R. M. Pytkowicz and D. R. Kester, *Amer. J. Sci.*, **267**, 217 (1969).

[h] D. R. Kester and R. M. Pytkowicz, *Limnol. Oceanogr.* **13**, 670 (1968).

[i] B. Elgquist, *J. Inorg. Nucl. Chem.*, **32**, 437 (1970).

[j] These constants are defined by

$$K_1' = [H_2CO_{3_T}^*]/\{H^+\}[HCO_{3_T}^-],$$

$$K_2' = [HCO_{3_T}^-]/\{H^+\}[CO_{3_T}^{2-}],$$

where $\{H^+\}$ is H^+ ion activity according to the pH definition of the National Bureau of Standards.

[k] C. Mehrbach et al., *Limnol. Oceanogr.*, **18**, 897 (1973).

[l] D. Dyrssen et al., *J. Chem. Educ.*, **46**, 252 (1969).

[m] K. Srinivasan and G. A. Rechnitz, *Anal. Chem.*, **40**, 509 (1968).

TABLE 6.7   INORGANIC COMPLEXES IN FRESH WATER[a]

Ion	HCO_3^-	CO_3^{2-}	SO_4^{2-}	Free Ion
Na^+	6.3	7.6	6.4	3.30 (100%)[b]
K^+	—	—	6.9	4.00 (100%)
Mg^{2+}	5.9	5.8	5.1	3.54 (96.3%)
Ca^{2+}	4.9	5.2	4.8	3.17 (95%)
H^+	—	—	9.6	8.0
Free ion[b]	2.70 (99% of C_T)	4.97	3.75 (88%)	—

[a] Composition: $C_T = 2.05 \times 10^{-3}$ M, $Ca(II)_T = 7 \times 10^{-4}$ M, $Mg(II)_T = 3 \times 10^{-4}$ M, $Na^+ = 5 \times 10^{-4}$ M, $K^+ = 1 \times 10^{-4}$ M, $SO_4^{2-} = 2 \times 10^{-4}$ M, pH = 8 (25°C), $I = 3 \times 10^{-3}$ M. Concentrations are given as $-\log M$.
[b] Percentage of total species present as free ions.

Major Inorganic Species in Seawater.   With the help of stability constants valid for seawater, a model for distribution of the most important dissolved species can be calculated. Garrels and Thompson [27] were the first to establish such a seawater model. Their calculations were based on stability constants (determined in simple electrolyte solutions and corrected or extrapolated to $I = 0$) and estimated activity coefficients of the individual ionic species in seawater. The mean ionic activity coefficients were assumed to be the same as those that would apply to a pure solution of the salt at the same ionic strength as seawater. This assumption is supported phenomenologically.

Operational concentration stability constants valid for seawater are now available (Table 6.6, II) that permit direct computation of the major inorganic species (without using individual activity coefficients).†

The distribution of the species in seawater according to the model suggested by Garrels and Thompson and as calculated on the basis of more recent information by Hanor and by Pytkowicz and Hawley is given in Table 6.8. According to these models, the major cations in seawater are predominantly present as free aquo metal ions. This is understandable because they are A-type metals and the concentration of major cations is much greater than the concentration of associating anions. Correspondingly, a significant fraction of the anions CO_3^{2-}, SO_4^{2-}, and HCO_3^- is associated with metal ions.

The use of recent values of activity coefficients and stability constants has modified views on the distribution of ligands more than of metal ions, but the pertinent qualitative features of the Garrels and Thompson model are still valid.

---

† The calculations consist essentially in solving 18 equations with 18 unknowns simultaneously. Ten stability constants are available and 8 mass balance relations, for example,

$$[SO_{4_T}] = [SO_4^{2-}] + [MgSO_4(aq)] + [CaSO_4(aq)] + [KSO_4^-] + [NaSO_4^-]$$

for Na_T, Mg_T, Ca_T, K_T, Cl_T, SO_{4_T}, HCO_{3_T}, CO_{3_T}. The calculation can be simplified if the assumption is made first that the cations are not significantly complexed (e.g., $[Ca^{2+}] = Ca_T$).

TABLE 6.8 CHEMICAL SEAWATER MODELS[a]

Percentage of Major Species at 25°C (1 atm)

Species	Na(I) G 0.4752 m	Na(I) H 0.4823 m	Na(I) P+H 0.4822 m	Mg(II) G 0.0540 m	Mg(II) H 0.05485 m	Mg(II) P+H 0.05489 m	Ca(II) G 0.0104 m	Ca(II) H 0.01062 m	Ca(II) P+H 0.01063 m	K(I) G 0.0100 m	K(I) H 0.01020 m	K(I) P+H 0.01062 m
Free ion	99	99.0	97.7	87	89.9	89.2	91	91.5	88.5	99	98.5	98.9
MSO_4	1.2	1.0	2.2	11	9.2	10.3	8	7.6	10.8	1	1.5	1.1
$MHCO_3$	0.01	0.0	0.1	1	0.6	0.1	1	0.7	0.3	—	0.0	—
MCO_3	—	0.0	0.0	0.3	0.3	0.1	0.2	0.2	0.3	—	0.0	—
Mg_2CO_3	—	—	—	—	—	0.0	—	—	—	—	—	—
$MgCaCO_3$	—	—	—	—	—	0.0	—	—	0.1	—	—	—

Species	SO_4^{2-} G 0.0284 m	SO_4^{2-} H 0.02909 m	SO_4^{2-} P+H 0.02906 m	HCO_3^- G 0.00238 m	HCO_3^- H 0.00186 m	HCO_3^- P+H 0.00213 m	CO_3^{2-} G 0.000269 m	CO_3^{2-} H 0.00011 m	CO_3^{2-} P+H 0.000171 m	F^- P+H 0.000080 m
Free ion	54	62.9	39.0	69	74.1	81.3	9	10.2	8.0	51.0
NaX	21	16.4	37.1	8	8.3	10.7	17	19.4	16.0	—
MgX	21.5	17.4	19.5	19	14.4	6.5	67	63.2	43.9	47.0
CaX	3	2.8	4.0	4	3.2	4.0	7	7.1	21.0	2.0
KX	1	0.5	0.4	—	0.0	0.4	—	0.0	—	—
Mg_2CO_3	—	—	—	—	—	—	—	—	7.4	—
$MgCaCO_3$	—	—	—	—	—	—	—	—	3.8	—

[a] Chlorinity 19.0‰ (G, H) or 19.375‰ (H + P). (G) R. M. Garrels and M. Thompson, *Amer. J. Sci.*, **260**, 57 (1962); (H) J. S. Hanor, *Geochim. Cosmochim. Acta*, **33**, 894 (1969); (P + H) R. M. Pytkowicz and J. E. Hawley, *Limnol. Oceanogr.*, **19**, 223 (1974).

Pressure and Temperature Dependence. In dilute electrolyte solutions, ion association is usually accompanied by a positive change in entropy. Thus, in dilute solutions, a decrease in the extent of ion association with decreased temperature is to be expected. The data given by Kester and Pytkowicz [28] for SO_4^{2-} association with Na^+, Mg^{2+}, and Ca^{2+} (Table 6.9), show that at high I or in seawater the tendency to form MSO_4 increases as the temperature decreases. These workers have calculated that a decrease in the water temperature causes a decrease in the free $[SO_4^{2-}]$, a marked increase in $[NaSO_4^-]$, a slight increase in $[MgSO_4^0]$, and no change in $[CaSO_4^0]$.

The effect of pressure on ion association can be estimated from partial molar volume data (Millero [29]). Various assumptions are involved in these calculations, and until direct measurements are made the effect of pressure on ion-pair formation will be uncertain. Kester and Pytkowicz have determined the pressure dependence of the $NaSO_4^-$ association. At 1.5°C the following dependence of $\log K_1$ ($I = 0.72$) was found:

$$\ln K_1 = 1.26 - 0.7 \times 10^{-3} p \text{ atm}$$

Hence association of $NaSO_4^-$ increases with pressure. Millero has estimated theoretically the effect of pressure on $CaSO_4^0$ and $MgSO_4^0$ association. With the exception of $NaSO_4^-$ at 1000 atm, little is known about the formation of other ion pairs in seawater at high pressures.

Association with Fluoride, Borate, Silicate, and Phosphate. F, B, and Si are often (usually in seawater) present at concentrations larger than 1 mg liter[-1] and may therefore be included among the major components. Fluorine has a high electronegativity and is the only one of the three bases that can form relatively stable complexes with Ca^{2+} and Mg^{2+}. The following distribution of total fluoride in seawater can be computed ($[F_T] = 8 \times 10^{-5}$ mol kg^{-1}): $[F^-] = 4.1 \times 10^{-5} M$ (51%), $[MgF^+] = 3.7 \times 10^{-5} M$ (47%), $[CaF^+] = 1.6 \times 10^{-6} M$ (2%), $[HF] = 3 \times 10^{-10} M$, $[HF_2^-] = 7.2 \times 10^{-14} M$.

TABLE 6.9 THE EFFECT OF TEMPERATURE AND PRESSURE ON SULFATE SPECIFICATION IN SEA WATER[a]

T (°C)	P (atm)	Free SO_4^{2-} (%)	$NaSO_4^-$ (%)	$MgSO_4^0$ (%)	$CaSO_4^0$ (%)
25	1	39.0	38.0	19.0	4.0
2	1	28.0	47.0	21.0	4.0
2	1000	39.0	32.0		
		42.0[b]	35.0[b]	19.0[b]	4.0[b]

[a] Given as percentage of the total sulfate as given species; results at 25 and 2°C (1 atm) are taken from Kester and Pytkowicz [28].
[b] Estimated theoretically by Millero [29].

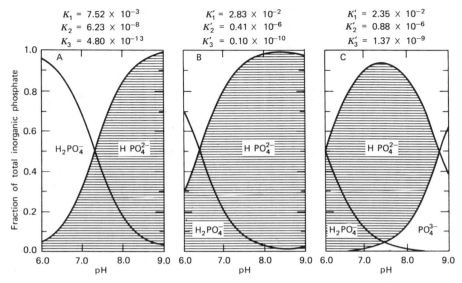

Figure 6.13   Distribution of phosphate species at 20°C. (*A*) Pure water. (*B*) 0.68 *M* NaCl. (*C*) Artificial sea water, 33‰ salinity. The shift in the distribution of phosphate species between (*A*) and (*B*) is due mainly to ionic strength effects; the shift between (*B*) and (*C*) is caused predominantly by ion association. (From Kester and Pytkowicz [30]. Reproduced with permission from *Limnology and Oceanography*.)

Borate ($B(OH)_4^-$) and silicate ($H_3SiO_4^-$) form weak complexes with major cations. Because the concentration of major cations is much larger than that of silicate, complexing of cations by silicate is negligible, but complexing of silicate by metal ions may be significant. For the complex formation $M + A \rightleftarrows MA$, the ratio of complex bound to unbound anion A is

$$\frac{[MA]}{[A]} = K_1[M] \tag{25}$$

Because [M] can be as large as 0.5 or 0.05, even a small K_1 value (e.g., $K_1 = 1$) indicates a marked degree of complexing of A.)†

Phosphate. The same considerations apply to phosphate where $[P] \ll [M]$. By comparing thermodynamic acidity constants of phosphoric acids with apparent constants determined in seawater [30], it is possible to distinguish between nonspecific interactions (effect of ionic strength) and specific interactions, that is, interactions, such as ion pair and complex formation, that depend on the specific constituents of the medium. This distinction is made very apparent by comparing the differences in distribution of phosphate species in pure water, in 0.68 *M* NaCl, and in artificial seawater (Figure 6.13). The shift of the peaks from a to b is caused mainly by ionic strength effects, whereas the shift from

---

† Similarly it may be noted that $[MgOH^+] > [OH^-]$ in seawater (cf. the appendix to this chapter.)

b to c is due to specific interactions. In seawater of pH $= 8$, 12% of the inorganic phosphate exists as $PO_{4_T}^{3-}$, 87% as $HPO_{4_T}^{2-}$, and 1% as $H_2PO_{4_T}^{-}$. Kester and Pytkowicz estimate that 99.6% of the PO_4^{3-} species and 44% of the HPO_4^{2-} species are complexed with cations other than Na^+ in seawater.

Water as a Ligand.   As has been mentioned, ions are associated with H_2O. In concentrated solutions such as seawater, the concentration of "free" water is considerably less than that of total water. Christenson and Gieskes [31] estimate that only about 2.5 mol of H_2O per kg are free. The activity of H_2O, a_{H_2O}, in seawater is given by the ratio of the vapor pressure (fugacity) of seawater, p, to that of pure water, p^0, at the same temperature:

$$a_{H_2O} = \frac{p}{p^0} \tag{26}$$

In seawater of $19.4\%_o$ chlorinity the activity of H_2O at $25°C$ is 0.981.

Aquo complexes of A-type metal cations are more stable than would be expected on the basis of their electrostatic ion–dipole interaction. This enhanced stability has been accounted for by assuming that interconnecting hydrogen bridges reinforce the first hydration sheath. F^- and O donors are able to react with these aquo A-type metal cations, but halogen ions, CN^-, and S^{2-} cannot compete with and successfully displace the H_2O dipoles. For example, $Al(H_2O)_5F^{2+}$ has a formation constant of $\sim 10^5$, but even in concentrated HCl no corresponding chloro complex can be formed; apparently, F^- is able to replace an H_2O of the coordinated water sheath; Cl^- cannot do this, partly for structural reasons, but also because it has little tendency to form hydrogen bridges [32]. The role of water structure in coordination chemistry has been reviewed by Krindel and Eliezen [33].

Measurements of partial molar volumes of dissolved electrolytes give information on structural interactions between ions and water and on ion–ion interactions. A dissolved salt appears to cause an effect on the water similar to that produced by a high external pressure [34,35]. Duedall and Weyl [36] have measured the partial molar volumes $\overline{V}^0$ (at infinite dilution in the ionic medium) for a number of salts in synthetic seawater. Millero (1969) has separated the $\overline{V}^0$ values in seawater and 0.725 M NaCl into their ionic components and analyzed these ionic $\overline{V}^0$ components using a simple model for ion–water interactions.

Coordination with Sulfide.   The tendency to form thio complexes is particularly strong with B-type metals. Traces of sulfides readily displace OH and Cl groups from the coordination sheath of Ag(I) and Hg(II) (Figure 6.14). Figure 6.7 illustrates that the tendency to coordinate with S^{2-} increases from Mn(II) to Co(II). As we have seen (Section 6.5), metal ions that form insoluble sulfides have a tendency to form thio complexes. Comparison of the sulfide solubility with the tendency to coordinate with O groups illustrates why, in nature, Mn

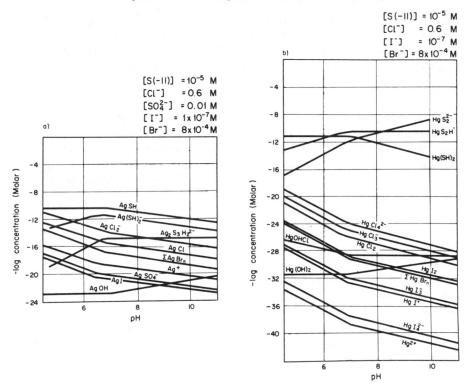

Figure 6.14 Solubility of Ag_2S and HgS in seawater (25°C, 1 atm pressure). (*a*) Species in equilibrium with solid HgS. (*b*) Species in equilibrium with solid HgS.

and Fe occur primarily as oxygen-bearing ores, whereas Co, Ni, Cu, Zn, are found as sulfides. Conceivably, the geochemical deposition of the sulfide ores of Co, Ni, Cu, and Zn occurred from solutions containing excess hydrogen sulfide, the metals being in the form of soluble $M(SH)_n^{(2-n)}$ complexes [32].

Mixed Complexes

There is the possibility that mixed (or ternary) complexes may form, for example, $HgOHCl$, $Fe(OH)_2SO_4^-$, $MgCaCO_3^{2+}$. So far few stability constants for mixed complex formation have been experimentally determined. Often it is possible to estimate the stability constant on the basis of statistical arguments [37–39].

The stability constant of a species $MA_m B_n$ can (in simple cases) be calculated from the stability of MA_{m+n} and MB_{m+n} by the equation

$$\log \beta_{MA_m B_n} = \frac{m}{m+n} \log \beta_{MA_{m+n}} + \frac{n}{m+n} \log \beta_{MB_{m+n}} + \log S \qquad (27)$$

where $\beta_{MA_m B_n} = \{MA_m B_n\}/\{M\}\{A\}^m\{B\}^n$ and S is a statistical factor that expresses the number of ways in which these species can be formed. For example, the statistical probability (in the presence of equal concentrations of ligands A and B) of forming MAB is $S = 2$ relative to the probability of forming the simple complex ML_2; similarly $S = 3$ for the probability of forming $MA_2 B$ relative to ML_3. In simple cases, S can be calculated by

$$S = \frac{(m + n)!}{m!\, n!} \tag{28}$$

For example, the stability constant of HgOHCl, β_{HgOHCl}, is obtained from equations 27 and 28 by

$$\beta_{HgOHCl} = 2(\beta_{Hg(OH)_2}\beta_{HgCl_2})^{1/2} \tag{29}$$

Figure 6.15a gives a predominance diagram for $HgOH_m Cl_n$ species as a function of pH and pCl. In most fresh waters the Hg species $Hg(OH)_2$, HgOHCl, and $HgCl_2$ are prevalent, while the predominant species in seawater is $HgCl_4$ (Figure 6.15b). As Figure 6.15a illustrates, mixed complexes predominate only under restricted conditions.

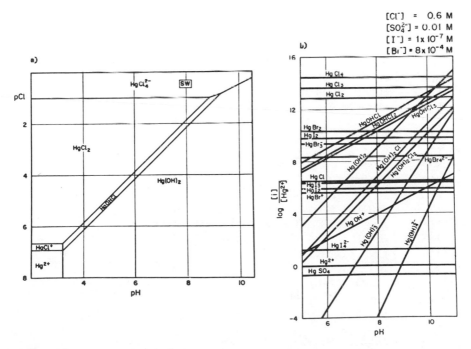

Figure 6.15 Hg(II) species. (a) Predominance of Hg species as a function of pCl and pH. In seawater (SW) $HgCl_4^-$ predominates. (b) Concentration ratio diagram for Hg(II) in seawater as a function of pH. This diagram gives the equilibrium concentrations of Hg(II) complexes relative to the concentration of Hg^{2+}.

Inorganic Species of Trace Metals and Trace Elements

Despite the biochemical importance of many of the trace elements, the chemical forms of these components are not always known. At concentrations of less than $10^{-6}\,M$, analytical techniques usually cannot provide information concerning the speciation. Table 6.10 gives a list of probable dissolved inorganic species. Complexation by organic material needs to be considered in addition to inorganic complex formation.

TABLE 6.10   PROBABLE   MAIN   DISSOLVED   INORGANIC   SPECIES   IN NATURAL WATERS (AEROBIC CONDITIONS)[a]

Element	Probable Main Species	Element	Probable Main Species
H	H_2O	Rb	Rb^+
Li	Li^+	Sr	Sr^{2+}
Be	$BeOH^+$	Y	$Y(OH)_3$
B	H_3BO_3, $B(OH)_4^-$	Zr	$Zr(OH)_n^{4-n}$
C	HCO_3^-	Mo	MoO_4^{2-}
N	N_2, NO_3^-	Ag	$AgCl_2^-$ (S), Ag^+ (F)
O	H_2O	Cd	$CdCl_2^0$ (S), Cd^{2+} (F), $CdOH^+$ (F)
F	F^-, MgF^+ (S)	Sn	$SnO(OH)_3^-$
Na	Na^+	Sb	$Sb(OH)_6^-$
Mg	Mg^{2+}	Te	$HTeO_3^-$
Al	$Al(OH)_4^-$	I	IO_3^-, I^-
Si	$Si(OH)_4$, MgH_3SiO_4 (S)	Cs	Cs^+
P	HPO_4^{2-}, $MgPO_4^-$ (S)	Ba	Ba^{2+}
S	SO_4^{2-}, $NaSO_4^+$ (S)	La	La^{3+}, $LaOH^{2+}$
Cl	Cl^-	Ce	Ce^{3+}, $CeOH^{2+}$
K	K^+	Pr	Pr^{3+}, $PrOH^{2+}$
Ca	Ca^{2+}	Nd	Nd^{3+}, $NdOH^{2+}$
Sc	$Sc(OH)_3^0$	Other	
Ti	$Ti(OH)_4^0$	rare	Me^{3+}, $MeOH^{2+}$
V	$H_2VO_4^-$, HVO_4^{2-}	earths	
Cr	$Cr(OH)_3^0$, CrO_4^{2-}	Lu	$LuOH^{2+}$
Mn	Mn^{2+}, $MnCl^+$ (S)	W	WO_4^{2-}
Fe	$Fe(OH)_2^+$	Re	ReO_4^-
Co	Co^{2+}, $CoCO_3^0$ (?)	Au	$AuCl_2^-$ (S), $Au(OH)_3^0$ (F)
Ni	Ni^{2+}, $NiCO_3^0$ (?)	Hg	$HgCl_4^{2-}$ (S), $Hg(OH)_2^0$ (F)
Cu	$CuCO_3^0$, $CuOH^+$		$HgOHCl$ (F)
Zn	$ZnOH^+$, Zn^{2+}, $ZnCO_3^0$	Tl	$TlCl^0$ (S), Tl^+ (F)
Ga	$Ga(OH)_4^-$	Pb	$PbCO_3^0$, $Pb(CO_3)_2^{2-}$
Ge	$Ge(OH)_4^0$	Bi	BiO^+, $Bi(OH)_2^+$
As	$HAsO_4^{2-}$, $H_2AsO_4^-$	Ra	Ra^{2+}
Se	SeO_3^{2-}	Th	$Th(OH)_n^{4-n}$, $Th(CO_3)_n^{4-2n}$ (?)
Br	Br^-	U	$UO_2(CO_3)_3^{4-}$

[a] S, species prevalent in seawater only; F, Species prevalent in fresh water.

One may approach the thermodynamics of the problem by extending the chemical model for the equilibrium distribution of the major species—such as the Garrels–Thompson model—to minor components [40,41]. The relative predominance of various possible species can also be elucidated from activity or concentration ratio diagrams from which relative proportions of the various species can be derived. The concentrations of the free ligands, that is, residual concentrations remaining after complexation with major cations (Table 6.8), enter into computation of the complexation equilibria of the trace metals. For inorganic ligands $[L_{free}] \gg [M_{trace}]$ and there is negligible competition in the complexing between individual trace metals. Furthermore, the quantitative degree of complex formation, ΔpM, is independent of the total metal-ion concentration, hence inorganic complex formation systems of the trace metals may be considered individually. The concentration ratio diagrams for Zn(II) and Cd(II) valid for seawater (Figure 6.16) may be compared with that for Hg(II) (Figure 6.15). These computations indicate that chloro complexes, $HgCl_4^{2-}$ and $CdCl_2^0$, are the most stable species in the Hg(II) and Cd(II) systems, whereas $ZnOH^+$ and $Zn(OH)_2^0$ predominate in the Zn(II) system. Determination of the

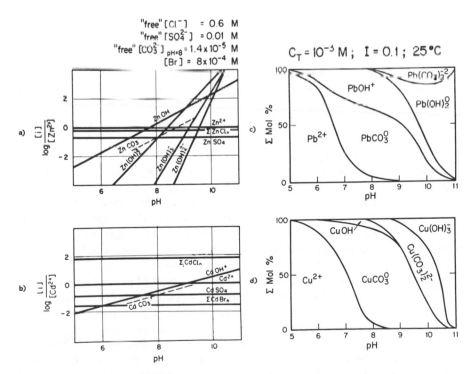

Figure 6.16 Relative concentrations of Zn, Cd, Pb, and Cu species (25°C, 1 atm). (*a* and *b*) Seawater conditions (activity ratio diagram); (*c* and *d*) freshwater conditions.

complexing of Cd(II) and Zn(II) is difficult, as there is some uncertainty concerning the stability of carbonato complexes. Using similar types of computations, one can establish that $CuCO_3$ and $PbCO_3$ should predominate in fresh and seawater. From a thermodynamic point of view, chloro complexes should predominate in seawater not only for Hg(II) and Cd(II) but also for Ag ($AgCl_2^-$) and Au ($AuCl_2^-$). Soluble hydroxo species are prevalent species in natural water systems: Al [$Al(OH)_4^-$]; Sc(III) [$Sc(OH)_3^0$], and some rare earth ions [$Ln(OH)_3^0$]; Fe(III) [$Fe(OH)_2^+$, $Fe(OH)_3^0$]; Zn ($ZnOH^+$); Th(IV). A few trace elements occur as free aquo ions: Rb^+, Cs^+, Ba^{2+}, Ra^{2+}, some rare earth ions, and probably Mn^{2+}, Ni^{2+}, and Co^{2+} (Table 6.10).

Clearly, the conclusions reached above depend on the values of the equilibrium constants used in the computations; for some of the species there are large variations among values reported by various authors, and for some of the species no thermodynamic data exist. Although calculated equilibrium species distributions remain tentative and are subject to corrections when better data become available, these calculations constitute a useful general framework that permits a better understanding of the observed behavior and functions of trace metals and of how their chemical forms are influenced by solution variables.

6.8   ORGANIC COMPLEXES IN NATURAL WATERS; PROBLEMS OF SPECIFICITY

Natural waters contain only small amounts of organic compounds (Chapter 8), but these often exert a pronounced influence on the inorganic constituents. Metal ions in waters often appear not to behave as predicted from the known chemistry of the metals in question. Frequently seawater shows an apparent supersaturation with respect to many inorganic compounds. A substantial proportion of the metal species present are not dialyzable or extractable with reagents known to be suitable for simpler aqueous systems. The addition of organic substances to water or their exudation by algae may increase or decrease biological activity, presumably because of increased or decreased rates of metal-ion uptake by organisms. Conversely, the reactions of the organic substances present in seawater are influenced by the presence of metal ions. Complex formation frequently renders the organic molecule more reactive toward nucleophilic reagents. Obviously, interactions between inorganic and organic substances must account for at least some of the observed phenomena enumerated above. In this section the evidence for organic complex formation or chelation in waters is considered, and an examination is made of whether knowledge of the degree of combination of organic compounds with metal ions can resolve some of the discrepancies between observed and predicted behavior.

Despite these arguments in favor of organic complexation, very little direct proof for the existence of soluble chelates in natural waters is available. It is very difficult to detect soluble chelates in natural waters by any direct method or to isolate them from natural waters, especially with the very small amounts of metal ions that are usually present. In this section we will investigate interactions between metal ions and organic constituents and inquire as to the type of metal organic associations found in natural waters.

The extent of complex formation of metal ions with organic constituents will be investigated in three speciation models representing the properties of an aerobic fresh water, an aerobic seawater, and an anaerobic sewage.

The Complex-Forming Properties of Organic Matter. Concentrations of dissolved organic matter in natural waters range from 0.1 to 10 mg liter^{-1}. The upper concentration is reached in lakes, streams, and estuaries. The lower concentration is more typically encountered in unpolluted and nonproductive fresh water and in seawater (0.1 to 0.5 ng C liter^{-1}). Sediments are the principal depositories of posthumous organic debris and in interstitial waters; in these, dissolved organic matter may reach concentrations of up to 100 mg C liter^{-1} and in a few instances of up to a few hundred mg C liter^{-1} [42]. As shown in Chapter 8, only a small fraction of the total dissolved organic carbon in surface and subsurface waters has been identified. The organic matter found includes compounds such as amino acids, polysaccharides, carbohydrates, amino sugars, fatty acids, organic phosphorus compounds, aliphatic and aromatic compounds containing $-OH$ and $-COOH$ functional groups, and porphyrins which contain donor atoms suitable for complex formation. The end product of the biological degradation of the organic matter appears to be humic material containing a variety of functional groups in addition to carboxylic groups (see Section 8.3). Natural waters contain primarily metals of Group I and Group II of the periodic table, but they also contain trace quantities of all naturally occurring metals and this includes most transition elements.

Although some exudation products of biota may have special steric arrangements of donor atoms that make them relatively selective toward individual trace metals most organic matter may not be present in the form of selective complex formers; the organic functional groups compete for association with inorganic cations and protons; cations and protons can satisfy their coordinative requirements with inorganic anions including OH^-, as well as with organic ligands. Most organic ligands in seawater appear to have, at best, complex-forming tendencies similar to those of acetate, citrate, amino acids, phthalate, salicylate, carbohydrates, and quinoline-carboxylate. An inspection of "Stability Constants of Metal-Ion Complexes" shows that these classes of compounds are able to form moderately stable complexes with most multivalent cations. Ethylenediamine tetraacetate and nitrilotri-acetate, both of which are able to form particularly stable 1 : 1 complexes, are probably not representative models for the complex-forming tendency of the organic matter normally encountered in seawater.

Competition by Ca^{2+} and Mg^{2+} Decreases the Tendency to Form Soluble Complexes

The ratio of inorganic to organic constituents is very high. How can trace concentrations of organic matter specifically interact with individual metal ions? Does not the presence of Ca^{2+} and Mg^{2+} at concentrations many orders of magnitude greater than that of the potential organic complex-forming species blur any complex-forming tendency of the organic functional groups? These questions are well justified, because in natural water systems the stability of a complex is influenced not only by H^+ and OH^- but also by other metal ions, especially Ca^{2+} and Mg^{2+}.

In order to illustrate the problem we select a system that appears to be especially favorable to complex formation with Fe(III). As we have seen in Figure 6.12, EDTA forms rather strong complexes with Fe(III). To such a system we now add Ca^{2+}. Simple complex formation of Fe(III) by EDTA

$$Fe^{3+} + Y^{4-} = FeY^- \qquad \log K_{FeY} = 25.1 \qquad (30)$$

is now replaced by

$$CaY^{2-} + Fe^{3+} = FeY^- + Ca^{2+} \qquad \log \frac{K_{FeY}}{K_{CaY}} = 14.4 \qquad (31)$$

where $\log K_{CaY} = 10.7$.

As equation 31 illustrates, FeY^- cannot be formed to any significant extent if the ratio $[Ca^{2+}]$ to $[Fe^{3+}]$ is sufficiently large. For an exact calculation the hydroxo complexes FeYOH and $FeY(OH)_2$ have to be considered in addition to CaY^{2-} and FeY^-. (Above pH 4 the complexes $CaHY^-$ and FeHY can be neglected.) For a given set of conditions in the system

$$M_1 = Ca^{2+} \quad M_2 = Fe^{3+} \quad M_3 = H^+ \quad L_1 = EDTA\ (Y^{4-}) \quad L_2 = OH^-$$

the calculated equilibrium concentrations of the more important species are plotted in Figure 6.17a. Figure 6.17b illustrates more explicitly how Ca^{2+} depresses the extent of Fe(III)–EDTA complexation. Such a strong complex former, even if available in excess of Fe(III), cannot in the presence of Ca^{2+} prevent Fe(III) from precipitating as $Fe(OH)_3(s)$ in alkaline solutions. The competing effect of other cations depends on the relative stability of the complexes that are formed. For example, in the case of L = citrate^{4-}, the equilibrium, $CaHL^- + Fe^{3+} = FeHL + Ca^{2+}$, is characterized by an equilibrium constant, $K = K_{FeHL}/K_{CaHL}$, of $10^{7.4}$. Correspondingly, the relative effect of Ca^{2+} upon Fe(III) complexing by citrate is much smaller than that encountered with EDTA (Figure 6.17c).

The examples given are not yet very realistic. In nature we encounter a multitude of complex-forming bases, each one most likely present at concentrations lower than $10^{-6} M$ (~ 0.1 mg liter^{-1}) organic matter. Furthermore,

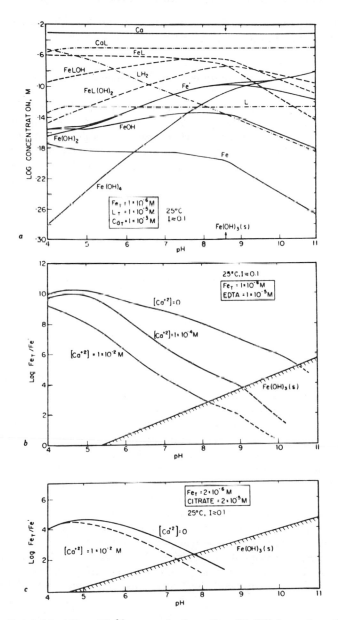

Figure 6.17 Competitive effect of Ca^{2+} on complex formation of Fe(III). In most practical systems, ions other than H^+ or OH^- compete as coordination partners, as illustrated here in a "simple" two-metal, two-ligand system: $M_1 = Fe^{3+}$, $M_2 = Ca^{2+}$, $(M_3 = H^+)$; $L_1 = $ EDTA or citrate, $L_2 = OH^-$. (*a*) Ca^{2+}, Fe(III)–EDTA equilibria and the relationship to Fe(OH)$_3$(s) $(K_{s0} = 10^{-38})$ formation. Fe(OH)$_3$(s) precipitates at pH > 8.6 (*b*) Effects of total Ca concentration upon Fe(III)–EDTA equilibrium. $Me' = \sum_0^k Me(OH)_n$; $Me_T = Me' + \sum MeHL + \sum MeOHL$. (*c*) Effect of Ca^{2+} on Fe(III)–citrate equilibrium.

other competing ions (Mg^{2+}, Al^{3+}, etc.) need to be considered. In Section 6.9, models of multimetal, multiligand systems will be analyzed.

Selectivity in Complex Formation

Compounds containing F and O donor atoms are rather general complex formers. On the other hand, the soft bases, for example, ligands with N, Cl, Br, I, and S are more selective; the selectivity increases with decreasing electronegativity of the element. Polyamines and compounds with sulfur are much more selective for B-type cations and transition metals than polycarboxylic acid. Although different ions may be held with a varying degree of strength, inspection of *Stability Constants of Metal-Ion Complexes* shows that no complete specificity for a single metal ion has been achieved with synthetic chelating agents. Thus it is not surprising that of the many valuable metals (e.g., 7×10^7 tons of gold and 7×10^9 tons of copper) known to be dissolved in seawater, none has yet been commercially extracted with the help of a chelating agent [43].

Concentration of Metal Ions by Organisms.   Although simple chelating agents cannot concentrate a single metal exclusively, living organisms are almost capable of achieving this result. Some extreme examples are cited by Bayer: Vanadium is a million times more concentrated in the blood cells of the tunicate *Phallusia mamillata* than in seawater. Similarly copper is concentrated by a factor of 10^5 in the blood of *Octopus vulgaris*. It appears that in the organism organic complexing agents are highly selective for one ion. What molecules have such a near-absolute specificity? Because functional groups containing donor atoms are not sufficiently selective, it is necessary to provide molecules with special geometric arrangements. A simple example is provided by 1,10-phenanthroline, which gives color reactions with Fe(II) and Cu(I). The reaction with Fe(II) can be suppressed by introducing alkyl groups into the 2- and 9-positions, giving a reagent that still reacts with Cu(I). Introducing bulky constituents into a chelate ligand usually does not aid complex formation but only hinders it from a kinetic point of view; but specificity can be improved by this method. It is conceivable that protein structures contain enclaves into which only special metal ions can fit, and that specificity in enzyme catalysis by metals arises in this way [44]. In line with this argument are the observations made by Bayer and co-workers on the hemocyanin derived from *Octopus vulgaris*. This hemocyanin has a molecular weight of 2.7×10^6. Its composition is that of a protein containing no prosthetic groups, that is, the complex-forming function is part of the molecular protein. Apparently, in addition to the donor atoms (e.g., mercapto groups of the cysteine linked by peptide bonds), the sequence of the links and the steric structure must be responsible for the specificity. Bayer [43] has been able to synthesize macromolecular chelating agents possessing high selectivity for rare-metal ions.

Accumulation of Trace Elements in Organisms.   The general problem of the enrichment of metals and trace elements in the marine biosphere has been treated by Lowman et al. [45] and reviewed by Brewer [46] (see Table 6.11). Martin and Knauer [47] have analyzed the elemental composition of phytoplankton and of mixed zooplankton populations. Very useful information about the distribution of chemical elements in the biosphere can be found in Bowen's book [48]. Goldberg [1] suggested that accumulation processes may be initiated by coordinative reactions at the external surface of the organism.

**TABLE 6.11   ACCUMULATION OF ELEMENTS BY PHY-
TOPLANKTON**

Concentration Factor[a]	Elemental Composition of Phytoplankton Median Values ($\mu g\ g^{-1}$ dry wt)[b]		
	I	II	III
Na	—	138×10^3	106×10^3	89×10^3
K	—	13×10^3	13×10^3	11×10^3
Mg	—	16×10^3	15×10^3	11×10^3
Ca	—	6.5×10^3	5.3×10^3	6.5×10^3
Si		147	119	697
Ba	—	33	19	287
Al	10^5	110	444	38
Fe	4×10^4	224	1510	231
Mn	4×10^3	6.1	13.3	7.7
Ti	—	N.D.	27	N.D.
Cr	2×10^3	N.D.	3.9	—
Cu	3×10^4	3.2	7.4	11
Ni	5×10^3	1.9	7.8	2.3
Zn	2×10^4	19	122	24
Ag	2×10^4	0.2	0.6	N.D.
Cd	—	1.5	1.6	3.9
Pb	4×10^4	N.D.	7.2	9.2
Hg	—	0.19	0.16	0.16
C	4×10^3	—	—	—
P	3×10^4	—	—	—

[a] Concentration factor is the ratio of the concentration of an element in an organism or its tissue to the concentration directly available to the organism from the organism's environment under equilibrium or steady state conditions. Lowman, et al. [45].

[b] Determined from samples collected with a 76-μm-mesh net (i.e., microplanktonic organisms were excluded. The data are classified on a chemical basis into samples containing no significant Ti I, samples containing Ti, II, and samples containing significant concentrations of Sr, III.

Polydentate and Macrocyclic Ligands. These ligands frequently occur in nature, where they are important in enzymatic and biological processes. An interesting model of a complexing system of iron(III) in nature is provided by *ferrichromes*. Ferrichromes represent a class of naturally occurring heteromeric peptides containing a trihydroxamate as an iron(III)-binding center. Figure 6.18a shows the structure of a ferrichrome (desferri-ferrichrome). Polyhydroxamic acid rather specifically forms strong soluble complexes with Fe(III) and Mn(III). Its structure is related to that of formaldoxime, which can be used analytically for the determination of iron and manganese. Ferrichromes appear to be the

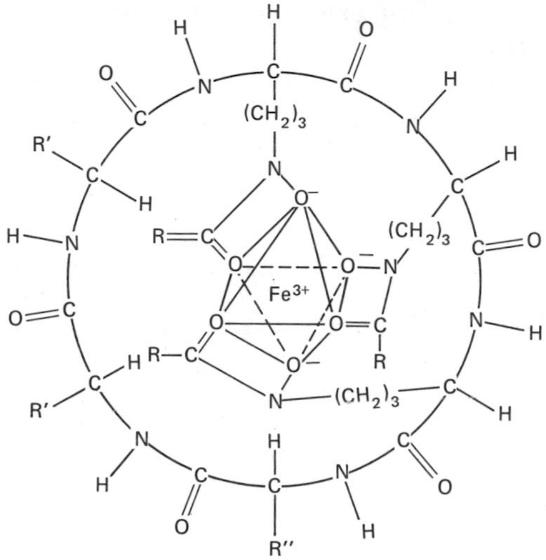

Figure 6.18 Macrocyclic complex formers. (*a*) Structure of a ferrichrome (desferriferrichrome). One of the strongest complex formers presently known for Fe(III). The iron-binding center is an octahedral arrangement of six oxygen donor atoms of trihydroxamate. It has been suggested that such naturally occurring ferrichromes play an important role in the biosynthetic pathways involving iron. (*b*) The vitamin B_{12} (or cyanocobalamin) as a macrocyclic multidentate complex of cobalt in the porphyrin-resembling part. The cyanide group CN^- can be exchanged for Cl^- or OH^-. Vitamin B_{12} is an essential growth factor for several bacteria and auxotrophic phytoplankton.

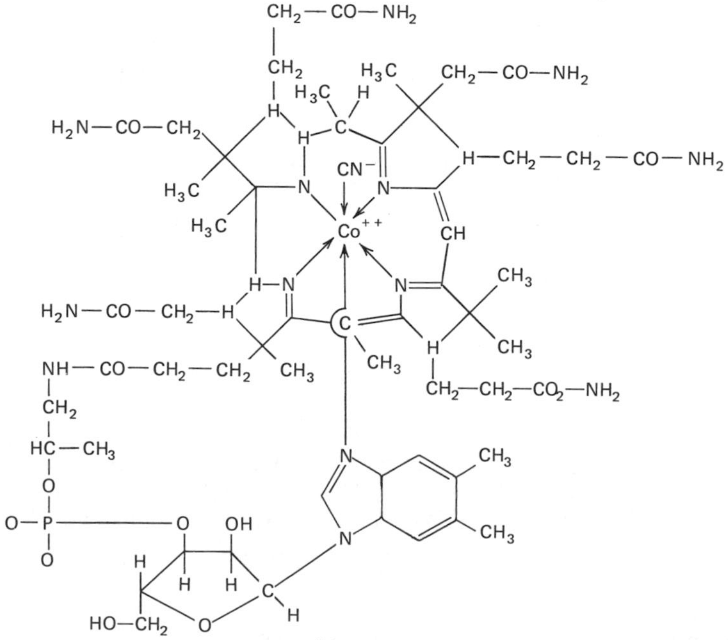

strongest iron(III) complex formers presently known. The strong affinity of Fe^{3+} for the trihydroxamate is plausible in view of the tendency of Fe^{3+} to coordinate preferentially with basic oxygen donor atoms. The binding center (Figure 6.18) consists of an octahedral arrangement of oxygen ligand atoms, thus satisfying all the coordinative requirements of Fe^{3+}. These natural trihydroxamates bind iron so highly that iron remains complex-bound in solution even at great dilutions and high pH. Furthermore, competition of earth alkali ions with iron(III) appears to be rather weak, thus making ferrichromes relatively specific for ferric iron.

Ferrichromes appear to be widely distributed in microorganisms. Fungi have been used for the routine preparation of ferrichromes in the laboratory [49]. Ferrichrome has been suggested as a cofactor in microbial iron metabolism. According to Neilands [49], the ferric ion, once coordinated as soluble trihydroxamate, can be transported to or into the cell and donated to the iron enzymes. Neilands has also shown that many microorganisms are capable of augmenting the biosynthesis of metal-free trihydroxamic acids during iron deficiency, and he infers that this may represent an evolutionary invention whereby the organism protects itself against iron deprivation.

There are many natural macrocyclic compounds such as porphyrins [e.g., vitamin B_{12}, a cobalt complex (Figure 6.18b)]; heme, an iron porphyrin molecule; and chlorophyll, a magnesium complex) and certain antibiotics. Naturally occurring antibiotics such as nonactin, monactin (cyclic polyethers containing 8 oxygens in a ring of 32 atoms), and valiomycin (a 36-membered ring that includes six ether oxygens and six peptide nitrogens) show highly selective action in complexing metal ions.† Recently, synthetic macrocyclic compounds have been prepared [50]. These macrocyclic molecules contain central hydrophilic cavities ringed with either electronegative or electropositive ligand atoms, and the external framework exhibits hydrophobic behavior. These substances function as multidentate complexing agents in living matter or at the interface between living matter and water.

Ternary Complexes. Since many different ligands are likely to compete for metal ions, mixed organic complexes may occur in natural waters. Ternary coordination complexes play an important role in biological processes as exemplified by many instances in which enzymes form enzyme–metal–substrate complexes. Mixed complexes may provide very specific structures of high selectivity [51,52]. Some general principles involved in assessment of the stability of mixed complexes have been discussed in Section 6.7 (cf. equations 27 and 28).

6.9 METAL BINDING BY POLYMERS AND AT THE SOLID–SOLUTION INTERFACE

In addition to simple solutes, the following categories of macromolecular, colloidal, or particulate matter may interact chemically with and play important roles in the transformation of metal ions:

1   Most polymeric organic substances contain a sufficient number of hydrophilic functional groups ($-COO^-$, $-NH_2$, $-R_2NH$, $-RS^-$, $-RO^-$) to allow them to remain

† Valiomycin, dissolved in *n*-hexane or octanol, can be placed in a liquid membrane electrode assembly (which is comparable in its selectivity to biological membranes of living cells). Because the membrane's permeability to K^+ is much higher than to Na^+, the electrode can be used as a K^+-selective electrode.

in solution despite their molecular size. These functional groups contain O, N, and S donor groups capable of complex formation. Polypeptides, certain lipids, polysaccharides, and many substances classified as humic acids, fulvic acids, and *Gelbstoffe* (see Section 8.3) belong to this category.

2   Colloidal or particulate organic material, either high-molecular-weight compounds and organic substances sorbed or chemically bound to inorganic colloids, or surfaces of biota and of biological debris.

3   Inorganic solids, especially hydrous oxides of Fe(III), Al(III), Si(IV), and Mn(IV).

The surfaces of these oxides are covered with OH groups that participate in H^+ transfer or metal-ion binding (see Figure 10.6).

The complex formation reactions can be generalized by

$$\{GH\} + Me^{z+} \rightleftharpoons \{GMe^{z-1}\} + H^+ \tag{32}$$

$$2\{GH\} + Me^{z+} \rightleftharpoons \{G_2Me^{z-2}\} + 2H^+ \tag{33}$$

where GH can be any polymeric or "crosslinked" organic or inorganic group such as $(R{-}COOH)_n$, $(ROH)_n$, $(ROHCOOH)_n$, $(RNH_2COOH)_n$, $(\equiv AlOH)_n$ $(\equiv FeOH)_n$, or $(\equiv SiOH)_n$.

Polymeric Complex Formers (e.g., Humic Substances)

As with monomeric complex formers the extent of complex formation depends on the tendency of the ligand groups to replace H^+ with Me^{2+}. Hence we must know both the acid–base behavior and the metal-ion binding tendency of the polymeric base (Figure 6.19a). Few metal complex equilibria involving macromolecules have been studied. The procedures vary and differ from those used for small organic ligands. The tendency of a given metal ion to interact with a particular ligand group of the macromolecule varies with the electrostatic free energy needed to change the electric charge of the macromolecule; this free energy is itself a function of all the ionic equilibria in which the various types of ligand groups in the macromolecule may take part [53].

Acid–Base Behavior.   Formally, the acidity constant K' of a polynuclear acid with identical acid groups can be derived [54,55] as

$$K' = [H^+]\frac{\bar{f}}{1 - \bar{f}} = [H^+]\frac{[\{G^-\}]}{[\{GH\}]} \tag{34}$$

or, in logarithmic form,

$$pK' = pH - \log\frac{\bar{f}}{1 - \bar{f}} \tag{35}$$

where $\bar{f}$ is the fraction of acid groups that has lost protons. In Figure 6.19, alkalimetric titration curves for a polyacrylic acid in the absence and presence of Cu^{2+} are given. Obviously, because of the electrostatic interaction, pK', the microscopic acidity constant (cf. Section 3.3), decreases with increasing $\bar{f}$. With every proton lost the charge of the polyelectrolyte becomes more negative; thus it becomes more difficult to dissociate H^+ from

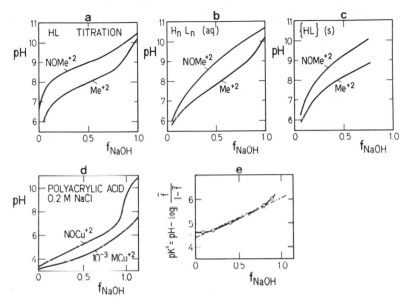

Figure 6.19 Acid–base properties of polyelectrolytes and hydrous oxides. (a–c) Schematic representation of the acid–base and complex formation behavior of a monoprotic, polyelectrolytic acid and a cross-linked (solid) polyacid (e.g., amorphous SiO_2). The shift in the titration curve is a measure of the extent of complex formation. [From W. Stumm, C. P. Huang, and R. Jenkins, *Croat. Chem. Acta,* **42**, 223 (1970).] (d) Alkalimetric titration of polyacrylic acid in the absence and presence of Cu^{2+}. (e) The microscopic acidity constant K' calculated from the titration curve is plotted as a function of $\bar{f}$. The value of K' decreases with increasing $\bar{f}$, because with increasing negative surface charge it becomes more difficult for the surface OH group to dissociate a H^+; an intrinsic acidity constant K'_{int}, for a hypothetically uncharged polyacid ($\bar{f} = 0$), can be calculated.

the acid groups. Conveniently the acidity constant can be interpreted to be composed of two factors [54,55], namely, an intrinsic acidity constant K'_{int}, which is independent of the degree of neutralization (for a neutral polynuclear acid, $K' = K'_{int}$ if $\bar{f} = 0$), and a factor that represents the electrostatic interaction, $e^{F\Psi/RT}$. Hence

$$pK' = pH - \log \frac{\bar{f}}{1 - \bar{f}} = pK'_{int} - \frac{F\Psi}{2.3RT} \qquad (36)$$

With increasing $\bar{f}$, Ψ becomes more negative and pK' increases, where Ψ is the potential difference (volts) between the surface of the polymer and the bulk of the solution, F is the Faraday (96,500 C eq^{-1}), and R is the gas constant. For a given $\bar{f}$ the charge density is fixed and the absolute value of Ψ decreases with increasing ionic strength. Accordingly, $pK'_{\bar{f}=0.5}$ (i.e., the average pK' value) decreases with increasing ionic strength.†

† An average equilibrium constant K_{av} quite representative of the free energy of the protolysis reaction per mole is obtained from

$$\ln K_{av} = \int_0^1 \ln K'd\bar{f}$$

Equation 36 can also be written as

$$\text{pH} - \log \frac{\bar{f}}{1 - \bar{f}} = pK'_{\text{int}} - 2\frac{1}{2.3}w\bar{Z} \tag{37}$$

where $\bar{Z}$ is the average molecular charge of the macromolecule and w is an electrostatic proportionality factor. The left-hand side of equation 37 is often found empirically to be a linear function of $\bar{f}$ (Figure 6.19e) or of $\bar{Z}$. The proportionality factor w can then be obtained from the slope of the line; extrapolation to $\bar{f} = 0$ gives the intrinsic constant. For various crude models of polyions theoretical expressions of w have been derived [54,55].

The extent of complex formation is given by the stability constant

$$K_{\text{M}} = \frac{[\{GMe^{z-1}\}]}{[\{G^-\}][Me^{z+}]} \tag{38}$$

The bonding of metal ions, similar to that of H^+, depends on the polymeric charge density, and thus experimentally determined (microscopic) K_{M} values are a function of $\bar{f}$ and pH. Intrinsic stability constants can be defined for a hypothetically uncharged polymer:

$$K_{\text{M}} = K_{\text{M(int)}} \exp(-2wz\bar{Z}) \tag{39}$$

Equation 38 can now be rewritten by considering equations 34, 37, and 38:

$$\frac{\theta_{\text{M}}}{1 - \theta_{\text{M}}} = \frac{K_{\text{M(int)}}[Me^{z+}] \exp(-2wz\bar{Z})}{1 + ([H^+]/K_{\text{int}})\exp(-2w\bar{Z})} \tag{40}$$

where θ_{M} is the fraction of bonding sites occupied by metal ions:

$$\theta_{\text{M}} = \frac{[\{GMe\}]}{[\{G^-\}] + [\{GH\}] + [\{GMe\}]}$$

Fulvic Acid. Characterization of humic substances and of fulvic acid is given in Section 8.3. The literature abounds with empirical evaluations of the complex-forming tendency of fulvic or humic acids. Much of this information cannot be generalized because the acid–base characteristics of the ligand were not described; often metal-ion hydrolysis [e.g., in case of Fe(III)] was not considered. Often no account was taken of whether the system was in true solution or in a colloidal state.†

Recently considerable progress has been made in attempting to characterize quantitatively the acid–base and metal-ion binding properties of fulvic acid [56–60].

These investigations show that the stability of fulvic acid–metal complexes is similar in magnitude to that of the corresponding complexes of monomeric units (e.g., carboxylic

† It is difficult to distinguish operationally between dissolved and colloidally dispersed substances. Statements frequently found in the literature that claim, for example, that decaying vegetable matter such as humic acids which carry carboxyl and hydroxyl functional groups form soluble chelates in natural waters must be taken with reservations. These substances can coordinate and form chelates with iron(III). Such substances undoubtedly can bind ferric iron and at high concentrations are capable of preventing the precipitation of ferric iron. Colored water usually contains higher iron(III) concentrations. It is not yet certain whether such substances can really keep ferric iron in *solution* at the pH values of interest. It might be more probable that the coordinative products formed between the color bases, OH^-, and Fe(III) are insoluble and are present as highly dispersed colloids. The diameter of $Fe(OH)_3$ can be smaller than 100 Å.

acids, salicylic acids, dihydroxybenzoic acid, phthalic acid) believed to contain functional groups similar to those in fulvic acid.

6.10   THE NEED FOR CHEMICAL SPECIATION

The chemical behavior of any element in the environment depends on the nature of its components. Historically, however, limnologists' and oceanographers' interests lay primarily in determining collective parameters and the elemental composition. This information alone is often inadequate for identifying the mechanisms that control the composition of natural waters and for understanding their perturbation. We need to know the form of the species in which the element is present in order to gain insight into the physical chemistry of natural waters and into the nature of pollutant interaction, as well as into the complexities of the biochemical cycle.

Dependence of Biological Availability and Toxicity of an Element on Speciation

It is especially important to recall that biological availability of metals and their physiological and toxicological effects depend on the species of the individual metal present; for example, $CuCO_3^0(aq)$, $Cu(en)_2$, and so on, affect the growth of algae in a different way than $Cu \cdot aq^{2+}$. This has been exemplified in some recent studies on the relationships between Cu^{2+} ion activity and the toxicity of copper to phytoplankton [61–64] (see Figure 6.20). In these experiments cupric

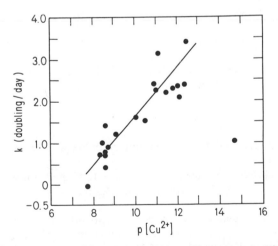

Figure 6.20   Effect of "free" $Cu \cdot aq^{2+}$ on growth of algae (*Chaetoceros socialis*) in seawater. The seawater contained pCu buffers made by appropriate additions of Cu(II) and of various chelators. pCu ($= -\log[Cu^{2+}]$) was calculated from the concentrations of total Cu(II) and complex formers present and from pH. (From Jackson and Morgan [63].)

ion activity was altered independent of total Cu(II) concentration by varying the complex-forming concentration and the pH (i.e., by using pCu buffers). Similarly, the nutritional supply of trace metals to phytoplankton is affected by complex formation, most likely also by lowering the concentrations of free metal ions [62,63].

The Methylation of Metals and Metalloids

Metals and metalloids that form alkyl compounds (e.g., methylmercury) deserve special concern because these compounds are volatile and accumulate in cells; they are poisonous to the central nervous system of higher organisms. Because methylmercury (or other metal alkyls) may be produced at a rate faster than it is degraded by other organisms, it may accumulate in higher organisms such as fish [65]. It is one of the few examples that demonstrate an increase in biological availability of a nonessential element in toxic concentrations through chemical transformation by the system. Metals and metalloids reported to be biomethylated (converted by bacteria under suitable conditions into methyl compounds) include Sn, As, Se, Te, Pd, Pt, Au, Hg, Tl, and Pb [66,67].

Methylmercury Species.   Because of the role of metal and metalloid alkyls in the toxicity of certain elements and the importance of such forms in the movement, bioaccumulation, and transformation of these elements, we briefly discuss here some properties of CH_3Hg^+ and its complexes. CH_3Hg^+ exists in aqueous solutions as an aquo complex $CH_3-Hg-OH_2^+$ with a covalent bond between Hg and O. The cation behaves as a soft acid and has a strong preference for the addition of only one ligand (Table 6.12). CH_3Hg^+ undergoes rapid coordination reactions with S, P, O, N, halogen, and C; the rate of the formation of Cl^-, Br^-, and OH^- complexes is extremely fast (diffusion-controlled). The CH_3Hg^+ unit itself, however, is kinetically remarkably inert toward decomposition. Therefore methylmercury compounds once formed—usually by biologically mediated methylation—are not readily demethylated. The neutral CH_3Hg^+ species are hydrophilic, lipophilic, and volatile; thus they can readily pass through boundaries. This, together with their broad tendency to form stable complexes quickly and the robustness of the CH_3Hg^+ unit, characterizes some of the far-reaching toxicological properties of methylmercury.

Example 6.7. The Thermodynamic Stability of CH₃Hg Species

Interpret the equilibrium constants given in Table 6.12 to predict under what conditions various CH_3Hg species are formed (thermodynamically) and which species predominate in natural waters.

A concentration ratio diagram may be constructed both for seawater ($[Cl^-] = 0.6\,M$) and for fresh water ($[Cl^-] = 2 \times 10^{-4}\,M$). A concentration of $[CH_4(aq)] = 10^{-4}\,M$ is assumed. The resulting diagrams (Figure 6.21) may be compared with Figure 6.15 which gives the relative abundance of inorganic Hg species. The concentration ratio diagram allows the following conclusions: (1) CH_3HgOH is the stable methylmercury species in fresh water. (2) Methylmercury compounds can—thermodynamically speaking—be

TABLE 6.12 STABILITY OF METHYLMERCURY AND ITS COMPLEXES^a

Reaction	Log K
(1) $Hg^{2+} + CH_3^- = CH_3Hg^+$	~50
(2) $CH_4(aq) = CH_3^- + H^+$	~−47
(3) $Hg^{2+} + CH_4(aq) = CH_3Hg^+ + H^+$	~3
(4) $CH_3Hg^+ + CH_3^- = (CH_3)_2Hg$	~37
(5) $CH_4(aq) + HgCl_2 = CH_3HgCl + H^+ + Cl^-$	−5.2
(6) $CH_3Hg^+ + Cl^- = CH_3HgCl$	5.25
(7) $CH_3Hg^+ + H_2O = CH_3HgOH + H^+$	−4.63
(8) $CH_3Hg^+ + CO_3^{2-} = CH_3HgCO_3^-$	6.1
(9) $CH_3Hg^+ + SO_4^{2-} = CH_3HgSO_4^-$	0.94
(10) $CH_3Hg^+ + S^{2-} = CH_3HgS^-$	21.02
(11) $CH_3Hg^+ + CH_3HgOH = (CH_3Hg)_2OH$	6.1
(12) $CH_3Hg^+ + CH_3HgS^- = (CH_3Hg)_2S$	16.34
(13) $HgS(s) + CH_4 = CH_3HgS^- + H^+$	~−26

^{a} Equlibrium constants (20°C, $I = 0$) are mostly from I. W. Erni, Ph.D., thesis, Chemistry Department, Swiss Federal Institute of Technology, Zürich. The more important inorganic Hg species are characterized by the following equilibrium constants (25°C, $I = 1$): $Hg(OH)_n^{2-n}$. $\log \beta_1 = 10.1, \log \beta_2 = 21.1; HgCl_n^{2-n}: \log \beta_1 = 6.72, \log \beta_2 = 13.23, \log \beta_3 = 14.2, \log \beta_4 = 15.3; \log {}^*\beta_{HgOHCl} (Hg^{2+} + H_2O + Cl^- = HgOHCl) = 3.67.$

formed in natural waters containing CH_4; these compounds should (thermodynamically) decompose at low pH:

$$CH_3HgOH + 2H^+ = CH_4 + Hg^{2+} \qquad \log K \approx 1.6$$
$$CH_3HgCl + H^+ + Cl^- = CH_4 + HgCl_2 \qquad \log K \approx 5.2$$

Because of the inertness of the CH_3Hg units, this decomposition occurs only very slowly. In other words, we can treat the CH_3Hg species in their coordination reactions (reactions 5 to 12 of Table 6.12) as metastable. Because CH_3Hg species can be formed in a micro-environment (i.e., inside a cell), they can be present in the bulk phase in concentrations that seem thermodynamically incompatible.

Methods for Species Identification. No single method presently available permits unequivocal identification of a species. Some of the principal methods in use are listed in Table 6.13. Usually, the evidence for a particular form of occurrence is circumstantial and is based on complementary evaluations together with kinetic and thermodynamic considerations. The identification of minor species has proved to be very difficult, as such species occur at concentrations smaller than 10^{-6} M in the presence of larger excesses of substances that often interfere with specific *in situ* sensing methods. Because of this, investigations

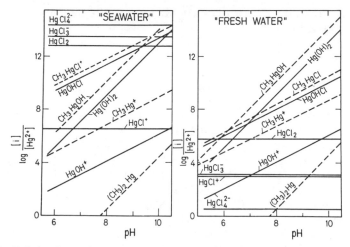

Figure 6.21   Relative thermodynamic stability of CH_3Hg^+ species. With the help of the equilibrium constants given in Table 6.13, concentration ratio diagrams have been constructed for the following conditions: seawater $[Cl^-] = 0.6\,M$, $[CH_4(aq)] = 10^{-4}\,M$; fresh water $[Cl^-] = 2 \times 10^{-4}\,M$, $[CH_4(aq)] = 10^{-4}\,M$ (see Example 6.7). CH_3Hg species once formed (usually by biomethylation) are inert with regard to demethylation; their coordination reactions, however, occur fast and in agreement with those predicted by equilibrium constants. Thus, while equilibria among the various CH_3Hg species and equilibria among the inorganic Hg species depicted prevail, the CH_3Hg species may not be in equilibrium with the inorganic Hg species.

using synthetic solutions in which the variables are known and can be controlled may frequently provide valuable clues to the types of species that exist in real water. Electrochemical methods helpful in species identification will be discussed in Section 7.9.

Size Fractionation.   The concept of chemical species as "dissolved" or "particulate" as defined by the pore size of a membrane filter can no longer be considered adequate. Among the size fractionation methods that seem to be particularly promising for characterizing the molecule-size distribution of soluble organic macromolecular material and of the metal ions associated with it is gel filtration (filtration and elution from columns containing gels of dextran, silica, or other molecular sieves). It has been used so far mostly for the fractionation of humic compounds and other coloured components of natural waters.

Multimetal, Multiligand Models

In order to understand the chemical properties of aquatic systems we wish to characterize different systems in terms of their dominant variables—major ions, oxidation–reduction status, acid–base components, minor ions, complexing

**TABLE 6.13   METHODS FOR ASSISTING IN THE SPECIFIC IDENTIFICA-
TION OF INDIVIDUAL SPECIES**

Method and Principle	Examples
Physical-mechanical separation Separation based on size (molecular weight), density, or charge	Membrane filtration, dialysis, electrodialysis, centrifugation, chromatography, gel filtration
Auxiliary equilibria A familar equilibrium system (e.g., a color-forming reaction or an ion-exchange system) is introduced to provide indication for the species	Effect of complex formation on acid-base equilibrium, adsorption, ion-exchange or redox reaction, or solubility equilibrium; solvent extraction
Equilibrium potentiometric methods Evaluation of an electrical potential difference related to the chemical potential (activity) of certain species.	Redox electrodes, ion-selective electrodes (metal, glass, hydrogen, solid-state, and membrane electrodes), electrodes of the second kind (e.g. Ag/AgCl).
Electrode kinetics Interdependence of current, potential, and time for a given electrode process; depends on the species participating	Polarography (square wave, pulse, inverse or anodic stripping), chronopotentiometry, chronoamperometry
Direct detection of electrode or atomic structure Measurement of properties based on electronic or atomic structure	Optical methods (spectrophotometry), magnetic properties (electron spin resonance), sound absorption
Catalytic effects and bioassays Many species, especially metal ions act as catalysts, growth (or inhibition) of organisms or rate of enzyme processes depends on species	Initiation of coordination or electron transfer reactions, batch or continuous-culture experiments with organisms, enzymatic reactions

components, and adsorbing surfaces. A systematic approach—from the abundant ions and their simple interactions with the rest of the system to the less abundant ions and their complex interactions with the rest of the system— should prove valuable in defining the relative importance of different variables in determining the stability of aquatic chemical systems. From the computation of chemical equilibrium models it is possible to decide, for example, in what complex form a given metal ion is likely to be found in a body of water containing a certain set of reacting ligands and competing metals.

The multimetal, multiligand models that will be used to exemplify this approach should aid in answering the following type of questions:

1 How do metal-ion speciations vary as a result of the addition of complex-forming organic matter to inorganic seawater or fresh water?

2 What is the effect of pH change (e.g., resulting from acid discharge or acid rain) on the free metal-ion concentrations of a fresh water?

3 What effect do particles, capable of adsorbing metal ions, have on metal-ion speciation? What is the competition between particles and soluble ligands for the metal ions?

4 What is the chemical speciation under anoxic conditions, for example, in anaerobic sewage?

A Hypothetical Experiment: Titration of Inorganic Fresh Water and Seawater with Organic Material.   An attempt is made to gain insight into the interaction of common organic material with metal ions. It is necessary to consider a system that contains, besides the major and minor ions typically found in natural waters, a variety of organic substances individually present at very small concentrations ($C < 10^{-5} M$). This is done by computing the effects of increments of organic matter added to an inorganic fresh water or seawater; after each addition the distribution of all the species is computed by considering all the equilibria involved.† Because of lack of information, the organic substances considered in the hypothetical system are not realistically representative of all detailed properties of natural waters. The organic substances selected for the model include amino acids (glycine and glutaminate), citrate as a representative of a substance of importance in the biochemical pathway, tartrate and phthalate which contain an arrangement of functional groups similar to that of humic acids and other natural color compounds. A few representative results are given in Table 6.14 and in Figure 6.22 [68].

Some significant implications can be derived from the results of these calculations: (1) Conventional complex-forming organic ligands at concentrations equal to or greater than those encountered in open surface waters affect the distribution of trace metal species only to a limited extent; for the examples taken, only the distribution of Cu species is affected markedly by organic complex formation. (2) About one-third of the organic complex-forming donor groups become bound to cations, mostly to Ca and Mg. (3) Concentrations of individual amino acids as high as 100 μg liter^{-1} are not sufficient to cause significant interactions with trace metals. (4) At the concentration levels considered, monodentate complexing agents like acetate are less efficient than chelate formers in binding trace metals (e.g., glycine to Zn, citrate to Cu); B-type cations

† The computer program, REDEQL, developed by F. Morel and J. J. Morgan, [*Environ. Sci. Technol.*, **6**, 58 (1972)] has been used.

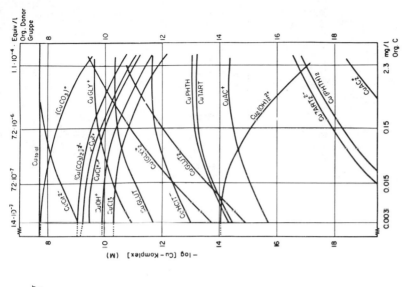

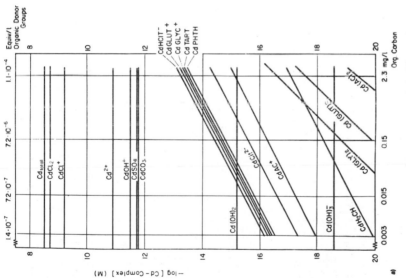

Figure 6.22 Distribution of Cd(II) and Cu(II) species as a function of the amount of complex-forming organic carbon added to seawater. For composition of organic material see Table 6.14. Whereas organic complex formation affects the distribution of Cu species (b), there is little effect on the distribution of inorganic species of most other trace metals, for example, distribution of Cd species (a). (*Note*: Citric acid behaves as a tetraprotic acid, H_4L.)

391

TABLE 6.14 EQUILIBRIUM MODEL; EFFECT OF COMPLEX FORMATION ON DISTRIBUTION OF METALS IN AEROBIC WATERS[a]

I. Seawater

Inorganic seawater, pH = 8.0, 25°C, free ligands: pSO4, 1.95; pHCO3, 2.76; pCO3, 4.86; pCl, 0.25

Inorganic seawater plus 7×10^{-6} mol liter^{-1} of each of the indicated organic ligands corresponding to 2.3 mg liter^{-1} of soluble organic carbon[b]; inorganic ligands remain unchanged; pSO4, 1.95; pHCO3, 2.76; pCO3, 4.86; pCl, 0.25

| | | Free | | Free | | Organic Complexes[d] | | | | | |
| | | | | | | Acetate, | Citrate, | Tartrate, | Glycinate, | Glutamate, | Phthalate, |
M	M_T^f	M	Major Species	M	Major Inorganic Species	Ligand → 5.21	14.7	5.41	6.96	6.89	5.2
Ca	1.97	2.03	CaSO4, 2.94 CaCO3, 3.50	2.03	CaSO4, 2.94 CaCO3, 3.50	7.41[d]	5.90	6.41	9.06	8.19	6.28
Mg	1.26	1.31	MgSO4, 2.25 MgCO3, 3.3	1.31	MgSO4, 2.25 MgCO3, 3.3	6.06	5.25	5.56	7.31	6.34	—[e]
Na	0.32	0.33	NaSO4, 1.97 NaHCO3, 3.3	0.33	NaSO4, 1.97 NaHCO3, 3.3	—	—	—	—	—	—
K	1.97	1.98	KSO4, 3.93	1.98	KSO4, 3.93	—	—	—	—	—	—
Fe(III)	8.0	18.9	Fe(OH)2, 8.3 FeSO4, 18.5	18.9	Fe(OH)2, 8.3 FeSO4, 18.5	20.7	8.6	—	15.9	13.7	—
Mn(II)	7.5	8.1	MnCl, 7.8[g] MnCl2, 8.3[g]	8.1	MnCl, 7.8 MnCl2, 8.3	12.8	11.4	—	13.1	12.2	—
Cu(II)	7.7	9.2	CuCO3, 7.7 Cu(CO3)2, 9.1	10.8	CuCO3, 9.4 Cu(CO3)2, 10.5	14.3	7.7	16.7	9.6	10.6	13.0
Cd	8.5	10.9	CdCl2, 8.7 CdCl, 9.2	10.9	CdCl2, 8.7 CdCl, 8.7	15.1	13.1	13.5	13.5	13.4	13.6
Ni	7.7	7.9	NiCl, 8.3 NiSO4, 8.7	8.0	NiSO4, 8.8	12.5	8.4	—	9.2	9.4	11.1
Pb	8.2	9.9	PbCO3, 8.6 PbOH, 8.7	9.9	PbOH, 8.7	13.2	11.34	11.5	11.8	—	11.7
Co(II)	8.3	8.5	CoCl, 9.0 CoSO4, 9.1	8.5	CoSO4, 9.1	12.7	26.5	11.9	10.8	10.8	14.9
Ag	8.7	13.1	AgCl2, 8.7 AgCl, 10.0	13.1	AgCl, 10.0	17.9	26.5	—	16.7	—	—
Zn	7.2	7.8	ZnOH, 7.4 ZnCl, 8.0	7.8	ZnCl, 8.0	11.7	11.3	10.9	8.8	9.7	10.9
					%[f]	13.0	98.6	44.9	0.7	6.6	7.5

II. Fresh Water

Inorganic fresh water, pH = 7.0, 25°C, free ligands: pSO₄ 3.4; pHCO₃ 3.1; pCO₃ 6.1; pCl 3.3 — Inorganic fresh water plus 7×10^{-6} mol liter^{-1} of each of the indicated organic ligands corresponding to 2.3 mg liter^{-1} of soluble organic carbon[b]. Inorganic ligands remain unchanged

M	M_T^c	Free M	Major Species	Free M	Major Inorganic Species	Acetate, 7.55	Citrate, 6.91	Tartrate, 8.02	Glycinate, 5.16	Glutamate, 5.16	Phthalate, 6.99
						(Free Ligand →)					
Ca	2.7	2.72	CaHCO₃, 4.6	2.72	CaHCO₃, 4.6	7.0	5.2	5.6	9.1	8.6	5.7
Mg	3.7	3.72	MgSO₄, 5.1	3.72	MgSO₄, 5.1	8.0	7.0	7.1	8.0	9.1	—
Fe(III)	Satd.	17.70	Fe(OH)₂, 8.7	17.73	Fe(OH)₂, 8.7	19.0	7.2	—	15.1	11.1	—
Mn(II)	7.0	7.04	MnSO₄, 8.5	7.04	MnSO₄, 8.5	11.3	9.7	—	11.5	11.1	—
Cu(II)	7.0	7.46	CuCO₃, 7.2	9.93	CuCO₃, 9.7	13.1	7.0	11.3	9.4	9.4	11.4
Zn(II)	6.7	6.72	ZnSO₄, 8.2	6.72	ZnSO₄, 8.2	10.3	10.5	8.9	9.6	8.6	9.1
Cd(II)	7.7	7.73	CdSO₄, 9.2	7.76	CdSO₄, 9.2	11.5	9.2	9.6	11.5	10.3	9.7
Pb(II)	7.0	8.02	PbCO₃, 7.1	8.04	PbCO₃, 7.1	11.0	8.9	8.8	10.1	—	9.2
Ag(I)	9.0	9.19	AgCl, 9.5	9.19	AgCl, 9.5	13.8	17.5	—	13.3	—	—
%^{f}						1.5	98.2	35.2	0.3	0.4	29.0

(Organic Complexes[d,e])

[a] All concentrations are given as −log (mol liter^{-1}). Charges of species are omitted.

[b] Organic matter of approximate composition $C_{13}H_{11}O_{12}N$.

[c] Total concentration of metal species. The concentrations of heavy metals given are higher than those typically found in unpolluted seawater or fresh water. The relative effects of complex formation as trace elements are independent of the total concentration of these elements; waters are in equilibrium with $Fe(OH)_3$(s).

[d] The concentrations refer to the sum of all complexes, for example, CuCit, CuHCit, CuCit₂.

[e] A dash means that no stability constants are available for such complexes.

[f] Percentage of each ligand bound to metal ions.

[g] Some uncertainty regarding stability constants for Mn(II)-chloro complexes; some computations show Mn^{2+} as a major species.

393

are relatively more strongly bound by molecules containing mixed N and O donor groups (amino acids) than A-type cations.

Figure 6.22a and b illustrates the different effects of organic matter upon the distribution of Cd and Cu in seawater. Under the conditions assumed for this model, the distribution of inorganic Cd species (and that of most other trace metals) is not affected by the presence of organic functional groups at the concentration levels normally found in seawater. The trend of the curves in Figure 6.22a indicates, and computations confirm, that higher concentrations of organic matter (e.g., 5×10^{-4} eq liter^{-1} and above) will influence the distribution of most trace metals. For Cu(II) however, organic ligands are able to compete successfully with inorganic ligands for the available coordinative sites of the Cu atom even at low concentrations. Because algal growth is strongly dependent upon free Cu^{2+}, organic matter, either incipiently present in the water or formed from exudates of algae, may indeed exert a pronounced influence on the physiological response of algae by regulating free Cu [62,63].

The model presented suggests that many soluble trace metals in surface waters are primarily present as inorganic species. Only if concentrations of metals and organic complex-forming components are higher, as they may be in interstitial waters [69], or if stronger and much more selective complex formers are present, will metal ions generally be present as organic complexes.†

Interdependence of Complex Formation.   In the example discussed, one notes that, even at a low concentration, organic matter becomes bound to Cu(II). This indicates that Cu(II), by tying up organic ligands, may regulate the chelation of other metals. A computation, for example, shows that increasing the total Cu(II) concentration (e.g., by pollution) by a factor of 100 will decrease the concentration of citrato complexes of all other metal ions by a factor of 10; this affects in turn—although to a lesser extent—the equilibrium concentrations of other metal complexes. Similar effects can be caused by the addition of other elements or by the addition of individual ligands. As Figure 6.22 suggests, metals, H^+, and ligands form a complicated network of interactions; because each cation interacts and equilibrates with all ligands and each ligand similarly equilibrates with all cations, the free concentration of metal ions and the distribution of both cations and ligands depend on the total concentrations of all the other constituents of the system [70,71]. The addition of Fe(III) (or of any other metal) to a water medium, for example, in a productivity experiment, produces significant reverberations in the interdependent "web" of metals and ligands and may lead to a redistribution of all trace metals (hence an observed change in productivity is not necessarily caused by a change in the availability

† These results are in agreement with those reported by R. F. C. Mantoura, A. Dickson, and J. P. Riley [*Estuarine Coastal Mar. Sci.*, **6**, 387 (1978)]. These authors determined conditional stability constants for Ca^{2+}, Mg^{2+}, and heavy-metal cations with humic compounds for pH = 8. Their model calculations show that humic compounds affect trace-metal speciation in such a way that only Cu(II) (seawater and fresh water) and Hg(II) (fresh water) are complexed by humic materials.

of Fe to the cell). Morel et al. [71] have illustrated how organic complex-forming substances mediate the many pathways of interdependence among the constituents in natural water systems and have elaborated on methods for quantifying such interactions.

The Sequestration of Metals by Particles. Particles—through their surfaces—are scavengers for reactive elements in their transport from land to rivers and lakes and from continents to the floor of the oceans. Turekian [72] writes, "The great particle conspiracy is active from land to sea to dominate the behavior of the dissolved species." As we have seen in Section 6.8, and as will be discussed further in Section 10.4, hydrous oxide surfaces, as well as organically coated and organic surfaces, contain functional surface groups ($\equiv$MeOH, $\equiv$ROH, R—COOH) that act as coordinating sites on the surface (equations 32 and 33). Since surface complex formation constants have been determined (especially for SiO_2 and Al_2O_3 surfaces) [73–76], with the help of these equilibrium constants simple models can be established to evaluate the competition between the coordination sites of soluble ligands and those of surfaces for metal ions.

Acidification of Lake Water. An example is illustrated in Table 6.15 where we estimate the proportion of free, complex-bound, and adsorbed metal ions in a hypothetical lake water. The pH of this water—which remains in equilibrium with the CO_2 pressure of the atmosphere—is lowered from 8 to 6 (e.g., simulating the input of acid rain into a lake of a crystalline rock region).

As Table 6.15 illustrates, H^+ ions can successfully compete with metal ions for the available ligands, for example,

$$CO_3^{2-} + H^+ \;\; = HCO_3^-$$
$$RCOO^- + H^+ = RCOOH$$
$$\equiv MeO^- + H^+ = \equiv MeOH$$
$$Fe(OH)_3(s) + H^+ = Fe(OH)_2^+$$

and thus the proportion of free metal ions increases markedly with lower pH, for example,

$$CuCO_3(aq) + H^+ = Cu^{2+} + HCO_3^-$$
$$RCOOZn^+ + H^+ = RCOOH + Zn^{2+}$$
$$\equiv MeOPb^+ + H^+ = \equiv MeOH + Pb^{2+}$$

This example also illustrates, like those given in Table 6.14, that organic complex formers (in our example, salicylate) change above all the speciation of Cu(II). The example illustrates that surfaces can tie up significant proportions of trace metals even in the presence of an organic chelate former at a relatively high concentration (see Cu at pH = 7). In a real lake or ocean, the effects of particles on the regulation of metal ions are enhanced because the continuously settling particles (phytoplankton and particles introduced by rivers) act like a conveyor belt in transporting reactive elements to the sediments.

Speciation of Trace Metals in Anoxic Water. Morel et al. [77] have calculated equilibrium models accounting for the chemical speciation of trace metals in primary sewage effluent. Many metals are found as solid sulfides (Zn, Hg, Ag, Cu, Cd, Pb) or solid oxides (Cr, Fe), while some (Ni, Co, Mn) are relatively soluble.

TABLE 6.15   LAKE ACIDIFICATION EQUILIBRIUM MODEL[a]

	Inorganic Lake Water, No Adsorbing Solids Surfaces Present			Inorganic Lake Water plus 10^{-5} M Salicylate plus Adsorbing Silica Surfaces with 10^{-5} M Suface Sites[b]		
	pH8	pH7	pH6	pH8	pH7	pH6
Ca	97[c]	99	99	89	92	99[d]
	1.5[c]	0	0	1.5	—	0[d]
				0	—	—[d]
				9	7.5	0.5[d]
Mg	97	99	99	95.5	97	99
	2	0	0	2	0	0
				—	0	0
				1.5	1	0
Fe(III)	0	0	0[e]	0	0	0
	0	0	0[e]	0	0	0
				0	0	0
				0	3	16[e]
Mn(II)	94	98	99	86	91	98
	4	0	0	3	0	0
				0	0	0
				9	8	1
Cu	5	76	98	3	15	73
	89	20	0	54	3	0
				17	7	4
				21	73	21
Cd	50	96	97	28	42	91
	47	1	0	24	1	0
				0	0	0
				47	56	7
Zn	54	97	99	18	25	85
	42	1	0	12	0	0
				0	0	0
				66	74	14

TABLE 6.15   *(continued)*

	Inorganic Lake Water, No Adsorbing Solids Surfaces Present			Inorganic Lake Water plus 10^{-5} M Salicylate plus Adsorbing Silica Surfaces with 10^{-5} M Surface Sites[b]		
	pH8	pH7	pH6	pH8	pH7	pH6
Ni	44	95	99	30	53	94
	48	2	0	33	1	0
				0	0	0
				31	44	5
Pb	1	48	97	1	7	55
	96	46	0	80	6	0
				0	0	0
				17	87	42
pCO_3^{2-}	5.5	7.5	9.5	5.5	7.5	9.5
$pHCO_3^-$	3.3	4.3	5.3	3.3	4.3	5.3

[a] $p_{CO_2}(g) = 10^{-3.5}$ atm, 25°C. Total concentrations expressed as $-\log M$: pCa $= 4.0$, pMg $= 4.3$, pFe(III) $= 6.0$, pMn(II) $= 6.7$, pCu $= 7.0$, pCd $= 8.0$, pZn $= 6.7$, pPb $= 8.0$, pNi $= 8.0$, pSO$_4$ $= 4.3$, pCl $= 3.0$, pNH$_3$ $= 5.5$, pPO$_4$ $= 6.3$, pH$_4$SiO$_4$ $= 4.7$, pNO$_3$ $= 5.5$. pCO_3^{2-} is controlled by $p_{CO_2}(g)$ and pH.
[b] 10^{-5} mol liter^{-1} SiO$_2$ surface sites correspond to 0.5 gl liter^{-1} of SiO$_2$ particles with a diameter of ~ 1 μm. The surface complex formation constants used in the calculation are those of Schindler et al. [73].
[c] The two entries listed for each pH are % of free metal and carbonato-bound metal.
[d] The four entries listed for each pH are percentages of free ion, carbonato-bound ion, Salicylate-bound ion and absorbed ion for each metal.
[e] Remainder is Fe(OH)$_3$(s).

   A set of organics representing various functional groups and complexing behaviors was included in the model (third column, Table 6.16). The only metals that show significant binding with the organic ligands are Co and Ni; the reactivity of the other metals is limited because of the extreme insolubility of the solid forms. The last column in Table 6.16 shows the results of including adsorbing surfaces in the model. Four metals were found to adsorb markedly: Cr, Co, Ni, Mn.

   Comparing the results of this model with those at higher redox potential makes it evident that most metals tend to be solubilized upon disposal in aerobic receiving waters.

TABLE 6.16 EQUILIBRIUM MODELS OF SEWAGE[a]

$-$Log Total Concentration	Inorganic Model $(\%)^b$	Addition of Organics $(\%)^{b,c}$	Addition of Adsorbing Surface $(\%)^{b,c,d}$
Fe, 3.7	$Fe_3O_4(s)$, 100	$Fe_3O_4(s)$, 100	$Fe_3O_4(s)$, 100
Cr, 4.8	$Cr(OH)_3(s)$, 97	$Cr(OH)_3(s)$, 97	$Cr(OH)_3(s)$, 52
	$Cr(OH)_4^-$, 3	$Cr(OH)_4^-$, 3	$Cr(OH)_4^-$, 3
			ADS, 45
Cu, 5.0	$CuS(s)$, 100	$CuS(s)$, 100	$CuS(s)$, 100
Cd, 6.5	$CdS(s)$, 100	$CdS(s)$, 100	$CdS(s)$, 100
Pb, 6.0	$PbS(s)$, 100	$PbS(s)$, 100	$PbS(s)$, 100
Zn, 4.5	$ZnS(s)$, 100	$ZnS(s)$, 99	$ZnS(s)$, 99
Ag, 6.7	$Ag_2S(s)$, 100	$Ag_2S(s)$, 100	$Ag_2S(s)$, 100
Hg, 8.3	$HgS(s)$, 100	$HgS(s)$, 100	$HgS(s)$, 100
Ni, 5.4	$NiS(s)$, 42	$NiS(s)$, 22	$NiS(s)$, 14
	Ni^{2+}, 2	Ni^{2+}, 2	Ni^{2+}, 2
	$Ni(CN)_4^{2-}$, 56	$Ni(CN)_4^{2-}$, 56	$Ni(CN)_4^{2-}$, 56
		GLY, 10	GLY, 10
		GLU, 10	GLU, 10
			ADS, 8
Co, 6.8	$CoS(s)$, 97	$CoS(s)$, 95	$CoS(s)$, 93
	Co^{2+}, 2	Co^{2+}, 2	Co^{2+}, 2
		GLU, 2	GLU, 2
			ADS, 2
Mn, 5.6	Mn^{2+}, 50	Mn^{2+}, 50	Mn^{2+}, 45
	$MnHCO_3^+$, 24	$MnHCO_3^+$, 24	$MnHCO_3^+$, 21
	$MnSO_4$, 19	$MnSO_4$, 19	$MnSO_4$, 17
	$MnCl^+$, 6	$MnCl^+$, 6	$MnCl^+$, 6
			ADS, 9

[a] From Morel et al. (1975) [77].
[b] Other inputs to the model: pH $= 7.7$ pX_T: pCa $= 2.75$, pMg $= 3.0$, pBa $= 5.0$, pAl $= 4.5$, $pCO_3 = 2.0$, $pSO_4 = 2.3$, $pS(-II)_T = 4.7$, pCl $= 1.8$, pF $= 4.0$, $pNH_3 = 2.2$, $pPO_4 = 3.3$, pCN $= 5.0$. $I = 0.01$, $T = 25°C$.
[c] Organic ligands added: pX_I: pAcetate $= 3.3$, pGlycine $= 3.8$, pTartrate $= 3.6$, pGlutamate $= 3.7$, pSalicylate $= 3.8$, pPhthalate $= 3.9$.
[d] Adsorbing surface: 10 m^2 liter^{-1}, constant potential. The computation was performed according to the James and Healy model [*J. Colloid Interface Sci.*, **40**, 42–65 (1972)] which considers coulombic, solvation, and specific chemical energy interactions. The chemical energy term is essentially a parameter and has been chosen arbitrarily in the model to represent high chemical affinity of the adsorbing surface for all metals (about 10 kcal mol^{-1}).

REFERENCES

1   E. D. Goldberg, *J. Geol.*, **62**, 249 (1954).

2   K. B. Krauskopf, *Geochim. Cosmochim. Acta*, **9**, 1 (1956).

3   L. G. Sillén, in *Oceanography*, M. Sears, Ed., American Association for the Advancement of Science, Washington, D.C., 1961.

4   R. M. Garrels, *Mineral Equilibria at Low Temperature and Pressure*, Harper and Row, New York, 1960.

5   R. M. Garrels and C. L. Christ, *Solutions, Minerals and Equilibria*, Harper and Row, New York, 1965.

6   C. F. Baes, Jr., and R. E. Mesmer, *The Hydrolysis of Cations*, Wiley-Interscience, New York, 1976.

7   H. Wendt, *Chimia*, **27**, 575 (1973).

8   P. Schindler, private communication.

9   G. Anderegg, in *Coordination Chemistry*, A. E. Martell, Ed., Van Nostrand Reinhold, New York, 1971.

10   S. Ahrland, S. J. Clatt, and W. R. Davies, *Quart. Rev. (London)*, **12**, 265 (1958).

11   H. Irving and R. J. P. Williams, *J. Chem. Soc.*, **1953**, 3192.

12   G. Schwarzenbach, *Chimia*, **27**, 1 (1973).

13   R. G. Pearson, *J. Amer. Chem. Soc.*, **85**, 3533 (1963); for general review see Pearson, R. G., *Science*, **151**, 172 (1966) and *J. Chem. Educ.*, **45**, 581 and 643 (1968).

14   H. B. Gray, *Chemical Bonds: An Introduction to Atomic and Molecular Structure*, W. A. Benjamin, Menlo Park, Calif., 1973.

15   C. F. Bell, *Metal Chelation: Principles and Applications*, Clarendon, Oxford, 1977.

16   R. G. Burns, *Mineralogical Applications of Crystal Field Theory*, Cambridge University Press, Cambridge, 1970.

17   H. L. Schläfer and G. Gliemann, *Einführung in die Ligandfeld-Theorie*, Akademische Verlagsgesellschaft, Frankfurt a/M, 1967.

18   F. A. Cotton and G. Wilkinson, *Advanced Inorganic Chemistry*, 3rd ed., Wiley-Interscience, New York, 1972.

19   M. Eigen and K. Tamm, *Z. Elektrochem.*, **66**, 93 (1962).

20   F. H. Fischer, *Science*, **157**, 823 (1967).

21   R. M. Garrels and M. Thompson, *Amer. J. Sci.*, **260**, 57 (1962).

22   R. M. Pytkowicz and J. E. Hawley, *Limnol. Oceanogr.*, **19**, 223 (1974).

23   G. Schwarzenbach and M. Widmer, *Helv. Chim. Acta*, **46**, 2613 (1963); **49**, 111 (1966).

24   A. Ringbom, *Complexation in Analytical Chemistry*, Wiley-Interscience, New York, 1963.

25   G. Schwarzenbach, *Complexometric Titrations*, Interscience, New York, 1959.

26   G. Schwarzenbach, in *Advances in Inorganic Chemistry and Radiochemistry*, Volume 3, H. J. Emeleus and A. G. Sharpe, Eds., Academic, New York, 1961, p, p. 257.

27   R. M. Garrels and M. E. Thompson, *Amer. J. Sci.*, **260**, 57 (1962).

28   D. R. Kester and R. M. Pytkowicz, *Geochim. Cosmochim. Acta*, **34**, 1039 (1970).

29   F. J. Millero, *Chem. Rev.*, **71**, 147 (1971).

30   D. E. Kester and R. M. Pytkowicz, *Limnol. Oceanogr.*, **12**, 243 (1967).

31   Christenson, P. G., and J. M. Gieskes, *J. Chem. Eng. Data*, **16**, 398 (1971).

32   G. Schwarzenbach, *Chimia*, **27**, 1 (1973).

33   P. Krindel and I. Eliezen, *Coord. Chem. Rev.*, **6**, 217 (1972).

34   F. J. Millero, *Chem. Rev.*, **71**, 147 (1971).

35   F. J. Millero, in *Water and Aqueous Solutions*, R. A. Horne, Ed., Wiley-Interscience, New York, 1972.

36   I. W. Duedall and P. Weyl, *Limnol. Oceanogr.*, **12**, 52 (1967).

37   J. L. Watters, J. Mason, and A. Aaron, *J. Amer. Chem. Soc.*, **82**, 1333 (1960).

38   V. S. Sharma and J. Schubert, *J. Chem. Educ.*, **46**, 506 (1969).

39   D. Dyrssen, D. Jagner, and F. Wengelin, *Computer Calculations of Ionic Equilibria and Titration Procedures*, Almqvist and Wiksell, Stockholm, 1968.

40   F. Morel and J. J. Morgan, *Environ. Sci. Technol.*, **6**, 58 (1972).

41   A. Zirino and T. Yamamoto, *Limnol. Oceanogr.*, **17**, 661 (1972).

42   F. S. Brown, M. J. Baedecker, I. R. Kaplan, and A. Nissenbaum, *Geochim. Cosmochim. Acta*, **32**, 397 (1972).

43   E. Bayer, *Angew. Chem.*, **3**, 325 (1964).

44   F. R. N. Gurd, *Chemical Specificity in Biological Interactions*, Academic, New York, 1954.

45   F. G. Lowman, T. R. Rice, and F. A. Richards, in *Radioactivity in the Marine Environment*, National Academy of Sciences, Washington, D.C., 1971.

46   P. G. Brewer, in *Chemical Oceanography*, Vol. 1, J. P. Riley and G. Skirrow, Eds., 2nd Ed., Academic, New York, 1975.

47   J. H. Martin and G. A. Knauer, *Geochim. Cosmochim. Acta*, **37**, 1639 (1973).

48   H. J. M. Bowen, *Trace Elements in Biochemistry*, Academic, New York, 1966.

49   J. B. Neilands, in *Essays in Coordination Chemistry*, W. Schneider, R. Gut, and G. Anderegg, Eds., Birkhäuser, Basel, 1964, p. 222.

50   J. J. Christen, J. O. Hill, and R. M. Izatt, *Science*, **174**, 459 (1971).

51   H. Sigel, Ed., *Metal Ions in Biological Systems*, Vol. 2, *Mixed Ligand Complexes*, Dekker, New York, 1973. The following chapters are of particular interest: R. P. Martin et al., "Mixed Ligand Metal Ion Complexes of Aminoacids and Peptides"; H. Sigel, "Structural Aspects of Mixed Ligand Complex Formation in Solution"; D. D. Perrin and R. P. Agarwol, "Multimetal-Multiligand Equilibria—A Model for Biological Systems."

52   H. Sigel, *Angew. Chem.*, **11**, 391 (1975).

53   L. G. Sillén and A. E. Martell, *Stability Constants*, Chemical Society, Special Publications Nos. 17 and 25, London 1964 and 1971.

54   E. J. King, *Acid-Base Equilibria*, Macmillan, New York, 1965. (Chapter 9 gives a lucid discussion of polyprotic acids and the interpretation of their alkalimetric titration curves.)

55   C. Tanford, *Physical Chemistry of Macromolecules*, Wiley, New York, 1961, p. 572.

56   D. S. Gamble and M. Schnitzer, in *Trace Metals and Metal-Organic Interactions*, P. C. Singer, Ed., Ann Arbor Science Publications, Ann Arbor, Mich., 1973.

57   D. E. Wilson and P. Kinney, *Limnol. Oceanogr.*, **22**, 281 (1977).

58   D. S. Gamble et al., *Can. J. Chem.*, **54**, 1239 (1976).

59   R. F. C. Mantoura and J. P. Riley, *Anal. Chim. Acta*, **78**, 193 (1975).

60   J. Buffle et al., *Anal. Chem.*, **49**, 216 (1977).

61   S. E. Manahan and M. J. Smith, *Environ. Sci. Technol.*, **7**, 829 (1973).

62   W. Sunda and R. L. Guillard, *J. Mar. Res.*, **34**, 511 (1976).

63   G. Jackson and J. J. Morgan, *Limnol. Oceanogr.*, **23**, 268 (1978).

64   R. T. Barber, in *Trace Metals and Metal Organic Interactions in Natural Waters*, P. C. Singer, Ed., Ann Arbor Science Publications, Ann Arbor, Mich., 1973.

65   F. M. D'Itri, *The Environmental Hg Problem*, CRC, Cleveland, 1972.

66   J. M. Wood, *Naturwissenschaften*, **62**, 357 (1975).

67   W. P. Ridley, L. J. Dizikes, and J. M. Wood, *Science*, **197**, 329 (1977).

68   W. Stumm and P. A. Brauner, in *Chemical Oceanography*, J. P. Riley and G. Skirrow, Eds. Vol. 1, 2nd ed., Academic, New York, 1975, p. 173.

69   A. Nissenbaum and D. Swaine, *Geochim. Cosmochim. Acta*, **40**, 809 (1976).

70   E. K. Duursma, in *Organic Matter in Natural Waters*, D. W. Hurd, Ed., University of Alaska, College, 1970.

71   F. Morel, R. E. McDuff, and J. J. Morgan, in *Trace Metals and Metal-Organic Interactions in Natural Waters*, P. C. Singer, Ed., Ann Arbor Science Publications, Ann Arbor, Mich., 1973.

72   K. Turekian, *Geochim. Cosmochim. Acta*, **41**, 1139 (1977).

73   P. W. Schindler, B. Fürst et al., *J. Colloid Interface Sci.*, **55**, 769 (1976).

74   H. Hohl and W. Stumm, *J. Colloid Interface Sci.*, **55**, 281 (1976).

75   W. Stumm, H. Hohl, and F. Dalang, *Croat. Chim. Acta*, **48**, 491 (1976).

76   J. A. Davis, R. O. James, and J. A. Leckie, *J. Colloid Interface Sci.*, **67**, 90 (1978).

77   F. M. Morel, J. C. Westall, C. R. O'Melia, and J. J. Morgan, *Environ. Sci. Technol.*, **9**, 757 (1975).

READING SUGGESTIONS

Introductory Texts

Basolo, F., and R. G. Pearson, *Mechanism of Inorganic Reactions, A Study of Metal Complexes in Solution*, 2nd ed., Wiley, New York, 1967.

Bell, C. F., *Metal Chelation, Principles and Application*, Clarendon, Oxford, 1977.

Cotton, F. A., and G. Wilkinson, *Basic Inorganic Chemistry*, Wiley, New York, 1976.

Earnshaw, A. and T. J. Harrington, *The Chemistry of Transition Elements*, Clarendon, Oxford, 1973.

Gray, H. B., *Chemical Bonds: An Introduction to Atomic and Molecular Structure*, W. A. Benjamin, Menlo Park, Calif., 1973.

Hunt, J. P., *Metal Ions in Solution*, W. A. Benjamin, Menlo Park, Calif., 1963.

Jørgensen, C. K., *Inorganic Complexes*, Academic, New York, 1963.

Nancollas, G. H., *Interactions in Electrolyte Solutions*, Elsevier, New York, 1966.

Pass, G., *Ions in Solution*, Clarendon, Oxford, 1973.

Quagliano, J. V., and L. M. Vallorino, *Coordination Chemistry*, D. C. Heath, Lexington, Mass., 1969.

Schneider, W., *Einfuhrung in die Koordinationschemie*, Springer, Berlin, 1968.

Quantitative Complex Formation: Stability of Complexes and Analytical Applications

Ahrland, S., "Thermodynamics of the Stepwise Formation of Metal Ion Complexes in Aqueous Solution," in *Structure and Bonding*, J. D. Dunitz, *et al.* eds., Springer, New York, vol. 15, p. 465, 1973.

Ringbom, A., *Complexation in Analytical Chemistry*, Wiley-Interscience, New York, 1963.

Rossotti, F. J. C., and H. Rossotti, *The Determination of Stability Constants*, McGraw-Hill, New York, 1961.

Schindler, P., "Heterogeneous Equilibria Involving Oxides, Hydroxides, Carbonates and Hydroxide Carbonates." In *Equilibrium Concepts in Natural Water Systems*, Advances in Chemistry Series, No. 67, American Chemical Society, Washington, D.C., 1967, p. 196.

Selective and Specific Formation of Complexes

Amphlett, C. B., *Inorganic Ion Exchangers*, Elsevier, New York, 1964.

Edwards, J. O., *Inorganic Reaction Mechanisms*, W. A. Benjamin, Menlo Park, Calif., 1964.

Schwarzenbach, G., "The General, Selective and Specific Formation of Complexes by Metallic Cations." In *Advances in Inorganic Chemistry and Radiochemistry*, Vol. 3, Academic, New York, 1961.

Schwarzenbach, G., "Koordinationsselektivität und die Thermodynamik der Komplexbildung in Lösung," *Chimia*, **27**, 1 (1973).

Smith, R. M., and A. E. Martell, *Critical Stability Constants*, Plenum, New York, 1976.

Wilkins, R. G., "The Kinetics of Complex Formation in Aqueous Solution." In *The Nature of Seawater*, E. D. Goldberg, Ed., Dahlem Konferenzen, Berlin, 1975, p. 397.

Schneider, W., "Kinetics and Mechanism of Metalloporphyrin Formation." In *The Nature of Seawater*, E. D. Goldberg, Ed., Dahlem Konferenzen, Berlin, 1975, p. 375.

Metal-Ion Hydrolysis

Baes, C. F., and R. E. Mesmer, *Hydrolysis of Cations*, Wiley-Interscience, New York, 1976. (This book presents a complete and critical review of all hydrolysis equilibria and provides insight into why metal ions provide such a wide variety of hydrolysis products.)

Metal Ions in Natural Waters

Brewer, P. G., "Minor Elements in Seawater." In *Chemical Oceanography*, J. P. Riley and G. Skirrow, Eds., Vol. 1, 2nd ed., Academic, New York, 1975, pp. 415–496. (Contains data on abundance and residence times of minor elements and discusses the interaction of metals with marine organisms.)

Dyrssen, D., and M. Wedborg, "Equilibrium Calculations of the Speciation of Elements in Seawater." In *The Sea*, Vol. 5, E. D. Goldberg, Ed., Wiley-Interscience, New York, 1974. pp. 181–196.

Foerstner, U. and G. T. W. Wittmann, *Metal Pollution in the Aquatic Environment*, Springer, Berlin, 1979.

Hem, J. D., *Study and Interpretation of the Chemical Characteristics of Natural Waters*, 2nd ed., Geological Survey Water Supply Paper 1473, Washington, D.C., 1975. (Provides an excellent introducion to water analysis interpretation and to processes that shape and control the chemical composition of fresh waters.)

Hoffmann, M. R., "Trace Metal Catalysis in Aquatic Environments," *Environ. Sci. Technol.*, **14**, 1061 (1980).

Jenne, E. A., "Controls on Mn, Fe, Co, Ni, Cu and Zn Concentrations in Soils and Water." In *Trace Inorganics in Water*, Advances in Chemistry Series, No. 73, American Chemical Society, Washington, D.C., 1968, p. 337.

Morel, F., R. E. McDuff, and J. J. Morgan, "Interactions and Chemostasis in Aquatic Chemical Systems: Role of pH, $p\varepsilon$, Solubility, and Complexation." In *Trace Metals and Metal Organic Interactions in Natural Waters*, P. C. Singer, Ed., Ann Arbor Science Publications, Ann Arbor, Mich., 1973, pp. 157–200.

Sillén, L. G., "Physical Chemistry of Seawater." In *Oceanography*, M. Sears, Ed., American Association for the Advancement of Science, Washington, D.C., 1961, pp. 549–581.

Stumm, W., and P. A. Brauner, "Chemical Speciation." In *Chemical Oceanography*, J. P. Riley and G. Skirrow, Eds., Vol. 1, 2nd ed., Academic, London, 1975, pp. 173–240.

Metals in Biological Systems

Eichhorn, G., "Organic Ligands in Natural Systems." In *The Nature of Seawater*, E. D. Goldberg, Ed., Dahlem Konferenzen, Berlin, 1975, (Describes organic ligands specifically designed in biological systems for metal binding.)

Lowman, F. G., T. R. Rice, and F. A. Richards, "Accumulation and Redistribution of Radionuclides by Marine Organisms." In *Radioactivity in the Marine Environment*, National Academy of Sciences, Washington, D.C., 1971, p. 161.

Morel, F., and J. G. Yeasted, "On the Interfacing of Chemical Physical and Biological Water Quality Models." In *Fate of Pollutants in the Air and Water Environments*, Part I, I. H. Suffet, Ed., Wiley-Interscience, New York, 1977, pp. 253–268.

Phipps, D. A., *Metals and Metabolism*, Clarendon, Oxford, 1976.

Williams, D. R., *The Metals of Life*, Van Nostrand Reinhold, New York, 1971.

PROBLEMS

6.1 Make a distribution diagram for the various Cr(VI) species as a function of pH at $Cr_T = 1$, 10^{-2}, and 10^{-4} M. Neglect activity corrections.

Equilibria:

$$CrO_4^{2-} + H^+ = HCrO_4^- \qquad \log K = 6.5$$
$$HCrO_4^- + H^+ = H_2CrO_4 \qquad \log K = -0.8$$
$$2\,HCrO_4^- = Cr_2O_7^{2-} + H_2O \qquad \log K = 1.52$$
$$Cr_2O_7^{2-} + H^+ = HCr_2O_7^- \qquad \log K = 0.07$$

6.2 Find the pH at which "$Fe(OH)_3$" begins to precipitate from a 0.1 M $Fe(ClO_4)_3$ solution. Consider the dimerization of $FeOH^{2+}$.

6.3 Derive an equation that permits the computation of the charge of a hydrous oxide (more exactly, that portion of the charge caused by direct or indirect binding of H^+ or OH^-) from an alkalimetric titration curve of the hydrous oxide.

6.4 What equilibria need to be considered in assessing the fish toxicity of a galvanic waste containing Zn(II) and cyanide?

6.5 An organic compound isolated from natural waters is found to contain phenolic and carboxylic functional groups. From an alkalimetric titration curve the intrinsic acidity constants $pK_1 \approx 4.1$ and $pK_2 \approx 12.5$ can be estimated. This compound forms colored complexes with Fe(III), the highest color intensity being observed in slightly acidic solutions for a mixture of equimolar concentrations of Fe(III) and the complex former. If to a $10^{-3} M$ Fe(III) solution at pH = 3 this complex former is added to attain a concentration of $5 \times 10^{-2} M$, a potentiometric electrode for Fe^{3+} (e.g., a ferro-ferri cell) registers a shift in $[Fe^{3+}]$ corresponding to a $\Delta pFe \simeq 3.5$. Provide a rough estimate for the stability of this complex. Constants for the hydrolysis of Fe(III) are given in Example 6.1.

6.6 In a series of phytoplankton culture experiments the growth rate is found to be affected by the copper chemistry of the medium. The constant pH medium contains a chelating agent, X, so that

$$TOT\ Cu = [Cu^{2+}] + [CuOH^+] + [CuX]$$

The experimental results show that when the chelating agent concentration is varied at constant TOT Cu, the growth rate increases with increasing TOT X. In the absence of Cu, variation of TOT X produces no effect on growth. Discuss these results with respect to:

(a) Equilibrium properties of the CuX chelate system;

(b) Kinetics of CuX formation and dissociation in comparison to rate of phytoplankton growth.

6.7 Sketch a titration curve for titrating Ca^{2+} with EDTA at a pH of 10. Show the equilibrium composition of CaY^{2-}, Ca^{2+}, and uncomplexed Y as a function of pCa. (H_4Y = EDTA.)

6.8 Is $BaSO_4$ appreciably soluble ($S > 10^{-3} M$) in a $10^{-1} M$ solution of citrate of pH = 8? *Information:* Acidity constants of citric acid (H_4L) $pK_1 = 3.0$, $pK_2 = 4.4$, $pK_3 = 6.1$, $pK_4 \simeq 16$. BaHL is the only complex to be considered.

$$\frac{[BaHL]}{[Ba^{2+}][HL]} = 10^{2.4}$$

$$[Ba^{2+}][SO_4^{2-}] = 10^{-9}$$

ANSWERS TO PROBLEMS

6.1 The problem may be approached in a fashion similar to that given in Example 6.1. The basic equation to start with is

$$Cr_T = [CrO_4^{2-}] + [HCrO_4^-] + 2[Cr_2O_7^{2-}] + 2[HCr_2O_7^-]$$

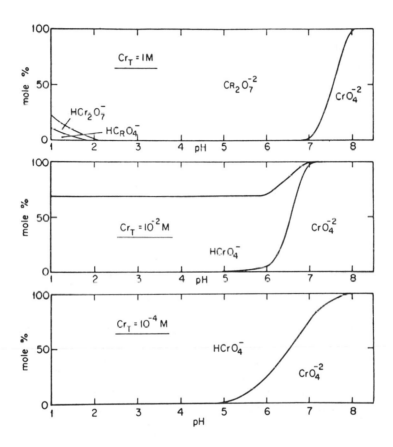

6.2 pH = 1.7.

6.3 $\dfrac{C_B - C_A + [H^+] - [OH^-]}{\text{mol Me(OH)}_n} = \text{mean charge}.$

6.4
$Zn^{2+} + H_2O = ZnOH^+ + H^+; *K_1$
$2Zn^{2+} + H_2O = Zn_2(OH)^{3+} + H^+; *\beta_{12}$
$Zn^{2+} + CN^- = Zn(CN)^+; \beta_1$
$Zn^{2+} + 2CN^- = Zn(CN)_2; \beta_2$
$Zn^{2+} + 3CN^- = Zn(CN)_3^-; \beta_3$
$Zn^{2+} + 4CN^- = Zn(CN)_4^{2-}; \beta_4$
$H^+ + CN^- = HCN$

Chemical species toxic to fish are primarily HCN, Zn^{2+}, $ZnOH^+$.

$$Zn_T = [Zn^{2+}] + \sum_{n=1}^{4} [Zn(CN)_n^{2-n}] + [ZnOH^+] + 2[Zn_2OH^{3+}]$$

$$CN_T = [HCN] + [CN^-] + \sum_{n=1}^{4} n[Zn(CN)_n^{2-n}]$$

6.8 No. The solubility is increased from $10^{-4.5}$ to $10^{-3.8}$ M. The calculation is made very simple by realizing that $[HL]$ is much larger than any other L-bound species; hence $\alpha_3 \simeq [HL]/L_T \simeq 1.0$, and $\alpha_{Ba(HL)} = [Ba^{2+}]/Ba_T = \{1 + 10^{2.4}[HL]^{-1} \simeq 10^{-1.4}$. The conditional solubility product for $BaSO_4$ becomes $P = [Ba_T][SO_4^{2-}] = K_{s0}\alpha_{Ba(HL)}^{-1} = 10^{-7.6}$.

APPENDIX: THE VARIOUS SCALES FOR EQUILIBRIUM CONSTANTS, ACTIVITY COEFFICIENTS, AND pH

In dealing with equilibria in natural waters, we wanted to give above all a feeling of the power of approach. In order not to overwhelm the reader with a large number of intricate details, we attempted to make nonideality corrections for electrolyte solutions simple and effective. The objective of this appendix is to review the various equilibria conventions usable for different natural water media—especially to compare the available conventions for describing (and measuring) pH and ionic equilibria in seawater—and to give an introduction to the ionic interaction theory, which is an expedient alternative and complementary approach to the ion association theory.

Procedures for finding the equilibrium distribution of species are based upon the principle that at equilibrium the total free energy of the system is at a minimum. This total free energy is the sum of the contributions from each of the constituent chemical species in the system; the contribution of each species depends on its standard free energy of formation, its activity, and the temperature and pressure of the system.

Equilibrium Constants and Activity Conventions

As discussed before (see Table 2.3 and Figure 3.2), three types of equilibrium constants are in common use:

1 Thermodynamic constants based on activities (rather than concentrations), the activity scale being based on the infinite dilution reference state.

2 Apparent or stoichiometric equilibrium constants expressed as concentration quotients and valid for a medium of given ionic strength.

3 Conditional constants that hold only under specified experimental conditions (e.g., at a given pH).

Equilibrium constants in the form of concentration quotients are just as thermodynamically valid as the traditional thermodynamic constants, the main difference being the choice of activity scale and reference state (see section 2.9).

For the reference state on the infinitely dilute solution scale the activity coefficient of a species approaches unity as the concentrations of all the solutes approach zero:

$$f_A \to 1 \quad \text{as} \quad \left(c_A + \sum_i c_i \right) \to 0$$

An alternative increasingly used convention is defined such that the activity coefficient of a species approaches unity as its concentration approaches zero in the medium of given ionic strength:

$$f'_A \to 1 \quad \text{as} \quad c_A \to 0$$

in a solution where the total concentration is still $\sum_i c_i$ and $\{H_2O\} = 1$. Activity coefficients of a species on this scale are also close to unity as long as the species concentration is small in comparison to those of the medium ions. Compilations of equilibrium constants [1,2] usually give data for both thermodynamic equilibrium constants and apparent constants valid for a medium of given ionic strength. Complicated ionic equilibria can only be studied quantitatively in an ionic medium of high total molarity or molality (solutions are adjusted by the addition of an indifferent electrolyte, e.g., $NaClO_4$) [3]. By maintaining an ionic medium with a concentration ca. 10 times larger than those of the reacting species, usually no correction terms for activity factor changes are necessary. (In addition, constant ionic media "swamp out" changes in the liquid junction potential of galvanic cells.)

For computational approaches it is possible to use thermodynamic equilibrium constants in conjunction with activities. In order to do this and to relate the constants to concentrations, the values of single-ion activity coefficients must be known. Alternatively, apparent equilibrium constants valid for the medium of particular interest (or a closely similar one) or constants that have been corrected for the medium under consideration can be used in conjunction with concentrations. Nonthermodynamic assumptions are involved in either case.

In our equilibrium calculations so far, we have favored the following approaches:

1   *In dilute solutions* ($I < 10^{-2}$ M), that is in *fresh waters*, our calculations are usually based on the infinite dilution activity convention and thermodynamic constants. In these dilute electrolyte mixtures, deviations from ideal behavior are primarily caused by *long-range electrostatic interactions*. The Debye–Hückel equation or one of its extended forms (see Table 3.3) is assumed to give an adequate description of these interactions and to define the properties of the ions. Correspondingly, individual ion activities are estimated by means of individual ion activity coefficients calculated with the help of the

Güntelberg or Davies equation [(3) and (4) of Table 3.3]; or it is often more convenient to calculate, with these activity coefficients, a concentration equilibrium constant valid at a given I,

$$K = \frac{\{AB\}}{\{A\}\{B\}}$$

$$= \frac{[AB]}{[A][B]} \frac{f_{AB}}{f_A f_B} \tag{2}$$

$$= K^c \frac{f_{AB}}{f_A f_B}$$

2   In more concentrated solutions and *in seawater*, our calculations were usually based on the ionic medium scale; that is, we used for the evaluation of the concentration of coexistent species at equilibrium apparent equilibrium constants (valid for the medium under consideration) expressed in concentration terms, for example, for seawater,

$$K_{s0}^{sw} = [Ca^{2+}][SO_4^{2-}] \tag{3}$$

$$K_{s0}^{'sw} = [Ca_T][SO_{4_T}^{2-}] \tag{4}$$

$$K_2^{'sw} = \frac{\{H^+\}[CO_{3_T}^{2-}]_T}{[HCO_{3_T}^-]} \tag{5}$$

The subscript T refers to total (free plus associated) ions, (e.g., $[HCO_{3_T}^-]$ = $[HCO_3^-] + [NaHCO_3^0] + [CaHCO_3^+] + [MgHCO_3^-] + 2[Mg(HCO_3^-)_2]$) [4].

By using "medium-bound" constants, we have bypassed activity coefficients in stoichiometric calculations. We can only do so if equilibrium constants valid for seawater or the medium of the interest have been determined. If such constants are not known, we are forced to use thermodynamic constants and estimated activity coefficients.†

Activity coefficients in most concentrated solutions reflect deviation from ideal behavior because of (1) the general electric field of the ions, (2) solute–water interactions, and (3) specific ionic interactions (association by ion-pair and complex formation). None of the major cations of seawater appears to interact significantly with chloride to form ion pairs; hence activity coefficients in these solutions appear to depend primarily upon the ionic strength modified by the extensive hydration of ions [5]. Thus synthetic solutions of these chlorides provide reference solutions in obtaining activity coefficients of the cations. The

---

† Pytkowicz writes: "Thermodynamic constants give a superficial appearance of convenience because they do not depend upon the composition of the medium, while apparent constants do. However, in the application of thermodynamic constants to the calculation of concentrations in concentrated multi-electrolyte solutions, activity coefficients have to be determined as a function of composition thereby cancelling what may have seemed to be an advantage."

single activity coefficients for free ions in seawater can be obtained from mean activity coefficient data in chloride solutions at the corresponding ionic strength by various ways.

In the *mean-salt method*,† the behavior of KCl in solution is the standard basis for obtaining individual ion activity coefficients [6]. Various lines of evidence indicate that f_{K^+} and f_{Cl^-} have similar values, that is, as an approximation,

$$f_{\pm(KCl)} - (f_K f_{Cl})^{1/2} = f_{K^+} = f_{Cl^-} \tag{6}$$

With the use of equation 6, a table of values for other ions can be built up from the appropriate mean ion activity coefficients, for example, for a monovalent chloride,

$$f_{\pm MCl} - (f_{M^+} f_{Cl^-})^{1/2} = (f_M \cdot f_{\pm KCl})^{1/2} \tag{7}$$

$$f_{M^+} = \frac{f_{\pm MCl}^2}{f_{\pm KCl}} \tag{8}$$

For a bivalent chloride,

$$f_{\pm MCl_2} = (f_{M^{2+}} f_{Cl^-}^2)^{1/3} = (f_{M^{2+}} f_{\pm KCl}^2)^{1/3} \tag{9}$$

$$f_{M^{2+}} = \frac{f_{\pm MCl_2}^3}{f_{\pm KCl}^2} \tag{10}$$

For salts of K^+ other than Cl^- the reverse relation can be used, for example,

$$f_{SO_4^{2-}} = \frac{f_{\pm K_2SO_4}}{f_{\pm KCl}^2} \tag{11}$$

For a salt like $CuSO_4$, $f_{Cu^{2+}}$ can be obtained, since

$$f_{\pm CuSO_4} = (f_{Cu^{2+}} f_{SO_4^{2-}})^{1/2} \tag{12}$$

From equations 11 and 12 one obtains

$$f_{Cu^{2+}} = \frac{f_{\pm CuSO_4}^2 f_{\pm KCl}^2}{f_{\pm K_2SO_4}^3} \tag{13}$$

The MacInnes convention has come under criticism recently, and other methods have been proposed [5]. Scales of ionic activity at ionic strengths above 0.1 have also been developed based on considerations of the degree of hydration of the ions involved [5].

Ion Association and Activity Coefficients. The activity coefficients for single free anions—SO_4^{2-}, HCO_3^-, CO_3^{2-}—cannot be estimated as well as those of the cations‡ because solutions of their salts show association with major seawater

---

† Often called the MacInnes convention [D. A. MacInnes, *J. Chem. Soc.*, **41**, 1068 (1919)].

‡ As Table 6.9 illustrates, most of the cations in seawater are present as free aquo cations. The concentration of associating SO_4^{2-}, HCO_3^-, and CO_3^{2-} is much smaller than that of major cations.

cations. In order to understand the properties of seawater, information on both the nature of the species and the extent of such associations is needed. Many difficulties are involved in evaluation of the degree of association between a cation and a ligand even in electrolyte solutions less complicated than seawater. It is difficult to separate the effects of ionic strength and of ion association. Either the ion association is known from other experiments or the activity coefficient effect is known from other experiments (both involve nonthermodynamic assumptions) [7]. The main activities of salts (NaCl, KCl, etc.) may be measured in mixed electrolytic solutions using ion-selective glass, membrane, or amalgam electrodes.† A decrease in activity caused by the addition of another electrolyte at given ionic strength (e.g., a marked decrease in the activity of NaCl consequent on the addition of carbonate or bicarbonate) may be explained in terms of a model in which ion pairs ($NaHCO_3$ or $NaCO_3^-$) are formed. Such a model is self-consistent when the formation constants are relatively independent of the composition at constant ionic strength [7].

Alternatively the effects of different ionic media on equilibria (such as carbonate protonation) can be determined. For example, the acidity constant of HCO_3^- is increased by the presence of Mg^{2+} ions. The lowering of the pK value in such solutions and in seawater is caused by the formation of carbonate ion pairs [6,8].

Because there is no method by which unequivocal structures of the species present may be obtained, ion association is a phenomenological concept. Ion-pair formation is involved to explain deviations from normal behavior.‡ However, the impossibility of knowing unambiguously the relevant activity coefficients in seawater implies that the concept of normal behavior is not clearly defined [9].

In Table A.6.1 activity coefficients measured in single salt solutions are compared with those measured in seawater and those calculated from an association model. We have to distinguish between total activity coefficients [4] (cf. equations 3 and 4)

$$^T f_A = \frac{\{A\}}{[A_T]} \tag{14}$$

and free activity coefficients

$$^F f_A = \frac{\{A\}}{[A_{free}]} \tag{15}$$

---

† For example, R. D. Lanier, *J. Phys. Chem.*, **69**, 3992 (1965); M. Thompson, *Science*, **153**, 966 (1966); R. M. Pytkowicz and D. R. Kester, *Limnol. Oceanogr.*, **14**, 686 (1969); J. N. Butler and R. Huston, *J. Phys. Chem.*, **24**, 2976 (1970); P. G. Christenson and J. M. Gieskes, *J. Chem. Eng. Data*, **16**, 398 (1971).

‡ Nevertheless ion pairs do indeed exist as shown for example by raman spectroscopy [F. P. Daly et al., *J. Phys. Chem.*, **76**, 3664 (1972)] and by sound attenuation (e.g., $MgSO_4^0$ in seawater) [F. H. Fisher, *Science* **157**, 823 (1967)].

TABLE A.6.1 COMPARISON OF ION ACTIVITY COEFFICIENTS AT 25°C (1 atm) AND EFFECTIVE IONIC STRENGTH $I = 0.7\ M$

Constituent	"Free" Activity Coefficient — Measured in Single-Salt Solutions[a,b]	"Total" Activity Coefficient in Seawater — Measured in Seawater	"Total" Activity Coefficient in Seawater — Calculated from Association Model	Specific Ion Interaction[c] (1)	(2)	(3)
Na^+	0.71	0.67[d], 0.70[f]	0.70[e]	0.68	0.65	0.64
K^+	—	0.60[g], 0.61[b]	0.62[e]	0.63	0.62	0.61
Mg^{2+}	—	0.26[i]	0.25[e]	0.23	0.22	0.22
Ca^{2+}	0.26	0.20[j]	0.24[e]	0.21	0.20	0.21
HCO_3^-	0.68	0.55[j]	0.51[e]	—	—	—
CO_3^{2-}	0.20	0.021[j]	0.021[e]	—	—	—
SO_4^{2-}	—	0.11[k]	0.068[e], 0.09[l]	0.1	0.12	0.13

[a] Garrels and Thompson (see footnote g) have used the following activity coefficients of individual species in sea water: Na^+, 0.76; K^+, 0.64; Mg^{2+}, 0.36; Ca^{2+}, 0.28; Cl^-, 0.64; CO_3^{2-}, 0.20; SO_4^{2-}, 0.12; $NaCO_3^- = NaSO_4^- = MgHCO_3^+$, etc., 0.68; $NaHCO_3^0 = MgCO_3^0 = CaCO_3^0 = MgSO_4^0 = CaSO_4^0$, 1.13. Pytkowicz and Kester (1971) have pointed out that uncharged ion pairs are dipolar ions, and they consider an activity coefficient of about 0.8 for $MgSO_4^0$ more appropriate.

[b] D. R. Kester and R. M. Pytkowicz, *Limnol. Oceanogr.*, **14**, 586 (1969).

[c] (1) Brønsted–Guggenheim model including interactions between ions of like charge [J. V. Leyendekkers, *Mar. Chem.*, **1**, 75 (1975)]; (2) Brønsted–Guggenheim model [M. Whitfield, *Mar. Chem.*, **1**, 251 (1973)]; (3) modified Pitzer equations [M. Whitfield. *Mar. Chem.*, **3**, 197 (1975)].

[d] R. F. Platford, *J. Mar. Res.*, **23**, 55 (1965).

[e] R. A. Berner, *Chemical Sedimentology*, McGraw-Hill, New York, 1971, p. 48.

[f] R. F. Garrels, in *Glass Electrodes*, Dekker, New York, 1967, p. 344.

[g] R. M. Garrels and M. Thompson, *Amer. J. Sci.*, **260**, 57 (1962).

[h] P. G. Mangelsdorf, Jr., and T. R. S. Wilson, *J. Phys. Chem.*, **75**, 1418 (1971).

[i] M. E. Thompson, *Science*, **153**, 866 (1966).

[j] R. A. Berner, *Geochim. Cosmochim. Acta*, **29**, 947 (1965).

[k] R. F. Platford and T. Dafoe. *J. Mar. Res.*, **23**, 68 (1965).

[l] N. van Breeman, *Geochim. Cosmochim. Acta*, **37**, 101 (1972).

411

The relation between ^Tf_A and ^Ff_A is the distribution coefficient

$$\alpha_A = \frac{[A_{free}]}{[A_T]} = \frac{^Ff_A}{^Tf_A} \tag{16}$$

which can be explained by the association model. For example, α_{Na} in seawater is given by

$$\alpha_{Na} = \frac{[Na_{free}^+]}{[Na^+]_{free} + [NaHCO_3] + [NaCO_3^-] + [NaSO_4^-]}$$

$$= (1 + K_{HCO_3}[HCO_3^-] + K_{NaCO_3}[CO_3^{2-}] + K_{NaSO_4}[SO_4^{2-}])^{-1} \tag{17}$$

The Specific Ionic Interaction Model as an Alternative and Complement to the Ion Association Model

As we have seen, the effect of long-range electrostatic interactions between the ions on the activity coefficients can be predicted fairly accurately in most cases for ionic strengths below 0.01 M. Deviations from the ideal Debye–Hückel theory at higher ionic strengths are attributed mostly to short-range interionic forces. The ion association model (often referred to as the Bjerrum ion association hypothesis) discussed above assumes that the deviations from the Debye–Hückel theory are primarily caused by the relatively strong binding of counterions to form ion pairs. An alternative procedure that can be employed in the evaluation of several thermodynamic properties of aqueous solutions, including the activity coefficients and osmotic coefficients of the individual ions in solution, is the Brønsted–Guggenheim ion interaction model [10]. In this model specific interactions among the various ions are considered and a thermodynamically mutually consistent set of formulas proposed to deduce—over a broad range of ionic strengths—the thermodynamic properties of mixed electrolytes from those of single electrolytes. These formulas contain parameters called interaction coefficients B_{MX} for each combination of a cation M and an anion X. The value of each interaction coefficient B_{MX} is determined from measurements on a solution containing only the electrolyte whose ions are M and X. The value of each interaction coefficient so determined can then be used for estimating the thermodynamic properties of mixed electrolytes. It is assumed that the multiple interactions with a specific ion are additive. Thermodynamic properties of mixed electrolytes and of seawater can be treated without explicitly considering ion association (ion pairs);† this approach has yielded reasonable results for the activity coefficients and other thermodynamic properties of seawater [11–14].

---

† "Thermodynamics is not compelled to take cognizance of the various molecular species which may exist in a system, particularly when the existence of such species cannot be absolutely demonstrated." (G. N. Lewis and M. Randall, *Thermodynamics*, revised by K. S. Pitzer and L. Brewer, McGraw-Hill, New York, 1961 p. 272).

The simplest procedure based on the Brønsted–Guggenheim hypothesis [10] gives rise to an equation for the mean ion activity coefficient that consists only of an electrostatic term and a statistical term linear in the salt concentration [13a]:

$$\log \gamma_{\pm MX} = \log \gamma_{EL} + \frac{\nu_M}{\nu_M + \nu_X} \sum_{X^*} B_{MX^*}[X]_T^* + \frac{\nu_X}{\nu_M + \nu_X} \sum_M B_{MX}^*[M]_T^* \quad (18)$$

where ν_M and ν_X are the number of cations and anions per "molecule" of electrolyte, $[M]_T$ and $[X]_T$ are the total molalities of cations and anions, B_{MX} is a coefficient of interaction between cation M and anion X, and $\log \gamma_{EL} = -|Z_M Z_X| A I^{1/2}/(1 + I^{1/2})$, where Z_M and Z_X are the charge numbers. The quantities with an asterisk are varied during the summation. Activity coefficients of individual ions in a multicomponent system can be formulated as follows [13a]:

$$\log \gamma_X = \left| \frac{Z_X}{Z_M} \right| \log \gamma_{EL} + \sum_M B_{M^*X}[M]_T^* \quad (19)$$

$$\log \gamma_M = \left| \frac{Z_M}{Z_X} \right| \log \gamma_{EL} + \sum_X B_{MX^*}[X]_T^* \quad (20)$$

That is, an interaction term (related to the molality of the solution) is added to the Debye–Hückel function. B values may be calculated from the tabulations in Pitzer and Brewer [14].

Pitzer and his co-workers [15,16] have proposed a more detailed, but at the same time more complex, approach. Whitfield [13,17] has applied these equations to seawater and has shown that this model gives good agreement with available experimental data for the osmotic coefficient and for the mean ion activity coefficient of the major electrolyte components. The results obtained yield numerical results similar to the predictions of the ion association model (see Table 6.A.1).

One advantage of the ion interaction theory is that it can be applied to solutions of different salinities, that is, to brines, seawaters with different salinities, and estuarine waters. While the ionic medium method provides a very simple solution to many problems, especially for the speciation of constituents in the open ocean, it cannot be readily applied to solutions of different salinity; that is, brines, seawater, and estuarine waters must be treated as separate solvents.

It is not possible at present to decide which model presents the most accurate description of seawater. The ion association model has been successful in dealing with the acid–base equilibria in seawater. The effects of the CO_2 system cannot be considered at present because data on the interaction coefficients are incomplete.†

---

† F. C. Cepeda, Ph.D. thesis, California Institute of Technology, Pasadena, 1977, has evaluated, $MHCO_3$ virial coefficients and determined the temperature dependence of the activity coefficient of the HCO_3^- ion.

Extension of the ion interaction model to minor components meets with some difficulty. Many heavy metals form relatively stable ion pairs or inorganic complexes (e.g., $CaHPO_4$, $CuCO_3$, $AgCl_2^-$, $CdCl_2$, $ZnCl^+$) whose existence may be of great biological and geological importance. Whitfield [17] has extended interaction models for seawater to encompass trace constituents. The two models (ion association and ion interaction) are best seen as providing complementary insights into the chemical nature of seawater, since each treats phenomena not covered by the other [13].

We should also be aware that our insight into the chemical speciation of minor components in natural waters is less hampered by the uncertainty of activity coefficients than by the uncertainty of hydrolysis and complex formation constants. For example, the percentage speciation of free Zn^{2+} varies from 68% [18] to 16% [19], depending on the stability constants selected in the speciation model.

pH Concepts in Seawater

The difficulties encountered in establishing activity coefficients also relate to the problem of defining pH in seawater. Accurate assessment of the thermodynamic properties of seawater depends on measurements of pH. At the present time, there is no general agreement on a single standard pH scale for seawater.

Conceptually we again depend on definitions related to the infinite dilution scale and the constant ionic medium scale. The NBS pH scale endorsed by IUPAC (see Sections 3.4 and 7.9) is based on the infinitely dilute aqueous reference state; the pH is determined relative to that of a standard buffer (whose pH has been estimated in terms of $-\log\{H^+\}$) from measurements in cells with a liquid junction. In dilute solutions ($I < 0.1$) the measured pH corresponds to $-\log\{H^+\}$ to within ± 0.02. This is not so in seawater. Because of our ignorance of liquid junction potentials, pH values measured in seawater by the NBS procedure do not approach $-\log\{H^+\}$ on an infinite dilution scale ($\{H^+\}/[H^+] \rightarrow 1$ when $[H^+] \rightarrow 0$ in pure water). The pH values so measured are on a different activity scale; they may be used to characterize and compare seawater samples and so serve as an index of acid–base balance and speciation.

The equilibrium constants K' given in Tables 4.8 to 4.10, originally determined by Lyman, for example,

$$K'_2 = \frac{\{H^+\}[CO_{3_T}^{2-}]}{[HCO_{3_T}^-]}$$

have been determined by measuring $\{H^+\}$ by the NBS procedure. These constants are internally consistent and can be used to calculate concentrations of carbonate species of a sample, but the pH should be measured with a cell containing the same reference electrode as that used by Lyman, that is, by the recommended procedure and calibrated with the same buffer.

The hydrogen ion concentration or molality in seawater, m_H, is a clearly defined concept free of the uncertainties attached to the use of single-ion activities [20]. But establishing an experimental method for the measurement of m_H in seawater is not without difficulties.

Using seawater as a constant ionic medium we can define pH by the relationship

$$pm_H = -\log m_H \qquad (21)$$

where m_H is the molality of free hydrogen ion in the seawater. $\{H^+\}/[H^+] \to 1$ when $[H^+] \to 0$ in the ionic medium. Correspondingly an electrode system can be calibrated in terms of equation 21 when a salt solution such as NaCl contains a known concentration of strong acid (e.g., HCl), the observed potentiometer reading being compared with m_{H^+}. If we attempt to carry out such a procedure with a seawater medium, let us say with a synthetic seawater free of carbonates and borates, we must consider that some of the H^+ ions become associated with SO_4^{2-}. In other words, we are still faced with the problem of distinguishing between free H^+ and total H^+:

$$[H_T^+] = [H^+] + [HSO_4^-] \qquad (22)$$

The electrode system can be calibrated in terms of $-\log[H_T^+]$. This is the *Hansson pH scale* [21]:

$$-\log[H_T^+] = -\log([H^+] + [HSO_4^-]) \qquad (23)$$

which Hansson defined on the M_w concentration scale (mol kg^{-1} seawater). $\{H\}/([H^+] + [HSO_4^-]) \to 1$ when $[H^+] + [HSO_4^-] \to 0$ in the seawater medium. Using this pH scale, Hansson [22] has determined a new set of acidity constants for carbonic acid and boric acid in seawater.

If the concentration of free sulfate ion in the medium does not change, the relationship between equations 21 and 23 can be established:

$$[H_T^+] = [H^+] + [HSO_4^-] = [H^+](1 + K_2^{-1}[SO_4^{2-}]) \qquad (24)$$

where K_2^{-1} is the acidity constant of HSO_4^- in seawater medium. The last term in equation 24 is ca. 1.4 for seawater in mildly alkaline solution.

Therefore pH values measured on Hansson's scale are ca. 0.15 ($=\log 1.4$) pH units lower than pm_H:

$$p[H_T^+] \cong pm_H - 0.15 \qquad (25)$$

pm_H values can in principle be measured if the calibration for a given m_{H^+} is carried out in a synthetic seawater medium free of SO_4^{2-}.

Obviously we encounter the same problems in defining $\{OH^-\}$: The total concentration of OH^- in seawater is given by

$$[OH_T^-] = [OH^-] + [MgOH^+] + [CaOH^+] \qquad (26)$$

where, for seawater at $-\log[H_T^+] = 8$,

$$p[OH_T^-] = p[OH^-] - 0.45.$$

Thus the formation of HSO_4^- and of $MgOH^+$ ($[CaOH^+]$ is negligible) explains the differences in ion products between 0.7 M_w NaCl and standard seawater (25°C) [21]:

$$0.7 \ M_w \ \text{NaCl: } p[H^+] + p[OH^-] = 13.77$$

$$\text{Standard seawater: } p[H_T^+] + p[OH_T^-] = 13.19$$

REFERENCES

1 L. G. Sillén and A. E. Martell, *Stability Constants*, Chemical Society, Special Publications Nos. 17 and 25, London, 1964 and 1971.

2 R. M. Smith and A. E. Martell, *Critical Stability Constants*, Plenum, New York, 1976.

3 G. Biedermann, in *The Nature of Seawater*, E. D. Goldberg, Ed., Dahlem Konferenzen, Berlin, 1975, p. 339.

4 R. M. Pytkowicz, *Geochem. J.*, **3**, 184 (1969).

5 R. G. Bates, B. R. Staples, and R. A. Robinson, *Anal. Chem.*, **42**, 867 (1970).

6 R. M. Garrels and C. L. Christ, *Solutions, Minerals and Equilibria*, Harper and Row, New York, 1965; p. 58.

7 J. N. Butler and R. Huston, *J. Phys. Chem.*, **24**, 2976 (1970).

8 R. M. Pytkowicz and J. E. Hawley, *Limnol. Oceanogr.*, **19**, 223 (1974).

9 J. C. Hindman and J. C. Sullivan, in *Coordination Chemistry*, A. E. Martell, Ed., Van Nostrand Reinhold, New York, 1971, p. 409.

10 E. H. Guggenheim and J. C. Turgeon, *Trans. Farad. Soc.*, **51**, 747 (1954).

11 R. A. Robinson and R. H. Wood, *J. Sol. Chem.*, **1**, 48 (1972).

12 J. V. Leyendekkers, *Mar. Chem.*, **1**, 75 (1973).

13a M. Whitfield, *Mar. Chem.*, **1**, 251 (1973).

13b M. Whitfield, *Mar. Chem.*, **3**, 197 (1975).

14 K. S. Pitzer and L. Brewer, in *Thermodynamics*, G. N. Lewis and M. Randall, Eds., McGraw-Hill, New York, 1961, p. 570.

15 K. S. Pitzer, *J. Phys. Chem.*, **77**, 268 (1973).

16 K. S. Pitzer and J. J. Kim, *J. Amer. Chem. Soc.*, **98**, 5701 (1974).

17 M. Whitfield, *Geochim. Cosmochim. Acta*, **39**, 1545 (1975).

18 W. Stumm and P. A. Brauner. In *Chemical Oceanography*, J. P. Riley and G. Skirrow, Eds., Vol. 1, 2nd ed., Academic, New York, 1975, p. 231.

19 D. Dryssen and M. Wedborg, in *The Sea*, Vol. 5, E. D. Goldberg, Ed., Wiley-Interscience, New York, 1977, p. 193.

20 R. G. Bates, in *The Nature of Seawater*, Dahlem Konferenzen, Berlin, 1975.

21 I. Hansson, *Deep Sea Res.*, **20**, 479 (1973).

22 I. Hansson, *Deep Sea Res.*, **20**, 461 (1973).

READING SUGGESTIONS

Bates, R. G., "pH-Scales for Seawater." In *The Nature of Seawater*, E. D. Goldberg, Ed., Dahlem Konferenzen, Berlin, 1975, pp. 313–338.

Hansson, I., "A New Set of pH Scales and Standard Buffers for Seawater." *Deep Sea Res.*, **20**, 479–491 (1973).

MacIntyre, F., "Concentration Scales: A Plea for Physico-chemical Data," *Mar. Chem.*, **4**, 205–224 (1976). (The four concentration scales molal, molar, mol fraction, and "mokal" (mol kg^{-1} seawater) are considered; conversion ratios are given for multicomponent ionic mixtures and as functions of salinity.)

Millero, F. J., "Thermodynamic Models for the State of Metal Ions in Seawater." In *The Sea*, Vol. 7, E. D. Goldberg et al., Eds., Wiley-Interscience, New York, 1977, pp. 653–693.

Pytkowicz, R. M., "Use of Apparent Equilibrium Constants in Chemical Oceanography, Geochemistry and Biochemistry," *Geochem. J.*, **3**, 181–184 (1969). (Illustrates that ion association and pH considerations make it often unadvisable to use thermodynamic constants in concentrated multielectrolyte solutions.)

Sillén, L. G., "Master Variables and Activity Scales." In *Equilibrium Concepts in Natural Water Systems*, W. Stumm, Ed., American Chemical Society, Advances in Chemistry Series, No. 67, Washington, D.C., 1967, p. 45. (Compares the merits of infinite dilution activity scale and ionic medium activity scale.)

Whitfield, M., "Seawater as an Electrolyte Solution," In *Chemical Oceanography*, J. P. Riley and G. Skirrow, Eds., Vol. 1, 2nd ed., Academic, New York, 1975, pp. 44–162.

Whitfield, M., "The Extension of Chemical Models for Seawater to Include Trace Components," *Geochim. Cosmochim. Acta*, **39**, 1545–1557 (1975).

7

Oxidation and Reduction

7.1 INTRODUCTION

A functional description of the causal relationships effective in a natural water system must include parameters that characterize the influence of electrons on the environment.

In this chapter we stress the stability relations of pertinent redox (oxidation–reduction) components in natural water systems. However, one must be aware that concentrations of oxidizable or reducible species may be far from those predicted thermodynamically, because many redox reactions are slow. In the sea or in a lake there is a marked difference in redox environment between the surface in contact with the oxygen of the atmosphere and the deepest layers of the sediments. In between are numerous localized intermediate zones, resulting from imperfections in mixing or diffusion and from varying biological activities, none of which is truly at equilibrium. The need for biological mediation of most redox processes encountered in natural waters means that approaches to equilibrium depend strongly on the activities of the biota. Moreover, quite different oxidation–reduction levels, different from those prevalent in the overall environment, may be established within biotic microenvironments; diffusion or dispersion of products from the microenvironment into the macroenvironment may give an erroneous view of redox conditions in the latter. Also, because many redox processes do not couple with one another readily, it is possible to have several different apparent oxidation–reduction levels in the same locale. Therefore detailed, quantitative exposition of redox conditions and processes will depend ultimately on understanding the dynamics of aquatic systems—the rates of approach to equilibrium—rather than on describing the total or partial equilibrium compositions. Elementary aspects of biological mediation and a few kinetic considerations of some redox reactions will be discussed in the latter part of this chapter.

Equilibrium considerations can greatly aid attempts to understand in a general way the redox patterns observed or anticipated in natural waters. In all circumstances equilibrium calculations provide boundary conditions toward which the system must be proceeding. Moreover, partial equilibria (those involving some but not all redox couples) are approximated frequently, even though total equilibrium is not reached. In some instances active poising† of particular redox couples allows us to predict significant oxidation–reduction

† Buffering with regard to oxidation or reduction.

levels or to estimate properties and reactions from computed redox levels. Valuable insight is gained even when differences are observed between computations and observations. The lack of equilibrium and the need for additional information or more sophisticated theory are then made clear.

Additional difficulties occur with attempts to measure oxidation–reduction potentials electrochemically in aquatic environments. Values obtained depend on the nature and rates of the reactions at the electrode surface and are seldom meaningfully interpretable. Even when suitable conditions for measurement are obtained, the results are significant only for those components whose behavior is electrochemically reversible at the electrode surface.

7.2   REDOX EQUILIBRIA AND THE ELECTRON ACTIVITY

There is a conceptual analogy between acid–base and reduction–oxidation reactions. In a similar way that acids and bases have been interpreted as proton donors and proton acceptors, reductants and oxidants are defined as electron donors and electron acceptors. e stands for the electron; since it is negatively charged we could also write e^-. Because there are no free electrons, every oxidation is accompanied by a reduction, and vice versa; or an oxidant is a substance that causes oxidation to occur while being reduced itself.

$$O_2 + 4H^+ + 4e = 2H_2O \qquad \text{reduction}$$
$$4Fe^{2+} = 4Fe^{3+} + 4e \qquad \text{oxidation}$$

$$\overline{O_2 + 4Fe^{2+} + 4H^+ = 4Fe^{3+} + 2H_2O \qquad \text{redox reaction}}$$

The Oxidation State.   As a result of the electron transfer (mechanistically the transfer may occur as a transfer of a group that carries the electron), there are changes in the oxidation states of reactants and products. Sometimes, especially in dealing with reactions involving covalent bonds, there are uncertainties in the assignment of electron loss or electron gain to a particular element. The oxidation state (or oxidation number) represents a hypothetical charge that an atom would have if the ion or molecule were to dissociate. This hypothetical dissociation, or the assignment of electrons to an atom, is carried out according to rules. The rules and a few examples are given in Table 7.1. In this book roman numerals are used to represent oxidation states, and arabic numbers represent actual electronic charge. The concept of an oxidation state may often have little chemical reality, but the concept is extremely useful in discussing stoichiometry—as a tool for balancing redox reactions—and in systematic descriptive chemistry.

Sometimes the balancing of redox reactions causes difficulties. One of various approaches is illustrated in the following example.

TABLE 7.1 OXIDATION STATE

Rules for Assigning Oxidation States:

(1) The oxidation state of a monoatomic substance is equal to its electronic charge.

(2) In a covalent compound, the oxidation state of each atom is the charge remaining on the atom when each shared pair of electrons is assigned completely to the more electronegative of the two atoms sharing them. An electron pair shared by two atoms of the same electronegativity is split between them.

(3) The sum of oxidation states is equal to zero for molecules, and for ions is equal to the formal charge of the ions.

Examples:

Nitrogen Compounds		Sulfur Compounds		Carbon Compounds	
Substance	Oxidation States	Substance	Oxidation States	Substance	Oxidation States
NH_4^+	$N = -III,\ H = +I$	H_2S	$S = -II,\ H = +I$	HCO_3^-	$C = +IV$
N_2	$N = 0$	$S_8(s)$	$S = 0$	$HCOOH$	$C = +II$
NO_2^-	$N = +III,\ O = -II$	SO_3^{2-}	$S = +IV,\ O = -II$	$C_6H_{12}O_6$	$C = 0$
NO_3^-	$N = +V,\ O = -II$	SO_4^{2-}	$S = +VI,\ O = -II$	CH_3OH	$C = -II$
HCN	$N = -III,\ C = +II,\ H = +I$	$S_2O_3^{2-}$	$S = +II,\ O = -II$	CH_4	$C = -IV$
SCN^-	$S = -I,\ C = +III,\ N = -III$	$S_4O_6^{2-}$	$S = +2.5,\ O = -II$	C_6H_5COOH	$C = -2/7$
		$S_2O_6^{2-}$	$S = +V,\ O = -II$		

Example 7.1. Balancing Redox Reactions

Balance the following redox reactions: 1. Oxidation of Mn^{2+} to MnO_4^- by PbO_2, 2. oxidation of $S_2O_3^{2-}$ to $S_4O_6^{2-}$ by O_2.

1   Reactants: $Mn(II)$, $Pb(IV)$; Products: $Mn(VII)$, $Pb(II)$
    Oxidation: $Mn(II) = Mn(VII) + 5e$
    Reduction: $Pb(IV) + 2e = Pb(II)$

Half-reactions:

$$Mn^{2+} = Mn(VII) + 5e$$

$$Mn(VII) + 4O(-II) = MnO_4^-$$

$$4H_2O = 4O(-II) + 8H^+$$

$$\overline{\phantom{Mn^{2+} + 4H_2O = MnO_4^- + 8H^+ + 5e}}$$

$$Mn^{2+} + 4H_2O = MnO_4^- + 8H^+ + 5e \qquad (i)$$

$$Pb(IV) + 2e = Pb^{2+}$$

$$PbO_2 = Pb(IV) + 2O(-II)$$

$$2O(-II) + 4H^+ = 2H_2O$$

$$\overline{\phantom{PbO_2 + 4H^+ + 2e = Pb^{2+} + 2H_2O}}$$

$$PbO_2 + 4H^+ + 2e = Pb^{2+} + 2H_2O \qquad (ii)$$

Adding reactions i and ii:

$$2Mn^{2+} + 5PbO_2 + 4H^+ = 2MnO_4^- + 5Pb^{2+} + 2H_2O$$

2   Reactants: $S(II)$, $O(0)$; Products: $S(+2.5)$, $O(-II)$
    Oxidation: $2S(II) = 2S(+2.5) + e$
    Reduction: $O(0) + 2e = O(-II)$

Half reactions:

$$2S(II) = 2S(+2.5) + e$$

$$S_2O_3^{2-} = 2S(II) + 3O(-II)$$

$$2S(+2.5) + 3O(-II) = \tfrac{1}{2}S_4O_6^{2-}$$

$$\overline{\phantom{S_2O_3^{2-} = \tfrac{1}{2}S_4O_6^{2-} + e}}$$

$$S_2O_3^{2-} = \tfrac{1}{2}S_4O_6^{2-} + e \qquad (iii)$$

$$\tfrac{1}{2}O_2 + 2e = O(-II)$$

$$O(-II) + 2H^+ = H_2O$$

$$\overline{\phantom{\tfrac{1}{2}O_2 + 2H^+ + 2e = H_2O}}$$

$$\tfrac{1}{2}O_2 + 2H^+ + 2e = H_2O \qquad (iv)$$

Adding half reactions iii and iv:

$$2S_2O_3^{2-} + \tfrac{1}{2}O_2 + 2H^+ = S_4O_6^{2-} + H_2O$$

Electron Activity and pε

Aqueous solutions do not contain free protons and free electrons, but it is nevertheless possible to define relative proton and electron activities. The pH

$$pH = -\log\{H^+\} \tag{1}$$

measures the relative tendency of a solution to accept or transfer protons. In an acid solution this tendency is low, and in an alkaline solution it is high. Similarly we can define an equally convenient parameter for the redox intensity

$$p\varepsilon = -\log\{e\} \tag{2}$$

pε gives the (hypothetical) electron activity at equilibrium and measures the relative tendency of a solution to accept or transfer electrons.† In a highly reducing solution the tendency to donate electrons, that is, the hypothetical "electron pressure," or electron activity, is relatively large. Just as the activity of hypothetical hydrogen ions is very low at high pH, the activity of hypothetical electrons is very low at high pε. Thus a high pε indicates a relatively high tendency for oxidation. In equilibrium equations H^+ and e are treated in an analogous way. Thus oxidation or reduction equilibrium constants can be defined and treated similarly to acidity constants as shown by the equations in Table 7.2 where the corresponding relations for pH and pε are derived in a stepwise manner. In order to relate pε to redox equilibria, we recall first the relationship derived for pH and acid–base equilibria (left-hand side of Table 7.2). An electron transfer reaction, in analogy with a proton transfer reaction, can be interpreted in terms of two reaction steps [(2) and (3) in Table 7.2].

In parallel to the convention of arbitrarily assigning $\Delta G^\circ = 0$ for the hydration of the proton [(3a) in Table 7.2], we also assign a zero free energy change for the oxidation of $H_2(g)$ [(3b) in Table 7.2]. Equations (5a) and (5b) in Table 7.2 show that pH and pε are measures of the free energy involved in the transfer of 1 mol of protons or electrons, respectively.

Any oxidation or reduction can be written as a half-reaction, for example

$$IO_3^- + 3H_2O + 6e = I^- + 6OH^- \qquad K_3 \tag{3}$$

Such a reduction is always accompanied by an oxidation, for example, $3H_2O = \frac{3}{2}O_2(g) + 6H^+ + 6e$. Even though there are no free electrons in solution, we can formulate an equilibrium expression for the half-reaction (equation 3) (cf. Table 7.2)

$$\frac{\{I^-\}\{OH^-\}^6}{\{IO_3^-\}\{e\}^6} = K_3 \qquad \log K_3 = 26.1 \ (25°C)$$

---

† As we shall see, pε is related to the equilibrium redox potential E_H (volts, hydrogen scale) by $p\varepsilon = -\log\{e^-\} = E_H/2.3RTF^{-1}$ where R = gas constant, T = absolute temperature, F = the faraday (96,490 C mol^{-1}) ($RTF^{-1} = 0.059$ V mol^{-1} at 25°C).

TABLE 7.2 pH AND pε

pH = $-\log\{H^+\}$		pε = $-\log\{e\}$	
Acid-base reaction: $HA + H_2O = H_3O^+ + A^-$; K_1	(1a)	Redox reaction: $Fe^{3+} + \tfrac{1}{2}H_2(g) = Fe^{2+} + H^+$ K_1	(1b)
Reaction (1a) is composed of two steps:		Reaction (1b) is composed of two steps:	
$HA = H^+ + A^-$ K_2	(2a)	$Fe^{3+} + e = Fe^{2+}$ K_2	(2b)
$H_2O + H^+ = H_3O^+$ K_3	(3a)	$\tfrac{1}{2}H_2(g) = H^+ + e$ K_3	(3b)
According to thermodynamic convention: $K_3 = 1$		According to thermodynamic convention: $K_3 = 1$	
Thus:		Thus:	
$K_1 = K_2 = K_2K_3 = \{H^+\}\{A^-\}/\{HA\}$	(4a)	$K_1 = K_2 = K_2K_3 = \{Fe^{2+}\}/\{Fe^{3+}\}\{e\}$	(4b)
or		or	
$pH = pK + \log[\{A^-\}/\{HA\}]$	(5a)	$p\varepsilon = p\varepsilon^\circ + \log[\{Fe^{3+}\}/\{Fe^{2+}\}]$	(5b)
Since $pK = -\log K = \Delta G^\circ/2.3RT$		Since $p\varepsilon^\circ = \log K = -\Delta G^\circ/2.3RT$	
$pH = \Delta G^\circ/2.3RT + \log[\{A^-\}/\{HA\}]$	(6a)	$p\varepsilon = -\Delta G^\circ/2.3RT + \log[\{Fe^{3+}\}/\{Fe^{2+}\}]$	(6b)
or for the transfer of 1 mole of H^+ from acid to H_2O:		or for the transfer of 1 mole of e to oxidant from H_2:	
$\Delta G/2.3RT = \Delta G^\circ/2.3RT + \log[\{A^-\}/\{HA\}]$	(7a)	$-\Delta G/2.3RT = -\Delta G^\circ/2.3RT + \log[\{Fe^{3+}\}/\{Fe^{2+}\}]$	(7b)
For the general case where n protons are transferred:		For the general case where n electrons are transferred:	
$H_nB + nH_2O = nH_3O^+ + B^{-n}$ β^*		$ox + (n/2)H_2 = red + nH^+$; $ox + ne = red$; K^*	
$pH = (1/n)p\beta^* + (1/n)\log[\{B^{-n}\}/\{H_nB\}]$	(8a)	$p\varepsilon = (1/n)\log K^* + (1/n)\log[\{ox\}/\{red\}]$	(8b)
$pH = \Delta G/n2.3RT$		$p\varepsilon = p\varepsilon^\circ + (1/n)\log[\{ox\}/\{red\}]$;	(9b)
$= \Delta G^\circ/n2.3RT + (1/n)\log[\{B^{-n}\}/\{H_nB\}]$	(10a)	$p\varepsilon = -\Delta G^\circ/n2.3RT + (1/n)\log[\{ox\}/\{red\}]$	(10b)
$\Delta G = -nFE$ (E = acidity potential)	(11a)	$\Delta G = -nFE_H$ (E_H = redox potential)	(11b)
$pH = -E/(2.3RTF^{-1})$		$p\varepsilon = E_H/(2.3RTF^{-1})$	
$= -E^\circ/(2.3RTF^{-1}) + (1/n)\log[\{B^{-n}\}/\{H_nB\}]$	(12a)ᵃ	$= E^\circ/(2.3RTF^{-1}) + (1/n)\log[\{ox\}/\{red\}]$	(12b)ᵃ
Acidity potential:		Redox potential (Peters–Nernst equation):	
$E = E^\circ + (2.3RT/nF)\log[\{H_nB\}/\{B^{-n}\}]$	(13a)	$E_H = E_H^\circ + (2.3RT/nF)\log[\{ox\}/\{red\}]$	(13b)

ᵃ At 25°C, $2.3RTF^{-1} = 0.059$ (V mol⁻¹). From (10) and (12): 25°C, $p\varepsilon = E/0.059$, $p\varepsilon^\circ = E_H^\circ/0.059$.

423

for which $p\varepsilon$ can be calculated:

$$p\varepsilon = \tfrac{1}{6}\log K_3 + \tfrac{1}{6}\log \frac{\{IO_3^-\}^{1/6}}{\{OH^-\}\{I^-\}^{1/6}} \tag{4}$$

This can be rewritten as

$$p\varepsilon = p\varepsilon^\circ + \tfrac{1}{6}\log \frac{\{IO_3^-\}^{1/6}}{\{OH^-\}\{I^-\}^{1/6}} \tag{5}$$

where $p\varepsilon^\circ = \tfrac{1}{6}\log K$ and K is defined in terms of the equilibrium constant of the reduction (electron acceptance) reaction.

Equation 5 can be expressed more generally for the reaction

$$\sum_i n_i A_i + ne = 0 \qquad K$$

where A_i designates the participating species and n_i their numerical coefficients, positive for reactants and negative for products, in terms of the relations

$$p\varepsilon = p\varepsilon^\circ + \frac{1}{n}\log\left(\prod_i \{A_i\}^{n_i}\right) \tag{6}$$

and

$$p\varepsilon^\circ = \frac{1}{n}\log K \tag{7}$$

The quantity $p\varepsilon^\circ$ is the (relative) electron activity when all species other than the electron are at unit activity.

Equilibrium Calculations

So far we have treated the electron like a ligand in complex formation reactions. Indeed in *Stability Constants of Metal–Ion Complexes* the first ligand considered in Section I (inorganic ligands) is the electron, and the equilibrium constants listed there are for redox reactions corresponding to equation 6.

In subsequent examples we illustrate a few basic equilibrium calculations. For simplicity we assume activities approximately equal concentrations.

Example 7.2. The Formal Computation of $p\varepsilon$ Values

Calculate $p\varepsilon$ values for the following equilibrium systems ($25°C$, $I = 0$):

(i) An acid solution 10^{-5} M in Fe^{3+} and 10^{-3} M in Fe^{2+}.

(ii) A natural water at pH $= 7.5$ in equilibrium with the atmosphere ($p_{O_2} = 0.21$ atm).

(iii) A natural water at pH $= 8$ containing 10^{-5} M Mn^{2+} in equilibrium with $\gamma\text{-}MnO_2(s)$.

Stability Constants of Metal–Ion Complexes give the following equilibrium constants:

$$Fe^{3+} + e = Fe^{2+} \qquad K = \frac{\{Fe^{2+}\}}{\{Fe^{3+}\}\{e\}} \qquad \log K = 13.01 \quad \text{(i)}$$

$$\begin{aligned}\tfrac{1}{2}O_2(g) + 2H^+ + 2e \\ = H_2O(1)\end{aligned} \qquad K = \frac{1}{p_{O_2}^{1/2}\{H^+\}^2\{e\}^2} \qquad \log K = 41.55 \quad \text{(ii)}$$

$$\begin{aligned}\gamma\text{-}MnO_2(s) + 4H^+ + 2e \\ = Mn^{2+} + 2H_2O(1)\end{aligned} \qquad K = \frac{\{Mn^{2+}\}}{\{H^+\}^4\{e\}^2} \qquad \log K = 40.84 \quad \text{(iii)}$$

By using equations 1 and 2 the following $p\varepsilon$ values are obtained for the conditions stipulated

(i) $p\varepsilon = 13.01 + \log \dfrac{\{Fe^{3+}\}}{\{Fe^{2+}\}} = 11.01$

(ii) $p\varepsilon = 20.78 + \tfrac{1}{2}\log(p_{O_2}^{1/2}\{H^+\}^2) = 12.94$

(iii) $p\varepsilon = 20.42 + \tfrac{1}{2}\log \dfrac{\{H^+\}^4}{\{Mn^{2+}\}} = 6.92$

Example 7.3. Equilibrium Composition of Simple Solutions

Calculate the equilibrium composition of the following solutions (25°C, $I = 0$), both in equilibrium with the atmosphere ($p_{O_2} = 0.21$ atm):

(i) An acid solution (pH $= 2$) containing a total concentration of iron, $Fe_t = 10^{-4}\ M$.

(ii) A natural water (pH $= 7$) containing Mn^{2+} in equilibrium with $\gamma\text{-}MnO_2(s)$. The equilibrium constants were given in Example 7.2. The redox equilibria are defined by the conditions given (p_{O_2} and pH). Hence $p\varepsilon$ for both can be calculated with the equation

$$p\varepsilon = 20.78 + \tfrac{1}{2}\log(p_{O_2}^{1/2}\{H^+\}^2)$$

The following values are obtained:

(i) $p\varepsilon = 18.43$
(ii) $p\varepsilon = 13.42$

Correspondingly we find

$$p\varepsilon = 13.01 + \log \frac{\{Fe^{3+}\}}{\{Fe^{2+}\}}$$

and

$$p\varepsilon = 20.42 + \tfrac{1}{2}\log \frac{\{H^+\}^4}{\{Mn^{2+}\}}$$

for solution i $\{Fe^{3+}\}/\{Fe^{2+}\} = 10^{5.4}$ or $\{Fe^{3+}\} = 10^{-4}$ and $\{Fe^{2+}\} = 10^{-9.4}$; and for solution ii $\{Mn^{2+}\} = 10^{-14}$ M.

There are, of course, other approaches in calculating the equilibrium composition; for example, we may first compute the equilibrium constants of the overall redox reactions:

$\tfrac{1}{2}O_2(g) + 2H^+ + 2e = H_2O(l)$	$\log K = 41.55$
$2Fe^{2+} = 2Fe^{3+} + 2e$	$\log K = -26.02$
$\tfrac{1}{2}O_2(g) + 2Fe^{2+} + 2H^+ = 2Fe^{3+} + H_2O(l)$	$\log K = 15.53$

$\tfrac{1}{2}O_2(g) + 2H^+ + 2e = H_2O(l)$	$\log K = 41.55$
$Mn^{2+} + 2H_2O(l) = \gamma\text{-}MnO_2(s) + 4H^+ + 2e$	$\log K = -40.84$
$\tfrac{1}{2}O_2(g) + Mn^{2+} + H_2O(l) = \gamma\text{-}MnO_2(s) + 2H^+$	$\log K = 0.71$

With the equilibrium constants defined for a given p_{O_2} and pH, the ratio $\{Fe^{3+}\}/\{Fe^{2+}\}$ and the equilibrium activity of Mn^{2+} can be calculated.

pε as a Master Variable. The logarithmic equilibrium expressions [e.g., (9b) in Table 7.2] lend themselves to graphical presentation in double logarithmic equilibrium diagrams. As we used pH as a master variable in acid–base equilibria, we may use pε as a master variable for the graphical presentation of redox equilibria.

For example, in an acid solution of Fe^{2+} and Fe^{3+} (hydrolysis is neglected) the redox equilibrium

$$\frac{\{Fe^{2+}\}}{\{Fe^{3+}\}\{e\}} = K \tag{8}$$

may be combined with the concentration condition

$$[Fe^{2+}] + [Fe^{3+}] = Fe_T \tag{9}$$

to obtain the relations

$$[Fe^{3+}] = \frac{Fe_T K^{-1}}{\{e\} + K^{-1}} \tag{10}$$

and

$$[Fe^{2+}] = \frac{Fe_T\{e\}}{\{e\} + K^{-1}} \tag{11}$$

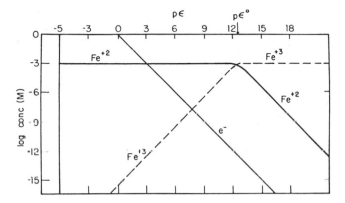

Figure 7.1   Redox equilibrium Fe^{3+}, Fe^{2+}. Equilibrium distribution of a $10^{-3} M$ solution of aqueous iron as a function of pε (Example 7.3).

which, in logarithmic notation, can be formulated in terms of asymptotes. For $\{e\} \gg 1/K$ or pε $<$ pε°,

$$\log[Fe^{3+}] = \log Fe_T + p\varepsilon - p\varepsilon° \qquad (12)$$

$$\log[Fe^{2+}] = \log Fe_T \qquad (13)$$

Similarly for $\{e\} \ll 1/K$ or pε $>$ pε°

$$\log[Fe^{3+}] = \log Fe_T \qquad (14)$$

$$\log[Fe^{2+}] = \log Fe_T + p\varepsilon° - p\varepsilon \qquad (15)$$

These relations can be conveniently plotted (Figure 7.1). The diagram shows how pε changes with the ratio of $\{Fe^{3+}\}$ and $\{Fe^{2+}\}$ for log $Fe_T = -3$.

For aqueous solutions at a given pH each pε value is associated with a partial pressure of H_2 and of O_2:

$$2H^+ + 2e = H_2(g) \qquad \log K = 0 \qquad (16)$$

or

$$2H_2O + 2e = H_2(g) + 2OH^- \qquad \log K = -28 \qquad (17)$$

and

$$O_2(g) + 4H^+ + 4e = 2H_2O \qquad \log K = 83.1 \qquad (18)$$

or

$$O_2(g) + 2H_2O + 4e = 4OH^- \qquad \log K = 27.1 \qquad (19)$$

The equilibrium redox equations in logarithmic form are

$$\log p_{H_2} = 0 - 2pH - 2p\varepsilon \qquad (20)$$

$$\log p_{O_2} = -83.1 + 4pH + 4p\varepsilon \qquad (21)$$

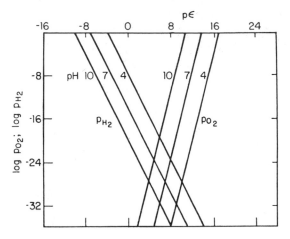

Figure 7.2   The stability of water $[H_2O(l) = H_2(g) + \frac{1}{2}O_2(g)]$. Partial pressure of H_2 and O_2 at various pH values in equilibrium with water.

Figure 7.2 gives a representation of equilibria 16 and 18 in logarithmic form. Thus, for example, a water of pH = 10 and of pε = 8 corresponds to $p_{O_2} = 10^{-11}$ atm and $p_{H_2} = 10^{-34}$ atm. Instead of using pε as a measure of oxidizing intensity, it is possible to characterize this intensity by specifying pH and p_{O_2} or p_{H_2}†
(Figure 7.2). As Figure 7-2 also illustrates, the pε range of natural waters (pH = 4 to 10) extends from approximately pε = −10 to 17; beyond these values water is reduced to H_2 or oxidized to O_2, respectively.

Example 7.4a. Redox Equilibrium SO_4^{2-}–HS^-

Construct a diagram showing the pε dependence of a 10^{-4} M SO_4^{2-}–HS^- system at pH = 10 and 25°C. The reaction is

$$SO_4^{2-} + 9H^+ + 8e = HS^- + 4H_2O(l) \tag{i}$$

and the redox equilibrium equation is

$$p\varepsilon = \tfrac{1}{8}\log K + \tfrac{1}{8}\log\frac{[SO_4^{2-}][H^+]^9}{[HS^-]} \tag{ii}$$

We may calculate the equilibrium constant from available data on the standard free energy of formation. The U.S. Bureau of Standards gives the following $\bar{G}_f^\circ$ values (kJ mol^{-1}) (cf. the appendix at the end of the book) SO_4^{2-}: −742.0;

---

† The resulting partial pressures are often mere calculation numbers; some correspond to less than one H_2 molecule in a space as large as the solar system. In 1921 Clark proposed a reduction intensity parameter defined by $r_H = -\log p_{H_2}$; fortunately r_H is no longer used; even Clark himself discouraged its use later.

HS^-: 12.6; $H_2O(l)$: -237.2. $\bar{G}_f^\circ$ for the aqueous proton and the electron are zero. Hence the standard free energy change in equation i is $\Delta G^\circ = -194.2$, and the corresponding equilibrium constant $(K = 10^{-\Delta G^\circ/2.3RT})$ is 10^{34}. Hence, substituting in equation ii we obtain

$$p\varepsilon = 4.25 - 1.125pH + \tfrac{1}{8}\log[SO_4^{2-}] - \tfrac{1}{8}\log[HS^-] \qquad \text{(iii)}$$

Or, for pH = 10,

$$p\varepsilon = -7 + \tfrac{1}{8}\log[SO_4^{2-}] - \tfrac{1}{8}\log[HS^-] \qquad \text{(iv)}$$

Equation iv has been plotted in Figure 7.3a for the condition $[SO_4^{2-}] + [HS^-] = 10^{-4}$ M. HS^- is the predominant S($-$II) species at pH = 10. The lines for $[SO_4^{2-}]$ and $[HS^-]$ intersect at $p\varepsilon = -7$; the asymptotes have slopes of 0 and ± 8, respectively. Lines for the equilibrium partial pressures of O_2 and H_2 are also given in the diagram. As the diagram shows, rather high relative electron activities are necessary to reduce SO_4^{2-}. At the pH value selected, the reduction takes place at $p\varepsilon$ values slightly less negative than for the reduction of water. In the presence of oxygen, only sulfate can exist; its reduction is only possible under very anaerobic conditions ($p\varepsilon < -6$; $p_{O_2} < 10^{-68}$ atm).

Example 7.4b. Activity Ratio Diagram; Equilibrium System:
SO_4^{2-}, S($\bar{0}$), S($-$II)

In the previous example, no consideration was given to solid sulfur as a possible intermediate state in the reduction of SO_4^{2-} to S($-$II).

1 Evaluate whether rhombic solid sulfur can be formed under the conditions specified (pH = 10, $[SO_4^{2-}] + [HS^-] = 10^{-4}$ M).
2 Establish a log concentration–$p\varepsilon$ diagram for the SO_4^{2-}–S(s)–H_2S(aq) system at pH = 4, assuming that the concentration of soluble sulfur species must not exceed 10^{-2} M.

In addition to the equilibrium constant for the reduction of SO_4^{2-} to HS^-, we need an equilibrium constant for a reaction with S(s). *Stability Constants of Metal–Ion Complexes* lists

$$S(s) + 2H^+ + 2e = H_2S(aq) \quad \log K = 4.8 \qquad \text{(i)}$$

This reaction may be combined with other pertinent equilibria as follows:

$$SO_4^{2-} + 9H^+ + 8e = HS^- + 4H_2O \qquad \log K = 34.0 \qquad \text{(ii)}$$

$$H_2S(aq) = S(s) + 2H^+ + 2e \qquad \log K = -4.8 \qquad \text{(iii)}$$

$$H^+ + HS^- = H_2S(aq) \qquad \log K = 7.0 \qquad \text{(iv)}$$

$$\overline{\rule{0pt}{1em}\hspace{3em}}$$

$$SO_4^{2-} + 8H^+ + 6e = S(s) + 4H_2O \qquad \log K = 36.2 \qquad \text{(v)}$$

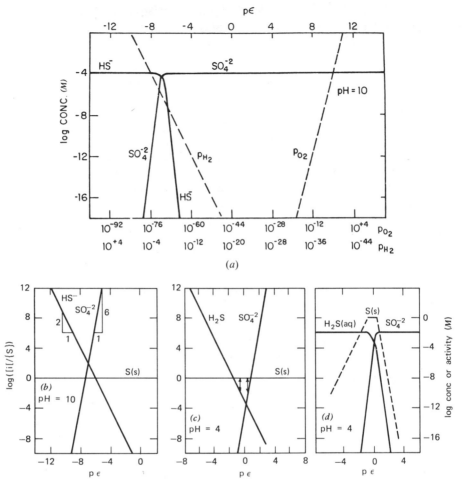

Figure 7.3 Equilibrium distribution of sulfur compounds. (*a*) As a function of pε at pH = 10 and 25°C. Total concentration of compounds is 10^{-4} M (Example 7.4a). (*b* and *c*) Activity ratio versus pε diagrams at pH = 10 and 4. Elemental sulfur is chosen as a reference. It is apparent that at pH = 10, elemental sulfur cannot exist thermodynamically, whereas at pH = 4, S(s) becomes the relatively most stable species in a narrow pε range (Example 7.4b). (*d*) Equilibrium diagram at pH = 4. Solid S is formed when its activity becomes unity. (It may be formed at lower activity if deposited as solid solution.) (7.4b)

For the S(s)–S(−II) system we can combine equations iii and iv to yield

$$HS^- = S(s) + H^+ + 2e \qquad \log K = 2.2 \qquad (vi)$$

1   In order to investigate whether sulfur can be formed under the conditions given, it might be most convenient to prepare an *activity ratio diagram* (see Example 5.6) to select the activity of S(s) as a reference and describe the dependence of $\log([SO_4^{2-}]/\{S\})$ and of $\log([HS^-]/\{S\})$ as a function of pε.

From equations v and vi we obtain the relations

$$\log \frac{[SO_4^{2-}]}{\{S\}} = -36.2 + 8pH + 6p\varepsilon \qquad \text{(vii)}$$

$$\log \frac{[HS^-]}{\{S\}} = -2.2 - pH - 2p\varepsilon \qquad \text{(viii)}$$

which can be plotted conveniently for pH = 10. As Figure 7.3b illustrates, over the entire pε range the relative activity of solid sulfur is always smaller than that of $[SO_4^{2-}]$ and $[HS^-]$; hence elemental sulfur cannot be formed and need not be considered for the construction of the diagram in Figure 7.3b.

2 For convenience, the pertinent reactions to be considered should be written in terms of the species prevailing under the given pH conditions. At pH = 4 the predominant species are SO_4^{2-} and $H_2S(aq)$. For the construction of an activity ratio diagram, equation vii may be used together with equation iii, whose redox equilibrium equation is

$$\log \frac{[H_2S]}{\{S\}} = 4.8 - 2pH - 2p\varepsilon \qquad \text{(ix)}$$

The activity ratio diagram (Figure 7.3c) reveals that elemental sulfur can be formed in a narrow pε range. Only between pε = −0.6 and +0.4 are the ratios of $[H_2S]/\{S\}$ and $[SO_4^{2-}]/\{S\}$ smaller than 10^{-2}. For this condition the log C–pε diagram can be constructed (Figure 7.3d). Outside the pε range −0.6 to +0.4, elemental sulfur acquires a (hypothetical) activity of less than unity (dashed line).

Combining equations i, ii, and iv gives

$$SO_4^{2-} + 3H_2S(aq) + 2H^+ = 4S(s) + 4H_2O \qquad \log K = 21.8 \qquad \text{(x)}$$

This equation indicates, as do Figures 7.3c and 7.3d, that the formation of elemental sulfur is only possible thermodynamically under acid conditions (pH < 7). Note that elemental sulfur (rhombic sulfur) may be formed as an intermediate and may persist as a metastable solid phase. Solid elemental sulfur occurs quite commonly in recent marine sediments.

Example 7.5. Stability Relations of Iron and Manganese Compounds at pH = 7

Discuss the stability relations of Fe and Mn compounds with the help of activity ratio diagrams for the conditions pH = 7, $[HCO_3^-] = 10^{-3}$ M, $[SO_4^{2-}] = 10^{-3}$ M. The equilibrium constants at 25°C can be obtained from the data on the free energy of formation in the appendix at the end of the book.

For construction of the activity ratio diagram we take $\{Fe^{2+}\}$ as a reference. The following equations can be derived:

$$\log \frac{\{\alpha\text{-Fe}_2O_3(s)\}^{1/2}}{\{Fe^{2+}\}} = -11.1 + 3pH + p\varepsilon \tag{i}$$

$$\log \frac{\{\alpha\text{-FeOOH(s)}\}}{\{Fe^{2+}\}} = -11.3 + 3pH + p\varepsilon \tag{ii}$$

$$\log \frac{\{am\cdot Fe(OH)_3(s)\}}{\{Fe^{2+}\}} = -17.1 + 3pH + p\varepsilon \tag{iii}$$

$$\log \frac{\{Fe_3O_4(s)\}^{1/3}}{\{Fe^{2+}\}} = -10.1 + \tfrac{8}{3}pH + \tfrac{2}{3}p\varepsilon \tag{iv}$$

$$\log \frac{\{FeCO_3(s)\}}{\{Fe^{2+}\}} = -0.2 + pH + \log\{HCO_3^-\} \tag{v}$$

$$\log \frac{\{Fe(OH)_2(s)\}}{\{Fe^{2+}\}} = -11.7 + 2pH \tag{vi}$$

$$\log \frac{\{\alpha\text{-FeS(s)}\}}{\{Fe^{2+}\}} = 38 - 8pH + \log\{SO_4^{2-}\} - 8p\varepsilon \tag{vii}$$

$$\log \frac{\{FeS_2(s)\}}{\{Fe^{2+}\}} = 86.8 - 16pH + 2\log\{SO_4^{2-}\} - 14p\varepsilon \tag{viii}$$

These equations are plotted for the conditions specified in Figure 7.4a. For the Mn species the relevant equations are:

$$\log \frac{\{\gamma\text{-MnO}_2(s)\}}{\{Mn^{2+}\}} = -43.0 + 4pH + 2p\varepsilon \tag{ix}$$

$$\log \frac{\{\delta\text{-MnO}_2(s)\}}{\{Mn^{2+}\}} = -43.6 + 4pH + 2p\varepsilon \tag{x}$$

$$\log \frac{\{\gamma\text{-MnOOH(s)}\}}{\{Mn^{2+}\}} = -25.3 + 3pH + p\varepsilon \tag{xi}$$

$$\log \frac{\{MnS(s)\}}{\{Mn^{2+}\}} = 34.0 - 8pH + \log\{SO_4^{2-}\} - 8p\varepsilon \tag{xii}$$

$$\log \frac{\{MnO_4^-\}}{\{Mn^{2+}\}} = -124.6 + 8pH + 5p\varepsilon \tag{xiii}$$

$$\log \frac{\{MnCO_3(s)\}}{\{Mn^{2+}\}} = -0.2 + pH + \log\{HCO_3^-\} \tag{xiv}$$

$$\log \frac{\{Mn(OH)_2(s)\}}{\{Mn^{2+}\}} = -15 + 2pH \tag{xv}$$

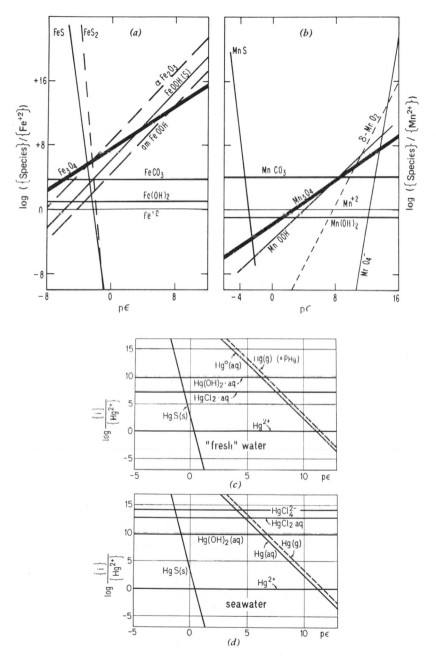

Figure 7.4 Activity ratio diagrams. (*a* and *b*) Fe and Mn for pH = 7.0, $C_T = 10^{-3}$ M, $\{SO_4^{2-}\} = 10^{-3}$ M (Example 7.5). (*c* and *d*) Hg(II) species in fresh water and seawater.

These equations are plotted in Figure 7.4b. Because of the uncertainty of free energy data, especially for the various iron oxides, the positions of the lines are not exact. Hematite is more stable than geothite† and geothite is more stable than amorphous $Fe(OH)_3$. There is considerable uncertainty about the $p\varepsilon$ values at which equilibrium between iron(III) oxide and Fe_3O_4 occurs. $FeCO_3(s)$ (siderite) is the most stable solid Fe(II) phase in the oxidative transition of FeS_2 to Fe(III) oxide. In the $p\varepsilon$ range 8 to 10, MnO_2, Mn_3O_4, and $MnCO_3$ are of similar relative activity. It appears from the diagram that under the specified conditions Mn_3O_4 may not occur as a stable phase. The diagrams, furthermore, suggest that iron and manganese sulfides start to be formed as one passes to $p\varepsilon$ values lower than -2 to -3. FeS(s) appears to be metastable with regard to pyrite. Even if sulfides are being formed, the sulfate concentration in many natural waters does not vary appreciably; hence the assumption of a constant sulfate concentration is justified. But if $p\varepsilon$ values drop further (below $p\varepsilon = -3$ at pH $= 7$), $[SO_4^{2-}]$ does not remain constant but decreases.

Example 7.6. Elemental Aqueous Mercury, Hg^0 (aq)

1. Estimate the water solubility of $Hg^0(aq)$ from the following information (25°C):

$$Hg^{2+} + 2e = Hg^0(aq) \qquad E_H^\circ = 0.659 \qquad p\varepsilon^\circ = 11.15 \qquad \text{(i)}$$

$$Hg^{2+} + 2e = Hg(l) \qquad E_H^\circ = 0.854 \qquad p\varepsilon^\circ = 14.4 \qquad \text{(ii)}$$

For the reaction

$$Hg(l) = Hg(aq) \qquad\qquad\qquad \text{(iii)}$$

the log equilibrium constant is given by log K_i − log K_{ii} (where log $K = np\varepsilon^\circ$). Hence log $K_{iii} = -6.5$. The solubility of aqueous mercury is $10^{-6.5}$ M ($3 \times 10^{-7} M$, 0.06 mg Hg liter^{-1}). This solubility corresponds to that given by D. N. Glew and D. A. Hames [*Can. J. Chem.*, **49**, 3114 (1971)].

2. How does the relative equilibrium contribution of elemental aqueous mercury to the total soluble inorganic mercury in fresh water and seawater and the volatility of $Hg^0(aq)$ depend on $p\varepsilon$? We assume for fresh water pH $= 8$, $[Cl^-] = 10^{-3}$ M, $[SO_4^{2-}] = 10^{-3}$ M; and for seawater, pH $= 8$, $[Cl^-] = 0.6$ M, $[SO_4^{2-}] = 10^{-2}$ M.

From Figure 6.15 we know that the major soluble Hg(II) species are $Hg(OH)_2(aq)$ in fresh water and $HgCl_4^{2-}$ in seawater. We may establish diagrams in which relative activities of $Hg^0(aq)$, $Hg^0(g)$, Hg^{2+}, $HgCl_2^0(aq)$,

---

† The stability depends on the mode of formation or precipitation, age, and crystal size. For large particle sizes goethite may be more stable than hematite (see Figure 5.23b).

$HgCl_4^{2-}$, $Hg(OH)_2(aq)$, and $HgS(s)$ are plotted as a function of $p\varepsilon$ for both fresh water and seawater (Figure 7.4c and d).

$$Hg^{2+} + 2e = Hg^0(aq) \qquad \log K = 22.3 \qquad \text{(iv)}$$

$$Hg^{2+} + 2Cl^- = HgCl_2(aq) \qquad \log K = 13.2 \qquad \text{(v)}$$

$$Hg^{2+} + 4Cl^- = HgCl_4^{2-} \qquad \log K = 15.1 \qquad \text{(vi)}$$

$$Hg^{2+} + 2OH^- = Hg(OH)_2(aq) \qquad \log K = 21.9 \qquad \text{(vii)}$$

$$Hg^{2+} + SO_4^{2-} + 8H^+ + 8e = HgS(s) + 4H_2O \qquad \log K = 70.0 \quad \text{(viii)}$$

$$Hg^0(aq) = Hg^0(g) \qquad \log K = 0.93 \qquad \text{(ix)}$$

Such activity ratio diagrams illustrate that $Hg^0(aq)$ becomes the major dissolved inorganic Hg species below the $p\varepsilon$ range 4 to 6 for both seawater and fresh water; above this $p\varepsilon$ range $Hg(OH)_2(aq)$ and $HgCl_4^{2-}$ are the preponderant species in fresh and seawater, respectively.

We may also note from equilibrium iii that the water solubility of Hg^0 (elemental mercury) is substantial and that the volatility of $Hg^0(aq)$ as characterized by equilibrium viii (Henry's law constant $p_{Hg}/\{Hg(aq)\} = 8.5$ atm M^{-1}) is relatively large; at normal temperatures the unionized mercury solute, especially in the not fully aerobic $p\varepsilon$ range, is readily lost by volatilization from its water solutions at normal temperatures (cf. A. Lerman, *Geochemical Processes*, Wiley-Interscience, New York, 1978, Chapter 4; R. Wollast, G. Billen, and F. T. Mackenzie, in *Ecological Toxicology Research*, A. D. McIntire and C. F. Mills, Eds., Plenum, New York, 1975, Chapter 7, p. 145).

Intensity and Capacity. $p\varepsilon$ is an intensity factor; it measures oxidizing intensity. Oxidation or reduction capacity must be expressed in terms of a quantity of system electrons that must be added or removed in order to attain a given $p\varepsilon$. This is analogous to the acid- or base-neutralizing capacity with respect to protons; for example, alkalinity and acidity are measured in terms of the proton condition. Thus the oxidative capacity of a system with respect to a given electron energy level will be given by the equivalent sum of all the oxidants below this energy level minus the equivalent sum of all the reductants above it. For example, the oxidative capacity of a solution with respect to an electron level corresponding to Cu(s) is

$$2[Cu^{2+}] + 2[I_3^-] + [Fe^{3+}] + 4[O_2] - 2[H_2] \qquad (22)$$

The Stability of Water.   There are reducing substances that are stronger reductants than H_2 and oxidizing substances that are stronger oxidants than O_2. Hence, thermodynamically speaking, for example, Cl_2 is unstable in H_2O and will acquire electrons from H_2O, that is, oxidize H_2O to O_2. The solvent exerts a leveling influence on the system and restricts the range of accessible pε values (Figure 7.2). As already given in Table 7.2, pε is related to the redox potential E_H (the suffix H denotes that this potential is on the hydrogen scale) by

$$p\varepsilon = \frac{F}{2.3RT} E_H \quad \text{and} \quad p\varepsilon^\circ = \frac{F}{2.3RT} E_H^\circ \tag{23}$$

7.3   THE ELECTRODE POTENTIAL;
THE PETERS–NERNST EQUATION

E_H as given in equation 23 is in accord with the IUPAC 1953 Stockholm Convention. Under certain conditions it is possible to measure E_H electrochemically. The thermodynamic relation of the potential E_H to the composition of the solution, as already given in Table 7.2, is generally known as the Nernst equation:

$$E_H = E_H^\circ + \frac{2.3RT}{nF} \log \frac{\prod_i \{ox\}^{n_i}}{\prod_j \{red\}^{n_j}} \tag{24}$$

for example, for the reaction

$$NO_3^- + 6H^+ + 5e = \tfrac{1}{2}N_2(g) + 3H_2O$$

$$E_H = E_H^\circ + \frac{2.3RT}{5F} \log \frac{\{NO_3^-\}\{H^+\}^6}{p_{N_2}^{1/2}}$$

where E_H°, the standard redox potential (i.e., the potential that would be obtained if all substances in the redox reaction were in their standard states of unit activity) is related to the free energy change for the cell reaction, ΔG°, or the equilibrium constant of the reduction reaction (cf. equations 7 and 23):

$$E_H^\circ = -\frac{\Delta G^\circ}{nF} = \frac{RT}{nF} \ln K = \frac{2.3RT}{nF} \log K \tag{25}$$

It is necessary to distinguish between the concept of a potential and the measurement of a potential. Redox or electrode potentials (quoted in tables in *Stability Constants of Metal–Ion Complexes* or by others [1–4]) have been derived from equilibrium data, thermal data, and the chemical behavior of a redox couple with respect to known oxidizing and reducing agents, and from direct measurements of electrochemical cells. Hence there is no a priori reason to identify the thermodynamic redox potentials with measurable electrode potentials.

We can express (relative) electron activity in pε or in volts. The use of pε, which is dimensionless, makes calculations simpler than the use of E_H, because every 10-fold change in the activity ratio causes a unit change in pε. Furthermore, because an electron can reduce a proton, the intensity parameter for oxidation might preferably be expressed in units equivalent to pε. Of course in making a direct electrochemical measurement of the oxidizing intensity an emf (volts) is being measured, but the same is true of pH measurements, and a few decades ago the "acidity potential" was used to characterize the relative H^+ ion activity.

Sign Convention. There have been controversies regarding sign conventions, and the one adapted here is based on IUPAC conventions. Tables 7.3 and 7.5 list for illustrative purposes a few representative redox potentials. According to IUPAC conventions, all half-reactions are written as reductions with a sign that corresponds to the sign of log K of the reduction reaction.

Useful reference works may use other sign conventions. For example, in Latimer's *Oxidation Potentials* the half-cell equation is written in the reverse direction. For example,

$$Zn(s) = Zn^{2+} + 2e \qquad E^\circ_{cell} = 0.76 \text{ V} \qquad (26)$$

corresponding to the overall reaction

$$Zn(s) + 2H^+ = Zn^{2+} + H_2(g) \qquad (27)$$

$E^\circ_{(Latimer)}$ corresponds to the emf measured in an electrochemical cell, E°_{cell}, where reaction 27 takes place. If we consider the reverse reaction

$$Zn^{2+} + H_2(g) = Zn(s) + 2H^+ \qquad (28)$$

TABLE 7.3   EQUILIBRIUM CONSTANTS FOR A FEW REDOX REACTIONS

Reaction	Log K at 25°C	Standard Electrode Potential (V) at 25°C
$Na^+ + e = Na(s)$	-46	-2.71
$Zn^{2+} + 2e = Zn(s)$	-26	-0.76
$Fe^{2+} + 2e = Fe(s)$	-14.9	-0.44
$Co^{2+} + 2e = Co(s)$	-9.5	-0.28
$V^{3+} + e = V^{2+}$	-4.3	-0.26
$2H^+ + 2e = H_2(g)$	0.0	0.00
$S(s) + 2H^+ + 2e = H_2S$	$+4.8$	$+0.14$
$Cu^{2+} + e = Cu^+$	$+2.7$	$+0.16$
$AgCl(s) + e = Ag(s) + Cl^-$	$+3.7$	$+0.22$
$Cu^{2+} + 2e = Cu(s)$	$+11.4$	$+0.34$
$Cu^+ + e = Cu(s)$	$+8.8$	$+0.52$
$Fe^{3+} + e = Fe^{2+}$	$+13.0$	$+0.77$
$Ag^+ + e = Ag(s)$	$+13.5$	$+0.80$
$Fe(OH)_3(s) + 3H^+ + e = Fe^{2+} + 3H_2O$	$+17.1$	$+1.01$
$IO_3^- + 6H^+ + 5e = \frac{1}{2}I_2(s) + 3H_2O$	$+104$	$+1.23$
$MnO_2(s) + 4H^+ + 2e = Mn^{2+} + 2H_2O$	$+43.6$	$+1.29$
$Cl_2(g) + 2e = 2Cl^-$	$+46$	$+1.36$
$Co^{3+} + e = Co^{2+}$	$+31$	$+1.82$

then the cell emf must have a sign opposite to that of reaction 27:

$$Zn^{2+} + 2e = Zn(s) \qquad E^\circ_{cell}(=E^\circ_H) = -0.76 \text{ V} \tag{29}$$

Whereas an emf can change signs depending on the direction of the reaction, an electrode potential or pε must not, according to IUPAC conventions, change signs because these parameters reflect electron activity or in an electrochemical cell reflect the electrostatic charge in the electrode. Hence we can define a standard electrode potential E°_H or a pε° of -0.76 V or -12.8, respectively, for the Zn^{2+}–Zn couple independent of the direction of writing the half-reaction. Note that multiplying or dividing the stoichiometric coefficients of a reaction does not affect pε° or E°_H but of course affects $\log K$ or ΔG°. The Nernst potential is independent of the number of moles of electrons transferred.

Effect of Ionic Strength and Complex Formation on Electrode Potentials; Formal Potentials

In dealing with redox equilibria, we are also confronted with the problem of evaluating activity corrections or maintaining the activities under consideration as constants. The Nernst equation rigorously applies only if the activities and actual species taking part in the reaction are inserted in the equation. The activity scales discussed before, the infinite dilution scale and the ionic medium scale, may be used. The standard potential or standard pε on the infinite dilution scale is related to the equilibrium constant for $I = 0$ of the reduction reaction by

$$\frac{F}{RT(\ln 10)} E^\circ_H = \frac{1}{n} \log K = p\varepsilon^\circ \tag{30}$$

and is usually obtained by either extrapolating the measured constant or measured potential to infinite dilution.

In a constant ionic medium the concentration quotient becomes the equilibrium constant K, and correspondingly one might define $^cE^\circ_H$ and $p^c\varepsilon^\circ$

$$\frac{F}{RT(\ln 10)} {}^cE^\circ_H = \frac{1}{n} \log {}^cK = p^c\varepsilon^\circ \tag{31}$$

Complex Formation. How the standard potential is influenced by complex formation may be illustrated by the dissolution of zinc

$$Zn(s) + 2H^+ = Zn^{2+} + H_2(g) \tag{32}$$

which is characterized by the half reaction

$$Zn^{2+} + 2e = Zn(s) \qquad \Delta G^\circ_1 = -RT \ln K = -2FE^\circ_{Zn^{2+}, Zn(s)} \tag{33}$$

If the dissolution of Zn occurs in a medium containing ligands that can displace coordinated H_2O from the zinc ions, that is, form complexes, such as in the reaction of Cl^- ions with Zn^{2+} to form $ZnCl_4^{2-}$

$$ZnCl_4^{2-} = Zn^{2+} + 4Cl^- \qquad \Delta G^\circ_2 = RT \ln \beta_4 \tag{34}$$

where β_4 is the formation constant $[\{ZnCl_4^{2-}\}/(\{Zn^{2+}\}\{Cl^-\}^4)]$, then we may characterize the overall half reaction by

$$ZnCl_4^{2-} + 2e = Zn(s) + 4Cl^- \qquad \Delta G_3^\circ = -RT \ln {}^+K = -2FE^\circ_{ZnCl_4^{2-}, Zn(s)} \quad (35)$$

This is the sum of equations 33 and 34. Correspondingly, we may write any one of the following relations

$$\Delta G_3^\circ = \Delta G_1^\circ + \Delta G_2^\circ \tag{36}$$

$$\log {}^+K = \log K - \log \beta_4 \tag{37}$$

$$p\varepsilon^\circ_{ZnCl_4^{2-}, Zn(s)} = p\varepsilon^\circ_{Zn^{2+}, Zn(s)} - \tfrac{1}{2}\log \beta_4 \tag{38}$$

$$E^\circ_{ZnCl_4^{2-}, Zn(s)} = E^\circ_{Zn^{2+}, Zn(s)} - \frac{RT(\ln 10)}{2F} \log \beta_4 \tag{39}$$

These equations show that Cl^- stabilizes the higher oxidation state, facilitating the dissolution of Zn. The Nernst equation may now be expressed either in terms of free Zn^{2+} or in terms of $ZnCl^{2-}$.

$$E = E^\circ_{Zn^{2+}, Zn(s)} + \frac{RT(\ln 10)}{2F} \log\{Zn^{2+}\}$$

$$= E^\circ_{ZnCl_4^{2-}, Zn(s)} + \frac{RT(\ln 10)}{2F} \log\{ZnCl_4^{2-}\} - \frac{2RT(\ln 10)}{F} \log\{Cl^-\} \tag{40}$$

Formal Potentials. As with conditional constants, that is, constants valid under specifically selected conditions, for example, a given pH and a given ionic medium, conditional or formal potentials are of great utility.

$$\frac{F}{RT(\ln 10)} {}^FE_H^\circ = \frac{1}{n} \log P = p^F\varepsilon^\circ \tag{41}$$

The measurement of a formal pε or formal electrode potential consists of measurement of the emf of an electrochemical cell in which, under the specified conditions, the analytical concentration of the two oxidation states is varied. For example, in a 0.1 M H_2SO_4 solution the formal electrode potential for Fe(III)–Fe(II) is 0.68 V in comparison to 0.77 V for the Fe^{3+}/Fe^{2+} ($I = 0$) system:

$$^FE^\circ_{(I = 0.1\,M\,H_2SO_4)} = 0.68 + \frac{RT}{F} \ln \frac{Fe(III)_T}{Fe(II)_T} \tag{42}$$

In this case the formal potential includes correction factors for activity coefficients, acid–base phenomena (hydrolysis of Fe^{3+} to $FeOH^{2+}$), complex formation (sulfate complexes), and the liquid junction potential used between the reference electrode and the half-cell in question. Although the correction is strictly valid only at the single concentration at which the potential has been determined, formal potentials may often lead to better predictions than standard potentials because they represent quantities subject to direct experimental measurement.

Example 7.7. Formal Potential of the Fe(III)–Fe(II) System in the Presence of F$^-$

Estimate the formal potential of the Fe(III)–Fe(II) couple for solutions of $[H^+] = 10^{-2}\ M$ and $[F^-] = 10^{-2}\ M$, $I = 0.1\ M$. The following constants are available:

$$Fe^{3+} + e = Fe^{2+} \qquad\qquad \log K(I = 0) = 13.0 \qquad\qquad \text{(i)}$$

$$Fe^{3+} + H_2O = FeOH^{2+} + H^+ \qquad \log K_H(I = 0.1) = -2.7 \qquad \text{(ii)}$$

$$Fe^{3+} + F^- = FeF^{2+} \qquad\qquad \log \beta_1(I = 0.1) = 5.2 \qquad\qquad \text{(iii)}$$

$$Fe^{3+} + 2F^- = FeF_2^+ \qquad\qquad \log \beta_2(I = 0.1) = 9.2 \qquad\qquad \text{(iv)}$$

$$Fe^{3+} + 3F^- = FeF_3 \qquad\qquad \log \beta_3(I = 0.1) = 11.9 \qquad\qquad \text{(v)}$$

In order to compute the formal potential we first consider the activity correction of the Fe^{3+}–Fe^{2+} electrode. Using the Güntelberg approximation, $K_{(i)}$ is corrected to $^cK_{(i)}$:

$$\frac{[Fe^{2+}]}{[Fe^{3+}]\{e\}} = {^cK} = K\frac{f_{Fe^{3+}}}{f_{Fe^{2+}}} = K\frac{0.083}{0.33} = 10^{12.4} \qquad p^c\varepsilon^\circ = 12.4 \qquad \text{(vi)}$$

from which $^cE_H^\circ$ is obtained as $+0.73$ V. It has been assumed that the liquid junction makes a negligible contribution. The same result is obtained if we consider that

$$^cE^\circ_{Fe^{3+},\,Fe^{2+}} = E^\circ_{Fe^{3+},\,Fe^{2+}} + \frac{RT}{F}\ln\frac{f_{Fe^{3+}}}{f_{Fe^{2+}}}$$

Next, the correction caused by hydrolysis and complex formation can be taken into account. For the conditions specified, $FeOH^{2+}$ is the only important hydrolysis species:

$$Fe(III)_T = [Fe^{3+}] + [FeOH^{2+}] + [Fe^{2+}] + [FeF_2^+] + [FeF_3] \qquad \text{(vii)}$$

Under the specified conditions, ferrous iron does not form complexes with F^- and OH^-, hence

$$Fe(II)_T = [Fe^{2+}] \qquad\qquad \text{(viii)}$$

Equation vii can be rewritten as

$$Fe(III)_T = [Fe^{3+}]\left(1 + \frac{K_H}{[H^+]} + \beta_1[F^-] + \beta_2[F^-]^2 + \beta_3[F^-]^3\right) \qquad \text{(ix)}$$

For the conditions specified,

$$\alpha_{Fe} = \frac{Fe(III)_T}{[Fe^{3+}]} = 9.5 \times 10^5 \qquad\qquad \text{(x)}$$

and equilibrium may be formulated as in

$$\frac{Fe(II)_T}{Fe(III)_T\{e\}} = \frac{^cK}{\alpha_{Fe}} = P \qquad\qquad \text{(xi)}$$

and

$$\log P = 6.4 = p^F\varepsilon^\circ \qquad\qquad \text{(xii)}$$

Correspondingly, the formal potential of the Fe(III)–Fe(II) electrode for the given conditions is $^{F}E^{\circ} = 0.38$ V. The potential of course is the same whether it is expressed in terms of actual concentrations or analytical concentrations:

$$E = 0.73 + \frac{RT}{F} \ln \frac{[Fe^{3+}]}{[Fe^{2+}]} = 0.38 + \frac{RT}{F} \ln \frac{Fe(III)_T}{Fe(II)_T} \qquad \text{(xiii)}$$

or

$$p\varepsilon = 12.4 + \log \frac{[Fe^{3+}]}{[Fe^{2+}]} = 6.4 + \log \frac{Fe(III)_T}{Fe(II)_T} \qquad \text{(xiv)}$$

Note that from an electrode kinetic point of view the Nernst equation does not give any information as to the actual species that establishes the electrode potential. In the case of a fluoride-containing solution it is very possible that one of the fluoro iron(III) complexes rather than the Fe^{3+} iron participates in the electron-exchange reaction at the electrode. Complexation usually stabilizes a system against reduction. In the example just considered, because complexation is stronger with Fe(III) than with Fe(II), the tendency for reduction of Fe(III) to Fe(II) is decreased. It is apparent that coordination with a donor group, in general, decreases the redox potential. In the relatively rare instances where the lower oxidation state is favored (e.g., complexation of aqueous iron with phenanthroline), the redox potential is increased as a result of coordination.

7.4 pε–pH DIAGRAMS

An attempt has been made thus far to describe the stability relationships of the distribution of the various soluble and insoluble forms through rather simple graphical representation. Essentially two types of graphical treatments have been used: first, equilibria between chemical species in a particular oxidation state as a function of pH and solution composition; second, equilibria between chemical species at a particular pH as a function of pε (or E_H). Obviously these diagrams can be combined into pε–pH diagrams. Such pε–pH stability field diagrams show in a comprehensive way how protons and electrons simultaneously shift the equilibria under various conditions and can indicate which species predominate under any given condition of pε and pH.

The value of a pε–pH diagram consists primarily in providing an aid for the interpretation of equilibrium constants (free energy data) by permitting the simultaneous representation of many reactions. Of course, such a diagram, like all other equilibrium diagrams, represents only the information used in its construction.

Natural waters are in a highly dynamic state with regard to oxidation–reduction rather than in or near equilibrium. Most oxidation–reduction reactions have a tendency to be much slower than acid–base reactions, expecially in the absence of suitable biochemical catalysis. Nonetheless equilibrium diagrams can greatly aid attempts to understand the possible redox patterns in natural waters.

The Construction of pε–pH Diagrams

The construction of a pε–pH diagram may be introduced by considering the redox stability of water. As we have seen (equations 16 to 21), H_2O can be oxidized to O_2 or reduced to H_2. Equations 20 and 21 can be rewritten as

$$p\varepsilon = 0 - pH - \tfrac{1}{2}\log p_{H_2} \tag{43}$$

$$p\varepsilon = 20.78 - pH + \tfrac{1}{4}\log p_{O_2} \tag{44}$$

These equations can be plotted in a pε–pH diagram (Figure 7.6, 7.7a, 7.8a). The lines of both equations have slopes $dp\varepsilon/dpH$ of -1, and they intersect the ordinate, at $pH = 0$, at $p\varepsilon = 20.78$ (for $p_{O_2} = 1$) and $p\varepsilon = 0$ (for $p_{H_2} = 1$), respectively. Above the upper line, water becomes an effective reductant (producing oxygen); below the lower line, water is an effective oxidant (producing hydrogen).

For any partial pressure of O_2, the equilibrium between water and oxygen is characterized by a straight line with a slope $dp\varepsilon/dpH = -1$; any decrease in p_{O_2} by 10^4 lowers the line by 1 pε unit.

Example 7.8. pε–pH Digram for the Sulfur System

Construct a pε–pH diagram for the SO_4^{2-}, S(s)–H_2S(aq) system assuming that the concentration of soluble S species is 10^{-2} M (corresponding to the conditions of Example 7.4b, where the necessary equilibrium constants were given).

The lines in the pε–pH diagram (Figure 7.5) are characterized

(1)   For the equilibrium

$$SO_4^{2-} + 8H^+ + 6e = S(s) + 4H_2O \qquad \log K = 36.2$$

$$p\varepsilon = \frac{36.2}{6} + \frac{1}{6}\log[SO_4^{2-}] - \frac{8}{6}pH \tag{i}$$

(2)   For the equilibrium

$$SO_4^{2-} + 10H^+ + 8e = H_2S(aq) + 4H_2O \qquad \log K = 41.0$$

$$p\varepsilon = \frac{41}{8} + \frac{1}{8}\log\frac{[SO_4^{2-}]}{[H_2S(aq)]} - \frac{10}{8}pH \tag{ii}$$

(3)   For the equilibrium

$$S(s) + 2H^+ + 2e = H_2S(aq) \qquad \log K = 4.8$$

$$p\varepsilon = \frac{4.8}{2} - pH - \frac{1}{2}\log[H_2S] \tag{iii}$$

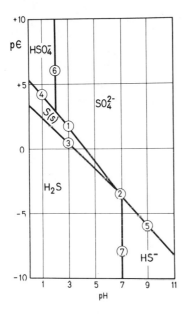

Figure 7.5 pε–pH diagram for the SO_4–$S(s)$–H_2S system. (The equations for the numbered lines are given in Example 7.7). Total dissolved S-species is 10^{-2} M.

(4)   For the reaction

$$HSO_4^- + 7H^+ + 6e = S(s) + 4H_2O \qquad \log K = 34.2$$

$$p\varepsilon = \frac{34.2}{6} + \frac{1}{6}\log[HSO_4^-] - \frac{7}{6}pH \qquad \text{(iv)}$$

(5)   For the reaction

$$SO_4^{2-} + 9H^+ + 8e = HS^- + 4H_2O \qquad \log K = 34.0$$

$$p\varepsilon = \frac{34.0}{8} + \frac{1}{8}\log\frac{[SO_4^{2-}]}{[HS^-]} - \frac{9}{8}pH \qquad \text{(v)}$$

(6)   For the equilibrium

$$HSO_4^- = SO_4^{2-} + H^+ \qquad \log K = -2.0$$

$$\log\frac{[SO_4^{2-}]}{[HSO_4^-]} - pH = -2.0 \qquad \text{(vi)}$$

(7)   For the equilibrium

$$H_2S(aq) = H^+ + HS^- \qquad \log K = -7.0$$

$$\log\frac{[HS^-]}{[H_2S]} - pH = -7.0 \qquad \text{(vii)}$$

Example 7.9. Chlorine Redox Equilibria

Summarize in a $p\varepsilon$–pH diagram the information contained in the equilibrium constants, $I = 0$, 25°C, of the following three reactions involving $Cl_2(aq)$, Cl^-, OCl^-, and $HOCl$. (For convenience, in addition to the equilibrium constant, the standard redox potential, E_H° (volts), is given.) Total soluble chlorine is assumed to be $Cl_T = 0.04\ M$.

$$HClO + H^+ + e = \tfrac{1}{2}Cl_2(aq) + H_2O \qquad \log K = 26.9,\ E_H^\circ = 1.59 \qquad (i)$$

$$\tfrac{1}{2}Cl_2(aq) + e = Cl^- \qquad\qquad\qquad \log K = 23.6,\ E_H^\circ = 1.40 \qquad (ii)$$

$$HClO = H^+ + ClO^- \qquad\qquad\qquad \log K = -7.3 \qquad\qquad\quad (iii)$$

We can write the following equilibrium equations for reactions i and ii:

$$p\varepsilon = 26.9 + \log \frac{[HClO]}{[Cl_2]^{1/2}} - pH \qquad (iv)$$

$$p\varepsilon = 23.6 + \log \frac{[Cl_2]^{1/2}}{[Cl^-]} \qquad (v)$$

Combining reactions i and ii gives the expression for the reduction of HOCl to Cl^-:

$$p\varepsilon = 25.25 + \frac{1}{2} \log \frac{[HClO]}{[Cl^-]} - 0.5\,pH \qquad (vi)$$

The reduction of ClO^- to Cl^- is given by

$$p\varepsilon = 28.9 + \frac{1}{2} \log \frac{[OCl^-]}{[Cl^-]} - pH \qquad (vii)$$

Since $Cl_T = [HOCl] + [ClO^-] + 2[Cl_2] + [Cl^-] = 0.04\ M$, and assuming unit atomic ratio for oxidants and reductants, we have the following concentrations: at the Cl_2–HOCl boundary: $[HOCl] = \tfrac{1}{2}Cl_T = 2 \times 10^{-2}\ M$, $[Cl_2] = \tfrac{1}{4}Cl_T = 10^{-2}\ M$; at the Cl_2–Cl^- boundary: $[Cl_2] = Cl_T/4 = 10^{-2}\ M$, $[Cl^-] = Cl_T/2 = 2 \times 10^{-2}\ M$. Finally, the line separating HOCl from OCl^- is given by (cf. equation iii)

$$\log \frac{[HClO]}{[ClO^-]} + pH = 7.3 \qquad (viii)$$

With these substitutions the resulting four equations are plotted in Figure 7.6. The line for equation v is pH-independent and thus plots as a horizontal line; it intersects with the line representing equation iv which has $dp\varepsilon/dpH$ of -1. Why do the lines discontinue at this intersection? To the right of pH = 1.9 HOCl is a more stable oxidant than $Cl_2(aq)$. (If any doubt should arise about which species predominates thermodynamically, an activity ratio diagram

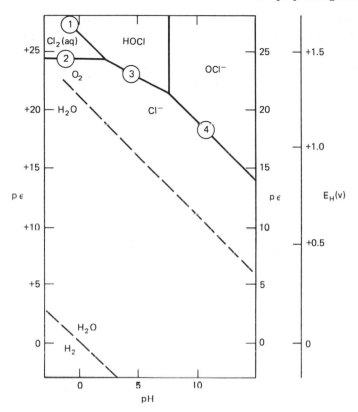

Figure 7.6 pε–pH diagram for the chlorine system. $Cl_T = 0.04\ M$; unit atomic ratios for oxidants and reductants at the boundaries. In dilute solutions $Cl_2(aq)$ exists only at low pH. Cl_2, OCl^-, and $HOCl$ are all unstable or metastable in water. Numbers 1–4 refer to the equilibria described by equations iv, v, vi, and vii, respectively, of Example 7.9.

either at a given pε or at a given pH may immediately clarify the stability relations.) Equations vi and vii have slopes of -0.5 and -1, respectively, in the graphical representation; the lines intersect at $pH = pK$ of the hypochlorous acid.

It is convenient to introduce the equations that define the stability limit of H_2O into the same diagram.

Figure 7.8 reveals the following pertinent features of the equilibria involving Cl at the oxidation state $-I$, $\bar{0}$, and $+I$:

1 $Cl_2(aq)$ in dilute solutions exists only at rather low pH. Addition of Cl_2 to water is accompanied by disproportionation into $HOCl$ and Cl^-. The disproportionation equilibrium is given by

$$Cl_2(aq) + H_2O = HClO + H^+ + Cl^- \qquad \log K = -3.3 \qquad \text{(ix)}$$

This reaction is obtained by combining equations i and ii.

2 Cl as Cl_2 or HOCl and ClO^- is a strong oxidant, stronger than O_2; in other words, Cl_2, HOCl, and ClO^- are thermodynamically unstable in water because these species oxidize water. (In the absence of suitable catalysts, however, this reaction is extremely slow.)

3 Within the entire $p\varepsilon$–pH range of natural waters, Cl^- is the stable Cl species; it cannot be oxidized by O_2.

Example 7.10. pε–pH Diagram for the Fe–CO₂–H₂O System

Construct a $p\varepsilon$–pH diagram (25°C) for the Fe–CO_2–H_2O system; delineate the stability conditions of the solid phases Fe, $Fe(OH)_2$, $FeCO_3$, and amorphous $Fe(OH)_3$; $C_T = 10^{-3}$ M, concentration of soluble Fe species $= 10^{-5}$ M. Free energy data and pertinent equilibrium constants are given in Table 5.1 and in the appendix at the end of the book. We also have to consider the following hydrolysis equilibria (25°C, 1 atm):

$$Fe^{3+} + 3H_2O = Fe(OH)_3(s) + 3H^+ \qquad \log K = -3.0 \qquad \text{(i)}$$

$$Fe^{3+} + H_2O = FeOH^{2+} + H^+ \qquad \log {}^*K_1 = -2.2 \qquad \text{(ii)}$$

$$Fe^{3+} + 2H_2O = Fe(OH)_2^+ + 2H^+ \qquad \log {}^*\beta_2 = -7.2 \qquad \text{(iii)}$$

$$Fe(OH)_3(s) + H_2O = Fe(OH)_4^- + H^+ \qquad \log {}^*K_s = -19.2 \qquad \text{(iv)}$$

$$Fe^{2+} + H_2O = FeOH^+ + H^+ \qquad \log {}^*K_1 = -9.5 \qquad \text{(v)}$$

$$Fe^{2+} + 2H_2O = Fe(OH)_2(s) + 2H^+ \qquad \log K = -11.7 \qquad \text{(vi)}$$

Table 7.4 gives the equations needed; the results are displayed in Figure 7.7a. Figure 7.7b gives a $p\varepsilon$–pH diagram for the corresponding Mn–CO_2–H_2O system.

Comparing Various Equilibrium Diagrams. The main advantage of a $p\varepsilon$–pH diagram is that it provides a good survey and a clear picture of the situation, but it cannot give too much detail, especially about the concentration dependence of the predominance areas. It is possible to draw the boundary lines for various assumed activities and to construct three-dimensional diagrams with activity as one of the axis. There is little limit to the combinations of variables and types of phase diagrams that can be constructed, but we must not forget that the main reason for making a phase diagram is to try to understand or to solve complicated equilibrium problems.

In the authors' opinion, activity ratio diagrams (at constant $p\varepsilon$ or constant pH) consisting of only straight lines and showing quantitatively the concentration dependence provide the most useful information for the small labor involved.

TABLE 7.4 EQUATIONS NEEDED FOR THE CONSTRUCTION OF A pε–pH DIAGRAM FOR THE Fe–CO$_2$–H$_2$O SYSTEM

$Fe^{3+} + e = Fe^{2+}$ (1)

$Fe^{2+} + 2e = Fe(s)$ (2)

$Fe(OH)_3 \text{ (amorph, s)} + 3H^+ + e = Fe^{2+} + 3H_2O$ (3)

$Fe(OH)_3 \text{ (amorph, s)} + 2H^+ + HCO_3^- + e = FeCO_3(s) + 3H_2O$ (4)

$FeCO_3(s) + H^+ + 2e = Fe(s) + HCO_3^-$ (5)

$Fe(OH)_2(s) + 2H^+ + 2e = Fe(s) + 2H_2O$ (6)

$Fe(OH)_3(s) + H^+ + e = Fe(OH)_2(s) + H_2O$ (7)

$FeOH^{2+} + H^+ + e = Fe^{2+} + H_2O$ (8)

$FeCO_3(s) + 2H_2O = Fe(OH)_2(s) + H^- + HCO_3^-$ (a)

$FeCO_3(s) + H^+ = Fe^{2+} + HCO_3^-$ (b)

$FeOH^{2+} + 2H_2O = Fe(OH)_3(s) + 2H^+$ (c)

$Fe^{3+} + H_2O = FeOH^{2+} + H^+$ (d)

$Fe(OH)_3(s) + H_2O = Fe(OH)_4^- + H^{+\,*}$ (e)

pε Functions

$$p\varepsilon = 13 + \log\{Fe^{3+}\}/\{Fe^{2+}\} \tag{1}$$

$$p\varepsilon = -6.9 + \tfrac{1}{2}\log\{Fe^{2+}\} \tag{2}$$

$$p\varepsilon = 16 - \log\{Fe^{2+}\} - 3pH \tag{3}$$

$$p\varepsilon = 16 - 2pH + \log\{HCO_3^-\} \tag{4}$$

where $\{HCO_3^-\} = C_T \alpha_1$

$$p\varepsilon = -7.0 - \tfrac{1}{2}pH - \tfrac{1}{2}\log\{HCO_3^-\} \tag{5}$$

$$p\varepsilon = -1.1 - pH \tag{6}$$

$$p\varepsilon = 4.3 - pH \tag{7}$$

$$p\varepsilon = 15.2 - pH - \log(\{Fe^{2+}\}/\{FeOH^{2+}\}) \tag{8}$$

pH Functions

$$pH = 11.9 + \log\{HCO_3^-\} \tag{a}$$

$$pH = 0.2 - \log\{Fe^{2+}\} - \log\{HCO_3^-\} \tag{b}$$

$$pH = 0.4 - \tfrac{1}{2}\log\{FeOH^{2+}\} \tag{c}$$

$$pH = 2.2 - \log(\{Fe^{3+}\}/\{FeOH^{2+}\}) \tag{d}$$

$$pH = 19.2 + \log\{Fe(OH)_4^-\} \tag{e}$$

447

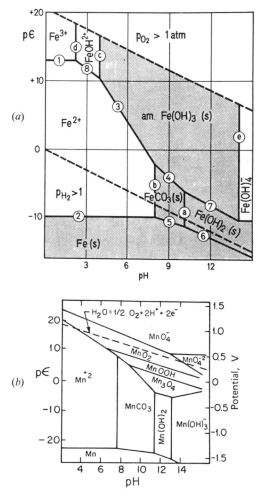

Figure 7.7 $p\varepsilon$–pH diagrams for the Fe, CO_2, H_2O, and Mn–CO_2 systems (25°C). (*a*) Solid phases considered: amorphous $Fe(OH)_3$, $FeCO_3$ (siderite), $Fe(OH)_2$, Fe, $C_T = 10^{-3}$ M, [Fe] $= 10^{-5}$ M. Equilibria and equations needed for construction of diagram are given in Table 7.4. (*b*) Solid phases considered: $Mn(OH)_2(s)$ (pyrochroite), $MnCO_3(s)$ (rhodochrosite), $Mn_3O_4(s)$ (hausmannite), γ-MnOOH (manganite), γ-MnO_2 (nsutite).

7.5 REDOX CONDITIONS IN NATURAL WATERS

Only a few elements—C, N, O, S, Fe, Mn—are predominant participants in aquatic redox processes. Table 7.5 presents equilibrium constants for several couples pertinent to consideration of redox relationships in natural waters and their sediments. Data are taken principally from the second edition of *Stability Constants of Metal-Ion Complexes*. A subsidiary symbol $p\varepsilon°(W)$ is convenient for considering redox situations in natural waters. $p\varepsilon°(W)$ is

TABLE 7.5 EQUILIBRIUM CONSTANTS OF REDOX PROCESSES PERTINENT IN AQUATIC CONDITIONS (25°C)

Reaction	$p\varepsilon^\circ$ ($\equiv \log K$)	$p\varepsilon^\circ$ (W)[a]
(1) $\frac{1}{4}O_2(g) + H^+ + e = \frac{1}{2}H_2O$	+20.75	+13.75
(2) $\frac{1}{5}NO_3^- + \frac{6}{5}H^+ + e = \frac{1}{10}N_2(g) + \frac{3}{5}H_2O$	+21.05	+12.65
(3) $\frac{1}{2}MnO_2(s) + \frac{1}{2}HCO_3^-(10^{-3}) + \frac{3}{2}H^+ + e$		
$\quad = \frac{1}{2}MnCO_3(s) + H_2O$	—	+8.9[b]
(4) $\frac{1}{2}NO_3^- + H^+ + e = \frac{1}{2}NO_2^- + \frac{1}{2}H_2O$	+14.15	+7.15
(5) $\frac{1}{8}NO_3^- + \frac{5}{4}H^+ + e = \frac{1}{8}NH_4^+ + \frac{3}{8}H_2O$	+14.90	+6.15
(6) $\frac{1}{6}NO_2^- + \frac{4}{3}H^+ + e = \frac{1}{6}NH_4^+ + \frac{1}{3}H_2O$	+15.14	+5.82
(7) $\frac{1}{2}CH_3OH + H^+ + e = \frac{1}{2}CH_4(g) + \frac{1}{2}H_2O$	+9.88	+2.88
(8) $\frac{1}{4}CH_2O + H^+ + e = \frac{1}{4}CH_4(g) + \frac{1}{4}H_2O$	+6.94	−0.06
(9) $FeOOH(s) + HCO_3^-(10^{-3}) + 2H^+ + e$		
$\quad = FeCO_3(s) + 2H_2O$	—	−0.8[b]
(10) $\frac{1}{2}CH_2O + H^+ + e = \frac{1}{2}CH_3OH$	+3.99	−3.01
(11) $\frac{1}{6}SO_4^{2-} + \frac{4}{3}H^+ + e = \frac{1}{6}S(s) + \frac{2}{3}H_2O$	+6.03	−3.30
(12) $\frac{1}{8}SO_4^{2-} + \frac{5}{4}H^+ + e = \frac{1}{8}H_2S(g) + \frac{1}{2}H_2O$	+5.25	−3.50
(13) $\frac{1}{8}SO_4^{2-} + \frac{9}{8}H^+ + e = \frac{1}{8}HS^- + \frac{1}{2}H_2O$	+4.25	−3.75
(14) $\frac{1}{2}S(s) + H^+ + e = \frac{1}{2}H_2S(g)$	+2.89	−4.11
(15) $\frac{1}{8}CO_2(g) + H^+ + e = \frac{1}{8}CH_4(g) + \frac{1}{4}H_2O$	+2.87	−4.13
(16) $\frac{1}{6}N_2(g) + \frac{4}{3}H^+ + e = \frac{1}{3}NH_4^+$	+4.68	−4.68
(17) $\frac{1}{2}(NADP^+) + \frac{1}{2}H^+ + e = \frac{1}{2}(NADPH)$	−2.0	−5.5[c]
(18) $H^+ + e = \frac{1}{2}H_2(g)$	0.0	−7.00
(19) Oxidized ferredoxin + e = reduced ferredoxin	−7.1	7.1[d]
(20) $\frac{1}{4}CO_2(g) + H^+ + e = \frac{1}{24}$ (glucose) $+ \frac{1}{4}H_2O$	−0.20	−7.20[e]
(21) $\frac{1}{2}HCOO^- + \frac{3}{2}H^+ + e = \frac{1}{2}CH_2O + \frac{1}{2}H_2O$	+2.82	−7.68
(22) $\frac{1}{4}CO_2(g) + H^+ + e = \frac{1}{4}CH_2O + \frac{1}{4}H_2O$	−1.20	−8.20
(23) $\frac{1}{2}CO_2(g) + \frac{1}{2}H^+ + e = \frac{1}{2}HCOO^-$	−4.83	−8.33

[a] Values for $p\varepsilon^\circ$ (W) apply to the electron activity for unit activities of oxidant and reductant in neutral water, that is, at pH = 7.0 for 25°C.
[b] These data correspond to $(HCO_3^-) = 10^{-3}$ M rather than unity and so are not exactly $p\varepsilon^\circ$ (W); they represent typical aquatic conditions more nearly than $p\varepsilon^\circ$ (W) values do.
[c] M. Calvin and J. A. Bassham, *The Photosynthesis of Carbon Compounds*, W. A. Benjamin, Menlo Park, Calif., 1962. NADP = Nicotinanide adenine dinucleotide phosphate.
[d] D. I. Arnon, *Science*, **149**, 1460 (1965).
[e] A. L. Lehninger, *Bioenergetics*, W. A. Benjamin, Menlo Park, Calif., 1965.

analogous to $p\varepsilon^\circ$ except that $\{H^+\}$ and $\{OH^-\}$ in the redox equilibrium equations are assigned their activities in neutral water. Values for $p\varepsilon^\circ$(W) for 25°C thus apply to unit activities of oxidant and reductant at pH = 7.00. $p\varepsilon^\circ$(W) is defined by

$$p\varepsilon^\circ(W) = p\varepsilon^\circ + \frac{n_H}{2} \log K_W \qquad (45)$$

where n_H is the number of moles of protons exchanged per mole of electrons.

The listing of $p\varepsilon^\circ(W)$ values in Table 7.5 permits an immediate grading of different systems in the order of their oxidizing intensity at pH $= 7$. Any system in Table 7.5 will tend to oxidize equimolar concentrations of any other system having a lower $p\varepsilon^\circ(W)$ value. For example, we see that, at pH $= 7$, NO_3^- can oxidize HS^- to SO_4^{2-}:

$$\tfrac{1}{8}NO_3^- + \tfrac{5}{4}H^+(W) + e = \tfrac{1}{8}NH_4^+ + \tfrac{3}{8}H_2O$$
$$p\varepsilon^\circ(W) = +6.15, \log K(W) = +6.15 \quad (46)$$

$$\tfrac{1}{8}HS^- + \tfrac{1}{2}H_2O = \tfrac{1}{8}SO_4^{2-} + \tfrac{9}{8}H^+(W) + e$$
$$p\varepsilon^\circ(W) = -3.75, \log K(W) = +3.75 \quad (47)$$

$$\tfrac{1}{8}NO_3^- + \tfrac{1}{8}HS^- + \tfrac{1}{8}H^+(W) + \tfrac{1}{8}H_2O = \tfrac{1}{8}NH_4^+ + \tfrac{1}{8}SO_4^{2-}$$
$$\log K(W) = +9.9 \quad (48)$$

or

$$NO_3^- + HS^- + H^+(W) + H_2O = NH_4^+ + SO_4^{2-} \quad \log K(W) = 79.2 \quad (49)$$

[$\log K(W)$ is the equilibrium constant for the redox reaction in neutral water, pH $= 7.00$ at $25°C$; note that $p\varepsilon^\circ$, because it is a measure of oxidizing intensity, maintains the same sign independent of the direction in which the reaction is

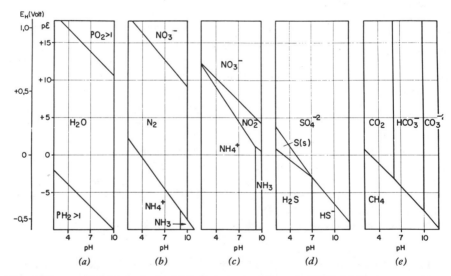

Figure 7.8 $p\varepsilon$–pH diagrams for biologically important elements ($25°C$). (a) The upper and lower lines represent equations 44 and 43, respectively, the oxygen and hydrogen equilibria with water. (b) The nitrogen system, considering only stable equilibria. The only oxidation states involved are (-III), the elemental state, and (V). (c) NH_4^+, NH_3, NO_3^- and NO_2^- are treated as species metastable with regard to N_2; that is, N_2 is treated as a redox inert component (cf. Figure 9.10c and text). (d) Sulfur species stable for assumed conditions are SO_4^{2-}, elemental sulfur, and sulfides (cf. Figure 7.5 and Example 7.7). (e) The thermodynamically possible existence of elemental C (graphite) is ignored.

written.] Since $\Delta p\varepsilon°(W)$ or $\log K(W)$ is positive, the reaction is thermo-dynamically possible in neutral aqueous solutions at standard concentrations. Figure 7.8 gives $p\varepsilon$–pH diagrams for biologically important elements.

Example 7.11. Oxidation of Organic Matter by SO_4^{2-}

Is the oxidation of organic matter, here CH_2O, thermodynamically possible under conditions normally encountered in natural water systems?

The overall process is obtained by combining (12) and (22) in Table 7.5.

$$\tfrac{1}{8}SO_4^{2-} + \tfrac{5}{4}H^+(W) + e = \tfrac{1}{8}H_2S(g) + \tfrac{1}{2}H_2O$$
$$p\varepsilon°(W) = -3.50, \quad \log K(W) = -3.50 \quad \text{(i)}$$

$$\tfrac{1}{4}CH_2O + \tfrac{1}{4}H_2O = \tfrac{1}{4}CO_2(g) + H^+(W) + e$$
$$p\varepsilon°(W) = -8.20, \quad \log K(W) = +8.20 \quad \text{(ii)}$$

$$\tfrac{1}{8}SO_4^{2-} + \tfrac{1}{4}CH_2O + \tfrac{1}{4}H^+(W) = \tfrac{1}{8}H_2S(g) + \tfrac{1}{4}CO_2(g) + \tfrac{1}{4}H_2O$$
$$\log K(W) = +4.70 \quad \text{(iii)}$$

This may also be written

$$SO_4^{2-} + 2CH_2O + 2H^+(W) = H_2S(g) + 2CO_2(g) + 2H_2O;$$
$$\log K(W) - 37.6$$

Hence at standard concentrations the reaction at pH = 7 is thermodyna-mically possible. The same results will also hold for any equal fractions of unit activity because the numbers of molecules of sulfur-containing and carbon-containing species do not change as a result of the reaction. Further-more, for assumed actual conditions a calculation of $\Delta G = RT \ln(Q/K)$ will show whether the oxidation can proceed thermodynamically. Thus, for a set of assumed actual conditions, such as $p_{CO_2} = 10^{-3.5}$ atm, $[CH_2O] = 10^{-6}$ M, $[SO_4^{2-}] = 10^{-3}$ M, $p_{H_2S} = 10^{-2}$ M, a value of $10^{-31.6}$ is obtained for Q/K; hence ΔG is clearly negative, indicating that SO_4^{2-} can oxidize CH_2O under these conditions.

Redox Intensity and the Biochemical Cycle

The maintenance of life resulting directly or indirectly from a steady impact of solar energy (*photosynthesis*) is the main cause of nonequilibrium conditions (Figure 7.9). Photosynthesis may be conceived as a process producing localized centers of highly negative $p\varepsilon$ and a reservoir of oxygen. Nonphotosynthetic organisms tend to restore equilibrium by catalytically decomposing the un-stable products of photosynthesis through energy-yielding redox reactions. Organisms, themselves a product of inorganic matter, are primarily built up from "redox elements," and their relatively constant stoichiometric composition $(C_{106}H_{263}O_{110}N_{16}P)$ and the cyclic exchange between chemical elements of

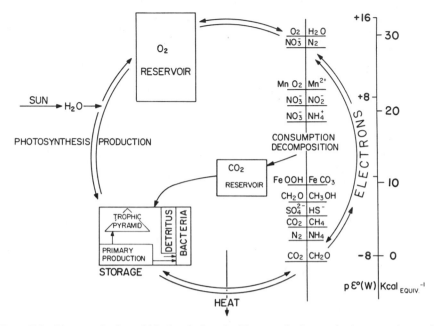

Figure 7.9 Photosynthesis and biochemical cycle. Photosynthesis may be interpreted as a disproportionation into an oxygen reservoir and reduced organic matter (biomass containing high-energy bonds made with hydrogen and C, N, S, and P compounds). The nonphotosynthetic organisms tend to restore equilibrium by catalytically decomposing the unstable products of photosynthesis through energy-yielding redox reactions. The $p\varepsilon°(W)$ scale on the right gives the sequence of the redox reactions observed in an aqueous system.

the water and the biomass have pronounced effects on the relative concentrations of the elements in the environment. The biologically active elements circulate in a different pattern than water itself or inactive (conservative) solutes.

Photosynthesis, by trapping light energy and converting it to chemical energy, produces reduced states of higher free energy (high-energy chemical bonds) and thus nonequilibrium concentrations of C, N, and S compounds. In a simplified context, photosynthesis may be conceived as a process producing localized centers of highly negative $p\varepsilon$. As shown in (20) of Table 7.5, the conversion of CO_2 to glucose at unit activities requires $p\varepsilon°(W) = -7.2$. Although this value may be modified somewhat for actual intracellular activities, it does represent approximately the negative $p\varepsilon$ level that must be reached during photosynthesis.

The NADP system [5], ubiquitous in living organisms and believed to play a major role in electron transport during photosynthesis, exhibits $p\varepsilon°(W) = +5.5$[†] Moreover, various ferredoxins, now widely considered to be the pri-

[†] Hydrogen resulting from the photolysis of water to be used eventually for the reduction of CO_2 is first bound to the coenzyme nicotinamide adenine dinucleotide phosphate (NADP).

mary electron receptors from excited chlorophylls, show $p\varepsilon°(W)$ values in the range -7.0 to -7.5 (Table 7.5). The coincidence of this range with the $p\varepsilon°(W)$ value for conversion of CO_2 to glucose is suggestive.

In contrast the respiratory, fermentative, and other nonphotosynthetic processes of organisms tend to restore equilibrium by catalyzing or mediating chemical reactions releasing free energy and thus increasing the mean $p\varepsilon$ level.

Water in solubility equilibrium with atmospheric oxygen has a well-defined $p\varepsilon = 13.6$ (for $P_{O_2} = 0.21$ atm, $E_H = 800$ mV at pH 7 and 25°C). Calculations from the $p\varepsilon°$ values of Table 7.5 show that at this $p\varepsilon$ all the other elements should exist virtually completely in their highest naturally occurring oxidation states: C as CO_2, HCO_3^-, or CO_3^{2-} with reduced organic forms less than 10^{-35} M; N as NO_3^- with NO_2^- less than 10^{-7} M; S as SO_4^{2-} with SO_3^{2-} or HS^- less than 10^{-20} M; Fe as $FeOOH$ or Fe_2O_3 with Fe^{2+} less than 10^{-18} M; and Mn as MnO_2 with Mn^{2+} less than 10^{-10} M. Even the N_2 from the atmosphere should be largely oxidized to NO_3^-.

Since in fact N_2 and organic matter are known to persist in waters containing dissolved oxygen, a total redox equilibrium is not found in natural water systems, even in surface films. At best there are partial equilibria, treatable as approximations to equilibrium either because of slowness of interaction with the other redox couples or because of isolation from the total environment as a result of slowness of diffusional or mixing processes.

The ecological systems of natural waters are thus more adequately represented by dynamic than by equilibrium models. The former are needed to describe the free energy flux absorbed from light and released in subsequent redox processes. Equilibrium models can only depict the thermodynamically stable state and describe the direction and extent of processes tending toward it.

When comparisons are made between calculations for an equilibrium redox state and concentrations in the dynamic aquatic environment, the implicit assumptions are that the biological mediations are operating essentially in a reversible manner at each stage of the ongoing processes or that there is a metastable steady state that approximates the partial equilibrium state for the system under consideration.

Microbial Mediation

As already pointed out, nonphotosynthetic organisms tend to restore equilibrium by catalytically decomposing the unstable products of photosynthesis through energy-yielding redox reactions, thereby obtaining a source of energy for their metabolic needs. The organisms use this energy both to synthesize new cells and to maintain the old cells already formed [6]. The energy exploitation is of course not 100% efficient; only a proportion of the free energy released can become available for cell use. It is important to keep in mind that organisms cannot carry out *gross* reactions that are thermodynamically not possible. From a point of view of overall reactions these organisms act only as redox

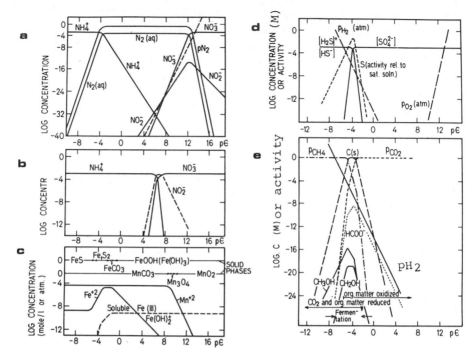

Figure 7.10 Equilibrium concentrations of biochemically important redox components as a function of pε at a pH of 7.0. (a) Nitrogen; (b) Nitrogen, with elemental nitrogen N_2 ignored; (c) Iron and Manganese; (d) Sulfur; (e) Carbon. These equilibrium diagrams have been constructed from equilibrium constants listed in Table 7.5 for the following concentrations: C_T (total carbonate carbon) = $10^{-3} M$; $[H_2S(aq)] + [HS^-] + [SO_4^{2-}] = 10^{-3} M$; $[NO_3^-] + [NO_2^-] + [NH_4^+] = 10^{-3} M$; $p_{N_2} = 0.78$ atm and thus $[N_2(aq)] = 0.5 \times 10^{-3} M$. For the construction of (b) the species NH_4^-, NO_3^-, and NO_2^- are treated as metastable with regard to N_2.

catalysts. Therefore organisms do not oxidize substrates or reduce O_2 or SO_4^{2-}, they only mediate the reaction or, more specifically, the electron transfer, for example, of the specific oxidation of the substrate and of the reduction of O_2 or SO_4^{2-}. Since, for example, SO_4^{2-} can be reduced only at a given pε or redox potential, an equilibrium model characterizes the pε range in which reduction of sulfate is not possible and the pε range in which reduction of sulfate is possible. Thus pε is a parameter that characterizes the ecological milieu restrictively. The pε range in which certain oxidation or reduction reactions are possible can be estimated by calculating equilibrium composition as a function of pε. This has been done for nitrogen, manganese, iron, sulfur, and carbon at a pH of 7. The results are shown in Figure 7.10.

Nitrogen System. Figure 7.10a shows relationships among several oxidation states of nitrogen as a function of pε for a total atomic concentration of nitrogen-containing species equal to $10^{-3} M$. The maximum $N_2(aq)$ concentration is therefore 5×10^{-4} M, corresponding to about p_{N_2} of 0.77 atm. For most of the aqueous range of pε, N_2 gas is the most stable species, but at quite negative pε values ammonia becomes predominant and nitrate dominates for pε greater than $+12$ and pH $= 7$. The fact that nitrogen gas has not been converted largely into nitrate under prevailing aerobic conditions at land and water surfaces indicates a lack of efficient biological mediation of the reverse reaction also, for the mediating catalysis must operate equally well for reaction in both directions. It appears then that denitrification must occur by an indirect mechanism such as reduction of NO_3^- to NO_2^- followed by reaction of NO_2^- with NH_4^+ to produce N_2 and H_2O.†

Because reduction of N_2 to NH_4^+ (N_2 *fixation*) at pH 7 can occur to a substantial extent when pε is less than about -4.5, the level of pε required is not as negative as for the reduction of CO_2 to CH_2O. It is not surprising then that blue-green algae are able to mediate this reduction at the negative pε levels produced by photosynthetic light energy. What is perhaps surprising is that nitrogen fixation does not occur more widely among photosynthetic organisms and proceed to a greater extent as compared with CO_2 reduction. Kinetic problems in breaking the strong bonding of the N_2 molecule probably are major factors here.

Because of this kinetic hindrance between "bound" nitrogen and N_2, it might also be useful to consider a system in which NO_3^-, NO_2^-, and NH_4^+ are treated as components metastable with respect to gaseous N_2. A diagram for such a system, Figure 7.10b, shows the shifts in relative predominance of the three species all within the rather narrow pε range from 5.8 to 7.2. That each of the species has a dominant zone within this pε range seems to be a factor contributing to the observed highly mobile characteristics of the nitrogen cycle.

Sulfur System. The reduction of SO_4^{2-} to H_2S or HS^- provides a good example of the application of equilibrium considerations to aquatic relationships. Figure 7.10d shows relative activities of SO_4^{2-} and H_2S at pH 7 and 25°C as a function of pε when the total concentration of sulfur is 1 mM. It is apparent that a significant reduction of SO_4^{2-} to H_2S at this pH requires p$\varepsilon < -3$. The biological enzymes that mediate this reduction with oxidation of organic matter must then operate at or below this pε. Because the system is dynamic rather than static, only an upper bound can be set in this way, for the excess driving force in terms of pε at the mediation site is not indicated by

---

†Because of the nonreversibility of the biological mediation of the $NO_3^- \rightleftharpoons N_2$ conversion, the NO_3^--N_2 couple cannot be used as a reliable redox indicator. For example, NO_3^- may be reduced to N_2 in an aquatic system even if the bulk phase contains some dissolved oxygen. The reduction may occur in a microenvironment with a pε value lower than that of the bulk, for example, inside a floc or within the sediments; the N_2 released to the aerobic bulk phase is not reoxidized although reoxidation is thermodynamically feasible.

equilibrium computations. Since, however, many biologically mediated reactions seem to operate with relatively high efficiency for utilizing free energy, it appears like that the operating $p\varepsilon$ value is not greatly different from the equilibrium value.

Combining (11) and (14) of Table 7.5 gives

$$SO_4^{2-} + 2H^+(W) + 3H_2S = 4S(s) + 4H_2O \qquad \log K(W) = 4.86 \quad (50)$$

This equation indicates a possibility of formation of solid elemental sulfur in the reduction of sulfate at pH 7 and standard concentrations. A concentration of 1 M sulfate is unusual, however. If the concentration of SO_4^{2-} is reduced to about 0.018 M (about 1600 mg liter^{-1}) and the H_2S activity is taken correspondingly as 0.09 atm (H_2S is about half-ionized at pH 7, and the solubility of H_2S at 1 atm is about 0.1 M; thus this condition means about 0.018 M total sulfide), then $\Delta G = RT \ln Q/K$ is positive in accordance with Figure 7.5; solid sulfur cannot form thermodynamically by this reaction at pH 7.

The solubility of $CaSO_4(s)$ is about 0.016 M at 25°C. According to the foregoing rough calculation, sulfur should form in the reduction of SO_4^{2-} in saturated $CaSO_4(s)$ only if the pH is somewhat below 7 (see Example 7.4). There are some indications that this conclusion agrees with the condition of natural sulfur formation. Elemental sulfur may be formed, however, as a kinetic intermediate or as a metastable phase under many natural conditions.

Iron and Manganese. In constructing Figure 7.10c, solid FeOOH ($\bar{G}_f^\circ = -98$ kcal mol^{-1}) has been assumed as the stable ferric oxide. Although thermodynamically possible, magnetite [$Fe_3O_4(s)$] has been ignored as an intermediate in the reduction of ferric oxide to Fe(II). As Figure 7.10 shows, in the presence of O_2, $p\varepsilon > 11$, aqueous iron and manganese are stable only as solid oxidized oxides. Soluble forms are present at concentrations less than 10^{-9} M. The concentration of soluble iron and manganese, as Fe^{2+} and Mn^{2+}, increases with decreasing $p\varepsilon$, the highest concentrations being controlled by the solubility of $FeCO_3(s)$ and $MnCO_3(s)$, respectively. ($[HCO_3^-] = 10^{-3}$ M has been assumed for construction of the diagram.)

Carbon System. A great number of organic compounds are synthesized, transformed, and decomposed—mostly by microbial catalysis—continually. For operation of the *carbon cycle* degradation is just as important as synthesis (Chapter 8). With the exception of CH_4, no organic solutes encountered in natural waters are thermodynamically stable. For example, the disproportionation of acetic acid

$$CH_3COOH = 2H_2O + 2C(s) \qquad \log K = 18$$
$$CH_3COOH = CH_4(g) + CO_2(g) \qquad \log K = 9$$

is thermodynamically favored, but prevented by slow kinetics. Similarly, formaldehyde is unstable with respect to its decomposition into carbon (graphite) and water

$$CH_2O(aq) = C(s) + H_2O; \quad \log K = 18.7$$

but there is no evidence that this reaction occurs.

Even though reversible equilibria are not attained at low temperatures, it is of considerable interest to compare the equilibrium constants of the various steps in the oxidation of organic matter. The compounds CH_4, CH_3OH, CH_2O, and $HCOO^-$ given in Figure 7.10e represent organic material with formal oxidation states of $-IV$, $-II$, $\bar{0}$, and $+II$, respectively. The diagram has been constructed for the condition $p_{CH_4} + p_{CO_2} = 1$ atm.† [The construction of such a diagram is facilitated by first drawing an activity ratio diagram using $\{CH_2O(aq)\}$ as a reference.] The major feature of the equilibrium carbon system is simply a conversion of predominant CO_2 to predominant CH_4 with a half-way point at $p\varepsilon = -4.13$. At this $p\varepsilon$ value, where the other oxidation states exhibit maximum relative occurrence, formation of graphite is thermodynamically possible.

Methane fermentation may be considered a reduction of CO_2 to CH_4; this reduction may be accompanied by oxidation of any one of the intermediate oxidation states.‡ Since all of the latter have $p\varepsilon°(W)$ values less than -6.2 (this for CH_3OH), each can provide the negative $p\varepsilon$ level required thermodynamically for reduction of CO_2 to CH_4 in its oxidation. Physiologically different organisms may typically be involved in methanogenesis. Certain organisms break down organic materials to organic acids and alcohols, producing eventually acetate, H_2, and CO_2 as intermediates [7]:

$$\text{complex organic material} \longrightarrow \begin{array}{c} \text{organic} \\ \text{acids} \end{array} \begin{array}{c} \nearrow H_2 + CO_2 \longrightarrow CH_4 \\ \searrow CH_3COO^- \longrightarrow CH_4 \end{array}$$
$$(51)$$

As indicated in equation 51 H_2—formed by redox disproportionation, for example, by β-oxidation of fatty acids such as $CH_3CH_2CH_2COO^- + H_2O = 2CH_3COO^- + 2H_2(g) + H^+$—acts as a reductant of CO_2:

$$4H_2(g) + CO_2(g) = CH_4 + 2H_2O \quad \Delta G° = -31 \text{ kcal} \quad \log K = 22.9$$
$$(25°C) \quad (52)$$

† This condition is not fulfilled in the $p\varepsilon$ range where $\{C(s)\} = 1$.
‡ This statement does not imply a mechanism. Methane may be formed directly, for example, from acetic acid:

$$CH_3COOH \longrightarrow CH_4 + CO_2$$

This reaction could be classified as the sum of the reactions $CH_3COOH + 2H_2O = 2CO_2 + 8H^+ + 8e$; $CO_2 + 8e + 8H^+ = CH_4 + 2H_2O$.

Alcohol fermentation may be exemplified by redox disproportionation of CH_2O (or $C_6H_{12}O_6$):

$$CH_2O + CH_2O + H_2O = CH_3OH + HCOO^- + H^+ \tag{53}$$

or

$$CH_2O + 2CH_2O + H_2O = 2CH_3OH + CO_2(g) \tag{54}$$

$$C_6H_{12}O_6 = 2C_2H_5OH + 2CO_2(g) \qquad \Delta G° = -58.3 \text{ kcal} \tag{55}$$

The reduction of CH_2O to CH_3OH can occur at $p\varepsilon < -3$. Because the concomitant oxidation of CH_2O to CO_2 has $p\varepsilon°(W) = -8.2$, there is no thermodynamic problem.

Microbially Mediated Oxidation and Reduction Reactions

Although, as stressed, conclusions regarding chemical dynamics may not generally be drawn from thermodynamic considerations, it appears that all the reactions discussed in the previous section, except possibly those involving $N_2(g)$ and $C(s)$, are biologically mediated in the presence of suitable and abundant biota. Table 7.6 surveys the oxidation and reduction reactions which may be combined to result in exergonic processes. The possible combinations represent the well-known reactions mediated by heterotrophic and chemoautotrophic organisms. It appears that in natural habitats organisms capable of mediating the pertinent redox reactions are nearly always found.

The Sequence of Redox Reactions.   In a closed aqueous system containing organic material—say CH_2O—oxidation of organic matter is observed to occur first by reduction of O_2 [$p\varepsilon(W) = 13.8$]. This will be followed by reduction of NO_3^- and NO_2^-. As seen in Figures 7.9 and 7.10, the succession of these reactions follows the decreasing $p\varepsilon$ level. Reduction of MnO_2 if present should occur at about the same $p\varepsilon$ levels as that of nitrate reduction, followed by reduction of $FeOOH(s)$ to Fe^{2+}. When sufficiently negative $p\varepsilon$ levels have been reached, fermentation reactions and reduction of SO_4^{2-} and CO_2 may occur almost simultaneously.

The described sequence would be expected if reactions tended to occur in order of their thermodynamic possibility. The reductant (CH_2O) will supply electrons to the lowest unoccupied electron level (O_2); with more electrons available, successive levels—NO_3^-, NO_2^-, $MnO_2(s)$ and so on—will be filled up. The described succession of reactions is mainly reflected in the vertical distribution of components in a nutrient-enriched (eutrophied) lake and in general also in the temporal succession in a closed system containing excess organic matter, such as a batch digester (anaerobic fermentation unit).

Since the reactions considered (with the possible exception of the reduction of $MnO_2(s)$ and $FeOOH(s)$) are biologically mediated, the chemical reaction

TABLE 7.6 REDUCTION AND OXIDATION REACTIONS THAT MAY BE COMBINED TO RESULT IN BIOLOGICALLY MEDIATED EXERGONIC PROCESSES (pH = 7)

Reduction	$p\varepsilon°(W) = \log K(W)$	Oxidation	$p\varepsilon°(W) = -\log K(W)$
(A) $\frac{1}{4}O_2(g) + H^+(W) + e = \frac{1}{2}H_2O$	$+13.75$	(L) $\frac{1}{4}CH_2O + \frac{1}{4}H_2O = \frac{1}{4}CO_2(g) + H^+(W) + e$	-8.20
(B) $\frac{1}{5}NO_3^- + \frac{6}{5}H^+(W) + e = \frac{1}{10}N_2(g) + \frac{3}{5}H_2O$	-12.65	(L-1) $\frac{1}{2}HCOO^- = \frac{1}{2}CO_2(g) + \frac{1}{2}H^+(W) + e$	-8.73
(C) $\frac{1}{2}MnO_2(s) + \frac{1}{2}HCO_3^- (10^{-3}) + \frac{3}{2}H^+(W)$ $+ e = \frac{1}{2}MnCO_3(s) + H_2O$	$+8.9$	(L-2) $\frac{1}{2}CH_2O + \frac{1}{2}H_2O = \frac{1}{2}HCOO^-$ $+ \frac{3}{2}H^+(W) + e$	-7.68
(D) $\frac{1}{8}NO_3^- + \frac{5}{4}H^+(W) + e = \frac{1}{8}NH_4^+ + \frac{3}{8}H_2O$	$+6.15$	(L-3) $\frac{1}{2}CH_3OH = \frac{1}{2}CH_2O + H^+(W) + e$	-3.01
(E) $FeOOH(s) + HCO_3^- (10^{-3}) + 2H^+(W)$ $+ e = FeCO_3(s) + 2H_2O$	-0.8	(L-4) $\frac{1}{2}CH_4(g) + \frac{1}{2}H_2O = \frac{1}{2}CH_3OH$ $+ H^+(W) - e$	$+2.88$
(F) $\frac{1}{2}CH_2O + H^+(W) + e = \frac{1}{2}CH_3OH$	-3.01	(M) $\frac{1}{8}HS^- + \frac{1}{2}H_2O = \frac{1}{8}SO_4^{2-}$ $+ \frac{9}{8}H^+(W) + e$	-3.75
(G) $\frac{1}{8}SO_4^{2-} + \frac{9}{8}H^+(W) + e = \frac{1}{8}HS^- + \frac{1}{2}H_2O$	-3.75	(N) $FeCO_3(s) + 2H_2O = FeOOH(s)$ $+ HCO_3^- (10^{-3}) + 2H^+(W) - e$	-0.8
(H) $\frac{1}{8}CO_2(g) + H^+(W) + e = \frac{1}{8}CH_4(g) + \frac{1}{4}H_2O$	-4.13	(O) $\frac{1}{8}NH_4^+ + \frac{3}{8}H_2O = \frac{1}{8}NO_3^- + \frac{5}{4}H^+(W) + e$	$+6.16$
(J) $\frac{1}{6}N_2 + \frac{4}{3}H^+(W) + e = \frac{1}{3}NH_4^+$	-4.68	(P) $\frac{1}{2}MnCO_3(s) + H_2O = \frac{1}{2}MnO_2(s)$ $+ \frac{1}{2}HCO_3^- (10^{-3}) + \frac{3}{2}H^+(W) + e$	8.9

sequence is paralleled by an *ecological succession* of microorganisms (aerobic heterotrophs, denitrifiers, fermentors, sulfate reducers, and methane bacteria). It is perhaps also of great interest from an evolutionary point of view that there appears to be a tendency for more energy-yielding mediated reactions to take precedence over processes that are less energy-yielding.

The quantity of biological growth resulting from inorganic and organic metabolism is related to the free energy released by the mediated reaction $[-\Delta G°(W)]$ and the efficiency of the microorganisms in capturing this energy

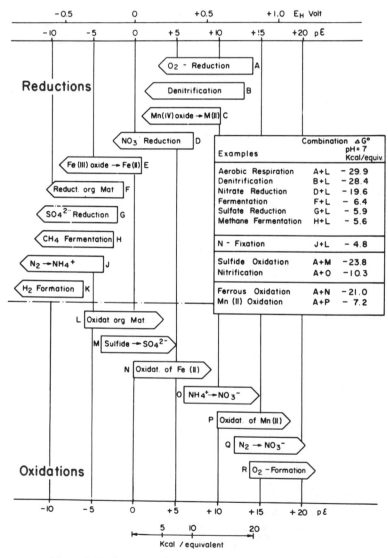

Figure 7.11 Sequence of microbially mediated redox processes.

[6]. Organic carbon compounds (with the exception of CH_4) are unstable over the entire $p\varepsilon$ range, but it is frequently assumed that anaerobic conditions are more favorable to the preservation of organic matter than aerobic conditions.

Another type of reaction sequence would be observed in a system of incipient low $p\varepsilon$ to which O_2 is added. This is the situation commonly encountered after a stream has become polluted with a variety of reducing substances. In such a case it is typically observed that aerobic respiration takes precedence over nitrification, that is, bacterially mediated nitrification is, at least partially, inhibited or represented in the presence of organic material.

It may also be noted (combination J and L in Figure 7.11) that there is a thermodynamic possibility for N_2 reduction accompanied by CH_2O oxidation. This is the gross mechanism mediated by nonphotosynthetic nitrogen-fixing bacteria.

The concern over so-called nonbiodegradable pollutants and the recovery from sediments of organic substances hundreds of millions of years old (see Chapter 11) may serve as a reminder that a state of equilibrium is not always attained, not even within geological time spans, and that microorganisms are not "infallible" in catalyzing processes toward the stable state. Equilibrium models can describe the conditions of stability of redox components in natural water systems, but more extended quantitative inferences must be made with great caution.

How Strong an Oxidant is Oxygen?

The plausible reaction steps in the reduction of oxygen are given in the following simplified scheme

$$O_2 \xrightarrow{+e^-} O_2^- \cdots {+H_2O\,(ads)} \tag{56}$$

with branches $\xrightarrow{+H^+} \searrow {-OH^-}$ leading to $HO_2 \xrightarrow{e^-} HO_2^- \xrightarrow{+H^+} H_2O_2$

$$H_2O_2 \xrightarrow{+e^-} OH + OH^- \tag{57}$$

with $\downarrow {+e^-}$ and $\downarrow {+H^+}$ leading to H_2O, then $OH^- \dashrightarrow H_2O$ ($+H^+$)

The overall reaction

$$O_2 + 4H^+ + 4e^- \rightleftharpoons 2H_2O \qquad \log K = 83.1, \; E_H^\circ = 1.229 \tag{58}$$

can be subdivided into two two-electron sequences:

$$O_2 + 2H^+ + 2e^- \rightleftharpoons H_2O_2 \qquad \log K = 23.1, \; E_H^\circ = 0.68 \tag{59}$$

and

$$H_2O_2 + 2H^+ + 2e^- \; \rightleftharpoons \; H_2O \qquad \log K = 60, E^\circ_H = 1.77 \qquad (60)$$

Oxygen is a strong oxidant if the reduction occurs in a more-or-less synchronous four-electron step ($\log K = 83$ for equation 58). However, O_2 is a much weaker oxidant if the first two-electron reduction sequence becomes operative only; if the second reduction sequence ($H_2O_2 \rightarrow H_2O$) is much slower (presumably because of the cleavage of the O—O bond) than the first one, the reaction $O_2 + 2H^+ + 2e^- \rightleftharpoons H_2O_2$ with a standard potential of only 0.68 V ($\log K = 23.1$) determines the oxidizing ability of O_2. This seems to be the case for the electronation of oxygen in many electrode systems. [8]†

On the other hand, the rate of oxygenation of many transition metals (Fe(II), Mn(II), Ti(III), V(III), U(IV)) is dependent on p_{O_2}; the reaction kinetics correspond to a mechanism in which the first step, that is, the reaction with the O_2 molecule itself, is rate-determining, and O_2 appears to be reduced all the way to H_2O and the full redox potential appears to be realized [9].

Aerobic life is surprisingly fast; O_2 reduction most likely does not take place through slow one-electron steps. Macromolecular electron transfer catalysts (enzymes) with fixed steric positions are presumably able to catalyze the more-or-less synchronous four-electron reduction of O_2. Thus, for most biologically mediated redox reactions, the O_2–H_2O system appears to be the operative redox couple.

The establishment of a carbon cycle was a necessary prerequisite for the evolution of higher forms of life. This could not have been achieved without the direct participation of oxygen in certain metabolic reactions. The controlled activation of oxygen is catalyzed by microbial oxygenases [10].

Can pɛ Be Defined for a Nonequilibrium System?

Obviously a conceptually meaningful pɛ cannot be defined for a nonequilibrium, that is, nonstable or nonmetastable, system. Based on some of the observed activities of redox components of a seawater system (atmosphere, hydrosphere, and sediments), different pɛ values can be calculated [11]. The examples in Figure 7.12 illustrate immediately that the various redox components are not in equilibrium with each other and that the real system cannot be characterized by a unique pɛ.

An Equilibrium Model for the Sea.   Abstracting from the complexity of nature, an idealized counterpart of the aerobic ocean (atmosphere, water, sediment) may be visualized. Oxygen obviously is the atmospheric oxidant that is most

† Breck has proposed that the O_2–H_2O_2 couple determines the pɛ of the sea. For critical discussions of this idea see W. Stumm [*Thalassia Jugosla.*, **14**, 197 (1978)] and S. Ben-Yaakov and I. R. Kaplan [*J. Mar. Res.*, **31**, 79 (1973)].

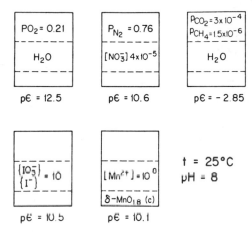

Figure 7.12 pε of seawater. Different pε values can be calculated for a few observed activities of redox components of an oxic seawater system (atmosphere, hydrosphere, and sediments). It is difficult to characterize the real system by a unique pε.

influential in regulating with its redox partner, water, the redox level of aerobic water. It is more abundant—within the time span of its atmospheric residence time—in the atmosphere than in the other accessible exchange reservoirs [12]. It is chemically and biologically reactive; its redox processes (photosynthesis and respiration) occur with a mean flux of ca. $40e$ m^{-2} surface per year. The other major abundant potential oxidant is N_2; because of its relative inertness in redox reactions (low electron flux), it does not appear to be a suitable candidate for controlling the redox composition of the atmosphere–hydrosphere interface. Thus the oxygen–water couple appears to be the predominant redox buffer that determines the redox level of other less abundant redox couples [13]. Biological mediation of the four-electron reduction of O_2 justifies modeling of the equilibrium system in terms of $O_2(g) + 4H^+ + 4e^- \rightleftharpoons 2H_2O(l)$; $\log K = 83.1$ (25°C).

This equilibrium system acquires at pH = 8, for $p_{O_2} = 0.2$ atm, a pε of 12.5. As we have seen, at this pε all biochemically important elements should exist virtually completely in their highest naturally occurring oxidation states.

7.6 KINETICS OF REDOX PROCESSES

Within the last decades, much progress has been made in understanding the mechanisms of redox reactions, especially how, during the reaction, the changes in the state of coordination are coupled to the changes in oxidation state [14]. Many, if not most, redox reactions involve substitutional changes as an integral part of the overall process.

That the addition of an electron can cause a dramatic change in structural geometry and lability may be exemplified by the reduction of $Cr(H_2O)_6^{3+}$ to

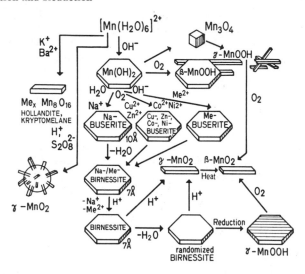

Figure 7.13 Various pathways in the Oxidation of Mn(II). In the nomenclature used here buserite corresponds to the 10-Å manganate(IV) (often also referred to as *cis*-todorokite), birnessite corresponds to 7-Å manganate(IV) (often referred to as *cis*-δ-MnO$_2$). In lake waters, γ-MnOOH(s) has been found to be the final product of the sequence of oxidation of Mn(II) by oxygen with Mn$_3$O$_4$ as an intermediate. A solid manganate(IV), that is, buserite (the structure of which may be stabilized by transition-metal cations that become incorporated into the lattice), may be formed at the sediment–water interface by oxidation of Mn^{2+} that diffuses (together with transition elements) to the sediments. See R. Giovanoli et al., Chimia, **29**, 517 (1975) and W. Stumm and R. Giovanoli, Chimia, **30**, 423 (1976). For a general review on the chemistry of manganese deposits see G. P. Glasby, Ed., *Marine Manganese Deposits*, Elsevier, Amsterdam, 1977.

$Cr(H_2O)_6^{2+}$. $Cr(H_2O)_6^{3+}$ exists as a symmetric octahedral structure of high stability. The half-time for water exchange between $Cr(H_2O)_6^{3+}$ and the solvent is rather large ($\sim 10^6$ sec). $Cr(H_2O)_6^{2+}$ is still octahedral, but its structure is distorted, that is, elongated along one axis; the complex is extremely labile, the half-time for water exchange between aquo complex and solvent being less than 10^{-9} sec.† Thus the change in oxidation state changes the rate of substitution by a factor of more than 10^{15} [14]. On the other hand, the oxidation of Cr(III) to Cr(VI) changes the coordination number to 4 (CrO_4^{2-} or $Cr_2O_7^{2-}$).

A better understanding of the redox kinetics of processes in aquatic systems depends on a better appreciation of the relationship between structural and dynamic aspects of chemical behavior. Figure 7.13 illustrates that often a

---

† The splitting of the d orbitals in an octahedral field permits one to rationalize some aspects of stereochemistry and lability. In the $+3$ oxidation state of chromium, $3s^2p^6d^3$, the three d electrons half fill the threefold degenerate t_{2g} orbitals. This symmetric electronic arrangement gives rise to a regular octahedral shape. In $Cr(H_2O)_6^{2+}$ the four d electrons have parallel spins, three occupying t_{2g} orbitals singly and the fourth an e_g orbital. The nonsymmetric filling of the e_g orbitals gives rise to a pronounced distortion of the regular octahedral structure.

multitude of reaction pathways—depending on solution variables and crystal-lographic conditions—are possible. Unfortunately, very little information is available on the reaction mechanisms of most chemical processes in natural waters. Many redox processes in natural waters are catalyzed by enzymes. Some kinetic aspects of microbial oxidation, especially the oxidation of organic substances, are discussed in Section 8.4.

We illustrate the rate laws of a few redox reactions pertinent to aquatic systems.

Oxidation of Fe(II) and Mn(II)

The rate of oxygenation of Fe(II) in solutions of pH ≥ 5 was found to be first-order with respect to the concentrations of both Fe(II) and O_2 and second-order with respect to the OH^- ion. Thus a 100-fold increase in the rate of reaction occurs for a unit increase in pH. The results of representative kinetic experiments are shown in Figure 7.14. Figures 7.14a and b show the course of Fe(II) and Mn(II) disappearance from the solution at different pH values. It is evident that the reaction rates are strongly pH-dependent. Oxidation of Fe(II) is very slow below pH 6. Figure 7.15 shows rate constants for Fe(II) oxidation by oxygen over the pH range 1–6. Catalysts (especially Cu^{2+} and Co^{2+}) in trace quantities, as well as anions that form complexes with Fe(III) (e.g., HPO_4^{2-}), increase the reaction rate significantly. The oxygenation kinetics follow the rate law [15a, 15b, 16]

$$\frac{-d[Fe(II)]}{dt} = k[Fe(II)][OH^-]^2 p_{O_2} \tag{61}$$

where $k = 8.0\,(\pm 2.5) \times 10^{13}$ min^{-1} atm^{-1} mol^{-2} l^2 at 20°C. Frequently it is more convenient to use the rate law in the form

$$\frac{-d[Fe(II)]}{dt} = \frac{k_H[O_2(aq)]}{[H^+]^2}[Fe(II)] \tag{62}$$

where, at 20°C $k_H = 3 \times 10^{-12}$ min^{-1} mol^1 liter^{-1}. For a given pH, the rate increases about 10-fold for a 15°C temperature increase (activation energy ≈ 23 kcal mol^{-1} at constant $[H^+]$). Kester et al. [17] found the same rate law based on measurements in natural seawater from Narragansett Bay (salinity $= 31.2\%_0$) and in surface Sargasso seawater (36.0$\%_0$). They report half-times nearly 100 times larger than those observed in fresh water at the same pH.†

By comparing Figure 7.14a and b, it is obvious that Mn(II) oxygenation does not follow the same rate law as Fe(II) oxygenation. The manner of the

---

† Recently, an explanation for the kinetics of Fe(II) oxidation in estuaries and seawater has been suggested by results of W. Sung and J. J. Morgan, *Environ. Sci. Technol.*, **14**, 561 (1980), who reported strong effects of ionic strength and medium anions (Cl^-, SO_4^{2-}) in slowing the reaction.

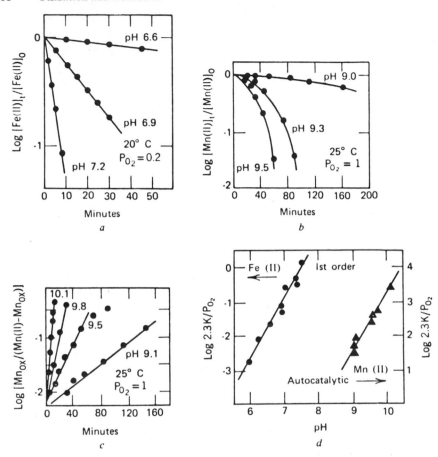

Figure 7.14 Oxidation of Fe(II) and Mn(II) by oxygen. All experiments were conducted with dissolved Fe(II) or Mn(II) concentrations of less than 5×10^{-4} M. In each series of experiments the pH was controlled by continuously bubbling CO_2- and O_2-containing gas mixtures through HCO_3^- solutions of known alkalinity. (a) Oxygenation of Fe(II) in bicarbonate solutions. (b) Removal of Mn(II) by oxygenation in bicarbonate solutions. (c) Oxidation of Mn(II) in HCO_3^- solutions; autocatalytic plot. (d) Effect of pH on oxygenation rates.

decrease in the Mn(II) concentration with time suggests an autocatalytic reaction [16].† The integrated form of the rate expression

$$\frac{-d[\text{Mn(II)}]}{dt} = k_0[\text{Mn(II)}] + k[\text{Mn(II)}][\text{MnO}_2] \tag{63}$$

was found to fit the experimental data well (Figure 7.14c), thus lending support to an autocatalytic model.

---

† A confirmation of the autocatalytic nature of this reaction and an investigation of the catalytic effect of various metal oxide surfaces have been provided by R. W. Coughlin and I. Matsui [*J. Catal.*, **41**, 108 (1976)].

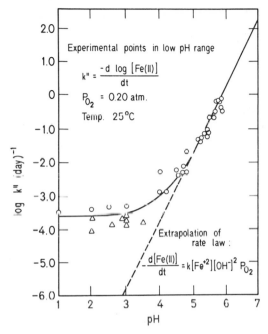

Figure 7.15 Oxidation rate of ferrous iron as a function of pH. At low pH the oxidation rate is independent of pH, while in the high pH range equation 47 (second-order dependence on $[OH^-]$) is fulfilled. (From Singer and Stumm [15b].)

The reaction might be visualized as proceeding according to the following pattern (reactions are not balanced with respect to water and protons):

$$Mn(II) + \tfrac{1}{2}O_2 \xrightarrow{\text{slow}} MnO_2(s) \tag{64}$$

$$Mn(II) + MnO_2(s) \xrightarrow{\text{fast}} Mn(II) \cdot MnO_2(s) \tag{65}$$

$$Mn(II) \cdot MnO_2(s) + \tfrac{1}{2}O_2 \xrightarrow{\text{slow}} 2MnO_2(s) \tag{66}$$

Although other interpretations of the autocatalytic nature of the reaction are possible, the following experimental findings are in accord with such a reaction scheme: (a) The extent of Mn(II) removal during the oxygenation reaction is not accounted for by the stoichiometry of the oxidation reaction alone; that is, not all the Mn(II) removed from the solution [as determined by specific analysis for Mn(II)] is oxidized (as determined by measurement of the total oxidizing equivalents of the suspension); (b) as pointed out before, the products of Mn(II) oxygenation are nonstoichiometric, showing various average degrees of oxidation ranging from ca. $MnO_{1.3}$ to $MnO_{1.9}$ (30 to 90% oxidation to MnO_2) under varying alkaline conditions; and (c) the higher-valent manganese oxide suspensions show large sorption capacities for Mn^{2+} in slightly alkaline solutions. The relative proportions of Mn(IV) and Mn(II) in the solid phase depend strongly on pH and other variables.

Both Mn(II) oxidation and removal rates follow the rate law of equation 63. The rate dependence on the O_2 concentration is the same as that of Fe(II). Thus k in equation 63 can be formulated as

$$k = k'[OH^-]^2 p_{O_2} \tag{67}$$

The pH dependences of Fe(II) and Mn(II) oxidation are compared in Figure 7.14. Metal ions (Cu^{2+}) and complex formers do not appear to have a marked effect upon the reaction rate, although catalytic effects of hydroxy carboxylic acids have been observed. It may be inferred from the catalytic influence of MnO_2, however, that surface catalysis by other active interfaces can influence the reaction rate. In many natural waters the oxidation of Mn(II) to γ-MnOOH or to manganate(IV) is catalyzed by manganese-oxidizing microorganisms.

Oxidation Mechanisms. It has been shown that the autoxidation (i.e., the spontaneous oxidation by free oxygen at room temperature) of transition-metal ions generally is strongly dependent on the reaction medium. Typically the rate increases with pH and is first-order with respect to the metal ion to be oxidized. Most likely the reactive species is a hydrolyzed species $MOH^{(\lambda-1)+}$. The scheme for the oxidation (cf. reactions 42 and 43) was first proposed by Haber and Weiss [18]:

$$MOH^{(\lambda-1)+} + O_2 + H^+ \longrightarrow MOH^{\lambda+} + HO_2{}^{\cdot} \tag{68}$$

$$M^{\lambda+} + HO_2{}^{\cdot} + H_2O \longrightarrow MOH^{\lambda+} + H_2O_2 \tag{69}$$

$$M^{\lambda+} + H_2O_2 \longrightarrow MOH^{\lambda+} + OH^{\cdot} \tag{70}$$

$$M^{\lambda+} + OH^{\cdot} \longrightarrow MOH^{\lambda+} \tag{71}$$

The reaction kinetics, with regard to metal concentration and O_2, found for the oxidation of most transition elements, correspond to a mechanism in which the first step (equation 68); that is, the reaction with the O_2 molecule itself is rate-determining. Thus the general redox reaction (reaction 68) involves an electron transfer from the valence electron shell of the metal ion to the O_2 molecule [9]. Certain possibilities have been advanced [9] for the geometric arrangement of the transition state of such a reaction and the effect of O ligand atoms (particularly in hydroxide and phenoxide anions) bound to the metal ion [Fe(II) and other metal ions with an incomplete d or f shell] in accelerating the electron transfer.

Reaction 68, however, would not predict a second-order dependence on pH as is observed for the oxidation of Fe(II) and Mn(II) (Figure 7.16). As suggested by Kester et al. [17], a reaction step such as

$$FeOH(H_2O)^+ + O_2 \longrightarrow Fe(OH)_2^+ + HO_2{}^{\cdot} \tag{72}$$

could possibly account for the second-order dependence of the rate on OH^-.

Effect of Organic Substances. A variety of organic substances, expecially those that contain hydroxy carboxylic functional groups (e.g., phenols and/or polyphenols, gallic acid, tanic acid), reduce both ferric iron and MnO_2 reasonably fast in synthetic solutions (minutes to hours). This is especially interesting in the case of iron, where the same type of substances that reduce ferric iron can also catalyze the oxygenation rate. This apparent contradiction may be explained by the following kind of reaction sequence:

$$Fe(II) + \tfrac{1}{4}O_2 + org \longrightarrow Fe(III)\text{–org complex} \tag{73}$$

$$Fe(III)\text{–org complex} \longrightarrow Fe(II) + \text{oxidized org} \tag{74}$$

$$Fe(II) + \tfrac{1}{4}O_2 + org \longrightarrow Fe(III)\text{–org complex} \tag{73}$$

Such a reaction pattern has been observed with phenols, "tannic acid," and cysteine. In these cases the ferrous–ferric system acts merely as a catalyst for the oxidation of organic material by oxygen. If conditions (pH, concentrations) are such that the rate of Fe(II) oxygenation is slow in comparison to the Fe(III) reduction by the organic material, a relatively high steady state concentration of Fe(II) can be maintained in the system as long as the organic material is not fully oxidized [19]. Under such circumstances organic material retards the overall oxidation of Fe(II) but hastens the specific Fe(II) oxygenation step. Light can modify such redox reactions; in the presence of certain organic complex formers, light tends to enhance the reduction of ferric iron.

Oxidation of Ferrous Iron in Acid Solutions. There are a number of instances in nature where iron-bearing waters of pH considerably below 5 are encountered. Of special concern are waters in coal-mining regions where pH values of 3 are not uncommon. Iron oxidation in such acidic systems is not characterized by the same kinetic relationships that describe the reaction in neutral waters. Studies in the low pH range (pH < 4) show the rate of oxidation to be independent of pH. By combining the experimental results obtained by Singer and Stumm [15b] for acidic systems with those obtained by Stumm and Lee [15a] for neutral waters, we can plot the rate of oxygenation of ferrous iron over the entire pH range as in Figure 7.15. The oxidation reaction is catalyzed by interfaces and by light. In the presence of light the reaction is approximately two to three times as fast as in its absence. Substantial surface area concentrations ($S > 100$ m^2 liter^{-1}) are necessary to enhance the oxidation reaction markedly.

Oxidation of ferrous iron is followed by hydrolysis to insoluble hydrous ferric oxide. The attachment of a hydroxo group to a ferric ion is very fast.

The Oxidation of Pyrite; the Release of Acidity into Coal Mine Drainage Waters

The sulfur-bearing minerals that predominate in coal seams are the iron sulfide ores pyrite and marcasite. Both have the same ratio of sulfur to iron, but their crystallographic properties are quite different. Marcasite has an orthorhombic structure, while pyrite is isometric. Marcasite is less stable and more easily decomposed than pyrite. The latter is the most widespread of all sulfide minerals and, as a result of its greater abundance in the eastern United States, pyrite is recognized as the major source of acid mine drainage. $FeS_2(s)$ is used here as a symbolic representation of the crystalline pyritic agglomerates found in coal mines.

During coal-mining operations pyrite is exposed to air and water; the following overall stoichiometric reactions may characterize the oxidation of pyrite:

$$FeS_2(s) + \tfrac{7}{2}O_2 + H_2O = Fe^{2+} + 2SO_4^{2-} + 2H^+ \tag{75}$$

$$Fe^{2+} + \tfrac{1}{4}O_2 + H^+ = Fe^{3+} + \tfrac{1}{2}H_2O \tag{76}$$

$$Fe^{3+} + 3H_2O = Fe(OH)_3(s) + 3H^+ \tag{77}$$

$$FeS_2(s) + 14Fe^{3+} + 8H_2O = 15Fe^{2+} + 2SO_4^{2-} + 16H^+ \tag{78}$$

The oxidation of the sulfide of the pyrite to sulfate (equation 75) releases dissolved ferrous iron and acidity into the water. Subsequently, the dissolved ferrous iron undergoes oxygenation to ferric iron (equation 76) which then hydrolyzes to form insoluble "ferric hydroxide" (equation 77), releasing more acidity to the stream and coating the streambed. Ferric iron can also be reduced by pyrite itself, as in equation 78, where sulfide is again oxidized and acidity is released along with additional ferrous iron which may reenter the reaction cycle via reaction 76.

The concentration of sulfate or acidity in the water can be directly correlated with the amount of pyrite that has been dissolved. The introduction of acidity into the stream arises from the oxidation of $S_2(-II)$ and the ensuing hydrolysis of the resulting Fe(III). The dissolution of 1 mol of iron pyrite leads ultimately to the release of 4 eq acidity—2 eq from the oxidation of $S_2(-II)$ and 2 from the oxidation of Fe(II). The decomposition of iron pyrite is among the most acidic of all weathering reactions because of the great insolubility of Fe(III).

A model describing the oxidation of iron pyrite in natural mine waters is proposed in Figure 7.16. The reactions shown are schematic and do not represent the exact mechanistic steps. The model is similar to and carries with it the same overall consequences as that suggested by Temple and Delchamps [20]. The rate-determining step is a reactive step in the specific oxidation of ferrous iron reaction b. As Figure 7.15 shows, the rate of oxidation of ferrous iron under chemical conditions analogous to those found in mine waters is very slow, indeed considerably slower than the oxidation of iron pyrite by ferric iron, reaction c. At pH 3, half-times for the oxidation of Fe(II) are on the order of 1000 days, while in the case of oxidation of pyrite by Fe(III), half-times on the order of 20 to 1000 min were observed. Similar rates have been observed for this latter reaction by Garrels and Thompson [21].

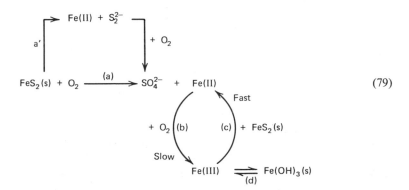

(79)

Figure 7.16   Model for the oxidation of pyrite [17].

To indicate the sequence, pyrite is oxidized directly by oxygen (*a*) or is dissolved and then oxidized (*a'*). The ferrous iron formed is oxygenated extremely slowly (*b*) and the resultant ferric iron is rapidly reduced by pyrite (*c*), releasing additional acidity and new Fe(II) to enter the cycle via reaction *b*. Once the sequence has been started, oxygen is involved only indirectly in the reoxidation of Fe(II) (*b*), the oxygenation of FeS$_2$ (*a*) being no longer of significance.

Precipitated ferric hydroxide deposited in the mine and the streams serves as a reservoir for soluble Fe(III) (*d*). If the regeneration of Fe(III) by reaction *b* is halted so that the concentration of soluble Fe(III) decreases, it will be replenished by dissolution of the solid Fe(OH)$_3$ and will be free to act again should it come in contact with additional FeS$_2$.

The following pertinent consequences of this model need to be emphasized.

1   Ferric iron cannot exist in contact with pyritic agglomerates. Fe(III) is rapidly reduced by iron pyrite. The exclusion of oxygen inhibits regeneration of Fe(III) by the oxidation of Fe(II).

2   The overall rate of dissolution of pyrite should be independent of its surface structure and its surface area concentration.

3   Microorganisms influence the overall rate of reaction by mediating the oxidation of ferrous iron, since it alone is the rate-determining step [22, 23]. Direct microbial oxidation of pyrite must be discounted. Microbial catalysis, as by the autotrophic iron bacteria *Thiobacillus* and *Ferrobacillus ferrooxidans*, seems to be ecologically significant as evidenced by the few field investigations conducted.

The same cycle is probably responsible for the dissolution and leaching of other mineral sulfides found in copper and uranium mines. Microbial leaching of these other minerals has always been demonstrated in the presence of iron, pyrite being the most abundant and widespread of all mineral sulfides.

Autoxidation of Sulfide

Sulfide and oxygen may, under certain conditions, coexist in aqueous solutions for relatively long periods of time. In seawater (pH = 8), saturated with O$_2$ half-times of sulfide removal are typically on the order of a few hours [24]. Despite various attempts to determine the kinetics and mechanisms for the autoxidation of sulfide, no unified results are available. Obviously the oxidation has complex kinetics, and a variety of oxidation products are formed [24–26]:

$$H_2S, HS^-, S_8, S_2O_3^{2-}, SO_3^{2-}, S_4O_6^{2-}, S_4^{2-}, S_5^{2-}, SO_4^{2-} \tag{80}$$

Under pseudo-first-order conditions ([O$_2$] $\gg$ [S($-$II)$_T$], [H$^+$], and [Catalyst] = constant) a generalized rate law can be written:

$$\frac{-d[S(-II)_T]}{dt} = k_{obs}[S(-II)_T] \tag{81}$$

where [S($-$II)$_T$] = [H$_2$S] + [HS$^-$] + [S^{2-}]

However, k_{obs} usually does not depend in a simple way on [O$_2$], [Catalyst], or [H$^+$]. The pH dependence observed (an increase in reaction rate between pH 4 and 8) indicates that HS$^-$ is the principal reactive sulfide species in solution, but the pH dependency is further complicated by the fact that the catalyzed autoxidation is sensitive to general base catalysis. Trace catalysis by metals, anions, and organic molecules affects the rate significantly.

Hoffmann and Lim [26] investigated the catalytic effect of metal phthalocyanine complexes. Phthalocyanines are macrocyclic tetrapyrole compounds that form square planar complexes which have been shown to be effective catalysts for the autoxidation of

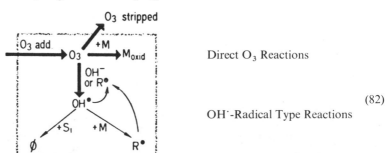

M = Co, Ni, Cu

hydroxylamine, hydrazine, and many organic substances. Their catalytic activity has been explained in terms of the relative capacity of the square planar complexes to bind O_2 reversibly. In the oxidation of HS^- with O_2 the catalyzed reaction is postulated [26] to proceed via the formation of a tertiary activated complex in which HS^- and O_2 bind reversibly with the phthalocyanine complex:

O_2 π acceptor

Co

SH σ donor

Addition of the Co(II) complex of tetrasulfophthalocyanine at μg liter^{-1} levels resulted in an increase in the rate of autoxidation by a factor of 10^4. Because of their relationship to naturally occurring porphyrin pigments, metalphthalocyanine complexes may be suitable models for studying catalytic effects of trace metals in aquatic systems.†

Ozone-Initiated Oxidation of Solutes

The primary oxidations initiated by ozone in water can be described by the following reaction sequences (Hoigné and Bader [27]):

O_3 stripped

Direct O_3 Reactions

(82)

OH·-Radical Type Reactions

† For a brief review on trace metal catalysis in aquatic environments see M. R. Hoffmann, *Environ. Sci. Technol.*, **14**, 1061 (1980).

On the one hand, part of the ozone (O_3) dissolved in water reacts directly with the solutes M. Such direct reactions are highly selective and often rather slow (minutes). On the other hand, part of the ozone added decomposes before it reacts with solutes; this leads to free radicals. Among these, the OH· radicals belong to the most reactive oxidants known to occur in water. OH· can easily oxidize all types of organic contaminants and many inorganic solutes (radical-type reactions). They are therefore consumed in fast reactions (microseconds) and exhibit little substrate selectivity. Only a few of their reactions are of specific interest in water treatment processes. Measured oxidations in model solutions indicate that up to 0.5 mol OH· formed per mole of ozone decomposed. The higher the pH, the faster the decomposition of ozone, which is catalyzed by hydroxide ions (OH^-). The decomposition is additionally accelerated by an autocatalyzed sequence of reactions in which radicals formed from decomposed ozone act as chain carriers. Some types of solutes react with OH· radicals and form secondary radicals (R·) which still act as chain carriers. Others, for instance bicarbonate ions, transform primary radicals to inefficient species (ϕ) and thereby act as inhibitors of the chain reaction. Therefore the rate of the decomposition of ozone depends on the pH of the water as well as on the solutes present. The overall effect is a superposition of the direct reaction and the radical-type reaction.

Kinetic Formulation. Following Hoigné's treatment [27], we may write for the direct reaction of molecular ozone with a solute (M):

$$O_3 + \eta M \xrightarrow{\ k\ } \eta M_{oxid} \tag{83}$$

The rates of these reactions are first-order with respect to ozone and, as a rule, nearly first-order with respect to solute concentration. Therefore the rate at which a solute is oxidized becomes

$$\frac{-d[M]}{dt} = -\eta \frac{d[O_3]}{dt} = \eta k[O_3][M] \tag{84}$$

and the elimination of M is given by

$$-\ln \frac{[M]_e}{[M]_0} = \eta k[\overline{O}_3]t \tag{85}$$

where $[M]_0$ and $[M]_e$ = initial and final concentration of M; η = yield factor of M elimination per ozone used (mol mol^{-1}); k = rate constant of reaction of ozone with M; $[\overline{O}_3]$ = mean O_3 concentration during reaction period t. The relative solute elimination by this direct reaction depends only on the mean concentration of ozone, the time the ozonation lasts, and the rate constant k. Some experimentally determined data for rate constants are presented in Figure 7.17a for illustration. The rate constants may be read on the left-hand scale.

Also in Figure 7.17a are elimination lines indicating eliminations down to 37, 1, and 0.01 %, respectively. The necessary ozonation time leading to such eliminations may be read off the horizontal time scale (compare equation 85).

Oxidation by OH· Radicals. The amount of OH· radicals available for oxidation of a solute M depends on the amount of OH· formed and the relative rate with which it reacts

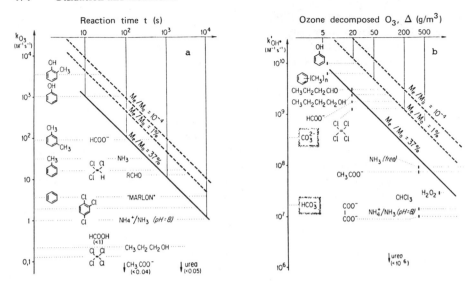

Figure 7.17   Rates of ozone-initiated reactions. (From Hoigné and Bader [27].) (a) Elimination via direct reaction. Second-order reaction rate constant of ozone and ozonation time for eliminations. See Equation 85 with: $\eta = 1.0$; $[\bar{O}_3] = 5$ g/m³. (b) Elimination via OH· radical reaction. Second-order reaction rate constant of OH· radicals and amount of ozone to be decomposed to OH· for solute eliminations for Lake Zürich (equations 88 and 89): $\eta'\eta'' = 0.5$; $\sum (k_i[S_i]) = 5 \times 10^5$ sec^{-1}.

with M when compared with the rate at which it is consumed by all other solutes, in accordance with the scheme:

Reaction: *Rate of OH· consumption:*

$$O_{3,\Delta} + \eta' OH^{\cdot}
\begin{cases}
\xrightarrow{+ M} & k'_M[M][OH^{\cdot}] \\
\xrightarrow{+ \sum_i S_i} & \sum (k'_i[S_i])[OH^{\cdot}]
\end{cases}$$

(86)

Thereby, $O_{3,\Delta} = O_3$ decomposed during process; η' = yield for OH· formation from $O_{3,\Delta}$; k' = second-order reaction rate constant for OH·; $\sum (k'_i[S_i])$ = rate of OH· scavenging by all solutes present including O_3 and M. The reactivity of OH· radicals is very high toward all organic solutes in water. Even free ammonia (NH₃), hydrogen peroxide, and ozone may interfere with OH· (compare Figure 7.17b). Only a fraction (η'') of the OH· radicals reacting with solutes will result in a solute elimination. If this yield factor η'' is also taken into account, then the rate of solute oxidations in the presence of competing scavengers becomes

$$\frac{-d[M]}{dt} = \eta'\eta'' \frac{dO_{3,\Delta}}{dt} = k'_M[M][\sum (k'_i[S_i])]^{-1}$$

(87)

Within a limited range of oxidation, where $\sum (k_i[S_i])$ depends on neither the degree of ozonation nor the ozone-concentration, integration of equation 87 yields

$$-\ln\left(\frac{[M]_e}{[M]_0}\right) = \eta'\eta''O_{3,\Delta}k'_M[\sum (k'_i[S_i])]^{-1} \tag{88}$$

when

$$k'_M[M] \ll \sum (k_i[S_i]) \tag{89}$$

The elimination of M is first-order with respect to the M concentration. The amount of ozone that must be decomposed in order to give a certain elimination factor is expected to increase linearly with the rate at which OH˙ radicals are consumed by the sum of all radical scavengers present. The amount of ozone required to be decomposed decreases linearly with the rate constant with which M itself reacts with the oxidant.

Rate constants for OH˙ radical reactions with hundreds of aqueous solutes are known. Figure 7.17b gives an approximation of the magnitudes of such constants.

Photo-redox Reaction

The conversion of light into chemical energy may be illustrated by the photo-redox reaction of the thionine–iron system. The reaction sequence is described schematically by an electronically excited state A*. Under the influence of light, A (thionine) is converted into

$$A \longrightarrow A*$$

photooxidation:

$$A* + Fe(II) \longrightarrow R + Fe(III)$$

$$A + Fe(II) \rightleftharpoons R + Fe(III)$$

dark reaction:

$$A + H \rightleftharpoons 2R$$

In the case of thionine, A is

$$TH^+ = \text{[structure]} \qquad R = TH \text{ or } TH_2^+$$

Similarly a photo-redox reaction can be formulated for the iodine–iron system:

$$I_3^- \longrightarrow (I_3^-)*$$

$$(I_3^-)* + 2\,Fe(II) \longrightarrow 3I^- + 2\,Fe(III)$$

$$3I^- + 2\,Fe(III) \rightleftharpoons I_3^- + 2\,Fe(II)$$

$$I_3^- \rightleftharpoons I_2 + I^-$$

† The possible importance of photochemistry in seawater has been discussed by O. C. Zafiriou, *Marine Chem.*, **5**, 497 (1977).

7.7   THE ELECTROCHEMICAL CELL

Galvanic and Electrolytic Cells.   Many chemical reactions can take place in an electrochemical cell. These reactions can occur spontaneously (galvanic cell) or can be driven by an applied external potential (electrolytic cell). Electrical terminals of a cell are labeled according to the type of electrode reactions associated with them. Reduction takes place at the *cathode* and oxidation at the *anode*. Hence the electrons enter an electrochemical cell at the cathode and leave at the anode. Terminals are also called plus or minus, but this depends on whether the cell is galvanic or electrolytic (Figure 7.18).

Electrochemical cells are described by a shorthand notation. For example, the cell reaction

$$Zn(s) + Cu^{2+} = Zn^{2+} + Cu(s) \tag{90}$$

is represented as

$$Zn|Zn^{2+}(a_1)\|Cu^{2+}(a_2)|Cu \tag{91}$$

in which a_1 and a_2 indicate activities, a single vertical line represents a phase boundary across which there is a potential difference, and a double vertical line signifies that the liquid junction is either ignored or considered to be eliminated by a salt bridge.† The following convention is used for the sequence in the diagram. The electrode reaction on the right is a reduction (making this electrode a cathode), and the electrode reaction on the left is an oxidation (anode).

The emf of the half-cell $Zn^{2+}|Zn$ corresponds to the emf of the cell:

$$Pt, H_2(p_{H_2} = 1)|H^+(a = 1)\|Zn^{2+}|Zn \tag{92}$$

where the electrode on the left is the standard hydrogen electrode. According to IUPAC conventions such a cell emf is the electrode potential. If both Zn and Zn^{2+} are at unit activity, we speak of the standard electrode potential of the Zn electrode.

Electrodes.   In the hydrogen electrode platinum acts as a catalyst for the reaction between H^+ and H_2 molecules and acquires a potential characteristic of this reaction. Similarly, electrodes may be characterized by the redox couple they represent, such as Cu^{2+}/Cu, Cl_2/Cl^-, H^+/H_2, O_2/H_2O, $AgCl/Ag$, and Hg_2Cl_2/Hg. The last two electrodes mentioned are also called electrodes of the second kind. These electrodes tend to maintain a constant electrode potential, because the concentration of the cations associated with the electrode

---

† Liquid junction potentials arise from the transfer of ionic species through the transition region that divides the test solution from the reference electrode solution. The liquid junction potential makes a contribution to the emf of the cell; it increases with increasing difference between the two solutions that form a single junction. The liquid junction potential can often be kept small by using a concentrated salt bridge.

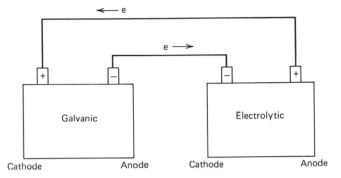

Figure 7.18   Plus and minus terminals in galvanic and electrolytic cells. A cathode is always the electrode where reduction takes place, an anode is where oxidation occurs. Electrons leave at the anode and enter at the cathode. The direction of electron flow is in the direction opposite to that of positive electricity. (After J. Waser, *Basic Chemical Thermodynamics*, W. A. Benjamin, Menlo Park, Calif., 1966.) Reproduced with permission from W. A. Benjamin, Inc.

metal is kept constant by buffering through the solubility product principle. Thus a AgCl/Ag electrode consists in principle of a silver electrode coated with AgCl immersed in a solution of high $[Cl^-]$. Because AgCl(s) is present, the activity of Ag^+ is given by

$$\{Ag^+\} = \frac{K_{s0(AgCl)}}{\{Cl^-\}} \tag{93}$$

Furthermore, if some Ag^+ is reduced to Ag(s), or some Ag(s) is oxidized to Ag^+, the dissolution or precipitation of AgCl, respectively, will keep $\{Ag^+\}$ constant. Hence the electrode potential of a AgCl/Ag electrode remains constant even if some current is flowing through the half-cell (nonpolarizable electrode). Of course the current must be small enough so that it does not exceed the exchange current of the reaction

$$AgCl(s) + e = Ag(s) + Cl^- \tag{94}$$

Such a nonpolarizable electrode is a convenient reference electrode. Another important reference electrode is the calomel electrode with the half-reaction

$$Hg_2Cl_2(s) + 2e = 2Hg(l) + 2Cl^- \tag{95}$$

Its electrode potential depends on the activity of the chloride ion.

Other electrodes represent redox couples in which oxidized and reduced forms are soluble species in the solution and the electron transfer occurs at the inert electrode (Pt, Au). For example, a platinum electrode immersed in an acid solution of Fe^{2+} and Fe^{3+} (Pt/Fe^{3+}, Fe^{2+}) may under favorable conditions acquire the potential characteristics of a Fe^{3+}/Fe^{2+} couple

$$Fe^{3+} + e = Fe^{2+} \tag{96}$$

When half-cells are combined in an electrochemical cell, the emf of the cell is given by

$$E_{cell} = E_{right\ cell} - E_{left\ cell} \tag{97}$$

$$E_{cell} = E_{H\ right} - E_{H\ left} \tag{98}$$

If the left electrode is a reference electrode, we may also write

$$E_{cell} = E_{H(ox\text{-}red)} - E_{ref} \tag{99}$$

Example 7.12. Electrode Potential of Fe^{3+}-Fe^{2+} System

Consider the following cell and compute its emf:

$$Hg\,|\,Hg_2Cl_2,\ KCl_{satd}\,\|\,HClO_4(1\ M),\ Fe^{3+}(10^{-3}),\ Fe^{2+}(10^{-2})\,|\,Pt$$

According to Table 7.7 the potential of the saturated calomel electrode (25°C) is 0.244 V. Hence

$$E_{cell} = E_{Fe^{3+},Fe^{2+}} - 0.244$$

and

$$E_{Fe^{3+},Fe^{2+}} = 0.771 + 0.059 \log \frac{10^{-3}}{10^{-2}} = 0.712\ V$$

$$E_{cell} = 0.468\ V$$

Example 7.13. Solution Composition from Measured emf

With an electrode pair consisting of an inert Pt and a saturated calomel reference electrode in a sample of a sediment–water interface of pH = 6.4

TABLE 7.7   POTENTIALS OF REFERENCE ELECTRODES[a]

Temperature °C	Calomel		AgCl, Ag	
	0.1 M KCl	Saturated	0.1 M	Saturated
12	0.3362	0.2528	—	—
20	0.3360	0.2508	—	—
25	0.3356	0.2444	0.2900	0.1988

[a] The values listed (V) included the liquid junction potential. Cell: Pt, H_2/H^+ ($a = 1$) ‖ reference electrode. Cf. R. G. Bates, in *Treatise in Analytical Chemistry*, Part I, Vol. I, I. M. Kolthoff and P. J. Elving, Eds., Interscience, New York, 1959, p. 319. Reproduced with permission from Interscience Publishers.

(25°C) a potential difference (emf) of 0.47 V is measured. The sediment contains solid amorphous $Fe(OH)_3$, and we assume that the measured potential corresponds to an oxidation–reduction potential of the aquatic environment.

(a) What is the E_H and $p\varepsilon$ of the sample? (b) What is the activity of Fe^{2+}? (c) Does the redox level found indicate aerobic or anaerobic conditions?

1 The emf has been measured in a cell

$$Hg, Hg_2Cl_2 | KCl \| ox, red/Pt$$

and

$$E_{cell} = E_{H|ox\text{-}red} - E_{ref}$$

Because $E_{ref} = 0.244$, we obtain $E_{H(ox\text{-}red)} = 0.47 + 0.244 = 0.714$ V. This is equivalent to $p\varepsilon = 0.714$ V/0.0916 V $= 12.1$

2 At equilibrium $Fe(OH)_3(s)$ is in equilibrium with Fe^{2+}, the redox reaction being

$$Fe(OH)_3(s) + 3H^+ + e = Fe^{2+} + 3H_2O \qquad \log K = 16 \qquad \text{(i)}$$

Thus $E_{H(ox\text{-}red)}$ and $p\varepsilon$, respectively, are given by (see equations 7, 24, and 25)

$$E_{H(ox\text{-}red)} = E^\circ_{Fe(OH)_3-Fe^{2+}} + 2.3\,\frac{RT}{nF}\log\left(\frac{\{H^+\}^3}{\{Fe^{2+}\}}\right) \qquad \text{(ii)}$$

$$p\varepsilon = p\varepsilon^\circ_{Fe(OH)_3-Fe^{2+}} + \log\left(\frac{\{H^+\}^3}{\{Fe^{2+}\}}\right) \qquad \text{(iii)}$$

The value of $E^\circ_{Fe(OH)_3-Fe^{2+}}$ or of $p\varepsilon^\circ_{Fe(OH)_3-Fe^{2+}}$ can be computed from data on free energy of formation for equation i. The reduction reaction at 25°C has $\Delta G^\circ = -91.4$ kJ, $\log K = 16.0$. From this we calculate (cf. equations 7 and 25)

$$E^\circ_{Fe(OH)_3-Fe^{2+}} = \frac{91,400 \text{ J}}{96,480 \text{ J V}^{-1}} = 0.947 \text{ V}$$

or

$$p\varepsilon^\circ_{Fe(OH)_3-Fe^{2+}} = 16.0$$

The activity of Fe^{2+} can be calculated by

$$p\varepsilon = p\varepsilon^\circ + \log\left(\frac{\{H^+\}^3}{\{Fe^{2+}\}}\right)$$

$$12.1 = 16 - 3pH + pFe^{2+}$$

This gives $pFe^{2+} = 15.3$ for $pH = 6.4$.

3 In order to decide whether this corresponds to aerobic or anaerobic conditions, p_{O_2} must be computed. For

$$O_2(g) + 4H^+ + 4e = 2H_2O \qquad p\varepsilon^\circ = 20.8$$

because

$$p\varepsilon = p\varepsilon° + \tfrac{1}{4}\log p_{O_2}\{H^+\}^4$$

$$12.1 = 20.8 - pH + \tfrac{1}{4}\log p_{O_2}$$

$$\log p_{O_2} = 4(12.1 - 20.8 + pH)$$

$$p_{O_2} = 10^{-9.2} \text{ atm}$$

This corresponds to ca. 10^{-12} M dissolved O_2. We may properly speak of anaerobic conditions.

Example 7.14. Standard Potential of the Cl_2/Cl^- Couple

G. Faita, P. Longhi, and T. Mussini [*J. Electrochem. Soc.*, **114**, 340 (1967)] determined the standard potential of the Cl_2/Cl^- electrode. They made emf measurements with the cell

$$Pt|Ag|AgCl|1.75 \ M \ HCl|Cl_2(=1 \text{ atm})|Pt-Ir(45\%), Ta|Pt \qquad (i)$$

The right-hand electrode consisted of a tantalum foil (attached to a Pt wire) coated with a platinum–iridium alloy. This alloy, containing 45% iridium, is used as a Cl_2 electrode because Pt is not fully appropriate since it is subject to corrosive attack by Cl_2 in the presence of HCl. Their results for the cell (i) are as follows: 25°C: $E_{cell(i)} = 1.13596$ V; 30°C: $E_{cell(i)} = 1.13309$ V; 40°C: $E_{cell(i)} = 1.12711$ V; 50°C: $E_{cell(i)} = 1.12110$ V

1 What is the chemical reaction taking place in the cell?
2 What is the standard potential of the Cl_2/Cl^- electrode?
3 Determine from the experimental data $\Delta G°$, $\Delta H°$, and $\Delta S°$ of the cell reaction of the reduction of Cl_2 to Cl^-, respectively.

The standard $Ag/AgCl/Cl^-$ electrode, $E°_{AgCl/Ag}$, has the following values (volts):

$E°_{AgCl/Ag}$: 25°C, 0.22234; 30°C, 0.21904; 40°C, 0.21208; 50°C, 0.20449

Furthermore, the concentration of HCl in cell i has been chosen such that $\{Cl^-\} = 1.0$.

1 The chemical reaction taking place in cell i is composed of the half-reactions

$$Ag(s) + Cl^- = AgCl(s) + e$$

$$\tfrac{1}{2}Cl_2(g) + e = Cl^-$$

$$Ag(s) + \tfrac{1}{2}Cl_2(g) = AgCl(s) \qquad (ii)$$

The emf of $E_{cell(i)}$ is a direct measure of the free energy of the reaction $\Delta G° = -nFE_{cell}°$. Hence at 25°C, $\Delta G° = -109.612$ J

2   The standard potential of the Cl_2/Cl^- electrode, $\frac{1}{2}Cl_2(g) + e = Cl^-$, is given by

$$E_{cell(i)} = E_{Cl_2/Cl} - E_{AgCl/Cl} \tag{iii}$$

Thus

$$E_{Cl_2/Cl^-} = E_{cell(i)} + E_{AgCl/Cl} \tag{iv}$$

and, at 25°C, $E_{Cl_2/Cl^-}° = 1.35830$ V. Thus $\Delta G°$ for the reaction $\frac{1}{2}Cl_2(g) + e = Cl^-$ [which is the same as for the reaction $\frac{1}{2}H_2(g) + \frac{1}{2}Cl_2(g) = H^+ + Cl^-$] is -131.068 J mol^{-1}, corresponding to a log K value of 22.97.

3   $\Delta S°$ and $\Delta H°$ for the cell reaction $Ag(s) + \frac{1}{2}Cl_2(g) = AgCl(s)$ can be calculated by considering that $\Delta S = nF(dE_{cell}/dT)$ and $\Delta H = -nFE + nFT(dE_{cell}/dT)$. The data given for $E_{cell(i)}$ have been fitted as a function of absolute temperature T (least square procedure) by $E_{cell(i)} = 1.28958 - (4.31562 \times 10^{-4})T - (2.7922 \times 10^{-7})T^2$. At 25°C the first derivative of this equation is 5.986×10^{-4} (V deg^{-1}). (This temperature dependence may also be obtained approximately by plotting $E_{cell(i)}$ versus $1/T$.) Hence $\Delta H_{(25°C)} = -126.82$ J mol^{-4}. Similarly $\Delta S°$ and $\Delta H°$ values at 25°C (or at other temperatures) may be obtained from the temperature dependence of $E_{Cl_2/Cl^-}°$. The latter is found (25°C) to be -1.246×10^{-3} V deg^{-1}. Correspondingly $\Delta S_{(25°C)}° = -120.2$ J deg^{-1} mol^{-1} and $\Delta H_{(25°C)} = -167$ J mol^{-1}.

7.8   THE POTENTIOMETRIC DETERMINATION OF INDIVIDUAL SOLUTES: pH MEASUREMENT

Methods that involve the introduction of electrodes into natural media without contamination are most appealing. The glass electrode has proved to be sufficiently specific and sensitive. More recently other electrodes specific for more than a dozen ions have joined the pH-type glass electrode in commercial production. Many of these electrodes are sufficiently specific and sensitive to permit measurement or monitoring of individual solution components.

An electrochemical cell can be used conveniently to study the properties of a solution quantitatively. For example, it is possible to determine $\{Ag^+\}$ with an Ag electrode, $\{H^+\}$ with a Pt/H_2(g) electrode or with a glass electrode, $\{Cl^-\}$ with an AgCl/Ag electrode, and $\{SO_4^{2-}\}$ with a PbSO$_4$/Pb electrode. Table 7.8 gives a survey of most of the ion-sensitive electrodes employed in potentiometric measurements.

In the case of metal electrodes the potential-determining mechanism is a fast electron exchange, given by the indicator reaction

$$Me^{n+} + ne = Me(s) \tag{100}$$

TABLE 7.8   ION SELECTIVE ELECTRODES

Ion	Electrode Material	Electrode Reaction
I. Metal Electrodes		
H^+	Platinized Pt, H_2	$H^+ + e = \frac{1}{2}H_2(g)$
Ag^+, Cu^{2+}, Hg^{2+}	Ag, Cu, Hg	$Me^{n+} + ne = Me(s)$
Zn^{2+}, Cd^{2+}	Zn(Hg), Cd(Hg)	$Me^{2+} + 2e + Hg(l) = Me(Hg)$
II. Electrodes of the Second Kind		
Cl^-	AgCl/Ag	$AgCl(s) + e = Ag(s) + Cl^-$
S^{2-}	Ag_2S/Ag	$Ag_2S(s) + 2e = 2Ag(s) + S^{2-}$
SO_4^{2-}	$PbSO_4$/Pb(Hg)	$PbSO_4 + Hg + 2e = Pb(Hg) + SO_4^{2-}$
H^+	Sb_2O_3/Sb	$Sb_2O_3 + 6H^+ + 6e = 2Sb(s) + 3H_2O$
III. Complex Electrodes		
H^+	Pt, quinhydrone	$C_6H_4O_2 + 2H^+ + 2e = C_6H_4(OH)_2$
$EDTA(Y^{4-})$	HgY^{2-}/Hg	$HgY^{2-} + 2e = Hg(l) + Y^{4-}$
IV. Glass Electrodes		
H^+	Glass	$Na^+(aq) + H^+_{Membrane} = Na^+_{Membrane} + H^+(aq)$
$Na^+, K^+, Ag^+,$ NH_4^+	Cation-sensitive glass	Ion exchange
V. Solid-State or Precipitate Electrodes		
$F^-, Cl^-, Br^-, I^-,$ $S^{2-}, Cu^{2+}, Cd^{2+},$ Na^+, Ca^{2+}, Pb^{2+}	Precipitate, impregnated or solid-state electrodes	Ion exchange
VI. Liquid–Liquid Membrane Electrodes		
$Ca^{2+}, Mg^{2+}, Pb^{2+},$ $Cu^{2+}, NO_3^-, Cl^-,$ ClO_4^-	Liquid ion exchange	Ion exchange

for which the equilibrium potential is

$$E_H = E^{\circ}_{H_{Me^{n+}, Me(s)}} + \frac{RT(\ln 10)}{nF} \log\{Me^{n+}\} \qquad (101)$$

If a potential of a given indicator electrode is to be controlled by a specific redox reaction, it is clear that the essential solid phases must be present in adequate amounts and that the electrode reaction must be characterized by a large enough exchange current. The magnitude of the latter determines the concentration at which the indicator solute must be contained in the solution. Obviously the Nernst equation must not be expected to hold for indefinitively decreasing activity of the potential-determining species. If the exchange current is too small, the electrode response becomes too slow, and reactions with impurities or spurious oxides or oxygen at the electrode surface influence the potential and render it unstable and indefinite. Two of the highest exchange current densities are for H^+ discharge at Pt and for reduction of Hg^{2+} at a Hg surface. It may be noted, however, that metal-ion electrodes often may show an unusually sensitive response because the electroactive species is actually a complex present at a larger concentration than the free metal ions. For example, if a slower electrode responds properly to $\{Ag^+\}$ in a sulfide-containing medium, the electrode response is caused by the two silver complexes usually present at much higher concentrations than the free Ag^+.

For many metals the system $Me^{n+} + ne = Me(s)$ is slow and the equilibrium potential is established too slowly in using such metals as indicator electrodes. Strongly reducing metals cannot be used as indicator electrodes because the potentials established with such metals are mixed potentials. Hence we cannot use an Fe electrode and a Zn electrode to measure $\{Fe^{2+}\}$ and $\{Zn^{2+}\}$, respectively. The Zn electrode is characterized by a mixed potential, because H_2O is reduced at potentials more positive than those at which Zn^{2+} is reduced (corrosion of Zn). If a zinc amalgam electrode is used instead of a zinc electrode, hydrogen discharge occurs at potentials more negative than those at the Zn electrode and the reduction of Zn^{2+} at more positive potentials. The reaction at the electrode is now $Zn^{2+} + Hg(l) + 2e = Zn(Hg)$, and the electrode potential is given by

$$E = E_0 + \frac{RT}{nF} \ln \frac{\{Zn^{2+}\}}{\{Zn(Hg)\}} \tag{102}$$

where $\{Zn(Hg)\}$ is the activity of the Zn in the amalgam.

Glass Electrodes

When a thin membrane of glass separates two solutions, an electric potential difference that depends on the ions present in the solutions is established across the glass. Glass electrodes responding chiefly to H^+ ions have become common laboratory tools, and a number of treatises are concerned with their properties [28–31]. More recently modification of the glass composition has led to the development of electrodes selective for a variety of cations other than H^+.

The origin of the glass electrode potential is not discussed here, but it may be helpful to mention that the glass membrane functions as a cation exchanger and that a Nernst potential is observed if such a membrane separates two solutions at two different concentrations:

$$E_{cell} = \frac{RT}{F} + \ln \frac{\{^1H^+\}}{\{^2H^+\}} \tag{103}$$

Because the glass electrode contains a solution (acid) of constant $\{^2H^+\}$ inside the glass bulb, the emf measured depends only on $\{^1H^+\}$ in the external solution:

$$E_{cell} = const + \frac{RT}{F}\ln\{^1H^+\} \tag{104}$$

Measurement of Activities by Potentiometric Electrodes. The standards may be chosen according to either one of the activity scales discussed before (Section 3.4 and the appendix to Chapter 6.1).

Other Ion-Selective Electrodes

In Table 7.8 other types of electrodes are enumerated.[†] In these the glass membrane is replaced by a synthetic single-crystal membrane (solid-state electrodes), by a matrix (e.g., inert silicone rubber) in which precipitated particles are imbedded (precipitate electrodes), or by a liquid ion-exchange layer (liquid–liquid membrane electrodes). The selectivity of these electrodes is determined by the composition of the membrane. All these electrodes show a response in their electrode potentials according to the Nernst equation.

No electrode is fully selective; other species often similar to that to be measured may affect the electrode response. The effect of an other (disturbing) ion, S, on the measured electrode potential can typically be expressed by

$$E = E_0 + \frac{2.3RT}{nF}\log(\{M^{n+}\} + \sum K_{M-S}\{S\}^{n_M/n_S}) \tag{105}$$

where K_{M-S} is a constant representative for the selectivity relation between M and S (e.g., if $K_{M-S} = 10^{-3}$, the electrode responds 1000 times more selectively to M than to S).

The so-called oxygen electrode is based on a different principle: An electrolysis cell is used, and the current measured at the inert cathode is, under standardized conditions, a function of the oxygen concentration (activity). Selectivity is enhanced by covering the cathode with a membrane permeable to molecules only.

pH Measurement

Nowadays most pH values are measured with glass electrodes. Before discussing the cell used for measurement with the glass electrode we discuss the measurement of pH under ideal conditions with a cell such as

$$\text{Pt, } H_2(g)|\text{solution with } H^+, Cl^- |AgCl(s), Ag \tag{106}$$

working electrode   sample   reference electrode

[†] For a more detailed discussion, see M. Whitfield, *Ion Selective Electrodes for the Analysis of Natural Waters,* Australian Marine Sciences Association, Handbook No. 2, Sydney, 1971; R. A. Durst, *Ion Selective Electrodes,* NBS, Special Publication No. 314, (1969); K. Cammann, *Das Arbeiten mit ionenselektiven Elektroden* 2nd revised ed, Springer, Berlin, 1977.

The Nernst equation for cell 106 with the cell reaction

$$H^+ + Ag(s) + Cl^- \; \rightleftharpoons \; \tfrac{1}{2}H_2(g) + AgCl(s)$$

is given, for $p_{H_2} = 1$ atm, by

$$E = E^\circ - k \log(\{H^+\}\{Cl^-\}) \qquad (107)$$

or

$$E = E^\circ - k \log([H^+][Cl^-]) - k \log f_{H^+} f_{Cl^-} \qquad (108)$$

where $k = RT \ln 10/F = 59.2$ mV at 25°C.

The cell can be calibrated by using a HCl solution of known concentration $C([H^+] = [Cl^-] = C)$. Then we have

$$E = E^\circ - 2k \log C \quad k \log f_{H^+} f_{Cl^-} \qquad (109)$$

For a series of solutions of known but variable C, E° and $\log f_{H^+} f_{Cl^-}$ can be determined. (Most conveniently $E + 2k \log C$ is plotted versus C and extrapolated to $C \to 0$; when $C = 0$, $k \log f_{H^+} f_{Cl^-} \to 0$; then, from E and C, the value of $\log f_{H^+} f_{Cl^-}$ can be calculated.) Using the infinite dilution activity scale, for dilute HCl solutions, $\log f_{H^+} f_{Cl^-}$ can be approximated by the Davies equation (Table 3.3):

$$\log f_{H^+} f_{Cl^-} = - \left(\frac{\sqrt{I}}{1 + \sqrt{I}} - 0.2I \right) \qquad (110)$$

where the ionic strength $I = C$.

If we now use a dilute solution of ionic strength I where $[H^+] \neq [Cl^-]$ (e.g., a solution of weak acid or of a buffer containing NaCl), we obtain from equation 108

$$-\log\{H^+\} f_{Cl^-} = \frac{1}{k}(E - E^\circ + k \log[Cl^-]) \qquad (111)$$

Everything on the right-hand side of equation 111 is experimentally accessible. If f_{Cl^-} can be calculated by the Davies equation

$$\log f_{Cl^-} = -0.5 \left(\frac{\sqrt{I}}{1 + \sqrt{I}} - 0.2I \right) \qquad (112)$$

$-\log\{H^+\}$ can be determined (Bates–Guggenheim convention). The approach outlined above corresponds to the method typically used to measure $-\log\{H^+\}$ of standard buffer solutions.

If the measurements are made in the presence of a swamping inert electrolyte, for example, $NaClO_4$ (constant ionic medium), we would find, instead of equation 110 (compare Figure 3.2),

$$\log \gamma_{H^+} \gamma_{Cl^-} \approx 0$$

and, instead of equation 111,

$$-\log[H^+] = \frac{1}{k}(E - E^{\circ\prime} + k \log[Cl^-]) \qquad (113)$$

where $E^{\circ\prime} = E^{\circ} +$ constant. (This correction is due to liquid junction effects.)

pH Measurement with the Glass Electrode.   Typically the cell used to measure pH can be represented† by

$$\text{Glass electrode} \mid \begin{array}{c} \text{standard S} \\ \text{or} \\ \text{sample X} \end{array} \mid \text{salt bridge (conc. KCl)} \mid \text{AgCl, Ag}   \quad (114)$$

The operational pH definition is

$$\text{pH(X)} = \text{pH(S)} + \frac{1}{k}\,[\text{E(X)} - \text{E(S)}] \quad (115)$$

The operational manual of a standard pH meter is based on equation 115. The temperature compensation is based on k $(=RT \ln 10/F)$. The buffers used in the standardization have pH values measured with cells of type 106 using equation 111 and the convention (112). If the measurement is carried out in the pH range 3 to 9 and in solutions of $I < 0.1$, the measurement represents a very good approximation of the conceptual definition pH $= -\log\{H^+\}$. If $I > 0.1$ and pH < 3 and pH > 9, the measured pH can not be conceptually interpreted.

For measurement of the H^+ *concentration*, $[H^+]$ in a solution of constant ionic medium, the calibration of a pH meter can be carried out with the help of a strong acid solution of known concentration, $[H^+] = C$, to which the electrolyte of the ionic medium has been added. The dial on the pH meter is set in accordance with the known $[H^+]$. For example, seawater has an ionic strength of ca. $I = 0.7$; a standard of $-\log[H^+] = 2.00$ may be prepared by adjusting a 1.00×10^{-2} M HCl solution with NaCl to an ionic strength of 0.70 (composition: $[H^+] = 0.01$ M, $[Na^+] = 0.69$ M, $[Cl^-] = 0.70$ M). In comparing this standard with ocean water, one may find $E(S) = 118$ mV; $E(X) = -249$ mV. Hence $-\log[H^+] = +2.00 + (1/59.2)\,[118 - (-249)] = 8.20$. (See the appendix to Chapter 6 for definition of different pH scales for seawater.)

Potentiometric Titration.   Potentiometry may be used to follow a titration and to determine its end point. The principles have already been discussed in connection with acid–base or complex formation titrations where pH or pMe is used as a variable. Any potentiometric electrode may serve as an indicator electrode which either indicates a reactant or a reaction product. Usually the measured potential will vary during the course of the reaction and the end point will be characterized by a "jump" in the curve of voltage versus amount of reactant added.

---

† Instead of an Ag,AgCl reference electrode, often a calomel electrode, Hg,Hg_2Cl_2, is used.

Although the potentiometric measurement permits determination of the concentration (activity) of a species, the potentiometric titration determines the total analytical concentration (capacity); for example, a Ca^{2+} electrode responds to free Ca^{2+} ions only and not to Ca^{2+} complexes present in the solution. Using the Ca electrode as an indicator electrode with a strong complex former (forming more stable complexes with Ca^{2+} than those already present in the solution) gives the sum of the concentrations of all Ca^{2+} species. In a similar way total F^- may be determined by titrating with $La(NO_3)_3$, using a F^- electrode, or total K^+ may be measured from titration with a specific K^+ electrode using Ca tetraphenylborate, $(Ca[B(C_6H_5)_4]_2)$, as a reagent.

The result obtained by a titration is usually more precise than that obtained in a direct potentiometric measurement. It is usually not too difficult to create an end point with a precision of better than 5%. In the direct potentiometric determination the relative error F_{rel} is given by

$$F_{rel} = \frac{\Delta C}{C} = 2.3\Delta \log C \tag{116}$$

In other words, if the measurement with a glass electrode can be reproduced within 0.04 pH units (2.5 mV), the relative error in $[H^+]$ is 10%. For a similar reproducibility with a Ca^{2+} electrode, $\Delta pCa = 0.08$ pCa units (the Nernst coefficient is half that for the pH electrode), the relative error in $[Ca^{2+}]$ is 20%.

Titration Curve. The morphology of the titration curve can be calculated using the same principles discussed earlier. However, for many systems the slopes of the theoretical titration curves do not conform to the experimental values because the indicator electrode does not respond with sufficient sensitivity or selectivity. A detailed example on a redox titration, given below, will serve for illustration. As with acid–base titrations, Gran plots (see the appendix to Chapter 4) can be used with advantage for a most precise end point detection. A good example is the titration of Cl^- with $AgNO_3$ using a AgCl/Ag electrode as an indicator electrode.

The stability of a redox system with respect to a change in pε as a result of the addition of a reductant, C_R

$$S = \frac{dC_R}{dpE} \tag{117}$$

may be called the *redox poising intensity* and is analogous to the buffer intensity in acid–base systems.

Example 7.15. Titration of Fe(II) with MnO_4^-

Construct a theoretical titration curve of redox potential ($E_{half-cell}$ versus H_2 electrode) for the titration of Fe^{2+} with MnO_4^- in a dilute H_2SO_4 solution using a Pt electrode as an indicator electrode. The stoichiometry of the reaction is given by

$$MnO_4^- + 8H^+ + 5e = Mn^{2+} + 4H_2O$$

$$5Fe^{2+} = 5Fe^{3+} + 5e$$

$$MnO_4^- + 8H^+ + 5Fe^{2+} = 5Fe^{3+} + 4H_2O + Mn^{2+} \qquad \text{(i)}$$

1. The *conceptual redox potential* is given by

$$E_H' = E_{H'Fe^{3+},Fe^{2+}}^\circ + 0.06 \log \frac{[Fe^{3+}]}{[Fe^{2+}]} \qquad \text{(ii)}$$

or

$$E_H' = E_{H'MnO_4^-,Mn^{2+}}^\circ + \frac{0.06}{5} \log \frac{[MnO_4^-]}{[Mn^{2+}]} \qquad \text{(iii)}$$

where E_H' and $E_H^{\circ\prime}$ are formal potentials.

  II. The *mass balances* are (neglecting dilution by titrant)

$$[Fe^{2+}] + [Fe^{3+}] = C_0 \qquad \text{(iv)}$$

$$5[MnO_4^-] + 5[Mn^{2+}] = C_0 f \qquad \text{(v)}$$

where f is the mole fraction of Fe(II) titrated (i.e., Fe(III)/[Fe(II) + Fe(III)] assuming stoichiometric completion of the reaction). (For example, at the end point $f = 1$, we have added $\frac{1}{5}$ mol of MnO_4^- per mole of Fe(II) originally present; correspondingly, at the end point we will have a concentration of Mn_T of one-fifth of that of $Fe_T : [Mn_T] = \frac{1}{5}[Fe_T]$ or $5[Mn_T] = Fe_T = C_0$.)

  III. The *electron balance* is

$$5[Mn^{2+}] = [Fe^{3+}] \qquad \text{(vi-}a\text{)}$$

that is, apply similar considerations as in proton balance. For every Mn^{2+} formed five electrons are lost, and for every ferric ion formed one electron is gained (or every MnO_4^- can accept five electrons and every Fe^{2+} can donate one electron).

The equations derived from *equilibrium constants* [i.e., redox potential– Nernst equation, (ii) and (iii)], from *mass balances* [(iv) and (v)] and the *electron balance* (vi-a) define E' as a *function of f* (five equations and five unknowns). This gives the exact solution.

Since operational E is not necessarily conceptual E, *approximate* solution may be more expedient if the assumption is made that the reaction goes virtually to completion.

1. *Before the end point*: All MnO_4^- added is reduced and oxidizes Fe^{2+} to Fe^{3+} quantitatively

$$[Fe^{3+}] = C_0 f = 5[Mn^{2+}] \qquad [Fe^{2+}] = C_0(1 - f)$$

According to equation ii

$$E_H' \simeq 0.7 + 0.06 \log \frac{f}{1 - f} \qquad \text{(vii)}$$

2. *Beyond the end point:* Essentially all the ferrous converted to ferric and $[Fe^{2+}]$ in equation iv can be neglected. Accordingly equation vi-a becomes $5[Mn^{2+}] \simeq C_0$, and the MnO_4^- added essentially stays as $[MnO_4^-]$; therefore (see also equation v)

$$5[MnO_4^-] \simeq C_0(f - 1) \qquad \text{(viii)}$$

and

$$E_H' \simeq 1.5 + \frac{0.06}{5} \log(f - 1) \qquad \text{(ix)}$$

At $f \simeq 2$, $E_H' = E_{H \cdot MnO_4^-, Mn^{2+}}^{\circ}$.

3. *At the end point:* Addition of equations ii and iii leads to

$$2E_H' = 0.70 + 1.5 + 0.06 \log \frac{[Fe^{3+}] [MnO_4^-]^{1/6}}{[Fe^{2+}] [Mn^{2+}]^{1/6}} \qquad \text{(x)}$$

At the end point $E_H' = E_H'(ep)$ and equation vi as well as the condition

$$5[MnO_4^-] = [Fe^{2+}] \qquad \text{(vi-}b)$$

is fulfilled.

From an operational point of view we may consider that the Pt electrode does not appear to respond to MnO_4^- and Mn^{2+} ions. Therefore the operational potential equals the conceptual potential only under conditions where a Pt electrode responds satisfactorily to the $Fe^{2+}-Fe^{3+}$ couple. If $[Fe^{3+}] \gg [Fe^{2+}]$ (i.e., beyond the end point) and if $[Fe^{2+}] \gg [Fe^3]$ (i.e., in the initial portion of the titration curve), the Pt electrode indicates a mixed potential that does not change appreciably with f. An illustration of the titration curve together with the schematic polarization curves are given in Figure 7.19.

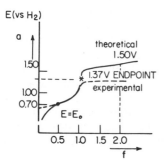

Figure 7.19 Titration of Fe(II) with MnO_4^- (Example 7.15). The experimental titration curve is compared with the calculated one.

7.9 MEASURING THE REDOX POTENTIAL, E_H OR $p\varepsilon$, IN NATURAL WATERS

It has already been pointed out that it is necessary to distinguish between the concept of a potential and the measurement of a potential. E_H measurements are of great value in systems for which the variables are known or under control. In this section we will discuss the measurement and indirect evaluation of redox potentials. The problems encountered in operationally measuring redox potentials have been reviewed extensively [11, 32–37].

Kinetic Considerations.   Some of the essential principles involved in the measurement of an electrode potential can be described qualitatively by consideration of the behavior of a single electrode (platinum) immersed in an acidified Fe^{2+}–Fe^{3+} solution. To cause the passage of a finite current at this electrode, it is necessary to shift the potential from its equilibrium value. We thus obtain a curve depicting the electrode potential as a function of the applied current (polarization curve). At the equilibrium potential, that is, at the point of zero applied current, the half-reaction

$$Fe^{3+} + e \rightleftharpoons Fe^{2+}$$

is at equilibrium, but the two opposing processes, the reduction of Fe^{3+} and the oxidation of Fe^{2+}, proceed at an equal and finite rate:

$$v_1 \text{ (rate of reduction)} = v_2 \text{ (rate of oxidation)} \qquad (118)$$

which is proportional to the rate of passage of electrons in both directions. Although the net rate of passage of electrons is equal in both directions and thus the (applied) current is zero, the passage of current in a single direction is not zero and is called the *exchange current* i_0 (see Figure 7.20).

At equilibrium, where no net current flows ($i_a = i_c$), concentrations at the electrode surface equal bulk concentrations, and $p\varepsilon$ values or potentials measured correspond to equilibrium conditions, yielding:

$$\frac{[Fe^{3+}]}{[Fe^{2+}]} = 10^{(p\varepsilon - p\varepsilon^\circ)} = \exp\frac{F}{RT}(E_H - E_H^\circ) \qquad (119)$$

or, in a more general sense,

$$\frac{[ox]}{[red]} = 10^{n(p\varepsilon - p\varepsilon^\circ)} = \exp\frac{nF}{RT}(E_H - E_H^\circ) \qquad (120)$$

Equations (119) and (120) are identical to the Nernst equation (equation 24).

The net current can be visualized as the summation of two opposing currents (cathodic and anodic). The rate of Fe^{3+} reduction (conventionally expressed as cathodic current) generally increases exponentially with more negative electrode potential values and is, furthermore, a function of the concentration of $[Fe^{3+}]$ and of the effective electrode area. Similar considerations apply to the rate of Fe^{2+} oxidation (anodic current), which is proportional to $[Fe^{2+}]$, electrode area, and the exponential of the potential. It is obvious from the schematic representation that an infinitesimal shift of the electrode potential from its equilibrium value will make the half-reaction proceed in either of the two opposing directions, provided the concentration of these ions is sufficiently large. Measurement of the equilibrium electrode potential in such a case is feasible; the amount the potential must

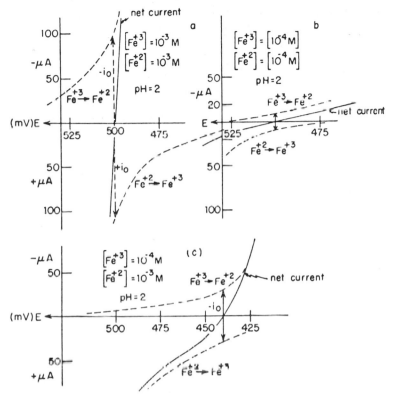

Figure 7.20 Polarization curves for various concentrations of Fe^{2+} and Fe^{3+}. Solid lines are polarization curves (electrode area = 1 cm^2); i_0, exchange current. Dashed lines are hypothetical cathodic ($-i$) and anodic ($+i$) currents. Curves are schematic but based on experimental data at significant points.

be shifted to obtain an indication with the measuring instrument, hence the sharpness and reproducibility of the measurement, is determined by the slope of the net current in the vicinity of the balance point. This slope is proportional to the exchange current i_0; i_0 depends upon the concentration of reactants and the electrode area.

Under favorable circumstances with modern instrumentation for which the current drain is quite low, reliable potential measurements can be made with systems giving i_0 greater than about 10^{-7} A. If 10-fold smaller concentrations of both ions are present, i_0 and the slopes are only $\frac{1}{10}$ as great; but with $i_0 = 10^{-5}$ A, reliable measurements can still be made, as shown in Figure 7.20b. If the concentration of only one ion is decreased, the drop in i_0 is not as great, and E_H, the potential corresponding to equal cathodic and anodic currents, is shifted, as shown in Figure 7.20c. If both Fe^{3+} and Fe^{2+} are at 10^{-6} M concentrations (ca. 0.05 mg liter^{-1}), i_0 is 10^{-7} A and measurements are no longer precise. Actually, because of other effects caused by trace impurities, it becomes difficult to obtain measurements in accord with simple Nernst theory when either Fe ion concentration is less than about 10^{-5} M.

Electrodes with large areas are advantageous, but they also tend to magnify the effects of trace impurities or other reactions on the electrode surface itself, such as adsorption of surface-active materials leading to a reduction in the electron-exchange rate or in the effective area A.

"Slow" Electrodes.   We might contrast such behavior with the conditions we would encounter in attempting the measurement of the electrode potential in water containing O_2. A schematic representation of the polarization curve for this case is given in Figure 7.21. The equilibrium electrode potential again should be located at the point where the net applied current (i.e., the sum of cathodic and anodic currents) is zero. The exact location becomes difficult to determine. Over a considerable span of electrode potentials the net current is virtually zero; similarly, the electron-exchange rate, or the exchange current reflecting the opposing rates of the-half reaction

$$H_2O \rightleftharpoons \tfrac{1}{2}O_2 + 2H^+ + 2e$$

is virtually zero. Operationally, a remarkable potential shift must be made to produce a finite net current, and the current drawn in the potentiometric measurement is very large compared with the exchange current. Even with modern instrumentation in which the current drain can be made extremely low, the experimental location of the equilibrium potential is ambiguous. It has been estimated [38] that for O_2 (1 atm) the specific exchange current i_0/A is 10^{-9} A cm^{-2}, far less than 10^{-7}. The current utilized by impurities may exceed the exchange current. It has been shown [39] that, under conditions of extremely high purity, an oxygen equilibrium potential of 1.23 V can be attained on a Pt electrode.

Mixed Potentials.   Another difficulty arises in E_H measurements. The balancing anodic and cathodic currents at the apparent "equilibrium" potential need not correspond to the same redox process and may be a composite of two or more processes. An example of this is shown in Figure 7.21b for a $Fe^{3+}-Fe^{2+}$ system in the presence of a trace of dissolved oxygen. The measured zero-current potential is the value where the rate of O_2 reduction at the electrode surface is equal to the rate of Fe^{2+} oxidation rather than the value of E_{eq},

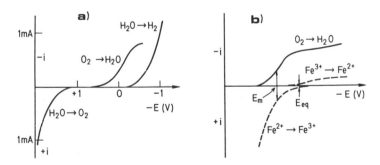

Figure 7.21 Electrode polarization curves for oxygen-containing solutions. (a) In otherwise pure water; (b) in the presence (nonequilibrium) of some Fe^{2+}. Curves are schematic but in accord with available data at significant points. Because the net current (a) is virtually zero over a considerable span of the electrode potentials, the exact location of the redox potential becomes difficult to determine or is determined by insidious redox impurities. A mixed potential (b) may be observed at the point where the anodic and cathodic currents are balanced; but because the various redox partners are not in equilibrium with each other, it is not amenable to quantitative interpretation.

since at the latter point simultaneous O_2 reduction produces excess cathodic current. In addition, because the net reaction at E_m converts Fe^{2+} to Fe^{3+}, the measured potential exhibits a slow drift. Such "mixed" potentials are of little worth in determining equilibrium E_H values. Many important redox couples in natural waters are not electroactive. No reversible electrode potentials are established for $NO_3^- - NO_2^- - NH_4^- - H_2S$ or $CH_4 - CO_2$ systems. Moreover, few organic redox couples yield reversible potentials. Unfortunately, many and perhaps most measurements of E_H (or pε) in natural waters represent mixed potentials not amenable to quantitative interpretation.

Indirect Evaluation of Redox Potentials. Of the standard redox potentials quoted in handbooks or summaries, relatively few have been determined by direct potential measurements; the others have been calculated from a combination of free energy data or from equilibrium constants. It is possible to evaluate the redox level in natural water systems by determining the relative concentrations of the members of one of the redox couples in the system and applying the electrochemical relations in reverse. As Figure 7.10 suggests, quantitative analytical information on any one of the species $O_2, Fe^{2+}, Mn^{2+}, HS^- - SO_4^{2-}$, and $CH_4 - CO_2$ gives a conceptually defined pε (or E_H), provided the system is in equilibrium. In practice, however, there are limitations even to this basically sound procedure. The system must be in equilibrium or must be in a sufficiently constant metastable state to make the concept of a partial equilibrium meaningful. For example, although most aqueous media are not in equilibrium with regard to processes involving N_2, the inertness of these reactions may allow us to ignore them while treating the equilibrium achieved by other species. A singular pε value can be ascribed to a system if equal values of E_H or pε are obtained for each of the redox couples in a multicomponent system; otherwise the couples are not in equilibrium and the concept of a singular redox potential becomes meaningless.

Within these limitations analytical determination of the oxidized and reduced forms of a redox couple can provide quite precise values for E_H. Depending on the number of electrons transferred in the process, determinations accurate within a factor of 2 will give E_H values within 5 to 20 mV or pε values within 0.1 to 0.3 units.

Example 7.16. Computation of pε and E_H from Analytical Information

Estimate pε and E_H for the aquatic habitats characterized by the following analytical information.

1 Sediment containing $FeOOH(s)$ and $FeCO_3(s)$ in contact with 10^{-3} M HCO_3^- at pH = 7.0.
2 Water from the deeper layers of a lake having a dissolved O_2 concentration of 0.03 mg O_2 liter^{-1} and a pH of 7.0.
3 An anaerobic digester in which 65% CH_4 and 35% CO_2 are in contact with water of pH = 7.0.
4 A water sample of pH = 6, containing $\{SO_4^{2-}\} = 10^{-3}$ M and smelling of H_2S.
5 A sediment–water interface containing $FeCO_3(s)$ with a crust of black $FeS(s)$ at pH = 8.

1 This corresponds to the conditions of (9) in Table 7.5. p$\varepsilon = -0.8$ and $E_H = -0.047$ V. This system, in equilibrium, has an infinite poising (redox-buffering) intensity. The redox level can be defined rather precisely.

2   This corresponds to (1) in Table 7.5, and $p\varepsilon = 13.75 + \frac{1}{4}\log p_{O_2}$; 10^{-6} mol O_2 liter^{-1} is equivalent to a partial pressure, $p_{O_2} \approx 6 \times 10^{-4}$, and $\frac{1}{4}\log p_{O_2} = -0.8$. $p\varepsilon$ for this system is 12.95 and $E_H = +0.77$ V. Note that E_H varies little with p_{O_2} or dissolved O_2 concentration. A reduction in O_2 from 10 (approximately air saturated) to 0.1 mg liter^{-1} will lower E_H by 30 mV.

3   According to (15) in Table 7-5

$$p\varepsilon = -4.13 + \tfrac{1}{8}\log\frac{p_{CO_2}}{p_{CH_4}} = -4.16$$

$$E_H = -0.25 \text{ V}$$

4   The equilibrium is characterized by

$$\tfrac{1}{8}SO_4^{2-} + \tfrac{5}{4}H^+ + e = \tfrac{1}{8}H_2S(g) + \tfrac{1}{2}H_2O$$

$p\varepsilon°$ can be calculated from the $p\varepsilon°(W)$ value given in Table 7.5. $p\varepsilon° = p\varepsilon°(W) - (n_H/2)\log K_W$ [see (30) of Section 7.5]; $p\varepsilon° = -3.5 + 8.75 = 5.25$. Hence

$$p\varepsilon = 5.25 - \tfrac{5}{4}pH + \tfrac{1}{8}\log[SO_4^{2-}] - \tfrac{1}{8}\log pH_2S$$

$$= 2.62 - \tfrac{1}{8}\log p_{H_2S}$$

It is reasonable to set p_{H_2S} between 10^{-2} and 10^{-8} atm. This puts $p\varepsilon$ between -1.6 and -2.4, corresponding to E_H of between -0.09 and -0.14 V.

5   We do not have all the information necessary, but we can make a guess. A reaction defining the most likely redox poising equilibrium could be formulated as

$$SO_4^{2-} + FeCO_3(s) + 9H^+ + 8e = FeS(s) + HCO_3^- + 4H_2O$$

With the use of the free energy values given in the appendix at the end of the book the equilibrium constant of the reaction given above is calculated (25°C) to be log $K = 38.0$. Correspondingly $p\varepsilon° = 4.75$, and $p\varepsilon$ can be calculated from

$$p\varepsilon = 4.75 - \tfrac{9}{8}pH + \tfrac{1}{8}(pHCO_3^- - pSO_4^{2-})$$

The term in parentheses has to be estimated; it is very unlikely that it is outside the range -2 to $+2$. This fixes $p\varepsilon$ for this equilibrium system at pH $= 8$ between -4.0 and -4.5. Correspondingly, $-0.27 < E_H < -0.24$ V.

Example 7.17. Redox Potential of S(s)-Bearing Sediments

R. A. Berner [*Geochim. Cosmochim. Acta*, **27**, 563 (1963)] has demonstrated that a Pt electrode inserted into H_2S-containing sediments gives electrode potentials in accord with the half-cell $S(s) + 2e = S^{2-}$ (S^{2-} was determined specifically with a Ag_2S/Ag electrode). Establish a relationship between electrode potential of measured E_H versus pS^{2-} (25°C). From equilibrium constants we can make the combinations:

$$S(s) + 2H^+ + 2e = H_2S(aq) \qquad \log K = 4.8$$

$$H_2S(aq) = H^+ + HS^- \qquad \log K = -7.0$$

$$HS^- = H^+ + S^{2-} \qquad \log K = -14$$

$$\overline{\hspace{3cm}}$$

$$S(s) + 2e = S^{2-} \qquad \log K = -16.2$$

Hence we can write

$$p\varepsilon = -8.1 + 0.5pS^{2-}$$

or

$$E_H = -0.48 + 0.030pS^{2-}$$

This relationship has been observed by Berner for natural and artificial H_2S-containing sediments.

There are two questions relevant in this context:

1   How does elemental sulfur participate in the electron exchange at the Pt surface? Sulfur probably does not participate in the electrode reaction; most likely soluble species that are in equilibrium with elemental sulfur, such as polysulfides $[S(s) + S^{2-} = S_2^{2-}; 4S(s) + S^{2-} = S_5^{2-}]$ are electroactive. As shown by Peschanski and Valensi and as quoted by Berner, because of such polysulfides the Pt electrode behaves with respect to the $S(s)$–S^{2-} system as a kinetically reversible system.

2   Do such measurements indicate the redox level of the sediments? Though the response of the Pt electrode is controlled by the $S(s)$–$S(-II)$ couple, the E_H measured is representative of the redox potential of the sediment only if the $S(s)$–$S(-II)$ couple is in equilibrium with all other redox couples present in the same locale. If different apparent redox levels coexist, the Pt electrode records the couple characterized by the highest exchange current. Consider for example the case where an organic redox system is not in equilibrium with the $S(s)$–$S(-II)$ couple has a lower apparent $p\varepsilon$. Because the organic species are not electroactive, the Pt electrode is not affected by this lower partial redox level.

In summary, because many redox reactions in natural waters do not couple with one another readily, different apparent redox levels exist in the same locale; thus an electrode or any other indicator system cannot measure a unique E_H or $p\varepsilon$. If the electrode (or the indicators) reached equilibrium with one of the redox couples, it would indicate the redox intensity of that couple only. A few conditions are necessary to obtain meaningful operational E_H values:

1   The measuring electrode must be inert. As shown by Whitfield [36] and others, the Pt electrode may form in aerobic milieu PtO and PtO_2 and in sulfide-bearing waters PtS. An oxide-coated electrode is—in the absence of high concentrations of other redox couples—primarily pH-responsive.

2   The exchange current should be large in comparison to the drain by the measuring electrode and to "impurity" currents. The measured potential is given by the system with the highest exchange rate. Considering the fact that, even in oxic waters, reductants (CO, H_2, CH_4) other than H_2O (or H_2O_2) are present at a level of 10^{-6} M, the impurity current, that is, the current utilized by the impurities, may exceed by a factor of 100 or 1000 that of the exchange current of the Pt–O_2 system [40].

3   In a system where various redox partners are not in equilibrium with each other, the balancing anodic and cathodic currents at an apparent equilibrium potential need not correspond to the same redox process and may be a composite of two or more processes; thus one may be observing a *mixed potential* not amenable to quantitative interpretation.

REFERENCES

1   W. M. Latimer, *Oxidation Potentials*, Prentice-Hall, Englewood Cliffs, N.J., 1952.

2   W. M. Clark, *Oxidation Reduction Potentials of Organic Systems*, William & Wilkins, Baltimore, 1960.

3   G. Charlot, *Selected Constants and Oxido-Reduction Potentials*, Pergamon, Elmsford, N.Y., 1958.

4   M. Pourbaix, *Atlas d'Equilibres Electrochimiques*, Gauthiers-Villars, Paris, 1963.

5   M. Calvin and J. A. Bassham, *The Photosynthesis of Carbon Compounds*, W. A. Benjamin, Menlo Park, Calif., 1962.

6   P. L. McCarty, in *Organic Compounds in Aquatic Environments*, S. D. Faust and J. V. Hunter, Eds., Dekker, New York, 1971, p. 495.

7   A. J. B. Zehnder, in *Water Pollution Microbiology*, Vol. 2, R. Mitchell, Ed., Wiley, New York, 1978.

8   W. G. Breck, in *The Sea*, Vol. 5, E. D. Goldberg, Ed., Wiley-Interscience, New York, 1974, p. 153.

9   S. Fallab, *Angew. Chem. Int. Ed.*, **6**, 496 (1967).

10 S. Dagley, *Naturwissenschaften*, **65**, 85 (1978).

11 W. Stumm, *Thalassia Jugosl.*, **14**, 197 (1978).

12 C. Junge, *Promet*, **2**, 19 (1975).

13 L. G. Sillén, *Ark. Kemi*, **24**, 431 (1965); **25**, 159 (1965).

14 H. Taube, *J. Chem. Educ.*, **45**, 452 (1968).

15a W. Stumm and G. F. Lee, *Ind. Eng. Chem.*, **53**, 143 (1961). [The rate constants given in Table II of [15a] are in error by a factor of $(1n\ 10)^2$.

15b P. C. Singer and W. Stumm, *Science*, **167**, 3921 (1970).

16 J. J. Morgan and W. Stumm, *Proceedings of the Second Conference on Water Pollution Research*, Pergamon, Elmsford, N.Y., 1964.

17 D. R. Kester, R. H. Byrne, and Y. J. Liang, in *Marine Chemistry in the Coastal Environment*, T. M. Church, Ed., American Chemical Society, Symposium Series, No. 18, Washington, D.C., 1975, p. 56.

18 F. Haber and J. Weiss, *Proc. Roy. Soc. London*, **A147**, 332 (1934).

19 T. L. Theis and P. C. Singer, *Environ. Sci. Technol.*, **8**, 569 (1974).

20 K. L. Temple and E. W. Delchamps, *Appl. Microbiol.*, **1**, 255 (1953).

21 R. M. Garrels and M. W. Thompson, *Amer. J. Sci.*, **258**, A57 (1960).

22 M. P. Silverman and D. G. Lundgren, *J. Bacteriol.*, **77**, 642 (1959); **78**, 325 (1959).

23 E. Stumm-Zollinger, *Arch. Mikrobiol.*, **83**, 110 (1972).

24 T. Almgren and I. Hagström, *Water Res.*, **8**, 395 (1974).

25 K. Y. Chen and J. C. Morris, *J. Sanit. Eng. Div. Amer. Soc. Civ. Eng.*, **98**, 215 (1972).

26 M. R. Hoffmann and B. C. Lim, in *Environ. Ser. Technol.*, **13**, 1406 (1979).

27 J. Hoigné and H. Bader, *Science*, **190**, 782 (1975); *Environ. Sci. Technol.*, **12**, 79 (1978); *Int. Assoc. Water Pollut. Res. Progr. Water Technol.*, **10**, 645 (1978).

28 G. Eisenmann, R. Bates, G. Mattock, and S. M. Friedman, *The Glass Electrode*, Wiley-Interscience, New York, 1965.

29 G. Eisenmann, Ed., *Glass Electrodes for Hydrogen and Other Cations*, Dekker, New York, 1973.

30 R. G. Bates, *Determination of pH: Theory and Practice*, 2nd ed., Wiley, New York, 1973.

31 M. Whitfield, in *Chemical Oceanography*, Vol. 4, 2nd ed., J. P. Riley and G. Skirrow, Eds., Academic, New York, 1975, p. 1.

32 C. E. Zobell, *Bull. Amer. Soc. Petrol. Geol.*, **30**, 477 (1946).

33 J. C. Morris and W. Stumm, in *Equilibrium Concepts Natural Water Systems*, Advances in Chemistry Series, No. 67, American Chemical Society, Washington, D.C., 270, 1967.

34 R. A. Berner, *Principles of Chemical Sedimentology*, McGraw-Hill, New York, 1971.

35 M. Whitfield, *Limnol. Oceanogr.*, **17**, 383 (1972).

36 M. Whitfield, *Limnol. Oceanogr.*, **19**, 857 (1974).

37 R. Parsons, in *The Nature of Seawater*, E. D. Goldberg, Ed., Dahlem Konferenzen, Berlin, 1975.

38 Bockris, J. O'M and A. K. M. S. Huq, *Proc. Roy. Soc. London*, **A237**, 277 (1956).

39 Watanabe, W and M. A. V. Devanathan, *J. Electrochem. Soc.*, **111**, 615 (1964).

40 J. O. M. Bockris and A. K. N. Reddy, in *Modern Electrochemistry*, Plenum, New York, 1970, Chapter 10.

READING SUGGESTIONS

Redox Equilibria

Garrels, R. M., and C. L. Christ, *Solutions, Minerals and Equilibria*, Harper and Row, New York, 1965.

Hem, J. D., *Study and Interpretation of Chemical Characteristics of Natural Waters*, Geological Survey Water Supply Paper No. 1473, Washington, D.C., 1975.

Latimer, W. M., *Oxidation Potentials*, Prentice-Hall, Englewood Cliffs, N.J., 1952.

Sillén, L. G. In *Treatise on Analytical Chemistry*, Part I, Vol. I, I. M. Kolthoff and P. J. Elving, Eds., Interscience, New York, 1959, p. 227; in *Chemical Equilibrium in Analytical Chemistry*, Wiley-Interscience, New York, 1965.

Sillén, L. G., "Master Variables and Activity Scales." In *Equilibrium Concepts in Natural Water Systems*, Advances in Chemistry Series, No. 67, American Chemical Society, Washington, D.C. 1967, p. 45.

Mechanisms of Oxidation–Reduction Reactions

Duke, F. R., "Mechnism of Oxidation-Reduction Reactions." In *Treatise on Analytical Chemistry*, Part I, Volume I, I. M. Kolthoff and P. J. Elving, Eds., Interscience, New York, 1959, Chapter 15.

Fallab, S., "Reactions with Molecular Oxygen," *Angew. Chem. Int. Ed.*, **6**, 496–507 (1967).

Marcus, R. A., "Electron Transfer in Homogeneous and Heterogeneous Systems." In *The Nature of Seawater*, E. D. Goldberg, Ed., Dahlem Konferenzen, Berlin, 1975, pp. 477–504.

Taube, H., Mechanism of Oxidation-Reduction Reactions, *J. Chem. Educ.*, **45**, 453 (1968).

Wilkins, R. G., *The Study of Kinetics and Mechanisms of Reactions of Transition Metal Complexes*, Allyn and Bacon, Boston, 1974.

Zafiriou, O. C., "Marine Organic Chemistry Previewed," *Marine Chem.*, **5**, 497 (1977).

Electrochemistry

Bard, A. J. and L. R. Faulkner, *Electrochemical Methods*, Wiley, New York, 1980.

Bockris, J. O. M., and A. K. N. Reddy, *Modern Electrochemistry*, Vols. 1 and 2, Plenum, New York, 1970.

Charlot, G., J. Badoz-Lambling, and B. Tremillon, *Electrochemical Reactions*, Elsevier, Amsterdam, 1962, Chapter 7.

Laitinen, H. A., and W. E. Harris, *Chemical Analysis*, 2nd ed., McGraw-Hill, New York, 1975.

Murray, R. W., and C. N. Reilley, *Electroanalytical Principles*, Wiley-Interscience, New York, 1963.

Tanaka, N., and R. Tamushi, "Kinetic Parameters of Electrode Reaction," *Electrochem. Acta*, **9**, 963 (1964).

Electrodes; pH Measurements

Bates, R. G., "pH Scales for Seawater." In *The Nature of Seawater*, E. D. Goldberg, Ed., Dahlem Konferenzen, Berlin, 1975, pp. 313–338.

Durst, R. A., "Mechanism of the Glass Electrode Response," *J. Chem. Educ.*, **44**, 175 (1967).

Eisenmann, G., Ed., *Glass Electrode for Hydrogen and Other Cations: Principles and Practice*. Dekker, New York, 1967.

Warner, T. B., "Ion Selective Electrodes in Thermodynamic Studies." In *The Nature of Seawater*, E. D. Goldberg, Ed., Dahlem Konferenzen, Berlin, 1975, pp. 191–218.

Whitfield, M., "The Electroanalytical Chemistry of Seawater." In *Chemical Oceanography*, J. P. Riley and G. Skirrow, Eds., Vol. 4, 2nd ed., Academic, New York, 1976. p. 1.

Redox Reactions and Organisms

Kirschbaum, J., "Biological Oxidations and Energy Conservation," *J. Chem. Educ.*, **45**, 29 (1968).

Klotz, I. M., *Energy Changes in Biochemical Reactions*, Academic, New York, 1967.

Lehninger, A. L., *Bioenergetics*, W. A. Benjamin, Menlo Park, Calif., 1965.

McCarty, P. L., "Energetics and Bacterial Growth." In *Organic Compounds in Aquatic Environments*, S. J. Faust and J. V. Hunter, Eds. M. Dekker, New York, 1971, pp. 495–532.

Redox Potential as an Environmental Parameter

Abelson, P. H., "Chemical Events on the Primitive Earth," *Proc. Natl. Acad. Sci. U.S.*, **55**, 1365 (1966).

Baas-Becking, L. G. M., I. R. Kaplan, and O. Moore, "Limits of the Natural Environments in Terms of pH and Oxidation-Reduction Potentials," *J. Geol.*, **68**, 243 (1960).

Berner, R. A., *Principles of Chemical Sedimentology*, McGraw-Hill, New York, 1971.

Bostrom, K., "Some pH Controlling Redox Reactions in Natural Waters." In *Equilibrium Concepts in Natural Water Systems*, Advances in Chemistry Series, No. 67, American Chemical Society, Washington, D.C., 1967, p. 286.

Morris, J. C., and W. Stumm, "Redox Equilibria and Measurements of Potentials in the Aquatic Environment." In *Equilibrium Concepts in Natural Water Systems*, Advances in Chemistry Series, No. 67, American Chemical Society, Washington, D.C., 1967, p. 270.

Sillén, L. G., "Oxidation State of Earth's Ocean and Atmosphere," *Ark. Kemi*, **24**, 431 (1965); **25**, 159 (1965).

Stumm, W., "What is the pε of the Sea?" *Thalassia Jugosl.*, **14**, 197–208 (1978).

Whitfield, M., "The Electrochemical Characteristics of Natural Redox Cells," *Limnol. Oceanogr.*, **17**, 383 (1972)

PROBLEMS

7.1 (a) Write balanced reactions for the following oxidations and reductions:
  (i) Mn^{2+} to MnO_2 by Cl_2
  (ii) Mn^{2+} to MnO_2 by OCl^-
  (iii) H_2S to SO_4^{2-} by OCl^-
  (iv) MnO_2 to Mn^{2+} by H_2S

  (b) Arrange the following in order of decreasing pε:
  (i) Lake sediment
  (ii) River sediment
  (iii) Seawater
  (iv) Groundwater containing 0.5 mg liter^{-1} Fe^{2+}
  (v) Digester gas (CH_4, CO_2)

7.2 The II, III, and IV oxidation states of Mn are related thermodynamically by these equilibria in acid solution:

$$Mn(III) + e = Mn(II) \quad \log K = 25$$

$$Mn(IV) + 2e = Mn(II) \quad \log K = 40$$

(a) What is the standard electrode potential of the couple

$$Mn(IV) + e = Mn(III)?$$

(b) Is Mn(III) stable with respect to simultaneous oxidation and reduction to Mn(II) and Mn(IV)? Explain.

7.3 (a) Under what pH conditions can NO_3^- be reduced to NO_2^- by ferrous iron? (Use constants available in Tables 7.4 and 7.5.)

  (b) What is the ratio of the concentrations of NO_3^- to NO_2^- in equilibrium with an aqueous Fe^{2+}–$Fe(OH)_3(s)$ [or $FeOOH(s)$] system that has a pH of 7 and a pFe^{2+} of 4?

7.4   (a) Can the oxidation of NH_4^+ to NO_3^- by SO_4^{2-} be mediated by micro-organisms at pH = 7?

(b) Is the oxidation of HS^- to SO_4^{2-} by NO_3^- thermodynamically possible at a pH of 9?

(c) Estimate the pε range in which sulfate-reducing organisms can grow.

7.5   In a closed tank, water and gases have been brought into equilibrium. The water at equilibrium has the following composition: Alkalinity $= 2 \times 10^{-2}$ eq liter^{-1}; $[Fe^{2+}] = 2 \times 10^{-5}$ M; pH = 6.0; Fe(III) = negligible but ferric hydroxide is precipitated.

(a) What is the partial pressure of CO_2?

(b) What is the pε of the solution?

(c) What partial pressure of O_2 corresponds to this pε?

The following information is available for the appropriate temperature:

$CO_2 + H_2O = H_2CO_3^*$	$\log K = -1.5$
$H_2CO_3^* = H^+ + HCO_3^-$	$\log K = -6.3$
$HCO_3^- = H^+ + CO_3^{2-}$	$\log K = -10.3$
$Fe^{2+} + CO_3^{2-} = FeCO_3(s)$	$\log K = +10.6$
$Fe^{3+} + 3OH^- = Fe(OH)_3(s)$	$\log K = +38.0$
$O_2 + 4e + 4H^+ + 2H_2O$	$E° = +1.23$ V
$H^+ + OH^- = H_2O$	$\log K = +14$
$Fe(OH)_3(s) + e = Fe^{2+} + 3OH^-$	$E° = -1.41$ V

7.6   What p_{O_2} cannot be exceeded so that a reduction of SO_4^{2-} to HS^- (pH = 7) can take place? (Use Table 7.5.)

7.7   (a) Can glucose $(C_6H_{12}O_6)$ be converted into CH_4 and CO_2 (e.g. in a digester)?

(b) What is the percent composition of the resulting gas?

7.8   Can $Fe_2SiO_4(s)$ in a primitive reducing atmosphere convert $CO_2(g)$ and $N_2(g)$ into alanine $(NH_2CH_3CHCOOH)$? [Compare M. E. Baur, "Thermodynamics of Heterogeneous Iron-Carbon Systems," *Chem. Geol.*, **22**, 189 (1978).] The following values for standard free energy of formation (kcal mol^{-1} may be used: $Fe_2SiO_4(s)$, -319; $Fe_3O_4(s)$, -242.4; $H_2O(l)$, -56.690; $NH_2CH_3CHCOOH$, -88.7; $SiO_2(s)$, -204.6; $CO_2(g)$, -94.26.

7.9   Separate each of the following reactions into its half-reactions and in each case write down the schematic representation of a galvanic cell in which the reaction would take place. Wherever possible devise a cell without liquid junction potentials.

(a) $H_2 + PbSO_4(s) = 2H^+ + SO_4^{2-} + Pb$

(b) $AgCl = Ag^+ + Cl^-$

(c) $3\,Mn^{2+} + 2\,MnO_4^- + 4\,OH^- = 5\,MnO_2 + 2\,H_2O$

(d) $6\,Cl^- + IO_3^- + 6\,H^+ = 3\,Cl_2 + I^- + 3\,H_2O$

For each reaction above compute E_{cell} for the corresponding cell and decide the direction in which each reaction would take place spontaneously if all substances were present at unit activity.

7.10 The solubility product of mercurous sulfate is 6.2×10^{-7} at 25°C, and the cell, $H_2(735 \text{ mm})H_2SO_4(0.001 \text{ } M)$, Hg_2SO_4/Hg has an emf of $+0.8630$ V at 25°C. Compute the standard potential of the half-reaction $Hg_2^{2+} + 2e = 2\,Hg$.

7.11 Potassium iodide (0.05 mol) and iodine (0.025 mol) are dissolved together in a liter of water at 25°C. By appropriate analysis it is found that only 0.00126 mol of the iodine in the solution has remained in the form of I_2, the remainder having reacted to form I_3^- according to the equation:

$$I_2(aq) + I^- = I_3^-$$

Calculate the equilibrium constant. Check your result on the basis of the following information:

$$I_2(s) + 2e = 2I^- \qquad\qquad p\varepsilon^\circ = 9.1$$

$$2I^- = I_2(s) + 2e \qquad\qquad E_H^\circ = -0.5355$$

$$I_3^- + 2e = 3I^- \qquad\qquad p\varepsilon^\circ = 9.1$$

$$3I^- = I_3^- + 2e \qquad\qquad E_H^\circ = -0.536$$

$$\Delta G^\circ \text{ for } I_2(aq) = +3.93 \text{ kcal}$$

7.12 What is the solubility of Au(s) in seawater [cf. L. G. Sillén, *Svensk Kem. Tidskr.*, **75**, 161 (1954)]? The following equilibrium constants (18 to 20°C) may be used:

$$AuCl_2^- + e = Au(s) + 2Cl^- \qquad \log K = 19.2$$

$$AuCl_4^- + 3e = Au(s) + 4Cl^- \qquad \log K = 51.3$$

$$Au(OH)_4^- + 4H^+ + 4Cl^- = AuCl_4^- + 4H_2O \qquad \log K = 29.64$$

7.13 Consider the equilibrium of an aqueous solution containing the solid phases FeOOH(s), $FeCO_3$(s). Cf. the reaction

$$FeOOH(s) + HCO_3^- + 2H^+ + e = FeCO_3(s) + 2H_2O$$

Does $p\varepsilon$ increase, decrease or stay constant upon addition of small quantities of the following: (a) FeOOH(s); (b) CO_2; (c) HCl; (d) EDTA at constant pH; (e) O_2 at constant pH; (f) NaOH.

7.14 The following reaction describes the reduction of *ferrate* iron [$Fe^{(VI)}$] to ferric iron:

$$FeO_4^{2-} + 3e + 8H^+ = Fe^{3+} + 4H_2O \qquad E_H^\circ = 2.2 \text{ V}$$

(a) What is $p\varepsilon^\circ$?

(b) Plot $p\varepsilon$ versus pH for the ferrate–ferric equilibrium in a system where $Fe_{T,aq} = 10^{-5} \text{ } M$ for each aqueous species.

(c) At pH $= 7$, which is the stronger oxidant, ferrate or HOCl?

(d) What change can you predict for long-term storage of ferrate salts in pure water?

7.15 The oxidation–reduction reaction

$$FeO(s) + \tfrac{1}{4}O_2(g) = \tfrac{1}{2}Fe_2O_3(s)$$

has been suggested as having regulated the oxygen content of the earth's atmosphere in earlier times. You may use the appendix at the end of the book.

(a) Discuss the stability of FeO(s) with respect to the two other reduced phases of iron, $Fe(OH)_s(s)$ and $FeCO_3$ (i.e., ignoring the oxidation–reduction reaction). Is FeO(s) a stable reduced phase in the absence of CO_2? Is $FeCO_3(s)$ a stable phase at $p_{CO_2} = 3 \times 10^{-4}$ atm in the absence of oxygen?

(b) Now consider the stability of reduced iron with respect to Fe_2O_3 in the presence of O_2.

(i) Write a reaction describing the conversion of $FeCO_3(s)$ to Fe_2O_3;

(ii) Calculate ΔG°;

(iii) Calculate K_{eq};

(iv) Discuss the stability relationships at $p_{CO_2} = 3 \times 10^{-4}$ atm and $p_{CO_2} = 0.21$ atm (present conditions).

ANSWERS TO PROBLEMS

7.1 (a) $Mn^{2+} + Cl_2 + 2H_2O = MnO_2(s) + 2Cl^- + 4H^+$; $Mn^{2+} + ClO^- + H_2O = MnO_2(s) + 2H^+ + Cl^-$; $H_2S + 4OCl^- = SO_4^{2-} + 4Cl^- + 2H^+$; $4MnO_2(s) + H_2S + 6H^+ = SO_4^{2-} + 4Mn^{2+} + 4H_2O$. (b) (iii) ($O_2$–$H_2O$, $p\varepsilon \geq 12$); (ii) (NH_4^+–NO_3^-, $p\varepsilon \simeq 5$–8); (iv) (10^{-5} M Fe^{2+}–FeOOH(s), $p\varepsilon \simeq -2$–0); (i) (HS^-–SO_4^{2-}, $p\varepsilon \simeq -2$ to -4); (v) (CH_4–CH_2O–CO_2, $p\varepsilon \simeq -3$ to -5).

7.2 (a) $E_H^\circ = +0.885$ V ($p\varepsilon^\circ = 15$); (b) No, $2Mn(III) = Mn(IV) + Mn(II)$, $\log K = 10$.

7.3 (a) For $[Fe^{2+}] \geq 10^{-4}$ M and $([NO_2^-]/[NO_3^-]) \geq 1$, pH ≥ 4.3 (consider the reaction $\tfrac{1}{2}NO_3^- + Fe^{2+} + 2\tfrac{1}{2}H_2O = \tfrac{1}{2}NO_3^- + 2H^+ + Fe(OH)_3(s)$; $\log K \simeq -4.6$). (b) $\sim 10^{-11}$.

7.4 (a) No, $\log K(W) \simeq -10$ for reaction $\frac{1}{8}NH_3^+ + \frac{1}{8}SO_4^- = \frac{1}{8}NO_3$ $+ \frac{1}{8}H^+ + \frac{1}{8}HS^-$ $([H^+] = 10^{-7}) + \frac{1}{8}H_2O.$ (b) Yes. (c) For pH of natural waters $(4.6 < pH < 9.4),$ $0 > p\varepsilon > -6.$

7.5 (a) 1 atm. (b) 4.8. (c) 10^{-40} atm.

7.6 $p_{O_2} < 10^{-70}$ atm.

7.7 (a) Yes. (b) $50\% CO_2, 50\% CH_4.$

7.8 Yes. $\frac{1}{2}N_2(g) + 3CO_2(g) + 9Fe_2SiO_4(s) + \frac{7}{2}H_2O(l) = NH_2CH_3CH$-COOH(aq) $+ 6Fe_3O_4 + 9SiO_2(s),$ $\Delta G^\circ(25°C) = -32$ kcal. If p_{N_2} is set at 0.01 atm and $p_{CO_2} = 0.03$ atm, alanine would be produced (thermodynamically). Baur shows that the formation of reduced carbon compounds is a thermodynamically spontaneous process in the heterogeneous system comprised of CO_2 and N_2

7.9 (a) $H_2(g) = 2H^+ + 2e;$ $Pb(s) + SO_4^{2-} = PbSO_4(s) + 2e;$ $Pb(s), PbSO_4(s), H_2SO_4/H_2SO_4, H_2(Pt); E_{cell}^\circ = +0.35$ V.

(b) $AgCl(s) + e = Ag(s) + Cl^-, Ag(s) = Ag^+ + e$ $Ag(s), AgCl(s), KCl/AgNO_3, Ag(s), E_{cell}^\circ = 0.58$ V.

(c) $MnO_4^- + 2H_2O + 3e = MnO_2(s) + 4OH^-;$ $Mn^{2+} + 4OH^- = MnO_2(s) + 2H_2O + 2e;$ $(Pt)MnO_2(s), Mn^{2+}, NaOH/MnO_4^-, MnO_2(s)(Pt); E_{cell}^\circ = 1.3$ V.

(d) $Cl_2(g) + 2e = 2Cl^-; IO_3^- + 6H^+ + 6e = I^- + 3H_2O$ $(Pt)Cl_2(g), NaCl/NaI, NaIO_3(Pt); E_{cell}^\circ = 0.46$ V.

7.10 For $Hg^{2+} + 2e = Hg(l),$ $E^\circ = +0.798$ V.

7.11 $K = 10^{2.88}.$

7.12 $\sim 10^{-7.2}$ M.

7.13 No change for (a), (d), and (e); increase for (b) and (c); decrease for (f).

8

Organic Carbon; Some Aspects of its Origin, Composition, and Fate

8.1 INTRODUCTION

After having devoted much attention to the discussion of mostly inorganic species, some consideration should be given to organic aquatic chemistry. Despite the fact that organic substances are thermodynamically unstable in natural waters, some aspects of organic chemistry may be discussed even in a book emphasizing chemical equilibria: Carbon forms the link in interactions between the inorganic environment and living organisms. The carbon cycle is interrelated with the cycles of all other elements. The composition of the inorganic environment thus depends to a large degree on the cyclic activity of life. Inorganic compounds may be considered intermediates in the continuous circuit of redox reactions.

Blumer [1] has defined the main goals of organic geochemistry and environmental organic chemistry in the following way: *Organic geochemistry* attempts to understand in terms of chemical, physical, and biological processes the formation, composition, and destruction of organic compounds in nature and

TABLE 8.1 GEOCHEMISTRY AND ENVIRONMENTAL CHEMISTRY OF ORGANIC COMPOUNDS[a]

Areas of Inquiry	Scope and Time Scale	
	Geochemistry	Environmental Chemistry
Sources	Biochemical and geochemical	Synthetic and fossil
Composition	Wide molecular-weight and compositional range	100–300 $\pm$ molecular weight, mainly C, H, halogen, metal
Interaction	All parts of the environment	Biosphere, food web, humans
Transformation	Long-term; slow reactions	Short-term; fast reactions
Fate	To $10^8 +$ years	Decades

[a] From Blumer [1].

their interaction with the environment during geological time spans. *Environmental organic chemistry* in a similar way attempts to understand the origin, composition, and fate, in today's environment, especially of organic compounds that affect living organisms and particularly humans in their best use of natural resources. Most of the activity in this field has followed the discovery of the environmental effects and of transmission of synthetic organics via the food web, and also the growing awareness of fossil fuels as environmental pollutants. The scope and time scale of the research on organic geochemistry and organic environmental chemistry overlap (Table 8.1).

8.2   THE CYCLE OF ORGANIC CARBON

Most of the carbon in the earth's crust has cycled through organisms and plants. Table 4.1 contains a survey of carbon in its various forms in the atmosphere and biosphere. Simple aspects of the pathways of C are illustrated in Figure 8.1. The flux reactions are initiated by photosynthesis (see Figure 7.9). Gross photosynthesis is estimated to fix ca. 54×10^{14} mol C year^{-1} [2,3]. Of the total carbon fixed annually 99.95% is respired and oxidized; only 0.05% ($= 0.025 \times 10^{14}$ mol C year^{-1}) escapes oxidation and becomes buried in the sediments. Most of the buried carbon will ultimately also be oxidized when it—subsequent to metamorphosis and uplifting of sediments—becomes weathered as fossil C.

The Production and Consumption of Organic Matter

Only a small fraction (4×10^{-4}) [4] of the solar energy supply is used for photosynthesis. Phytoplankton use a fraction of this energy (ca. one-sixth) for their own respiration. The remaining energy is bound biochemically and becomes available to heterotrophic consumers and decomposers (see Figure 11.9). The principle of the *food chain* may be illustrated by an example given by Cole [4]. Of 1000 cal fixed by the algae of Cayuga Lake, 150 cal can be transferred to biochemical bonds in their predators, small aquatic animals. Only 30 cal can be incorporated into the next step of the food chain, for example, into protoplasm of smelt. Then, 6 cal of the 30 cal are further transferred to biochemical bonds in the fat and muscle tissue of the person who eats the smelt. If trout is a link between smelt and humans, the yield of energy to humans shrinks to 1.2 cal. For each link in the food chain energy is lost to the environment, organic carbon converted into CO_2, and some organic intermediates are released to the water as by-products of metabolic reactions.

Odum [5], Ryther [6], and Fogg [7] have summarized production and respiration data for several ecological systems. A few illustrative data are presented in Table 8.2. Gross primary production (productivity) is the total rate of photosynthesis. Net primary productivity is the rate of synthesis of organic matter in excess of its respiratory utilization during the period of measurement.

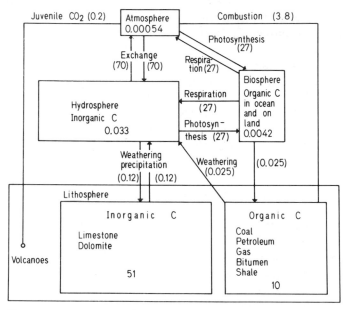

Figure 8.1 The circulation of carbon in nature is determined primarily by biochemical reactions. Amounts of carbon on the earth's surface in 10^{20} mol (geomoles). Parentheses indicate rates of turnover of carbon in 10^{14} mol year^{-1} (microgeomol year^{-1}). In 1933 Goldschmidt made quantitative estimates of the amount of carbon in the different parts of the cycle. His data have been revised by various workers. The data given here are mostly from Garrels and Perry [2]. (For a recent review see [3].) Gross photosynthesis (54×10^{14} mol C year^{-1}) has been equally divided between land and ocean (7.5 mol C m^{-2} year^{-1} in the ocean and 18 mol C m^{-2} year^{-1} on land). R. H. Whittaker and G. E. Likens in [3] give for total net primary production 44×10^{14} mol C year^{-1} for land and 21×10^{14} mol C year^{-1} for oceans. The C reservoir in the sediments is based on an average C content of 0.5% by weight and a sedimentary reservoir of 25×10^{23} g. Although the estimates given for the annual turnover are quite uncertain, they clearly illustrate that most of the carbon in the earth's surface has been cycled through organisms.

The concentration of organic matter in a natural body of water results from an interplay of net productivity, exudation of organic substances by phytoplankton, and import and export (inflow, outflow, dissolution, sedimentation, etc.) of organic matter. Productive lakes have a gross primary production of organic matter on the order of a few g m^{-2} day^{-1}. Oceanic waters show gross productions of up to 1 g m^{-2} day^{-1}. Production in flowing waters is usually very high. Odum has suggested that streams are among the most productive biological environments. The highest primary production rates occur in the recovery zones of streams polluted with sewage; this is a consequence of the supply of fertilizing elements and of additional organic nutritive sources possibly contained in sewage. High production is not limited, however, to polluted water. Salt marshes and estuaries (ca. 0.35% of the world surface area) are estimated [8] to produce ca. 2% of the net world primary production.

TABLE 8.2 METABOLISM OF SOME ECOSYSTEMS

	Gross Photosynthesis (g dry wt m^{-2} day^{-1})^a	Total Respiration (g dry wt m^{-2} day^{-1})^a	Net Production (g dry wt m^{-2} year^{-1})
Land			
Wheat (growing season, world average)	3	0.7	
Sugarcane (growing season, world average)	6.1	1.4	
Forest, tropical	13	3	
Forest, deciduous	6.7	1.6	
Average land			350–500
Marine			
Average hydrosphere			150–290
Coral reef maxima	18	—	
Continental shelf	0.55	0.25	
Long Island Sound (annual average)	2.1	1.2	
Open ocean	0.2–0.6	0.1–0.3	
Polluted systems:			
Galveston Bay	20–58	36–87	
Corpus Christi and Houston ship channels	0–31	0–51	
Potomac estuary, polluted	8	12	
Open estuaries			1000–2000
Salt marshes			1000–2000
Fresh Water			
White River, Indiana:			
Immediately below sewage outfall	0.15	21	
30 mi below sewage outfall	16	16	
60 mi below sewage outfall (zone of recovery)	41	29	
Lakes			
Clear Deep Lake, Wisconsin	0.7	—	
Lake Erie, summer	9.0	4.0	
Lake Tahoe, California–Nevada	0.2	0.1	
Laurence Lake, Michigan			430

^{a} The examples given here are based on results of various investigators and are not precisely comparable; the data are meaningful in a semiquantitative way only; 2.5 g dry weight corresponds to ca. 1 g C.

The Concentration of Aqueous Organic Carbon

In view of the considerable variety of sources of organic material from outside the water basin, a great diversity in the concentration of organic matter must be expected, even in the absence of human contamination. In freshwater bodies one typically encounters concentrations of a few mg C liter^{-1}; waters low in Ca^{2+} and Mg^{2+} often contain humic substances (see Section 8.3), occasionally (e.g., in bog or swamp waters) concentrations may be as high as 50 mg C liter^{-1}. In the oceans the concentrations of organic carbon range from 0.5 to 1.2 mg C liter^{-1}, with the higher values occurring in the surface waters [9,10]. Particulate organic carbon, including planktonic organisms, generally accounts for about 10% of the total organic carbon in the surface waters and for about 2% in the deep waters. Over much of the ocean we have, according to Wangersky, a "soluble" organic carbon concentration of about 500 μg C liter^{-1} and a total particulate organic carbon concentration of about 50 μg C liter^{-1}, all ultimately derived from a surface phytoplankton population equivalent to a concentration of about 5 μg C liter^{-1}. These figures are significantly higher for waters on the continental shelf, especially for regions where nutrients are returned to the surface by upwelling. Concentrations in surface layers are generally greater than in deeper waters. Below a depth of 400 to 600 m the concentration variations in organic carbon become very slight. This suggests that materials present in deep waters decompose extremely slowly [11,12]. Presumably much of the readily degradable organic matter has already been decomposed in the upper layers. Furthermore, life processes appear to be slower in the deep sea than in shallow water [12].

Organic Aggregates

The nonliving particulate organic matter in waters has usually been assumed to be fecal pellets and minute plankton remains in various stages of disintegration [13]. However, in recent decades evidence has accumulated that many of these particles are delicate platelike aggregates a few micrometers to several millimeters in diameter. They are amorphous and contain both organic and inorganic nonliving material with inclusions of microorganisms [13–17]. It has furthermore been shown that the nuclei of such particles can be formed by adsorption of dissolved organic compounds on air bubbles and other surfaces. The initial aggregates tend to increase in size by further aggregation and adsorption.

The process of particle formation on air bubbles from dissolved carbon seems to be restricted to organic molecules with a molecular weight of more than 100,000. In light of this observation, the lytic products of microorganisms and polymeric excretions of plankton deserve special consideration.

The *water–air* interface is a major site of biological activity. The surface microlayer is dominated by a 0.1-μm-thick microlayer of wet surfactants below the sea interface that behaves like a polysaccharine–protein complex [17]. There is an enrichment of dissolved organic carbon (DOC) (estimated effective C concentration of 0.5 to 2 g liter^{-1} of which ca 30% is carbohydrate) and neuson (microorganisms living in this microlayer). Thus the surface microlayers are largely heterotrophic microcosms whose nutrient and microorganism concentrations approach those in laboratory cultures [18].

Pollution. Pollution by humans adds organic matter to the receiving waters. Modern chemical industry has synthesized and processed a great variety of organic chemicals. These stream pollutants need to be distinguished from biogenic organic constituents even though they may also occur "naturally." The method of carbon dating can be used to distinguish between petrochemical and natural sources of organic matter. Since petrochemical wastes are derived from fossil carbon sources (coal, petroleum), they do not contain ^{14}C. Biogenic materials (products of photosynthesis and their decay and organic waste constituents of domestic sewage) contain essentially contemporary levels of ^{14}C.

Sediments. The quantity of organic matter carried in natural waters is small compared to the total amount of organic matter found in sediments. The latter are the principal depositories of posthumous organic debris accumulated throughout the earth's history. Most of the organic matter occurs in a finely disseminated state and is associated with fine-grained sediments. In order to illustrate more vividly the ratio of inorganic to organic matter, Degens [19] states "From an evaluation of the total thickness of sedimentary materials that have been formed and deposited over the last 3 to 4 billion years, it can be assumed that a sediment layer of approximately 1000 m around the earth's surface has been laid down. Roughly 2 percent of this layer, namely ~20 m, is organic, the rest is inorganic in nature. Of this 20-m organic part, coal comprises ~5 cm and crude oil only a little more than 1 mm. The rest represents the finely disseminated organic matter in shales, limestones and sandstones." Surface waters and sediments are the principal sites of biodegradation and bioconversion. It can be inferred from the published literature that most of the chemical alteration of organic matter takes place in the environment of deposition and during early stages of diagenesis.

Example 8.1. Steady-State Distribution of Organic Matter in the Ocean

Estimate from the data given in Figure 8.1 the residence time of organic C in the ocean.
    The organic carbon reservoir of the ocean (3.6×10^{14} m^2 surface area, 3.8×10^3 m mean depth) is 1.4×10^{21} dm$^3 \times 0.5$ mg C liter$^{-1} = 7 \times 10^{17}$ g C. The input by photosynthesis is 27×10^{14} mol C year^{-1} or 3.2×10^{16} g C year^{-1}.

Other imputs by rain and rivers (2 to 3 $\times$ 10^{14} g C year^{-1}) are negligible. Hence the residence time (=amount in reservoir per input) is ca. 20 years. This residence time is meaningful only with regard to the organic C in the surface waters of the oceans. As we have seen, little biodegradation of deep sea organic C occurs. We can estimate the residence time of the deep sea C by considering the output by sedimentation, which according to Figure 8.1 is ca. 3 $\times$ 10^{13} g C year^{-1}. Since the deep ocean contains only about two-thirds of the organic C of the ocean, a residence time of 4.7 $\times$ 10^{17} g C $\div$ 3 $\times$ 10^{13} g C year^{-1} or $\sim$15,000 years for organic C is obtained. Williams, Oeschger, and Kinney [20] have determined by ^{14}C measurements the apparent "age" of organic matter at 2000 m in the northeast Pacific to be 3400 years. From this age a fraction of the primary production actually entering the deep ocean can be estimated: input into deep ocean = 4.7 $\times$ 10^{17} g C/3400 years = 1.4 $\times$ 10^{14} g C year^{-1}; that is, according to this estimate only about 0.4% of primary production enters the deep ocean.

These estimations underline the fact that organic matter in the deep waters of the ocean is chronologically old and chemically or biochemically rather inert.

Analytical Considerations. Organic matter in natural waters includes a great variety of organic compounds, usually present in minute concentrations, many of which elude direct isolation and identification. *Collective parameters*, such as chemical oxygen demand (COD), biological oxygen demand (BOD), and total organic carbon (TOC) or dissolved organic carbon (DOC), are therefore often used to estimate the quantity of organic matter present. COD is obtained by measuring the equivalent quantity of an oxidizing agent, usually permanganate or dichromate in acid solution, necessary for oxidation of the organic constituents; the amount of oxidant consumed is customarily expressed in equivalents of oxygens. The BOD test measures the oxygen uptake in the microbiologically mediated oxidation of organic matter directly. In both tests not all the organic matter reacts with the oxidant. In determinations of total organic carbon, we measure the CO_2 produced in the oxidation or combustion of a water sample (from which carbonate has been removed). TOC, DOC, and COD are capacity terms; the last measures the reduction capacity of organic matter. If COD is expressed in mol O_2 liter^{-1} and TOC in mol C liter^{-1}, the "average" oxidation state of the organic carbon present can be obtained from (Figure 8.2)

$$\frac{4(\text{TOC} - \text{COD})}{\text{TOC}} = \text{oxidation state} \qquad (1)$$

The COD of a compound is very nearly proportional to its heat of formation, and a crude value of its energy of formation can often be obtained. Thus a rough estimate of the potential energy available for the metabolic needs of organisms can be made from COD data. Although measurements of TOC, DOC, BOD, and COD are conveniently obtained, these collective parameters lack physiological meaning. The rates of microbial growth and the overall use of organic

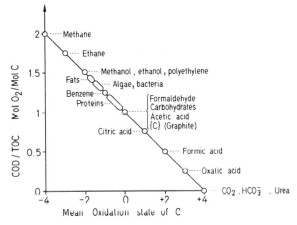

Figure 8.2 Oxygen demand and mean oxidation state of organic C.

matter in multisubstrate media depend in a complex way on the activities of a great variety of different enzymes and on various mechanisms by which these activities are ecologically interrelated and physiologically coordinated.

8.3 ORGANIC SUBSTANCES IN NATURAL WATERS

Organic life substances result either directly or indirectly from photosynthesis. Many of the organic chemicals found in natural waters can be regarded as products of both biosynthesis and biodegradation. Organic compounds span a wide range in size, from CH_4 to polymers, and in functionality, from hydrocarbons to complex multifunctional compounds [21]. Knowledge of their structures reveals information about the processes in which they have been involved.

Biochemicals on Their Way to Graphite

Information on the catabolic pathways of organic substances permits prediction of the type of organic substances to be found in natural waters. Table 8.3 gives a greatly condensed, simplified survey of the decomposition products of life substances and includes an abbreviated list of specific organic compounds reportedly found and identified in natural waters. For more details various reviews [1,20,11,22–25] may be consulted.

Much of the *thermodynamic stabilization* occurs in diagenesis. (Diagenesis refers to changes that take place within the sediment during and after burial.) The least stable and most reactive components *or their substituents* are gradually eliminated. This process leads with increasing age and depth of burial to a

TABLE 8.3 NATURALLY OCCURRING ORGANIC SUBSTANCES

Life Substances	Decomposition Intermediates	Intermediates and Products Typically Found in Nonpolluted Natural Waters
Proteins	Polypeptides → RCH(NH$_2$)COOH → $\begin{cases} RCOOH \\ RCH_2OHCOOH \\ RCH_2OH \\ RCH_3 \\ RCH_2NH_2 \end{cases}$ amino acids	NH$_4^+$, CO$_2$, HS$^-$, CH$_4$, HPO$_4^{2-}$, peptides, amino acids, urea, phenols, indole, fatty acids, mercaptans
Polynucleotides	Nucleotides → purine and pyrimidine bases	
Lipids		CO$_2$, CH$_4$, aliphatic acids, acetic, lactic, citric, glycolic, malic, palmitic, stearic, oleic acids, carbohydrates, hydrocarbons
Fats Waxes Oils	RCH$_2$CH$_2$COOH + CH$_2$OHCHOHCH$_2$OH → $\begin{cases} RCH_2OH \\ RCOOH \\ \text{shorter chain acids} \\ RCH_3 \\ RH \end{cases}$ fatty acids · · · · · · · · · glycerol	
Hydrocarbons		
Carbohydrates		HPO$_4^{2-}$, CO$_2$, CH$_4$, glucose, fructose, galactose, arabinose, ribose, xylose
Cellulose Starch Hemicellulose	$\Big\}$ C$_x$(H$_2$O)$_y$ → $\begin{cases} \text{monosaccharides} \\ \text{oligosaccharides} \\ \text{chitin} \end{cases}$ → $\begin{cases} \text{hexoses} \\ \text{pentoses} \\ \text{glucosamine} \end{cases}$ →	
Lignin	(C$_2$H$_2$O)$_x$ → unsaturated aromatic alcohols → polyhydroxy carboxylic acids	
Porphyrins and Plant Pigments		Phytane Pristane, carotenoids
Chlorophyll Hemin	Chlorin → pheophytin → hydrocarbons	Isoprenoid, alcohols, ketones, acids Porphyrins
Carotenes and Xantophylls	$\Big\}$	
Complex Substances Formed from Breakdown Intermediates, e.g.,		
	Phenols + quinones + amino compounds →	Melanins, melanoidin, gelbstoffe
	Amino compounds + breakdown products of carbohydrates →	Humic acids, fulvic acids, "tannic" substances

512

gradual stabilization, not necessarily of each individual compound but of the sedimentary organic matter as a whole [25]. In terms of structures the transformation of open chains to saturated rings and finally to aromatic networks is favored; hydrogen becomes available for inter- or intramolecular reduction processes [25]. Eventually highly ordered, stable structures of graphite may be formed. The approach to thermodynamic equilibrium is very slow; unstable compounds may survive from time spans equivalent to the age of the earth [25].

Blumer [2] has pointed out that the most characteristic feature of organic diagenesis is the appearance of extreme structural complexity and disorder at an intermediate stage, interposed between the high degree of biochemical order in the starting material and the even greater crystallographic order in graphite, the end product of diagenesis. Organic geochemistry is critically important to the understanding of many environmental problems. For instance, the severe effect of spilled fossil fuels is directly related to their composition and to the diagenetic processes that have produced them in the earth's interior. Our understanding of the chemistry of fossil fuels and of the changes they undergo during diagenesis is directly applicable also to an understanding of their biological and chemical degradation in the environment [25].

Naturally Occurring Water Soluble Compounds

Dawson [24] has given a synopsis of the average concentrations of organic components in seawater. The synopsis is based on literature reports and on some of his own data (Table 8.4).

TABLE 8.4   AVERAGE CONCENTRATIONS OF ORGANICS IN SEAWATER

Components	Concentration in Seawater (μg C liter^{-1})
Free amino acids	20
Combined amino acids	50 (to 100?)
Free sugars	20
Fatty acids	10
Phenols	2
Sterols	0.2
Vitamins	0.006
Ketones	10
Aldehydes	5
Hydrocarbons	5
Urea	20
Uronic acids	18
Approximate total	340 μg C liter^{-1}

In Lake Zürich waters [26] amino acids are typically present in concentrations between $10^{-10} M$ $(10^{-3} \mu g$ C liter$^{-1})$ (methionine) and 1 to $4 \times 10^{-8} M$ (0.5 to 2 μg C liter^{-1}) (valine, alanine, glycine, lysine, serine). Acetate and pyruvate are also present in the concentration range of 1 to $3 \times 10^{-8} M$ (0.3 to 1 μg C liter^{-1}). Obviously a significant fraction of the naturally occurring dissolved organic carbon remains uncharacterized.

Polymeric organic substances which contain a sufficient number of hydrophilic functional groups ($-COO^-$, $-NH_2$, R_2NH, $-RS^-$, ROH, RO^-) remain in solution despite their molecular size. Polypeptides, certain lipids, polysaccharides, and many of the substances classified as humic acids, fulvic acids, and *Gelbstoffe* belong to this category.

Humic and Related Substances

Transformation of biogenic substances within the soil, water, and sediment environment leads to humic substances which are believed to represent a significant fraction of the bulk of organic matter in most soils and waters [27–31]. Humic substances constitute between 6 and 30 % of the total organic matter in seawater [30]; they are a major component of the organic matter reservoir in recent marine sediments [31]. The yellow substances referred to as *Gelbstoffe* fall into this group.

Humic substances may be described as polymers (300 to 30,000 molecular weight) containing phenolic OH and carboxylic groups with a lower number of aliphatic OH groups. Based on their solubility in alkaline and acid solutions, humic substances are usually divided into three fractions: (1) *humic acid*, which is soluble in alkaline solution but is precipitated by acidification; (2) *fulvic acid*, which is the humic fraction that remains in the aqueous acidified solution; that is, it is soluble over the entire pH range; and (3) *humin*, the fraction that cannot be extracted by acid or base. Structurally the three fractions are believed to be similar; they appear to differ in molecular weight and functional group content [27]. Fulvic acid has probably a lower molecular weight but more hydrophilic functional groups than humic acid and humin. Schnitzer has suggested on the basis of degradative studies and nondegradative fractionation that fulvic acid is made up of phenolic and benzene carboxylic acids joined by hydrogen bonds to form a polymeric structure of considerable stability (Figure 8.3).

Stuermer and Payne [30] have shown (with the help of ^{13}C and proton nuclear magnetic resonance and infrared spectroscopy) that there are significant structural differences between marine and terrestrial fulvic acids. Marine fulvic acids are of less aromatic and more aliphatic character and have lower molecular weights than terrestrial fulvic acids. There is also a significant difference in the N content (6.4 % for marine fulvic acids, 0.5 % for terrestrial fulvic acids). Since the marine environment lacks abundant aromatic precursors (lignin is not abundant in marine plants), the humic substances in seawater—genetically related to degraded planktonic material [31]—apparently are formed from the

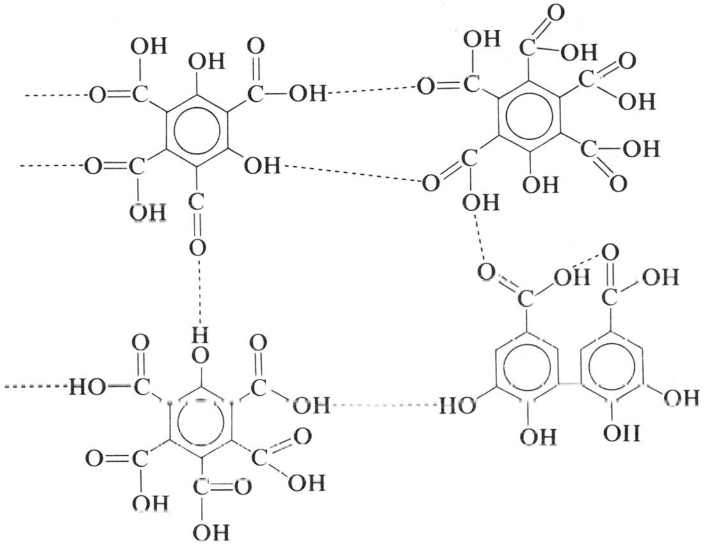

Figure 8.3 Structure of fulvic acid (after Schnitzer [27]).

precursors present, such as amino acids, carbohydrates, lipids and pigments [30].

Tannins (derivatives of multimeric gallic acid),

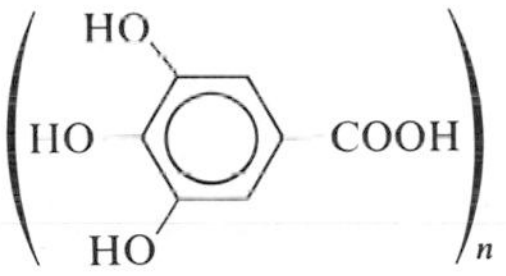

are sometimes found in fresh waters as a result of leaching of bark and leaf litter. In some regards (complex formation, adsorbability, color) tannins have properties similar to those of humic substances.

Properties. The conclusions about the structure of fulvic acid made by Schnitzer and Khan Christman and Ghassemi are helpful in explaining many properties of fulvic acids and related compounds.

1 *Attachment by and to other substances.* Presumably because of the voids and holes in their structure and because of the tendency of the functional groups to participate in H bonding, fulvic acids can adsorb or trap other organic substances, such as alkanes, fatty acids, phthalates, and possibly also carbohydrates, peptides, and pesticides [27].† Humic acids may operate as

---

† Fulvic acids isolated from sewage sludges contain appreciable amounts of sulfonates resulting from partially degraded detergent-sulfonates (G. Sposito, personal communication).

concentration and transport agents for trace toxic materials. Humic substances also have a strong tendency to become adsorbed on hydrous oxides, clays, and other surfaces. This adsorption may often be interpreted in terms of a ligand exchange (see Section 10.4), for example,

$$R-\overset{\overset{\displaystyle O}{\|}}{C}-O^- + HOAl\equiv \;\rightleftharpoons\; R-\overset{\overset{\displaystyle O}{\|}}{C}-OAl\equiv + OH^- \qquad (2)$$

where $\equiv$AlOH represents the surface of a hydrous aluminum oxide. Colloidal iron oxide is stabilized in a similar way by humates.

2 *Tendency to form complexes with metal ions.* Humic and fulvic acids constitute polyelectrolytic weak acids which tend to form complexes with metal ions [27], for example,

$$+ Cu^{2+} \;\rightleftharpoons\; + H^+ \qquad (3)$$

They are believed to play a very important role in the organometallic geochemistry of sediments [32]. Some aspects of this complex formation have been discussed in Section 6.9. Humates, tannates, and related compounds impart color to waters. Iron(III) often enhances the color intensity.

3 *Precipitation; colloid-chemical behavior.* Like most polycarboxylic acids, humates precipitate in the presence of Ca^{2+} and Mg^{2+}. Many of the humic substances are present in natural waters as colloids. Humates are able to form negatively charged colloids. Often these colloids (e.g., stabilized iron(III) oxide) are highly dispersed so that they pass through a 0.45-μm-diameter membrane filter. They are readily coagulated by Mg^{2+} and Ca^{2+}. Correspondingly, surface waters with high Ca^{2+} and Mg^{2+} (e.g., $>10^{-3}$ M) contain almost no humic substances (<1 mg C liter^{-1}). Estuaries are particularly efficient in removing humic substances, probably by coagulation (by seawater cations) of humates typically bound to hydrous iron oxides [33].

8.4 KINETIC CONSIDERATIONS OF THE OXIDATION OF ORGANIC MATTER

The enormous turnover of carbon in nature depends partly upon catabolic reactions catalyzed by microorganisms (see Section 7.5). Synthetic compounds may persist in the environment if their chemical structures preclude attack by these microbial enzymes. Aerobic microorganisms make use of most of the biochemical sequences and cycles generally found in most living forms. As pointed out by Dagley [23c], they possess in addition the unique biochemical asset of being able to catalyze the oxidation of numerous natural products using molecular oxygen, thereby initiating reaction sequences that enter the central pathways of metabolism (Krebs cycle or fatty acid cycle). The ability to initiate catabolism is a most important biochemical aptitude. An alkane, for example, remains biochemically inert until a terminal carbon has been oxidized. The fatty acids so formed then undergo β-oxidation. An unsubstituted benzene nucleus is rather inert, but enzymatic ring fission may take place when the benzene ring is suitably substituted [23c].

The rate of oxidation of organic matter, S (for substrate), is related to the rate of growth of microorganisms:

$$\frac{dB}{dt} = \mu B \tag{4}$$

where B is the concentration of organisms (e.g., dry weight or number concentration of cells) or a measure of enzymatic activity at time t and μ is the specific growth rate constant (time^{-1}). During growth part of the substrate is converted into microbial cells:

$$-\frac{dB}{dS} = y \tag{5}$$

where y is the yield and S is the concentration of substrate present in the solution. The combination of equations 4 and 5 describes the relationship between substrate utilization and microbial growth:

$$-\frac{dS}{dt} = \frac{\mu}{y} B \tag{6}$$

In accordance with a simple enzyme model μ varies with the substrate concentration in the following way

$$\mu = \frac{\mu_0 S}{K + S} \tag{7}$$

where μ_0 is the maximum specific growth rate constant and is a characteristic for a particular substrate and microorganism ($S = K$ when $\mu = 0.5\mu_0$).†

Most natural waters provide a nutritionally complex habitat for a multiplicity of microorganisms of large biological and physiological diversity. The cellular functions of each organism are susceptible to stimulation and retardation by various mechanisms. The extemely low levels of energy sources, essential nutrients, and microorganisms present in many natural waters present difficulties in interpretation, since it is often not possible to extrapolate from experimental data collected at higher concentrations [36,37].

As equation 6 indicates, there is a quantitative relationship between the rate of elimination of organic substances in a natural water and the quantity and

---

† The agreement of experimental data with equation 7 does not necessarily imply that the growth or substrate utilization rate is controlled by active enzymatic transport processes.

quality of the biomass in continuous contact with the water. This has been quantitatively confirmed in studies on the biological self-purification (rate of removal of organic substances) of rivers where it has been shown [38] that the rate of transformation of an organic substance is dependent on the structure of the microphytic community (which is in turn a function of the nature and concentration of organic substances present in the aquatic habitat) and on the standing crop of biomass in intimate contact with the water.

Bioassays.   For estimation of the concentrations of a substrate S, various known quantities of a substrate, usually ^{14}C-labeled, of concentration A^* are added to aliquot samples containing unknown quantities of the same substrate. In the procedure adapted by Vaccaro et al. [39], a pure bacterial culture whose uptake constant is well established is added to aliquots of membrane-filtered seawater. When equations 6 and 7 are combined, the following relations between the measurable rate of substrate utilization, $V = -dS/dt$, and the substrate originally present, S, is obtained:

$$V = \frac{\mu_0 B(S + A^*)}{y(K + S + A^*)} \tag{8}$$

Because K, μ_0, y, and B are assumed to remain constant during the test period (B is constant when $B > S$ and when the uptake time is very brief), a set of measurements of B for various A^* values permits the evaluation of S. The method is nearly specific and quite sensitive; with bacterial isolates from natural waters K values, that is, substrate concentrations that allow half-maximum uptake velocities, on the order of 10^{-8} to 10^{-5} M have been determined for various substrates. The method has been applied to both fresh waters [40] and seawater [39]. Vaccaro et al. [39] used the method to measure the glucose concentration along a transoceanic section. The concentrations varied from a high of 10^{-6} M to values below 10^{-8} M.

Table 8.5 gives a few representative examples of organic substances determined by bioassays.

Equations 4 to 8 apply in principle to all biologically mediated redox processes. It is generally held that anaerobic conditions are more favorable to the preservation of organic material than aerobic conditions, but the reason for this difference is not very well understood. The fact that more organic compounds are found under anaerobic conditions may also be explained by the fact that an organically enriched environment becomes by necessity anaerobic. Although anaerobic microorganisms multiply more slowly than aerobic ones, their efficiency in biodegrading organic matter is not necessarily lower; because the smaller growth rate of anaerobes is accompanied by a smaller yield [41], the specific rate of substrate utilization (expressed in terms of oxidation equivalents) of anaerobes is about equal to that of aerobes (for phenomenological evidence see [42]). However, other factors may play a significant role. Multicellular organisms cannot exist under anaerobic conditions; furthermore, conditions

TABLE 8.5   REPRESENTATIVE EXAMPLES OF KINETICS OF SUBSTRATE BIOASSAY[a]

Substance	S (10^{-9} mol liter^{-1})	K (10^{-9} mol liter^{-1})	V (10^{-9} mol liter^{-1} hr^{-1})	V_0 (10^{-9} mol liter^{-1} h^{-1})	Waters	Reference[b]
Gly	1.2	0.07	0.03	0.032	York River, estuary, September 1967, $t = 23°C$	(1)
Ser	0.51	8.55	0.014	0.25		
Ala	0.13	0.68	0.008	0.05		
Glu	0.14	8.8	0.003	0.19		
Asp	0.25	9.2	0.007	0.26		
Tyr	0.13	41.6	0.002	0.06		
Arg	0.10	27.5	0.001	0.27		
Glucose	240	970	850	4300	South Atlantic 1–35 m off South America, $T = 25.5°C$	(2)
Glucose	<27	~5.5	0.12	0.143	Lake Erken(s), mesotrophic, December 1964, $T = 3°C$	(3)
Acetate	132	33	0.48	0.6		
Glucose	5	ca. 10	3.3	11	Lake Lötsjön (S), eutrophic, August 1965, $T = 18°C$, total soluble organic $C = 10$ mg liter^{-1}	(4)
Acetate	300	30	300	330		

[a] Modified from K. Wuhrmann, *Pathol. Microbiol.*, **39**, 55 (1973). V and V_0 are measured and maximum substrate utilization rates, respectively. K is the half-rate constant (equations 6 and 8).

[b] (1) J. E. Hobbie et al., *Science*, **159**, 1463 (1968); (2) R. F. Vaccaro et al. [39]; (3) R. T. Wright and J. E. Hobbie, *Ecology*, **47**, 477 (1976); (4) H. L. Allen, *Int. Ges. Hydrobiol.*, **54**, 7 (1969).

toxic to organisms (H_2S, high or low pH) are more likely to occur in anaerobic than in aerobic environments.

Aquatic scientists are often primarily interested in the overall elimination rate of organic matter. Equations 4 to 8 apply strictly to the oxidation of individual compounds. In reality, one is dealing with many different simultaneous reactions. The superimposition of many processes—each one characterized by the rate law according to equations 4 and 6—leads to an overall rate of decomposition of total organic carbon that may reasonably be approximated by a first-order rate law:

$$-\frac{dC_T}{dt} = k\bar{C}_T \qquad (9)$$

where k is the rate constant (time^{-1}) and $\bar{C}_T$ is the concentration of potentially metabolizable organic carbon.

A *synthetic compound* will be biodegradable if it is susceptible to attack by the enzymatic apparatus acquired by microbes during the course of evolution. This in turn depends upon two factors: first, the ability of microbial enzymes to accept as substrates compounds having chemical structures similar to, but not identical with, those found in nature; and second, the ability of these novel substrates, when in the presence of microbes, to induce or repress the synthesis of the necessary degradative enzymes [23c]. Synthetic compounds, for example, pesticides, can be rendered less biodegradable by incorporating into the molecule structural features not typically encountered in natural products.

Chemical Degradation. Data on the stability of alanine obtained by studying the (purely chemical) degradation of ^{14}C-tagged alanine in the laboratory may serve as an illustration [50]. Decomposition of alanine occurs at rates following first-order kinetics ($-dC/dt = kC$), where the rate constant k can be formulated as

$$k = Ae^{-E_a/RT} \qquad (10)$$

At a temperature of 100°C, an activation energy $E_a = 44$ kcal mol^{-1} and a frequency factor of 3×10^{13} are obtained. At 100°C the experimental findings indicate, for example, a half-time of ca. 10^4 years. If extrapolation to lower temperatures were allowable, half-times of more than 10^{10} years at temperatures below 50°C would result. (Extrapolation to low temperatures is not justified in this case, because at low temperatures a second-order reaction involving O_2 appears.) Abelson [50] has surveyed the stability of other amino acids, and the laboratory findings are generally in agreement with observations on fossil materials. Bonds between adjacent carbon atoms in hydrocarbons and in aromatic structures are very strong. With activation energies higher than 50 kcal mol^{-1} for such bonds, half-times of more than 10^{10} years are calculated at temperatures of 100°C or lower.

REFERENCES

1   M. Blumer, *Angew. Chem.*, **14**, 507 (1975).

2   R. M. Garrels and E. A. Perry, Jr., in *The Sea*, Vol. 5, E. D. Goldberg, Ed., Wiley-Interscience, New York, 1974, p. 303.

3   G. M. Woodwell and E. V. Pecan, Eds., *Carbon and the Biosphere*, U.S. Atomic Energy Commission, Symposium Series, Washington, D.C., 1973; see especially the Appendix prepared by W. A. Reiners, p. 368.

4   L. C. Cole, *Sci. Amer.*, **198**, 83 (1958).

5   H. T. Odum, *Limnol. Oceanogr.*, **1**, 102 (1956).

6   J. H. Ryther, *Science*, **130**, 602 (1959).

7   G. E. Fogg, in *Chemical Oceanography*, J. P. Riley and G. Skirrow, Eds., Vol. 2, 2nd ed., Academic, New York, 1975.

8   G. M. Woodwell, P. H. Rich, and C. A. S. Hall, in *Carbon and the Biosphere*, G. M. Woodwell and E. V. Pecan, Eds., U.S. Atomic Energy Commission, Symposium Series, Washington, D.C., 1973.

9   P. J. Wangersky, *Amer. Sci.*, **53**, 358 (1965); *Chimia*, **26**, 559 (1972).

10 P. M. Williams in *Organic Compounds in Aquatic Environments*, S. D. Faust and J. V. Hunter, Eds., Dekker, New York, 1971, p. 145.

11 P. J. Le B. Williams, in *Chemical Oceanography*, J. P. Riley and G. Skirrow, Eds., Vol. 2, 2nd ed., Academic Press, New York, 1975, p. 301.

12 H. W. Jannasch and C. D. Wirsen, *Sci. Amer.*, 42 (1977).

13 D. Lal, *Science*, **198**, 997 (1977).

14 G. A. Riley, *Limnol. Oceanogr.*, **8**, 372 (1963); G. A. Riley, in *Carbon and the Biosphere*, G. M. Woodwell and E. V. Pecan, Eds, U.S. Atomic Energy Commission, Symposium Series, Washington, D.C., 1973, p. 204.

15 R. T. Barber, *Nature*, **211**, 257 (1966).

16 R. W. Sheldon, T. P. T. Evelyn, and T. R. Parsons, *Limnol. Oceanogr.*, **12**, 367 (1967).

17 G. Cauwet, *Organic Chemistry of Sea Water Particulates* and N. Ogura, Particulates in Symposium on Concepts in Marine Organic Chemistry, University of Edinburgh, 1976.

18 J. M. Sieburth et al., *Science*, **194**, 1415 (1976).

19 E. T. Degens, *Geochemistry of Sediments*, Prentice-Hall, Englewood Cliffs, N.J., 1965.

20 P. M. Williams, H. Oeschger, and P. Kinney, *Nature*, **224**, 256 (1969).

21 D. J. Faulkner et al., in *The Nature of Seawater*, E. D. Goldberg, Ed., Dahlem Konferenzen, Berlin, 1975, p. 623.

22 D. Hood, Ed., *Organic Matter in Natural Waters*, University of Alaska, College, 1970.

23a P. A. Cranwell, Environmental Organic Chemistry of Rivers and Lakes, Both Water and Sediment. *Environmental Chemistry*, **1**, Chem. Soc. (1975).

23b R. J. Morris and F. Culkin, in *Environmental Chemistry*, Vol. 1, Chemical Society, London 1975.

23c S. Dagley, *Naturwissenschaften*, **65**, 85 (1978).

24 R. Dawson, *Water Soluble Organic Compounds in Seawater* (*the Characterized Fraction*), Symposium on Concepts in Marine Oceanic Chemistry, Edinburgh, 1976.

25 M. Blumer, *Chemical Fossils: Trends in Organic Geochemistry*.

26 Wuhrmann, Personal Communication. 1976.

27 M. Schnitzer and S. U. Khan, *Humic Substances in the Environment*, Dekker, 1972.

28 M. Ghassemi and R. F. Christman, *Limnol. Oceanogr.*, **13**, 583 (1968).

29 C. Steelink, *J. Chem. Educ.*, **54**, 599 (1977).

30 D. H. Stuermer and J. R. Payne, *Geochim. Cosmochim. Acta*, **40**, 1109 (1976).

31 A. Nissenbaum and I. R. Kaplan, *Limnol. Oceanogr.*, **17**, 570 (1972).

32 A. Nissenbaum and D. S. Swaine, *Geochim. Cosmochim. Acta*, **40**, 809 (1976).

33 E. A. Boyle, J. M. Edmond, and E. R. Sholkovitz, *Geochim. Cosmochim. Acta* **4**, 1313 (1977).

36 H. W. Jannasch, *Mitt. Int. Ver. Limnol.*, **17**, 7 (1969).

37 E. Stumm-Zollinger and R. H. Harris, in *Organic Compounds in Aquatic Environments*, S. Faust and J. V. Hunter, Eds., Dekker, New York, 1971, p. 555.

38 K. Wuhrmann, *Mitt. Int. Ver. Limnol.*, **20**, 324 (1974).

39 R. F. Vaccaro, S. E. Hicks, H. W. Jannasch, and F. C. Carey, *Limnol. Oceanogr.*, **13**, 356 (1968).

40 J. E. Hobbie and R. T. Wright, *Limnol. Oceanogr.*, **10**, 471 (1965).

41 P. L. McCarty, in *Organic Compounds in Aquatic Environments*, S. D. Faust and J. V. Hunter, Eds., Dekker, New York, 1971, p. 495.

42 F. A. Richards, in *Organic Matter in Natural Waters*, D. W. Hood, Ed., University of Alaska, College, 1970, p. 399.

50 P. H. Abelson, in *Researches in Geochemistry*, P. H. Abelson, Ed., Wiley, New York, 1959.

9

The Regulation of the Chemical Composition of Natural Waters

9.1 INTRODUCTION

Natural waters acquire their chemical characteristics by dissolution and by chemical reactions with solids, liquids, and gases with which they have come into contact during the various parts of the hydrological cycle. Waters vary in their chemical composition, but these variations are at least partially understandable if the environmental history of the water and the chemical reactions of the rock–water–atmosphere systems are considered. Dissolved mineral matter originates in the crustal materials of the earth; water disintegrates and dissolves mineral rocks by weathering. Gases and volatile substances participate in these processes. As a first approximation, seawater may be interpreted as the result of a gigantic acid–base titration—acid of volcanoes versus bases of rocks (oxides, carbonates, silicates) [1]. The composition of fresh water similarly may be represented as resulting from the interaction of the CO_2 of the atmosphere with mineral rocks.

Figure 1.1 gave the major solutes contained in seawater and in average river water. A survey of the frequency distribution of various constituents of terrestrial waters by Davies and DeWiest [2] (see Figure 9.1) makes it apparent that many of the aquatic constituents show little natural variation in their concentrations. For example, according to Figure 9.1, 80% of the water analyses for dissolved silica show concentrations between $10^{-3.8}$ and $10^{-3.2}$ M. The range of $[H^+]$ of naturally occurring bodies of water is generally $10^{-6.5}$ to $10^{-8.5}$ M. The composition of seawater is remarkably constant, and the waters of different oceans differ very little in chemical analysis. Perhaps more interesting is the hypothesis supported by geological records that seawater has remained relatively constant in its chemical composition for at least the past 100 million years.

In this chapter an attempt at the following will be made:

1   To discuss the processes involved in the acquisition of chemical substances by natural waters. In this context some of the principles of chemical weathering of rocks and of the formation of soils will be treated.

2   To give attention to the mechanisms that regulate and control the mineral composition of natural waters. Equilibrium models facilitate identification of the many variables and establish chemical boundary conditions toward which aquatic environments must proceed.

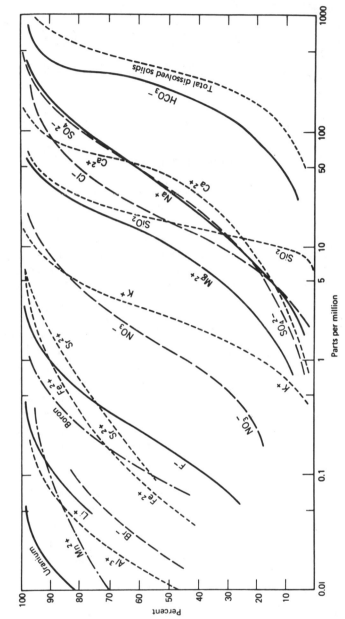

Figure 9.1 Cumulative curves showing the frequency distribution of various constituents in terrestrial water. Data are mostly from the United States from various sources [2]. (Reproduced with permission from J. Wiley and Sons.)

3 To consider the interrelation and interaction of the chemical environment with organisms and to illustrate how biological activity can influence the temporal and spatial distribution of aquatic constituents. In a balanced ecological system a balance between photosynthetic activity and respiratory activity seems to be maintained.

9.2   THE ACQUISITION OF SOLUTES

In order to better understand how natural waters acquire their chemical composition, the processes involved in weathering of rocks and in formation of soils must be considered. Weathering reactions—essentially caused by the interaction of water and atmosphere with the crust of the earth—take place because the original constituents of the crust, the igneous rocks, are thermo-dynamically unstable in the presence of water and the atmosphere. The igneous rocks have been formed under physical and chemical conditions entirely different from those presently existing at the earth's surface. The biosphere is also important to most weathering phenomena because living matter increases CO_2 as a result of respiration and, organic matter produced in the biosphere may serve as a reducing agent and may produce organic solutes that can complex cations and enhance their mobility.

The Cycle of Rocks and the Course of Weathering

All matter on the earth's surface and in the uppermost layers of the lithosphere participate in a slow and rather complicated migration. According to Rankama and Sahama [3] and others the migration of matter may be divided into two parts: *the minor (or exogenic) cycle* and *the major (or endogenic) cycle*. The minor cycle takes place under the direct influence of water and atmospheric agents; it starts in solid crystalline rocks and ends in sedimentary rocks; reactions take place primarily in one direction only. This exogenic cycle forms but part of the major cycle of nature which takes place at deeper levels in the crust. In this endogenic cycle, sedimentary rocks that have been deposited undergo further transformations (metamorphism) into new types of rocks (schist, gneiss, lime-silicate rock, quartzite, marble, graphite) by the action of heat, pressure, and migrating fluids. Parallel with this metamorphism are changes in the level of the earth's crust (folding, faulting, thrusting, formation of mountains). Gradually a rock melt, magma, is produced. Such a magma contains volatile components. Different solid phases (igneous rocks) crystallize out sequentially from magma, producing at the same time gases, which may be released as volcanic emanations and possibly as a water solution. This endogenic cycle is not closed; it receives primary magma and heat from the interior.

Figure 9.2 shows some of the pertinent features of part of the cycle of matter. The hydrosphere participates in the migration of matter (a) as a *conveyor* of matter in suspended and dissolved form, and (b) as a *reactant* in chemical

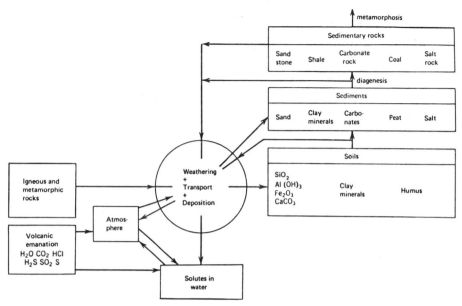

Figure 9.2   Interaction of the cycle of rocks with that of water.

transformations of matter. As Figure 9.2 illustrates, the atmosphere, whose composition is directly or indirectly influenced by volcanic emanations, participates in these chemical transformations by providing acids (CO_2) and oxidants (O_2).

The compositions of inland waters are related primarily to the following types of reactions of water and atmospheric gases with rock-forming minerals: (a) congruent dissolution reactions, (b) incongruent dissolution reactions, and (c) redox reactions. Furthermore, evaporation of water occurs, and organisms absorb on or loose certain constituents. Table 9.1 lists some of the more important rock-forming minerals and Table 9.2 gives some typical examples of weathering reactions. Highly schematic and idealized formulas are used in these tables to represent the complex mineral phases that occur in nature; furthermore, some of the reactions listed may go through intermediates not shown in the stoichiometric equations.

Weathering Processes

The weathering of *carbonate rocks* can be readily understood with the principles discussed in Chapter 5, where we have shown that the solubility of $CaCO_3$ (calcite) or any other $MeCO_3$ depends on the partial pressure of CO_2:

$$CaCO_3(s) + CO_2(g) + H_2O = Ca^{2+} + 2HCO_3^- \qquad {}^*K_{ps0} \qquad (1)$$

Some *silicates* such as fosterite, chrysotile, and talc can dissolve congruently (see Table 9.2), and in principle the equilibrium composition of waters in contact

TABLE 9.1 IMPORTANT ROCK-FORMING MINERALS

Silicon oxides	α-Quartz	SiO_2	Trigonal; densely packed array of SiO_2 tetrahedra
	α-Tridymite	SiO_2	Orthorhombic, open structure
	α-Cristobalite	SiO_2	Tetragonal sheets of six-membered rings of $[SiO_4]$ tetrahedra
	Opal	SiO_2	Hydrous, cryptocrystalline form of cristobalite
Aluminum oxides	Corundum	$\alpha\text{-}Al_2O_3$	Oxygen in hexagonal closest packing
	Gibbsite	$Al_2O_3 \cdot 3H_2O$	Monoclinic. A layer of Al ions sandwiched between two sheets of closely packed hydroxide ions
	Boehmite	$\gamma\text{-}AlOOH$	Orthorhombic. Double sheets of octahedra with Al ions at their centers
	Diaspore	$\alpha\text{-}AlOOH$	Orthorhombic Al^{3+} in octahedrally coordinated sites
Orthosilicates (olivines)	Forsterite	Mg_2SiO_4	SiO_4 tetrahedra linked by bivalent cations in octahedral coordination
	Fayolite	Fe_2SiO_4	
Serpentines	Chrysotile	$Mg_3[Si_2O_5](OH)_4$	Two-layered sheet silicate; SiO_4 tetrahedra and a modified brucite layer
Two-layer clays (kaolinites)[a]	Kaolinite	$Al_4[Si_4O_{10}](OH)_8$	Sheet consisting of two layers: (1) SiO_4 tetrahedra in a hexagonal array and (2) a layer of Al in 6 coordination
	Halloysite	$Al_4[Si_4O_{10}](OH)_8 \cdot 2H_2O$	

(continued)

527

TABLE 9.1 (*continued*)

Three-layer minerals	Micas		A layer of octahedrally coordinated cations (usually Al) is sandwiched between two identical layers of $[(Si,Al)O_4]$ tetrahedra
	Muscovite	$K_2Al_4[Si_6Al_2O_{20}](OH,F)_4$	
	Biotite	$K_2(Mg,Fe)_6[Si_6Al_2O_{20}](OH,F)_4$	
	Glauconite		
	Illite	$K_xAl_4[Si_{1-x}Al_xO_{20}](OH)_4$	
	Talc	$Mg_6[Si_8O_{20}](OH)_4$	
Expandable three-layer clays	Smectites (montmorillonite)	$(Na,K)_{x+y}(Al_{2-x}Mg_x)_2[(Si_{1-y}Al_y)_8O_{20}](OH)_4 \cdot nH_2O$	Octahedral Al on Mg sheets, tetrahedral Si sheets. Al partially replaced by Mg and occasionally by Fe, Cr, Zn. In tetrahedral sheet occasional replacement of Si by Al
	Vermiculite	$(Ca,Mg)(Mg_{3-x}Fe_x)_2[(Si_6Al_2)_8O_{20}](OH)_4 \cdot 8H_2O$	
	Chlorite	$(Mg,Al)_{12}[(Si,Al)_8O_{20}](OH)_{16}$	
Sulfides	Pyrite	FeS_2	Cubic, octahedral coordination of Fe by S
	Marcasite	FeS_2	Orthorhombic, octahedral coordination of Fe by S
	Pyrrhotite	FeS	Monoclinic pseudohexagonal
	Galena	PbS	Cubic
Sulfates	Baryte	$BaSO_4$	Orthorhombic
	Gypsum	$CaSO_4 \cdot 2H_2O$	Monoclinic
	Anhydrite	$CaSO_4$	Orthorhombic; more stable than gypsum above 42°C

Group	Mineral	Formula	Remarks
Carbonates	Calcite	$CaCO_3$	Trigonal. Mg- and Mn-calcite
	Rhodochrosite	$MnCO_3$	Similar to calcite
	Magnesite	$MgCO_3$	Similar to calcite
	Siderite	$FeCO_3$	Similar to calcite
	Dolomite	$MgCa(CO_3)_2$	One layer calcite combined with one layer magnesite
	Huntite	$Mg_3Ca(CO_3)_4$	
	Aragonite	$CaCO_3$	Orthorhombic
	Strontianite	$SrCO_3$	Similar to aragonite
Iron oxides	Goethite	$\alpha\text{-FeOOH}$	Similar to diaspore
	Lepidocrocite	$\gamma\text{-FeOOH}$	Similar to boehmite
	Limonite	$FeOOH \cdot nH_2O$	Hydrated oxides of iron with poorly crystalline character
	Magnetite	$\alpha\text{-Fe}_2O_3$	Trigonal, occurs in sediments. Spinel type
		Fe_3O_4	$8\ Fe^{2+}$ in 4 coordination; $16\ Fe^{3+}$ in 6 coordination
Titanium oxide	Rutile	TiO_2	Tetragonal; band of octahedra
Magnesium hydroxide	Brucite	$Mg(OH)_2$	Trigonal, two sheets of OH parallel to basal plane with sheet of Mg ion between them
Phosphates	Apatite	$Ca_5(OH,F,Cl)(PO_4)_3$	
	Carbonate–apatite	$Ca_5(PO_4,OH,CO_3)_3(F,OH)$	Hexagonal

ᵃ A brief introduction to clay structures is given in the Appendix to this Chapter.

529

TABLE 9.2 EXAMPLES OF TYPICAL WEATHERING REACTIONS

I. Congruent Dissolution Reactions

$$SiO_2(s) + 2H_2O = H_4SiO_4$$
quartz

$$CaCO_3(s) + H_2O = Ca^{2+} + HCO_3^- + OH^-$$
calcite

$$CaCO_3(s) + H_2CO_3^* = Ca^{2+} + 2HCO_3^-$$

$$Al_2O_3 \cdot 3H_2O(s) + 2H_2O = 2Al(OH)_4^- + 2H^+$$
gibbsite

$$Mg_2SiO_4(s) + 4H_2CO_3^* = 2Mg^{2+} + 4HCO_3^- + H_4SiO_4$$
forsterite

$$Fe_2SiO_4(s) + H_2CO_3^* = 2Fe^{2+} + 4HCO_3^- + H_4SiO_4$$
fayalite

$$Mg_6Si_8O_{20}(OH)_4(s) + 12H^+ + 8H_2O = 6Mg^{2+} + 8H_4SiO_4$$
talc

$$Mg_3Si_2O_5(OH)_4(s) + 6H^+ = 3Mg^{2+} + 2H_4SiO_4 + H_2O$$

II. Incongruent Dissolution Reactions

$$MgCO_3(s) + 2H_2O = HCO_3^- + Mg(OH)_2(s) + H^+$$
magnesite — brucite

$$Al_2Si_2O_5(OH)_4(s) + 5H_2O = 2H_4SiO_4 + Al_2O_3 \cdot 3H_2O(s)$$
kaolinite — gibbsite

$$NaAlSi_3O_8(s) + \tfrac{11}{2}H_2O = Na^+ + OH^- + 2H_4SiO_4 + \tfrac{1}{2}Al_2Si_2O_5(OH)_4(s)$$
albite — kaolinite

$$NaAlSi_3O_8(s) + H_2CO_3^* + \tfrac{9}{2}H_2O = Na^+ + HCO_3^- + 2H_4SiO_4 + \tfrac{1}{2}Al_2Si_2O_5(OH)_4(s)$$

$$CaAl_2O_8(s) + 3H_2O = Ca^{2+} + 2OH^- + Al_2Si_2O_5(OH)_4(s)$$
anorthite — kaolinite

$$CaAl_2Si_2O_8(s) + 2H_2CO_3^* + H_2O = Ca^{2+} + 2HCO_3^- + Al_2Si_2O_5(OH)_4(s)$$

$$4Na_{0.5}Ca_{0.5}Al_{1.5}Si_{2.5}O_8 + 6H_2CO_3^* + 11H_2O$$
plagioclase (andesine)
$$= 2Na^+ + 2Ca^{2+} + 4H_4SiO_4 + 6HCO_3^- + 3Al_2Si_2O_5(OH)_4(s)$$
kaolinite

$$3KAlSi_3O_8(s) + 2H_2CO_3^* + 12H_2O$$
K-feldspar (orthoclase)
$$= 2K^+ + 2HCO_3^- + 6H_4SiO_4 + KAl_3Si_3O_{10}(OH)_2(s)$$
mica

$$7NaAlSi_3O_8(s) + 6H^+ + 20H_2O$$
albite
$$= 6Na^+ + 10H_4SiO_4 + 3Na_{0.33}Al_{2.33}Si_{3.67}O_{10}(OH)_2(s)$$
Na$^+$ montmorillonite

$$KMg_3AlSi_3O_{10}(OH)_2(s) + 7H_2CO_3^* + \tfrac{1}{2}H_2O$$
biotite
$$= K^+ + 3Mg^{2+} + 7HCO_3^- + 2H_4SiO_4 + \tfrac{1}{2}Al_2Si_2O_5(OH)_4(s)$$
kaolinite

$$Ca_5(PO_4)_3F(s) + H_2O = Ca_5(PO_4)_3(OH)(s) + F^- + H^+$$
fluoroapatite — hydroxyapatite

530

$$KAlSi_3O_8(s) + Na^+ = K^+ + NaAlSi_3O_8(s)$$
orthoclase   albite

$$CaMg(CO_3)_2(s) + Ca^{2+} = Mg^{2+} + 2CaCO_3(s)$$
calcite

III. Redox Reactions

$$MnS(s) + 4H_2O = Mn^{2+} + SO_4^{2-} + 8H^+ + 8e$$

$$3Fe_2O_3(s) + H_2O + 2e = 2Fe_3O_4(s) + 2OH^-$$
hematite   magnetite

$$FeS_2(s) + 3\tfrac{3}{4}O_2 + 3\tfrac{1}{2}H_2O = Fe(OH)_3(s) + 4H^+ + 2SO_4^{2-}$$
pyrite

$$PbS(s) + 4Mn_3O_4(s) + 12H_2O = Pb^{2+} + SO_4^{2-} + 12Mn^{2+} + 24OH^-$$
galena

---

with these minerals may be calculated from solubility products or stability relations [4]. (During the early phase of the dissolution of these minerals an initial exchange of surface Mg^{2+} ions with H^+, that is a noncongruent pattern, is usually observed.)

The presence of Al_2O_3 (and of FeO) affects the weathering of ultramafic, basaltic, and granite rocks. The structural breakdown of *aluminum silicates* is accompanied by a release of cations and of silic acid. Often this process may be represented schematically [4] by

$$\text{Cation Al silicate(s)} + H_2CO_3^* + H_2O$$
$$= HCO_3^- + H_4SiO_4 + \text{cation} + \text{Al silicate(s)} \qquad (2)$$

As a result of such reactions alkalinity is imparted to the dissolved phase from the bases of the minerals. In most silicate phases Al is conserved during the reaction, the solid residue being higher in Al than the original silicates. Because the alkalinity of the solution increases during the weathering process, the solid residue has a higher acidity than the original aluminum silicate. Minerals of the kaolinite group are the main alteration products of weathering of feldspar. Besides kaolinites, smectites and micas are possible products or intermediates. Mica (illite) has been identified as an intermediate in the decomposition of K feldspar (orthoclase).

Example 9.1. Synthesizing Natural Waters Stoichiometrically

1. Estimate the gross composition of a water that results from the reaction of 1 mmol of CO_2 upon 1 liter of water in contact with

> (i) calcite
> (ii) anorthite
> (iii) andesine
> (iv) biotite

Considering the stoichiometry of the equations given in Table 9.2, the resulting compositions are:

(i) $pCa = 3.0$, $pHCO_3 = 2.7$
(ii) $pCa = 3.3$, $pHCO_3 = 3.0$
(iii) $pNa = 3.48$, $pCa = 3.48$, $pHCO_3 = 3.0$, $pH_4SiO_4 = 3.18$
(iv) $pK = 3.85$, $pMg = 3.37$, $pHCO_3 = 3.0$, $pH_4SiO_4 = 3.54$

Note that the weathering of anorthite leads to solutions having the same mole ratio of $[Ca^{2+}]/[HCO_3^-]$ as waters resulting from the dissolution of calcite.

2. Estimate the gross composition of a spring water that results from the weathering of the following quantities of source minerals per liter of water: 0.2×10^{-4} mol K-feldspar, 0.15×10^{-4} mol biotite, and 1.2×19^{-4} mol andesine. Further, 0.40 mol NaCl and 0.26 mol Na_2SO_4 as atmospheric salt particles from the sea have become dissolved per liter of water.

The composition of the resulting water is:

	Concentration $M \times 10^4$							
Reaction	Na^+	Ca^{2+}	Mg^{2+}	K^+	HCO_3^-	SO_4^{2-}	Cl^-	H_4SiO_4
K feldspar → kaolinite	—	—	—	0.20	0.20	—	—	0.40
Biotite → kaolinite			0.45	0.15	1.05			0.30
Andesine → kaolinite	0.60	0.60			1.8			1.2
NaCl, Na_2SO_4	0.92					0.26	0.40	
Spring water	1.52	0.60	0.45	0.35	3.05	0.26	0.40	1.9

Feth et al. [6] and Garrels and Mackenzie [7], using data from the California Sierras, originally showed that stream water chemistry could be related to the chemistry of the source rock and to equilibria among secondary minerals formed by weathering. In other weathering studies in the Absaroka Mountains in Wyoming [8] mass balance calculations relating stream water chemistry to rock alteration indicated that the dominant weathering process that controls the water chemistry is slight alteration of large volumes of rock rather than the development of chemical equilibria involving secondary phases in the soil zone.

An incongruent dissolution reaction consists essentially of an exchange of lattice ligands. Thus the conversion of feldspar into kaolinite can be visualized as a hydrolysis reaction, that is, a partial exchange of silica for OH^-. The reaction may go through intermediates. Virtually all bonds must be broken in the tetrahedric Al framework structure before rearrangement of the lattice with Al in 6 coordination (as in kaolinite) becomes possible. Understandably the rate of reaction is very slow.

Similarly in a fluoroapatite, the F^- may be exchanged for OH^-; even some of the phosphate groups in the apatite might under suitable conditions be replaced by CO_3^{2-}. The exchange of cations, in an analogous way, reflects a change in the coordinative relations. Typically clay minerals have the ability to exchange interlayer and surface cations with cations from the solution, and these processes are also of great importance in modifying the solution composition of natural waters.

The dislocation of structural constituents of the minerals depends in a more general way upon the prevailing environment. The dissolution processes may become modified by complex formation or by changes in the oxidation state. Table 9.2, III, illustrates that naturally occurring redox processes can have a pronounced pH controlling action. For example, the oxidation of pyrite yields ferric oxide and acid (4 mol of acidity per mole of pyrite; Table 9.2). The acid in turn may then react with rocks. Alternatively, the reduction of metal oxides such as Mn_3O_4 and Fe_2O_3 may produce large quantities of OH^- ions. Complex formers may preferentially "leach" out certain lattice constituents and thus enhance the rock disintegration. Naturally occurring organic matter may act as chelate complex formers. The influence of organisms and plants on the weathering reaction may be interpreted to result (in addition to pH variation) from the effect of organic chelates released by the cells. Under the influence of complex formation, for example, the formation of soluble AlY, an otherwise incongruent dissolution may become congruent:

$$Al_2Si_2O_5(OH)_4 + 2Y^- = 2AlY + 2H_4SiO_4 + H_2O \qquad (3)$$
$$\text{kaolinite} \qquad\qquad\qquad \text{soluble}$$

Water-soluble anions of organic acids of low molecular weight, that is, malic, malonic, acetic, succinic, tartaric, vanillic, and p-hydroxybenzoic acids are, assumed to be the important complexers. The presence of such complex formers has been demonstrated in top soils in concentrations as high as 10^{-5} to 10^{-4} M, oxalate being the most abundant [8b].

Kinetics. When feldspars are exposed to aqueous solutions, one observes after an initial rapid exchange of alkali ions for H^+ ions a slow dissolution in which the elements of the feldspar framework are released to the solution or to the formation of a secondary solid phase. The time dependence of the latter process of the reaction of feldspar with the solution has been reported to take the form of a parabolic rate law

$$\frac{dc_i}{dt} = k_i t^p \qquad (4)$$

where $p = -0.5$ and k_i is the rate constant for the change in concentration of the ith aqueous species. Various models of incongruent feldspar dissolution have been proposed: (1) The dissolution rate is controlled by the reaction of the unaltered feldspar with the solution at the interface between these two phases. (2) Feldspar weathering is controlled largely by diffusion through and between the grains of an aluminosilicate surface on an amorphous or crystalline precipitate layer [9–13]. On the basis of X-ray photoelectron spectroscopy on feldspar grains, subjected to dissolution at 82°C for 12–17 days, Petrović et al. [13a,b], however, rule out the presence of a continuous precipitate layer or of a leached layer more than one feldspar unit cell thick.

The complexing by oxalate, F^- and other ligands speeds weathering of soil minerals; it tends to raise the Al (and Fe) concentration to levels that allow significant mass transfer in the soil.

Furthermore the adsorption of these ligands to the surface of the aluminum oxide (surface coordination) modifies the surface charge and its surface acidity and enhances access of H^+.

9.3 SOME CHARACTERISTICS OF RIVER WATERS AND GROUNDWATERS

The water carried in streams may be considered to consist of a fraction made up of subsurface water and groundwater which reenters (or infiltrates into) the surface water, and a (direct) surface runoff fraction which enters the drainage system during and soon after precipitation periods (Figure 9.3c).† The relative proportions of these components and the concentration of dissolved species in each, as influenced by the interactions of rainwater with minerals and vegetation and by evaporation and transpiration from plants, largely determine the composition of river waters. The direct (surface) runoff has had only a short contact with soil or vegetation; reactions in the ground are often extensive enough for rainwater that penetrates into the subsurface to impart to this base flow an increased dissolved solids content. Two extreme cases may be considered: (1) The direct runoff is small (e.g., by evaporation) in comparison to the subsurface and groundwater flow; the latter has a relatively constant solute concentration, for example, by nearly reaching solid–solution equilibria. Then the solute concentration in the river is relatively independent of river flow. (2) If the direct runoff is larger than the flow that has penetrated into the subsurface, the solute concentrations of the subsurface and groundwater become "diluted" by the direct runoff. Thus the solute concentrations of river waters tend to be inversely related to river flow. Superimposed are the influences of pollution and waste disposal. Holland [14a] has analyzed the effect of runoff on the concentration and composition of dissolved components of river waters.

The uppermost part of the earth's rocks constitutes a porous medium in which water is stored and through which it moves. Up to a certain level these rocks are saturated with water that is free to flow laterally under the influence of gravity. Subsurface water in this saturated zone is *groundwater*, and the uppermost part of the zone is the water table. The chemistry of the groundwater is influenced by the composition of the aquifer and by the chemical and biological

---

† The term runoff is usually considered synonymous with stream flow and is the sum of surface runoff and groundwater flow that reaches the streams. Surface runoff equals precipitation minus evaporation, surface retention, and infiltration.

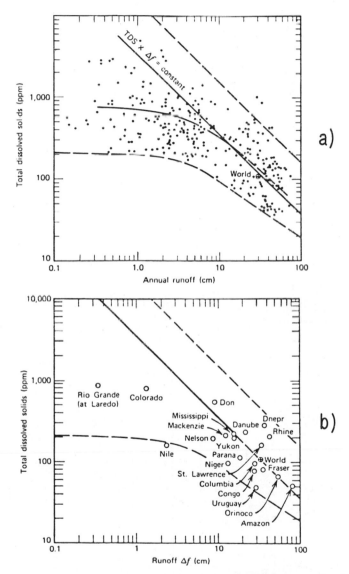

Figure 9.3 Dissolved solids of rivers as a function of runoff. (*a*) River waters of the United States; (*b*) major rivers. [From Holland [14*a*]. Reproduced with permission of Wiley and Sons, New York.)

535

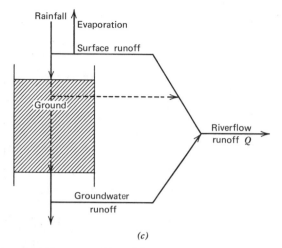

(c)

Figure 9.3 *(Continued)* *(c)* The composition of river waters depends on the relative proportions of the surface runoff which enters the river during and soon after precipitation and the groundwater and subsurface flow that reenters (or infiltrates into) the river.

processes occurring in the *infiltration*. The soil profile (Figure 9.4) is developed in response to chemical weathering, organic activity, and time. In a well-defined profile the A horizon is more permeable than the B horizon. The B horizon is usually composed of masses of soil particles with clay and colloids. In the C horizon permanent material is weathered.

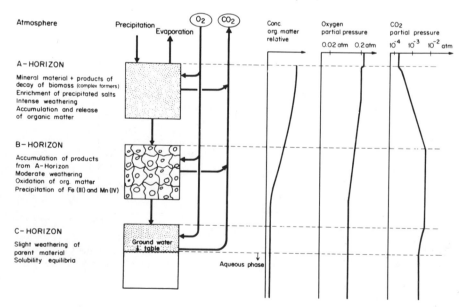

Figure 9.4 Effect of soil profile on the acquisition of solutes during infiltration. (A similar figure has been given by E. Eriksson.)

TABLE 9.3 THE PROVENANCE OF SOLUTES IN AVERAGE RIVER WATER

Source	Anions (meq kg^{-1})			Cations (meq kg^{-1})				Neutral Species (mmol kg^{-1})
	HCO_3^-	SO_4^{2-}	Cl^-	Ca^{2+}	Mg^{2+}	Na^+	K^+	SiO_2
Atmosphere[b]	0.58[c]	0.05[d]	0.06	0.01	≤0.01	0.05	≤0.01	≤0.01
Weathering or solution of								
Silicates	0	0	0	0.14	0.20	0.10	0.05	0.21
Carbonates	0.31	0	0	0.50	0.13	0	0	0
Sulfates	0	0.07	0	0.07	0	0	0	0
Sulfides	0	0.07	0	0	0	0	0	0
Chlorides	0	0	0.16	0.03	≤0.01	0.11	0.01	0
Organic Carbon	0.07	0	0	0	0	0	0	0
Sum	0.96	0.23	0.22	0.75	0.35	0.26	0.07	0.22

[a] From Holland (1978) [14a]. Reproduced with permission of John Wiley and Sons, New York.
[b] These figures do not include soil-derived material.
[c] Largely as atmospheric CO_2.
[d] Much of this is apparently balanced by H^+.

A general relationship between the composition of the water and that of the solid minerals with which the water has come in contact during infiltration and in the aquifer can be expected. Biological activity, especially in the organic layer above the mineral part and in the A horizon, has a pronounced effect on the acquisition of solutes. Complex formers that are released—apparently to a large extent by fungi [8a]—enhance the weathering of the primary minerals. Because of microbial respiration the CO_2 pressure is increased. (The relative change in CO_2 is much larger than that of O_2 because CO_2 is present in a much smaller concentration in the atmosphere than O_2.) The increased CO_2 pressure tends to increase the alkalinity and the concentration of Ca^{2+} and other solutes (see reactions in Table 9.2). When dissolved organic matter enters the groundwater, the latter may become anaerobic; iron and manganese are present under these conditions as soluble Fe^{2+} and Mn^{2+}.

Concentrations of important solutes in river waters of the world were given in Figure 1.1. We have tried to illustrate the various weathering reactions that cause the acquisition of solutes. Holland [14a] has provided a detailed analysis on the variations of the major constituents of the river waters. The most important results—as summarized by Holland—are given in Table 9.3. According to this table nearly two-thirds of the C in river HCO_3^- is derived from the atmosphere either directly as CO_2 or via photosynthesis followed by plant decay. The remaining HCO_3^- is derived largely from the weathering of carbonates. The source of SO_4^{2-}, according to Holland, is still somewhat uncertain; roughly 40% is cycled through the atmosphere, and 60% is derived from the weathering of sulfides and the solution of sulfate minerals (gypsum and anhydrite). Atmospheric chloride contributes less than half of the average chloride in river water; chlorides mostly from evaporitic halite contribute to the bulk. Most of the Ca^{2+} in river waters is derived from the solution of carbonates. However, the weathering of silicates contributes about 50% more Mg^{2+} than the solution of carbonates. NaCl (from evaporites and atmospheric recycling) provides most of the Na^+ in rivers; only 35% of the Na^+ owes its origin to the weathering of silicates. K^+ and dissolved SiO_2 are largely contributed by silicate weathering.†

9.4 SOLUBILITY OF MINERALS

The earth's surface may be considered a huge chemical laboratory in which a quantitative rock analysis is being made [3]. In this rock analysis, products are formed incipiently by congruent and incongruent dissolution of the rocks; the products then are subjected to chemical separations and redepositions (sedimentation) in new surroundings followed by aging, recrystallization, and formation of new minerals (diagenesis into sedimentary rocks).

---

† J. M. Martin and M. Meybeck [*Marine Chem.*, **7**, 173 (1979)] have recently reassessed the elemental mass balance of material carried by major world rivers.

Evaluation of the solubility of minerals combined with field observations contribute to a better understanding of the functional interrelationship between minerals and their environment. Overall weathering reactions of course are characterized by a decrease in free energy. The latter may be relatively small; repeated leaching with fresh water tends to decrease the reaction quotient Q, hence $\Delta G = RT \ln (Q/K)$, that is, to "drive" the weathering reaction to the right. The thermodynamic weathering sequence could be specified if the stability relations of the various rocks could be defined reliably. Unfortunately, information on the free energy of formation of aluminum silicates is insufficient, and thermochemical data for many minerals of significance are not known with any precision; conflicting values have been reported by various authors. Thermochemical data for aluminum silicate minerals at low temperature and pressure may be obtained by extrapolation from investigations on mineralogical equilibria at elevated temperatures and pressures or from tedious low-temperature solubility studies. Free energies of formation may also be estimated by an ionic bonding approach, that is, essentially from the combination of free energies of the oxide and, hydroxide components of the silicate minerals [14b–17].

There are additional uncertainties: (1) We do not know the exact composition of the solid phases; (2) since cations in many minerals may replace each other to varying degrees, the solid phase may not be invariant of the solution composition thus we are not sure whether we should treat the solid as a "pure" solid or as a solid solution; (3) many aluminum silicates are metastable; and (4) the free energy and solubility of a mineral vary with its crystallinity. These difficulties make the establishment of a detailed thermodynamic weathering sequence uncertain. By using tentative values of equilibrium constituents, the principles of the chemical behavior of rocks in water can be circumscribed and stability diagrams that may help in understanding weathering transformations can be constructed.

Carbonates

Experience has shown that equilibrium models involving $CaCO_3$ and some other carbonates can often be applied, especially in groundwater systems. The equilibrium expression of the reaction

$$CaCO_3(s) + CO_2(g) + H_2O = Ca^{2+} + 2 HCO_3^-; \qquad {}^+K_{ps0} \qquad (1)$$

can be given as

$$\frac{[Ca^{2+}][HCO_3^-]^2}{p_{CO_2}} = {}^+K_{ps0} \qquad (5)$$

Because of the electroneutrality condition, $2[Ca^{2+}] \simeq [HCO_3^-]$, equation 5 can be rewritten (cf. Example 5.5) as

$$[Ca^{2+}] \simeq 0.63 {}^+K_{ps0}^{1/3} p_{CO_2}^{1/3} \qquad (6)$$

where the equilibrium constant $^+K_{ps0}$ can be calculated from the solubility product K_{s0}, the first and second acidity constants of $H_2CO_3^*$, K_1 and K_2, and the Henry constant K_H:

$$^+K_{ps0} = K_{s0} K_1 K_H K_2^{-1} \tag{7}$$

For a water in equilibrium with dolomite we can write instead of equation 6

$$[Ca^{2+}] = [Mg^{2+}] \simeq {^+K_{ps0(\text{dolomite})}^{1/3}} p_{CO_2}^{1/3} \tag{8}$$

The solubility of dolomite and of calcite increases with the cube root of the CO_2 pressure. This CO_2 pressure in natural waters, especially in soil–water systems, is often larger than that of the atmosphere mostly because of respiration.

The Chemistry of Aqueous Silica

The solubility of SiO_2 can be characterized by the following equilibria:

$$SiO_2(s, \text{quartz}) + 2H_2O = Si(OH)_4 \qquad \log K = -3.7 \tag{9}$$

$$SiO_2(s, \text{amorphous}) + 2H_2O = Si(OH)_4 \qquad \log K = -2.7 \tag{10}$$

$$Si(OH)_4 = SiO(OH)_3^- + H^+ \qquad \log K = -9.46 \tag{11}$$

$$SiO(OH)_3^- = SiO_2(OH)_2^{2-} + H^+ \qquad \log K = -12.56 \tag{12}$$

$$4Si(OH)_4 = Si_4O_6(OH)_6^{2-} + 2H^+ + 4H_2O \qquad \log K = -12.57 \tag{13}$$

The equilibrium constants given are valid at 25°C. Data for equations 9 to 11 are those given by Lagerstrom [18] as valid for 0.5 M $NaClO_4$.

In these equations silicic acid is written as $Si(OH)_4$ (rather than as H_4SiO_4) in order to emphasize that metalloids (Si, B, Ge) similar to metal ions have a tendency to coordinate with hydroxo and oxo ligands. Like multivalent metal ions, such metalloids tend to form multinuclear species.

The rate of crystallization of quartz is so slow in the low-temperature range that the solubility of amorphous silica represents the upper limit of dissolved aqueous silica (Siever [19]); we first consider the solubility of amorphous silica. The equilibrium data of equations 2 to 5 permit computation of the solubility of amorphous SiO_2 for the entire pH range and the relative concentration of the species in equilibrium with amorphous SiO_2 (Figure 9.5). Only $Si(OH)_4$ occurs within neutral and slightly alkaline pH ranges. Under alkaline conditions the solubility of SiO_2 becomes enhanced because of the formation of monomeric and multimeric silicates. Although there is some uncertainty concerning the exact nature of the multimeric species, the experimental data of Lagerstrom [18] and Ingri [20] leave no doubt about the existence of multinuclear species. Even if multinuclear species other than $Si_4O_2(OH)_6^{2-}$ were present, the solubility characteristics of SiO_2 would not be changed markedly.

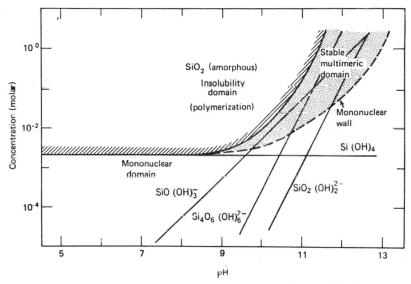

Figure 9.5 Species in equilibrium with amorphous silica. Diagram computed from equilibrium constants (25°C.) The line surrounding the shaded area gives the maximum soluble silica. The mononuclear wall represents the lower concentration limit below which multinuclear silica species are not stable. In natural waters the dissolved silica is present as monomeric silicic acid.

The Solubility of Aluminum Silicates

Example 9.2. Solubility of Kaolinite

Calculate the solubility of kaolinite, $Al_2Si_2O_5(OH)_4(s)$, in pure water. Construct a predominance diagram in the system $Al_2O_3-SiO_2-H_2O$ for the solid phases gibbsite ($Al_2O_3 \cdot 3H_2O$), amorphous SiO_2, and kaolinite.

The following information is available:

$$Al_2Si_2O_5(OH)_4(s) + 5H_2O = Al_2O_3 \cdot 3H_2O(s)$$
$$+ 2H_4SiO_4 \qquad \log K = -9.4 \qquad \text{(i)}$$

$$\tfrac{1}{2}Al_2O_3 \cdot 3H_2O(s) = Al^{3+} + 3OH^- \qquad \log K = -34.0 \qquad \text{(ii)}$$

$$\tfrac{1}{2}Al_2O_3 \cdot 3H_2O(s) + OH^- = Al(OH)_4^- \qquad \log K = -1.0 \qquad \text{(iii)}$$

$$SiO_2(\text{amorph}) + 2H_2O = H_4SiO_4 \qquad \log K = -2.7 \qquad \text{(iv)}$$

[The equilibrium constant of equation i is from R. M. Garrels, *Am. Mineral.*, **42**, 789 (1957).]

Furthermore, we can combine equations i and ii as well as equations i and iii in order to obtain equilibria for equations v and vi:

$$\tfrac{1}{2}Al_2Si_2O_5(OH)_4(s) + 2\tfrac{1}{2}H_2O = Al^{3+} + H_4SiO_4 + 3OH^-$$
$$\log K = -38.7 \quad \text{(v)}$$

$$\tfrac{1}{2}Al_2Si_2O_5(OH)_4(s) + 2\tfrac{1}{2}H_2O + OH^- = Al(OH)_4^- + H_4SiO_4$$
$$\log K = -5.7 \quad \text{(vi)}$$

If kaolinite is introduced into pure H_2O, it dissolves incongruently (reaction i). Its solubility is very small; the electroneutrality of the solution will essentially be governed by $[H^+] \simeq [OH^-]$, or pH $\approx$ 7. While the existence of $Al(OH)_4^-$ and its formation constant have been well established, there is a great deal of uncertainty about the composition and stability of other aluminum hydrolysis species. Because these polynuclear hydroxo complexes do not materially alter the solubility characteristics of gibbsite and kaolinite, we carry out our calculations by considering $Al(OH)_4^-$ and Al^{3+} as the predominant equilibrium species.

We may first consider congruent solubility characteristics for the three solid phases, gibbsite, SiO_2(amorph), and kaolinite. The solubility of SiO_2 has been described in Figure 9.5. Figure 9.6a and b shows the pH dependence of the solubility of gibbsite and kaolinite, respectively. For the calculation of the kaolinite solubility, a fixed value of $[H_4SiO_4] = 10^{-2.7}$ M has been assumed.

With the information at hand we proceed to construct a predominance diagram (Figure 9.6c) valid for the dissolution of kaolinite in pure water (pH $\simeq$ 7). In this diagram the ordinate and abscissa represent molar concentrations of the "base" $Al(OH)_3$ and of the acid $[H_4SiO_4]$, respectively, which make up the "salt" kaolinite.

For our conditions of pH $\simeq$ 7 we obtain the maximum solubility of gibbsite at $[Al(OH)_4^-] \simeq 10^{-8}$ M (A axis in Figure 9.6c). The maximum solubility of amorphous SiO_2 is given by equation iv as $[H_4SiO_4] = 10^{-2.7}$ M. The transition between gibbsite and kaolinite is defined by equation i as $[H_4SiO_4] = 10^{-4.7}$. The boundary between kaolinite and the solution phase is given by equation vi:

$$\log \frac{[Al(OH)_4^-][H_4SiO_4]}{[OH^-]} = -5.7$$

As this diagram suggests, the hypothetical maximum solubility of kaolinite at pH = 7 falls into the stability region of gibbsite at C(incongruent) $\simeq 10^{-6.5}$ M (this point must be at the $A = B$ axis). Hence kaolinite dissolves incongruently according to equation i at equilibrium and coexists with gibbsite; $Al_T \simeq [H_4SiO_4] = 10^{-4.7}$ M.

As Figure 9.6 suggests, kaolinite converts into gibbsite if, as for example, through extensive leaching, $[H_4SiO_4]$ is kept below $10^{-4.7}$ M.

Example 9.3. Solubility of Albite

Estimate the dependence of the solubility of $NaAlSi_3O_8$ on the partial pressure of CO_2. Assume the following equilibrium constant (25°C):

$$NaAlSi_3O_8(s) + H^+ + 4\tfrac{1}{2}H_2O$$
$$\underset{\text{albite}}{}$$
$$= Na^+ + 2H_4SiO_4 + \tfrac{1}{2}Al_2Si_2O_5(OH)_4(s) \qquad \log K = -1.9 \quad \text{(i)}$$
$$\underset{\text{kaolinite}}{}$$

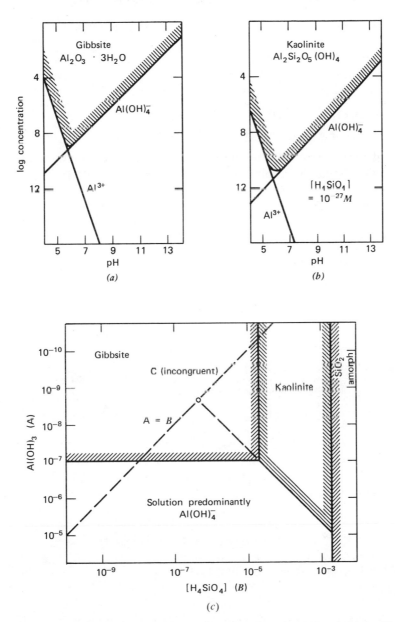

Figure 9.6 Incongruent dissolution of kaolinite (Example 9.2). (*a*) Solubility of gibbsite. (*b*) Hypothetical (component) solubility of kaolinite for $[H_4SiO_4] = 10^{-2.7}\ M$. (*c*) Predominance diagram for the dissolution of kaolinite in pure water (pH = 7).

Reaction i may be combined with the protolysis reaction of $CO_2(g)$

$$CO_2(g) + H_2O = HCO_3^- + H^+; \quad \log K = -7.8 \tag{ii}$$

to give

$$\text{albite(s)} + CO_2(g) + 5\tfrac{1}{2}H_2O$$
$$= Na^+ + HCO_3^- + 2H_4SiO_4 + \tfrac{1}{2}\text{kaolinite(s)} \qquad \log K = -9.7 \quad \text{(iii)}$$

If hypothetically pure albite dissolves under the influence of CO_2 according to equation iii, the resulting solution is characterized by the solutes HCO_3^-, Na^+, H_4SiO_4, and H^+. Because the solutions are near neutrality, CO_3^{2-} and silicates need not be considered. The equilibrium composition as a function of p_{CO_2} can be computed because, in addition to the constants defining the equilibrium of equations ii and iii, the charge balance

$$[Na^+] \simeq [HCO_3^-] \tag{iv}$$

and the stoichiometric relation

$$[Na^+] = \tfrac{1}{2}[H_4SiO_4] \tag{v}$$

are known. Combining equations iv and v with the equilibrium relation of equation iii gives

$$\frac{4[HCO_3^-]^4}{p_{CO_2}} \simeq 10^{-9.7} \tag{vi}$$

Thus, for example, for a $p_{CO_2} = 10^{-2}$, $pH_4SiO_4 = 2.78$. Then $[H^+]$ can be calculated with the equilibrium of equation iii, and $pH = 6.72$ is obtained.

The resulting equilibrium composition is in accord with the assumptions made in equations iv and v and the assumed presence of kaolinite as a stable phase. In Figure 9.7 the solubility of albite, expressed as $[HCO_3^-]$, is plotted as a function of p_{CO_2}.

Weathering Sequence and Solubility

With the help of equilibrium constants given in Table 9.4, the equilibrium solubility of a few individual minerals is considered in Figure 9.7. The left part of this figure is based on calculations such as the one given in the preceding example; that is, systems are made by introducing a mineral into pure water and exposing the solution to a selected partial pressure of CO_2; the individual curves have been calculated by assuming thermodynamic coexistence of the two solid phases specified for each reaction. In Figure 9.7b the solubility of a few congruently soluble minerals is indicated as a function of pH in the neutral region. The data given in Figure 9.7 suggest the following thermodynamic order of succession of minerals in the weathering sequence: gypsum, calcite, Ca feldspar, K feldspar, Na feldspar, Ca and Na montmorillonite, quartz, K mica,

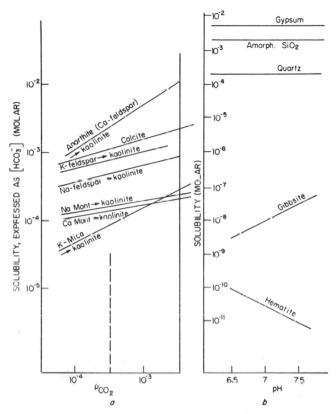

Figure 9.7 Solubility of minerals. (*a*) p_{CO_2}-dependent solubility of "pure" minerals. Equilibrium $[HCO_3^-]$ in reactions such as: albite(s) + $CO_2(g)$ + $5\frac{1}{2}H_2O$ − Na^+ + HCO_3^- + $2H_4SiO_4$ + $\frac{1}{2}$kaolinite(s); and calcite(s) + $CO_2(g)$ + H_2O = Ca^{2+} + $2HCO_3^-$; is used to express tendency for dissolution (see Example 8.3). (*b*) Congruent solubility of some minerals in the neutral pH range.

gibbsite, kaolinite, hematite. Petrographic observations appear to be consistent with this stability sequence. The weathering sequence proposed by Jackson et al. [21] is very nearly in accord with this stability sequence; in their sequence dark minerals such as hornblende, augite, and biotite are grouped between calcite and the feldspars.

Mobility of Individual Elements. Relative mobility has been used as an index of the final redistribution during the alteration of rock into soil. Most consistent in these considerations is the evidence that Ca^{2+} and Na^+ are the most mobile elements, whereas aluminum and iron belong to the group of least mobile elements, and magnesium, manganese, and silicon show an intermediate mobility. Although this order of relative mobility could be predicted qualitatively from the coordinating tendency of these elements (or from the stability

TABLE 9.4 EQUILIBRIUM CONSTANTS USED TO ESTABLISH STABILITY RELATIONS AMONG MINERALS[a]

	Log K at 25°C, 1 atm
(1) Na feldspar(s) + H$^+$ + $4\frac{1}{2}$H$_2$O = $\frac{1}{2}$ kaolinite(s) + 2H$_4$SiO$_4$ + Na$^+$	−1.9
(2) 3 Na montmorillonite(s) + H$^+$ + $11\frac{1}{2}$H$_2$O = $3\frac{1}{2}$ kaolinite(s) + 4H$_4$SiO$_4$ + Na$^+$	−9.1
(3) Ca feldspar(s) + 2H$^+$ + H$_2$O = kaolinite(s) + Ca^{2+}	+14.4
(4) 3 Ca montmorillonite(s) + 2H$^+$ + 23H$_2$O = 7 kaolinite(s) + 8H$_4$SiO$_4$ + Ca^{2+}	−15.4
(5) Kaolinite(s) + 5H$_2$O = 2 gibbsite(s) + 2H$_4$SiO$_4$	−9.4[b]
(6) SiO$_2$(amorph) + 2H$_2$O = H$_4$SiO$_4$	−2.7
(7) CaCO$_3$(s) (calcite) = Ca^{2+} + CO$_3^{2-}$	−8.3
(8) HCO$_3^-$ = H$^+$ + CO$_3^{2-}$	−10.3
(9) CO$_2$(g) + H$_2$O = H$^+$ + HCO$_3^-$	−7.8
(10) CO$_2$(g) + H$_2$O = H$_2$CO$_3^*$	−1.5

[a] Ca feldspar (Anorthite) = CaAl$_2$Si$_2$O$_8$; Na feldspar (Albite) = NaAlSi$_3$O$_8$; Na montmorillonite = Na$_{0.33}$Al$_{2.33}$Si$_{3.67}$O$_{10}$(OH)$_2$; Ca montmorillonite = Ca$_{0.33}$Al$_{4.67}$Si$_{7.33}$O$_{20}$(OH)$_4$; Kaolinite = Al$_2$Si$_2$O$_5$(OH)$_4$; Gibbsite = Al$_2$O$_3$ · 3H$_2$O; H$_2$CO$_3^*$ = CO$_2$(aq) + H$_2$CO$_3$.
[b] Cf. R. M. Garrels, *Amer. Mineralogist*, **42**, 789 (1957).

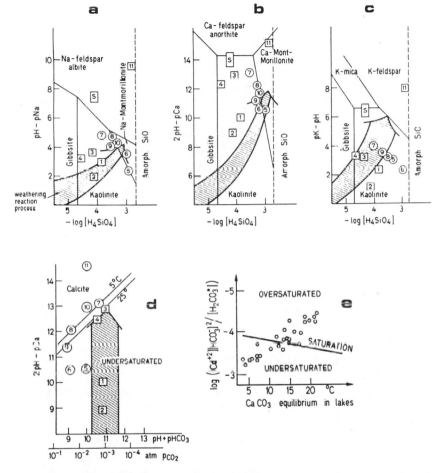

Figure 9.8 Predominance diagrams illustrating the stability relations of some mineral phases according to the equilibrium constants given in Table 9.4. The arrows indicate typical progress in the weathering reaction. The dissolution of feldspars (a–c) by a CO_2-enriched water is accompanied by an increase in $[Na^+]/[H^+]$, $[Ca^{2+}]/[H^+]^2$, or $[K^+]/[H^+]$ and of $[H_4SiO_4]$; initially the release of Al^{3+} leads to an oversaturation with respect to gibbsite; then kaolinite is formed and progressive accumulation of H_4SiO_4 can lead to conversion of kaolinite into Na and Ca montmorillonites; eventually even calcite may be precipitated. (d) A $CaCO_3(s)$ solubility equilibrium $[CaCO_3 + 2H^+ = Ca^{2+} + CO_2(g) + H_2O]$; the approach toward calcite saturation upon $CaCO_3$ dissolution at constant values of p_{CO_2} is given. For comparison, squares and circles represent analyses of Table 9.5 for surface and groundwaters, respectively; S represents seawater. (e) $CaCO_3$ equilibrium in lakes. The water layers at the surface tend to be oversaturated and the bottom waters tend to be undersaturated with respect to $CaCO_3$. The comparison is made with the help of the reaction $CaCO_3(s) + H_2CO_3^* = Ca^{2+} + 2HCO_3^-$. Data are for Lake Zürich.

547

constants of the respective solid phases), knowledge of the thermodynamic stability relations alone does not always suffice to understand thoroughly the geochemical behavior and the manner of occurrence of an element. Kinetic and crystal chemical properties must be considered in order to interpret more quantitatively the incorporation and distribution of elements in the various solid phases and in the solution phase.

Stability Relations

Predominance diagrams have been calculated with the help of the equilibrium constants given in Table 9.4, and they are shown in Figure 9.8. The method of graphical representation is identical to that described by Garrels and Christ [22] and Feth, Roberson, and Polzer [6]. For Figure 9a and b, however, equilibrium constants or solids with mineral formulas different from those used by Feth et al. and Garrels and Christ have been used. By choosing different constants and formulas, remarkably different predominance diagrams are obtained. It is important to realize that such stability diagrams represent estimates based on limited data or even tentatively assumed data; the objective is to gain qualitative or semiquantitative impressions of possible actual situations [24].

9.5   REACTION PROGRESS MODELS

Reaction progress diagrams are useful for showing the effects of initial reactions on the reaction path, on the appearance or disappearance of stable or metastable solid phases, and on the redistribution of aqueous species [5, 24–26]. Usually these diagrams are based on the concept of *partial equilibrium.* Partial equilibrium describes a state in which a system is in equilibrium with respect to at least one process (or reaction) but out of equilibrium with respect to others.†
For example, in considering the effect of a CO_2-bearing rainwater reacting with calcite, we assume that equilibrium among $H_2CO_3^*$, HCO_3^-, and H^+ is attained relatively rapidly, while $CaCO_3$ dissolution is slower. The pathway of the irreversible $CaCO_3$ dissolution may thus be indicated by considering the progress of the dissolution reaction on sequential states of homogeneous solution equilibrium among H_2CO_3, HCO_3^-, and H^+ until $CaCO_3$ saturation equilibrium is attained.

Reaction Paths for Calcite Dissolution.   How does the water composition change during the dissolution of $CaCO_3$ (calcite) by a CO_2-enriched water at

---

† An irreversible process involving a series of successive partial equilibrium states may result in a state of *local equilibrium* for the system; that is, a state in which no mutually incompatible phases are in contact even though the system as a whole is not in equilibrium (J. B. Thompson, Jr., in *Researches in Geochemistry*, P. H. Abelson, Ed., Wiley, New York, 1959, p. 427).

constant partial pressure of CO_2? This problem has already been treated in Example 5.4a, but we use here a different reaction path diagram. The extent of $CaCO_3$ dissolution is given by

$$[Ca^{2+}] = \tfrac{1}{2}[Alk] \simeq \tfrac{1}{2}[HCO_3^-] \tag{14}$$

The homogeneous equilibrium

$$CO_2(g) + H_2O = HCO_3^- + H^+ \qquad K_H K_1^{-1} \tag{15}$$

is assumed to be always attained during the $CaCO_3$ dissolution process. (The notation of the equilibrium constants is that of Table 5.2). The $CaCO_3$ dissolution equilibrium can be characterized by

$$CaCO_3(s) + 2H^+ = Ca^{2+} + CO_2(g) + H_2O \qquad K_{so} K_H^{-1} K_2^{-1} \tag{16}$$

or

$$CaCO_3(s) + 2H^+ = Ca^{2+} + H^+ + HCO_3^- \qquad K_{so} K_2^{-1} \tag{17}$$

Accordingly, $[Ca^{2+}]/[H^+]^2$ can be plotted as a function of either $p_{CO_2}^{-1}$ or $([H^+][HCO_3^-])^{-1}$ (Figure 9.8d). For any p_{CO_2} the approach to calcite saturation can be represented by a vertical line.

Feldspar Dissolution

The reaction progress for the dissolution of feldspars by a CO_2-enriched water can be calculated similarily by considering the various subsequent reaction steps. The dissolution of a mixture of albite and anorthite, for example, causes initially the release of Na^+, Ca^{2+}, Al^{3+}, and H_4SiO_4 and the consumption of H^+ (see Figure 9.8):

$NaAlSi_3O_8(albite) + 4H^+ + 4H_2O \longrightarrow$
$$Al^{3+} + Na^+ + 3H_4SiO_4 \tag{18}$$

$CaAl_2Si_2O_8(anorthite) + 8H^+ \longrightarrow$
$$2Al^{3+} + Ca^{2+} + 2H_4SiO_4 \tag{19}$$

After an initial congruent dissolution the released Al^{3+} (or hydroxo Al) causes the precipitation of gibbsite:

$$albite + H^+ \longrightarrow Al(OH)_3(s) + Na^+ + 3H_4SiO_4 \tag{20}$$

$$anorthite + 2H^+ \longrightarrow 2Al(OH)_3(s) + Ca^{2+} + 2H_4SiO_4 \tag{21}$$

From then on the dissolution of feldspars is incongruent; with increasing $[H_4SiO_4]$ the gibbsite kaolinite boundary will be reached.

$$Al(OH)_3(s) + H_4SiO_4 \longrightarrow \tfrac{1}{2}Al_2Si_2O_5(OH)_4(kaolinite) + 5H_2O$$

Only after all the gibbsite has been consumed will the solution composition leave the boundary given by the partial equilibrium between gibbsite and kaolinite and traverse the kaolinite field in the direction of Na^+ and Ca-mont-morillonites:

$$3\tfrac{1}{2}\text{kaolinite} + 4H_4SiO_4 + Na^+ \longrightarrow Na_{0.33}Al_{2.33}Si_{3.67}O_{10}(OH)_2$$
$$\text{(monmorillonite)} + H^+ + 11\tfrac{1}{2}H_2O \tag{22}$$

$$7\text{ kaolinite} + 8H_4SiO_4 + Ca^{2+} \longrightarrow 3Ca_{0.33}Al_{4.67}Si_{7.33}O_{20}(OH)_4$$
$$\text{(montmorillonite)} + 2H^+ + 23H_2O \tag{23}$$

Eventually $CaCO_3$ may be precipitated. During these dissolution processes the reactions $H_2CO_3^* = H^+ + HCO_3^-$ and $CO_2(g) + H_2O = H^+ + HCO_3^-$ are assumed to be in equilibrium. Considering these equilibria and the stoichiometry of the reaction, detailed reaction paths and the concomitant changes in solution composition and quantities of secondary minerals formed can be computed for various assumptions (ratio of albite/anorthite, p_{CO_2}, initial $[H_2CO_3]$ and temperature). A lucid example of such calculations, carried out with a computer program, were given by Drouby et al. [27a].

9.6   THE COMPOSITION OF FRESH WATERS IN COMPARISON TO EQUILIBRIUM MODELS

A few examples of *analyses of waters* of various types are compared in Table 9.5. It is tempting to try to relate the composition of the waters to the different geological environments. Because such relations depend upon so many interdependent variables, and because there is substantial variance in each genetic type of water, no reliable generalizations are possible with such a small number of examples. The publications by White, Hem, and Waring [27b] and Hem [27c] should be consulted for a discussion of the composition of a large number of waters in relation to their genesis. A few qualitative characteristics are apparent from Table 9.5. As expected, groundwaters have a higher ionic strength than surface waters. This results from the difference in CO_2 partial pressure to which these two types of waters are exposed. In comparing waters Nos. 5, 6 and 7, the influence of calcium plagioclase, which is less stable than the silicic rocks, becomes apparent. Sandstone beds are important aquifers; sandstones—as sedimentary rocks—may contain admixtures of carbonates besides grains of quartz. Thus the pH of sandstone waters tends to be higher than that of igneous rocks. Shales are laminated sediments whose constituent particles are of the clay grade. Most of these fine-grained sediments were deposited in saline environments. The relatively high $[HCO_3^-]$ of slate groundwaters may be due to high CO_2 concentrations resulting from oxidation of organic carbon. Waters from limestone and dolomite contain substantial quantities of Ca^{2+}, Mg^{2+}, and HCO_3^-; the concentrations of these solutes are usually controlled by carbonate equilibria.

TABLE 9.5 EXAMPLES OF NATURAL WATERS[a]

	1	2	3	4	5	6	7	8	9	10	11
	Stream	Stream	Stream	Lake Erie	Ground-water	Ground-water	Ground-water	Ground-water	Ground-water	Ground-water	Closed-Basin Lake
						Gabbro					
	Granite	Quartzite	Sandstone[b]		Granite	Plagioclase	Sandstone	Shale	Limestone	Dolomite	Soda Lake
pH	7.0	6.6	8.0	7.7	7.0	6.3	8.0	7.3	7.0	7.9	9.6
pNa	4.0	4.6	4.3	3.4	3.4	3.0	3.3	2.6	3.0	3.5	0.0
pK	4.7	5.1	4.8	4.3	4.0	4.5	4.0	4.2	3.7	—	1.7
pCa	4.0	4.3	3.1	3.0	3.5	3.1	3.0	2.5	2.7	2.8	4.5
pMg	4.6	5.1	4.0	3.4	3.8	3.2	3.5	2.5	3.4	2.8	4.6
pH_4SiO_4	3.8	4.2	4.1	4.7	3.2	3.0	3.9	3.5	3.7	3.4	2.8
$pHCO_3$	3.6	4.0	2.9	2.7	2.9	2.5	2.6	2.1	2.3	2.2	0.4
pCl	5.3	5.8	5.3	3.6	4.0	3.5	3.7	4.0	3.2	3.3	0.3
pSO_4	4.5	4.7	3.7	3.6	4.2	4.0	3.2	2.2	3.4	4.7	2.0
−log (ionic strength)	3.5	3.8	2.7	2.5	2.8	2.4	2.4	1.7	2.2	2.2	0.0

[a] 1–3: "Small Streams in New Mexico," *U.S. Geol. Surv. Bull.*, **1535F** (1961). 4: J. Kramer, *Geochim. Cosmochim. Acta*, **29**, 921 (1965). Types 5–10 are from *U.S. Geol. Surv. Bull.*, **440F** (1963). 5: Granite McCormick Co. (Table 1). 6: Harrisburg (Table 2). 7: Home Wood (Table 3). 8: Cuyahoga (Table 5). 9: Edwards limestone (Table 6). 10: Precambrian dolomite (Table 7). 11: "Albert Summer Lake Basin, Oregon," *North. Ohio Geol. Soc.*, J. L. Rau, Ed., 1966, p. 181.
[b] With slate and limestone beds.

Composition and Silicate Equilibria.   In Figure 9.8*a–c* values of $[Na^+]/[H^+]$, $[K^+]/[H^+]$, and $[Ca^{2+}]/[H^+]^2$ of the waters in Table 9.5 are plotted on the silicate diagrams as functions of $[H_4SiO_4]$. Most of the resulting points fall into the stability field of kaolinite. These waters, if they were in equilibrium, would be in equilibrium with kaolinite. They are not in equilibrium with feldspars; or, in other words, feldspars at the CO_2 pressures encountered are unstable and are degraded. At that stage, kaolinite apparently tends to be converted to montmorillonite, accounting for the limitation of silica content to about 60 mg liter^{-1} at the two-phase kaolinite–montmorillonite boundary.

If $[Ca^{2+}]$ and $[HCO_3^-]$ become sufficiently large, calcite precipitates; solubility equilibrium represents an upper limit for soluble carbonic constituents and calcium. The upper limits of the soluble silica content of natural waters are far less than saturation with amorphous silica ($\sim 2 \times 10^{-3} M$); $[H_4SiO_4]$ appears to be controlled by equilibrium between the waters and various silicate phases [5, 27*d*].

Equilibrium with Calcite.   Figure 9.8*d* illustrates that the waters in contact with $CaCO_3$ (sandstone, shale, limestone, dolomite, Table 9.5) are close to $CaCO_3$ saturation equilibrium or are oversaturated with respect to calcite. Lake Erie water falls into the same category.

In many *lake waters* precipitation and dissolution of calcite play a major role in buffering pH and water composition. In Figure 9.9 saturation of the water of Lake Zürich (Switzerland) with respect to calcite is represented as a function of temperature [24, 28]; that is, concentration quotients of analytical data for the reaction

$$CaCO_3(s) + H_2CO_3^* = Ca^{2+} + 2HCO_3^- + H_2O$$

are compared with the equilibrium constant (corrected for ionic strength) as a function of temperature. The figure shows that the lake water is oversaturated with $CaCO_3$ at high temperature (typically near surface waters) and undersaturated at low temperature (waters at greater depths). Correspondingly, in the upper layers calcite is precipitated, while in the lower layers it is partly dissolved. The variance from equilibrium is caused in large part by photosynthesis (elevation of pH in upper layers) and by respiration (decrease in pH) in the deeper layers (Figure 9.10). Kelt and Hsu [29] have given a detailed discussion on the chemistry of freshwater carbonate sedimentation in lakes.

Equilibrium with CO_2 of the Atmosphere.   As illustrated by Figure 9.8, all the waters of Table 9.4 contain more CO_2 than that corresponding to equilibrium with the atmosphere. Understandably, groundwaters influenced by a soil atmosphere enriched in CO_2 are further away from equilibrium with the atmosphere than are surface waters. Accordingly aeration of fresh waters leads frequently to an increase in pH when equilibrium conditions are approached.

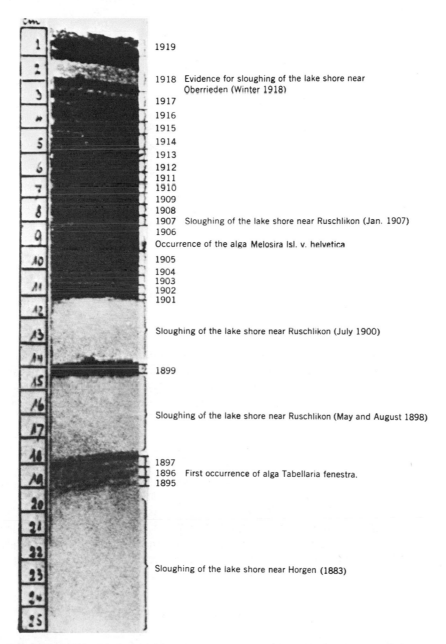

Figure 9.9 Sediment layers from Lake Zürich (1895–1919). The alternative sequence of layers result from deposition of $CaCO_3$ during the summer (oversaturation with respect to $CaCO_3$ because of pH increase due to photosynthesis) and accumulation of black iron(II) sulfide containing sludge during winter (anaerobic conditions). [After F. Nipkow, *Z. Hydrol.*, **1**, 101 (1920)]. Reproduced with permission from Birkhäuser Verlag, Basel, Switzerland.]

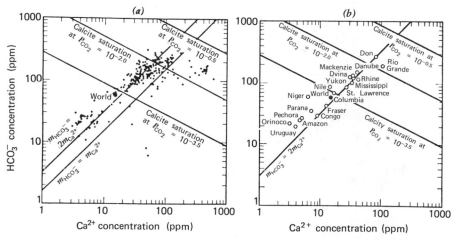

Figure 9.10 The relationship between the HCO_3^- and Ca^{2+} concentrations. (a) Rivers in the United States. Data largely from U.S. Geological Survey Water Supply Papers; world mean after Livingston (U.S. Geological Survey Paper No. 440G, 1963). (b) Major rivers in the world. Data from Livingston (U.S. Geological Survey Paper No. 440, 1963). (From Holland [14a]. Reproduced with permission from Wiley and Sons, New York.]

The conclusions to be drawn from this are: (a) Equilibration with atmospheric CO_2 is rather slow, and (b) reactions that tend to depress the pH of natural waters, such as repiration, ion exchange, and $CaCO_3$ deposition, tend to be more rapid than the CO_2 exchange process between these waters and the atmosphere.

River Water Composition and Carbonate Equilibria.   Figure 9.10a and b from Holland [14a] indicate that many rivers contain a concentration of CO_2 in excess of the concentration in equilibrium with the partial pressure of CO_2 in the atmosphere. The $HCO_3^-–Ca^{2+}$ plots correspond to the equilibrium $CaCO_3(s) + CO_2(g) = Ca^{2+} + 2 HCO_3^-$. Solutions saturated with respect to calcite at a given temperature and p_{CO_2} pressure have slopes of $-\frac{1}{2}$ in a log $[HCO_3^-]$ versus $\log[Ca^{2+}]$ diagram. Apparently most dilute river waters are undersaturated with respect to calcite, but in arid areas saturation is reached at the CO_2 pressures commonly encountered in river water.

Isothermal Evaporation

As we have seen, water composition is also affected by concentration resulting from evaporation (and evapotranspiration). Example 9.4 will illustrate the principles, procedures, and calculations of the effect of concentrating natural waters by isothermal evaporation for a few simplified systems. These calculations will illustrate how the reaction path of natural waters during evaporation depends on the Ca^{2+}/HCO_3^- ratio.

Example 9.4. Isothermal Evaporation of a Natural Water Leading to CaCO$_3$ Precipitation

Compute pH and solution composition during isothermal evaporation (25°C) for three types of calcium bicarbonate containing waters (1) $2[Ca^{2+}] = [Alk]$; (2) $2[Ca^{2+}] < [Alk]$; and (3) $[Ca^{2+}] > [Alk]$. All waters are assumed to remain in equilibrium with atmospheric CO_2 (log $p_{CO_2} = -3.5$ atm).

During evaporation the ionic strength continuously changes, and corrections for ionic strength should be made. For simplicity we make our calculations for $I = 5 \times 10^{-2}$. The values obtained without further correction must be regarded as approximations that may contain uncertainties of up to 0.3 logarithmic units for a concentration factor of 100. (An iterative procedure may be used to obtain more exact solutions: One first obtains approximate pH values and molarities by using constants valid at $I = 0$; with these tentative values I can be estimated as a function of the concentration factor in order to correct the equilibrium constants used). At $I = 5 \times 10^{-2}$ and 25°C equilibrium constants (corrected with the Davies equation) with the following numerical values are used (notation as in Table 5.2) $pK'_H = 1.47$, $pK'_1 = 6.27$, $pK'_2 = 10.16$, $pK'_{s0} = 7.72$; we choose the following initial conditions:

(1) $[Ca^{2+}]_0 = 5 \times 10^{-6} M$ $[Alk]_0 = 10^{-5}$ eq liter^{-1}

(2) $[Ca^{2+}]_0 = 2 \times 10^{-4} M$ $[Alk]_0 = 6 \times 10^{-4}$ eq liter^{-1}

(3) $[Ca^{2+}]_0 = 3 \times 10^{-4} M$ $[Alk]_0 = 4 \times 10^{-4}$ eq liter^{-1}

Initially all three waters are undersaturated with respect to calcite. Initially the alkalinity is present as HCO_3^-. The solution pH is given by the equilibrium

$$CO_2(g) + H_2O = HCO_3^- + H^+ \qquad K_H K_1$$

that is,

$$[H^+] = \frac{K_H K_1 p_{CO_2}}{[HCO_3^-]} \qquad (i)$$

At high pH values (above pH ~ 9) CO_3^{2-} may become a more important species; instead of equation i we may then write (compare equation 11 Section 4.3)

$$[Alk] = \left(\frac{K_H p_{CO_2}}{\alpha_0}\right)(\alpha_1 + 2\alpha_2) + \frac{K_w}{[H^+]} - [H^+] \qquad (ii)$$

The relationship between [Alk] and pH has been calculated before for Figure 4.3.

We may now proceed in steps and concentrate the solution successively by various factors (concentration factor = R).

In waters of type 1, $[Ca^{2+}]$ and $[Alk]$ (or $[HCO_3^-]$) will increase with R: $[Ca^{2+}] = R[Ca^{2+}]_0$ and $Alk = R[Alk]_0$ until the $CaCO_3$ solubility equilibrium is reached. Saturation with calcite and the atmosphere may be characterized by the equilibrium $CaCO_3(s) + CO_2(g) + H_2O = Ca^{2+} + 2HCO_3^-$; $^+K_{ps0}$ (where $^+K_{ps0} = K_{s0}K_1K_HK_2^{-1}$); that is,

$$[Ca^{2+}] = \frac{^+K_{ps0}\,p_{CO_2}}{[HCO_3^-]^2} \tag{iii}$$

or more generally (valid also at high pH) (cf. equation 39 in Section 5.3)

$$[Ca^{2+}] = \frac{K_{s0}\alpha_0}{K_H\alpha_2\,p_{CO_2}} \tag{iv}$$

$[Ca^{2+}]$ can be calculated from equation iii as a function of pH; this has already been done in Figure 5.7. As soon as solubility equilibrium is attained, the composition of the solution is constant (Figure 9.11a).

Solution of type 2 is characterized by the electroneutrality condition (Figure 9.11b)

$$[X] = Alk - 2[Ca^{2+}] \tag{v}$$

where X may be thought of as any (nonprecipitating) cation. During evaporation the left-hand side is given by $R[X]_0$; the terms on the right-hand side can be expressed as a function of pH (equations i or ii and iii or iv). These calculations are most conveniently carried out with a programmable calculator or a computer; otherwise, approximate results may be read from the diagrams in Figures 4.3 and 5.7. Most conveniently, for a given pH, $[Alk]$ and $[Ca^{2+}]$ are first calculated; then R is obtained from X in equation v. The result is given in Figure 9.11b. At a concentration factor of ca. 4, calcite starts to precipitate. During the bulk precipitation H^+, Ca^{2+} and Alk are somewhat buffered. After most calcite has been precipitated, $[Ca^{2+}]$ decreases while pH, $[Alk]$, and $[CO_3^{2-}]$ increase, and eventually the solution composition approaches that of a solution of soda.

Waters of type 3 are characterized by the electroneutrality condition

$$Y = 2[Ca^{2+}] - [Alk] \tag{vi}$$

where Y may be thought of as any (nonprecipitating) anion. The extent of concentration can be measured by $R[Y_0]$, and the terms on the right-hand side of equation vi can be expressed as before. As Figure 9.11c shows, pH starts to decrease after precipitation of the bulk of $CaCO_3$.

Garrels and Mackenzie [27d] have calculated the change in composition resulting from the evaporation of typical Sierra Nevada spring water coupled with sequential precipitation of calcite sepiolite and gypsum. In the early stage of evaporative concentration the relatively insoluble carbonates sulfates and silicates are precipitated. In the later stages of the evolution of saline lakes and brines the very soluble saline minerals may also be precipitated. For a review of

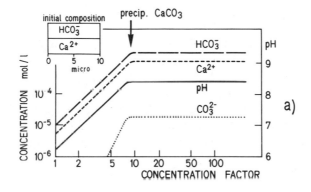

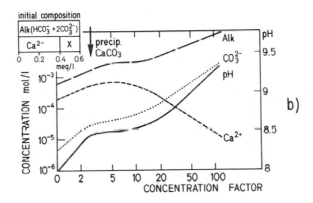

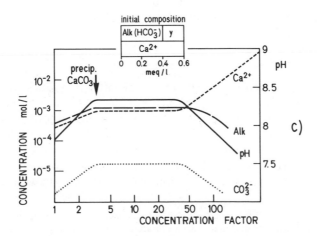

Figure 9.11 Isothermal (25°C) evaporation of natural waters leading to CaCO$_3$ precipitation. (*a*) 2[Ca^{2+}] = Alk, (*b*) 2[Ca^{2+}] < Alk, (*c*) 2[Ca^{2+}] > Alk (p_{CO_2} = 10$^{-3.5}$ atm).

the literature on evaporative concentration and brine evolution see Eugster and Hardie [30].

As shown in Example 9.4 and in Figure 9.11, the ratio Alk/[Ca²⁺] is of great significance in the genesis of natural waters and evolutionary paths during evaporative concentration. Waters in which $2[Ca^{2+}] \approx [HCO_3^-]$ do not change their [Ca]/[HCO₃⁻] ratio markedly upon evaporative concentration. Many river waters of the world are along the $2[Ca^{2+}] = [HCO_3^-]$ line in Figure 9.10. Upon extensive evaporation such waters may precipitate magnesite.

Weathering of silicate minerals usually supply cations in addition to Ca²⁺ and Mg²⁺. Such waters with $[HCO_3^-] > 2[Ca^{2+}]$ (residual alkalinity) tend upon evaporation to increase their pH values and concentrations of HCO₃⁻ and CO₃²⁻ and to decrease [Ca²⁺]; that is, in Figures 9 and 10a and b, these waters tend to move upward to the left along the lines of CaCO₃ saturation at constant p_{CO_2}. Upon extensive evaporation such waters acquire a composition similar to that found in natural soda lakes; eventually alkaline brines of the Na–CO₃–SO₄–Cl type may be formed.

Waters with a negative residual alkalinity ($2[Ca^{2+}] > [HCO_3^-]$; see Figure 9.11c), as a consequence of evaporation, tend in Figures 9 and 10a and b to move downward to the right along the lines of CaCO₃ saturation. Such waters tend to increase their calcium hardness, while their alkalinity and pH decrease;† eventually, upon extensive evaporation, they may reach saturation with respect to gypsum and become Ca–Na–SO₄–Cl brines.‡

Buffering

Buffering of pH of natural waters is not solely caused by the CO_2–HCO_3^-–CO_3^{2-} equilibrium. Heterogeneous equilibria are the most efficient buffer systems of natural waters. In Section 3.9 the pH buffer intensity, $\beta_{C_j}^{C_i}$, was defined for the incremental addition of C_i to a closed system of constant C_j at equilibrium

$$\beta_{C_j}^{C_i} = \frac{dC_i}{d\text{pH}} \tag{24}$$

We can now compare, for example, the buffer intensities, with respect to strong acid, of the following systems (Figure 9.12): (a) a carbonate solution of constant C_T, β_{C_T}; (b) an aqueous solution in equilibrium with calcite, β_{CaCO_3}; (c) a carbonate solution in equilibrium with a gas phase of constant p_{CO_2}, $\beta_{p_{CO_2}}$; and (d) a solution in equilibrium with both kaolinite and anorthite, $\beta_{an-kaol}$. As explained earlier (Section 3.9), the buffer intensity is found analytically by differentiating the appropriate function of C_i for the system with respect to pH.

---

† The conclusion of Figure 9.11b and c, that residual alkalinity increases upon evaporation (concomitant with pH increase) and that negative residual alkalinity decreases (concomitant with pH decrease) applies generally even if alkalinity is made up of anions other than HCO₃⁻ and CO₃²⁻.
‡ See also A. Al-Droubi, B. Fritz, J. V. Gac and Y. Tardy, *Amer. J. Sci.*, **280**, 560 (1980).

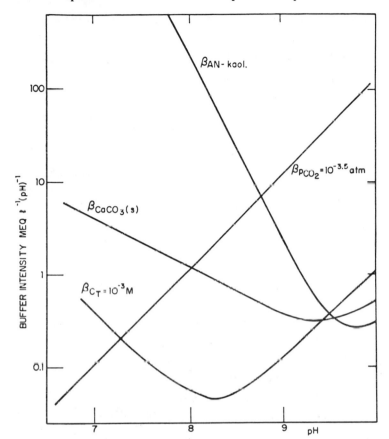

Figure 9.12 Buffer intensity versus pH for some heterogeneous systems and for the homogeneous dissolved carbonate system. Buffer intensities: β_{C_T} (dissolved carbonate, $C_T = 10^{-3}$ M), β_{CaCO_3} (carbonate solution in equilibrium with calcite), $\beta_{p_{CO_2}}$ (carbonate solution in equilibrium with $p_{CO_2} = 10^{-3.5}$ atm) and $\beta_{an\text{-}kaol}$ (solution in equilibrium with anorthite and kaolinite).

The buffer intensity of a homogeneous carbonate system has been derived before (see Chapter 4, Table 4.5). For heterogeneous systems the buffer intensities are derived in a similar way. Example 9.5 serves as an illustration.

Example 9.5   Buffer Intensity of a Heterogeneous System

Calculate the buffer intensity with respect to strong acid (or strong base), $\beta_{CaCO_3(s)}$, of a solution in equilibrium with calcite.

It is convenient to start out with the charge condition of a closed aqueous solution of $CaCO_3(s)$:

$$2[Ca^{2+}] + [H^+] = [HCO_3^-] + 2[CO_3^{2-}] + [OH^-] \qquad (i)$$

After addition of a strong acid, say HCl (i.e., $[Cl^-] = C_A$), the charge balance becomes

$$C_A = 2[Ca^{2+}] + [H^+] - [HCO_3^-] - 2[CO_3^{2-}] - [OH^-] \qquad \text{(ii)}$$

The buffer intensity is obtained after differentiation of equation ii with respect to pH. Because

$$[Ca^{2+}] = C_T = \left(\frac{K_s}{\alpha_2}\right)^{0.5} \qquad \text{(iii)}$$

(compare Section 5.3), the titration curve of equation ii can be calculated as a function of $[H^+]$ or pH. Instead of a cumbersome numerical differentiation, it may be more convenient to consider the pH dependence of the individual terms on the right-hand side of equation ii. The variation of $\log[Ca^{2+}]$, $\log[HCO_3^-]$, and $\log[CO_3^{2-}]$ with pH has been illustrated in Figure 5.5. In the pertinent pH region, $pK_1 < pH < pK_2$:

$$\frac{d\log[Ca^{2+}]}{-dpH} = 0.5 \qquad \text{(iv)}$$

$$\frac{d\log[HCO_3^-]}{-dpH} = 0.5 \qquad \text{(v)}$$

$$\frac{d\log[CO_3^{2-}]}{-dpH} = -0.5 \qquad \text{(vi)}$$

From these relations we readily obtain

$$\frac{d[Ca^{2+}]}{-dpH} = \frac{d\log[Ca^{2+}]}{-dpH}\frac{d[Ca^{2+}]}{d\log[Ca^{2+}]} = (0.5)2.3[Ca^{2+}] \qquad \text{(vii)}$$

Similarly,

$$\frac{d[HCO_3^-]}{-dpH} = 1.15[HCO_3^-] \qquad \text{(viii)}$$

$$\frac{d[CO_3^{2-}]}{-dpH} = -1.15[CO_3^{2-}] \qquad \text{(ix)}$$

$$\frac{d[OH^-]}{-dpH} = -2.3[OH^-] \qquad \text{(x)}$$

Thus with the help of Figure 5.5 the individual terms in the derivative of equation ii are obtained for every pH value (Figure 9.11). The following relationship holds in good approximation for the pH range stipulated:

$$\beta_{CaCO_3(s)}^{C_A} \simeq \left(\frac{1}{-dpH}\right)(2d[Ca^{2+}] - d[HCO_3^-])$$

$$= \frac{d[Ca^{2+}]}{-dpH}$$

$$= 1.15\left(\frac{K_s}{\alpha_2}\right)^{0.5} \qquad \text{(xi)}$$

In the case of the heterogeneous equilibria with silicates, highly simplified models are considered where, for example, an acid (or base) is added to an aqueous system of kaolinite and anorthite. The electroneutrality would be

$$[H^+] + 2[Ca^{2+}] = [OH^-] + C_A \tag{13}$$

where C_A is the concentration of strong acid added. $[Ca^{2+}]$ is given by $K[H^+]^2$, where K is the equilibrium constant for the anorthite–kaolinite system [(3) of Table 9.4, hence $\beta_{an\ kaol}$ can be evaluated. As Figure 9.12 shows, the homogeneous buffer intensity β_{L_T} is relatively small. For freshwater systems β_{CaCO_3} and $\beta_{p_{CO_2}}$ are of considerable practical interest.

$CaCO_3(s)$ is an efficient buffer in the neutral and acid pH range. If, for example, a large quantity of acid is discharged into a natural water system containing $CaCO_3(s)$, an initially large decrease in the pH of that system occurs. The extent of this decrease depends largely upon the magnitude of the fraction of total buffer intensity attributable to dissolved buffer components. Ultimately, however, the decrease in pH resulting from addition of the acid leads to dissolution of solid calcium carbonate and the establishment of a new position of equilibrium. Thus the final change in pH is much less than the initial decrease which is resisted only by the intensity contribution of dissolved buffer components. The addition of a strong base in large quantity, conversely, leads to a deposition of calcium carbonate, thus reducing the pH shift that would occur in the absence of dissolved calcium.

The aluminum silicates provide considerably more resistance toward pH changes. The equilibrium system anorthite–kaolinite at pH = 8 has a buffer intensity a thousand times higher than that of a 10^{-3} M carbonate solution. As has been shown, equilibrium systems consisting of a sufficient number of coexisting phases attain, in principle, an infinite buffer intensity.

What actually protects water from pH changes is also dependent upon the kinetics of the heterogeneous reactions. Reactions with solid carbonates and ion-exchange processes are faster than alteration reactions of solid silicates. Investigations on the rate of the buffering reactions are very much in need.

From a kinetic point of view we must also consider that biochemical processes effect pH regulation and buffer action in natural water systems. Photosynthetic activities decrease CO_2, whereas respiratory activities contribute CO_2.

For fresh waters there is a further restraint on pH rise: the CO_2 reservoir of the atmosphere. For a given p_{CO_2} the pH is a function of alkalinity. In order to raise the pH of a water in equilibrium with the atmosphere from 8 to 9, alkalinity must increase by nearly 5 meq liter^{-1} (either by base addition or by evaporation) (see Figure 4.3). Hence only soda lakes, that is, lakes containing substantial amounts of soluble carbonates and bicarbonates, can attain high pH values (Example 11 of Table 9.5); for example, Sierra Nevada spring waters discharged to the east of the Sierra and evaporated in a plaza of the California desert. As the example given below illustrates, generally a nearly neutral Na–Ca–HCO_3^- water is converted by evaporation into an alkaline Na–HCO_3^-–CO_3^{2-} water (soda lake). Buffering of pH (see Figure 9.10) occurs during the precipitation of solid phases.

9.7.   INTERACTION BETWEEN ORGANISMS AND ABIOTIC ENVIRONMENT

Aquatic organisms influence the concentration of many substances directly by metabolic uptake, transformation, storage, and release. In order to understand the chemistry of an aquatic habitat, the causal and reciprocal relationship between organisms and their environment must be taken into consideration.

Photosynthesis and Respiration

Energy-rich bonds are produced as a result of photosynthesis, thus distorting the thermodynamic equilibrium. Bacteria and other respiring organisms catalyze the redox processes that tend to restore chemical equilibrium. In a simplified way we may consider a stationary state between photosynthetic production $P = dp/dt$ (rate of production of organic material; p = algal biomass) and heterotrophic respiration R (rate of destruction of organic material) (Figure 9.13) and chemically characterize this steady state by a simple stoichiometry [31, 32]

$$106\,CO_2 + 16\,NO_3^- + HPO_4^{2-} + 122\,H_2O + 18\,H^+ \text{ (+ trace elements and energy)}$$

$$P \big\|\, R$$

$$\{C_{106}H_{263}O_{110}N_{16}P_1\} + 138\,O_2 \tag{24}$$
$$\text{algal protoplasm}$$

Algal protoplasm may conveniently be also expressed as

$$\{(CH_2O)_{106}(NH_3)_{16}(H_3PO_4)\}.$$

The flux of energy through the system is accompanied by cycles of nutrients and other elements. Like every ecosystem, lakes and their surroundings and oceans contain a biological community (primary producers, various trophic levels of decomposers, and consumers) in which the flow of energy is reflected in the trophic structure and in material cycles.

Although the stoichiometry of equation 24 may vary from one aquatic habitat to another, it is remarkable that the summation of the complicated processes of the P–R dynamics, in which so many organisms participate results in such simple relations. To a first approximation parcels of water differ in their P, N, and C concentrations to the extent that organisms have removed or added in the fixed biomass ratio (equation 24).

A vertical segregation of nutritional elements occurs in lakes (during the stagnation period) and in the ocean (Figure 9.13b). During photosynthesis nitrogen (NO_3^- or NH_4^+) and phosphate are taken up together with carbon (CO_2 or HCO_3^-) in the proportion C/N/P $\simeq$ 106:16:1. As a consequence of respiration (oxidation) of these organism-produced particles after settling, these elements are released again in approximately the same proportions.

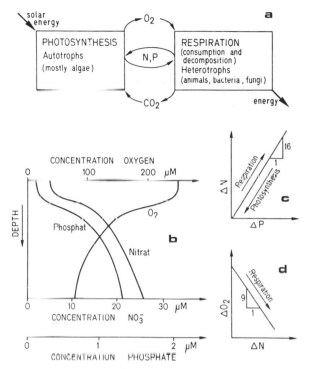

Figure 9.13 Photosynthesis and respiration. A well-balanced ecosystem may be characterized by a stationary state between photosynthetic production, *P* (rate of production of organic material) and heterotrophic respiration, *R* (rate of destruction of organic matter) (*a*). Photosynthetic functions and respiratory functions may become vertically segregated in a lake or in the sea. In the surface waters the nutrients become exhausted by photosynthesis. The subsequent destruction (respiration) of organism-produced particles after settling leads to enrichment of the deeper water layers with these nutrient elements and a depletion of dissolved oxygen [schematic representation in (*b*)]. The relative compositional constancy of the aquatic biomass and the uptake (**P**) and release (**R**) of nutritional elements in relatively constant proportions (see equation 24) is responsible for a co-variance of carbon, nitrate, and phosphate in lakes (during stagnation periods) and in the ocean; an increase in the concentration of these elements is accompanied by a decrease in dissolved oxygen. The constant proportions $\Delta C : \Delta N : \Delta P : \Delta O_2$ typically observed in these waters are caused by the stoichiometry of the **P**–**R** processes (*c* and *d*).

Respiration is accompanied by a respiration quotient $\Delta O_2/\Delta C \simeq -1.3$ (or $\Delta O_2/\Delta N \simeq -9$).

As shown in Figure 11.13 and in Figure 9.14, simple correlations in ΔN, ΔP, and ΔO_2 are typically observed in lakes and in the ocean. The stoichiometric formulation of equation 24 reflects in a simple way *Liebig's law of the minimum.* It follows from Figure 9.14*a* and *b* that seawater becomes exhausted simultaneously in dissolved phosphorus and nitrogen as a result of photosynthetic assimilation. We infer that nitrogen and phosphorus together determine the extent of organic production if temporary and local deviations are not considered. We might consider the possibility that originally phosphorus (e.g., from

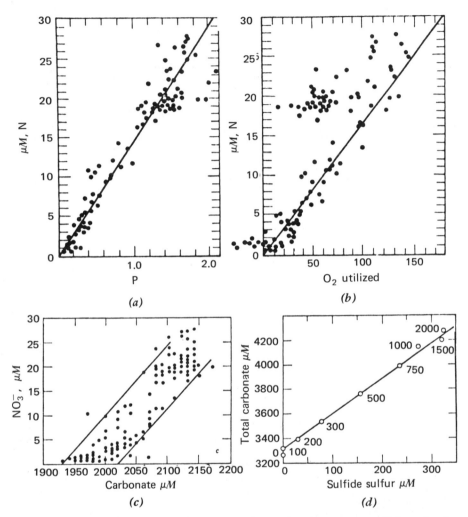

Figure 9.14 Stoichiometric correlations among nitrate, phosphate, oxygen, sulfide, and carbon. The correlations can be explained by the stoichiometry of reactions such as equation 24; concentrations are in micromolar. (*a*) Correlation between nitrate nitrogen and phosphate phosphorus corrected for salt error in waters of the western Atlantic. (*b*) Correlation between nitrate nitrogen and apparent oxygen utilization in same samples. The points falling off the line are for data from samples above 1000 m. (Redfield, A. C. *James Johnstone Memorial Volume* 1934, p. 177. Liverpool) (*c*) Correlation between nitrate nitrogen and carbonate carbon in waters of the western Atlantic. (*d*) Relation of sulfide sulfur and total carbonate carbon in waters of the Black Sea. Numbers indicate depth of samples. Slope of line corresponds to $\Delta S^{2-}/\Delta C = 0.36$. (From data of Skopintsev et al. (1958), as quoted in [31].) [(*a–c*) after Redfield [34]. Reproduced with permission from Wiley-Interscience Publishers.]

apatite) was the sole minimum nutrient, but that the concentration of nitrogen has been adjusted in the course of evolution as a result of nitrogen fixation and denitrification to the ratio presently found. Alternatively, we could also argue that the stoichiometric composition of the organisms as a result of evolution has become the same as that in the sea.

There is also a horizontal segregation of biologically utilized elements in the deep Atlantic and the deep Pacific. The water circulation pattern in the ocean is responsible for the enrichment of biologically active elements in the deep Pacific relative to the deep Atlantic. Such effects are lucidly explained by Broecker [33].

Anoxic Conditions.   The usual sequence of various redox reactions with organic matter is observed where the accumulation of organic matter is great. As shown in Table 7.6, oxygen dissolved in water first becomes exhausted, and then the oxidation of organic matter continues with nitrate serving as the oxidant; subsequently organic fermentation reactions and redox reactions with SO_4^{2-} and CO_2 as electron acceptors occur. The reduced products of the oxidation of organic matter accumulate in the water in addition to the products of the oxidation of organic matter. The formal equations given in Table 7.6 can be modified to develop stoichiometric models that predict how the components of the anoxic water will change as a result of the mineralization of settled plankton. Such stoichiometric models have been developed by Richards [34], who corroborated the validity of these models for a large number of anoxic basins and fjords. For example, under conditions of sufficiently low pε, planktonic

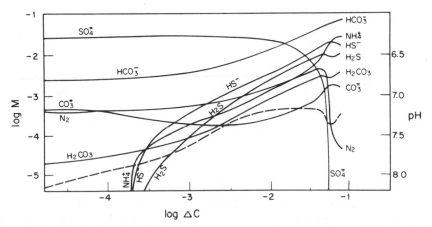

Figure 9.15   Distribution of predominant dissolved species during the decomposition of alanine ($C_3H_7O_2N$) in seawater at 25°C. The molar concentrations of dissolved species are given as a function of ΔC, the number of moles of organic carbon reacted per liter of solution. The dashed line represents pH. (From Thorstenson [26]. Reproduced with permission from *Geochimica et Cosmochimica Acta*.)

material becomes oxidized by SO_4^{2-} which is reduced to $S(-II)$. For the oxidation of one $\{C_{106}H_{263}O_{110}N_{16}P_1\}$, approximately 424 electrons are necessary. (Note that the carbon in the plankton formula has a formal oxidation state of approximately 0.) In the reduction of SO_4^{2-} to $S(-II)$, 53 SO_4^{2-} ions can provide the 424 electrons necessary; hence one would expect a ratio $\Delta S(-II)/\Delta C = 0.5$. As Figure 9.14d shows, this ratio is somewhat less for the waters of the Black Sea.

Using the concept of partial equilibria Thorstenson [25] has predicted in a reaction path calculation the compositional changes in the aqueous phase as a function of the amount of organic matter decomposed. Figure 9.15 (from Thorstenson) gives the distribution of predominant dissolved species at hypothetical equilibrium as a function of increments of organic matter decomposed, represented in this example by $C_3H_7O_2N$ ($=$ alanine).

9.8   THE OCEANS; CONSTANCY OF THE COMPOSITION AND CHEMICAL EQUILIBRIA

Figure 1.1 gives the concentrations of the most abundant constituents of seawater and average river water. The salt dissolved in seawater has remarkably constant major constituents (Table 9.6). Cl^-, SO_4^{2-}, Mg^{2+}, K^+, Ca^{2+}, and Na^+ dominate sea salt. Their ratios one to another are very constant. This constancy does not extend to all the trace components (Table 9.7) especially not to the biolimiting elements that are removed from the surface seawater by organisms. For bio-unlimited elements the ratio of the element to the total salt (e.g., chlorinity) in both surface and deep seawater samples is unchanged.

The relative proportions of the major ions in seawater and average river water are quite different. Concentration of river water does not produce ocean water. If average river water were concentrated by evaporation, a variety of minerals would precipitate and the ratio among the elements would change. As we have seen in Example 9.4 and Figure 9.11b, the composition of evaporated river water would resemble the composition of carbonate-bearing nonmarine evaporite brines much more than the composition of seawater. Furthermore, the dissolved material present in the ocean is only a small fraction of that delivered to the ocean by the rivers over geological time. Obviously, ions must be removed from the ocean approximately as fast as they are supplied by rivers. Thus the removal processes exert the major control on the chemical composition of seawater.

Essentially two major concepts regarding control of the composition of seawater complement each other: (1) control by chemical equilibria between seawater and oceanic sediments, and (2) kinetic regulation by the rate of supply of individual components and the interaction between biological cycles and mixing cycles.

Obviously the sea is an open, dynamic system with variable inputs and outputs of mass and energy for which the state of equilibrium is a construct. As we have seen, the concept of free energy, however, is not less important in dynamic

TABLE 9.6   MAJOR COMPOSITION OF SEAWATER

Constituent	Seawater at $S = 35\%_{00}$ $(g\ kg^{-1})^a$	$(g\ kg^{-1})\div$ Chlorinity[b]	$(mol\ kg^{-1})\div$ Chlorinity	Residence Time in Oceans, $\log \tau$ (years)[d]
Na^+	10.77	0.556	0.0242	7.7
Mg^{2+}	1.29	0.068	0.0027	7.0
Ca^{2+}	0.4121	0.02125	0.000530	5.9
K^+	0.399	0.0206	0.000527	6.7
Sr^{2+}	0.0079	0.00041	0.0000047	6.6
Cl^-	19.354	0.9989	0.0282	7.9
SO_4^{2-}	2.712	0.1400	0.0146	6.9
$HCO_3^{-\ c}$	0.1424	0.00735	0.00012	4.9
Br^-	0.0673	0.00348	0.000044	8
F^-	0.0013	0.000067	0.0000035	5.7
B	0.0045	0.000232	0.0000213	7.0
	$\Sigma = 35$	$\Sigma = 1.82$	$\Sigma = 0.058$	

[a] Salinity $S(\%_{00})$ is defined as the weight in grams of the dissolved inorganic matter in 1 kg of seawater after all Br^- and I^- have been replaced by the equivalent quantity of Cl^- and all HCO_3 and CO_3^{2-} converted to oxide. In over 97% of the seawater in the world, the salenity S is between 33 and 37‰. The total grams of major constituents (sea salt), g_T, for 1 kg of solution is related to the chlorinity by $g_T = 1.81578\ Cl\ (\%_{00})$. Salinity $S(\%_{00})$, is defined as $S(\%_{00}) = 1.80655\ (Cl\%_{00})$; thus $g_T = 1.00511\ S(\%_{00})$.

[b] The chlorinity, $Cl\ (\%_{00})$, is determined by the titration of seawater with $AgNO_3$. It was defined as the chlorine equivalent of the total halide concentration in $g\ kg^{-1}$ seawater; it is now defined as the mass in grams of Ag necessary to precipitate the halogens (Cl^- and Br^-) in 328.5233 g of seawater. It has been adequately demonstrated that the relative composition of the major (greater than $1\ mg\ kg^{-1}$ seawater) components of seawater is nearly constant. By measuring one constituent of seawater, the composition of other components can be characterized. The constituent normally selected is the chlorinity, $Cl\ (\%_{00})$.

[c] The results given for HCO_3^- are actually values of the carbonate alkalinity expressed as though it were all HCO_3^-.

[d] Residence times were computed by $\tau = M/Q$, where M for a particular constituent is equal to its concentration in seawater times the mass of the oceans and Q is equal to the concentration of the constituent in average river water times the annual flux of river water to the ocean.

systems. In considering equilibria and kinetics in ocean systems, it is useful to recall that different time scales need to be identified for the various processes. When a particular reaction of a phase or species has—within the time scale of consideration—a negligible rate, it is permissible to define a metastable equilibrium state. Similarly, in a flow system the time-invariant condition of a well-mixed volume approaches chemical equilibrium when the residence time is

TABLE 9.7 MINOR ELEMENTS IN SEAWATER[a]

Element	Concentration (−log M)	Residence time, [log τ (years)]	Element	Concentration (−log M)	Residence time, [log τ (years)]	Element	Concentration (−log M)	Residence time, [log τ (years)]
He	8.8	—	V	7.3	5	Rb	5.85	—
Li	4.6	6.3	Cr	8.2	—	Mo	7	5
B	3.39	7.0	Mn	8.4	4	Ag	9.4	5
N	1.97	6.3	Fe	7.5	2	Cd	9	4.7
F	4.17	5.7	Co	9.1	4.5	Sn	10	—
Ne	8.2	—	Ni	7.6	4	Sb	8.7	4
Al	7.1	2	Cu	8.1	4	I	6.3	6
Si	4.1	3.8	Zn	4.9	4	Xe	9.4	—
P	5.7	4	As	7.3	5	Cs	8.5	5.8
Ar	6.96	—	Kr	8.6	—	Ba	6.8	4.5
						W	9.3	—
						Au	10.7	5
						Hg	9.8	5
						Pb	9.7	—
						Th	10.4	—
						U	7.9	3.3

[a] Concentrations mostly from P. G. Brewer, in *Chemical Oceanography*, J. P. Riley and G. Skirrow, Eds., Vol. 1, 2nd ed., Academic, New York, 1975. For the calculation of residence times see Table 9.6, footnote *d*.

sufficiently large relative to the appropriate time scale of the reaction. We will first discuss some equilibrium concepts.

Equilibrium Models

Sillén, in a classic, most influential paper in 1961 [1], outlined an imaginary equilibrium model for the ocean sediment system. Before discussing some of the features of his model, we should be aware of some of the qualifications he expressed [35] concerning the merits of such an equilibrium approach:

> "First it is well known that there is no true equilibrium in the ocean. Why then bother to talk about this equilibrium model? I agree that there is no complete equilibrium in the oceans and that everything interesting we observe is caused by lack of equilibrium. Nevertheless, it may be worthwhile to try to determine what the equilibrium model would look like—i.e., what the solid phases and composition of the solution would be. Perhaps the equilibrium model will be a useful first approximation, the next step would then be to discuss how it is disturbed by various processes: radiation, life, transport of matter etc. Finally the model may be proved useless, but this has not yet been done, and perhaps one would learn many interesting things in refuting it."

The Phase Rule for Organizing Equilibrium Model

In considering equilibria of a water with various solid phases and a gas phase, the Gibbs phase rule may be used as a basis for organizing and interpreting such models.

A few simple equilibrium systems are considered in Table 9.8. They represent simple models for natural waters and, as illustrated in Section 5.8, are constructed by incorporating the specific components into a closed system and by specifying the phases to be included. Recall that the phase rule restricts the number of independent variables F to which we can assign values according to the number of components C and phases P:

$$F = C + 2 - P$$

For comparison Table 9.8a lists CO_2 and $CaCO_3$ equilibrium systems discussed before (Section 5.9). In the $CaCO_3$ solubility model (No. 3, Table 9.8a) a closed system containing the phases calcite, aqueous solution and gas (CO_2) can be constructed with the components H_2O, CO_2 and CaO. The system can be described with two independent variables such as temperature and pressure.

In Table 9.8b models containing the same five components—SiO_2, CaO, CO_2, H_2O, Al_2O_3—but with different numbers of phases, are compared with each other; an increase in P must be accompanied by a decrease in F. For model 4, in addition to temperature and p_{CO_2}, a concentration condition must be specified in order to define the equilibrium composition. On the other hand, the composition of model 5 can be calculated with two independent variables (e.g., temperature and p_{CO_2}).

TABLE 9.8 EQUILIBRIUM MODELS; APPLICATION OF PHASE RULE

	(a) CO_2 and $CaCO_3$ Solubility Models			(b) Aluminum Silicates and $CaCO_3$		
	1	2	3	4	5	6
Phases	Aqueous solution $CO_2(g)$	Aqueous solution Calcite(s)[a]	Aqueous solution $CO_2(g)$ Calcite(s)	Aqueous solution $CO_2(g)$ Kaolinite Ca montmorillonite	Aqueous solution $CO_2(g)$ Kaolinite Ca montmorillonite Calcite	Aqueous solution $CO_2(g)$ Kaolinite Ca montmorillonite Calcite Ca feldspar
P	2	2	3	4	5	6
Components	H_2O, CO_2	H_2O, CO_2, CaO	H_2O, CO_2, CaO	H_2O, CO_2, CaO,	Al_2O_3, SiO_2	
C	2	3	3	5	5	5
F	2	3	2	3	2	1
Variables[b]	$t = 25°C$ $-\log p_{CO_2} = 3.5$	$t = 25°C$ $-\log p = 0^a$ $[Ca^{2+}] = C_T^c$	$t = 25°C$ $-\log p_{CO_2} = 3.5$	$t = 25°C$ $-\log p_{CO_2} = 3.5$ $8[Ca^{2+}] = [H_4SiO_4]^c$	$t = 25°C$ $-\log p_{CO_2} = 3.5$	$t = 25°C$ $-\log p_{CO_2} = 4.5$
Composition						
pH	5.7	9.9[d]	8.3	7.4	8.3	9.0
$pHCO_3$	5.7	4.1	3.0	3.9	3.0	3.4
pCa		3.9	3.3	4.2	3.3	3.7
pH_4SiO_4				3.2	3.6	3.7

[a] $H_2CO_3^*$ is treated as a nonvolatile acid. The system is under a total pressure of 1 atm.

[b] By specifying p_{CO_2}, the total pressure P is determined ($P = p_{CO_2} + p_{H_2O}$). For the calculation, constants valid at $P = 1$ atm were used.

[c] This additional constraint is necessary for defining the system; other conditions could be specified.

[d] $pCO_3 = 4.4$.

Example 9.5. Equilibrium Composition in the System
H_2O–CO_2–CaO–Al_2O_3–SiO_2

1   Compute the equilibrium composition (25°C) of a model system containing the phases Ca montmorillonite, calcite, kaolinite, aqueous solution, and a gas phase with $p_{CO_2} = 10^{-3.5}$ atm (model 5, Table 9.8).
2   Add anorthite to the system considered under 1).
The solutes are interrelated by the electroneutrality condition

$$2[Ca^{2+}] + [H^+] = [HCO_3^-] + 2[CO_3^{2-}] + [OH^-] \qquad \text{(i)}$$

We are justified in neglecting the protolysis of H_4SiO_4 in formulating equation i. In equation i, $[Ca^{2+}]$ as well as $[HCO_3^-]$ and $[CO_3^{2-}]$ can be expressed as a function of $[H^+]$ and p_{CO_2}. $[Ca^{2+}]$ can be formulated through the solubility product of calcite, K_{s0},

$$[Ca^{2+}] = \frac{K_{s0}}{[CO_3^{2-}]} = \frac{K_{s0}}{C_T \alpha_2} \qquad \text{(ii)}$$

and C_T can be expressed with the help of Henry's law, K_H,

$$C_T = \frac{[HCO_3]}{\alpha_1} = \frac{[CO_3^{2-}]}{\alpha_2} = \frac{K_H p_{CO_2}}{\alpha_0} \qquad \text{(iii)}$$

After substitution of equations ii and iii into equation i, the latter can be solved for $[H^+]$, for example, by trial and error; subsequently C_T, $[HCO_3]$, $[CO_3^{2-}]$, and $[Ca^{2+}]$ are obtained. Thus far the computation is the same as that for the $CaCO_3$–CO_2–H_2O system (model 3 in Table 9.8). $[H_4SiO_4]$ can now be computed from the equilibrium constant of the Ca montmorillonite–kaolinite equilibrium (K in Table 9.4).

$$[H_4SiO_4]^8 = \frac{K[H^+]^2}{[Ca^{2+}]} \qquad \text{(iv)}$$

The result is given in Table 9.8 (model 5). Note that $[H_4SiO_4]$ varies with $p_{CO_2}^{1/8}$. Hence waters of this type have nearly constant $[H_4SiO_4]$.

We now add anorthite (model 6) to the system already considered. After specifying the temperature, no other degree of freedom remains for the given number of components and phases; then p_{CO_2} in the gas phase of the model will be determined by the equilibria and cannot be varied.†

Infinite Buffer Intensity.   The activities in a system such as model 3 or 5 (Table 9.8) remain constant and independent of the concentration of the components as long as the phases coexist in equilibrium. The composition of the

† The phase rule with special reference to the coexistence of magnesium calcites and dolomite interaction is discussed by R. Wollast and R. M. Pytkowicz, *Geochem. J.*, **12**, 199 (1978).

solution does not change if water (isothermal dilution) or the base $Ca(OH)_2$ is added. Such coexistence in equilibrium of the appropriate number of phases constitutes a chemostat or pH stat.

CO_2 Manostat.   Model 6 in Table 9.8 illustrates the possibility that the CO_2 content of the atmosphere is regulated at the sea–sediment interface by equilibria of reactions in which various aluminum silicates and $CaCO_3$ participate.

Sillén's Oceanic Model.   The classical geochemical material balance assumes that all sediments were ultimately derived from igneous rocks: primary rocks + volatile substances $\rightleftharpoons$ sediments + seawater + air. Goldschmidt's [36] balance suggested that, for each liter of present seawater, 600 g of igneous rocks reacted with about 1 kg of volatile substances [H_2O, HCl, CO_2, etc.); during the process about 600 g of sediment and 3 liters of air were also formed. This balance became the basis of Sillén's [1] equilibrium model. In an imaginary experiment the components of the real system (as given by Goldschmidt's values) were brought to equilibrium. Sillén's formal approach in constructing his model consisted of adding the components in sequence. The models given in Table 9.8 can be enlarged with the addition of each additional component to an equilibrium system resulting in either a new phase or an additional degree of freedom. Sillén has proposed equilibrium systems of different complexity as models for the oceans. His *nine-component system* contains HCl, H_2O, CO_2, SiO_2, Al_2O_3, NaOH, KOH, MgO, and CaO. The first three components correspond to the volatile substances originating in the interior of the earth which, together with the other components, are contained in igneous rocks and participate in forming the sea. The nine components are distributed in nine phases, aqueous solution, quartz (Si), kaolinite (Al), chlorite (Mg), mica-illite (K), montmorillonite (Na), phillipsite or some other zeolite (Ca), calcite (CO_2), and a gas phase. (The items in parentheses are the components whose activities Sillén believes may become fixed by that phase.) By fixing the temperature and [Cl^-], the composition of the sea and p_{CO_2} of the atmosphere will be determined.

The implications of these equilibrium models for the history of the mean water were discussed by Holland [37]. Mackenzie and Garrels [38, 39] suggested that the important reactions were not so much transformation of one clay mineral into another as a transformation of amorphous aluminum silicate resulting from weathering into crystalline clay minerals:

amorphous Al silicate + H_4SiO_4 + HCO_3^- + cations $\longrightarrow$
$$\text{cation Al silicate} + CO_2$$

These reverse weathering reactions would prevent the accumulation of alkali metals and HCO_3^- from accumulating in the oceans. Attempts to measure directly the chemical changes that take place when clay minerals pass from fresh waters into the oceans [40] indicated, however, that the types of reactions suggested by Sillén and by Garrels and Mackenzie could not have taken place to a significant extent.†

Nevertheless the *carbonate chemistry* of seawater is clearly constrained by thermodynamics. Although oceans were probably never precisely at saturation with respect to calcite, the presence of limestone in sedimentary rocks of all ages implies near saturation of mean water with respect to calcite during the past 3 billion years [41]. Holland [37, 41] has used the sedimentary record to define the limits of possible variations in the concentrations of Ca^{2+} and HCO_3^-. The approach used is illustrated in Example 9.6.

Example 9.6. p_{CO_2} Control during Recent Geological Past

Holland [42] shows that the CO_2 pressure in the atmosphere, at least in the recent geological past, has been between two boundaries. The lower boundary in the CO_2 pressure of seawater is given by

$$CaCO_3(s) + SO_4^{2-} + 2H^+ + H_2O = CaSO_4 \cdot 2H_2O(s) + CO_2(g) \quad \text{(i)}$$

Because gypsum is not a typical constituent of normal marine sediments, it seems unlikely that p_{CO_2} has been less than that given by equation i. The upper boundary of p_{CO_2} is given by the conversion of calcite into dolomite.

$$CaMg(CO_3)_2(s) + 2H^+ = CaCO_3(s) + Mg^{2+} + CO_2(g) + H_2O \quad \text{(ii)}$$

Since in areas of active dolomite precipitation $[Mg^{2+}]/[Ca^{2+}] - 20$, the CO_2 pressure concordant with this condition seems to be near the maximum p_{CO_2} to be expected.

Estimate the range of CO_2 pressure in the atmosphere. The pH is buffered by clays and constant (pH = 8.1).

The equilibrium constant of equations i and ii can be computed from the following information valid at 25°C.

$pK_{so}(CaCO_3) = 8.4; pK_{so}(CaMg(CO_3)_2) = 16.7; pK_{so}(CaSO_4 \cdot 2H_2O) = 4.6; pK(CO_2(g) + H_2O = CO_3^{2-} + 2H^+) = 15.9$, for example, for equation i:

$CaCO_3(s) = Ca^{2+} + CO_3^{2-}$		$\log K_{so} = -8.4$
$Ca^{2+} + SO_4^{2-} + 2H_2O = CaSO_4 \cdot 2H_2O$		$\log(1/K_{so}) = 4.6$
$CO_3^{2-} + 2H^+ = CO_2(g) + H_2O$		$\log K = 18.3$

$CaCO_3(s) + SO_4^{2-} + 2H^+ + H_2O = CaSO_4 \cdot 2H_2O(s) + CO_2(g)$
$$\log K_{(i)} = 14.5$$

---

†For a reassessment of Sillén's model see R. E. McDuff and F. M. Morel: "The Geochemical Control of Seawater (Sillén Revisited)," *Environ. Sci. Technol.*, **14**, 1182 (1980).

Correcting for ionic strength:

$$K'_{(i)} = \frac{p_{CO_2}}{[SO_4^{2-}]\{H^+\}^2} = K_{(i)} f_{SO_4^{2-}}$$

with $f_{SO_4^{2-}}$ in the water $= 0.12$ log $K'_{(i)}$ becomes 13.6. For $[SO_4^{2-}] = 3.8 \times 10^{-2}$ a CO_2 pressure (atm) of log $p_{CO_2} = -4.0$ obtains. Reaction ii has an equilibrium constant of log $K_{(ii)} = 9.7$. Corrected for activity $(K'_{(ii)} = p_{CO_2}[Mg^{2+}]/\{H^+\}^2)$, log $K'_{(ii)} = 10.14$ is obtained. Without oversaturation for $[Mg^{2+}] = 5.4 \times 10^{-2}$ M and pH $= 8.1$, an equilibrium $p_{CO_2} = 10^{-3.7}$ atm is obtained for equation ii. Because of the sluggishness of dolomite precipitation a considerable oversaturation in the reaction

$$2CaCO_3(s) + Mg^{2+} = CaMg(CO_3)_2 + Ca^{2+} \tag{iii}$$

is necessary to form dolomite. The equilibrium constant of equation iii is $K_{(iii)} = 3.1$. Hence the stipulated condition $[Ca^{2+}]/[Mg^{2+}] = 0.05$ represents a 60-fold oversaturation. The CO_2 pressure related to this oversaturation is obtained by considering that $[Ca^{2+}]$ is $\frac{1}{20}$ of the seawater concentration of magnesium $([Mg^{2+}] = 5.4 \times 10^{-2}$ $M)$. p_{CO_2} for this $[Ca^{2+}]$ is then given by the equilibrium

$$CaCO_3(s) + 2H^+ = Ca^{2+} + CO_2(g) + H_2O \qquad \log K = 10.2 \tag{iv}$$

with $[Ca^{2+}] = 2.7 \times 10^{-3}$ M and $f_{Ca^{2+}} = 0.28$, the CO_2 pressure (atm) becomes log $p_{CO_2} = -2.9$. CO_2 pressures of this magnitude or larger would lead to the precipitation of essentially all the Mg^{2+} brought into the sea by rivers as dolomite. These two boundary conditions suggest that p_{CO_2} has been less than $10^{-2.9}$ and more than $10^{-3.7}$ atm.

9.9 CONSTANCY OF COMPOSITION: STEADY STATE

Much of the chemistry of the oceans and of freshwater systems depends on the kinetics of various physical and chemical processes and on biochemical reactions rather than on equilibrium conditions. The simplest model describing systems open to their environment is the time-invariant steady-state model. Because the sea has remained constant for the recent geological past, it may be well justified to interpret the ocean in terms of a steady-state model.

Input is balanced by output in a steady-state system. The concentration of an element in seawater remains constant if it is added to the sea at the same rate that it is removed from the ocean water by sedimentation. Input into the oceans consists primarily of (a) dissolved and particulate matter carried by streams, (b) volcanic hot spring and basalt material introduced directly, and (c) atmospheric inputs. Often the latter two processes can be neglected in the mass balance. Output is primarily by sedimentation; occasionally emission into the atmosphere may have to be considered. Note that the system considered is a single box model of the sea, that is, an ocean of constant volume, constant temperature and pressure, and uniform composition.

The concept of the *residence time* (or the passage time) of an element in the sea, τ, is defined as [43, 44, 45]

$$\tau = \frac{\text{amount in the sea}}{\text{amount supplied per unit time or amount removed per unit time}} \quad (25)$$

or in shorter notation

$$\tau = \frac{M}{\sum J_i} \text{ (time)} \quad (26)$$

where the amount M may be given in the same units (e.g., moles) as the input or removal fluxes J_i are in the same units per unit of time (e.g., mol year^{-1}).

Each of the input or removal fluxes corresponds to a fractional mean residence time τ defined with respect to the particular process:

$$\tau_i = \frac{M}{J_i} \quad \text{(time)} \quad (27)$$

The mean residence time is then given by

$$\frac{1}{\tau} = \frac{1}{\tau_1} + \frac{1}{\tau_2} + \cdots + \frac{1}{\tau_n} \quad (28)$$

Equation 28 shows that the stronger fluxes (smaller fractional residence times) make the contributions of the weaker fluxes insignificant.

If J_{in} and J_{out} represent the input rate and removal rate (fluxes) of an element in the sea, respectively, then

$$\frac{dM}{dt} = J_{in} - J_{out} \quad (29)$$

If one assumes as a first approximation that the removal rate is proportional to the total amount of the element in the sea, that is, $J_{out} = kM$, where k is the rate constant, then at steady state equation 29 becomes

$$0 = J_{in} - kM$$

or

$$k^{-1} = MJ_{in}^{-1} = \tau \quad (30)$$

The inverse of the residence time is equal to the removal rate constant of an element in the sea.

The rate of sedimentation is controlled largely by the rate at which an element is converted (uptake by organisms, precipitation, coprecipitation, ion exchange) into an insoluble, settleable form. Hence the reactivity of the elements influences the time the elements spend, on the average, as constituents of the seawater.†‡ For most elements residence times have been determined on the basis of estimates of the input by runoff from the land or from calculations of sedimentation times. Remarkably similar results are obtained by these two methods. Residence times calculated from the river input)§ are given in Tables 9.6 and 9.7. Elements that are highly oversaturated (e.g., Al, Fe) have short residence times, that is, times that are smaller than those necessary for ocean mixing. On the other hand, elements with low reactivity such as Na and Li have very long residence times.

Steady-State Composition of the Marine Redox System

Kinetics are clearly of overwhelming importance in the control of the marine redox system. An analogy between the earth's surface geochemical system and a giant chemical engineering plant has been suggested by Lotka [46], Siever [47], and others. Figure 9.16 gives Siever's summary of the main regulatory processes. As he points out, the provocative feature of this hypothetical chemical engineering plant is that one can readily see the multiplicity of valves and switches that control the system and the ease with which it may be subject to some kind of perturbation if there are violent movements of any one of the switches. Of particular concern are the gas regulators of the CO_2 and O_2 tanks. The O_2 regulator is governed by photosynthesis and by weathering; the CO_2 regulator is probably controlled much more by the weathering system than by photosynthesis.

Broecker [48] has pointed out that atmospheric oxygen is used up in weathering reactions on a time scale of a few million years. Since the atmosphere has been reasonably constant for a few million years [49], the oxygen used by weathering reactions must have been replaced by the net supply of oxygen produced via photosynthesis. The servomechanism that balances O_2 use and O_2 production is not fully understood, but Broecker [50] has suggested that marine phosphate may be a critical link.

Table 9.9 gives the concentrations of major and some minor redox components in the atmosphere. Nearly 10 % of the electron flux caused by photosynthesis is shunted off for the production of CO and CH_4. The residence times

---

† M. Whitfield [*Mar. Chem.*, **8**, 101 (1979)] has shown that the mean oceanic residence time can be related to an ocean–rock partition coefficient, which in turn is related to the electronegativities of the elements on the assumption that seawater composition is controlled by general adsorption–desorption reactions at surfaces having oxygen donor groups.

‡ A. C. Lasaga [*Geochem. Cosmochim. Acta*, **44**, 815 (1980)] provides a systematic approach to the kinetic treatment of geochemical cycles by extending the concepts of residence times and response times. This extension is particularly useful when complex cycles are involved.

§ Ions such as Cl^- and Na^+ are cycled through the atmosphere—maritime aerosols from wind spray and bursting seawater bubbles are transported to the continents—and returned via rivers. If corrections are made for atmospheric cycling, residence times for Na^+ and Cl^- are obtained that are several times greater than the values given in Table 9.6.

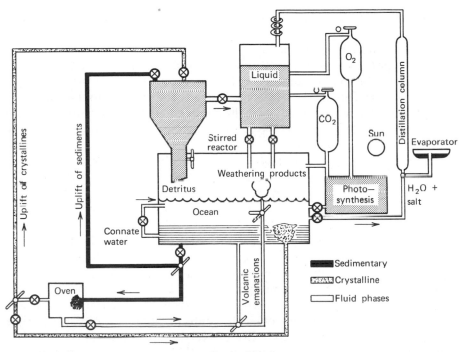

Figure 9.16 Steady-state model of the ocean. The earth's surface geochemical cycle is likened to a chemical engineering plant. (From Siever [47]. Reproduced with the permission of the author and Elsevier Publishing Company.)

TABLE 9.9 COMPOSITION OF ATMOSPHERE AND ESTIMATIONS OF RATE OF VARIOUS REDOX PROCESSES[a]

Atmosphere	Partial Pressure (atm)	Residence Time (Redox) (years)	Electron Flux [eq year^{-1} m^{-2}(sea surface)]^b
N_2	0.78	54×10^6	0.06
O_2	0.21	7000	42
CO_2	0.0003	10	42
CO	10^{-7}	0.06	1.3
CH_4	3×10^{-6}	3	2.7
N_2O	2×10^{-7}	40(?)	~0.01
NH_3	10^{-8}	0.3	~0.01
H_2	5×10^{-7}		

[a] Because the mass transfer of these gases has been many times the current content of this reservoir, these redox components are at a steady state. The residence time of the gases in the atmosphere is estimated by considering redox processes (photosynthesis and respiration for O_2 and CO_2, fixation of N_2 by organisms and civilization) only.

[b] eq = mol electrons

of the gases in the atmosphere—with respect to redox processes—are small in comparison to geological time scales and small in comparison to the time span of an atmosphere with reasonably constant composition ($\sim 600 \times 10^6$ years). Hence all these gases go through—mostly biospheric—redox cycles, and their mass transfer through the atmosphere has been many times the currect content of this reservoir. This implies a reasonable steady-state composition of the redox components of the atmosphere and of the ocean, and that therefore the oxidation states of the materials leaving the ocean are on the average equal to those of material undergoing weathering.

The minor gas components (less than 0.01 %) are by many orders of magnitude less abundant in the atmospheric reservoir than in the other exchange reservoirs. Hence their concentrations in the atmosphere are most likely to be "buffered" by (probably) biological processes in the ocean and soil rather than vice versa. CO_2 is also more abundant (as C) in the hydrosphere and lithosphere than in the atmosphere; its atmospheric concentration is regulated by biochemical (photosynthesis–respiration) and geochemical (weathering) processes, by chemical equilibria in the sea, and more recently also by humans. N_2 has the longest residence time in the atmosphere where it has accumulated with a much higher abundance than in other exchange reservoirs. N_2 is chemically remarkably inert, presumably because of kinetic problems in breaking its strong triple bond. Lightening discharges convert it to NO_3^-, but this process is probably a negligible part of the N cycle. Reduction to NH_3 by microbial N_2 fixation (and by industrial N_2 fixation) and microbial denitrification (reduction of NO_3^- and NO_2^- to N_2) regulate the output and input of N_2 in the atmospheric reservoir.

Global Cycling

One of the possible models of the joint cycle of CO_2 and O_2 (Garrels et al. [49]) is shown in Figure 9.17. This model serves to illustrate the complex chemical interconnections between the major geochemical reservoirs and that they all have an important influence in the atmosphere. The diagram presents the more important geochemical reservoirs that have played a role in the CO_2–O_2 cycle at least during the last 600 million years. As shown in the figure, the oceanic biomass (reservoir 3) takes up CO_2 in photosynthesis; C goes into the building of the biota, and O_2 is released to the atmosphere. (The terrestrial biomass, although it exceeds the standing crop of the oceanic biota by a factor of 100 to 400, is considered a closed cycle on the geological time scale; all the terrestrial organic matter being formed oxidizes and decays, returning CO_2 to the atmosphere. The major avenues of removal of CO_2 and O_2 from the atmosphere are the settling and burial of oceanic biota (F_{34} and F_{45}) and precipitation of $CaCO_3$ (F_{47}). The ocean (reservoir 4) is assumed to be in equilibrium with the $CaCO_3$ sediments (reservoir 7). This reservoir is capable of adding and substracting CO_2.

$$Ca^{2+} + 2HCO_3^- \rightleftharpoons CaCO_3 + H_2O + CO_2 \qquad (31)$$

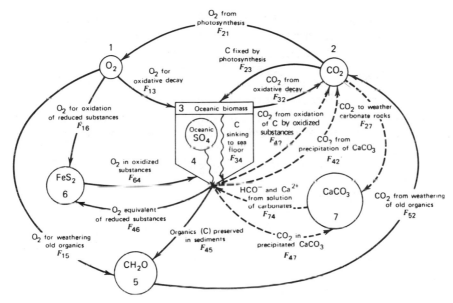

Figure 9.17 Major reservoir and fluxes of the exogenic cycles of CO_2 and O_2. Dashed arrows indicate equilibria in the CO_2 system. The C and O_2 contents of the geochemical reservoirs and the fluxes and residence times (as assessed by Garrels et al. [49]) are indicated below the figure. (From Garrels, Lerman, and Mackenzie [49]. Reproduced by permission of *American Scientist*.)

Reservoir Masses, Fluxes, and Residence Times

Reservoir (number)	Content M_i (mol)	Flux F_{ij} (10^{12} mol year^{-1})	Residence time τ_{ij} (years)
1. Atmosphere O_2	3.8×10^{19}	$F_{13} = 2496.5$	$\tau_{13} = 1.52 \times 10^4$
		$F_{15} = 2.5$	$\tau_{15} = 1.52 \times 10^7$
		$F_{16} = 1.0$	$\tau_{16} = 3.8 \times 10^7$
2. Atmosphere CO_2	5.5×10^{16}	$F_{23} = F_{21} = 2500$	$\tau_{23} = 22$
3. Oceanic biota C	6×10^{14}	$F_{32} = 2496.5$	$\tau_{32} = 0.24$
		$F_{34} = 3.5$	$\tau_{34} = 170$
4. Ocean water C			
C (as CO_2 and HCO_3^-)	3.234×10^{18}		
O_2 (as SO_4^{2-})	8.4×10^{19}		
Ca	1.4×10^{19}		
		(CO_2) $F_{42} = 1.0$	
		(O_2 equivalent) $F_{46} = 1.0$	
		(C) $F_{45} = 2.5$	
5. Sedimentary organics C	1×10^{21}	$F_{52} = 2.5$	$\tau_{52} = 4 \times 10^8$
6. Reduced substances (FeS$_2$ as O_2 demand equivalent)	4×10^{20}	$F_{64} = F_{16} = 1.0$	$\tau_{64} = 4 \times 10^8$

579

A major part of the oceanic biomass is oxidized (F_{32}), returning CO_2 to the atmosphere, but a small fraction sinks through the deep ocean (F_{34}). Some of the settling and settled material is oxidized (bacterial mediation) by SO_4^{2-}:

$$2CH_2O + SO_4^{2-} + 2H^+ \rightleftharpoons 2CO_2 + 2H_2O + H_2S \qquad (32)$$

The CO_2 formed is returned to the atmosphere (F_{42}). The remaining fraction of organic material becomes part (F_{45}) of the sedimentary record. The sulfide produced in equation 32 is added to the reservoir of reduced substances, collectively represented in Figure 9.17 as pyrite (FeS_2, reservoir 6). Finally, atmospheric O_2 is consumed in the oxidation of reduced minerals (F_{16}, F_{64}) and sedimentary organic matter (F_{15} and F_{53}).

The fluxes listed describe a balanced CO_2–O_2 cycle. Garrels et al. [49] has considered mathematically various perturbations of the steady state [increased rate of erosion, increased rates of photosynthesis, and cessation of productivity (doomsday)]. These investigations demonstrate that, in the ecosphere, effective feedback mechanisms may have protected the earth's surface environment from severe perturbations.

Lakes

In many regards the situation in lakes is even more complicated than in the sea [51]. Some of the factors regulating the concentration of lake components are given in Figure 9.18.

As we have seen, the residence time of an element or a compound E (equation 26) is defined by

$$\tau_E = \frac{[E]V}{[E]_{in} q} \qquad (33)$$

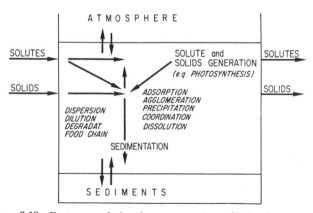

Figure 9.18 Factors regulating the concentrations of Lake Components.

where [E] and $[E]_{in}$ are the total concentration (e.g., mol liter^{-1}) in the system and in the inflow, respectively, V is the volume of the lake (e.g., liters), and q is the rate of inflow (e.g. liters year^{-1}). In lakes it is often convenient to define under idealized conditions for a one-box model of a well-mixed lake a relative residence time τ_{rel} that is a time relative to the residence time of water, $\tau_W = qV^{-1}$,

$$\tau_{rel} = \frac{\tau_E}{\tau_W} = \tau_W^{-1} \frac{[E]V}{\text{amount supplied per time}} = \frac{[E]}{[E]_{in}} \qquad (34)$$

where the amount supplied per unit of time ($=$ amount removed per unit of time by sedimentation, outflow, and other reactions) is given by $q[E]_{in}$. The relative residence time of an element or compound decreases with increasing reactivity (incorporation into biomass, suspended solids, or precipitates [31,52]).

Most substances entering lakes are nonconservative (i.e., have a residence time different from that of water). All these substances that are strongly adsorbed on suspended matter or become incorporated into a settling biomass have small relative residence times. The residence time of a substance or element is regulated by the loss through the outlet and the loss through nonhydrological processes of elimination (e.g., sedimentation):

$$\frac{1}{\tau_E} = \frac{1}{\tau_W} + \frac{1}{\tau_S} \qquad (35)$$

where τ_S is the residence time for the nonconservative pathway (e.g., the residence time of the settling constituent). Relative residence times for Na$^+$ and Cl$^-$ are very close to 1.0. The rates of removal of phosphorus to lake floor sediments are of the order of 25 to 50% of the input load ($\tau_{rel} = 0.5$ to 0.75) under aerobic conditions. ^{137}Cs and heavy-metal ions behave in a nonconservative way; they are strongly adsorbed on suspended materials and have residence times approaching those of suspended matter ($\tau_{rel} = 0.1$ to 0.6). The flux of materials into and out of the sediments usually accounts for a significant fraction of the lake budget [51–53].

The equations for mass balance and steady state for one-box lake models are summarized in Table 9.10. Application of such a model to the removal of heavy metals will be illustrated in Example 11.8 and Table 11.13. Application of one-box and two-box models (stratified lakes) to nutrients has been reviewed by Imboden and Lerman [52].

In most lakes the input rates of many substances have increased, and consequently concentrations of many constituents are not time-invariant. Nevertheless steady-state models give useful limits for comparison purposes, that is, concentrations that would be obtained if the input rate were kept constant for a time period of ca. $3\tau_E$. The shorter the residence time, the faster the system adjusts to steady state after the onset of a new constant input rate.

TABLE 9.10 MASS BALANCE AND STEADY STATE FOR ONE-BOX LAKE MODELS[a]

Changes in the chemical composition of a lake

$$= \text{input rates} - \frac{\text{removal rate}}{\text{through outflow}} - \frac{\text{removal rate into sediments}}{\text{or through other reactions}}$$

$$V\frac{d[E]}{dt} = Q_{in}[E]_{in} - Q_{out}[E] - BV \quad (\text{mol year}^{-1}) \tag{1}$$

or, if $Q_{in} = Q_{out}$:

$$\frac{d[E]}{dt} = \frac{1}{\tau_W}([E]_{in} - [E]) - B \quad (M \text{ year}^{-1}) \tag{2}$$

at steady state:

$$[E]_{ss} = [E]_{in} - B\tau_W \quad (M) \tag{3}$$

The retention factor r (ratio of removal rate by sedimentation or other reactions to total input rate) is defined by

$$r = \frac{BV}{[E]_{in}Q_{in}} \tag{4}$$

which at steady state becomes

$$r_{ss} = 1 - [E]_{ss}[E]_{in}^{-1} = B\tau_E[E]_{in}^{-1} \tag{5}$$

If the removal function B is assumed to be a linear function of $[E]$:

$$B = k[E] + B_0 \quad (M \text{ year}^{-1}) \tag{6}$$

Equation 2 becomes

$$\frac{d[E]}{dt} = \left(\frac{[E]_{in}}{\tau_W} - B_0\right) - \left(k + \frac{1}{\tau_W}\right)[E] \tag{7}$$

For constant coefficients the solution is

$$[E]_t = ([E]_0 - [E]_\infty)\exp\left[-\left(k + \frac{1}{\tau_W}\right)t\right] + [E]_\infty \tag{8}$$

$$[E]_\infty = \frac{[E]_{in}\tau_W^{-1} - B_0}{k + \tau_W^{-1}} \tag{9}$$

---

[a] V = Volume of lake (liters); Q_{in}, Q_{out} = rate of total water inflow or outflow, respectively (liters year^{-1}); $[E]$, $[E]_{in}$, $[E]_{ss}$, $[E]_\infty$ = concentration of an element, (M) in the lake, in the inflow, at steady state and at infinite time, respectively; ($[E]_{in}$ may represent a mean concentration: $[E]_{in} = (\Sigma \text{ input rates}) Q_{in}^{-1}$); B = net removal function $(M \text{ year}^{-1})$ for incorporation into sediments, e.g., by gas exchange, radioactive decay or chemical reactions; τ_W = residence time of water = VQ_{in}^{-1} (years); r, r_{ss} = retention factor and retention factor at steady state as defined by (4) and (5); k = first-order rate constant (year^{-1}), B_0 = zero-order rate constant $(M \text{ year}^{-1})$.

582

9.10   THE SEDIMENT–WATER INTERFACE

The sediments are not just depositories for the material removed from the ocean or lake water.

The flux of constituents from the sediments into the water, and vice versa, is important in controlling the composition of oceans and lakes [51–63].

The diagenetic† chemical reactions occurring within the sediments consist of abiotic and biogenic reactions [55]. Because of these reactions sediments exert a significant effect on the overlying ocean and lake waters (Figure 9.19). Most biogenic reactions depend on the decomposition of organic matter. The sequence of redox reactions occurring in the sediments is the same as that already discussed (see Table 7.6 in Section 7.5) for the interaction of excess organic matter with O_2, NO_3, SO_4^{2-}, and HCO_3^-, that is, the removal of dissolved O_2, the reduction of NO_3, SO_4^{2-}, and HCO_3^-, and the production of CO_2, NH_4^+, phosphate, HS , and CH_4 (Figure 9.19a).

The products of these reactions may in turn bring further changes in sediment chemistry such as the solubilization of iron and manganese after O_2 has been removed. HS^- resulting from SO_4^{2-} reduction, on the other hand, may react with detrital iron minerals to form iron sulfides. Excess HCO_3^- is produced by SO_4^{2-} reduction and NH_4^+ formation; eventually $CaCO_3$ may be precipitated. Build-up of dissolved phosphate may under suitable conditions bring about the precipitation of apatite. Mg^{2+} may be precipitated as a result of the removal of clay minerals of iron which reacts with HS^- to form iron sulfides. Reactions that are not biogenically controlled include the dissolution of opaline silica, $CaCO_3$, and feldspars, and various ion-substitution reactions in sediment minerals and ion-exchange processes on clay minerals.

The possible importance of reactions between seawater and the underlying basalt has been recognized [58,59,61,62]. The alteration of deeply buried volcanic rocks may also make a contribution of dissolved material to the ocean, at least for Ca^{2+}. This alteration procees may also form a sink for Mg^{2+} and Na^+ and possibly K^+ and CO_2 [58].

Fluxes of Solids, Water, and Solutes

Sedimentation of solid particles and entrapment of water in the pore spaces are two major fluxes of materials across the sediment–water interface.

Further processes responsible for transport across the sediment–water interface are upward flow of (pore) water caused by hydrostatic pressure gradients of groundwater in aquifers or land; molecular diffusional fluxes in pore water; and mixing of sediment and water at the interface (bioturbation and water turbulence). The rates of sediment deposition vary from mm per 1000 years in the pelagic ocean up to cm year^{-1} in lakes and near-shore oceanic

† Diagenesis refers to changes that take place within a sediment during and after burial [cf. Berner].

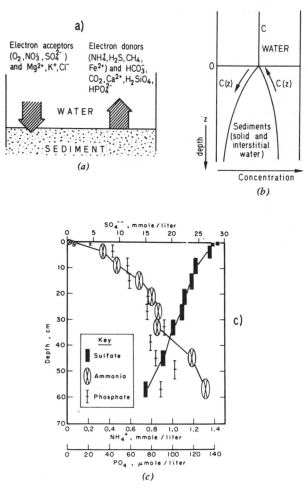

Figure 9.19 The sediment–water interface. (*a*) Direction of fluxes expected for dissolved constituents between sediment pore waters and the overlying waters (oceans and lakes). (*b*) For sediments and pore water, the one-dimensional distribution of concentrations is time- and depth-dependent. Arrows indicate fluxes at the sediment–water interface depending on the concentration gradient in pore water. The overlying water (ocean or lakes) is assumed to be well mixed (cf. Imboden and Lerman [52]). (*c*) Sulfate, phosphate, and ammonia versus depth in pore waters from Santa Barbara Basin, California. [From Sholkovitz, *Geochim. Cosmochim. Acta*, **37**, 2043 (1973).]

areas. A sediment can be thought of as made up of equal volumes of water and solid particles (sediment volume fraction = 0.5) of density 2.5 g cm^{-3} [44].

Since ocean water contains ca. 35 g liter^{-1} and fresh water ca. 0.01 g liter^{-1} dissolved material, the mean fluxes to the sediments are in the following range: solids = 6 × 10^{-4} to 6 × 10^{-1} g cm^{-2} year^{-1}; dissolved materials = 0.1 × 10^{-4} to 0.1 × 10^{-1} cm^{-2} year^{-1} [44]. In freshwater lakes dissolved material fluxes are closer to the lower value.

The net flux of a chemical species across the sediment–water interface, $F_{z=0}$, is due to (molecular) diffusion in pore water, $F_d = -\phi D(dC/dz)$, the flux due to advection in pore water, $F_a = \phi UC$, and the flux due to deposition of solid particles, $F_s = \phi U_s C_s$:

$$F_{z=0} = \phi\left(-D\frac{dC}{dz} + UC + U_s C_s\right) \qquad (36)$$

where C and C_s = concentration in solution and in solids, respectively, (units of mass per unit of pore water), z = depth (positive and increasing downward from the sediment–water interface), and U and U_s = rates of pore water advection and rate of sedimentation, respectively (cm year^{-1}). U is positive when the sediment and water flow are downward relative to $z = 0$; ϕ = porosity (volume fraction of sediment occupied by water).†

Two simplifying assumptions can be introduced into equation 36. (1) If the sediment porosity ϕ changes little with depth, then the sediment particles and pore water do not move relative to each other in a growing sediment column and we can set U equal to the sedimentation rate $U = U_s$. (2) A relationship between the concentration in pore water, C, and the concentration in the solids, C_s can be established. Under simplifying assumptions a linear relationship $C_s = KC$ may be assumed, where K is a (dimensionless) distribution coefficient. Thus in noncompacting sediments the general flux equation 55 may be written

$$F_{z=0} = \phi\left[-D\frac{dC}{dz} + U(K + 1)C\right]_{z=0} \qquad (37)$$

The larger K is, the more important the flux of a chemical species or settling particles, UKC, in comparison to the flux in pore water, UC. In compacting sediments the total flux become [55]

$$F_{z=0} = \phi\left[-D\frac{dC}{dz} + UC\left(\frac{K\phi(1 - \phi_\infty)}{\phi_\infty(1 - \phi)} + 1\right)\right] \qquad (38)$$

where ϕ_∞ is the porosity at a given depth when it has attained a steady value smaller than ϕ, the porosity at the interface.

For the calculation of the fluxes the following parameters are needed: porosity, sedimentation rate, diffusion coefficients, distribution constants, and the concentration gradients for each species $(dC_i/dz)_{z=0}$. In assessing diffusion coefficients coupling effects due to electroneutrality between coions and counterions may have to be considered. Serious errors may be introduced by using wrong values of (dC_i/dz), because it is very difficult to collect undisturbed sediment

---

† An observer balanced on the sediment–water interface ($z = 0$) as sediment particles continue to arrive from above and pile up will see the particles and pore water flow by in the downward direction. In this sense one can always speak of the fluxes of solids, waters, and solutes as moving up or down [44].

samples near the interface. Sharp gradients in the top few centimeters may be overlooked. Some of these problems may be circumvented by measuring the fluxes directly either *in situ* [64] or in the laboratory using carefully obtained cores overlain by water of controlled composition and hydrodynamic conditions.

In seawater where the sedimentation rates are small (< 5 cm/10^3 year) the advective or burial term in equations 37 and 38 for most species (with the possible exception of Na^+) is small in comparison to the diffusive term. Manheim's [57] results indicate that the diffusive fluxes between deep sea sediments and the ocean are on the same order of magnitude as river inputs. This is confirmed by the evaluation of Sayles [63] who shows that the fluxes of the major cations and HCO_3^- are similar in magnitude to those of river inputs. In the case of Mg^{2+} and K^+ these processes are a major part of the geochemical cycles of these elements. For Ca^{2+}, Mg^{2+}, K^+, and HCO_3^- 90 % of the fluxes across the interface are due to reactions at sediment depths of less than 100 cm.

9.11 BIOLOGICAL REGULATION OF THE COMPOSITION

We have already shown in the previous section that life—by cycling and transporting chemical elements—plays an important role in maintaining the earth in a homeostatic† state. Alfred J. Lotka [46] wrote more than 50 years ago

If we are satisfied to omit innumerable details, we can trace, for each of the most important chemical elements concerned, the broad outline of its cycle in nature. The elements and simple compound principally concerned are Carbon (CO_2), Oxygen (O_2), Nitrogen (N_2, NH_3, NO_2^-, NO_3^-), Water (H_2O), Phosphorus (PO_4^{3-} etc.).

Brief consideration will presently be given to each of these cycles in turn. For the drama of life is like a puppet show in which stage, scenery, actors and all are made of the same stuff. The players, indeed, 'have their exits and their entrances', but the exit is by way of translation into the substance of the stage; and each entrance is a transformation scene. So stage and players are bound together in the close partnership of an intimate comedy; and if we would catch the spirit of the piece, our attention must not all be absorbed in the characters alone, but must be extended also to the scene, of which they are born, on which they play their part, and with which, in a little while, they merge again.

A Kinetic Model for the Chemical Composition of Seawater

Obviously the composition of natural waters is markedly influenced by the growth, distribution, and decay of phytoplankton and other organisms. The dominant role of organisms in regulating the oceanic composition and its variation with depth of some of the important sea salt components (i.e., C, N, P, Si) will be illustrated here by introducing certain aspects of Broecker's kinetic

---

† The term homeostatic has been used to indicate constancy maintained by negative feedback.

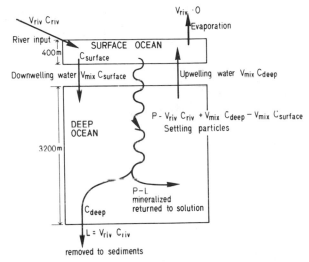

Figure 9.20 Broecker's [33] idealized kinetic model for the marine cycles of biologically fixed elements. V_{riv}, volume of river water entering the ocean per year expressed as volume per unit sea area (m³ m⁻² year⁻¹ or m year⁻¹) — 0.1 m year⁻¹; C_{riv}, concentration of an element in average river water (mol m⁻³); $V_{riv} \times C_{riv}$, input flux (mol m⁻² year⁻¹); V_{mix}, volume of water sinking into deep water box — volume of water rising to surface water box (volume m⁻² year⁻¹) = 20 m year⁻¹; $C_{surface}$, C_{deep}, concentration in surface ocean, in deep ocean (e.g., mol m⁻³); $V_{mix} \times C_{surface}$, flux (mol m⁻² year⁻¹) of an element that sinks from the ocean surface into the deep ocean (mol m⁻²).

model for the chemical composition of seawater [50].† We summarize Broecker's line of arguments.

The one process that yields geographic and depth variation in the chemical composition of sea salt components is uptake of dissolved constituents by organisms. The remains of these organisms sink under the influence of gravity and are gradually destroyed by oxidation. The superposition of this particular cycle upon the ordinary mixing cycle in the sea accounts for the present distribution of chemical properties (Figure 9.13).‡

Broecker's model is shown in Figure 9.20 (cf. Figure 4.11 and Example 4.9). The ocean is divided into two boxes—an upper one of a few hundred meters depth and a lower one of 3200-m depth. The zone separating the upper warmer water body from the lower cooler water body is a density gradient (the thermocline) which provides an obstruction to mixing. It is furthermore assumed that the only way an element is added is by river runoff from the continents and the

† The reader will find a clear and simple access to this model and a broad discussion on the interaction between mixing cycles and biological cycles in Broecker [33].
‡ Recent papers concerned with removal mechanism of elements of elements from the ocean include: Y. H. Li, "Mass Balance and Geochemical Cycles of Elements," in *Chemical Cycles in the Evolution of the Earth* (in press); D. W. Spencer, et al., "Chemical Fluxes from a Sediment Trap Experiment in the Deep Sargasso Sea," *J. Mar. Res.*, **36**, 493 (1978).

only way an element is removed from the ocean is by the fall of organism-produced particles to the sea floor, which are then permanently buried in the sediments.

The ocean and its two compartments are assumed to be at steady state. The warm surface ocean receives its supply of any given element (1) from the water entering from the rivers and (2) from the deep ocean which is steadily being exchanged with surface water. Downwelling and particulate settling match these two inputs, so that the concentration of any element remains constant in the surface ocean. The amount of a given element entering the surface reservoir each year is then $V_{mix} C_{deep} + V_{riv} C_{riv}$. The return flow to the deep sea carries away an amount of the element equal to $V_{mix} C_{surf}$. Material balance requires that the remainder, $P = V_{mix} C_{deep} + V_{riv} C_{riv} - V_{mix} C_{surf}$, be carried by falling particles (see Figure 9.20).

The yearly amount of runoff from the continents, V_{riv}, is equal in volume to a layer 10 cm thick over the entire ocean surface. As illustrated in Example 4.9, the mean residence time of water in the deep ocean—as calculated from the distribution of ^{14}C—is on the order of 1600 years. Since the depth of the deep ocean is ca. 3200 m, the upwelling rate (which must equal the downwelling rate) is ca. 2 m year^{-1}. Thus $V_{mix}/V_{riv} \simeq 20$ (for an explanation of symbols see Figure 9.20).

Defining the fraction of an element removed from the surface ocean by these particles, g,

$$g = \frac{\text{particle flux of an element into deep ocean}}{\sum \text{fluxes of this element into surface ocean}}$$

we have

$$g = \frac{V_{riv} C_{riv} + V_{mix} C_{deep} - V_{mix} C_{surf}}{V_{riv} C_{riv} + V_{mix} C_{deep}}$$

$$= 1 - \frac{V_{mix} C_{surf}/V_{riv} C_{riv}}{1 + V_{mix} C_{deep}/V_{riv} C_{riv}} \tag{39}$$

For an element such as phosphorus C_{deep}/C_{riv} is about 5 and C_{surf}/C_{riv} is 0.25. The corresponding value of g is 0.95; that is, 95% of the phosphate introduced into surface ocean is carried away by falling particles.

The fraction f of a given element carried to the deep sea by the particulate flux P surviving destruction,

$$f = \frac{\text{export } (L) \text{ into sediments}}{\text{import of particles } (P) \text{ into deep sea}}$$

must be equal to the river input $fP = V_{riv} C_{riv}$:

$$f = \frac{L}{P} = \frac{V_{riv} C_{riv}}{V_{riv} C_{riv} + V_{mix} C_{deep} - V_{mix} C_{surf}}$$

$$= \frac{1}{1 + V_{mix}/V_{riv}(C_{deep}/C_{riv} - C_{surf}/C_{riv})} \tag{40}$$

For phosphorus, for example, $f = 0.01$; that is, about 99% of the particulate phosphorus reaching the deep sea is oxidized to phosphate and is recycled; 1% is buried in the sediments.

fg gives the fraction of an element that is removed per oceanic mixing cycle.

$$fg = \frac{\text{export } (L) \text{ into the sediments}}{\sum \text{fluxes into surface ocean}}$$

$$fg = \frac{1}{1 + V_{mix} C_{deep}/V_{riv} C_{riv}} \tag{41}$$

For phosphorus, $fg \simeq 0.01$. Thus, only about 1% of the phosphorus entering the surface of the sea is removed to the sediments.

τ is the residence time of an element; that is, the time required to remove an amount of an element equal to that stored in the sea today, while $T_{mix} = (V_{mix}/\text{depth of the deep sea}) \simeq 1600$ years. Thus

$$\tau = \frac{T_{mix}}{fg} \tag{42}$$

Since phosphate is removed at the rate of 1% during every mixing cycle ($fg = 0.01$), the average life time of a P atom in the ocean must be about 1600 years. Broecker [33] describes this in the following way:

In words, a typical P atom, upon release from some sedimentary rock by erosion, is carried by rivers to the sea. It then goes through an average of 100 oceanic mixing cycles: 100 times it is fixed by an organism in surface water and becomes part of a particle that sinks and is destroyed in the deep sea. Each time, it waits in the dark abyss about 1600 years before being sent back to the surface. On the average, during the hundredth cycle, the particle bearing the P atom survives destruction and is trapped in the sediment. So our P atom makes 100 round trips of 1600 years each during its stay in the ocean. It then becomes part of the sediment, where it remains for several hundred million years until it is uplifted and exposed again to continental erosion. The life of a typical P atom is indeed bleak. It spends 99.9 percent of its time trapped in the sedimentary rocks of the earth; that is, out of every 200,000,000 years it has only one 160,000 year stint in the ocean. Since the warm layer of surface water is very thin and the time required for particulate loss is small, the P atom spends most of its time in the ocean in the cold dark abyss; for every 1600-year mixing cycle, it spends about four years in the surface water!

Table 9.11 gives a calculation summary for some examples of biolimiting, biointermediate, and biounlimited elements. For biolimiting elements g is close to unity and $f < 0.1$; for biounlimiting elements g must be very small ($g < 0.01$).

The mass balance considerations and calculations illustrate how concentrations of these elements are controlled. Rearranging equation 40, we have

$$\frac{C_{deep}}{C_{riv}} = \frac{1 - fg}{fg} \frac{V_{riv}}{V_{mix}} \tag{43}$$

TABLE 9.11   CALCULATION SUMMARY FOR THE ELEMENTS PHOSPHORUS, SILICON, BARIUM, CALCIUM, SULFUR, AND SODIUM[a]

Category	Element	$\dfrac{C_{\text{surface}}}{C_{\text{river}}}$	$\dfrac{C_{\text{deep}}}{C_{\text{river}}}$	g	f	$f \times g$	τ (years)
Biolimiting	P	0.25	5	0.95	0.01	$\dfrac{1}{100}$	2×10^5
	Si	0.05	1.6	0.97	0.03	$\dfrac{1}{300}$	6×10^5
Biointermediate	Ba	0.20	0.60	0.70	0.11	$\dfrac{1}{13}$	3×10^4
	Ca	30.0	30.3	0.01	0.16	$\dfrac{1}{500}$	1×10^6
Biounlimited	S	5000	5000	—	—	$\dfrac{1}{10,000}$	2×10^7
	Na	50,000	50,000	—	—	$\dfrac{1}{100,000}$	2×10^8

[a] From Broecker [33].

or, for a biolimiting element,

$$\frac{C_{\text{deep}}}{C_{\text{riv}}} \simeq \frac{1}{f}\frac{V_{\text{riv}}}{V_{\text{mix}}} \simeq \frac{1}{20f} \tag{44}$$

and, for a biounlimiting element,

$$\frac{C_{\text{deep}}}{C_{\text{riv}}} \simeq \frac{1}{fg}\frac{V_{\text{riv}}}{V_{\text{mix}}} \tag{45}$$

and

$$\frac{C_{\text{surf}}}{C_{\text{deep}}} \simeq 1 \tag{46}$$

As shown by equation 44, the phosphorus concentration dissolved in the sea is controlled by (1) the upwelling rate of deep seawater, (2) the fraction of particles falling to the deep sea that survive oxidation, (3) the phosphorus content of average river water, and (4) the rate of continental runoff.

CaCO₃ Precipitation and Dissolution.   Ca may serve as example for application of the Broecker model of a biointermediate element. The warm surface seawater is several times oversaturated with respect to calcite and aragonite (cf. Example

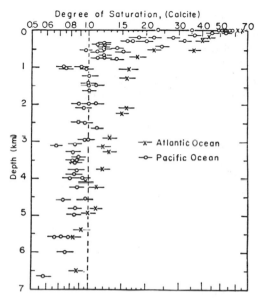

Figure 9.21   Degree of saturation of calcite as a function of depth in the Atlantic and Pacific Oceans. [From Y. H. Li, T. Takahashi, and W. S. Broecker, *J. Geophys. Res.*, **74**, 5507 (1969).

5.3). Spontaneous nucleation does not occur under these conditions in the sea (e.g., inhibition by Mg^{2+}, Section 5.9).

Unlike that in lakes, removal is entirely by organisms. Particular $CaCO_3$ becomes dissolved in the deep ocean because of the CO_2 released by the oxidation of biogenic debris (Figure 9.21). The ratio of $[Ca^{2+}]$ in surface seawater to that in the deep sea is ca. 0.99. Table 9.11 shows the value for f and g. The average Ca atom resides in the sea for 500 mixing cycles or 8×10^5 years. Oceanic sediments found at depths below 4500 m are nearly free of $CaCO_3$, while those at depths less than 3000 m are usually more than 70% by weight $CaCO_3$. Although there is some kinetic inhibition of $CaCO_3$ dissolution in waters undersaturated with respect to $CaCO_3$, this difference can be explained plausibly. The $CaCO_3$ fragments falling into the deepest part redissolve, while those falling into shallower parts are preserved.

REFERENCES

1   L. G. Sillén, in *Oceanography*, M. Sears, Ed., American Association for the Advancement of Science, Washington, D.C., 1961, p. 549.

2   S. N. Davies and R. C. M. DeWiest, *Hydrogeology*, Wiley, New York, 1966.

3   K. Rankama and Th. G. Sahama, *Geochemistry*, University of Chicago Press, Chicago, 1950.

4   O. P. Bricker, N. H. Nesbitt, and W. D. Gunten, *Amer. Mineral.*, **58**, 64 (1973).

5   O. P. Bricker and R. M. Garrels, in *Principles and Applications of Water Chemistry*. S. D. Faust and S. V. Hunter, Eds., Wiley, New York, 1967, p. 449.

6   J. H. Feth, C. E. Roberson, and W. L. Polzer, U.S. Geological Survey Water Supply Papers, No. 1535I, Washington, D.C., 1964.

7   R. M. Garrels and F. T. Mackenzie, in *Equilibrium Concepts in Natural Water Systems*, Advances in Chemistry Series, No. 67, American Chemical Society, Washington, D.C., 1967.

8a W. R. Miller and J. I. Drever, *Geochim. Cosmochim. Acta*, **41**, 1963 (1977).

8b W. C. Graustein, K. Kromack, and P. Sollins, *Science*, **198**, 1252 (1977); *Abstr. Geol. Soc. Amer.*, **8**, 891 (1976).

9   R. Wollast, *Geochim. Cosmochim. Acta*, **37**, 2641 (1973).

10 T. Paces, *Geochim. Cosmochim. Acta*, **31**, 635 (1967).

11 H. C. Helgeson, *Geochim. Cosmochim. Acta*, **35**, 421 (1971).

12 E. Busenbeng, and C. V. Clemency, *Geochim. Cosmochim. Acta*, **40**, 41 (1976).

13a R. Petrović, R. A. Berner and M. B. Goldhaber, *Geochim. Cosmochim. Acta*, **40**, 1509 (1976).

13b G. R. Holdren and R. A. Berner, *Geochim. Cosmochim. Acta*, **43**, 1161 (1979); R. A. Berner and G. R. Holdren, *ibid.*, **43**, 1173 (1979).

14a H. D. Holland, *The Chemistry of the Atmosphere and Oceans*, Wiley-Interscience, New York, 1978.

14b Y. Tardy and R. M. Garrels, *Geochim. Cosmochim. Acta*, **40**, 1051 (1976); **41**, 87 (1977).

15 S. V. Mattigod and G. Sposito, *Geochim. Cosmochim. Acta*, **42**, 1753 (1978).

16 J. O. Nriagu, *Amer. Mineral.*, **60**, 834 (1975).

17 C. H. Chen, *Amer. J. Sci.*, **275**, 801 (1975).

18 G. Lagerstrom, *Acta Chem. Scand.*, **13**, 722 (1959).

19 R. Siever, *Amer. Mineral.*, **42**, 826 (1957).

20 N. Ingri, *Acta Chem. Scand.*, **13**, 758 (1959).

21 M. L. Jackson, S. A. Tyler, A. C. Willis, G. A. Burbeau and R. P. Pennington, *J. Phys. Colloid Chem.*, **52**, 1237 (1948).

22 R. M. Garrels and C. L. Christ, *Solutions, Minerals, and Equilibria*, Harper and Row, New York, 1965

23 W. Stumm and E. Stumm-Zollinger, *Chimia*, **22**, 325 (1968).

24 H. C. Helgeson, *Geochim. Cosmochim. Acta*, **32**, 853 (1968).

25 D. C. Thorstenson, *Geochim. Cosmochim. Acta*, **34**, 745 (1970).

26 B. Fritz and Y. Tardy, *Sci. Geol. Bull.*, **26**, 339 (1973).

27a A. Drouby et al., *Sci. Geol. Bull.*, **29**, 45 (1976).

27b D. E. White, J. D. Hem, and G. A. Waring, *Chemical Composition of Subsurface Waters*, U.S. Geological Survey Professional Paper, No. 440-F, Washington, D.C., 1963.

27c J. D. Hem, *Study and Interpretation of Chemical Characteristics of Natural Waters*, U.S. Geological Survey Water Supply Paper No., 1473, G.P.O., Washington, D.C., 1970.

27d R. M. Garrels and F. T. Mackenzie, in *Equilibrium Concepts in Natural Water Systems*, Advances in Chemistry Series, No. 67, American Chemical Society, Washington, D.C., 1967.

28   A similar analysis has been made for the water of the Great Lakes by J. R. Kramer (in *Equilibrium Concepts in Natural Water Systems*, Advances in Chemistry Series, No. 67, American Chemical Society, Washington, D.C., 1967), p. 243.

29   K. Kelts and K. J. Hsü, in *Lakes: Chemistry, Geology, Physics*, A. Lerman, Ed., Springer, New York, 1978, p. 295.

30   II. P. Eugster and L. A. Hardie, in *Lakes: Chemistry, Geology, Physics*, A. Lerman, Ed., Springer, New York, 1978; p 237.

31   Redfield, A. C., B. H. Ketchum and F. A. Richards, in *The Sea*, Vol. II, M. N. Hill, Ed., Wiley-Interscience, New York, 1966, p. 26.

32   Stumm, W., and J. J. Morgan, in *Transactions of the 12th Annual Conference on Sanitary Engineering*, University of Kansas Press, Lawrence, 1962, p. 16.

33   W. S. Broecker, *Chemical Oceanography*, Harcourt Brace Jovanovich, New York, 1974.

34   F. A. Richards, in *Chemical Oceanography*, J. P. Riley and G. Skirrow, Eds., Academic, New York, 1965, Chapter 13.

35   L. G. Sillen, in *Equilibrium Concepts in Natural Water Systems*, Advances in Chemistry Series, No. 67, American Chemical Society, Washington, 1967, p. 57.

36   V. M. Goldschmidt, *Fortschr. Mineral. Krist. Petrol.*, **17**, 112 (1933).

37   H. D. Holland, *Proc. Nat. Acad. Sci. U.S.*, **53**, 1173 (1965).

38   F. T. Mackenzie and R. M. Garrels, *Amer. J. Sci.*, **264**, 507 (1966).

39   R. M. Garrels and F. T. Mackenzie, *Evolution of Sedimentary Rocks*, Norton, New York, 1971.

40   K. L. Russell, *Geochim. Cosmochim. Acta*, **34**, 893 (1970).

41   H. D. Holland, *Geochim. Cosmochim. Acta*, **36**, 637 (1972).

42   H. D. Holland, *Proc. Nat. Acad. Sci. U.S.*, **53**, 1173 (1965).

43   J. F. Barth, *Theoretical Petrology*, Wiley, New York, 1952.

44   A. Lerman, *Geochemical Processes*, Wiley-Interscience, 1979.

45   Y. H. Li, *Geochim. Cosmochim. Acta*, **41**, 555 (1977).

46   J. A. Lotka, *Elements of Mathematical Biology*, 1924, reprinted by Dover, New York, 1956.

47   R. Siever, *Sedimentology*, **11**, 5 (1968).

48   W. S. Broecker, *Science*, **168**, 1537 (1970).

49   R. M. Garrels, A. Lerman, and F. T. Mackenzie, *Amer. Sci.*, **64**, 306 (1976).

50   W. S. Broecker, *Quatern. Res.*, **1**, 188 (1971).

51   A. Lerman, *Hydrol. Sci. Bull.*, **19**, 25 (1974).

52   D. M. Imboden and A. Lerman, in *Lakes: Chemistry, Biology, Physics*, A. Lerman, Ed., Springer, New York, 1978; p. 341.

53   C. H. Mortimer, *Limnol. Oceanogr.*, **16**, 387 (1971).

54   R. A. Berner, *Principles of Chemical Sedimentology*, McGraw-Hill, New York, 1971.

55   R. A. Berner, in *The Benthic Boundary Layer*, I. N. McCave, Ed., Plenum, New York, 1976, p. 33.

56   F. L. Sayles and F. T. Manheim, *Geochim. Cosmochim. Acta*, **39**, 103 (1971).

57   F. T. Manheim, in *Chemical Oceanography*, J. P. Riley and G. Skirrow, Eds., Vol. 3, 2nd ed., Academic, New York, 1977, p. 115.

58   J. M. Gieskes, *Ann. Rev. Earth Planet. Sci.*, **3**, 433 (1975).

59   J. B. Maynard, *Geochim. Cosmoch. Acta*, **40**, 1253 (1976).

60   S. Emerson, *Geochim. Cosmochim. Acta*, **40**, 925 (1976); S. Emerson and G. Widmer, *Geochim. Cosmochim. Acta*, **42**, 1307 (1978).

61   R. H. Hart, *Can. J. Earth Sci.*, **10**, 799–816 (1973).

62   J. L. Bischoff and F. W. Dickson, *Earth Planet. Sci. Lett.*, **25**, 358 (1975).

63   F. L. Sayles, *Geochim. Cosmochim. Acta*, **43**, 527 (1979).

64   R. O. Hallberg et al., *Ambio*, **1**, 71 (1972).

READING SUGGESTIONS

Rocks Mineral and Sediments

Deer, W. A., R. A. Howie, and J. Zussman, *An Introduction to the Rock Forming Minerals*, Wiley, New York, 1966.

Marshall, C. E., *The Physical Chemistry of Minerals and Soils*, Vol. II, *Soils in Place*, Wiley-Interscience, New York, 1977.

Weathering

Berner, R. A., *Principles of Chemical Sedimentology*, McGraw-Hill, New York, 1971.

Bolt, G. H., and M. G. M. Bruggenwert, Eds., *Soil Chemistry*, Elsevier, Amsterdam, 1976.

Busenberg, E., and C. V. Clemency, "The Dissolution Kinetics of Feldspars at 25°C and 1 atm CO_2 Partial Pressure," *Geochim. Cosmochim. Acta*, **40**, 41–50 (1976).

Carroll, D., "Rainwater as a Chemical Agent of Geologic Processes—A Review," *U.S. Geological Survey Water Supply Paper*, No. 1535-G, 1962.

Cronan, D. S., "Authigenic Minerals in Deep Sea Sediments." In *The Sea*, Vol. 5, E. D. Goldberg, Ed., Wiley-Interscience, New York, 1974, pp. 491–525.

Feth, J. H., C. Robertson, and W. Polzer, "Sources of Mineral Constituents in Water from Granitic Rocks, Sierra Nevada California and Nevada," *U.S. Geological Survey Water Supply Paper*, No. 1535-I, 1964.

Garrels, R. M., and F. T. Mackenzie, *Evolution of Sedimentary Rocks*, Norton, New York, 1971.

Holdren, G. R., Jr., and R. A. Berner, "Mechanism of Feldspar Weathering," *Geochim. Cosmochim. Acta*, **43**, 1161–1186 (1979).

Jackson, M. L., "Weathering Sequence of Clay-Sized Minerals in Soils and Sediments: II. Chemical Weathering of Layer Silicates," *Soil Sci.*, **16**, 3–6 (1948).

Marshall, C. E., *The Physical Chemistry of Minerals and Soils*, Vol. II, *Soils in Place*, Wiley-Interscience, New York, 1977.

Miller, W. R., and J. A. Drever, "Chemical Weathering and Related Controls on Surface Water Chemistry in the Absaroka Mountains, Wyoming," *Geochim. Cosmochim. Acta*, **41**, 1693–1702 (1977).

Paces, T., "Steady-State Kinetics and Equilibrium between Ground Water and Granitic Rocks," *Geochim. Cosmochim. Acta*, **37**, 2641–2663 (1973).

Equilibrium Models and Evolution of Ocean and Atmosphere

Bricker, O., and R. M. Garrels, "Mineralogical Factors in Natural Water Equilibria." In *Principles and Applications of Water Chemistry*, S. D. Faust and J. V. Hunter, Eds., Wiley, New York, 1967, p. 449.

Goldberg, E. D., "Chemistry in the Oceans," In *Oceanography*, M. Sears, Ed., American Association for the Advancement of Science, Washington, D.C., 1961, p. 583.

Holland, H. D., "The History of the Ocean Water and Its Effect on the Chemistry of the Atmosphere," *Proc. Nat. Acad. Sci. U.S.*, **53**, 1173 (1965).

Holland, H. D., "The Geological History of Seawater: An Attempt to Solve the Problem," *Geochim. Cosmochim. Acta*, **36**, 637–651 (1972).

Kramer, J. R., "History of Sea Water. Constant Temperature-Pressure Equilibrium Models Compared to Liquid Inclusion Analyses," *Geochim. Cosmochim. Acta*, **29**, 921 (1965).

Lasaga, A. C., "The Kinetic Treatment of Geochemical Cycles," *Geochim. Cosmochim. Acta*, **44**, 815 (1980).

Pitman, J. I., "Carbonate Chemistry of Groundwater from Chalk, Givendale, East Yorkshire," *Geochim. Cosmochim. Acta*, **42**, 1885–1897 (1978).

Rubey, W. W., "Geological History of Sea Water," *Bull. Geol. Soc. Amer.*, **62**, 1111–1148 (1951). (Reprinted in *The Origin and Evolution of Atmospheres and Oceans*, P. J. Brancazio and A. G. W. Cameron, Eds., Wiley, New York, 1964).

Siever, R. "Sedimentological consequences of a Steady-State Ocean-Atmosphere," *Sedimentology*, **11**, 5 (1968).

Sillén, L. G., "The Physical Chemistry of Sea Water," In *Oceanography*, M. Sears, Ed., American Association for the Advancement of Science, Washington, D.C., 1961, p. 549.

Sillén, L. G., "Gibbs Phase Rule and Marine Sediments." In *Equilibrium Concepts in Natural Water Systems*, Advances in Chemistry Series, No. 64, Washington, D.C., 1967, pp. 57–69.

Chemical Composition of Waters

Brewer, P. G., "Minor Elements in Seawater." In *Chemical Oceanography*, J. P. Riley and G. Skirrow, Eds., 2nd ed., Academic, New York, 1975, pp. 415–497.

Drever, J. I., *The Geochemistry of Natural Waters*, Prentice Hall, Englewood Cliffs, 1981.

Freeze, R. A., and J. A. Cherry, *Groundwater*, Prentice Hall, Englewood Cliffs, N.J., (1979).

Garrels, R. M., and F. T. Mackenzie, "Origin of the Chemical Composition of Some Springs and Lakes." In *Equilibrium Concepts in Natural Water Systems*," Advances in Chemistry Series, No. 67, American Chemical Society, Washington, D.C., 1967, pp. 222–242.

Hem, J. D., *Study and Interpretation of the Chemical Characteristics of Natural Water*, U.S. Geological Survey Water Supply Paper, No. 1473, Washington, D.C., 1970.

Holland, H. D., *The Chemistry of the Atmosphere and Oceans*, Wiley–Interscience, New York, 1978.

Martin, J. M. and M. Meybeck, "Elemental Mass-Balance of Material Carried by Major World Rivers," *Marine Chem.*, **7**, 173 (1979).

The Sediment–Water Interface

Berner, R. A., "The Benthic Boundary Layer from the Viewpoint of a Geochemist." In *The Benthic Boundary Layer*, I. N. McCave, Ed., Plenum, New York, 1976, pp. 33–55.

Berner, R. A., *Early Diagenesis: A Theoretical Approach*, Princeton University Press, Princeton, 1980.

Gieskes, J. M., "Chemistry of Interstitial Waters of Marine Sediments," *Ann. Rev. Earth Planet. Sci.*, **3**, 433–453 (1975).

Lerman, A., *Geochemical Processes; Water and Sediment Environments*, Wiley-Interscience, New York, 1979.

Manheim, F. T., "Interstitial Waters of Marine Sediments." In *Chemical Oceanography*, J. P. Riley and R. Chester, Eds., Vol. 6, 2nd ed., Academic, New York, 1976, pp. 115–186.

Murray, J. W., V. Grundmanis, and W. M. Smethie, Jr., "Interstitial Water Chemistry in the Sediments of Saanich Inlet," *Geochim. Cosmochim. Acta*, **42**, 1011–1026 (1978).

Price, N. B., "Chemical Diagenesis in Sediments." In *Chemical Oceanography*, J. P. Riley and R. Chester, Eds., Vol. 6, 2nd ed., Academic, New York, 1976, pp. 1–58.

Sayles, F. L., and F. T. Manheim, "Interstitial Solutions and Diagenesis in Deeply Buried Marine Sediments: Results from the Deep Sea Drilling Project, *Geochim. Cosmochim. Acta*, **39**, 103–107 (1975).

Sayles, F. L., The Composition and Diagenesis of Interstitial Solutions—Fluxes across the Sea-water–Sediment Interface, *Geochim. Cosmochim. Acta*, **43**, 527–545 (1979).

Influence of Biota upon Composition of Natural Waters

Broecker, W. S., *Chemical Oceanography*, Harcourt Brace Jovanovich, New York, 1974.

Broecker, W. S., "A Kinetic Model for the Chemical Composition of Seawater, *Quatern. Res.*, **1**, 188–207 (1971).

DiToro, D. M., R. V. Thomann, and D. J. O'Connor, A Dynamic Model of Phytoplankton Population in the Sacramento San Joaquin Delta. In *Non-Equilibrium Systems in Natural Water Chemistry*, Advances in Chemistry Series, No. 106 American Chemical Society, Washington, D.C., 1971, pp. 131–180.

Imboden, D. M., and A. Lerman, "Chemical Models of Lakes." In *Lakes: Chemistry, Geology, Physics*, A. Lerman, Ed., Springer, New York, 1978, pp. 341–359.

Redfield, A. C., B. H. Ketchum, and F. A. Richards, "The Influence of Organisms on the Composition of Sea Water." In *The Sea*, Vol. II, M. N. Hill, Ed., Wiley-Interscience, New York, 1966.

Estuaries

Burton, J. D., and P. S. Liss, Eds., *Estuarine Chemistry*, Academic, New York, 1976.

APPENDIX: CLAY MINERALS†

Layer Structure

Clay minerals are primarily crystalline aluminum or magnesium silicates with stacked-layer structures.‡ Each unit layer is in turn a sandwich of silica and

---

† This appendix draws on a report prepared by Joel Gordon (Harvard University).
‡ R. E. Grim, *Clay Mineralogy*, 2nd ed., McGraw-Hill, New York, 1968; H. van Olphen, *An Introduction to Clay Colloid Chemistry*, Wiley-Interscience, New York, 1963.

gibbsite or brucite sheets. In the *silica* or *tetrahedral* (*T*) *sheet*, silicon atoms are each surrounded by four oxygen atoms in a tetrahedral arrangement; these tetrahedra are connected in an open hexagonal pattern in a continuous two-dimensional array. The *gibbsite* or *brucite* layer, or *octahedral* (*O*) *sheet*, consists of two layers of oxygen atoms (or hydroxyl groups) in a hexagonal closest packed arrangement with aluminum or magnesium atoms, respectively, at the octahedral sites.

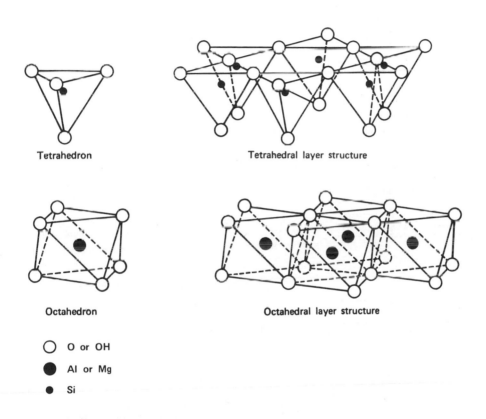

Tetrahedron   Tetrahedral layer structure

Octahedron   Octahedral layer structure

○   O or OH

●   Al or Mg

•   Si

These sheets are stacked into two- or three-layer units (T-O or T-O-T) in which the oxygen atoms at the vertices of the tetrahedra in the T sheet also form the basis hexagonal pattern of the O sheet (so that a T-O layer is five atoms thick and a T-O-T layer, seven atoms). The unshared oxygens in the O sheet are hydroxyl groups. Isomorphous substitution of Al(III) for Si(IV) in the T sheet [or Mg(Fe, Zn(II) for Al(III) in the O sheet] may lead to negatively charged layers. In clay minerals, the T-O or T-O-T units are stacked one on another, with layers of water and/or interlayer or surface cations (to compensate the negative charge) beyond the units. These cations may be exchangeable, and interlayer water may be absorbed by a dry clay, causing it to swell.

Types and Examples

Structure	Remarks	Names
A. Two-Layer Clays T O T O T O	Little isomorphous substitution Small cation exchange capacities (CEC) Nonexpanding	Kaolinite Dickite Nacrite Halloysite (Interlayer water)

B. Three-Layer Clays
 	
 T
 O
 T
 	
 M^{+m}, nH_2O
 	
 T
 O
 T
 	
 M^{+m}, nH_2O
 	
 T
 O
 T
	1. Expanding (Smectites or
	   Montmorillonites)
	 Substitution of a small amount of
	   Al for Si in T-sheet and of Mg,
	   Fe, Cr, Zn, Li for Al or Mg in
	   O-sheet
	 Large CEC (M^{m+} = Na^+, K^+,
	   Li^+, Ca^{2+}, ...)
	 Swell in water or polar organic
	   compounds
	 2. Nonexpanding (Illites)
	 About $\frac{1}{4}$ of Si in T-sheet replaced
	 by Al, similar O-sheet substitutions
	 Small CEC
	 M^{m+} = K^+
		Montmorillonite
		 Nontronite
		 Volkhonskyite
		 Hectorite
		 Saponite
		 Sauconite
		 Vermiculite
		
		 Poorly crystallized
		 Micas (muscovite,
		   biotite, phlogopite)

C. Chlorites
 	
 T
 O
 T
 	
 O (Brucite)
 	
 T
 O
 T
	Three-layer alternating with brucite
	 Brucite layer positively charged
	   [some Al(III) replacing M(II)],
	   partially balances negative charge
	   on T-O-T (mica) layer
	 Low CEC, nonswelling	

D. Fibrous Clays	Different type of structural units
	   consisting of double silica chains
	   (tetrahedral) joined to one-
	   dimensional O-layers and
	   containing interstitial water
		Attapulgite
		 Palygorskyite
		 Sepiolite

In addition, there are *mixed-layer clays* in which unit layers of different clay minerals are stacked more or less regularly, and poorly crystallized or amorphous clays known as *allophanes*.

10

The Solid–Solution Interface

10.1 INTRODUCTION

Most chemical reactions that occur in natural waters take place at phase discontinuities, that is, at atmosphere hydrosphere and lithosphere–hydrosphere interfaces (Table 10–1). In this chapter the discussion deals primarily with the solid–solution interface. This is still a very broad topic; even in single and reasonably well-defined systems the interfacial and colloidal properties involve so many variables and represent such complicated physical and chemical interactions that it is difficult to describe such systems as a unified subject.

The significance of the solid–solution interface in natural waters becomes apparent when one considers the state of subdivision of the solids typically present in natural waters. The dispersed phase in a natural body of water consists predominantly of inorganic colloids, such as clays, metal oxides, metal hydroxides, and metal carbonates, and of organic colloidal matter of detrital origin, as well as living microorganisms (algae and bacteria). Most of the clay minerals have physical dimensions smaller than 1 to 2 μm ($1~\mu$m $= 10^{-6}$ m). Montmorillonite, for example, has plate diameters of 0.002 to 0.2 μm and plate thicknesses in the range of 0.002 to 0.02 μm. Many solid phases have specific *surface energies* of the order of a few tenths J m^{-2}; hence, for substances that have specific surface areas of a few hundred m^2 g^{-1}, the total surface energy is of the order of 100 J g^{-1} (24 cal g^{-1}).

A description of the causal relationships effective in natural water systems must include surface chemical reactions, since these, in addition to those of biological activity, profoundly influence the temporal and spatial distribution of constituents. Because of the large extent of interfaces available, metal oxides and clays are of relevance in regulation of the water composition.

We will consider a few pertinent interfacial phenomena in simple systems and indicate ways in which the ideas that account for the properties of the single systems may be used for a qualitative understanding of some of the processes occurring in nature. Since a comprehensive survey of solid–solution phenomena would essentially cover all the divisions of colloid and surface chemistry, only a few areas have been selected for treatment.

Forces acting at interfaces are composed of extensions of forces acting within the two phases. Solids in natural waters have electrically charged surfaces. One side of the interface assumes a net electrostatic charge, either positive or negative, and an equivalent number of counterions of opposite charge form a counterlayer in the aqueous phase. Such an electric double layer exists at all

TABLE 10.1 SOME IMPORTANT NATURAL PROCESSES OCCURRING AT INTERFACES

Process	Interface	Examples
Weathering	Solid–liquid (rock–water)	Congruent and incongruent dissolution of rocks, erosion; soil formation
Gas exchange	Liquid–gas (water–atmosphere)	Aeration of water; evaporation of water; absorption of CO_2; loss of volatile substances from water to the atmosphere
Crystallization, precipitation	Liquid–solid (e.g., water–sediment)	Formation of sediments and rocks; nucleation and precipitation of calcite; nucleation of ice
Adsorption	Liquid–solid	Adsorption of H^+ and OH^- and of cations, anions, and weak acids on solid surfaces (e.g., oxides, clays, and biota); regulation of water composition by adsorption equilibria [e.g., interaction of Fe(III) oxides with phosphates; adsorption of surfactants and of nonionic difficultly water soluble substances (such as many pesticides, chlorinated hydrocarbons) on suspended particles
Absorption	Liquid–liquid	Dissolution of lipophilic substances in an oil or lipid phase; bioaccumulation of lipophilic substances in the food chain
Aerosol formation	Solid–gas	Emission of industrial and smoke particles; erosion of soil dust

interfaces in natural waters. Often particles tend to remain dispersed, that is, in a colloidal state; frequently, colloid stability results from electrostatic repulsion of the charged particle. Conceptual approaches for analyzing reactions at interfaces and the effect of these reactions may be subdivided into two categories:

1 The older, *chemical theory* assumes that colloids are aggregates of defined chemical structure, that the primary charge of surfaces arises from the ionization of complex ionogenic groups present on the surface of the particles, and that the destabilization of colloids is due to such chemical interactions as complex formation and proton transfer.

2 The newer, *physical theory* emphasizes the concept of the electrical double layer and the significance of predominantly physical factors, such as counter-ion adsorption, reduction of zeta potential, and ion-pair formation in the destabilization of colloids.

The physical or double-layer theory has been developed in great detail and has, in its various forms of simplification, found wide acceptance. It has become a most effective tool in the interpretation of interfacial phenomena. This theory has virtually superceded the older chemical theory.

These two theories are not, however, as mutually exclusive as they might appear to be at first sight. We want to call attention to the fact that purely chemical factors must be considered in addition to the theory of the double layer in order to explain, in a more quantitative way, many phenomena pertinent to solid–solution interfaces in natural water systems.

10.2 FORCES AT INTERFACES

Atoms, molecules, and ions exert forces upon each other. Molecules at a surface are subject to an inward attraction normal to the surface. This is explained in part by the fact that surface molecules have fewer nearest neighbors and, as a consequence, fewer intermolecular interactions than bulk molecules. Ideally, the energy of interaction at an interface can be interpreted as a composite function resulting from the sum of attractive and repulsive forces, but our insight into intramolecular and interatomic forces is far from satisfactory. At a qualitative—and necessarily introductory—level we briefly enumerate the principal types of forces involved.

Generally speaking, *chemical forces* extend over very short distances; a covalent bond can be formed only by a merging of electron clouds. *Electric forces* extend over longer distances. Basically, these forces are defined by Coulomb's law. The electrostatic force of attraction or repulsion between two point charges, q_1 and q_2, separated by distance x is given by

$$F = (\text{const})\frac{1}{\varepsilon}\frac{q_1 q_2}{x^2}$$

where ε is the dielectric constant relative to a vacuum (dimensionless) $(\varepsilon_{H_2O} \approx 80)$

The potential energy $E(x)$ referred to infinite separation is then given by

$$E(x) = \int_{\infty}^{x} F\,dx = -(\text{const})\frac{1}{\varepsilon}\frac{q_1 q_2}{x} \tag{1}\dagger$$

There are other forces, principally of electrical nature, present in molecular arrays whose constituents possess a permanent dipole (H_2O, NH_3). The energy resulting from such dipole–dipole interaction is also called the *orientation energy*. A charged species can also induce a dipole (induction energy).

London (1930) has shown that there is an additional type of force between atoms and molecules which is always attractive. Its origin can be qualitatively

---

† In the cgs system, where charges are expressed in esu, x in centimeters, and F in dynes, the proportionality constant (const) = 1. In the SI system (cf. the appendix to Chapter 1), the charge is measured in coulombs (C), x in meters (m), F in newtons (N), and the proportionality constant (or conversion factor), for historical reasons written as (const) = $1/4\pi\varepsilon_0$, takes the numerical value 8.99×10^9 N m^2 C^{-2} and ε_0, the permittivity (in vacuum) is 8.854×10^{-12} C^2 N^{-1} m^{-2} (or C^2 m^{-1} J^{-1}, or CV^{-1} m^{-1}).

explained by considering that neutral molecules or atoms constitute systems of oscillating charges producing synchronized dipoles that attract each other; hence this force is also, basically, an electric force. It is known as the *dispersion force* or the *London–van der Waals force*. This force is one of the reasons for the departure of real gases from ideal behavior and for the liquefaction of gases. The energy of interaction resulting from these dispersion forces (10 to 40 kJ mol^{-1}) is small compared to that of a covalent bond or electrostatic (ion-pair) bond ($\gg 40$ kJ mol^{-1}), but large compared to that of orientation or induction energies (< 10 kJ mol^{-1}). The van der Waals attraction energy between two atoms is inversely proportional to the sixth power of the separating distance over small distances. The total interaction between two semiinfinite flat plates is obtained by summing the pairwise interactions of all the constituent atoms. The resulting attractive energy per unit area between two semiinfinite plates is inversely proportional to the second power of the distance x:

$$V_A = - \frac{A}{12\pi x^2}$$

where A is the Hamaker constant. Its value depends on the density and polarizability of the material and is typically of the order of 10^{-19} to 10^{-20} J. This equation is valid for distances up to approximately 200 Å; for larger separating distances corrections must be made that make the attractive energy decay as $1/x^3$.

Because of the very small size of the hydrogen ion, its highest coordination number is 2. In a *hydrogen bond*, H^+ accommodates two electron pair clouds in order to bind two polar molecules. While van der Waals interactions are principally spherically symmetric, hydrogen bonding occurs at a preferred molecular orientation. Hydrogen bonding, however, is in the same energy range (10 to 40 kJ mol^{-1}) as van der Waals interactions. The unusually high boiling point of water, for example, in comparison to that of $H_2S(l)$, presents evidence for hydrogen bonding in addition to van der Waals association.

The Hydrophobic Effect.   Hydrophobic substances, for example, hydrocarbons, are readily soluble in many nonpolar solvents but only sparingly soluble in water; thus these substances tend to reduce the contact with water and seek relatively nonpolar environments. Many organic molecules (soaps, detergents, long-chain alcohols) are of a dual nature; they contain a hydrophobic part and a hydrophilic polar or ionic group; they are amphipathic. At an oil–water interface both parts of these molecules can satisfy their compatibility with each medium. Such molecules tend to migrate to the surface or interface of an aqueous solution; they also have a tendency toward self-association (*formation of micelles*). This is considered to be the result of *hydrophobic bonding*. This term may be misleading because the attraction of nonpolar groups to each other arises not primarily from a particular affinity of these groups for each other, but from the strong attractive forces between H_2O molecules which must be disrupted when any solute is dissolved in water. The hydrophobic effect is perhaps the single most important factor in the organization of the constituent molecules of living matter into complex structural entities such as cell membranes and organelles [1].

Surface Tension

Molecules in the surface or interfacial region are subject to attractive forces from adjacent molecules, which result in an attraction into the bulk phase. The attraction tends to reduce the number of molecules in the surface region (increase in intermolecular distance). Hence work must be done to bring molecules from the interior to the interface and, in turn, to increase the area of the interface. The minimum work required to create a differential increment in surface $d\bar{A}$ is $\gamma d\bar{A}$, where $\bar{A}$ is the interfacial area and γ is the surface tension or interfacial tension. γ is thus the interfacial Gibbs free energy for the condition of constant temperature, pressure, and composition:

$$\gamma = \left(\frac{\delta G}{\delta \bar{A}}\right)_{T, p, n} \tag{2}$$

γ is expressed in the units N cm^{-1} or J cm^{-2}; [1 N (newton) = 10^5 dyne].

In water the intermolecular interactions producing surface tension are essentially composed of (a) London–van der Waals dispersion interactions, $\gamma_{H_2O(L)}$ and (b) hydrogen bonds, $\gamma_{H_2O(H)}$:

$$\gamma = \gamma_{H_2O(L)} + \gamma_{H_2O(H)} \tag{3}$$

In water about one-third of the interfacial tension is due to van der Waals attraction, and the remainder is due to hydrogen bonding (Table 10.2). Similarly, for other liquids, the surface tension can be divided into two parts: the part due to dispersion forces and the part due to more specific chemical forces (e.g., hydrogen bonding, pi electron interactions in aromatic solvents, metallic bonds in metals) [2]. For liquid hydrocarbons the intermolecular attraction is almost entirely due to London dispersion forces; for Hg(l), on the other hand, the contribution due to metallic bonds is significant.

As shown by Fowkes [2], the interfacial tension between two phases (whose surface tensions—with respect to vacuum—are γ_1 and γ_2) is subject to the resultant force field made up of components arising from attractive forces in the bulk of each phase and the forces, usually the London dispersion forces, operating across the interface itself. Then the interfacial tension between two phases, γ_{12}, is given by

$$\gamma_{12} = \gamma_1 + \gamma_2 - 2(\gamma_{1(L)}\gamma_{2(L)})^{1/2} \tag{4}$$

The geometric mean of the dispersion force components $\gamma_{1(L)}$ and $\gamma_{2(L)}$ may be interpreted as a measure of the interfacial attraction resulting from dispersion

TABLE 10.2 ATTRACTIVE FORCES AT INTERFACES—SURFACE ENERGY, γ, AND LONDON–VAN DER WAALS DISPERSION FORCE COMPONENT OF SURFACE ENERGY, $\gamma_{(L)}$[a]

	γ $\times 10^5$ N cm^{-1}	$\gamma_{(L)}$ $\times 10^7$ J cm^{-2}
Liquids[b]		
Water	72.8	21.8
Mercury	484	200
n-Hexane	18.4	18.4
n-Decane	23.9	23.9
Carbon tetrachloride	26.9	—
Benzene	28.9	—
Nitrobenzene	43.9	—
Glycerol	63.4	37.0
Solids[c]		
Paraffin wax	—	25.5
Polyethylene	—	35.0
Polystyrene	—	44.0
Silver	—	74
Lead	—	99
Anatase (TiO$_2$)	—	91
Rutile (TiO$_2$)	—	143
Ferric oxide	—	107
Silica	—	123 (78)
Graphite	—	110

[a] Based on information provided by Fowkes [2]. A dash indicates that no value is available 20°C.
[b] $\gamma_{(L)}$ values for water and mercury have been determined (Fowkes) by measuring the interfacial tension of these liquids with a number of liquid saturated hydrocarbons. The intermolecular attraction in the liquid hydrocarbons is entirely due to London–van der Waals dispersion forces for all practical purposes. $\gamma_{(L)}$ was derived (Fowkes) from contact angle measurements.
[c] $\gamma_{(L)}$ of solids were derived (Fowkes) from contact angle measurements or from measurements of equilibrium film pressures of adsorbed vapor on the solid surface.

forces between adjacent dissimilar phases. $\gamma_{(L)}$ values of solid phases (Table 10.2) may be used to compare adsorbent properties of various materials, since $\gamma_{(L)}$ of the solid phase is a quantitative measure of the available energy of the solid surface for interaction with adjacent media [2]. Of the materials listed, graphite and oxides (rutile, ferric oxide) are strong adsorbents for hydrocarbons. Low-energy organic solids are weak adsorbents. The much weaker intermolecular forces in soft, organic solids of low molecular weight are reflected in their lower surface free energies; hence the dispersion force contribution to surface energy can also be used to estimate the long-range attraction forces between solid bodies. Values for the Hamaker constant have been estimated this way [2]. Adsorption of organic films onto a solid converts the surface into a low-energy surface. Similarly, adsorbed water greatly decreases the surface energy of glasses, silica, alumina, and metals [3]. As more than a monolayer of H_2O is adsorbed, the surface energy of the solid may approach the surface tension of a bulk water surface (73×10^{-7} J cm^{-2} at 20°C). Similarly, as pointed out by Baier et al. [3], a subtle change in the surface composition of a biological membrane can greatly decrease or increase its wetability and also its adhesiveness. An outermost substituent which decreases the interfacial tension, such as $-CH_3$ or $-CH_2-$, would decrease adhesiveness, whereas one which increases γ (such as $-C_6H_5$, $-OH$, $-SH$, $-COOH$ or $-NH_2$) would increase adhesiveness [3].

Adsorption from Solution

Adsorption of a solute molecule on the surface of a solid can involve removing the solute molecule from the solution, removing solvent from the solid surface, and attaching the solute to the surface of the solid. The net energy of interaction of the surface with the adsorbate may result from short-range chemical forces (covalent bonding, hydrophobic bonding, hydrogen bridges, steric or orientation effects) and long-range forces (electrostatic and van der Waals attraction forces). For some solutes, solid affinity for the solute can play a subordinate role in comparison to the affinity of the aqueous solvent.

In a simplified way the adsorption of organic matter, X, at a hydrated surface, S, may be viewed as

$$S(H_2O)_m + X(H_2O)_n \rightleftharpoons SX(H_2O)_p + (m + n - p)H_2O \qquad \Delta G^\circ_{ads}$$

(5)

This equation may be interpreted as the summation of the equations for the "hydration" of X:

$$X + n(H_2O) \rightleftharpoons X(H_2O)_n \qquad \Delta G^\circ_{solv}$$

the adsorption of unhydrated X at the hydrated surface

$$S(H_2O)_m + X \rightleftharpoons SX(H_2O)_m \qquad \Delta G^\circ_1$$

and the "hydration" of the product

$$SX(H_2O)_m + (p - m)H_2O \rightleftharpoons SX(H_2O)_p \qquad \Delta G^\circ_2$$

The free energy of adsorption is then given by

$$\Delta G^\circ_{ads} = \Delta G^\circ_1 + \Delta G^\circ_2 - \Delta G^\circ_{solv}$$

(6)

where ΔG_{solv}° is the free energy of solvation. If ΔG_2° is assumed to be relatively small, ΔG_{ads}° will be composed of ΔG_1° (affinity of the surface) and of $-\Delta G_{solv}^{\circ}$; $-\Delta G_{solv}^{\circ}$ is equivalent to the free energy of displacement of X from the water. Hydrophobic substances that are sparingly soluble in water tend to be adsorbed at solid surfaces. Organic dipoles and large organic ions are preferentially accumulated at the solid–solution interface, primarily because their hydrocarbon parts have a low affinity for the aqueous phase. Simple inorganic ions (e.g., Na^+, Ca^{2+}, Cl^-), even if they are specifically attracted to the surface of the colloid, may remain in solution because they are readily hydrated. Less hydrated ions (e.g., Cs^+, $CuOH^+$, and many anions) seek positions at the interface to a larger extent than easily hydrated ions.

Such considerations are also contained in the qualitative rule that a polar adsorbent adsorbs the more polar component of nonpolar solutions preferentially, whereas a nonpolar surface prefers to adsorb a nonpolar component from a polar solution. In accord with these generalizations and a consequence of the hydrophobic effect is *Traube's rule*, according to which the tendency to adsorb organic substances from aqueous solutions increases systematically with increasing molecular weight for a homologous series of solutes. Thus the interaction energy due to adsorption increases rather uniformly for each additional CH_2 group. Unless there is a special affinity of the surface for an organic solute (e.g., electrostatic attraction), the hydrophobic effect dominates the accumulation of organic solutes at high-energy surfaces.

The lipophilicity of a substance, that is, the tendency of a substance to become dissolved in a lipid, is often measured by the tendency of a substance to become dissolved in a nonpolar solvent, for example, by the *n*-octanol–water distribution coefficient. The lipophilicity of a substance is inversely proportional to its water solubility.

Adsorption Isotherms.   Adsorption is most often described in terms of isotherms which show the relationship between the bulk activity of adsorbate and amount adsorbed at constant temperature. Adsorption isotherms and two of the simpler equations commonly used to characterize adsorption equilibria are given in Figure 10.1 and Table 10.3.

By means of the *Gibbs equation* a quantitative relation between adsorption and the variation of interfacial tension is established. In a qualitative sense the equation illustrates that dissolved material which lowers the surface or interfacial tension of a solution tends to accumulate at the interface because less work is then required to increase (isothermally and reversibly) the area of the interface. While simple electrolytes in water raise the interfacial tension, the interfacial tension is reduced by many nonelectrolytes and especially by polar-nonpolar substances which work against the adhesive forces of the water molecules. Almost all organic substances found in natural waters fall into the latter category

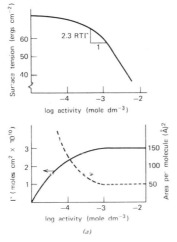

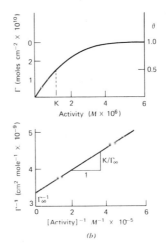

Figure 10.1 Adsorption isotherms. (*a*) Gibbs adsorption equation. The surface excess, Γ, hence the area occupied per molecule or ion of adsorbed substance, can be computed from the slope of a semilogarithmic plot of surface (or interfacial) tension versus log activity of sorbate [(4) of Table 10.3]. (*b*) Langmuir adsorption isotherm. From the adsorption isotherm [plotted in accordance with (5) and (6) of Table 10.3], the equilibrium constant K and the adsorption capacity, Γ_∞, obtain by plotting Γ^{-1} versus the reciprocal activity of the sorbate [(7) of Table 10.3].

and are thus concentrated or adsorbed at the surface or at the solid–liquid interface. In the absence of specific chemical or electrostatic attraction forces, small ions cannot be adsorbed at interfaces because they tend to increase the interfacial tension.

Molecules and ions that possess a hydrophobic and hydrophilic part tend to orient at interfaces. Detergents are notable examples of such "surfactants" with a dual constitution; the polar group is immersed in the water, and the hydrophobic group exposed to the air, oil, or solid phase.

Many experimental adsorption isotherms are satisfactorily described by the *Langmuir equation* (Figure 10.1). However, a satisfactory fit of the experimental points to the equation does not necessarily imply that the conditions that form the basis of the theoretical Langmuir model are fulfilled.

Water–Air Interface.   The surface studies of Baier et al. [4] tend to support the conclusion that all naturally adsorbed air-natural water films are of essentially the same macromolecular net-negatively-charged composition dominated by degraded proteinaceous components. MacIntire and Liss [5,6] have pointed out that, in the sea surface microlayer, microscopic chemical and hydrodynamic processes exert a profound influence over macroscopic geochemical and geophysical phenomena.

TABLE 10.3 ADSORPTION EQUILIBRIA

	Gibbs[b,c]		Langmuir[b,c]	
Adsorbed substance per unit area $\Gamma = n_i/A$ $= \Gamma_\infty \theta$ (for monolayer adsorption)	$\Gamma_i = -\left(\dfrac{\delta\gamma}{\delta\mu_i}\right)_{T,\,P,\,\text{all }\mu\text{'s except }\mu_i\text{ and }\mu_{H_2O}}$	(1)	$\dfrac{\theta}{1-\theta} = K^{-1}a_i$	(5)
	$\Gamma_i = -\dfrac{1}{RT}\left(\dfrac{\delta\gamma}{\delta \ln a_i}\right)$	(2)	$\Gamma_i = \dfrac{\Gamma_{i\infty}a_i}{K + a_i}$	(6)
Example: Adsorption of a hydrocarbon (HC)	$d\gamma = -RT\Gamma_{HC}\,d\ln\{HC\}$	(3)	$\dfrac{1}{\Gamma_i} = \dfrac{1}{\Gamma_{i\infty}} + \dfrac{K}{\Gamma_{i\infty}}\dfrac{1}{\{HC\}}$	(7)
Example: Adsorption of a hydrocarbon (HC) and of an alcohol (ROH)	$d\gamma = -RT\Gamma_{HC}\,d\ln\{HC\}$ $\quad\quad - RT\Gamma_{ROH}\,d\ln\{ROH\}$	(4)	$\Gamma_{ROH} = \dfrac{\Gamma_{ROH\infty}K_{ROH}^{-1}\{ROH\}}{1 + K_{ROH}^{-1}\{ROH\} + K_{HC}^{-1}\{HC\}}$	(8)
Basis for application: $\mu_{i(\text{interface})} = \mu_{i(\text{bulk phase})}$	Reversible equilibrium		(1) Reversible equilibrium up to monolayer ($\theta = 1$).[a] K is related to the free energy of adsorption: $K = \exp[\Delta G°/RT]$ (2) Fixed-site adsorption (immobility of adsorbate)[e] (3) Homogeneity in surface (energy of adsorption independent of θ)	

[a] μ = chemical potential; γ = interfacial tension, $(J\ cm^{-2})$; a_i = activity of species $i(M)$; θ = fraction of surface covered with adsorbate; R = gas constant 8.314 J mol^{-1} K^{-1}; K = equilibrium constant, (M) (if $K = a_i$, $\theta = 0.5$); Γ_i = adsorption density $(mol\ cm^{-2})$.

[b] In most cases, concentrations may be used instead of activities; however, the bulk concentration can no longer be defined in terms of quantities added if the solutes have associated to form micelles. Similar difficulties arise when the adsorbates in the interface undergo irreversible changes, such as denaturation reactions.

[c] Γ_i is defined in such a way that, in dividing the interface, the adsorption density of water is zero (Gibbs convention).

[d] A similar adsorption model for multilayer adsorption has been set forth in the *Brunauer, Emmett and Teller (BET)* equation which has the form:

$$X = \frac{X_\infty Ba_i}{(a_{is} - a_i)[1 + (B - 1)a_i/a_{is}]}$$

where X and X_∞ are the observed and ultimate quantity of substance adsorbed per unit area, B, an energy related constant, and a_{is} is the saturation activity of the solute.

[e] Varying energies of adsorption can arise because of heterogeneity of the surface, that is, the free energy decreased with coverage. The Tempkin isotherm is based on such a model. In the *Tempkin isotherm*, θ is a function of log a_i at intermediate coverages $(0.2 < \theta < 0.8)$.

609

10.3   THE ELECTRIC DOUBLE LAYER

A charged or electrified interface reflects an unequal distribution of charges (usually ions) at the phase boundary; that is, it results from a localized disturbance of electroneutrality. The interfacial system as a whole (the region from one bulk phase to the other) is of course electrically neutral.

The electric state of a surface depends on the spatial distribution of free (electronic or ionic) charges in its neighborhood. This distribution is usually idealized as an *electrochemical double layer*; one layer of the double layer is envisaged as a *fixed charge* or surface charge attached to the particle or solid surface, while the other layer is distributed more or less diffusely in the liquid in contact (Figure 10.2). This layer contains an excess of *counter ions*, opposite in sign to the fixed charge and usually a deficit of *coions* of the same sign as the fixed charge. We consider a negative surface with one type of cations as counterions. These counterions (1) are electrostatically attracted by the surface (while anions are depleted from the surface; (2) tend, because of thermal motion, to become more evenly distributed through the solution; and (3) may be attracted to the surface by other than electrostatic forces (specific adsorption). Various models attempt to describe schematically the spatial distribution of charges at the surface (Figure 10.2). In the simplest case, the electrified surface consists

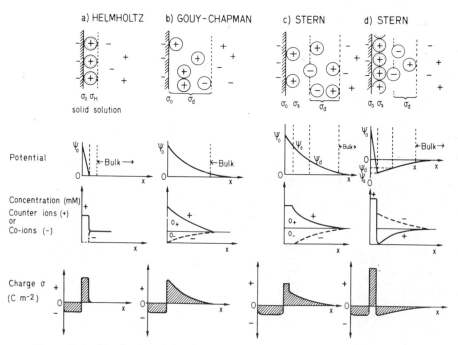

Figure 10.2   Distribution of charge, ions, and potential at a solid–solution interface.

of two sheets of charge, one on the surface and one in the solution (*Helmholtz model*, Figure 10.2*a*). If the charges in the solution are exposed to the forces of thermal motion, a balance between electrostatic and thermal forces is attained (*Gouy–Chapman diffuse charge model*, Figure 10.2*b*). Stern (1924) has divided the region near the surface into two parts, the first consisting of a layer of ions specifically adsorbed at the surface forming the compact *Stern layer*, and the second consisting of the diffuse layer or the *Gouy layer* (Figure 10.2*c*). Ions adsorbed in the Stern layer are subject to electrostatic and specific interaction. If the specific interaction outweighs the electrostatic one, the charge of the Stern layer may become more positive than that of the surface (superequivalent adsorption, Figure 10.2*d*).

Because of electroneutrality, the sum of the charges is zero:

$$\sigma_0 + \sigma_s + \sigma_d = 0 \qquad (7)$$

where σ_0 is the (fixed) surface charge density (C m^{-2}), and σ_s and σ_d are the charge densities in the Stern layer and in the diffuse (Gouy) layer. If large surface-active ions or molecules (soaps, polyelectrolytes) adsorb to the surface, the Stern layer may be thick.

Ψ Potential, ζ Potential, Capacitance

The potential $\Psi(x)$ at any point x is defined in such a way that $e\Psi(x)$ is the reversible isothermal electric work required to bring an elementary charge e from infinity to x, where x is the distance from the surface of the solid.

The Electrokinetic Potential (Zeta Potential), ζ, is the potential drop across the mobile part of the double layer that is responsible for electrokinetic phenomena, for example, electrophoresis (=motion of colloidal particles in an electric field). It is assumed that the liquid adhering to the solid (particle) surface and the mobile liquid are separated by a shear plane (slipping plane). The electrokinetic charge is the charge on the shear plane.

Often Ψ_d, the potential drop across the diffuse part of the double layer, is taken as identical to ζ. Hence the electrokinetic charge σ_ζ may also be set approximately equal to the diffuse charge σ_d. Unfortunately, the division between the fixed surface charge of the particle and the diffuse charge of the solution does not necessarily coincide with the shear plane.

Electrophoresis refers to the movement of charged particles relative to a stationary solution in an applied potential gradient, whereas in *electroosmosis* the migration of solvent with respect to a stationary charged surface is caused by an imposed electric field. The *streaming potential* is the opposite of electroosmosis and arises from an imposed movement of solvent through capillaries;

conversely a *sedimentation potential* arises from an imposed movement of charged particles through a solution. Operationally the zeta potential can be computed from electrophoretic mobility and other electrokinetic measurements. For example, for nonconducting particles whose radii are large when compared with their double-layer thicknesses, the zeta potential ζ is related to the electrophoretic mobility m_e [=velocity per unit of electric field (e.g., nm sec^{-1} V^{-1} cm^{-1})] by

$$m_e = \frac{\zeta \varepsilon \varepsilon_0}{\eta} \tag{8}$$

where ε and ε_0 are relative dielectric permittivity and permittivity in vacuum, respectively and η is the viscosity of the solution. Frequently many corrections that are difficult to evaluate must be considered in the computation of ζ.

The position of the shearing plane is not known, but generally ζ is smaller in magnitude than Ψ_0. The theory of electrokinetic phenomena has been treated by Dukhin and Derjaguin [7]. The concept of the zeta potential has been reviewed by Sennett and Olivier [8]. The measurement of electrophoretic mobility of particulate matter in fresh water and in seawater has been discussed by Black [9] and by Neihof and Loeb [10], respectively.

Charge and potential are related by the *capacitance*:

$$\sigma_0 = K\Psi_0; \qquad K = \text{integral capacitance} \tag{9a}$$

$$d\sigma_0 = Cd\Psi_0; \qquad C = \text{differential capacitance} \tag{9b}$$

Capacitances can also be defined for the Stern part and the diffuse part of the double layer

$$C_{\text{compact}} = \frac{d\sigma_0}{d(\Psi_0 - \Psi_d)} = \frac{\varepsilon_m \varepsilon_0}{\delta} \tag{10}$$

and

$$C_{\text{diffuse}} = \frac{d\sigma_d}{d\Psi_d} \tag{11}$$

where $C_{\text{total}}^{-1} = C_{\text{compact}}^{-1} + C_{\text{diffuse}}^{-1}$ (12)

where ε_m is the relative dielectric permittivity in the Stern layer and δ is the thickness of the compact (Stern) layer. In other words, the interface is treated as a two-layered electric condenser with the two capacitances in series.

Origin of Surface Charge

Figure 10.3 shows that many suspended and colloidal solids encountered in natural waters have surface charge, and that the charge may be strongly

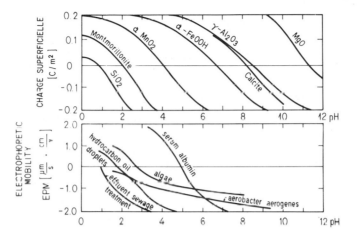

Figure 10.3 Effect of pH on charge and electrophoretic mobility. In the neutral pH range the suspended solids typically encountered in natural waters are negatively charged. These simplified curves are based on results by different investigators whose experimental procedures are not comparable and may depend upon solution variables other than pH. The curves are meant to exemplify trends and are meaningful in a semiquantitative way only.

affected by pH. There are three principal ways in which the surface charge may originate.

1 The charge may arise from *chemical reactions* at the surface. Many solid surfaces contain ionizable functional groups: $-OH$, $-COOH$, $-OPO_3H_2$. The charge of these particles becomes dependent on the degree of ionization (proton transfer) and consequently on the pH of the medium. For example, the electric charge of a silica surface in water can be explained by the acid–base behavior of the silanol ($-Si-OH$) groups found on the surface of hydrated silica

$$\equiv Si-OH_2^+ \xrightleftharpoons{K_1} \equiv Si-OH \xrightleftharpoons{K_2} \equiv Si-O^-$$

Most oxides and hydroxides exhibit such amphoteric behavior; thus the charge is strongly pH-dependent, being positive at low pH values. Similarly, for an organic surface, for example, that of a bacterium, one may visualize the charge as resulting from protolysis of functional amino and carboxyl groups, for example,

$$R\begin{matrix} COOH \\ \\ NH_3^+ \end{matrix} \xrightleftharpoons{K_1} R\begin{matrix} COO^- \\ \\ NH_3^+ \end{matrix} \xrightleftharpoons{K_2} R\begin{matrix} COO^- \\ \\ NH_2 \end{matrix}$$

At low pH a positively charged surface prevails; at high pH, a negatively charged surface is formed. At some intermediate pH (the isoelectric point) the net surface charge will be zero.

Charge can also originate by processes in which solutes become coordinatively bound to solid surfaces, for example,

$$S(s) + S^{2-} = S_3^{2-}(surface)$$

$$Cu(s) + 2H_2S = Cu(SH)_2^{2-}(surface) + 2H^+$$

$$AgBr(s) + Br^- = AgBr_2^-(surface)$$

$$FeOOH(s) + HPO_4^{2-} = FeOHPO_4^-(surface) + OH^-$$

$$R(COOH)_n + mCa^{2+} = R - [(COO)_nCa_m]^{-n+2m}(surface) + nH^+$$

$$MnO_2 \cdot H_2O(s) + Zn^{2+} = MnOOHOZn^+(surface) + H^+$$

The alteration in surface charge results from chemical reactions; these phenomena are frequently referred to as specific adsorption. Colloid chemists often assign these coordinating ions to the Stern layer; from a coordination chemistry point of view these ions—like H^+ and OH^-—may be included with the (fixed) surface charge.

2 Surface charge at the phase boundary may be caused by lattice imperfections at the solid surface and by *isomorphous replacements* within the lattice. For example, if in any array of solid SiO_2 tetrahedra an Si atom is replaced by an Al atom (Al has one electron less than Si), a negatively charged framework is established:

Similarly, isomorphous replacement of the Al atom by Mg atoms in networks of aluminum oxide octahedra leads to a negatively charged lattice. Clays are representative examples where such atomic substitution causes the charge at the phase boundary. Sparingly soluble salts also carry a surface charge because of lattice imperfections. An introduction to the structure of clays is given in the appendix to Chapter 9.

3 A surface charge may also be established by *adsorption of a surfactant ion*. Preferential adsorption of one type of ion on the surface can arise from London–van der Waals interactions and from hydrogen or hydrophobic bonding.

Constant-Charge and Constant-Potential Surfaces

We can distinguish two ideal cases: (1) the charge on the surface is fixed and remains independent of solution composition; and (2) the surface potential remains constant and its magnitude is not affected by the presence of indifferent electrolytes.

A constant charge is established, for example, on a mineral containing isomorphous substitutions in its lattice. Thus the plate of the clay surface is characterized by constant charge. Similarly, a solid surface covered with a monolayer of fatty acids has a fixed number of charges at constant pH.

Alternatively, the potential of a solid surface may be controlled externally. We may keep the potential of a metal electrode constant by means of an external potentiostatic arrangement, and the potential of a surface can be fixed by keeping constant the bulk activities of certain solutes whose specific chemical reaction with the surface gives rise to the potential.

Potential-determining ions are by definition [11,12] species of ions which by virtue of their equilibrium distribution between the two phases (or by their equilibrium with electrons in one of the phases) determine the interfacial potential difference, that is, the difference in galvanic potential between these phases.

It is readily seen that the electrode potential (or the potential difference between the electrode and the solution) of a Ag electrode depends on the Ag^+ ion activity of the solution (Nernst equation)

$$E = E_0 + \frac{RT}{F} \ln \frac{\{Ag^+\}}{\{Ag(s)\}}$$
(13)

and for an Ag/AgCl electrode on the activity of Ag^+ or Cl^- ions. Accepting the establishment of potential-determining equilibria between two phases, we may apply the concept of potential-determining species also to dispersed particles. It has been shown that the Nernst equation holds for silver–silver halogenide colloids, for example, for AgBr(s),

$$\Psi_0 = \frac{RT}{F} \ln \frac{\{Ag^+\}}{\{Ag_0^+\}} = -\frac{RT}{F} \ln \frac{\{Br^-\}}{\{Br_0^-\}}$$
(14)

where Ψ_0 is called the surface potential (i.e., the difference in potential between the surface and the solution) and $\{Ag_0^+\}$ and $\{Br_0^-\}$ are the activities of Ag^+ and Br^-, respectively, at a reference point, that is, at a point of zero potential. While a standard reference electrode (e.g., a H_2 electrode) is a standard reference state for electrode potentials, the point of zero potential or the *zero point of charge* may serve as a reference state in surface chemistry. As there are oxide electrodes, for example, the antimony oxide electrode, that show a potential dependence on $\{H^+\}$ or the interrelated $\{OH^-\}$ in accordance with the Nernst equation,

$$E = E_0 + \frac{RT}{F} \ln \frac{\{H^+\}}{\{H_0^+\}}$$
(15)

one expects that H^+ and OH^- determine the potential of hydrous oxide surfaces:

$$\Psi_0 = \frac{RT}{F} \ln \frac{\{H^+\}}{\{H_0^+\}}$$

or

$$\Psi_0 = \frac{2.3RT}{F} (pH_0 - pH) = \frac{2.3RT}{F} (pOH - pOH_0)$$
(16)

where $\{H^+\}$ and $\{OH^-\}$ are the H^+ and OH^- activities, respectively at the point of zero potential or at the *point of zero charge* and pH_0 or pH_{zpc} is the pH of zero point of charge. The experimental observation with most oxides investigated are not in accord with equation 16, the Nernst equation [12]. H^+ and OH^- nevertheless have a significant influence on the surface potential and the surface charge of oxides. Even if the Nernst equation is not fulfilled, we may speak of H^+ and OH^- ions as potential-determining ions;† hence it is also appropriate to speak of a pH of zero point of charge. pH_{zpc} is different for different metal oxides; it is a measure of the acidity and basicity of the hydrated surface oxide groups. pH_{zpc} is constant for a given oxide and is independent of the concentration of indifferent (nonspecifically adsorbed) solutes in the solution.

Thus, from a coordination chemistry point of view, most generally the following type of chemical equilibria between the solid and the solution or between the surface phase and the solution may be considered potential-determining equilibria:

$$Ag(s) = Ag^+ + e$$

$$AgBr(s) + e = Ag(s) + Br^-$$

$$AgBr(s) + Ag^+ = Ag_2Br^+(\text{surface})$$

$$CuS(s) + HS^- = Cu(S_2H^-)(\text{surface})$$

$$FePO_4(s) + H_2PO_4^- = Fe(HPO_4)_2^-(\text{surface})$$

$$FeOOH(s) + H^+ = Fe(OH)_2^+(\text{surface})$$

$$FeOOH(s) + OH^- + H_2O = Fe(OH)_4^-(\text{surface})$$

$$SiO_2(s) + OH^- = SiO_2(OH)^-(\text{surface}) \tag{17}$$

In each case the ion, species, or electron‡ (italicized in equation 17) that crosses or reacts at the interphase establishes the chemical equilibrium at the interface (equality of the electrochemical potential between solid or surface phase and solution) and thus determines (or influences) the interfacial potential difference. However, without further information, the validity of the Nernst equation may not be inferred.

Potential and Charge Distribution in Double Layers

In the *Gouy–Chapman model*, ions are treated as point charges that do not interact specifically (other than through coulombic forces) with each other.

---

† The concept of potential-determining ions has been restricted by Lyklema [12] to constituent ions of the sorbent. However, O^{2-} ions that are components of the hydrous oxides are in reversible equilibrium with OH^- ($O^{2-} + H_2O \rightleftharpoons 2OH^-$); hence OH^- is in equilibrium with both phases.
‡ The electron (equation 17) may stand for a suitable reductant.

Table 10.4 gives a brief derivation of the equations. This discussion attempts to provide *a qualitative understanding* of the most relevant feature of this model.†

The theory shows that the space charge density resulting from the distribution of ions in the solution decreases rapidly with distance from the surface. The net double-layer charge in the solution is reflected in the disturbance of the electroneutrality of the ions close to the surface; it is represented by the total area between the two concentration curves in Figure 10.2b. This figure also illustrates the approximately exponential decline of the potential Ψ as a function of distance.

In a medium of low ionic strength the disturbance in the electroneutrality extends further into the solution than in the case of a solution with high ionic strength. The center of charge falls at a plane of distance $1/\kappa$, referred to as the double-layer thickness, which depends on the electrolyte concentration. For water at 20°C, (8) of Table 10.4 gives

$$\kappa^{-1} \sim \frac{2.8 \times 10^{-8}}{\sqrt{I}} \text{ (cm)} \tag{18}$$

where I is the ionic strength (molarity).

To a first approximation, the double layer may be visualized as a parallel plate condenser of variable distance κ^{-1} between the two plates and with its capacitance (cf. equation 10):

$$C = \frac{\varepsilon\varepsilon_0}{\kappa^{-1}} \tag{19}$$

Accordingly, an increase in ionic strength results in a decrease in double-layer thickness. This is equivalent to an increase in capacitance, hence an increase in the σ/Ψ_0 ratio. The response of the two types of double layers (constant charge and constant potential) to an increase in electrolyte concentration (I) will be different:

$$\begin{aligned} &\Psi_0 \text{ decreases if } \sigma \text{ is constant} \\ &\sigma \text{ increases if } \Psi_0 \text{ is constant} \end{aligned} \tag{20}$$

These relationships are evident from Figure 10.4. Equations (12a) and (14b) of Table 10.4 are applicable only for small double-layer potentials ($\Psi_0 < 25$ mV), otherwise (12) and (14a) of Table 10.4 have to be used. Equation (13) of Table 10.4 shows that the total surface charge is reflected in the initial slope of the potential distance curve.

For fresh water κ^{-1} is typically in the range of 5 to 20 nm. In seawater κ^{-1} is on the order of 0.4 nm; this is a magnitude similar to that of the radius of a hydrated ion. Thus double layers in seawater are nondiffuse.

† For a more detailed discussion see [13–17].

TABLE 10.4 GOUY–CHAPMAN THEORY OF SINGLE FLAT DOUBLE LAYER[a]

I. Variation of Charge Density in Solution

Equality of electrochemical potential, $\bar{\mu}\ (=\mu + zF\Psi)$ of every ion, regardless of position.

Electrochemical Potential:

$$\bar{\mu}_{+(x)} = \bar{\mu}_{+(x=\infty)}, \qquad \bar{\mu}_{-(x)} = \bar{\mu}_{-(x=\infty)} \tag{1}$$

$$zF(\Psi_{(x)} - \Psi_{(x=\infty)}) = -RT \ln \frac{n_{+(x)}}{n_{+(x=\infty)}} = RT \ln \frac{n_{-(x)}}{n_{-(x=\infty)}} \tag{2}$$

Cations

$$n_+ = n_{+(x=\infty)} \exp\left(\frac{-zF\Psi_{(x)}}{RT}\right) \tag{3}$$

Anions

$$n_- = n_{-(x=\infty)} \exp\left(\frac{zF\Psi_{(x)}}{RT}\right) \tag{4}$$

Space Charge Density

$$q = zF(n_+ - n_-) \tag{5}$$

(if $z_+ = z_-$)

II. Local Charge Density and Local Potential

Ψ and q are related by Poisson's equation:

Poisson's Equation:

$$\frac{d^2\Psi}{dx^2} = -\frac{q}{\varepsilon\varepsilon_0} \tag{6}$$

Combining (3), (4), and (5) with (6) and considering that $\sinh x = (e^x - e^{-x})/2$ gives the *Double-Layer Equation:*

$$\frac{d^2\Psi}{dx^2} = \frac{\kappa^2 \sinh(zF\Psi/RT)}{(2F/RT)} \tag{7}$$

where κ is the reciprocal thickness of the double layer (the reciprocal Debye length)

$$\kappa = \left(\frac{e^2 \sum_i n_i z_i^2}{\varepsilon\varepsilon_0 kT}\right)^{1/2} \quad \text{(nm)} \tag{8}$$

For convenience the following substitutions can be made:

$$y = \frac{zF\Psi}{RT}; \qquad \bar{z} = \frac{zF\Psi_d}{RT}; \qquad \xi = \kappa x \tag{9}$$

Considering (9), (8) becomes the

Substituted Double-Layer Equation:

$$\frac{d^2y}{d\xi^2} = \sinh y \tag{10}$$

618

For boundary conditions, if $\xi = \infty$, $dy/d\xi = 0$ and $y = 0$

first integration:

$$dy/d\xi = -2\sinh(y/2), \text{ or} \tag{11}$$

$$d\Psi/dx = -\frac{RT}{zF}2\kappa\sinh(y/2) \tag{11a}$$

and for boundary conditions, if $\xi = 0$, $\Psi = \Psi_d$ or $y = \bar{z}$

second integration:

$$e^{y/2} - \frac{e^{\bar{z}/2} + 1 + (e^{\bar{z}/2} - 1)e^{-\xi}}{e^{\bar{z}/2} + 1 - (e^{\bar{z}/2} - 1)e^{-\xi}} \tag{12}$$

Simplified Equations for $\Psi_d \ll 25$ mV:

Instead of (7)

$$\frac{d^2\Psi}{dx^2} = \kappa^2\Psi \tag{7a}$$

Instead of (12)

$$\Psi = \Psi_d \exp(-\kappa x) \tag{12a}$$

III. Diffuse Double-Layer Charge and Ψ_d

Surface Charge Density:

$$\sigma_d = -\int_0^\infty q\,dx = \varepsilon\varepsilon_0 \int_0^\infty \left(\frac{d^2\Psi}{dx^2}\right)dx$$

$$= -\varepsilon\varepsilon_0 \left[\frac{d\Psi}{dx}\right]_{x=0} \tag{13}$$

Inserting (11a),

$$\sigma_d = (8\varepsilon\varepsilon_0 n_s \mathbf{k}T)^{1/2} \sinh\left(\frac{zF\Psi_d}{2RT}\right) \tag{14}$$

$$= 11.72\, C_s^{1/2} \sinh\left(\frac{zF\Psi_d}{2RT}\right) (\mu C\ cm^{-2}) \qquad (\text{at } 25°C) \tag{14a}$$

If $\Psi_d \ll 25$ mV, $\sigma_d \approx \varepsilon\varepsilon_0 \kappa\Psi_d$ \hfill (14b)

---

[a] μ = chemical potential; $\bar{\mu}$ = electrochemical potential; Ψ = local potential (V); Ψ_d = diffuse double-layer potential (V); q = (volumetric) charge density (C cm^{-3}); σ_d = diffuse surface charge (C cm^{-2}); x = distance from surface (cm); n_+ = local cation concentration (mol cm^{-3}); n_- = local anion concentration (mol cm^{-3}); $n_{(x=\infty)}$ = bulk ion concentration (mol cm^{-3}); n_i = number of ions i per cm^3 $[=Nn_{(x=\infty)}]$ (cm^{-3}); n_s = number of ion pairs (cm^{-3}); c_s = salt concentration (mol liter^{-1} or M); N = Avogadro's number, 6.03×10^{23} mol^{-1}; I = ionic strength (mol liter^{-1}); z = valence of ion; κ = reciprocal thickness of double layer (nm); e = charge of electron (elementary charge), 1.6×10^{-19} C; $\mathbf{k}$ = Boltzmann constant, 1.38×10^{-23} J K^{-1}; $\mathbf{k}T$ = Boltzmann constant times absolute temperature, 0.41×10^{-20} V C at 20°C; $RT = N \times \mathbf{k}T = 2.46 \times 10^3$ V C mol^{-1} at 20°C; ε = relative dielectric permittivity (dimensionless) ($\varepsilon = 80$ for water); ε_0 = permittivity in vacuum, 8.854×10^{-14} C V^{-1} cm^{-1}; F = Faraday = $6 \times 10^{23} \times e = 96,490$ C eq^{-1}; $F\Psi/RT = e\Psi/\mathbf{k}T = 1$ for $\Psi = 25$ m V at 20°C.

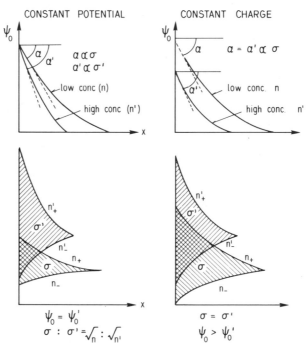

Figure 10.4 Effect of electrolyte concentration upon double-layer structure as predicted from Gouy theory.

Simple Applications

The Gouy theory is only applicable quantitatively in dilute solutions and for small potentials. In the presence of multivalent cations or anions specific adsorption effects occur. One of the pertinent applications of double-layer theory—the stability of colloids—will be taken up in Section 10.6. Example 10.1 illustrates how to handle the equations numerically.

Example 10.1. Surface Potential of Kaolinite

Consider a kaolinite mineral whose negative face charge is attributable to atomic substitutions in 0.05 % of all Si atoms. The surface area determination by the Brunauer–Emmett–Teller (BET) has yielded a specific surface area of 25 m^2 g^{-1}. Assume that the Gouy–Chapman theory is valid; $t = 20°C$.

1   Estimate the surface potential for a suspension in freshwater (ionic strength $= 10^{-3}$ M). For lack of any simple theory applicable to mixed electrolytes, it may be assumed that a water of ionic strength I is comparable to a solution

of a monovalent symmetric electrolyte of equal strength. [See, for example, K. M. Joshi and R. Parsons, *Electrochim. Acta*, **4**, 129 (1961).]

2 Estimate how much the pH in the immediate proximity of the surface differs from that in the bulk of the solution.

1. We first compute the total surface charge density. One mole of kaolinite $[Al_2Si_2O_5(OH)_4]$ equals ca. 200 g and contains 2 mol of silicon atoms. For an atomic substitution of five Si atoms in 10,000, 1 g kaolinite contains 5×10^{-6} mol of substitutions. The surface charge is readily computed by dividing the amount of substitutions by the surface area

$$\sigma = \frac{5 \times 10^{-6} \text{ eq g}^{-1} \times 9.65 \times 10^{4} \text{ C eq}^{-1}}{25 \times 10^{4} \text{ cm}^{2} \text{ g}^{-1}} \tag{i}$$

$$= 1.95 \times 10^{-6} \text{ C cm}^{-2}$$

We then estimate κ^{-1} for the condition specified. We may use equation 18 to obtain

$$\kappa^{-1} = 8.8 \times 10^{-7} \text{ cm}$$

The potential Ψ_0 $(= \Psi_d)$ can now be calculated using (14a) of Table 10.4; we obtain, assuming that $\sigma_0 = \sigma_d$,

$$\sinh\left(\frac{F\Psi_0}{2RT}\right) = \frac{\sigma_0}{(8\varepsilon\varepsilon_0 n_s kT)^{1/2}} \tag{ii}$$

$$= \frac{-1.95 \times 10^{-6} \text{ C cm}^{-2}}{(8 \times 80 \times 8.85 \times 10^{-14} \text{ C V}^{-1} \text{ cm}^{-1} 10^{-6} \text{ mol cm}^{-3} \times 6}{\times 10^{23} \text{ mol}^{-1} \times 0.41 \times 10^{-20} \text{ V C})^{0.5}}$$

$$= -5.22.$$

From tables on hyperbolic functions

$$\frac{F\Psi_0}{2RT} = -2.36$$

and

$$\Psi_0 = \frac{-2.36}{96490 \text{ C mol}^{-1}/(2 \times 2.46 \times 10^3 \text{ V C mol}^{-1})} = -0.12 \text{ V}$$

2. The accumulation of H^+ by the negatively charged surface may be estimated by considering (3) and (4) of Table 10.4 which predict that

$$2.3 \log \frac{[H^+]_{x=0}}{[H^+]_{x=\infty}} = -\frac{F\Psi_0}{RT} \tag{iii}$$

$$pH_{surf} = pH_{bulk} + 16.9\Psi_0 \qquad \text{(at 20°C)}$$

Hence $pH_{surface} = pH_{bulk} - 2.0$

This calculation overestimates ΔpH, because the theory treats ions as point charges. More realistically the calculation should be made for Ψ of the plane of closest approach of the hydrated hydrogen ion.

The Stern Layer

In his theory Stern (1924) has accounted for the finite size of ions and the possibility of their specific adsorbability. Stern assumes that the fraction of available sites occupied by ions is related to the unoccupied fraction by Boltzmann expressions [namely, (3) of Table 10.4]. If n_c is the number of counterions of valence z adsorbed from a solution of bulk mole fraction x_0 in the solute and n_s is the number of possible adsorption sites at the surface, the Stern relation can be expressed as

$$\frac{n_c}{n_s - n_c} = x_0 e^{(zF\Psi_\delta + \Phi)/RT} \tag{21}$$

where Ψ_δ is the potential at the plane of adsorption and Φ is the energy (per mole) of specific adsorption due to nonelectrostatic forces. The left-hand side of equation 21 denotes the actually occupied fraction of sites. For a dilute solution, the corresponding ratio is the mole fraction of the solute which, for water as a solvent, can be replaced (dilute solution, 25°C) by $x_0 = c/55.4$, where c is the molar concentration. Hence, for a single ionic species, equation 21 can be rewritten in the form of a Langmuir equation [see equation (5) of Table 10.3]

$$\frac{\theta}{1 - \theta} = \frac{c}{55.4} e^{(zF\Psi_\delta + \Phi)/RT} \tag{22}$$

where $(zF\Psi_\delta + \Phi)$ has the significance of a standard free energy of adsorption. If Φ is larger than $zF\Psi_\delta$, a reversal of the total surface charge occurs (Figure 10.2d) because $zFn_c = \sigma_c$ exceeds the original charge density σ_s on the surface.

As we have seen, the compact Stern layer can be treated as a flat plate condenser of thickness δ in which the potential drops linearly with distance from Ψ_0 to Ψ_δ:

$$\sigma_s = \frac{\varepsilon_{Stern}\varepsilon_0}{\delta}(\Psi_0 - \Psi_\delta) \tag{23}$$

The relative dielectric permittivity ε_{Stern} in this Stern layer must be assumed to be considerably lower than the ε of bulk water.

Adsorption Against Electrostatic Repulsion

Following the assumptions of Stern, the overall standard free energy of adsorptions equals the sum of the total specific ("chemical") adsorption energy and the electrochemical work involved in the adsorption:

$$\Delta \bar{G}^\circ = \Phi + zF\Psi_\delta \tag{24}$$

Experimentally it is difficult or impossible to separate the energy of adsorption into its chemical and coulombic components. Despite this operational restriction, it is instructive to formulate a numerical example. For the adsorption of an ion on a surface of similar charge, equation 24 indicates that the chemical contribution to the overall standard free energy of adsorption opposes the electrostatic contribution. For the adsorption of a monovalent organic ion to a surface of similar charge, and against a potential drop (Ψ_δ) of 100 mV, the electrostatic term in equation 24, $zF\Psi_\delta$, is 10 kJ mol^{-1}. The standard chemical adsorption energy for typical sorbable monovalent organic ions is on the order of -8 to -35 kJ mol^{-1}, thus indicating that the electrostatic contribution to adsorption easily can be smaller than the chemical contribution. It is therefore at least qualitatively understandable that the addition of a suitable counterion decreases the surface potential and may produce a reversal of the surface charge at higher concentrations.

Somasundaran, Healy, and Fuerstenau [18] have investigated the effects of alkyl chain length on the adsorption of alkylammonium cations on negative silica surfaces at pH 6.5 to 6.9. These authors consider that, in addition to electrostatic effects, surfactant adsorption is influenced by the van der Waals energy of interaction or hydrophobic bonding between CH_2 groups on adjacent adsorbed surfactant molecules. The van der Waals cohesive energy per CH_2 group is found to be approximately $1\,RT$ (at 25°C, $RT = 2.5$ kJ mol^{-1}). For a 12-carbon alkylamine, the van der Waals cohesive energy is therefore on the order of 30 kJ mol^{-1} and exceeds the electrostatic contribution.

Davies and Rideal [19] have reviewed the results of several investigators on the specific interaction energy arising from polarization (complex formation) and van der Waals forces for a number of surfaces. For example, they give for the specific energy of interaction on a carboxylate surface for Na^+, Mg^{2+}, Ca^{2+}, and Cu^{2+} values of ~ -2.4, ~ -5.4, ~ -10.9, and -24 kJ mol^{-1}, respectively.

Polymers.   That van der Waals forces and hydrophobic bonding may contribute significantly to the specific interaction energy is apparent from the observation that macromolecules have a strong tendency to accumulate at interfaces. For example, on an incipiently negatively charged silica surface, polystyrene sulfonate (Table 10.5) is adsorbed readily, while monomeric p-toluenesulfonate ($CH_3-C_6H_4-SO_3^-$) does not adsorb from 10^{-4} M solutions [20].

Each polymer molecule or polymeric ion (Table 10.5) can have many groups or segments that potentially can be adsorbed; these groups are often relatively free of mutual interaction [20,21]. Usually the extent of adsorption, but not necessarily its rate, increases with increasing molecular weight and is affected by the number and type of functional groups in the polymer molecule. Hydroxyl, phosphoryl, and carboxyl groups can be particularly effective in causing adsorption; the significance of such functional groups directs attention to the specificity of the chemical interactions involved in the adsorption process. Natural polymers, such as starch, cellulose, tannins, humic acids (Section 8.3),

**TABLE 10.5   EXAMPLES OF SYNTHETIC POLYMERS AND POLYELECTRO-
LYTES**

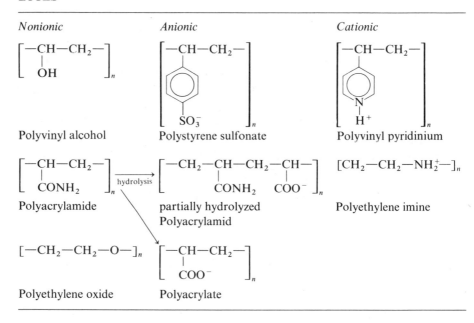

Nonionic

$$\left[-CH-CH_2-\atop\quad\ |\atop\quad OH\right]_n$$

Polyvinyl alcohol

$$\left[-CH-CH_2-\atop\quad\ |\atop\quad CONH_2\right]_n \xrightarrow{\text{hydrolysis}}$$

Polyacrylamide

$$[-CH_2-CH_2-O-]_n$$

Polyethylene oxide

Anionic

$$\left[-CH-CH_2-\atop\quad\ |\atop\quad\bigcirc\atop\quad\ |\atop\quad SO_3^-\right]_n$$

Polystyrene sulfonate

$$\left[-CH_2-CH-CH_2-CH-\atop\qquad\ |\qquad\qquad\ |\atop\qquad CONH_2\qquad COO^-\right]_n$$

partially hydrolyzed
Polyacrylamid

$$\left[-CH-CH_2-\atop\quad\ |\atop\quad COO^-\right]_n$$

Polyacrylate

Cationic

$$\left[-CH-CH_2-\atop\quad\ |\atop\quad\bigcirc\atop\quad\ |\atop\quad N\atop\quad\ |\atop\quad H^+\right]_n$$

Polyvinyl pyridinium

$$[CH_2-CH_2-NH_2^+-]_n$$

Polyethylene imine

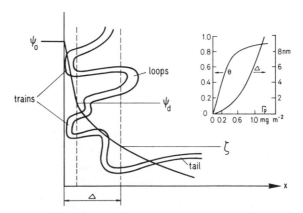

Figure 10.5   Effect of adsorbed polymer on the double-layer. Because of the presence of adsorbed
train segments, the Stern layer is modified. The surface charge is reduced because the adsorption
of polymer segments causes a displacement of specifically adsorbed counterions and a reduction in
the capacitance. The assumption of $\Psi_d \approx \zeta$ does not apply any more, because the adsorbed polymer
displaces the plane of shear. The parameters for describing adsorbed polymers are the fraction of the
first layer covered by segments, θ, and the effective thickness, Δ, of the polymer layer. The inset
(from Lyklema [21]) gives the distribution of segments over trains and loops for polyvinyl alcohol
adsorbed on silver iodide. Results obtained from double layer and electrophoresis measurements.
(After Lyklema [21].)

624

and synthetic polymers (Table 10.5) are of interest. Because many segments of the polymer can be in contact with the surface, a low bonding energy per segment may suffice to render the affinity of several segments together so high that their adsorption is virtually irreversible [19]. Figure 10.5 provides an illustration of the adsorption of a neutral polymer, polyvinyl alcohol, on a silver iodide surface, and the resulting effects on the double layer properties. Adsorption of anionic polymers on negative surfaces—especially in the presence of Ca^{2+} or Mg^{2+} which may act as coordinating links between the surface and the functional groups of the polymer—is not uncommon.

As will be discussed in Section 10.6, the adsorption of polymers may either increase or decrease colloid stability.

Hydrolyzed Metal Ions.   As pointed out in Chapter 6, the hydrolysis products of multivalent ions, that is, polynuclear polyhydroxo species, are adsorbed more readily at particle–water interfaces than nonhydrolyzed metal ions. In addition to metal-ion hydrolysis products, polysilicates, polyphosphates, and heteropoly ions (e.g., phosphotungstate and silicotungstate) have a tendency to become adsorbed.

10.4   THE SURFACE CHEMISTRY OF OXIDES, HYDROXIDES, AND OXIDE MINERALS

Oxides, especially those of Si, Al, and Fe, are abundant components of the earth's crust. Hence most of the solid phases in natural waters contain such oxides or hydroxides. Interactions of cations and anions with hydrous oxide surfaces are of importance in natural water systems, in geochemical processes, and in colloid chemistry.

In the presence of water, metal or metalloid oxides are generally covered with surface hydroxyl groups (Figure 10.6).

A hydroxylated oxide particle can up to a certain degree be understood as a polymeric oxoacid or -base. As already discussed in a preliminary way in Section 6.9 (see also Figure 6.19), we treat the specific adsorption of H^+ and OH^- and of cations, anions, and weak acids in terms of *surface coordination reactions* at the oxide–water interface (Figure 10.7).

The *pH-dependent charge* of metal or metalloid hydrous oxides results from proton transfers at the amphoteric surface:

$$\equiv MeOH_2^+ \;\rightleftharpoons\; \equiv MeOH + H^+; \qquad K_{a_1}^s = \frac{\{\equiv MeOH\}[H^+]}{\{\equiv MeOH_2^+\}} \quad (25)$$

$$\equiv MeOH \;\rightleftharpoons\; \equiv MeO^- + H^+; \qquad K_{a_2}^s = \frac{\{\equiv MeO^-\}[H^+]}{\{\equiv MeOH\}} \quad (26)$$

where [ ] and { } indicate concentrations of species in the aqueous phase (mol liter^{-1}) and concentrations of surface species (mol kg^{-1}), respectively.

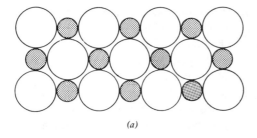

(a)

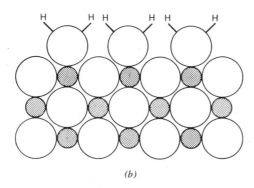

(b)

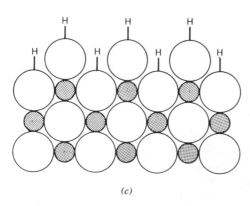

(c)

Figure 10.6   Schematic representation of the cross section of the surface layer of a metal oxide. ●, Metal ions; ○, oxide ions. The metal ions in the surface layer (a) have a reduced coordination number. They thus behave as Lewis acids. In the presence of water the surface metal ions may first tend to coordinate H_2O molecules (b). For most of the oxides dissociative chemisorption of water molecules (c) seems energetically favored. Oxide surfaces carry typically 4 to 10 hydroxyl groups per square nanometer. From P. Schindler, in *Adsorption of Inorganics at the Solid/Liquid Interface*, Anderson, N. and Rubin, A., Eds., Ann. Arbor. Science, Ann. Arbor., 1981.

The portion of the charge due to specific interaction with H^+ and OH^- ions corresponds to the difference in protonated and deprotonated ≡MeOH groups.

Operationally, there is a similarity between H^+ and metal ions (Lewis acids) and OH^- and other bases (Lewis bases). The OH group on a hydrous oxide surface has a complex-forming O-donor group like an OH^- or an OH group attached to another element (phosphate, silicate, polysilicate). Protons and metal ions compete with each other for the available coordinating sites on the surface:

$$\equiv MeOH + M^{z+} \rightleftharpoons \ \equiv MeOM^{(z-1)} + H^{+}$$

$$*K_1^s = \frac{\{\equiv McOM^{(z-1)}\}[H^{+}]}{\{\equiv MeOH\}[M^{z+}]} \quad (27)$$

$$\begin{matrix} =Me-OH \\ | \\ =Me-OH \end{matrix} + M^{a+} \longrightarrow (=MeO)_2M^{(a-2)} + 2H^{+}$$

$$*\beta_2^s = \frac{\{(\equiv MeO)_2M^{(a-2)}\}[H^{+}]^2}{\{\equiv Me_2(OH)_2\}[M^{z+}]} \quad (28)$$

The extent of coordination is related to the exchange (or displacement) of H^{+} by M^{z+} ions. Similarly, *ligand exchange* with coordinating anions leads to a release of OH^{-} from the surface:

$$\equiv Me-OH + A^{z-} \rightleftharpoons \ MeA^{(z-1)} + OH^{-} \qquad K_1^s \qquad (29)$$

$$\begin{matrix} =Me-OH \\ | \\ =Me-OH \end{matrix} + A^{z-} \rightleftharpoons (=Me)_2A^{(z-2)} + 2OH^{-} \qquad \beta_2^s \quad (30)$$

For protonated anions the ligand exchange may be accompanied by a deprotonation of the ligand at the surface. For example, in the case of HPO_4^{2-}:

$$\equiv MeOH + HPO_4^{2-} \rightleftharpoons \ =MeHPO_4 + OH^{-}$$

$$\rightleftharpoons \ \equiv MePO_4^{2-} + H_2O \qquad (31)$$

It is also conceivable that, under high pH conditions, adsorption of a metal ion may be accompanied by hydrolysis.

As Figure 10.7 illustrates qualitatively, the specific binding of H^{+} and cations increases and the specific binding of OH^{-} and anions decreases the net charge of the particle surface. The equilibrium constants that characterize the processes described can be used to quantify the extent of adsorption and the resultant net charge of the oxide particle surface as a function of pH and solute activity.

Amphoteric Properties and the pH of Zero Point of Charge

The principles involved in experimental determination of the fixed surface charge as a function of pH can be explained with the help of Figure 10.8. A model oxide with a large specific surface area (preferably some $m^2 \ g^{-1}$) is dispersed in a solution of an inert (nonspecifically adsorbable cations and anions) salt solution, for example, $NaClO_4$. Aliquots of the dispersion are titrated with standard base (NaOH) and with standard acid ($HClO_4$). The resultant titration curve may be compared with that obtained in the absence of the oxide,

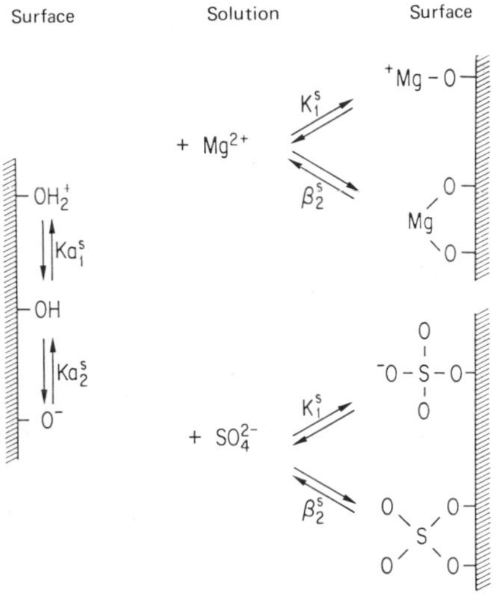

Figure 10.7 Interaction of hydrous oxides with acids and bases and with cations and anions.

and the quantity of H^+ or OH^- bound $(\Gamma_H - \Gamma_{OH})$ is calculated (see also Example 10.2). The point of zero proton condition, the pH_{zpc}, $(\{\equiv MeOH_2^+\} = \{\equiv MeO^-\})$ corresponds to the pH where the surface is uncharged and is, in the absence of specifically adsorbable ions other than H^+ and OH^-, identical with the *isoelectric point* (iep). The isoelectric point is the pH where the particle is electrokinetically uncharged.

The acidity constants in equations 25 and 26 are microscopic equilibrium constants (see Section 6.9) because each loss of a proton reduces the charge on the solid poly-acid and thus affects the acidity of the neighbor groups. The free energy of deprotonation consists of the *dissociation* as measured by an intrinsic acidity constant $K_a^s(intr)$ and the *removal* of the proton from the site of the dissociation into the bulk of the solution as expressed by the Boltzman factor; thus

$$K_{a1}^s = K_{a1}^s(intr) \exp\left(\frac{F\Psi'}{RT}\right) \tag{32}$$

where Ψ' is the effective potential difference between the binding site and the bulk solution and $K_a^s(intr)$ is the acidity constant of an acid group in a hypo-thetically completely chargeless surrounding. There is no direct way to obtain Ψ' theoretically or experimentally. It is possible, however, to determine the microscopic constants experimentally and to extrapolate these constants to zero surface charge in order to obtain intrinsic constants.

As shown in Figure 10.8, the intrinsic values for the acidity constants can be obtained by linear extrapolation to the zero charge condition. The proton condition where the surface charge caused by binding of H^+ or OH^- is zero is called the *zero point of charge* or pH_{zpc}. pH_{zpc} is given by

$$pH_{zpc} = \tfrac{1}{2}[pK_{a1}^s(intr) + pK_{a2}^s(intr)] \tag{33}$$

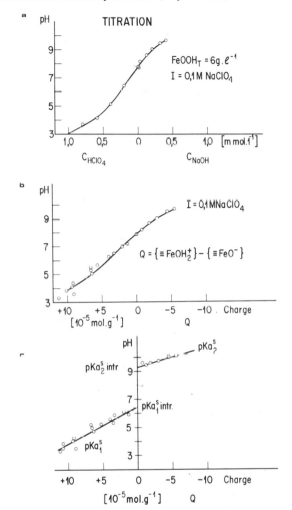

Figure 10.8 Titration of a suspension of α-FeOOH (goethite) in absence of specifically adsorbable ions. Data from Sigg and Stumm. (*a*) Acidimetric–alkalimetric titration in the presence of an inert electrolyte. (*b*) Charge calculated from the titration curve (charge balance). (*c*) Microscopic acidity constants calculated from (*a*) and (*b*). Extrapolation to charge zero gives intrinsic pK_{a1}^s and pK_{a2}^s.

The pH$_{zpc}$ obtained in this way should agree with the common intersection point of the titration curves obtained with different concentrations of inert salt (Figure 10.9*b*).

It is evident that oxides formed by metal ions that are strong acids (see Figure 10.6) have low pH$_{zpc}$ values and that less acidic metal ions have high pH$_{zpc}$ values. We can predict the pH$_{zpc}$ of metal oxides from electrostatic considerations. Parks [22] has shown that the pH$_{zpc}$ of a simple oxide is related to the appropriate cationic charge and radius of the central ion. As shown by Parks

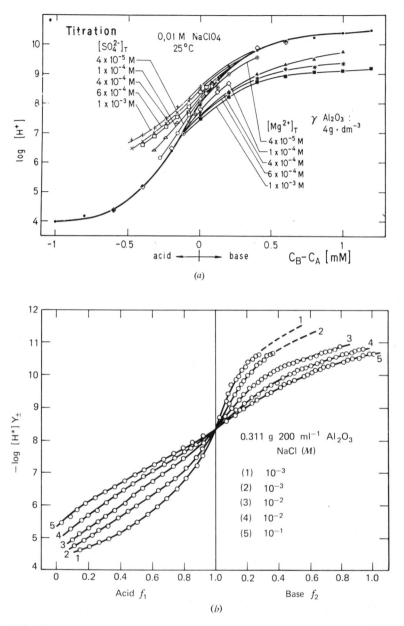

Figure 10.9 Characterization of the surface. Chemical properties of γ-Al_2O_3 in the presence of Mg^{2+} and SO_4^{2-}. (*a*) Alkalimetric–acidimetric titration curves in the absence and presence of surface-coordinating Mg^{2+} and SO_4^{2-}, respectively. (*b*) Titration curves at various ionic strengths (f = equivalent fraction of titrant added). pH of zero point of charge of the suspension is independent of electrolyte concentration.

and illustrated by a few examples in Table 10.6, the pH_{zpc} of a composite oxide is approximately the weighted average of the pH_{zpc} values of its components. Predictable shifts in pH_{zpc} occur in response to state of hydration, cleavage habit, and crystallinity [22].

pH_{zpc} of Minerals Other Than Oxides or Hydroxides.   The pH_{zpc} of salt-type minerals depends, sometimes in a complicated way, upon pH and on the concentration (activities) of all potential determining ions. Thus in the case of calcite, possible potential-determining species, in addition to H^+ and OH^-, are H_2CO_3, HCO_3^-, CO_3^{2-}, Ca^{2+}, and $CaOH^+$; various mechanisms of charge development are possible. It is not meaningful to refer to a pH_{zpc} of such non-oxides unless the solution composition is also specified.

In the absence of complications such as those caused by structural or adsorbed impurities, the zero point of charge of the solid should correspond to the pH of charge balance (electroneutrality) of potential-determining ions.

TABLE 10.6   ZERO POINT OF CHARGE[a]

Material	pH_{zpc}
α-Al_2O_3	9.1
α-$Al(OH)_3$	5.0
γ-$AlOOH$	8.2
CuO	9.5
Fe_3O_4	6.5
α-$FeOOH$	7.8
γ-Fe_2O_3	6.7
"$Fe(OH)_3$"(amorph)	8.5
MgO	12.4
δ-MnO_2	2.8
β-MnO_2	7.2
SiO_2	2.0
$ZrSiO_4$	5
Feldspars	2–2.4
Kaolinite	4.6
Montmorillonite	2.5
Albite	2.0
Chrysotile	> 12

[a] The values are from different investigators who have used different methods and are not necessarily comparable. They are given here for illustration.

Specific Chemical Interaction with Cations and Anions
at the Oxide Surface

The surface chemical properties of the oxide surface are sensitive to the composition of the aqueous phase because specific adsorption or binding of solutes to the surface may increase, decrease, or reverse the effective surface charge on the solid. The titration curve for a hydrous oxide dispersion in the presence of a coordinatable cation is shifted toward lower pH values (Figure 10.9) in such a way as to lower the pH of the zero proton condition at the surface. At this point the portion of the charge due to H^+ and OH^- or their complexes† becomes zero. Because of the binding of M^{z+} to the surface ($\Gamma_{M^{z+}}$), the fixed surface charge increases or becomes less negative‡ and, at the pH where the fixed surface charge becomes zero, the iep is shifted to higher pH values. Correspondingly, specifically adsorbable anions increase the pH of the zero proton condition but lower the pH of the iep (Figures 10.9 and 10.10, Example 10.2).

If the fixed surface charge of an oxide particle, σ (C m^{-2}), arises from the specific adsorption of H^+, OH^-, and cations and anions by the hydrous oxide surface, it is in principle possible to determine its value by determining experimentally (analytically) the extent of adsorption of charged species:

(i) in the absence of specifically adsorbable cations and anions:

$$\sigma = F(\Gamma_H - \Gamma_{OH}) \tag{34}$$

(ii) in the presence of a specifically adsorbable cation M^{z+}:

$$\sigma = F(\Gamma_H - \Gamma_{OH} + z\Gamma_{M^{z+}}) \tag{35}$$

(iii) in the presence of a specifically adsorbable anion $H_n A^{(z-n)-}$:

$$\sigma = F(\Gamma_H - \Gamma_{OH} - z\Gamma_{A^{z-}}) \tag{36}$$

or, generally,

$$\sigma = F\left(\Gamma_H - \Gamma_{OH} + \sum_{ji} z_j \Gamma_{M_i^{z+}} - \sum_{ji} z_j \Gamma_{A_i^{z-}}\right) \tag{37}$$

where F = Faraday constant (C mol^{-1}), z = magnitude of the charge of the nonhydrolyzed cation or deprotonated anion; Γ_H, Γ_{OH}, and Γ_{M^+}, $\Gamma_{A^{z-}}$ are the adsorption densities (mol m^{-2}) of H^+ (and its complexes), of OH^- (and its complexes), of cations, and of the deprotonated anion§) respectively.

---

† If a hydrolyzed metal ion is adsorbed, its OH^- will be included in the proton balance; similarly, in case of adsorption of protonated anions, their H^+ will be included in the proton balance.
‡ Some colloid chemists often place these specifically bound cations and anions in the Stern layer. From a coordination chemistry point of view it does not appear very meaningful to assign a surface-coordinating ion to a layer different than H or OH in a $\equiv$MeOH group.
§ If 1 mol of an ion like $H_2PO_4^-$ is adsorbed, the surface gains the equivalent of 2 mol of protons (to be included in the Γ_H) and 1 mol of PO_4^{3-}.

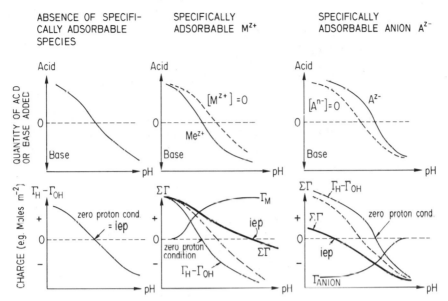

Figure 10.10   The net charge at the hydrous oxide surface is established by the proton balance (adsorption of H or OH⁻ and their complexes at the interface and specifically bound cations or anions. This charge can be determined from an alkalimetric–acidimetric titration curve and from a measurement of the extent of adsorption of specifically adsorbed ions. Specifically adsorbed cations (anions) increase (decrease) the pH of the isoelectric point (iep) but lower (raise) the pH of the zero proton condition.

Example 10.2.   Evaluation of Surface Charge from Alkalimetric and Acidimetric Titration Curves

1. We will first demonstrate how the surface charge of a hydrous oxide (γ-Al$_2$O$_3$) can be calculated from an experimental titration curve (e.g., Figure 10.9).

2. We will then illustrate how the titration curve is affected by Mg^{2+}, which becomes surface-coordinated at the oxide surface, and how this surface co-ordination (specific adsorption) affects the surface charge of the oxide and the pH of the isoelectric point, pH$_{\text{iep}}$.

1. In titrating a suspension of γ-Al$_2$O$_3$ in an inert electrolyte (10^{-2} M NaClO$_4$) with NaOH (C_B and C_A = concentration of base and acid, respectively, added per liter), we can write for any point on the titration curve (cf. Sections 3.8 and 6.3)

$$C_A - C_B + [\text{OH}^-] - [\text{H}^+] = [\equiv\text{AlOH}_2^+] - [\equiv\text{AlO}^-] \qquad \text{(i)}$$

where [ ] indicates concentrations of solute and surface species per unit volume solution (mol liter^{-1}). Equation i can also be derived from a charge balance. The right-hand side gives the net number of moles per liter of OH$^-$ ions bound to γ-Al$_2$O$_3$. The mean surface charge (i.e., the portion of the charge due to

OH^- or H^+) can be calculated as a function of pH from the difference between total added base or acid and the equilibrium OH^- and H^+ ion concentration for a given quantity a (kg liter^{-1}) of oxide used:

$$\frac{C_A - C_B + [OH^-] - [H^+]}{a} = \{\equiv AlOH_2^+\} - \{\equiv AlO^-\} = Q \qquad \text{(ii)}$$

where { } indicates the concentration of surface species in mol kg^{-1} (e.g., $[\equiv AlO^-]/a = \{\equiv AlO^-\}$). If the specific surface area S(cm^2 kg^{-1}) of the aluminum oxide used is known (in this case 1.2×10^9 cm^2 kg^{-1}), the surface charge σ (C cm^{-2}) can be calculated:

$$\sigma = QFS^{-1} = F(\Gamma_{H^+} - \Gamma_{OH^-}) \qquad \text{(iii)}$$

where F is the Faraday constant (96,490 C mol^{-1}) and Γ_{H^+} and Γ_{OH^-} are the "adsorption" densities of H^+ and OH^- (mol cm^{-2}).

The *zero point of charge* pH$_{zpc}$ corresponds to the *zero proton condition* at the surface:

pH$_{zpc}$ (zero proton condition = zero point of charge):

$$\{\equiv AlOH_2^+\} = \{\equiv AlO^-\}; \qquad \Gamma_{H^+} = \Gamma_{OH^-} \qquad \text{(iv)}$$

In this case pH$_{zpc}$ = 8.0.

2. In presence of Mg^{2+}, the titration curve is shifted in such a way that at a given pH (Figure 10.9a) more base is consumed. This results from the coordination reactions (cf. Figure 6.19)

$$\equiv AlOH + Mg^{2+} \quad \rightleftharpoons \quad \equiv AlOMg^+ + H^+$$

$$2\equiv AlOH + Mg^{2+} \quad \rightleftharpoons \quad (\equiv AlO)_2Mg + 2H^+ \qquad \text{(v)}$$

Considering the proton condition (also derivable from the electroneutrality condition) we can write for the titration curve (concentrations with an asterisk are those in the presence of Mg^{2+})

$$Q_H = \frac{\overset{*}{C}_A - \overset{*}{C}_B + [OH^-] - [H^+]}{a}$$

$$= \{\equiv Al\overset{*}{OH_2^+}\} - \{\equiv Al\overset{*}{O^-}\} - \{\equiv AlOMg^+\} - 2\{(\equiv AlO)_2Mg\} \qquad \text{(vi)}$$

where Q_H gives the portion of the charge caused by bound H^+ or OH^-

$$Q_H = \{\equiv Al\overset{*}{OH_2^+}\} - \{\equiv Al\overset{*}{O^-}\} - \{\equiv AlOMg^+\} - 2\{(\equiv AlO)_2Mg\} \qquad \text{(vii)}$$

The zero proton balance at the surface, $Q_H = 0$ or $\Gamma_H - \Gamma_{OH} = O$, is now at a pH considerably lower than that in the absence of Mg^{2+}. A charge balance of the surface species gives the fixed surface charge:

$$Q = \{\equiv AlOMg^+\} + \{\equiv AlOH_2^+\} - \{\equiv AlO^-\} \qquad \text{(viii)}$$

Q is experimentally accessible from the charge caused by the surface proton balance, Q_H (determined from the titration curve, $Q_H = (\overset{*}{C}_A - \overset{*}{C}_B + [OH^-] - [H^+])/a$), and the (analytically determined) quantity of Mg^{2+} bound to the surface (cf. equations vi and viii):

$$Q = \frac{\overset{*}{C}_A - \overset{*}{C}_B + [OH^-] - [H^+]}{a} + 2\{\equiv AlOMg^+\} + 2\{(\equiv AlO)_2Mg\} \quad (ix)$$

or

$$\sigma = F(\Gamma_H - \Gamma_{OH} + 2\Gamma_{Mg^{2+}}) \quad (x)$$

A comparison of equation ix with equation vi shows that $Q > Q_H$. Correspondingly the pH at which $Q = 0$ (or $\sigma = 0$) is considerably higher than the pH at which $Q_H = 0$. And $Q = 0$ corresponds most nearly to the condition where the particle is electrokinetically uncharged. Thus $Q = 0$ (or $\sigma = 0$ corresponds to the pH_{iep}. As Figure 10.9a shows, surface coordinate binding of a cation increases pH_{iep} but lowers pH_{zpc}.

Figure 10.11a illustrates the binding (adsorption) of Pb^{2+} on hydrous γ-Al_2O_3; although positively charged under these pH conditions, the Al_2O_3 surface removes Pb^{2+} cations from the solution effectively. In the pH range considered Pb^{2+} does not hydrolyze to any substantial extent. The pH-dependent binding of metal ions to amorphous SiO_2 is illustrated in Figure 10.11b.

The extent of adsorption of a metal ion is strongly pH-dependent. The fact that pH is a most pertinent variable governing the extent of adsorption was first recognized by Kurbatov et al. [23]. Later complex formation equilibrium models were developed by Maatman and his collaborators [24]. James and Healy [25] and others have postulated that metal hydroxo species are more strongly adsorbed than free metal ions. In the coordination model presented here, this pH dependence can be explained by the pH dependence of the activity of the $\equiv MeO^-$ groups and the affinity of this group for the metal ion.†

O'Connor and Kester [26] obtained the same type of pH dependence for the adsorption of Cu(II) and Co(II) on illite from fresh and marine systems. At high pH these ions desorb, most likely because of the formation of anionic hydroxo complexes.

Complex formers in solution may compete with surface ligands if they are able to occupy all the sites in the coordination shell of the metal ion in solution. Thus EDTA and NH_3, if present in stoichiometric excess, can prevent specific adsorption on the hydrous oxide surface. If the dissolved ligands occupy only part of the coordination shell of the metal, the adsorption may proceed by formation of ternary surface complexes, for example, $\equiv Si$—O—Cu—L or $\equiv Si$—L—Cu [27].

---

† It is important to distinguish between mononuclear and polynuclear metal-ion hydrolysis species. Many investigations on the adsorption of hydrolyzable metal ions have been carried out under solution conditions where, often unknowingly, polynuclear metal species are usually adsorbed strongly to surfaces.

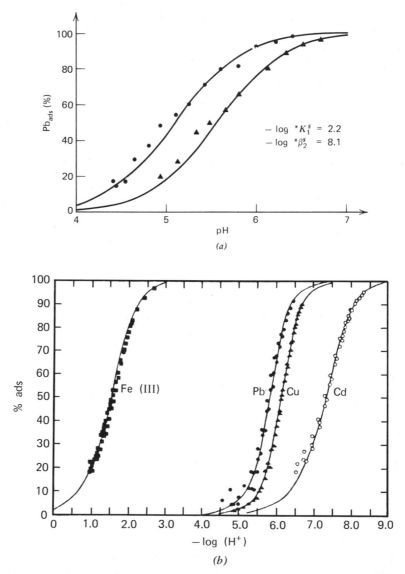

Figure 10.11 pH dependence of the binding of metal ions to hydrous oxide surfaces. (*a*) pH dependence of the adsorption of Pb(II) to γ-Al₂O₃ (alon). Percent adsorption (relative to Pb(II) in solution) measured in batch experiments (48-hour equilibration time). The lines are theoretically determined (i.e., calculated with the help of the complex formation constants $*K_1^s$ and $*\beta_2^s$). ●, 11.72 g alon dm⁻³, Pb(II)$_T$ = 2.94 × 10⁻⁴ M; ▲, 3.18 g alon dm⁻³, Pb(II)$_T$ = 9.8 × 10⁻⁵ M. [From Hohl and Stumm, *J. Colloid Interfacial Sci.*, **55**, 281 (1976).] (*b*) Adsorption of metal ions on amorphous silica as a function of −log[H⁺]. [From Schindler et al., *J. Colloid Interfacial Sci.*, **55**, 469 (1976).]

636

Anions and Weak Acids.   As for cations, the extent of adsorption of anions and of weak acids and its pH-dependence can be explained by considering the affinity of the surface sites for the ligands and the acid–base properties of the surface sites and those of the ligands. Figure 10.12 illustrates the effect of silicic acid upon the titration curve of goethite (α-FeOOH) Dissolved silica is primarily present as H_4SiO_4 below pH $= 9$. The shift in the titration curve caused by silica reflects a release of protons, which can be explained with the reactions

$$\equiv FeOH + H_4SiO_4 = \equiv FeOSi(OH)_3 + H_2O$$

$$\equiv FeOSi(OH)_3 = \equiv FeOSiO(OH)_2^- + H^+$$

The adsorption of dissolved silica decreases the fixed charge of the goethite surface (Figure 10.12). With the help of calculations using the equilibrium constants, the extent of adsorption as a function of pH can be predicted.

Adsorption of *polar organic substances* often gives results similar to those illustrated for silicate. The adsorption of the organic acid is accompanied by deprotonation of the adsorbate.

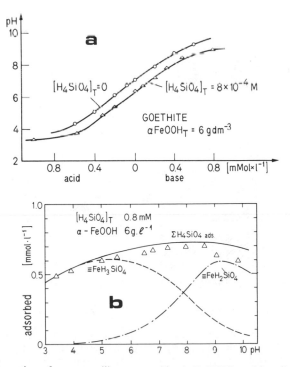

Figure 10.12   Adsorption of aqueous silica on goethite (α-FeOOH) and its effect on the alkalimetric titration curve and surface charge (in the presence of $0.1\,M$ $NaClO_4$). The adsorption of silicic acid on α-FeOOH tends to release protons (a) and causes a decrease in surface charge. The extent of adsorption as a function of pH can be predicted by an equilibrium model (b). (From L. Sigg and W. Stumm, 1979.)

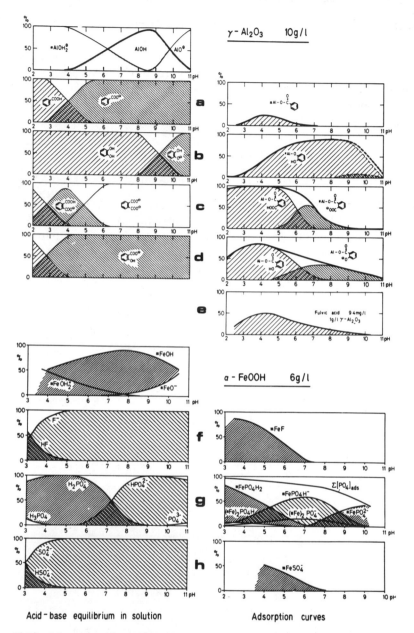

Figure 10.13 Adsorption of organic acids on γ-Al_2O_3 (a–e) and of anions on α-FeOOH (f–h). Curves (a–d) and (f–h) are calculated from experimentally derived stability constants (organic acids and anions 0.1 to 1 mmol liter^{-1}, 22°C, 0.1 M $NaClO_4$). Adsorption of fulvic acids (collected from lake sediments at high pH) (e) was experimentally determined. [Modified from Stumm, Kummert, and Sigg, *Croat. Chim. Acta* (1980).]

Figure 10.13 illustrates that the adsorption (binding) of simple weak acids or their anions is greatest at about $pH = pK_a$. Weak acids with several protons adsorb over a wider pH range with the possibility of deprotonating at the surface. The models help explain the adsorption properties of natural organic substances (such as fulvic acids) in fresh surface waters (Figure 10.13e).

The binding of *phosphate* to hydrous oxides, especially Al_2O_3 and $FeOOH$, is also characterized by a proton release and a shift of the isoelectric point to lower pH values.

Comparable Tendency of Ligands to Form Complexes in Solution as at the Oxide– Water Interface. We may, for example, compare the surface coordination of the acid H_2A at an iron oxide surface:

$$\equiv FeOH + H_2A \rightleftharpoons \equiv FeHA + H_2O; \qquad *K_1^s \qquad (38)$$

with the corresponding reaction in solution

$$FeOH^{2+} + H_2A \rightleftharpoons FeHA^{2+} + H_2O; \qquad *K_1 \qquad (39)$$

Comparable equations can be written for the interaction with monoprotic acids. Figure 10.14 illustrates that the tendency of ligands to displace OH^- from $\alpha\text{-FeOOH}$ and $\gamma\text{-Al}_2O_3$ surfaces is the same as the tendency of these

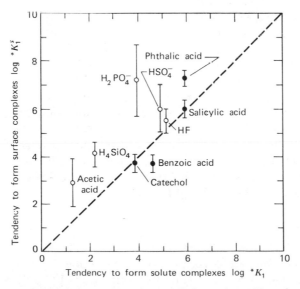

Figure 10.14   Relationship between coordination in solution and at oxide surfaces. The tendency of various ligands to form metal complexes in solution is similar to that of ligands to form surface complexes at the corresponding metal oxide surface (compare equations 38 and 39). This relationship can be utilized to estimate surface coordination equilibrium constants from the corresponding complex formation constants in solution. $\bigcirc$, $\alpha\text{-FeOOH}$; $\bullet$, $\gamma\text{-Al}_2O_3$.

ligands to displace OH^- from $FeOH^{2+}$(aq) and $AlOH^{2+}$(aq), respectively. Thus strong complexes in solution correspond to strong complexes at the surface. Since the solute complexes considered in this figure are inner-sphere complexes, we may infer that the surface complexes formed are of the inner-sphere type [28].

10.5   ION EXCHANGE

In every double layer one type of counterion can be exchanged for another type of ion of the same sign. Mixed oxides can be prepared in which a second cation of charge different from that of the parent cation is introduced into the structure, often resulting in remarkable alterations of the surface chemical behavior of these oxides. As we have seen, atomic isomorphic substitution and other lattice defects produce a fixed number of charges (anionic sites in the solid phase). Substitutions also strongly affect the acidity of structural OH groups. For example, the acid strength in the sequence $Si(OH)_4$, $PO(OH)_3$, $SO_2(OH)_2$, and $ClO_3(OH)$ increases from very weak (silicic acid) to very strong (perchloric acid), as can be predicted qualitatively from electrostatic and structural considerations. As Pauling has pointed out, an aluminum tetrahedron with corners shared with silicon tetrahedra is similar to the perchlorate ion, hence, we should expect to obtain a strong acid by replacing the K^+ ion of mica with hydrogen ion. Most clays behave as weak acids.

The cation-exchange capacity of a negative double layer may be defined as the excess of counterions that can be exchanged for other cations. This ion-exchange capacity corresponds to the area marked σ_+ in Figure 10.2. In the Gouy double layer the surface charge density is given by the sum of σ_+ (charge due to surplus of cations) and σ_- (charge due to deficiency of anions).

It can be shown that σ_+/σ_- remains independent of electrolyte concentration only for constant-potential surfaces. However, for constant-charge surfaces such as double layers on clays, σ_-/σ_+ increases with increasing electrolyte concentration, the deficiency of anions and the surplus of cations decreasing with increasing electrolyte concentration. Hence the cation-exchange capacity (as defined by σ_+) increases with dilution and becomes equal to the total surface charge at great dilutions [14].

Preferential Concentration of Multivalent Ions. The Gouy theory predicts that multivalent ions are concentrated in the double layer to a much larger extent than monovalent ions. Thus the ratio $[Ca^{2+}]/[Na^+]$ is much higher in the proximity of a negatively charged surface than in the medium. The fact that selectivity for a greater concentration of bivalent cations decreases with increasing ionic strength can also be derived from the Gouy model. Experimental data are in accord with these generalized predictions.

The surface chemical properties of clay minerals may often be interpreted in terms of the surface chemistry of the structural components, that is, sheets of tetrahedral silica, octahedral aluminum oxide (gibbsite), or magnesium hydroxide (brucite).

Zeolites are important mineral exchangers in marine sediments. A zeolite is a crystalline aluminosilicate with a tetrahedral framework enclosing cavities occupied by cations and H_2O molecules, both of which have enough freedom of movement to permit cation exchange and reversible dehydration [29]. Zeolites exhibit ion-exchange selectivity that correlates well with structural properties and chemical composition [30]. An enormous range of cation-exchange behavior is exhibited by inorganic phosphates [31]

Simple Models. Some of the pertinent characteristics of the ion-exchange process can be visualized from the model depicted in Figure 10.15a [32]. Here the fixed charges are indicated as anionic sites. The ion-exchange framework is held together by lattice forces or interlayer attraction forces. These forces are represented in the model by mechanical springs. They may be very elastic in organic exchangers and quite nonelastic in zeolites. In clays such as montmorillonites, because of the difference in osmotic pressure between solution and interlayer space, water penetrates into the interlayer space. Depending upon the hydration tendency of the counterions and the elasticity of the interlayer forces, different interlayer spacings may be observed. A composite balance among electrostatic, van der Waals, and osmotic forces influences the swelling pressure in the ion exchange phase and in turn also the equilibrium position of the ion-exchange equilibrium. This model illustrates that the extent of swelling increases with the extent of hydration of the counterion. Furthermore, the swelling is much less for a bivalent cation than for a monovalent counterion. Because only half as many Me^{2+} are needed as Me^+ to neutralize the charge of the ion exchanger, the osmotic pressure difference between the solution and the ion-exchange framework becomes smaller. Thus, if strongly hydrated Na^+ replaces less hydrated Ca^{2+} and Mg^{2+} in soils, the resulting swelling adversely affects the permeability of soils. Similarly, in waters of high relative $[Na^+]$, the bottom sediments are less permeable to water. Complex formation between the counterions and the anionic sites (oxo groups) also reduces the swelling pressure.

From a geometric point of view, clays can be packed rather closely. Muds containing clays, however, have a higher porosity than sand. The higher porosity of the clays is caused in part by the high water content (swelling), which in turn is related to the ion-exchange properties.

The Distribution of Charge. Figure 10.15c illustrates schematically the charge densities of ions on and surrounding a one-component ion exchanger as a function of pH. Such an exchanger has a finite number of "built-in" anionic sites, σ_{-I}. For the interpretation of the distribution of exchangeable cations in

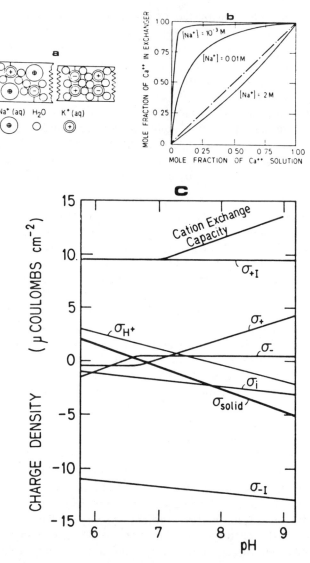

Figure 10.15 Ion exchange. (*a*) Buser's model [*Chimia*, **9**, 73 (1955)]. Incorporation of well-hydrated Na^+ ions leads to a larger volume and a higher swelling pressure than incorporation of less hydrated K^+ ions. (Reproduced with permission from *Schweizerische. Chemikerverband*, Zürich, Switzerland.) (*b*) Typical exchange isotherms for the reaction $Ca^{2+} + 2\{Na^+R^-\} \rightleftarrows \{Ca^{2+}R_2\} + 2Na^+$. In dilute solutions the exchanger shows a strong preference for Ca^{2+} over Na^+. This selectivity decreases with increasing ion concentration. The 45° line represents the isotherm with no selectivity. (*c*) Schematic representation of distribution of charge components as a function of pH on the surface of a one-component exchanger.

the proximity of the solid surface, it is convenient to distinguish between cations within the ion-exchange framework (interlayer cations and complex-bound cations) and cations in the diffuse double layer. A large part of the negative fixed-charge density σ_{-I} is compensated for by cations within the ion-exchange framework, σ_{+I}. The extent of this compensation, hence the magnitude of σ_{+I}, depends on the type and concentration of cation and on the pH of the solution. There is a slight pH dependence of σ_{-I}, because progressively less H$^+$ becomes dissociated from structural OH groups as the pH is lowered. A *net* intrinsic structural charge σ_i,

$$\sigma_i = \sigma_{+I} + \sigma_{-I} \tag{40}$$

and the charge imposed by potential-determining H$^+$ and OH$^-$, $\sigma_H = F(\Gamma_H - \Gamma_{OH})$, make up the total charge density σ_{solid} of the ion-exchange structure. This is now balanced by a charge in the diffuse part of the double layer, $\sigma_d = \sigma_+ + \sigma_-$. Combining the charge density of the structural cations with that of the cations in the diffuse double layer gives the total cation-exchange capacity (CEC):

$$CEC = \sigma_{+I} + \sigma_+ \tag{41}$$

As Figure 10.15c illustrates, the CEC increases with increasing pH above pH$_{zpc}$ but stays nearly constant at lower pH values. This is in accord with experimental observations. Some anion-exchange capacity exists as well on the acid side of pH$_{zpc}$. Because of the net intrinsic (negative) charge σ_i, pH$_{zpc}$ is shifted to a value lower than that observed in the absence of a structural charge.

Ion-Exchange Equilibria

The double-layer theory predicts qualitatively correctly that the affinity of the exchanger for bivalent ions is larger than that for monovalent ions and that this selectivity for ions of higher valency decreases with increasing ionic strength of the solution. However, according to the Gouy theory, there should be no ionic selectivity of the exchanger between different equally charged ions.

Relative affinity may be defined quantitatively by *formally* applying the mass law to exchange reactions:

$$\{Na^+R^-\} + K^+ = \{K^+R^-\} + Na^+ \tag{42}$$

$$2\{Na^+R^-\} + Ca^{2+} = \{Ca^{2+}R_2^{2-}\} + 2Na^+ \tag{43}$$

where R$^-$ symbolizes the negatively charged network of the cation exchanger. A selectivity coefficient Q can then be defined by

$$Q_{(NaR \to KR)} = \frac{X_{KR}\,[Na^+]}{X_{NaR}\,[K^+]} \tag{44}$$

$$Q_{(NaR \to CaR)} = \frac{X_{CaR}\,[Na^+]^2}{X_{NaR}^2\,[Ca^{2+}]} \tag{45}$$

where X represents the equivalent fraction of the counterion on the exchanger (e.g., $X_{CaR} = 2[CaR^{2+}]/(2[CaR^{2+}] + [NaR^+])$). The selectivity coefficients may be treated as mass-law constants for describing, at least in a semiquantitative way, equilibria for the interchange of ions, but these coefficients are neither constants nor are they thermodynamically well defined. Because the activities of the ions within the lattice structure are not known and vary depending on the composition of the ion-exchanger phase, the coefficients tend to deviate from constancy. Nevertheless, it is expedient to use equation 45 for illustrating the concentration dependence of the selectivity for more highly charged ions.

In Figure 10.15b the equivalent fraction of Ca^{2+} on the exchanger is plotted as a function of the equivalent fraction of Ca^{2+} in the solution. For a hypothetical exchange with no selectivity, the exchange isotherm is represented by the dashed line. In such a case the ratio of the counterions is the same for the exchanger phase as in the solution. The selectivity of the exchanger for Ca^{2+} increases markedly with increased dilution of the solution, but in solutions of high concentration the exchanger loses its selectivity. The representation in this figure makes it understandable why a given exchanger may contain predominantly Ca^{2+} in equilibrium with a fresh water; in seawater, however, the counterions on the exchanger are predominantly Na^+. Figure 10.15b also illustrates why, in the technological application of synthetic ion exchangers such as water softeners, Ca^{2+} can be selectively removed from dilute water solutions, whereas an exhausted exchanger in the Ca^{2+} form can be reconverted into a {NaR} exchanger with a concentrated brine solution or with undiluted seawater.

Table 10.7 gives the experimentally determined distribution of Ca^{2+} and K^+ for three clay minerals at various equivalent concentrations of Ca^{2+} and K^+. The results demonstrate the concentration dependence of the selectivity; they also show that marked differences exist among various clays.

TABLE 10.7   ION EXCHANGE OF CLAYS WITH SOLUTIONS OF CaCl₂ AND KCl OF EQUAL EQUIVALENT CONCENTRATION[a]

		Ca^{2+}/K^+ Ratios on Clay			
Clay	Exchange Capacity (meq g^{-1})	Concentration of Solution $2[Ca^{2+}] + [K^+]$ (meq liter^{-1})			
		100	10	1	0.1
Kaolinite	0.023	—	1.8	5.0	11.1
Illite	0.162	1.1	3.4	8.1	12.3
Montmorillonite	0.810	1.5	—	22.1	38.8

[a] From L. Wiklander, *Chemistry of the Soil*, F. E. Bear, Ed., 2nd ed., Van Nostrand Reinhold, New York, 1964. Reproduced with permission from Reinhold Publishing Corp.

The dependence of selectivity upon electrolyte concentration has important implications for analytical procedures for determining ion-exchange complements and the composition of interstitial waters. Sayles and Mangelsdorf [33] have shown in a well-documented report that the rinsing of marine samples with distilled water shifts the exchange equilibria away from true seawater conditions; the influence of rinsing increases the bivalent/monovalent ratio, especially the Mg^{2+}/Na^+ ratio. As Table 10.8 illustrates, Na^+ rather than Mg^{2+} is the major exchangeable cation of the clays studied. Sayles and Mangelsdorf [33] and Russel [34] have shown that the net reaction of fluvial clays and seawater is primarily an exchange of seawater Na^+ for bound Ca^{2+}. This process is of importance in the geochemical budget of Na^+.

Thermodynamic Relationships

Equilibrium relationships for ion-exchange reactions have been discussed by many writers, and various approaches have been used [35–38]. For example, equations for ion-exchange equilibria have been derived from the concept of Donnan equilibria, but the resulting formulations (essentially the same as those given in equations 44 and 45) contain terms for the activities in the ion exchanger phase (but the activity coefficients for these species cannot be evaluated).

Thomas and his collaborators [37] have recognized that the ion-exchange equilibria between clay minerals and solutions of salts cannot be described, even roughly, by a mass-law expression in terms of stoichiometric concentrations. If the clay phase is considered as a separate electrolyte phase, a wet montmorillonite is at the simplest a 3 M solution in which the anions are fixed in position and in which the motion of the cations is otherwise highly inhibited [36].

Affinity Series. With the help of ion-exchange constants, a general order of affinity can be given. Eisenmann [39] has worked out the consequences of the assumption that cation-exchange equilibria are dominated by (a) Coulombic interactions between the counterions (in various states of hydration) and the fixed groups of the exchanger, and (b) ion–dipole and ion-induced dipole interactions between the counterions and water molecules (ionic hydration). Where (a) is weak compared to (b) the normal affinity sequence (Hofmeister series)

$$Cs^+ > K^+ > Na^+ > Li^+$$
$$Ba^{2+} > Sr^{2+} > Ca^{2+} > Mg^{2+}$$

(46)

is observed, that is, the ion with the larger hydrated radius tends to be displaced by the ion of smaller hydrated radius. As interactions of the first kind predominate over those of the second, the selectivity may be reversed; that is, the affinity increases with decreasing crystal radii ($Li^+ > Na^+ > K^+ > Rb^+ > Cs^+$). For most clays the normal series prevails, while for some zeolites and glasses the reversed selectivity may be observed. For a more refined interpretation the reader is referred to Eisenmann [38]. The buffering effect of clays on $[H^+]$ is evident from alkalimetric titration curves.

TABLE 10.8 INTERACTION OF FLUVIAL CLAYS WITH SEAWATER[a]

	Σ cations (meq 100 g^{-1})	Original Exchange Composition (equivalent fractions)				Equilibrium Composition in Seawater (equivalent fractions)[b]			
		X_{Na}	X_{Mg}	X_{Ca}	X_K	X_{Na}	X_{Mg}	X_{Ca}	X_K
Clays originally equilibrated with river water[c]									
Dakota montmorillonite	54.3	0.03	0.28	0.66	0.002	0.50	0.22	0.26	0.02
Texas montmorillonite	80.9	0.03	0.23	0.71	0.007	0.55	0.22	0.19	0.03
Bath kaolinite	6.2	0.06	0.18	0.51	0.008	0.38	0.32	0.24	0.06
Dry clay minerals									
Wyoming montmorillonite	70.2	0.46	0.18	0.27	0.02	0.53	0.39	0.04	0.04
Mixed layer	38.5	0.01	0.23	0.39	0.08	0.56	0.32	—	0.07
Illite	16.8	0.01	0.01	0.74	0.15	0.47	0.24	0.11	0.17

[a] Based on data presented by F. L. Sayles and P. C. Mangelsdorf, *Geochim. Cosmochim. Acta*, **41**, 951 (1977).
[b] After equilibration for 7 to 12 days.
[c] Artificial "mean world river water."

The preference of clays and most other natural ion exchangers for K^+ over Na^+ provides a ready explanation of why Na^+/K^+ ratios in natural waters and especially in seawater are much larger than unity although K^+ is only slightly less abundant than Na^+ in igneous rocks. Ion-exchange processes continuously remove K^+ from solution and return it to the solid phase. Some of the K^+, however, is also removed from water by the conversion of clays, for example, montmorillonite into illite and chlorite.

10.6   AGGREGATION OF COLLOIDS

Figure 10.16 gives a size spectrum of water-borne particles. Particles with diameters less than 10 μm are usually called colloids. Colloids usually remain suspended in waters because their gravitational settling is less than 10^{-2} cm s^{-1}.

In dealing with colloids, the term *stability* has an entirely different meaning than in thermodynamics. A system containing colloidal particles is said to be

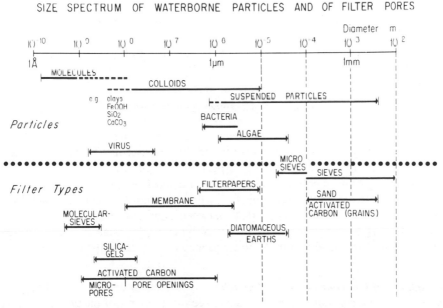

Figure 10.16   Suspended particles in natural and wastewaters vary in diameter from 0.005 to about 100 μm (5×10^{-9} to 10^{-4} m). For particles smaller than 10 μm, terminal gravitational settling will be less than about 10^{-2} cm sec^{-1}. Filter pores of sand filters, on the other hand, are typically larger than 500 μm. The smaller particles (colloids) can become separated either by settling if they aggregate or by filtration if they attach to filter grains. Particle separation is of importance in the following processes: aggregation of suspended particles (clays, hydrous oxides, phytoplankton, biological debris) in natural waters; coagulation (and flocculation) in water supply and wastewater treatment; bioflocculation (aggregation of bacteria and other suspended solids) in biological treatment processes; sludge conditioning (dewatering, filtration); filtration, ground water infiltration, removal of precipitates. [From W. Stumm, *Environ. Sci. Technol.*, **11**, 1066 (1977).]

stable if during the period of observation it is slow in changing its state of dispersion. The times for which sols are stable may be years or fractions of a second. The large interface present in these systems represents a substantial free energy which by recrystallization or agglomeration tends to reach a lower value; hence, thermodynamically, the lowest energy state is attained when the sol particles have been united with large crystals. The term stability is also used for particles having sizes larger than those of colloids; thus the stability of sols and suspensions can often be interpreted by the same concepts.

Historically, two classes of colloidal systems have been recognized: hydrophobic and hydrophilic colloids. In colloids of the second kind there is a strong affinity between the particles and water; in colloids of the first kind this affinity is negligible.

There exists a gradual transition between hydrophobic and hydrophilic colloids. Gold sols, silver halogenides, and nonhydrated metal oxides are typical hydrophobic colloid systems. Gelatin, starch, gums, proteins, and so on, as well as all biocolloids (viruses, bacteria), are hydrophilic. Hydrophobic and hydrophilic colloids have a different stability in the same electrolyte solution. Macromolecular colloids and many biocolloids are often quite stable. Many colloid surfaces relevant in water systems contain bound H_2O molecules at their surfaces. Amorphous silica and metal oxide surfaces are at least partially solvated and are thus intermediates between hydrophobic and hydrophilic colloids.

Kinetics of Particle Agglomeration

The rate of particle agglomeration depends on the frequency of collisions and on the efficiency of particle contacts (as measured experimentally, for example, by the fraction of collisions leading to permanent agglomeration).

The aggregation of colloids is known as *coagulation* or *flocculation*. The two terms are often used interchangeably when discussing chemical aspects of aggregation; we prefer to distinguish between aggregation due to electrolytes (coagulation) and aggregation due to polymers (flocculation). We address ourselves first to a discussion of the frequency of particle collision.

Frequency of Collisions between Particles.   Particles in suspension collide with each other as a consequence of at least three mechanisms of particle transport:

1   Particles move because of their thermal energy (Brownian motion). Coagulation resulting from this mode of transport is referred to as *perikinetic*.
2   If colloids are sufficiently large or the fluid shear rate high, the relative motion from velocity gradients exceeds that caused by Brownian (thermal) effects (*orthokinetic* agglomeration).
3   In settling, particles of different gravitational settling velocities may collide (agglomeration by differential settling).

The time-dependent decrease in the concentration of particles (N = number of particles per cubic centimeter) in a *monodisperse* suspension due to collisions by Brownian motion can be represented by a second-order rate law

$$- \frac{dN}{dt} = k_p N^2 \tag{47}$$

or

$$\frac{1}{N} - \frac{1}{N_0} = k_p t \tag{48}$$

As given by Von Smoluchowski (1917), k_p can be expressed as

$$k_p = \alpha_p 4 D \pi d \tag{49}$$

where D is the Brownian diffusion coefficient and d is the diameter of the particle. α_p is the fraction of collisions leading to permanent agglomeration and is an operational parameter for the stability ratio. The diffusion coefficient in equation 49 can be expressed as (Einstein–Stokes) $D = kT/3\pi\eta d$, where η is the absolute viscosity. With this substitution we obtain

$$- \frac{dN}{dt} = \alpha_r \frac{4kT}{3\eta} N^2 \tag{50}$$

The rate constant k_p is on the order of 2×10^{-12} cm^3 sec^{-1} for water at 20°C and for $\alpha_p = 1$. Thus, for example, a turbid water containing 10^6 particles cm^{-3} will reduce its particle concentration by half within a period of ca. 6 days (5×10^5 sec) provided that all particles are completely destabilized and that the particles are sufficiently small so that collisions result from Brownian motion only.

Agitation may accelerate the aggregation. The velocity of the fluid may vary both spatially and temporally. The spatial changes in velocity are characterized by a velocity gradient (see Figure 10.17). Since particles that follow the fluid motion will also have different velocities, opportunities exist for interparticle contact. The rate of decrease in particles due to agglomeration of particles

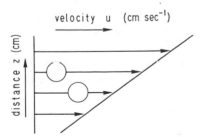

Figure 10.17 Particle collision in an idealized shear field of velocity gradient du/dz.

(having uniform particle size) under the influence of a mean velocity gradient G (time^{-1}) can be described by

$$-\frac{dN}{dt} = \frac{4}{3}\alpha_0 Gd^3 N^2 \tag{51}$$

where α_0 is (defined in the same way as α_p) the fraction of collisions leading to permanent agglomeration and d is the particle diameter. It is useful to consider the ratio of the rate at which contacts occur by orthokinetic agglomeration to the rate at which contacts occur by perikinetic agglomeration (equations 50 and 51); assuming

$$\alpha_0 = \alpha_p, \tag{52}$$

$$\frac{(dN/dt)_{\text{ortho}}}{(dN/dt)_{\text{peri}}} = \frac{Gd^3}{2kT} \tag{53}$$

In water at 25°C containing colloidal particles having a diamerer of 1 μm, this ratio is unity when the velocity gradient is 10 sec^{-1}.

If the volume of solid particles is conserved during agglomeration, the volume fraction of colloidal particles, the volume of colloids per unit volume of suspension, ϕ, can be expressed as

$$\phi = \frac{\pi}{6} d_0^3 N_0 \tag{54}$$

where N_0 is the initial number of particles and d_0 is the initial particle diameter; equation 51, that is, the reaction rate for a homogeneous colloid, may then be expressed as a pseudo-first-order reaction:

$$-\frac{dN}{dt} = -\frac{4}{\pi}\alpha_0 \phi GN \tag{55}$$

A numerical example might again illustrate the meaning of this rate law. For 10^6 particles of diameter $d = 1$ μm, ϕ becomes approximately 5×10^{-7} cm^3 cm^{-3}. For $\alpha_0 = 1$ and for a turbulence characterized by a velocity gradient $G = 5$ sec^{-1} (this corresponds to slow stirring in a beaker—about one revolution per second), the first-order rate constant $[(4/\pi)\phi G]$ is on the order of 3×10^{-6} sec^{-1}. Hence a period of ca. 3.7 days would elapse until the concentration of particles is halved as a result of orthokinetic agglomeration.

The overall rate of decrease in concentration of particles of any size is given by equations 51 and 55 by assuming additivity of the separate mechanisms [40]

$$-\frac{dN}{dt} = \alpha_p \frac{4kT}{3\eta} N^2 + \alpha_0 \frac{4\phi G}{\pi} N \tag{56}$$

The first term usually becomes negligible for particles with a diameter $d > 1$ μm, whereas the second term is less important than the first term, at least incipiently, for particles with a diameter $d < 1$ μm.

Heterodisperse Suspensions.   The rate laws given above apply to mono-disperse colloids. In polydisperse systems the particle size and the distribution of particle sizes have pronounced effects on the kinetics of agglomeration [41,42]. For the various transport mechanisms (Brownian diffusion, fluid shear, and differential settling), the rates at which particles come into contact are given in Table 10.9. These rate constants are compared for two cases in Figure 10.18. It follows that heterogeneity in particle size can significantly increase agglomeration rates.

Example 10.3. Effects of Particle Size on Agglomeration Rate

Compare the agglomeration rate of an aqueous suspension containing 10^4 virus particles per cubic centimeter ($d = 0.01$ μm) with that of a suspension containing, in addition to the virus particles, 10 mg liter^{-1} bentonite (number conc. $= 7.35 \times 10^6$ cm^{-3}; $d = 1$ μm). The mixture is stirred, $G - 10$ sec^{-1}, and

TABLE 10.9 AGGLOMERATION KINETICS OF COLLOIDAL SUSPEN-SIONS[a]

Transport Mechanism	Rate Constant for Heterodisperse Suspensions		Rate Constant If $d_1 = d_2$	
Brownian diffusion	$k_b = \dfrac{2}{3}\dfrac{kT}{\eta}\dfrac{(d_1 + d_2)^2}{d_1 d_2}$	(1)	$k_p = \dfrac{4kT}{3\eta}$	(4)
Laminar shear	$k_{sh} = \dfrac{(d_1 + d_2)^3}{6} G$	(2)	$k_0 = \tfrac{2}{3}d_p^3 G$	(5)
Differential settling	$k_s = \dfrac{\pi g(\rho - 1)}{72v}(d_1 + d_2)^3 (d_1 - d_2)$	(3)	$k_s = 0$	(6)

[a] The rate at which particles of sizes d_1 and d_2 come into contact by the jth transport mechanism is given by $F_j = k_j N_{d_1} N_{d_2}$. $F_j =$ collision rate in collisions per unit volume (cm^{-3} sec^{-1}); $k_j =$ bimolecular rate constant (cm^3 sec^{-1}) for the jth mechanism; N_{d_1} and $N_{d_2} =$ number concentrations of particles of size d_1 and d_2, respectively (cm^{-3}); $k =$ Boltzmann constant (1.38 $= 10^{-23}$ J K^{-1}; $\eta =$ absolute viscosity (g cm^{-1} sec^{-1}, or N cm^{-2} sec); $v =$ kinematic viscosity (cm^2 sec^{-1}); $\rho =$ specific gravity of the solids (g cm^{-3}); g $=$ gravity acceleration (cm sec^{-1}); $G =$ mean velocity gradient (sec^{-1}); $T =$ absolute temperature (K).

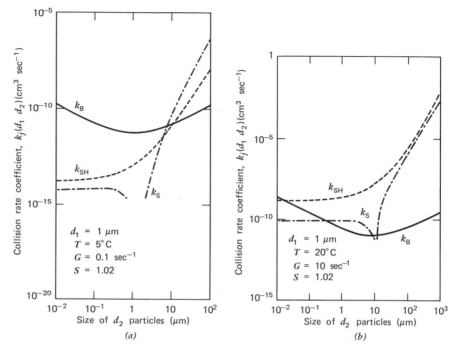

Figure 10.18   Effects of particle size on collision rate constants for agglomeration. Left: $d_1 = 1\ \mu m$; right: $d_1 = 10\ \mu m$.

the temperature is 25°C. Complete destabilization, $\alpha = 1$, may be assumed. (This example is from O'Melia [42].)

Let us neglect bentonite–bentonite particle interactions. We calculate from the equations given in Table 10.11 the following rate constants:

(1) $$k_p = 2 \times 10^{-12}\ cm^3\ sec^{-1}$$

(2) $$k_b = 3.1 \times 10^{-10}\ cm^3\ sec^{-1}$$

(3) $$k_{sh} = 1.7 \times 10^{-12}\ cm^3\ sec^{-1}$$

(4) $$k_s = 7.8 \times 10^{-13}\ cm^3\ sec^{-1}$$

According to equation 4 (Table 10.9), the time required to halve the concentration of the virus particles in the suspension containing the virus particles only would be almost 200 days. In the presence of bentonite ($k_b = 3.1 \times 10^{-10}$ $cm^3\ sec^{-1}$ and $N_{d_2} = 7.35 \times 10^6\ cm^{-3}$) we find after integrating that the free virus concentration after 1 hr of contact is only 2.6 particles cm^{-3}. This example illustrates that the presence of larger particles may aid significantly in the removal of smaller ones, even when Brownian diffusion is the predominant transport mechanism.

Particle Size Distribution

Particles may be sorted into size fractions (diameter, volume, mass, or number). The analysis of agglomeration, sediment transport, and dissolution processes of particles depends on information about their particle size distribution. Junge [43] and Friedlander [41] have advanced the understanding of atmospheric sciences by characterizing and explaining particle size distributions in atmospheric aerosols. In aquatic sciences similar developments should be forthcoming. Perhaps the simplest type of graphical representation is a bar histogram in which the particle number concentrations Δ (number) (vol^{-1}), found in each class interval, Δ (volume), or Δ (diameter), are plotted (discrete particle size distribution, Figure 10.19b). By measuring a large number of particles and making the class intervals approach zero, the smooth curve resulting represents the *continuous particle size distribution*. Usually the particle size distributions of natural materials (suspended, sedimentary, or airborne particles) are not normally distributed; they are often very asymmetric. (The normal frequency distribution as defined by Gauss is produced when an infinite number of factors cause independent variations of equal magnitude.)

Often it is convenient to plot the results of experimental measurements, for example, those obtained with a Coulter counter, as a *cumulative particle size distribution* (Figure 10.19a).

Here N (number cm^{-3}) denotes the total concentration of particles with a volume equal to or less than v (μm^3); the total concentration of all particles is given by N_∞. The slope of this curve, $\Delta N/\Delta v$ or dN/dv, is called *a particle size distribution* and is represented as $n(v)$. In this case $n(v)$ has units of number cm^{-3} μm^{-3}.

Particle volume is one of three common measures of particle size.

Surface area(s) and diameter (d_p) are also used, so that three particle size distribution functions can be defined:

$$\frac{\Delta N}{\Delta v} = \frac{dN}{dv} = n(v) \qquad \text{(number cm}^{-3}\ \mu m^{-3}) \qquad (57a)$$

$$\frac{\Delta N}{\Delta s} = \frac{dN}{ds} = n(s) \qquad \text{(number cm}^{-3}\ \mu m^{-2}) \qquad (57b)$$

$$\frac{\Delta N}{\Delta d_p} = \frac{dN}{d(d_p)} = n(d_p) \qquad \text{(number cm}^{-3}\ \mu m^{-1}) \qquad (57c)$$

These functions can be measured and used in both conceptual and empirical studies on coagulation and other particle transport processes.

The volume concentration of all particles in the interval between size 1 and size 2 may be written as

$$V_{1-2} = \int_1^2 dV$$

† In writing this section, the authors enjoyed the assistance of Charles R. O'Melia.

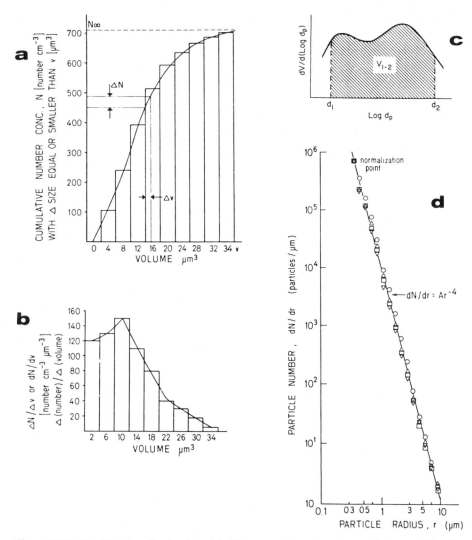

Figure 10.19 Particle size distribution. (*a*) Discrete and continuous *cumulative* particle size distribution. (*b*) Discrete and continuous particle size distribution. (*c*) Volume distribution plotted in accordance with equation 58. (*d*) Particle size distributions at four depths in a calcareous sediment from west equatorial Pacific Ocean, 1°6.0′S, 161°36.6′E, box core no. 136, water depth 3848 m. (From A. Lerman, *Geochemical Processes*, Wiley-Interscience, New York, 1979.)

Dividing and multiplying the right-hand side by $d(\log d_p)$ yields

$$V_{1-2} = \int_1^2 \frac{dV}{d(\log d_p)} \, d(\log d_p) \qquad (58)$$

A plot of $dV/d(\log d_p)$ versus $\log d_p$ is illustrated in Figure 10.19*c*. The total volume concentration provided by particles in the size interval from d_1 to d_2

is equal to the integrated (shaded) area in this figure. In preparing such plots from field measurements, it is frequently assumed that the particles are spherical. If this assumption is not true, the area is proportional (not equal) to the volume concentration.

For a plot of $dS/d(\log d_p)$ versus $\log d_p$ that is similar to Figure 10.19c, the integrated area under the resulting curve from d_1 to d_2 would represent the total concentration of surface area in a suspension provided by particles in the size interval from d_1 to d_2. A plot of $dN/d(\log d_p)$ versus $\log d_p$ would provide similar information about the total number concentration.

The size distribution of atmospheric aerosols and of aquatic suspensions is often found to follow a power law of the form

$$n(d_p) = A d_p^{-\beta} \tag{59}$$

in which A is a coefficient related to the total concentration of particulate matter in the system. As noted below, the exponent β has been determined experimentally and has also been shown on theoretical grounds to result from the interaction of various physical processes such as coagulation and settling.†

Measurements indicate that, for particles above about 1 μm in size (i.e., those detectable by present electronic or optical measurements), values of β range from 2 to 5. Lerman [44] reports measurements of size distributions at four locations in the North Atlantic. Fifty-three size distributions derived from samples taken at depths ranging from 30 to 5100 m yielded a mean value of $\beta = 4.01 \pm 0.28$.

Friedlander [45] has examined the effects of flocculation by Brownian diffusion and removal by sedimentation on the shape of the particle size distribution function as expressed by equation 59. The examination is conceptual; the predictions are consistent with some observations on atmospheric aerosols. For small particles, where flocculation by Brownian diffusion is predominant, β is predicted to be 2.5. For larger particles, where removal by settling occurs, β is predicted to be 4.75. Hunt [46] has extended this analysis to include flocculation by fluid shear (velocity gradients) and by differential settling. For these processes, β is predicted to be 4 for flocculation by fluid shear and 4.5 when flocculation by differential settling predominates. These theoretical predictions are consistent with the range of values for β observed in aquatic systems.

Thus the particle size distribution that occurs in natural waters may reflect physical and chemical processes in these systems. Very significant effects of

† Equation 59 has some interesting characteristics: A particle size distribution that follows an empirical power law with $\beta = 1$ has an equal number of particles in each logarithmic size interval. Similarly, for $\beta = 3$, the concentration of surface is uniformly distributed in each logarithmic size interval. Finally, for $\beta = 4$, the volume of solids is equally distributed in each logarithmic size interval, while the surface area and number concentrations are primarily in the smaller sizes.

Some pollutants can be characterized in terms of mass or volume concentrations. Examples include oil, suspended solids, and certain precipitates. Other pollutants are concentrated at surfaces. Examples include DDT adsorbed on detritus and trace metals adsorbed on clays. For materials such as these, the surface concentration of the particulate phase is of interest. Still other pollutants (e.g., pathogenic organisms) are best considered in terms of their number concentration.

particle size distributions on the performance of water treatment processes have been proposed by Lawler et al. [47].

A Physical Model for Colloid Stability

To what extent can theory predict the collision efficiency factor? Two groups of researchers, Verwey and Overbeek and Derjagin and Landau, independently of each other, have developed such a theory (the DLVO theory) [48] by quantitatively evaluating the balance of repulsive and attractive forces that interact when particles approach each other. Their model has undoubtedly become a most effective tool in the interpretation of many empirical facts in colloid chemistry. The reader is referred to some recent reviews [49, 50] and to van Olphen [51] for a detailed discussion on the stability of clays.

The DLVO theory considers van der Waals' attraction and diffuse double-layer repulsion as the sole operative factors. It calculates the interaction energy (as a function of interparticle distance) as the reversible isothermal work (i.e., Gibbs free energy) required to bring two particles from distance ∞ to distance d. Physically the requirement is that at any instant during interaction the two double layers are fully equilibrated [48, 49]. The mathematics of the interaction are different for the interaction of constant-potential surfaces and for the interaction of constant-charge surfaces. As long as the interaction is not very strong, that is, as long as the surfaces do not come too close, it does not make too much difference [49]. Table 10.10 gives the approximate equations for the constant-potential case. In order to illustrate the use of SI units, we calculate the interaction energy of two flat plates in Example 10.4.

TABLE 10.10 COLLOID STABILITY AS CALCULATED FROM VAN DER WAALS ATTRACTION AND ELECTROSTRATIC DIFFUSE DOUBLE-LAYER REPULSION[a, b]

Additive interactions of repulsive interaction energy V_R and attraction energy V_A:

$$V_T = V_R + V_A \tag{1}$$

Repulsive interaction per unit area
between flat plates:

$$V_R = \frac{64 n_s \mathbf{k} T}{\kappa} \left[\tanh\left(\frac{z e \Psi_d}{4 \mathbf{k} T} \right) \right]^2 e^{-\kappa d} \tag{2a}$$

for spherical particles:

$$V_R = \frac{64 \pi n_s z \mathbf{k} T}{\kappa^2} \frac{(a + \delta)^2}{R} \tanh\left(\frac{z e \Psi_d}{4 \mathbf{k} T} \right)^2 e^{-\kappa(H - 2\delta)} \tag{2b}$$

Van der Waals attraction per unit area[c]
for flat plates

$$V_A = -\frac{A_{11(2)}}{12\pi(d + 2\delta)^2} \tag{3a}$$

for spherical particles:

$$V_A = -\frac{A_{11(2)}}{6}\left(\frac{2}{s^2 - 4} + \frac{2}{s^2} + \ln\frac{s^2 - 4}{s^2}\right) \tag{3b}$$

where

$$s = \frac{R}{a}$$

For very short particle distances (3b) may be replaced by

$$V_A = -\frac{A_{11(2)}}{12\Pi} \tag{3v}$$

Electrolyte concentration required to just coagulate the colloids (25°C):[d]

$$c_s = 8 \times 10^{-36}\frac{[\tanh(ze\Psi_d/4kT)]^4}{A_{11(2)}^2 z^6} \tag{4}$$

Valence effect on stability:

$$c_s = \frac{3.125 \times 10^{-38}}{A_{11(2)}^2 z^2}\left(\frac{e\Psi_d}{kT}\right)^4 \tag{5}$$

Stability Ratio:

$$W = 2a\int_0^\infty \exp\left[\left(\frac{V_T}{kT}\right)\frac{dd}{(d + 2a)^2}\right] \tag{6}$$

---

[a] a = particle radius (nm); A = Hamaker constant (J); c_s = concentration of "salt" molecules (liter^{-1}); d = interaction distance between two surfaces (nm); e = elementary charge, 1.6×10^{-19} C; H = shortest interaction distance between two spherical particles (nm); k = Boltzmann constant, 1.38×10^{-23} J K^{-1}; n_s = number of "molecules" or ion pairs (cm^{-3}); R = distance between centers of two spheres (nm); W = stability ratio ($1/\alpha$) (in the kinetic equations, k_j can be replaced by k_j/W); T = absolute temperature (K); V = interaction energy (J m^{-2}); z = charge of ion (valence); δ = thickness of the Stern layer (nm); κ = reciprocal thickness of double layer (nm^{-1}); Ψ_d = potential (mV) at the plane where the diffuse double-layer begins.
[b] Cf. J. Lyklema, in *The Scientific Basis of Flocculation*. K. J. Ives, Ed., Sijthoff and Nordhoff, Alphen aan den Rijn, The Netherlands, 1978. Most equations below stem from DLVO theory. As exact solutions do not exist, recourse must be made to approximations.
[c] $A_{11(2)}$ is the Hamaker constant [dimension (energy)] that applies for the interaction between particles 1 in medium 2. This quantity can be related to the corresponding constants for attraction in vacuum between particles 1 or between particles 2 by $A_{11(2)} = (\sqrt{A_{11}} - \sqrt{A_{22}})^2$.
[d] Calculated for the condition that $V_T \leq 0$ and $\partial V_T/\partial d = 0$.

Example 10.4

Calculate the interaction energy of two flat Gouy plates of 25-mV surface potential (assumed to be constant), at 25°C in a 10^{-3} M NaCl solution, at a distance of 10 nm. A Hamaker constant $A_{11(2)} = 10^{-19}$ J may be used.

Using (1) to (3) of Table 10.10, and considering that, for a 10^{-3} M NaCl solution $\kappa^{-1} = 9.5 \times 10^{-9}$ m, we calculate

$$V_R = \frac{64 \text{ mol m}^{-3} \times 6.02 \times 10^{23} \times 1.38 \times 10^{-23} \text{ J K}^{-1} \times 298.13\kappa}{1.05 \times 10^{8} \text{ m}^{-1}}$$

$$\times \tanh \left(\frac{1.6 \times 10^{-19} \text{ C} \times 25 \times 10^{-3} \text{ V}}{4 \times 1.38 \times 10^{-23} \text{ J K}^{-1} \times 298.13\kappa} \right)^2$$

$$\times \exp - (1.05 \times 10^8 \text{ m}^{-1} \times 10^{-8} \text{ m})$$

$$= 1.5 \times 10^{-3} \text{ J m}^{-2} \times 5.5 \times 10^{-2} \times 0.35$$

$$= 2.9 \times 10^{-5} \text{ J m}^{-2} \text{ (or } 2.9 \times 10^{-7} \text{ erg cm}^{-2})$$

$$V_A = - \frac{10^{-19} \text{ J}}{12 \times 3.14 \times 10^{-16} \text{ m}^2}$$

$$= 2.65 \times 10^{-5} \text{ J m}^{-2} \text{ (or } 2.65 \times 10^{-7} \text{ erg cm}^{-2})$$

$$V_T = 2.5 \times 10^{-6} \text{ J m}^{-2} \text{ or } 2.5 \times 10^{-8} \text{ erg cm}^{-2}$$

In Figure 10.20a the energies of interaction (double-layer repulsion, V_R, and van der Waals attraction, V_A, and net total interaction, V_T) are plotted as a function of distance of the separation of the surfaces. As (2a) of Table 10.10 shows, V_R decreases in an exponential fashion with increasing separation. V_R increases roughly in proportion with Ψ_d^2 (for small Ψ_d, tanh $u \approx u$). The distance characterizing the repulsive interaction is similar in magnitude to the thickness of a single double layer (κ^{-1}). Thus the range of repulsion depends primarily on the ionic strength. The energy of attraction due to van der Waals attractive dispersion forces is plotted in the lower part of Figure 10.20a as a function of separation. This curve varies little for a given value of the Hamaker constant A which depends on the density and polarizability of the dispersed phase but is essentially independent of the ionic makeup of the solution. Often values for A between 10^{-19} and 10^{-20} J are adopted. Summing up repulsive and attractive energies gives the total energy of interaction (Figure 10.20a). Conventionally, the repulsive potential is considered positive and the attractive potential negative. At small separations, attraction outweighs repulsion, and at intermediate separations repulsion predominates. This energy barrier is usually characterized by the maximum (net repulsion energy) of the total potential energy curve, V_{max} (Figure 10.20b). The potential energy curve shows, under certain conditions, a secondary minimum at larger interparticle distances ($d \approx 10^{-6}$ cm). This secondary minimum depends among other things on the choice of the Hamaker constant and on the dimensions of the particles involved;

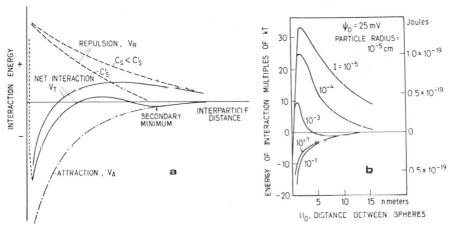

Figure 10.20   Physical Model for colloid stability. (*a*) Schematic forms of the curves of interaction energies (electrostatic repulsion V_R, van der Waals attraction V_A, and total (net) interaction V_I) as a function of the distance of surface separation. Electrolyte concentration c_s is smaller than c'_s. At very small distances a repulsion between the electronic clouds (Born repulsion) becomes effective. Thus, at the distance of closest approach, a deep potential energy minimum reflecting particle aggregation occurs. A shallow so-called secondary minimum may cause a kind of aggregation that is easily counteracted by stirring. (*b*) Net energy of interaction for spheres of constant potential surface for various ionic strengths (1:1 electrolyte) (cf. Verwey and Overbeek).

it is seldom deep enough to cause instability but might help in explaining certain loose forms of adhesion or agglomeration.

The Stability Ratio.   A stable colloidal dispersion is characterized by a high-energy barrier, that is, by a net repulsive interaction energy. Fuchs has defined a stability ratio W that is related to the area enclosed by the resultant curve of the energy of interaction versus separation distance. W is the factor by which agglomeration is slower than in the absence of an energy barrier. The conceptually defined W should correspond to the operationally determined α ($W = \alpha^{-1}$). In a first approximation the stability ratio W is related to the height of the potential energy barrier V_{max}. The latter is conveniently expressed in units of kT. If V_{max} exceeds the value of a few kT, relatively stable colloids will be found. For example, if $V_{max} \simeq 15kT$, only 1 out of 10^6 collisions will be successful ($\alpha = 10^{-6}$). Figure 10.20*b* shows how V_{max} typically decreases with an increase in electrolyte concentration.

Some of the pertinent interactions that affect colloid stability are readily apparent from these figures. The main effect of electrolytes is to reduce Ψ_d and to compact the double layer (reducing κ^{-1}). Experimentally it is known that the charge of the counterion plays an important role. The critical electrolyte concentration required just to agglomerate the colloids is proportional to $z^{-6}A_{11(2)}^{-2}$ for high surface potentials [(4) in Table 10.10]. This is the theoretical basis for the qualitative *valency rule of Schulze and Hardy*. Under simplifying conditions it is possible to derive the relation that the critical coagulation concentrations, *ccc*, of mono-, di-, and trivalent ions are in the ratio $(1/z)^6$ or $100:1.6:0.13$. At small surface potentials c_s is proportional to $\Psi_d^4 z^{-2} A_{11(2)}^{-2}$ [(5) in Table 10.10]. Ions with $|z| > 1$ are often, at least partially, specifically adsorbed; this leads to a reduction in Ψ_d. Because of the fourth power exponent of Ψ_d, c_s is often more dependent upon z.

The influence of the particle radius a on the kinetics of agglomeration is very complex. W is approximately proportional to a [(7) in Table 10.10]. Experiments show, however, that W is often independent of particle size. Two other factors influencing the rate of agglomeration should be considered: The theory (Table 10.10) assumes that complete equilibrium exists between the particle surface and the surrounding liquid at any separation between the particles. In the agglomeration of particles the time necessary for the adjustment of equilibrium may be longer than the time involved in the collision [49]. Furthermore, the last part of the approach of two particles is slowed down because it is difficult for the liquid to flow away from the narrow gap between the particles. This effect on the collision frequency has been calculated [52, 53].

As indicated, clays may carry different double-layer structures on the same particle: a negatively charged plate and a positively charged edge.† Therefore when a suspension of platelike particles aggregates, different modes of particle association may occur: face to face (FF), edge to face (EF), and edge to edge (EE). Hence the electric energies of interaction for the three types of associations are governed by three different combinations of the double layers. Furthermore, because of geometric factors, the van der Waals interaction energy will be different for the three types of associations. Especially important is the fact that, owing to the opposite sign of the charge of the edge and face double layers, edge-to-face association (self-coagulation) can take place in dilute electrolytes. Obviously, theoretical stability calculations for clays, simply based on considerations of the plate charge, thus may have little meaning concerning the state of aggregation. The different forms of particle association prevalent at different electrolyte concentrations cause very striking and technologically important variations in the rheological properties of clay–water systems. For a comprehensive discussion the reader is referred to van Olphen [51].

Chemical Factors and Colloid Stability

Purely chemical factors must be considered in addition to the theory of the double layer in order to explain the stability and the agglomeration properties of most colloids in natural waters. In Table 10.11 an attempt is made to classify

---

† According to van Olphen [51], at the edges of the plates the silica and alumina sheets are disrupted and primary bonds are broken. On such surfaces, similar to the surfaces of silica and alumina particles, the electrical double layer is created by potential-determining ions, predominantly by OH^- or H^+.

TABLE 10.11 MODES OF DESTABILIZATION AND THEIR CHARACTERISTICS[a]

Phenomena	Aggregation by "inert" Electrolytes Coagulation	Aggregation by Adsorbable Species Adsorption Coagulation[b]	Chemical Bridging Model Flocculation
Electrostatic interaction	Predominant	Important	Subordinate
Chemical interactions and adsorption	Mostly absent	Important	Predominant
Zeta potential for optimum aggregation	Near zero	Not necessarily zero	Usually not zero
Addition of an excess of destabilizing species	No effect	Restabilization usually accompanied by charge reversal	Restabilization due to complete surface coverage
Relationship between optimum dosage of destabilizing species and the concentration of colloid (or the concentration of colloidal surface)	ccc virtually independent of colloid concentration	Stoichiometry usual	Stoichiometry, a linear relationship between flocculant dose and surface area
Physical properties of the aggregates produced	Dense, great shear strength, but poor filtrability in cake filtration	Flocs of widely varying shear strength and density	Flocs of three-dimensional structure; low shear strength but excellent filtrability in cake filtration

[a] Modified from W. Stumm and C. R. O'Melia, *J. Amer. Water Works Assoc.*, **60**, 514 (1968).
[b] Aggregation by hydrolized metal ions fits into this category.

various modes of destabilization. If coagulation and flocculation have a scientific basis, their roots have to be found in the physical and chemical properties of the particle surfaces.

Hydrophilic Colloids.   The influence of electrolytes upon the stability of hydrophilic colloids is in a qualitative way similar to that in hydrophobic systems. By their screening effect the added indifferent salts lessen the mutual electrostatic interaction of the charged groups. Polyvalent cations effectively screen the negatively charged groups of hydrophilic particles, but other factors besides those of the valency and concentration play a role. It has been shown that the stability of hydrophilic colloids is strongly influenced by the pH of the medium and the composition of the ionized group of the hydrophilic surface. Remarkable differences in coagulative behavior toward multivalent cations are observed for colloids having different functional groups. This is caused by coordination of the multivalent cations to the ionized group, usually accompanied by a reduction in the effective charge of the particles and by an alteration of their surface solvation.

Complex formation with cations may further lead to a reversal of the charge of the particles. Charge reduction or charge reversal may be visualized by the following schematic example:

$$[R(COO^-)_n]^{n-} + mMe^{2+} = [R(COO_nMe_m)]^{(2m-n)+} \tag{60}$$

Multivalent metal ions are able to form soluble or insoluble complexes with substances containing carboxyl, phenolic, hydroxyl, sulfato, or phosphato groups. The marked difference in the coagulation response of colloids with carboxyl, sulfate, or phosphate functional groups toward a given multivalent cation, as well as the different sequence of cations with regard to the relative position of their critical concentration for each type of a colloid, are explainable if we consider the great variation in affinity of inorganic cations for the different functional groups. For example, 10 to 100 times smaller concentrations of Pb^{2+}, Cd^{2+}, Cu^{2+}, and Zn^{2+} are needed for coagulation of phosphate colloids than for coagulation of carboxyl colloids of similar charge density. For the coagulation of sulfate colloids, 5 to 10 times larger concentrations of the same bivalent metals are needed than for the coagulation of carboxyl colloids [48]. Such a sequence is analogous to the relative solubilities of the metal phosphates, acetates, and sulfates.

The effect of anions that can form complexes with colloidal surface groups is similar to the cases discussed; for example, phosphates and organic substances containing functional hydroxyl or carboxyl groups interact chemically with the aluminum edge of a clay particle. The resultant alteration of the edge charge (conversion from positive to negative double layer) modifies the stability of the clay (cf. Section 10.4).

Adsorption and Colloid Stability.   Sorbable species are observed to destabilize colloids at much lower concentrations than nonsorbable ions. The extent to which a species is sorbable is reflected in the concentration necessary to produce aggregation. Figure 10.21 gives schematic curves of residual turbidity as a function of coagulant dosage for a natural water treated with Na^+, Ca^{2+}, and Al^{3+}, with hydrolyzed metal ions, or with polymeric species. It is evident that there are dramatic differences in the coagulating abilities of simple ions (Na^+, Ca^{2+}, Al^{3+}), hydrolyzed metal ions (highly charged multimeric hydroxo metal species), and species of large ionic or molecular size. For species having equal counterion charges it is apparent that the critical coagulation concentration decreases with increasing ion sorbability.

Specifically sorbable species that coagulate colloids at low concentrations may restabilize these dispersions at higher concentrations. When the destabilization agent and the colloid are of opposite charge, this restabilization is accompanied by a reversal of the charge of the colloidal particles. Purely coulombic

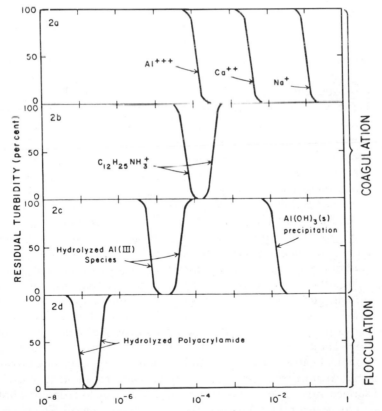

Figure 10.21   Schematic agglomeration curves for several different destabilizing agents. (From C. R. O'Melia, in *Physicochemical Processes for Water Quality Control*, W. J. Weber, Jr., Ed., Wiley-Interscience, New York, 1972.)

attraction would not permit an attraction of counterions in excess of the original surface charge of the colloid. Traces of hydrolyzed metal ions or surfactants alter the colloid stability. Simple adsorption models have been developed that depict colloid stability as a function of concentration of destabilizing species and of the surface concentration of the dispersed phase (Figure 10.22).

Polymers.   Natural and synthetic macromolecules have been used successfully as aggregating agents in water and waste treatment and for sludge conditioning. Natural (anionic) polymers are of great importance in sorptive stabilizing and destabilizing reactions of particles in natural waters and in biological systems; for example, the cohesion of tissue cells and the aggregation of microorganisms have been interpreted in terms of colloid–polymer interactions. As we have seen, polymers may adsorb readily on solid surfaces. Colloid stability can be either increased or decreased because of polymer adsorption. An increase in colloid stability may result from (1) an increase in the electrical repulsion between the particles (e.g., adsorption of anionic surfactants or physical displacement of counterions from the Stern layer by adsorbed (nonionic) polymers), (2) a decrease in the van der Waals attraction because the adsorbed layer contributed to the dispersion interaction, or (3) an increasing in a steric component of repulsion. [On collision the adsorbed layers might be compressed; this would reduce the volume available to adsorbed molecules, hence restrict the number of possible configurations for adsorbed polymer chains; a reduced number of configurations implies decreased entropy, and so an increase in free energy, hence a repulsion between the particles (volume restriction or elastic effect). The interpenetration of adsorbed layers increases locally the concentration of polymer segments, again increasing the free energy, hence the interparticle repulsion (osmotic effect)].

The decrease in colloid stability caused by adsorbed polymers is often called "sensitization." Polymers found to be good flocculants are usually linear homopolymers of high molecular weight (Table 10.5). Often they have ionizable groups along the chain, that is, they are *polyelectrolytes*. LaMer and coworkers and others have developed a chemical bridging theory that provides a model for understanding the ability of polymers to destabilize colloidal suspensions. In its simplest form the chemical bridging theory proposes that a polymer molecule can attach itself to the surface of a colloidal particle at one or more adsorption sites with the remainder of the molecule extending into the solution. These extended segments can then interact with vacant sites on another colloidal particle. Failing to find a suitable adsorption site on another particle, the extended segments can eventually adsorb at other sites on the original surface. Adsorption of anionic polymers (e.g., by hydrophobic bonding or through hydrogen bridges) on negative surfaces is common. Flocculation of a polymer–sol system of like charge results only if an appropriate concentration of a salt is present in the solution [54–56]. However, flocculation usually

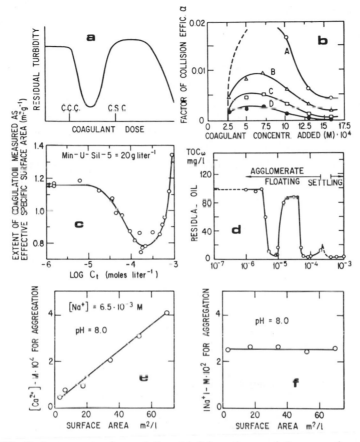

Figure 10.22 Various aspects of coagulation. (*a*) Schematic coagulation curve for specifically adsorbable coagulant. Coagulation starts at C.C.C. (the critical coagulation concentration). Adding an excess of adsorbable coagulant causes a charge reversal; at the C.S.C. (the critical stabilization concentration), the colloid is restabilized as oppositely charged sol. Upon further addition of coagulant, coagulation occurs by indifferent electrolytes. If hydrolyzing metal ions are used as coagulants, metal hydroxides are precipitated at high dosages. These hydroxides enmesh the colloidal particles originally present in the water (this is often called "sweep flocculation"). C.C.C. and C.S.C. are dependent on the sol concentration, or more precisely on their surface area concentration. (*b*) Effects of pH and Al(III) concentration upon relative colloid stability of amorphous silica expressed as collision efficiency factor. Colloidal silica: Ludox LS 0.3 g liter^{-1}; (*A*) pH = 5.75; (*B*) pH = 5.5; (*C*) pH = 5.25; (*D*) pH = 5.0 [cf. H. Hahn and W. Stumm, *J. Colloid Interfacial Sci.*, **28**, 134 (1968)]. (Reproduced with permission from Academic Press.) (*c*) Effective specific surface area of silica (Min-U-Sil) suspensions (20 g liter^{-1}) as a function of the concentration of hydrolyzed iron(III) applied. Effective surface areas are calculated from refiltration time measurements based on the Kozeny equation. [After C. R. O'Melia and W. Stumm, *J. Colloid Interfacial Sci.*, **23**, 437 (1967). Reproduced with permission from Academic Press.] (*d*) Coagulation of an oil dispersion with Al(III), pH = 6.3, NaCl = 10^{-3} *M* (25°C). At the first destabilization (zeta potential ≈ 0), the coagulated or coalesced oil ascends to the surface. At higher Al(III) dosage, the oil droplets become enmeshed in settling "Al(OH)$_3$" flocs (residual density larger than water). Note the difference between destabilization by soluble polyhydroxoaluminum species and by "sweep flocculation" caused by Al(OH)$_3$ flocs. (From Thuer and Stumm [71].) (*e* and *f*) Relationship between MnO$_2$ colloid surface area concentration and C.C.C. of Ca^{2+} and Na$^+$. Because of the specific interaction of the oxide surface with Ca^{2+}, a stoichiometric relationship exists between C.C.C. and the surface area concentration; in case of Na$^+$, however, this interaction is weaker, so that primarily compaction of the diffuse part of the double layer causes destabilization. [From W. Stumm, C. P. Huang, and S. R. Jenkins, *Croat. Chim. Acta*, **42**, 223 (1970).]

occurs at electrolyte concentrations much smaller than those necessary in the absence of polymers [56].

Many bacteria and algae have a tendency to adhere to interfaces and to each other (*bioflocculation*). Because of the hydrophilic surface, the stability of a microbial dispersion does not depend primarily on the electrostatic repulsive forces between the cells. Reduction in surface potential is not a prerequisite for bioflocculation; some bacterial suspensions can form stable dispersions at the pH_{iep}. Flocculation of microorganisms is effected by an interaction (bridging mechanism) of polymers excreted by the microorganisms or exposed at the microbial surface under suitable physiological conditions [55].

Naturally occurring aggregates that contain much of the nonliving organic matter in seawater commonly consist of pale yellowish or brownish amorphous matrices with inclusions of bacteria, silt particles, and sometimes phytoplankton (organic agregates).

Filtration Compared with Coagulation

Filtration is analogous to coagulation in many respects. This is illustrated by juxtaposing the basic kinetic equations on particle removal:

$$\text{orthokinetic coagulation:} \quad -\frac{dN}{dt} = \frac{4}{\pi}\,\alpha G\phi N \qquad (61)\dagger$$

$$\text{packed-bed filtration:} \quad -\frac{dN}{dL} = \frac{3}{2}\frac{(1-f)}{d}\,\alpha\eta N \qquad (62)$$

where t the time, $1-f$, the volume of filter media per unit volume of filter bed where f is porosity, η, a "single-collector efficiency" that reflects the rate at which particle contacts occur between suspended particles and the filter bed by mass transport, L, the bed depth, d, the diameter of the filter grain. The effectiveness of particle aggregation in coagulation and particle removal in filtration (attainable on integration of equations 61 and 62) depends on a dimensionless product of the variables shown in Table 10.12, which have comparable meanings for both processes. Particle transport, that is, by forces of fluid-mechanical origin, is required in both processes either to move the particles toward each other or to transport them to the surface of the filter grain or to the surface of a previously deposited particle [57, 58]. The contact opportunities of the particles for collisions with one another or with filter grains depend on Gt or $\eta(L/d)$. The detention time t is somewhat related to L/d, the ratio of bed depth to medium diameter [57].

---

† The rate of coagulation for particles of small diameter ($d < 1\ \mu m$) occurring mainly by Brownian interparticle contact is given in equation 50.

TABLE 10.12 COMPARISON BETWEEN COAGULATION AND FILTRATION[a]

Coagulation	α	ϕ	G	t
	collision efficiency	volumetric concentration of suspended particles	velocity gradient	time
Filtration	α	$1 - f$	η	$\dfrac{L}{d}$
		volumetric concentration of filter medium	single-collector efficiency (v,d)	number of collectors
			Contact opportunities	
Design and operational variables	Chemicals	Coagulation aids, media size, sludge recirculation	Energy input, mass transport	Residence time filter lengths and media diameter

[a] From W. Stumm, *Environ. Sci. Technol.*, **11**, 1065 (1977). The table is based at least partially on ideas expressed by C. R. O'Melia [57]. The effectiveness of coagulation and filtration depends on a related dimensionless product.

Figure 10.23a illustrates how coagulation in natural systems and in water and waste treatment systems depends on the variables in Table 10.12. In natural waters long detention times may provide sufficient contact opportunities despite very small collision frequencies (small G and small ϕ). In fresh water the collision efficiency is usually also low ($\alpha \sim 10^{-3}$ to 10^{-6}; that is, only 1 out of 10^3 to 10^6 collisions leads to a successful agglomeration). In seawater, colloids are less stable ($\alpha \cong 0.1$ to 1) because of the high salinity; κ^{-1} for seawater is about 0.36 nm, hence seawater double layers are nondiffuse [59]. Estuaries with their salinity gradients and tidal movements represent gigantic natural coagulation reactors where much of the dispersed colloidal matter of rivers settles. In water and waste treatment systems we can reduce detention time (volume of the tank) by adding coagulants at a proper dosage ($\alpha \rightarrow 1$) and by adjusting the power input (G). If the concentration ϕ is too small, it can be increased by adding additional colloids as so-called coagulation aids. O'Melia has shown that similar trade-offs in operation and design exist for optimizing particle removal in filtration [42, 57].

Figure 10.23b shows how, in the development from very slow filtration in groundwater percolation to ultrahigh-rate contact filtration, a relatively constant efficiency in particle removal (constant product $\alpha\eta(L/d)$) is maintained despite a dramatic increase in filtration rate. This is achieved by counterbalancing decreased contact opportunities (decreasing η and L and increasing d) by improving the effectiveness of particle attachment (through natural release or addition of suitable chemical destabilizing agents that increase α to 1).

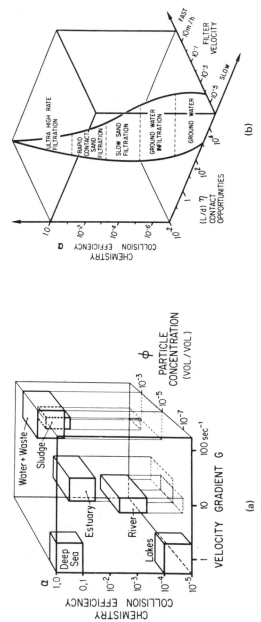

Figure 10.23 Variables that typically determine the efficiency of coagulation in natural waters and in water and waste treatment systems. (*a*) How the variables determine the coagulation efficiency. (*b*) Marked increase in filtration rate can be achieved by counterbalancing a reduction in contact opportunities by chemically improving the contact efficiency, with similar efficiency in particle removal. (Compare with Table 10.12.) [From W. Stumm, *Environ. Sci. Technol.,* **11**, 1065 (1977).

10.7   THE ROLE OF COAGULATION IN NATURAL WATERS

The aggregation of colloids is of great importance in the transport and distribution of matter in natural waters. Although dissolved substances tend to be distributed by convective mass transfer, the distribution of suspended matter is also influenced by the forces of gravity. Whether a particle will settle depends on its density, its size, and the water movement. Aggregation processes not only affect the distribution of suspended matter but they also play a role in the transformation of solutes because many dissolved substances, especially metal ions and organic material, adsorb onto colloids or react chemically with colloid surface groups [61–63]. It is estimated that all the rivers on the earth bring ca. 183×10^{14} g [60] (~ 8 km^3) of sediment into the oceans per year [62]. The majority of runoff water first collects in rivers and lakes; in estuaries much of the dispersed matter of rivers coagulates and settles [61]. Only very fine-grained material, predominantly clays, fine silt, and some organic matter, is carried into the sea. Suspended particulate matter plays a vital role in ocean chemistry [62].

Obviously colloid chemical reactions can profoundly influence the temporal and spatial distribution of dissolved and suspended constituents. In order to describe the various transportation processes of colloids and their effect on the ecological interdependence of natural water systems and on the ultimate distribution of suspended matter, one needs answers to the following types of questions [63]:

(1) What is the rate of coagulation of various naturally occurring colloids under specified conditions of pH and ionic strength? (2) How do particle concentration, particle size, and particle size distribution affect coagulation rates in natural waters? (3) How fast does suspended matter settle when the solution parameters are changed in such a way as to effect aggregation of the colloidal phase, and what is the rate of decrease in the turbidity (naturally occuring colloids or pollutants) of a body of water at given flow velocities or detention times? (4) How far are discharged colloidal pollutants transported by the movement of natural waters before they are removed through sedimentation, and what is the variation in the composition of colloidal aggregates found in the sediments as it depends on the distance from the point of discharge? (5) In what sequence are various colloids of different stability removed from a water of changing composition, such as is the case in the transition from a river water to the open sea; what are the effects upon the composition of the sediments in that transition zone?

Colloid chemistry can help to answer these questions, but we do not have adequate theories or sufficient empirical information concerning these complicated processes to provide quantitative and generally valid answers.

Colloid Stability in Natural Waters.   It is important to emphasize that coagulation phenomena in natural waters are quite specific. Because of the great variety of possible colloid–solute interactions the computation of a stability ratio on the basis of a simplified double-layer model might give unrealistic estimates of colloid stability. For example, for clays in fresh waters stability ratios on the order of 10^{100} or higher may be computed for a given surface charge density (e.g., 10 $\mu C\ cm^{-2}$ in a $10^{-3}\ M$ solution) [63]. According to such stability ratios, clays would remain in colloidal dispersion for geological time spans, but such predictions do not incorporate the effects of face-to-edge aggregation or destabilization by adsorption of traces of a hydrolyzed metal ion. Experimentally determined α values for various colloids encountered in natural water systems should be useful for predicting agglomeration rates [63–66].

Double-layer theory alone also cannot explain the colloid chemical behavior of most other inorganic colloids in natural waters.

Theory (see Figure 10.23a) predicts that colloid stability decreases with increasing ionic strength; hence stability ratios calculated for seawater are close to 1. Most colloids are less stable in *seawater* than in fresh water. Figure 10.24 shows experimental results for coagulation of kaolinite particles in solutions of different ionic strengths, simulating variations of salinity in an estuary. Nevertheless the collision efficiency is relatively small for most hydrophilic colloids (organic aggregates, biocolloids, and inorganic colloids well hydrated or coated with organic "colloid-protective" material).

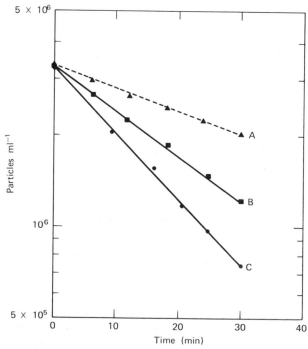

Figure 10.24 Rates of coagulation of kaolinite in synthetic solutions simulating estuarine waters of different salinities. Ionic strengths, uncorrected for ion pairing, were: *A*, 0.036; *B*, 0.087; *C*, 0.343. (From Edzwald et al. [65].)

Example 10.5. Effect of Detention Time on Residual Particle Concentration

Estimate the decrease in particle concentration resulting from coagulation in a reservoir as a function of detention time. Assume completely mixed conditions (negligible settling), spatially uniform composition, and turbidity to be proportional to the number concentration of aggregates.

In a steady-state model inflow into the reservoir will be balanced by outflow plus the decrease in concentration of aggregates resulting from coagulation. Using equations 50 and 55 for perikinetic and orthokinetic coagulation, respectively, equations for the steady-state balance are obtained:

$$\frac{Q}{V} N_0 - \frac{Q}{V} N - k_p N^2 - 0 \tag{i}$$

$$\frac{Q}{V} N_0 - \frac{Q}{V} N - k_0 N = 0 \tag{ii}$$

where Q = inflow or outflow rate (m^3 sec^{-1}), V = volume of the reservoir (m^3), V/Q = theoretical detention time τ (sec). By solving equations i and ii for N, the steady-state concentration of aggregates can be estimated:

perikinetic:
$$N = \frac{\tau^{-1} \pm \sqrt{\tau^{-2} + 4k_p N_0 \tau^{-1}}}{2k_p} \tag{iii}$$

orthokinetic:
$$N = \frac{\tau^{-1} N_0}{\tau^{-1} + k_0} \tag{iv}$$

We recall (from equations 50 and 55) that $k_p = \alpha_p 4kT/\eta$ and $k_0 = 4\alpha_0 \phi G/\pi$.

Fresh Waters. High turbidity, high amounts of filtrable iron (III), and humic substances[†] are typically observed in waters with a small detention time and of low ionic strength, for example, waters in crystalline rock areas relatively devoid of Ca^{2+} and Mg^{2+}. In limestone areas, because of the higher ionic strength ($I \geq 10^{-3}$ M) and the larger concentrations of Ca^{2+} and Mg^{2+}, turbidity and humic organic carbon are readily removed from the water body by coagulative processes.[‡] Ca^{2+} and Mg^{2+} increase α, not only by their contribution to the ionic strength but also because of their tendency to coordinate with the carboxyl and OH$^-$ functional groups of the humic substances and of the hydrous oxides and clays.

[†] In natural waters high concentrations of organic material are frequently associated with high concentrations of iron; many organic hydrophilic colloids form (often highly colored) complexes with ferric iron.

[‡] For example, in Switzerland where waters are (because of ubiquitous $CaCO_3$) high in Ca^{2+} and Mg^{2+} ($[Ca^{2+}] + [Mg^{2+}] \geq 2 \times 10^{-3}$ M), the concentration of humic acid organic matter is so low that a water pollution control bylaw stipulates that surface waters must not contain more than 2 mg liter^{-1} dissolved organic carbon. In soft-water areas humic substances are present in much higher concentrations; the world average is approximately 10 mg C liter^{-1}.

Infiltration and permeation through soils is affected by aggregate size. According to the Kozeny equation for flow through porous media, the filtration rate is inversely related to the square of the effective particle surface area and thus depends on the extent of coagulation. Trace constituents may affect the state of aggregation. Organic matter, especially polysaccharides excreted by microorganisms, and amorphous precipitates such as iron(II) sulfide tend to disperse the particles in the medium. On the other hand, electrolytes (Ca^{2+}), polyelectrolytes, and certain metal-ion hydrolysis products may, under suitable conditions, increase the permeability because of particle agglomeration and a decrease in swelling. A simple laboratory experiment may illustrate the effect of minute quantities of hydrolyzed Fe(III) upon the filtration rate through a porous medium of silica particles (Figure 10.22c).

Waste Disposal.   Suspended material is distributed in a different fashion than dissolved material. Most of the organic pollutants discharged from an effluent of a sewage treatment plant are in the form of colloidal solids. While dissolved matter is usually transported away from the outfall, suspended wastes may be deposited in places not far removed from their point of origin. The spatial distribution depends not only on water movements but also on the state of dispersion of the colloidal material, which in turn affects its sedimentation and flotation. Accumulation of trace metals, radionuclides, organic matter, and toxic material on colloidal interfaces is reversible; thus adsorbed materials may be released into solution as the particles are carried from one environment to another. Uptake or loss is accompanied by coagulation or redispersion and occurs as a result of changes in adsorption and ion-exchange equilibria [66].

Estuaries.   In a typical *estuarine circulation* [65], particles carried down with the river water will, in the stratified area, sink from the upper to the lower water layer—the gradient in electrolyte concentration probably enhances this settling because of agglomeration—and be carried back upstream. This process may repeat itself. Thus concentrations of suspended matter in an estuary may be many times higher than those found in the river or the open sea. The mechanism by which fresh water flows freely seaward while suspended particles are retained is operative even if colloid stability were not affected by electrolyte concentration. The destabilization by electrolytes can make the retention process more efficient because it influences the settling velocity of these materials. One size of material may be trapped more efficiently than another.

Colloid chemistry can aid in explaining the variation in sediment composition that parallels the increase in electrolyte content near a river inflow into the sea? A good example of such variation in abundance of different clays has been described by Grim and Johns [67]. We might argue that different clays become sequentially destabilized in meeting progressively increasing salinities. Although such effects may plausibly influence the spatial distribution of colloids in sediments, insufficient theoretical and experimental justification is available to establish this effect as the principal explanation for the observed variation in clay distribution. Chemical transformation of montmorillonite into illite and chlorite under the influence of the increasing salinity is an alternative explanation, but numerous other factors such as sorting by current circulation must be considered.

Metals present as colloidal humic complexes maintaining the semblance of a dissolved load are coagulated as their freshwater carrier is mixed with seawater at the continental boundary [68]. Field evidence for the coagulation of Fe and humic substances has been extensively documented, and laboratory experiments give insights into the processes operating [69, 70]. The river-borne "dissolved" iron consists almost entirely of mixed iron oxide–organic matter colloids of diameter less than 0.45 μm stabilized by the dissolved organic matter [70]. Coagulation occurs on mixing, because the seawater cations, especially Mg^{2+} and Ca^{2+}, destabilize the negatively charged iron-bearing colloids.

Seawater. Concentration of suspended materials in the open ocean generally decreases from the surface (tens to a few hundred μg liter^{-1} [dry weight]) down to about 10 to 20 μg liter^{-1}; the decrease in concentration is more or less exponential with depth. The most abundant material consists of particles with a radius of 1 to 7 μm [44].

Lerman [44] has analyzed the various sinks for settling particle assemblages. Obviously sedimentation to the floor of the ocean is a major sink that removes suspended materials from water. In a steady-state situation, the rate of input of particles at the water surface either by local production or inflow from outside is balanced by the rate of sedimentation on the bottom and production or destruction in the water column. As Lerman points out, if sedimentation were the only sink (i.e., nothing affected the particle mass and size in the water column), the particle concentration and size distribution would not change with depth as particles of different sizes continuously settle through the water column. The number concentration of particles decreases with depth (while the particle size distribution is rather independent of depth); thus an additional sink for the settling particles must occur. A sink for settling particles of any size class means that particles are removed from that size class. According to Lerman the sinks include the following processes: dissolution or biodegradation of mineral or organic particles, agglomeration and attachment (coagulation), and ingestion by organisms. Undoubtedly, dissolution and biodegradation is of great significance. Coagulation is more likely to be of importance in areas of high number concentrations of suspended materials (e.g., coastal waters). Furthermore, aggregation in the gut of filter-feeding animals—another form of coagulation— may contribute to the removal of suspended matter.

As discussed before, particle size distributions in the oceans are usually well represented by the relation [44, 46]

$$n(d_p) = A d_p^{-\beta} \tag{59}$$

with values of β often (and independent of location) close to 4. As discussed in Section 10.6, Hunt [46] has shown that the observed oceanic particle size distribution may be based on particle coagulation and gravitational settling. He has arrived at predicted particle size distributions in reasonable agreement with those observed in oceanic waters.

10.8   THE OIL–WATER INTERFACE

Approximately 0.2 % (~ 2 to 5×10^6 tons year^{-1}) of the mineral oil transported is lost to the hydrosphere. Pollution of inland waters is even greater than that of the oceans. We consider here some of the background necessary for understanding some of the phenomena associated with oil spills. The concepts discussed can usually be applied equally well to other interfaces. The term "oil" may be applied to any water-immiscible phase. The spreading and dispersion of mineral oil introduced into water can be found in the following forms: (1) as a separate phase (supernatant layer or film), (2) dispersed colloidal droplets, emulsion, or agglomerated coagulates, (3) (molecularily) dissolved hydrocarbons, and (4) hydrocarbons adsorbed on suspended particulate matter.

Partial dissolution, dispersion, adsorption, agglomeration, and flotation or sedimentation of oil components and loss of hydrocarbons by volatilization and by injection into the air, concomitant with bubble breaking, can be readily demonstrated in the laboratory by adding mineral oil to a stirred aquatic suspension of suspended matter [71]. Initially, an oil film is formed at the water surface, but some of the oil, especially the aromatic and low-molecular-weight hydrocarbons, are dissolved in the water phase; some of these molecules, especially the high-molecular-weight ones, tend to be adsorbed at solid–liquid interfaces. Some of the oil in the film will sooner or later—depending on the energy input (velocity gradient), emulsifiers present in the oil, and microbial activity (exudation of natural emulsifiers)—be broken up into colloidal oil droplets. Some of these oil droplets will tend to aggregate with other suspended solids; depending on the density of these aggregates they will settle or ascend. Bubbling of the suspension provides an efficient mechanism not only for volatilization of low-molecular-weight hydrocarbons but also for removal of oily interfacial films by aerosolization of film fragments by tiny jet droplets [72]. Fractionation and preferential ejection of polar organics into the aerosol phase have been observed [2]. This might explain the findings [73] that surface-active material occurs in marine aerosols in quantities in excess of that in the bulk seawater.†

The *work of adhesion* W_{OW} between two liquid phases, oil, O, and water, W, is the work necessary to separate 1 cm² of interface OW into two separate

† For a general discussion on sea surface processes see F. MacIntyre [5a], and on the impact of oil pollution on the sea–air interface see W. D. Garrett, in *The Changing Chemistry of the Ocean*, D. Dyrssen and D. Jagner, Eds., Almqvist and Wiksell, Stockholm, 1972, p. 75.

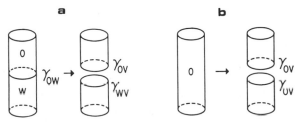

Figure 10.25   Work of adhesion (*a*) and of cohesion (*b*). Compare equations 69 and 70.

liquid–air (or liquid–vapor) interfaces, OV and WV (Figure 10.25*a*),

$$W_{OW} = \gamma_{OV} + \gamma_{WV} - \gamma_{OW} \tag{63}$$

where the γ are the respective interfacial tensions. The *work of cohesion* (Figure 10.25*b*) for the single liquids O and W is

$$W_{OO} = 2\gamma_{OV}; \qquad W_{WW} = 2\gamma_{WV} \tag{64}$$

The work of adhesion is influenced by the orientation of the molecules at the interface. For example, with the help of Table 10.2 and equation 10.6, the work of adhesion of *n*-decane–water (corresponding to a paraffinic oil–water system) and of glycerol–water can be computed to be 40 ergs cm^{-2} (40 × 10^{-3} J m^{-2}) and 56 ergs cm^{-2} (56 × 10^{-3} J m^{-2}), respectively. It requires more work to separate the polar glycerol molecules (oriented with the OH groups toward the water) from the water phase than the nonpolar hydrocarbon molecules. For paraffinic oils W_{OO} is about 44 ergs cm^{-2}, for water W_{WW} is 144 ergs cm^{-2}, and for glycerol W_{OO} is 127 ergs cm^{-2}.

Spreading of Oil on Water.   If an insoluble liquid, O, is placed on water, W, either a nonspreading lens is formed or a thin film spreads uniformly over the surface.†

By computing the surface free energy change $-dG/dA_B$, the tendency of O to spread over W can be estimated [74]. At constant temperature and pressure, a small change in the surface is given by

$$dG = \frac{\partial G}{\partial A_O} dA_O + \frac{\partial G}{\partial A_{OW}} dA_{OW} + \frac{\partial G}{\partial A_W} dA_W \tag{65}$$

Considering that $dA_W = -dA_B = dA_{BW}$ and that $(\partial G/\partial A_O) = \gamma_{OV}$, etc. (Section 10.4), one obtains

$$-\frac{dG}{dA_B} = S_{O/W} = \gamma_{WV} - \gamma_{OV} - \gamma_{OW} \tag{66}$$

† Benjamin Franklin observed (1765) that olive oil spread over water to a thickness of 25 Å. The length of the hydrocarbon chain is 25 Å. Because it was not possible to expand the film further, this result was proof of the ultimate indivisibility of matter and of the atomic theory.

or (substituting equations 63 and 64)

$$-\frac{dG}{dA_B} = S_{O/W} = W_{OW} - W_{OO} \tag{67}$$

$S_{O/W}$ is the *spreading coefficient* (positive if O spreads over W). It is seen from equation 67 that the spreading coefficient is the difference between the work of adhesion of O to W and the work of cohesion of O. Spreading occurs when the oil adheres to the water more strongly than it coheres to itself; this is generally the case when a liquid of low surface tension is placed on one of high surface tension. Thus mineral oil spreads on water, but water cannot spread on this oil. The initial spreading coefficient does not consider that the two liquids will, after contact, become mutually saturated. For example, for the benzene–water system (γ are expressed in units of 10^{-7} J cm^{-2}):

$$S_{B/W} = \gamma_{WV} - \gamma_{BV} - \gamma_{BW}$$

initially:

$$S_{B/W} = 72.8 - 28.9 - 35.0 = 8.9 > 0 \qquad \text{(spreading)}$$

after saturation:

$$S'_{B'/W'} = 62.4 - 28.8 - 35.0 = -1.4 < 0 \qquad \text{(no spreading)}$$

Low-surface-tension liquids in contact with water will have a positive initial spreading coefficient but a near-zero or negative final one. Pure hydrocarbons spread on water only if their molecular weight is low (*n*-alkanes up to octane). Even then, the initial spreading may stop or be reversed because the partial dissolution of small quantities of hydrocarbons in the water lowers γ_{OW} whereas γ_{WV} is not affected markedly.

Mineral oils often contain surface-active substances that—after immersion in water—accumulate at the oil–water interface; thereby the oil–water interfacial tension is lowered more than the surface tension of the water, thus enhancing the spreading tendency. The spreading tendency can be reduced by adding surfactants to the water that lower γ_{WV} more than γ_{OW}. Detergents have often been applied to prevent oil spills in the sea from reaching the shores; this technique, however, is of doubtful ecological value.

The spreading rate of many polar oils is about 10 cm sec^{-1}.

Monolayers. Surface films can be formed not only by spreading sparingly soluble liquids on a water surface but also by the migration of amphipathic molecules to the surface of the aqueous medium. One experimental technique for studying *monomolecular films* is to measure the *surface pressure* by means of a film balance. Such a film balance was originally constructed by Langmuir. In it the film is contained between a movable barrier and a float attached to a torsion wire arrangement. The surface pressure of the film is measured directly in terms of the horizontal force it exerts on the float, and the area of the film is varied by means of the movable barrier.

The film pressure Π is defined as the difference between the surface tension of the pure liquid and that of the film-covered surface

$$\Pi = \gamma_{W_0 V} - \gamma_{WV(film)} \tag{68}$$

Therefore any measurement of surface tension can also be used. The variation of surface pressure with the area available to the material in the monolayer (force–area curve) is a particularly useful characteristic of the monomolecular layer (film compressibility and limiting area per molecule, film coherence, etc.) (Figure 10.26).

Reducing Evaporation Rates through Monomolecular Films.

It has been estimated [75] that about 20 million tons of water are lost by evaporation annually from reservoirs in the western United States alone. The possibility of retarding evaporation by covering the water surface with a film is of obvious importance. Attempts to use monolayers to reduce evaporation have been made in Australia and in the United States. Ethyl alcohol has been used because it has a high rate of spreading and it can form (if spread using a solution in petroleum ether rather

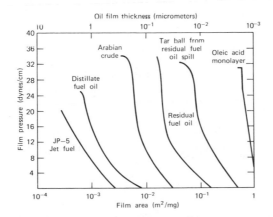

Figure 10.26 Film pressure as a function of oil film area and oil layer thickness for four potential pollutants, a tarlike residue, and a pure monomolecular layer of oleic acid. The upper abscissa is labeled as oil film thickness and is accurate only to the extent that the oil density is assumed to be unity. The oleic acid, a pure surface-active compound, formed a true water-insoluble monomolecular film which attained a maximum film pressure at 31.0 dyn cm^{-1}. A water surface of 0.6 m^2 is covered by 1 mg of this compound at its maximum film pressure. [From W. D. Garrett, *J. Rech. Atmos.* **8**, 555 (1974)].

than in benzene) a highly compressed film which gives high evaporation resistance.†

Dispersion of a Supernatant Oil Layer.

Oil layers resulting from mineral oil spills are eventually broken up. The formation of dispersed oil droplets from a coherent supernatant oil layer primarily depends on the work E_1 that is done in creating a new solid–liquid interfacial area A:

$$E_1 \approx A\gamma_{ow} \tag{69}$$

The oil volume that can be dispersed to create the interfacial area A depends on the diameter d of the (spherical) droplets, $V = A(d/6)$. The turbulence of the water, expressed as velocity gradient G (sec^{-1}) which in turn is a function of the energy input per time ε [$G = (\varepsilon/\mu)^{1/2}$, where μ is the dynamic viscosity] provides the energy necessary for dispersion per unit time:

$$E_2^* = \beta\varepsilon \tag{70}$$

where β is the fraction of energy input available for dispersion work. This energy must equal the energy required for creating new interfacial area per time $E_2^* = A^*\gamma_{ow}$. (Symbols with an asterisk are per unit of time and volume). Thus the dispersed oil volume V_{ow}^* (m^3 oil m^{-3} water sec^{-1}) can be estimated by

$$V_{ow}^* = A^* \frac{d}{6} = \frac{\beta\mu G^2}{\gamma_{ow}} \frac{d}{6} \tag{71}$$

Accordingly, it is proportional to the square of the velocity gradient and the diameter of the droplets formed, and inversely proportional to the interfacial tension; the latter depends on the composition of the oil and on the presence of surface-active substances in the spilled oil or in the water [71].

Equation 71 may be exemplified by estimating the amount of oil that can become dispersed in a typical subalpine river:

River characteristics: flow rate = 100 m^3 sec^{-1}, slope = 1%, velocity = 1 m sec^{-1}; this from potential energy ε = 100 W m^{-3}; $\mu = 10^{-3}$ kg m^{-1} sec^{-1}; velocity gradient $G = (\varepsilon/\mu)^{1/2} = 100$ sec^{-1}

Interfacial tension for commercial fuel oil:

$$\gamma_{ow} = 20 \text{ ergs cm}^{-2} = 20 \times 10^{-3} \text{ J m}^{-2}$$

$d = 10 \times 10^{-6}$ m; $\beta = 10^{-3}$

---

† The rate of evaporation may be considered to be governed by a steady-state transfer through the surface region and air space, the diffusion through the film being the rate-determining step. The flux F (g cm^{-2} sec^{-1}) is given by

$$F = -D\left(\frac{dc}{dx}\right) = -D\left(\frac{\Delta c}{\Delta x}\right)$$

where Δc is the gradient in water activity (i.e., the concentration of vapor above the surface minus the concentration of vapor in equilibrium with the substrate water), D is the diffusion coefficient of H_2O, and Δx may correspond to the thickness of the surface region.

Hence $V^*_{\overline{O}w} \approx 10^{-6} \text{ m}^3 \text{ m}^{-3} \text{ sec}^{-1}$ or 1 ml oil m^{-3} water sec^{-1}; that is, in a stretch of 1 km this river may carry 100 liters of oil in a dispersed form. This estimate is in accordance with observations made on accidental oil spills. In lakes, velocity gradients will be significantly smaller than in rivers, but somewhat larger oil droplets may be formed.

Agglomeration of Dispersed Oil with Suspended Solids.   Substantial quantities of water-polluting mineral oil components find their way into the sediments of natural water bodies. Hydrocarbons at the sediment–water interface are quite resistant to biological oxidation, especially under anaerobic conditions. These hydrocarbons affect chemical and biological processes and impair bottom fauna [76].

Mineral oil can be transferred to the sediments only if it is attached to settling particulate material (biological debris, $CaCO_3$, etc.) by adsorption and/or by agglomeration (coagulation) of colloidal oil droplets with suspended matter. The negative charge on dispersed oil colloids originating from ionization at the surface and the preferential adsorption of ions, especially surfactants, at the interface, determine the state of dispersion of the oil droplets. Most natural colloids carry, similarly to the oil dispersion, a negative surface charge. Nevertheless, some natural colloids are positively charged; furthermore, substances such as clays, although carrying a net negative charge, contain regions (e.g., Al_2O_3 edges on aluminosilicates) of positive surface charge. Furthermore, agglomeration of oil doplets with other colloids may be aided in natural waters by polymeric excretion products of algae and bacteria (bioflocculation).

Oil removal in waste treatment often cannot be achieved effectively by gravity separation (skimming tanks or separators). Coagulation is necessary to obtain coalescence (Figure 10.21*f*). Contact filters, using appropriate chemical destabilization of oil colloids, are particularly effective in oil removal [77].

REFERENCES

1   C. Tanford, *The Hydrophobic Effect; Formation of Micelles and Biological Membranes.* Wiley-Interscience, New York, 1973.

2   F. M. Fowkes, in *Chemistry and Physics at Interfaces,* American Chemical Society, Washington, D.C., 1965. This comprehensive presentation unifies several fields of surface chemistry. It gives a background useful in deciding what molecules or chemical groups should bond most strongly to a given substrate and also what contaminants are most likely to be strongly attached to surfaces or are capable of penetrating into interfaces. A detailed discussion on work of adhesion by the same author is in *J. Colloid Interface Sci.,* **28**, 493 (1968).

3   R. E. Baier, E. G. Shaffrin, and W. A. Zisman, *Science,* **162**, 1360 (1968).

4   R. E. Baier, D. W. Goupil, S. Perlmutter, and R. King, *J. Rech. Atmos.,* **8**, 571 (1974).

5   F. MacIntyre, In *The Sea,* Vol. 5, E. D. Goldberg, Ed., Wiley-Interscience, New York, 1974, p. 245.

6   P. S. Liss, in *Chemical Oceanography*, J. P. Riley and G. Skirrow, Eds., Vol. 2, 2nd ed., Academic, New York, 1975, Chapter 10.

7   S. S. Dukhin and B. V. Derjaguin, in *Surface and Colloid Science*, Vol. 1, E. Matijevic, Ed., Wiley, New York, 1974, Chapter 1.

8   P. Sennett and J. P. Olivier, in *Chemistry and Physics of Interfaces*, American Chemical Society, Washington, D.C., 1965, p. 73.

9   A. P. Black, *J. Amer. Water Works Assoc.*, **52**, 492 (1960).

10   R. A. Neihof and G. I. Loeb, *Limnol. Oceanogr.*, **17**, 7 (1972).

11   International Union of Pure and Applied Chemistry (IUPAC), *Pure Appl. Chem.*, **31**, (4) (1972).

12   Lyklema, J., in *Physical Chemistry: Enriching Topics from Colloids and Surface Science*, H. van Olphen and K. J. Mysels, Eds., Theorex, La Jolla, Calif., 1975, p. 281.

13   A. W. Adamson, *Physical Chemistry of Surfaces*, 3rd ed., Wiley-Interscience, New York, 1976.

14   H. van Olphen, *An Introduction to Clay Colloid Chemistry*, 2nd ed., Wiley-Interscience, New York, 1977.

15   M. J. Spaarnaay, *The Electric Double Layer*, Pergamon, Elmsford, N.Y., 1972.

16   D. C. Grahame, *Chem. Rev.*, **41**, 441 (1947).

17   A. L. Loeb, J. Th. G. Overbeek, and P. H. Wiersma, *The Electric Double Layer around a Spherical Particle*. MIT Press, Cambridge, Mass., 1960. (Contains spherical double-layer tables).

18   P. Somasundaran, T. W. Healy, and T. W. Fuerstenau, *J. Phys. Chem.*, **68**, 3562 (1964).

19   J. T. Davies and E. K. Rideal, *Interfacial Phenomena*, Academic, New York, 1961.

20   J. Th. G. Overbeek, *Pure Appl. Chem.*, **46**, (1976), p. 91.

21   J. Lyklema, in *The Scientific Basis of Flocculation*, K. J. Ives, Ed., Sijthoff and Noordhoff, Alphen aan den Rijn, The Netherlands, 1978.

22   G. A. Parks, *Chem. Rev.*, **65**, 177 (1965); in *Equilibrium Concepts in Natural Water Systems*, Advances in Chemistry Series, **67**, American Chemical Society, Washington, D.C., 1967, p. 121; in *Chemical Oceanography*, 2nd ed., J. P. Riley and G. Skirrow, Eds., Academic, New York, 1975; p. 241.

23   M. H. Kurbatov, G. B. Wood, and J. D. Kurbatov, *J. Phys. Chem.*, **55**, 258 (1951).

24   J. Stanton and R. W. Maatman, *J. Colloid Sci.*, **13**, 132 (1963); D. L. Dugger et al., *J. Phys. Chem.*, **68**, 757 (1964).

25   R. O. James and T. W. Healy, *J. Colloid Interfacial Sci.*, **40**, 42, 53, 65 (1972).

26   T. P. O'Connor and D. T. Kester, *Geochim. Cosmochim. Acta*, **39**, 1531 (1975).

27   A. C. M. Bourg and P. W. Schindler, *Chimia*, **32**, 166 (1978).

28   W. Stumm, R. Kummert, and L. Sigg, *Croat. Chim. Acta.* **52**, 291 (1980).

29   J. V. Smith, Special Paper No. 1, Mineralogical Society of America, 1963.

30   H. S. Sherry, in *The Nature of Seawater*, E. D. Goldberg, Ed., Dahlem Konferenzen, Berlin, 1975, p. 523.

31   A. Clearfield, in *The Nature of Seawater*, E. D. Goldberg, Ed., Dahlem Konferenzen, Berlin, 1975, p. 555.

32   W. Buser, *Chimia*, **9**, 73 (1955).

33   F. L. Sayles and P. C. Mangelsdorf, Jr., *Geochim. Cosmochim. Acta*, **41**, 951 (1977); **43**, 767 (1979).

34 K. L. Russell, *Geochim. Cosmochim. Acta*, **34**, 893 (1970).

35 F. Helfferich, *Ion Exchange*, McGraw-Hill, New York, 1962.

36 R. M. Garrels and C. L. Christ, *Solutions, Minerals and Equilibria*, Harper and Row, New York, 1965.

37 C. A. Faucher and H. C. Thomas, *J. Chem. Phys.*, **22**, 258 (1954); G. R. Frysinger and H. C. Thomas, *J. Phys. Chem.*, **64**, 224 (1960).

38 R. G. Gast, *Proc. Soil Sci. Soc. Amer.*, **33**, 37 (1969); R. G. Gast and W. D. Klobe, *Clays Clay Minerals*, **19**, 311 (1971).

39 G. Eisenmann, *Biophys. J.*, **2**(2), 259 (1962).

40 D. L. Swift and S. K. Friedlander, *J. Colloid Sci.*, **19**, 621 (1964).

41 S. K. Friedlander, *Smoke, Dust, and Haze*, Wiley Interscience, New York, 1977.

42 C. R. O'Melia, in *The Scientific Basis of Flocculation*, K. J. Ives, Ed., Sijthoff and Noordhoff, Alphen aan den Rijn, The Netherlands, 1978, p. 101.

43 C. E. Junge, *Air Chemistry and Radioactivity*, Academic, New York, 1964.

44 A. Lerman, *Geochemical Processes*, Wiley-Interscience, New York, (1979).

45 S. K. Friedlander, *J. Meteorol.*, **17**, 479 (1960).

46 J. R. Hunt, in *Particulates in Water: Characterization, Fate, Effects, and Removal*, M. C. Kavanaugh and J. O. Leckie, Eds., Advances in Chemistry Series, No.189, American Chemical Society, Washington, D.C., 1980.

47 D. F. Lawler, C. R. O'Melia, and J. E. Tobiason, in *Particulates in Water: Characterization, Fate, Effects, and Removal*, M. C. Kavanaugh and J. O. Leckie, Eds., Advances in Chemistry Series, No. 189, American Chemical Society, Washington, D.C., 1980.

48 E. J. W. Verwey and J. Th. G. Overbeek, *Theory of the Stability of Lyophobic Colloids*, Elsevier, Amsterdam, 1948.

49 H. Lyklema, in *The Scientific Basis of Flocculation*, K. J. Ives, Ed., Sijthoff and Noordhoff, Alphen aan den Rijn, The Netherlands, 1978, p. 3.

50 R. H. Ottewill, *J. Colloid Interfacial Sci.*, **58**, 357 (1977).

51 J. H. van Olphen, *An Introduction to Clay Colloid Chemistry*, 2nd ed., Wiley-Interscience, New York, 1977.

52 L. A. Spielman, *J. Colloid Interface Sci.*, **33**, 562 (1970).

53 E. P. Honig, G. J. Roebersen, and P. H. Wiersma, *J. Colloid Interface Sci.*, **36**, 97 (1971).

54 A. Sommerauer, D. L. Sussman and W. Stumm, *Kolloid-Z. Z. Polym.*, **225**, 147 (1968).

55 P. L. Busch and W. Stumm, *Environ. Sci. Technol.*, **2**, 49 (1968).

56 J. Gregory, in *The Scientific Basis of Flocculation*, K. J. Ives, Ed., Sijthoff and Noordhoff, Alphen aan den Rijn, Nertherlands, 1978, p. 3.

57 C. R. O'Melia, Report of the U.S. Environmental Agency, EPA 670/2-74-032, Washington, D.C., 1974.

58 L. A. Spielman, *Ann. Rev. Fluid Mech.*, **9**, 297 (1977).

59 J. Lyklema, in *The Nature of Seawater*, E. D. Goldberg, Ed., Dahlem-Konferenzen, Berlin, 1975, p. 579.

60 R. M. Garrels and F. T. Mackenzie, *Evolution of Sedimentary Rocks*, Norton, New York, 1971.

61 J. D. Burton, in *Estuarine Chemistry*, J. D. Burton and P. S. Liss, Eds., Academic, New York, 1976, p. 1.

62   D. Lal, *Science*, **198**, 997 (1977).

63   H. H. Hahn and W. Stumm, *Amer. J. Sci.*, **268**, 354 (1970).

64   H. H. Hahn and B. Eppler, *Colloid Interfacial Sci.*, **4**, 125 (1976).

65   J. K. Edzwald, J. B. Upchurch, and C. R. O'Melia, *Environ. Sci. Technol.*, **8**, 58 (1974).

66   H. Postma, in *Pollution and Marine Ecology*, T. A. Olson and F. J. Burgess, Eds., Wiley-Interscience, New York, 1967.

67   R. E. Grim and W. D. Johns, *Clays and Clay Minerals*, Proceedings of the 2nd National Conference, National Academy of Sciences—National Research Council, Washington, D.C., 1953, p. 81.

68   K. K. Turekian, *Geochim. Cosmochim. Acta*, **41**, 1139 (1977).

69   E. R. Sholkovitz, *Geochim. Cosmochim. Acta*, **40**, 831 (1976); *Earth Planet. Sci. Lett.*, **41**, 77 (1978).

70   E. A. Boyle, J. M. Edmond, and E. R. Sholkovitz, *Geochim. Cosmochim. Acta*, **41**, 1313 (1977).

71   M. Thuer and W. Stumm, *Prog. Water. Technol.*, **9**, 183 (1977).

72   R. E. Baier, *J. Geophys. Res.*, **77**, 5062 (1972).

73   W. R. Barger and D. W. Garrett, *J. Geophys. Res.*, **75**, 4561 (1970).

74   A. W. Adamson, *Physical Chemistry of Surfaces*, 3rd ed., Wiley, New York, 1976.

75   V. K. LaMer, Ed., *Retardation of Evaporation by Monolayers*, Academic, New York, 1962.

76   M. Blumer and J. Sass, *Mar. Poll. Bull.*, **3**, 92 (1972).

77   L. A. Spielman and S. L. Goren, *Funda.* **11**, 66 (1972).

READING SUGGESTIONS

Electric Double Layer and Surface Chemistry

Adamson, A. W., *Physical Chemistry of Surfaces*, 3rd ed., Wiley-Interscience, New York, 1976.

Aveyard, R. and D. A. Haydon, *An Introduction to the Principles of Surface Chemistry*, Cambridge University Press, Cambridge, 1973. (An introduction to the physical chemistry of surfaces).

Bockris, J. O'M., and A. K. N. Reddy, *Modern Electrochemistry*, Vols. 1 and 2, Plenum, New York, 1970.

Hiemenz, P. C., *Principles of Colloid and Surface Chemistry*, Dekker, New York, 1977.

van Olphen, H., *An Introduction to Clay Colloid Chemistry*, 2nd ed., Wiley-Interscience, New York, 1977.

Sparnaay, M. J., *The Electrical Double Layer*, Pergamon, Elmsford, N.Y., 1972.

Overbeek, J. Th. G., "Electrochemistry of the Double Layer." In *Colloid Science*, Vol. I, H. R. Kruyt, Ed., Elsevier, Amsterdam, 1952, Chapter IV.

Interfaces in Natural Waters

Bolt, G. H., "Surface Interaction between the Soil Solid Phase and the Soil Solution" and "Adsorption of Anions by Soil." In *Soil Chemistry*, G. H. Bolt and M. G. M. Bruggenwent, Eds., Elsevier, Amsterdam, 1976, pp. 43–53 and 91–95.

Bolt, G. H., M. G. M. Bruggenwent, and A. Kemphorst, Adsorption of Cations by Soil. In *Soil Chemistry*, G. H. Bolt and M. G. M. Bruggenwent, Eds., Elsevier, Amsterdam, 1976, pp. 54–90.

Lal, D., "The Oceanic Microcosm of Particles," *Science*, **198**, 997–1009 (1977).

Liss, P. S., "Chemistry of the Sea Surface Microlayer." In *Chemical Oceanography*, J. P. Riley and G. Skirrow, Eds., Vol. 2, 2nd ed., Academic, New York, 1975, pp. 193–243.

Lyklema, J., "Interfacial Electrochemistry of Hydrophobic Colloids." In *The Nature of Seawater*, E. D. Goldberg, Ed., Dahlem Konferenzen, Berlin, 1975, pp. 579–597.

Lyklema, J., "Inference of Polymer Adsorption from Electric Double Layer Measurements," *Pure Appl. Chem.*, **46**, 149–156 (1976).

Lyklema, J., "On the Relation between Surface Charge and Sol Stability," *Croat. Chim. Acta*, **48**, 565–571 (1976).

Parks, G. A., "Adsorption in the Marine Environment." In *Chemical Oceanography*, J. P. Riley and G. Skirrow, Eds., Vol. 1, 2nd ed., Academic, New York, 1975, pp. 241–308.

Colloid Stability

Gregory, J., "Effects of Polymers on Colloid Stability." In *The Scientific Basis of Flocculation*, K. J. Ives, Ed., Sijthoff and Noordhoff, Alphen aan den Rijn, The Netherlands, 1978, pp. 3–36.

Lyklema, J., "Surface Chemistry of Colloids in Connection with Stability." In *The Scientific Basis of Flocculation*, K. J. Ives, Ed., Sijthoff and Noordhoff, Alphen aan den Rijn, The Netherlands, 1978, pp. 3–36.

Matijević, E., "Colloid Stability and Complex Chemistry," *J. Colloid Interface Science*, **43**, 217–245 (1973).

Ottewill, R. H., "Stability and Instability in Disperse Systems," *J. Colloid. Interface Sci.*, **58**, 357–373 (1977).

Overbeek, J. Th. G., "Polyelectrolytes, Past, Present and Future," *Pure Appl. Chem.*, **46**, 91–101 (1976).

Overbeek, J. Th. G., "Recent Developments in the Understanding of Colloid Stability," *J. Colloid Interface Science*, **58**, 408–422 (1977).

Coagulation and Filtration

S. K. Friedlander, *Smoke, Dust and Haze*, Wiley-Interscience, New York, 1977.

D. F. Lawler, C. R. O'Melia, and J. E. Tobiason, "Internal Water Treatment Plant Design: From Particle Size to Plant Performance." In *Particulates in Water: Characterization, Fate. Effects, and Removal*, M. C. Kavanaugh and J. O. Leckie, Eds. Advances in Chemistry Series, No. 189 American Chemistry Society, Washington, D.C., 1980.

O'Melia, C. R., "Coagulation and Flocculation." In *Physicochemical Processes for Water Quality Control*, W. J. Weber, Jr., Ed., Wiley-Interscience, New York, 1972, pp. 61–110.

O'Melia, C. R., "Coagulation in Waste Water Treatment." In *The Scientific Basis of Flocculation*, K. J. Ives, Ed., Sijthoff and Noordhoff, Alphen aan den Rijn, The Netherlands, 1978, pp. 101–130.

O'Melia, C. R. "Aquasols: The Behavior of Small Particles in Aquatic Systems," *Environ. Sci. Technol.*, **14**, 1052 (1980).

Stumm, W., and C. R. O'Melia, "Stoichiometry of Coagulation," *J. Amer. Water Works Assoc.*, **60**, 514–539 (1968).

Surface Chemistry of Hydrous Oxides

Bowden, J. W., M. D. A. Bolland, A. M. Posner, and J. P. Quirk, "Generalized Model for Anion and Cation Adsorption at Oxide Surfaces," *Nature*, **245**, 81 (1973).

Davis, J. A., R. O. James, and J. O. Leckie, "Surface Ionization and Complexation at the Oxide/Water Interface." *J. Colloid Interface Sci.*, part I, **63**, 480–499 (1978); part II, **67**, 90–107 (1978); part III, **74**, 32–43 (1980).

James, R. O., and T. W. Healy, "Ions at the Oxide Water Interface," *J. Colloid Interface Sci.*, **40**, 42–81 (1972).

Lyklema, J., "The Electric Double Layer on Oxides," *Croat. Chim. Acta*, **43**, 249–260 (1971).

Schindler, P., "Surface Complexes at Oxide/Water Interfaces." In *Adsorption of Inorganics at the Solid/Liquid Interfaces*, Marc A. Anderson and Alan J. Rubin, Eds., Ann Arbor Science Publications, Ann Arbor, Mich., 1981.

Stumm, W., H. Hohl, and F. Dalang, "Interaction of Metal Ions with Hydrous Oxide Surfaces," *Croat. Chem. Acta*, **48**, 491–504 (1976).

Stumm, W., R. Kummert, and L. Sigg, "A Ligand Exchange Model for the Adsorption of Inorganic and Organic Ligands at Hydrous Oxide Interfaces," *Croat. Chim. Acta*, **52**, 291 (1980).

11

Some Concepts on Water Pollution
and its Control;
An Ecological Perspective

11.1 INTRODUCTION

At the bottom of many of our fresh waters, at the sediment–water interface, there lives a little worm; biologists call it tubifex. This worm has an interesting anatomy; its mouth is at the front end, but it breathes oxygen at the rear end. This presents the worm with a conflict situation: In order to eat, it tends to penetrate deeper into the food-bearing sludge where oxygen is depleted; in order to breathe, however, it tends to ascend into the higher water layers that contain dissolved oxygen but where there is little food.

This example of an antinomy—an organism being suspended between layers of nutrition and respiration—illustrates a basic property, a constitutive property of all organisms including humans: Conditions of life preservation are unavoidably interrelated with conditions of life endangerment. This constitutive element is particularly evident with humans in their conflict between exploitation of natural resources and protection of the environment. Humans have always built up their own order at the expense of some order in the environment. They have been forced to multiply destruction of environmental order to build up the increasingly intricate structure of cultural and technical civilization. Humans as physiological beings play a minor role in the ecosystem; domestic waste and garbage represent a small fraction of the detritus produced by all forms of life. Within the biosphere, the energy involved in human metabolism, ca. 4×10^{11} W (4×10^9 inhabitants with a food intake of ca. 100 W = 2100 kcal day^{-1} = 8800 kJ day^{-1}) may be compared with primary productivity—that is, the energy fixed by all plants—ca. 10^{14} W (based on photosynthesis of 6×10^{15} mol C year^{-1} and an energy equivalent of 100 kcal (418 kJ) mol^{-1} of C photosynthesized). If evenly distributed over the world, human biological wastes would have an almost negligible effect on the energy transfer of the ecosphere. Domestic wastes cause localized or temporary unfavorable environmental alteration only where they are discharged in high concentrations.

On the other hand, humans have developed from relatively unimportant physiological consumers to inventive intellectual beings and geochemical manipulators who use external flows of energy and materials to build up their civilization and increase their dominance. Humans dissipate 10 to 20 times (in the Northern Hemisphere 50 to 100 times) as much energy as they require for their metabolism. The disturbance of the aquatic environment caused mostly indirectly by this energy dissipation may outweigh that caused by the disposal of excreta (Figure 11.1). Most of the energy utilized by an industrial society for its own advantage (heat production, manipulation of the landscape, urban construction, agriculture, forestry, geological exploration, construction of dams)

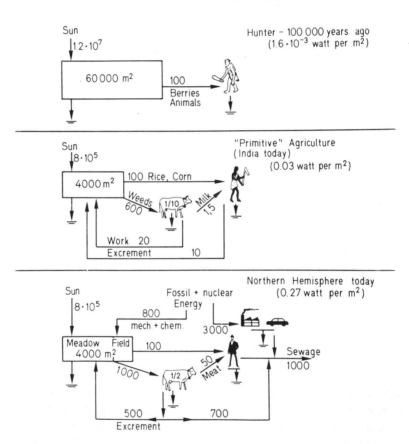

Figure 11.1 Energy and land needs of modern and primitive humans. In modern times, humans require for environmental manipulation and ecological domination much more energy than for mere metabolic activity. About 10^5 years ago, a hunter required about 60,000 m² land to provide basic food needs. With primitive agriculture, the productivity of the land is increased by a factor of almost 15. Cattle ($\frac{1}{10}$ cattle per person) provides a minimum of protein (milk). Intensive agriculture as it now prevails in the Western world has led to a further enormous increase in productivity. But because the increased productivity is paralleled by improved food quality (primary plant products have been replaced by meat production: $\frac{1}{2}$ head of cattle per person), humans today still require ca. 4000 m² land for food production. The system is no longer closed. The consequences of energy dissipation cause an impact on the environment that goes far beyond that of excreta disposal.

ultimately affects and perturbs ecosystems. Humans—as terrestrial beings—directly interfere primarily with the terrestrial environment; but, because of the interdependence of the land and water ecosystems and because of the sensitivity of the latter, the stress imposed upon the ecology by civilization is primarily reflected in the aquatic ecosystems. Many of the consequences of energy dissipation are also evident in the atmosphere which acts as an efficient conveyor belt for many pollutants. Human power to disturb the environment tends to run ahead of the technological possibilities for response to the environmental impact.

An *ecosystem* may be defined as a unit of the environment in which, as a result of solar energy input, a biological community (primary producers, consumers, and decomposers) is maintained; the flow of energy is used to organize the system and is accompanied by cycles of water, nutrients, and other elements and by cycles of life through various food users at different trophic levels (cf. [1]). The members of an ecosystem are interlocked by various feedback loops (homeostasis) and thus adapted to coexistence for mutual advantage. The network of checks and balances includes a multiplicity of transfers of inorganic and organic substances—the food web, nutrients, allelochemicals—and gives the ecosystem its functional unity.

The second law of thermodynamics demands that any spontaneous process be accompanied by an increase in entropy—that is, dS (source, sink) + dS (ecosystem) ≥ 0. Because dS (source, sink) > 0, the entropy of the ecosystem may decrease: $-dS$ (ecosystem) $\leq dS$ (source, sink). This decrease is reflected in the ordering of the ecosystem and the presence of such highly improbable aggregations of energy as living beings [2]. Their organization has been acquired at the expense of an increase in entropy in the environment.

Objectives.   This chapter consists essentially of three parts. We will first exemplify pollution and how it affects water quality; we will emphasize that we have experienced in the last decades not only a marked increase in pollutional load but that the character of pollution has changed. In the second part we will discuss the response of the aquatic ecosystem to human impact, especially to stress caused by chemical perturbations. Third, it will illustrate that pollution is no longer just a local or regional problem, but that human influence on global chemical cycles may change the global environment and affect the quality of natural waters.

So far we have endeavored to approach the various subjects in a quantitative and rigorous way. In this chapter, however, we often have to depend on rather qualitative, occasionally even on speculative arguments because our understanding of the sensitivity and the resilience of aquatic ecosystems is seriously hampered by lack of theoretical and sufficient empirical information on the effects of physical and chemical perturbations of aquatic ecosystems.

Although this chapter stresses the impact of human activities on the environment, the reader should not be left with the impression that technology and energy and resource utilization in themselves are bad. For the maintenance of our civilization and culture and for enhancement of the quality of life, and especially for food production for the increasing world population, we will have to continue to depend on technology and energy utilization. Social criteria and growing pressures for social equality must codetermine scientific and technological development.

11.2 POLLUTION AND WATER QUALITY

In a broad sense, pollution has been characterized as an alteration of our surroundings in such a way that they become unfavorable to us and to our life. This characterization implies that pollution is not synonymous with the addition of contaminants or pollutants to the environment but can also result from other direct or indirect consequences of human action.

Interdependence of Land and Water Ecosystems. Humans interfere more and more with cycles that connect land, water, and atmosphere. In Figure 11.2 we illustrate schematically with the help of a few selected physical, chemical, and biological attributes some interdependences of land and water. For example, a technological energy input into a terrestrial ecosystem (e.g., deforestation, conversion of grassland into cropland, intensification of agricultural production, land amelioration) reduces the quantity of biomass (decrease in vegetation cover); this in turn affects the microclimate and reduces evapotranspiration, thus increasing rain runoff and rates of erosion and siltation. Accelerated nutrient cycles and faster transport of soil constituents increase sedimentation rates and lead to enrichment with nutrients of surface waters followed by changes in the chemical and biological composition of aquatic habitats. Obviously, every aquatic ecosystem has physical, chemical, biological, and geological inputs and outputs. A detailed example of an analysis of the interdependence of a terrestrial and water ecosystem can be found in the studies on the Hubbard Brook ecosystem [3].

The ratio of pollutant fluxes to natural fluxes increases with increasing activity of civilization. The quality of water bodies thus generally reflects the range of human activity within their catchment area. In a broad sense, potential perturbation of lakes, rivers, estuaries, and coastal areas may be related to population density and energy dissipation in the drainage area of these water bodies. Figure 11.3 shows the relationships among per capita energy consumption, population density, and energy consumption per unit area for various countries [4]. These data may be compared with the mean biotic energy flux (energy fixed by photosynthesis) and the input of solar energy to the surface of the earth. As this figure shows, in most countries of the Northern Hemisphere energy flux by civilization exceeds biotic energy flux markedly.

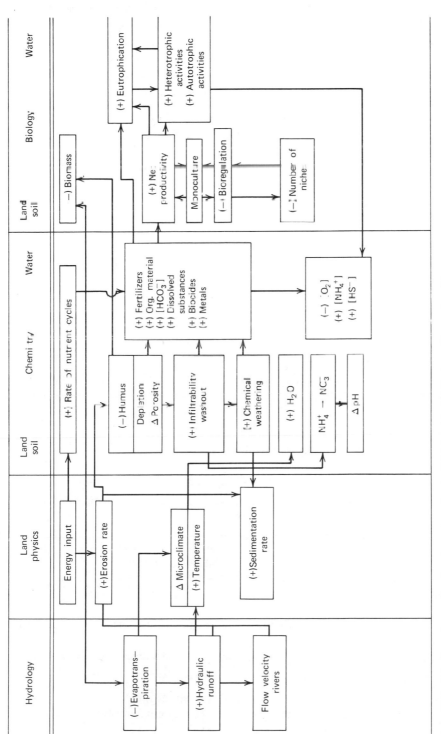

Figure 11.2 Interdependence of land–water systems (see text). (+) and (−) indicate positive and negative feedbacks, respectively.

689

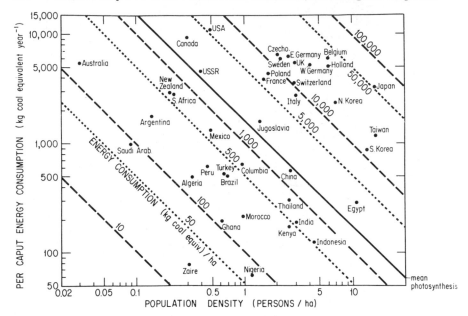

Figure 11.3 The relationships among per person energy consumption, population density, and energy consumption per unit area (or pollution potential), for different countries in 1972. [Note: The area of each country includes only FAO-defined agricultural areas. For comparison, solar input at earth surface is ca. 10^6 kg coal eq year^{-1} ha^{-1}. (1 kg coal eq year^{-1} ha^{-1} $\cong$ 1.1 $\times$ 10^{-4} W m^{-2}).] (cf. Li, [4]).

Gross Loading Parameters

Pollutional loading may be related to the population density and to the per capita waste production in a drainage area. The potential loading J of various rivers and estuaries may be estimated by

$$J = \frac{\text{inhabitants}}{\text{drainage area}} \times \frac{\text{drainage area}}{\text{runoff}} \times \frac{\substack{\text{waste production or} \\ \text{energy consumption or} \\ \text{gross national product}}}{\text{capita}} \times (1 - \eta) \quad (1a)$$

where η is the effectiveness of environmental protection measures (recycling, waste retention, waste treatment).

Similarly, for comparative purposes, the loading of a lake, J', could be formulated as

$$J' = \frac{\text{inhabitants}}{\text{drainage area}} \frac{\text{drainage area}}{\text{lake area}} \frac{1}{\text{lake depth}} \frac{\substack{\text{waste production or} \\ \text{energy consumption}}}{\text{capita}} (1 - \eta) \quad (1b)$$

The gross national product per time within the drainage area may be used to estimate the extent of potential waste production because it measures economic production, that is, the value of material goods and services for private and public consumption. Table 11.1 compares loading parameters for some rivers. Obviously the surface waters draining areas of high population density and extensive industrialization belong to the more heavily loaded aquatic systems in Table 11.1.

At the turn of the century a rule was established by the State Health Department of Massachusetts on the basis of data collected from different streams that objectionable conditions were unlikely to result when less than 5800 inhabitants per m^3 sec^{-1} live in the drainage area of a stream. Considering the fact that the American economy and correspondingly the production of wastes have increased by 1.5% per year per person, the limit today would be ca. 2000 inhabitants per m^3 sec^{-1} (cf. Figure 11.5c).

The six lakes at the top of the list in Table 11.2 are—or have been eutrophied prior to treatment or waste diversion.

Chemical Dynamics of a Polluted Watershed

A watershed responds to precipitation, weathering of rocks, agricultural runoff, and municipal and industrial wastes. The chemical mass flux of constituents reflects the extent of pollution and the many other processes acting in the drainage area. By using information on the geology, rainwater composition, land use, and density of population and domestic animals, and by considering the dependence of concentration on flow, the various processes contributing to the mass flux may be identified. For example, Ceasar et al. [5] have constructed a dynamic chemical flux model for the Merrimac River in northern New England.

Waste Discharge and Composition of the Receiving Waters.   Figure 11.4 depicts schematically the relationship among load (input), water quality, and objectives of water use (output). Water pollution consists of a variety of material flows that depend on population density, life style, and cultural activities. The various emissions are measured as flows (capacity factors, e.g., kg $time^{-1}$, watt). The resulting water composition is determined by the entity of interacting chemical, physical, and biological factors which are intensity factors (activity, concentration, redox potential, temperature, velocity gradient). These intensive variables, above all the activities of the chemical constituents, determine primarily the type of community of organisms present in the water.

Water quality criteria are scientifically established requirements about intensity factors; these criteria form the bases for judgments with respect to the compatibility of a water composition with ecological objectives or designated

TABLE 11.1 COMPARISON OF LOADING PARAMETERS OF SOME RIVERS

Rivers	Population Density (Inhabitants per km^2)	Inhabitants per Runoff (Inhabitants per m^{-3} sec^{-1})	Gross National Product per Unit Flow (Dollars per m^3)
Rhine	140	15,000	3.4
Danube	83	10,400	1.1
Ohio	76	5,800	1.3
Mississippi	19	3,300	0.75
Rhone	63	3,700	0.55
Kemijoki	2.5	250	0.03
All world rivers	27	3,000	0.15
All U.S. rivers	28	4,200	1.0
All European rivers	66	6,500	—
All Asian rivers	45	4,877	—
All African rivers	11	2,500	—
All North American rivers	15	1,600	—
All South American rivers	10	560	—

[a] Based on average runoff; it may be more appropriate to select runoff during a low flow period. The difference between average runoff and runoff during a dry year is especially large in the United States. The 95 % dry year (flow exceeded on the average in 19 out of 20 years) for Mississippi and Ohio is about half that of the average year.

TABLE 11.2 COMPARISON OF SOME LOADING PARAMETERS OF SOME LAKES

Lake	Country	Surrounding Factor[a]	Mean Depth (m)	Inhabitants per km^{-2}	Inhabitants per m^3 Lake Volume	Energy Consumption per Lake Volume (W m^{-3})
Greifensee	Switzerland	15	19	441	348	1.81
Plattensee	Hungary	10	3	~60	200	0.97
Lake Washington	United States	~15	18	~50	42	0.48
Lake Constance	Switzerland–Germany–Austria	19	90	114	24	0.12
Lake Lugano	Switzerland–Italy	11	130	264	22.3	0.11
Laka Biwa	Japan	4.5	41	~150	16	0.07
Lake Winnipeg	Canada	35	13	~3	8.1	0.07
Lake Titicaca	South America	14	~100	~40	5.6	0.001
Lake Victoria	Africa	3	40	~70	5.1	0.002
Lake Baikal	USSR	17	730	~5	0.6	0.0005
Lake Tanganjika	Africa	4	572	~50	0.3	0.0001
Laka Inari	Lapland	12	~50	0.5	0.1	0.0005
Lake Superior	Canada–United States	1.5	145	~5	0.05	0.0005

[a] Drainage area/lake area.

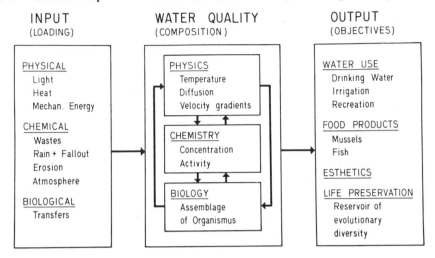

Figure 11.4   Transfer functions between loading, water quality, and water use.

water uses. *Standards* are tolerance levels established by governmental authorities as a program for water pollution abatement.

Transfer functions between load (immission†) and water composition (concentrations of chemical species) depend on a multitude of processes (adsorption; bioaccumulation, sedimentation, self-purification, etc. (Compare Figure 9.18). Only with soluble conservative immissions does a simple dilution model give the solute concentration. For a stream with a rate of flow Q, the chemical mass flux L is often composed of a Q-independent mass flux A (e.g., from relatively constant waste discharges) and a Q-dependent part bQ (e.g., from background fluxes such as equilibration with rocks and elution of soil).

$$L = CQ = A + bQ \qquad (2)$$

This equation must not be applied uncritically; its validity depends, among other things, on the time period over which Q is averaged.

Example 11.1. Dependence of Concentration on Runoff

Estimate the chemical mass flow of dissolved organic carbon from the data given in Figure 11.5a and estimate the proportion of this mass flux that results from an anthropogenic input.

Most conveniently the data can be plotted in accordance with equation 2. $C = A/Q + b$ (Figure 11.5b) where we obtain the intercept and slope; $A = 460$ g DOC sec^{-1} and $b = 2.06$ mg DOC liter^{-1}. A, the Q-independent load, generally is the portion of the load that may be characterized as being caused by

† It is convenient to distinguish between imission (input) and emission (output).

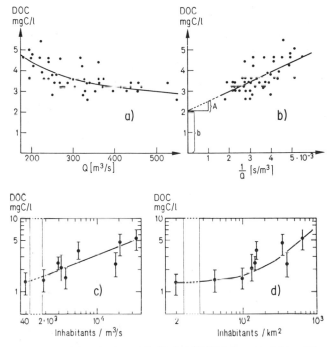

Figure 11.5   Dependence of dissolved organic Carbon (DOC) on rate of flow, Q (a and b) and on population density (c and d) (Example 11.1). In (a) and (b) data from the river Aare (Switzerland) are plotted. In (c) and (d) DOC data with a mean range between 5 and 95 % values (i.e., values observed in more than 5 % and less than 95 % of all samples) of various rivers are given (cf. J. Zobrist, J. Davis, and Hegi, *Gas Wasser Abwasser* **57**, 402 (1977) and J. Davis and Zobrist, *Progr. Water Technol.*, **10**, 65 (1978).

waste discharge, while bQ may be interpreted as the load related to natural processes (e.g., washout of any material from the earth's crust and soil). The correlation coefficient between DOC and $1/Q$ is 0.65. The geometric mean of the DOC data (data are normally distributed) is 3.56 mg C liter^{-1}.

11.3   WATER QUALITY CRITERIA

Objectives for major water use are listed as "output" in Figure 11.4. The preservation of fresh water as a supply of potable water and the maintenance of most natural waters as life preservation systems (production of aquatic food, reservoir for genetic diversity) are among the most important goals of water pollution control.

It is difficult to evaluate objectively and to codify water quality because (1) the effect of water composition on the various ecological consequences is little understood and very difficult to quantify, and (2) it is difficult to define a reference state, that is, a hypothetical pristine state of the water.

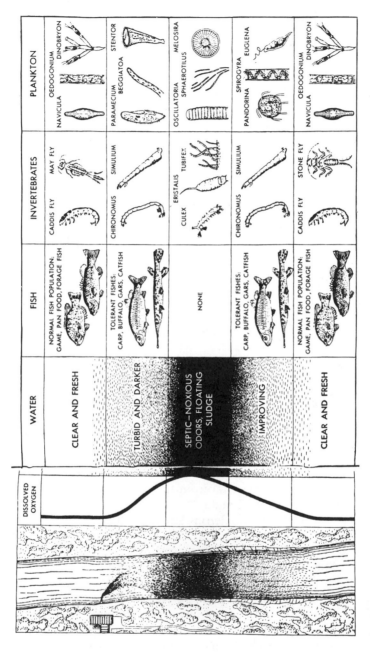

Figure 11.6 Diagrammatic presentation of the zones of pollution showing dissolved oxygen and types of organisms associated with each zone. Because the character of pollution has changed in the last decades, these quality parameters alone do no longer suffice for water quality evaluation. (From Eliassen [6] reproduced with permission from *Scientific American.*)

Two water quality criteria have been used most commonly in previous decades:

1   The concentration of dissolved oxygen or the oxygen saturation deficit as a pollution index and the biochemical oxygen demand as a loading parameter.
2   "Indicator organisms" which are indicative of the existence of certain pollution conditions. At the turn of the century, a saprobic index was developed in Germany by classifying organisms according to the environment that seemed to favor them.

The concept is illustrated in Figure 11.6 [6]. However, it *alone* can no longer suffice for the evaluation and codification of water quality for the following reasons:

1   The character of pollution, as we shall see, has changed over the last decades, especially in industrialized and highly settled regions. Dissolved oxygen concentrations and typical indicator organisms are primarily indicative of one type of pollution (putrescible substances that affect the heterotrophic response of organisms). The activity of oxygen is obviously an ecologically important parameter, but many substances that may impair the ecology of receiving waters or adversely affect the potability of water have no effect on dissolved oxygen.
2   In interpreting the presence of indicator organisms, we are often not aware of a cause-and-effect relationship. Chemical activities (together with physical factors) determine causally the biological characteristics of a water. Indicator organisms are a meaningful tool only with respect to the water quality for which they initially were shown to be indicative. Today, modern analytical chemistry can determine the numerous water quality parameters that affect water quality and the biocenosis more directly than a bioassay relying on the presence of certain organisms [7].

Chemical Context of Water Quality

Implicit in water quality criteria is a chemical model with the help of which tolerance levels compatible with designated water uses can be rationalized and quantified. In measuring and quantifying the diverse chemical variables (including pollutants), one encounters first the analytical problems of sensitivity and specificity (Figure 11.7). Analytical chemistry has made remarkable progress in improving the sensitivity of detection. With new gas chromatographic and electrochemical sensors or by using flameless atomic absorption or neutron activation, some components can routinely be detected down to concentrations of less than 10^{-3} μg liter^{-1}. The environmental effect of a substance (biological availability, physiological and toxicological effects, chemical and geochemical reactivity) is structure-specific; for example, $Cu(CO_3)(aq)$ affects the growth of

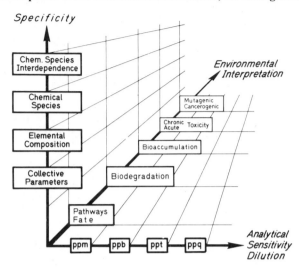

Figure 11.7 Sensitive analytical techniques and a knowledge of individual chemical species and their interdependence are prerequisites for water quality interpretations.

algae in a different way than Cu^{2+}(aq); organic isomers usually differ in their toxicological effects. While high-resolution gas chromatography is able to identify a large number of individual substances at high sensitivity (Figure 11.7), usually no single method presently available permits unequivocal identification of many inorganic species, and we often have to depend on chemical models for species recognition (e.g. cf. Figure 6.10) and for determination of species interdependence (Table 6.8).

Environmental interpretation depends, furthermore, on knowledge of pathways and of biogeochemical parity (stoichiometric relations in rocks, sediments, biota, etc.). Knowledge of the physiological and toxicological responses of individual organisms to factors in their environments is a prerequisite for understanding the distribution and abundance of aquatic organisms.

Max Blumer [8] points out: "It appears important to strive for analyses that provide the greatest possible resolution and the most complete insight into all components of natural mixtures. We must remain cautious in adopting tolerance levels as long as our analyses are so incomplete, and we should suggest and demand safety factors that are adequate to protect against the unanticipated effects of yet unsuspected biologically active compounds in the environment."

Trends in Quantity and Quality of Pollutional Load

Over the years the gross pollutional load has increased, and its character has changed. While a few decades ago most of our wastes were predominantly catabolic (excreta of humans and animals and other biogenic components), they are now more and more composed of discards of modern industrial society [synthetic chemicals, mining products (phosphates, metals), by-products of fossil fuel combustion and of energy production (metals, oxides of S, N, and H, heat, radioactive isotopes)]. The change in waste composition is evident in comparing annual growth figures for populations with industrial activities:

	Percentage year^{-1}
Population, world	2
Population, Northern Hemisphere	1.1
Gross national product, Northern Hemisphere	3–4
Production of chemicals, United States	5–7
Phosphate mining, world	7

Many industrial chemicals reach receiving waters indirectly (via households, agricultural drainage, the atmosphere). The changes in the character of the

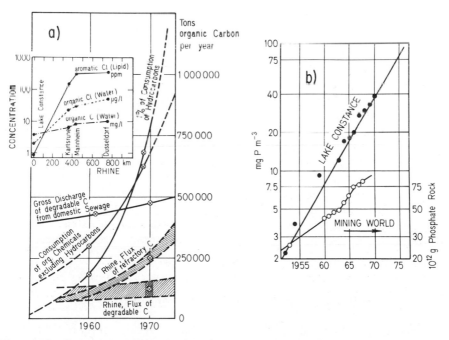

Figure 11.8 Change in pollution load of the Rhine River (organic carbon) (a) and Lake Constance (phosphate) (b). (a) Increase in consumption of organic chemicals in the catchment area of the Rhine River in comparison to the gross load of the Rhine with degradable organic carbon (excreta). The flux of refractory organic carbon C (calculated and measured for Dusseldorf) reflects the increase in consumption of synthetic chemicals. The flux of biodegradable C remains constant, however (data from Stumm and Roberts). The inset illustrates the increased downstream accumulation of chlorinated organic compounds in the water and in the biota (from data of H. Sontheimer). [Reproduced from *IAWR* **3**, 69 (1973)]. (b) The increase in concentration of P in Lake Constance (measured during overturn) is compared with the quantity of phosphate rocks mined globally.

pollution load are especially apparent in areas of high population and industrial density. Examples for the Rhine River and Lake Constance are given in Figure 11.8. In the Rhine River the mass flux of refractory† substances exceeds that of biodegradable substances. Conventional primary and secondary waste treatment for municipalities and industries within the catchment area is unable to prevent the progressive accumulation of refractory substances in the water, since self-purification and biological waste treatment are not very effective in eliminating these relatively persistent chemicals. These substances interfere markedly with the biocenosis of the Rhine water; in certain stretches the number of species has been reduced drastically.

11.4 DISTURBANCE OF BALANCE BETWEEN PHOTOSYNTHESIS AND RESPIRATION

As we have seen in Section 9.7, the causal and reciprocal relationship between organisms and the environment must be considered in understanding an aquatic habitat. Temporal or local disturbances of the stationary state of photosynthesis and respiration (Figure 11.9a) lead to chemical and biological changes that often may be interpreted as pollution.

Terrestrial and aquatic ecosystems differ in the organization of their food chain. Terrestrial ecosystems, which have productivities similar to those of aquatic systems, typically contain a large biomass containing few tropic levels. Aquatic systems on the other hand are characterized by a relatively small biomass and a more intricate foodweb. The plant biomass dominates on land, while in lake waters the animal biomass is similar in magnitude to the plant biomass. Primary production is consumed to a much larger extent in aquatic systems than in terrestrial systems; in the latter most of the biomass is decomposed by microorganisms. The delicate balance between photosynthetic and heterotrophic organisms, especially in the ocean, appears to be controlled to a large extent by herbivores (Steele [9]); phytoplankton may be eaten as fast as it is produced. The products of photosynthesis become very diluted in aquatic systems.

Sensitivity of Aquatic Systems to Chemical Perturbation

Bacteria and algae occur in aquatic habitats under varying conditions. The wide spectrum of ecological conditions (temperature, pressure, pH, redox intensity) under which microorganisms grow reflects the great capability of physiological regulatory mechanisms. The microbial transformations of a diversity of substrates distributed more or less uniformly throughout the water masses have to

---

† Refractory substances are those—frequently of synthetic origin—that resist biodegradation; since the persistence of a substance is relative, we may arbitrarily define a substance to be refractory if its half-life in an aerobic aquatic environment exceeds a certain time period (cf. Wuhrmann [7]).

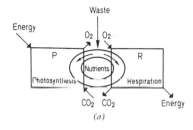

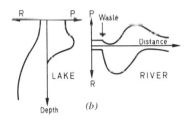

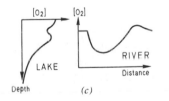

Figure 11.9   Ecological balance between photosynthesis and respiration. Photosynthetic organisms (mostly algae and plants) synthesize biomass (P = rate of photosynthesis) which is consumed by animals (consumers, herbivores) and biodegraded by bacteria and fungi (decomposers); respiration (R = rate of respiration) consists of biodegradation and decomposition. A steady state between P and R (a) is a prerequisite for the maintenance of a constant chemical composition and thus for a relatively constant structure of the population of organisms. This ecological balance is disturbed if a natural water receives an excess of organic bacterial nutrients or an excess of inorganic algal nutrients.

In stratified *lakes* a vertical separation of P and R organisms results from the fact that algae remain photosynthetically active only in the euphotic upper layers; algae that have settled under the influence of gravitation serve as food for the R organisms in the deeper layers of the lake (b). Organic biomass that has been synthesized in the upper layers of the lake becomes biochemically oxidized (respiration and consumption) in the deeper layers. Most of the photosynthetic oxygen escapes to the atmosphere and does not become available in the deeper water layers where, as a result of the oxidation of the settling biological debris, less aerobic or anaerobic conditions develop (c).

In *rivers* P and R organisms become separated more longitudinally. The introduction of wastes enhances primarily respiration. After wastes have become partially mineralized and suspended solids have settled out, photosynthesis becomes more important.

occur at concentrations that are typically at levels of below $10^{-5}\,M$. The resorption of an organic substrate by aquatic microorganisms (alga, bacteria)—typically by a mixed biocenosis (community of organisms)—occurs according to the equation

$$V = \frac{V_{max}\,S}{K_S + S} \qquad (3)$$

where V and V_{max} are, respectively, the observed rate of resorption and the maximum rate of resorption, S is the concentration of substrate, and K_S is the concentration of substrate, where $V = \frac{1}{2}V_{max}$ [10–13]. The half-rate concentration K_S for many substrates, such as orthophosphate, amino acids, and carbohydrates, is between 10^{-5} and $10^{-10}\,M$ [12–13].

Aquatic systems are particularly sensitive to many trace elements. It has been suggested that some metal ions, such as Cu, can be toxic to phytoplankton in naturally occurring concentrations, for example, in upwelling waters [14]. It has been shown that the growth rate of algae is affected by free Cu^{2+} even at concentrations as low as 10^{-10} to $10^{-12}\,M$ [15–17], while Cd concentrations as low as $10^{-8.5}$ are toxic to daphnia and other fish food [18]. The impact of trace elements on aquatic organisms depends on the chemical species distribution (Table 11.3). The factors that affect chemical speciation thus also influence the physiological and toxicological effects on organisms.

As we have seen (equation 24 in Chapter 9), solar energy recirculates C, N, P, S, and other elements in approximately fixed proportions:

$$106\,CO_2 + 16\,NO_3^- + HPO_4^{2-} + 122\,H_2O + 18\,H^+\,(+\text{trace elements: energy})$$

$$P \parallel R$$

$$\{C_{106}H_{263}O_{110}N_{16}P_1\} + 138\,O_2 \tag{4}$$
$$\text{algal protoplasm}$$

These proportions reflect the biogeochemical parity of the biomass. The rates of cycling are geared to preservation of the ecosystem and the relative proportions of chemical elements representative of the biota (e.g., an approximate 106:16:1 atomic ratio of carbon, nitrogen, and phosphorus in aquatic systems). Deviation from these requirements of ecosystem inputs and outputs, or acceleration of the cycling of one element, may produce ecological maladjustments globally, regionally, and locally.

The delicate balance between photosynthesis and respiration in a receiving water may be perturbed by adding either an excess of organic compounds or an excess of inorganic algal nutrients (e.g., phosphorus, nitrogen). (See Table 11.4.) In the first case, heterotrophic processes (decomposition) tend to dominate, and dissolved oxygen may become exhausted (biochemical oxygen demand). In the

TABLE 11.3 SOME FACTORS AFFECTING CHEMICAL SPECIATION AND THUS INFLUENCING THE TOXICITY OF TRACE METALS TO AQUATIC ORGANISMS

Variables	Explanation
pH, alkalinity, organic and inorganic ligands	Changes species distribution and $[Me^{n+}]$; influences formation of hydroxo, carbonato and other complexes; changes adsorbability of metal ions and resorptivity of cell organisms
Density of organisms	Reduces available $[Me]_T$ and changes species distribution because of adsorption on cell surfaces and/or by complexation by exudates of organisms
Concentration of particles and colloids	Metals are sequestered by particles oxides or iron and manganese and other organic colloids are particularly effective
Redox potential	Affects oxidation state of metal; methylation often occurs more readily at low redox potentials

TABLE 11.4 CHEMICAL PERTURBATION OF AQUATIC ECOSYSTEMS[a]

Type of Pollutant	Effect
Fertilizing substances (P and N compounds)	$P/R > 1$; promotion of phototrophic organisms, especially in stratified waters
Biologically degradable organic substances	$P/R < 1$; promotion of heterotrophic organisms
Nondegradable organic substances and biologically active inorganic substances (metals)	Disturbance of interspecies relationships, sociological changes in biocenosis

[a] P and R are the rates of photosynthetic production and of respiration, respectively.

second case, the immediate result is progressive accumulation of algae and plants. In either case the initial perturbation is followed by a readjustment that ultimately leads to a new balance (Figure 11.9).

Considerable effort has been made by sanitary engineers to quantify the three most important kinetic processes governing the dissolved oxygen in river waters—exchange with the atmosphere, total respiration, and photosynthesis. The dynamics of oxygen and CO_2 exchange, that is, the air water exchange rate constant and the rate of photosynthesis and respiration can in principle be estimated from a continuous record of dissolved oxygen, temperature, and an idealized light intensity curve [19].

Pollution by Algal Nutrients

In a stratified lake excessive production of algae and oxygen in the upper layers ($P \gg R$) may be paralleled by anaerobic conditions at the bottom ($R \gg P$), because most of the photosynthetic oxygen escapes to the atmosphere and does not become available to the deeper water layers while the algae eventually sink to the bottom of the lake (see Figure 11.9).

Limiting Nutrients.   For most inland waters phosphorus is the limiting nutrient in determining productivity. In some estuaries and in many marine coastal waters nitrogen appears to be more limiting to algal growth than phosphorus [20]. Deficiency in trace elements occurs usually only as a temporal or spatial transient.

Because of the complex functional interactions in lake ecosystems, the limiting factor concept needs to be applied with caution. We should distinguish between rate-determining factors (an individual nutrient, temperature, light, etc.) that determine the rate of biomass production and a limiting factor where a nutrient determines in a stoichiometric sense (equation 4) the maximum possible biomass standing crop.

Schindler [21] used evidence from whole-lake experiments to show convincingly the phosphorus limitation in lakes; he also demonstrated that natural mechanisms compensated for nitrogen and carbon deficiencies in eutrophied lakes. In experimentally fertilized lakes, the invasion of atmospheric carbon dioxide supplied enough carbon to support and maintain phytoplankton populations proportional to P concentrations over a wide range of values. There was a strong tendency in every case for lakes to correct carbon deficiencies —obviously the rate of CO_2 supply from the atmosphere was sufficiently fast [22]—maintaining concentrations of both chlorophyll and carbon that were proportional to the P concentration (Figure 11.10a).

Schindler [21] also demonstrated that biological mechanisms were in many cases capable of correcting algal nitrogen deficiencies (Figure 11.10b). While a sudden increase in the P input may cause algae to exhibit symptoms of limitation by either N or C or both, there are long-term processes at work which appear to correct the deficiencies eventually, once again leaving phytoplankton growth proportional to the P concentration. As Schindler points out, this "evolution" of appropriate nutrient ratios in fresh waters involves a complex series of interrelated biological, geological, and physical processes, including photosynthesis, the selection of species of algae that can fix atmospheric nitrogen, alkalinity, nutrient supplies and concentrations, rates of water renewal, and turbulence. Various authors have observed shifts in algal species with changing N/P ratios; low N/P ratios appear to favor N-fixing blue-green algae, whereas high N/P ratios, achieved by controlling P input by extensive waste treatment, cause a shift from a "water bloom" consisting of blue-green algae to one containing forms that are less objectionable. Shapiro [23] concludes that a lower pH (or increased CO_2) gives green algae a competitive advantage over blue-green algae.

Typically, N/P ratios change in passing from the land to the sea. If one compares field data with the average ratio in phytoplankton, one sees that fresh waters in most instances receive an excess of N over that needed. Agricultural drainage contains relatively large concentrations of bound nitrogen, because nitrogen is washed out more readily from fertilized soil than phosphorus. In estuaries denitrification is frequently encountered, because NO_3-bearing waters may come into contact with organically enriched water layers; the N/P ratio can also be shifted by differences in the circulation rate of these two nutrients.

Example 11.2. O_2 Consumption Resulting from Increased Productivity

Estimate the P loading of a lake that may be tolerated without causing anaerobic conditions as a function of the depth of the hypolimnion (water layer below the thermocline).

As is the case in the sea, the deeper portions of a lake receive P in two forms: (1) preformed P, that is, P that enters the lake as such (or adsorbed on clays),

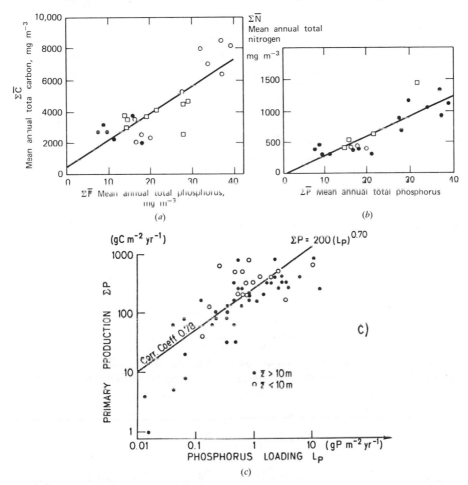

Figure 11.10   Phosphorus limitations in lakes. (*a*) and (*b*) (from Schindler [21]) illustrate that in experimentally fertilized lakes of the experimental lakes area of the Freshwater Institute in Winnipeg (Environment Canada), ratios of mean annual concentrations C/P and N/P tend to become constant. (*a*) shows that the C content increases as a consequence of P addition to the lakes, while (*b*) illustrates that the N content of a lake increases when the P input is increased, even when little or no nitrogen is added with fertilizer. Each point represents the results of a different lake. For details see [21]. (*c*) gives the primary productivity as a function of P loading per surface area for various North American and European lakes. [From D. M. Imboden and R. Gächter, in *Biological Aspects of Fresh Water Pollution*, O. Ravera, Ed., Pergamon, Elmsford, N.Y., 1978.]

and (2) P in the form of biogenic debris. Most of the latter is oxidized to form phosphates of oxidative origin, P_{ox}. For every P atom of oxidative origin found, a stoichiometric equivalent of oxygen atoms—276 oxygen atoms per P atom, or 140 g oxygen per 1 g P (cf. equation 4)—have been consumed. Accordingly a flux of P_{ox} is paralleled by a flux of O_2 utilization, as indicated schematically in Figure 11.11. The tolerable phosphorus loading of a lake may be related to the

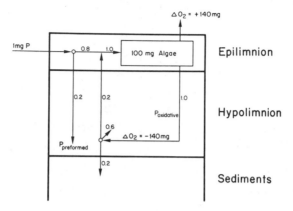

Figure 11.11   Simplified scheme of typical transformation of phosphorus in a stratified lake. One milligram of phosphorus introduced into a lake during the stagnation period may lead to the synthesis of 100 mg algae (dry mass) which upon mineralization cause an oxygen consumption in the hypolimnion of 140 mg O_2; from the organic P mineralized in the hypolimnion 0.6 mg are assumed to accumulate during the stagnation period, while 0.2 mg are assumed to be adsorbed [e.g., on iron(III) oxide] and transferred into the sediments; another 0.2 mg reaches the surface waters as phosphate by eddy diffusion.

hypolimnetic oxygen consumption. The annual P loading per lake surface, L_t (mg P m^{-2} year^{-1}) causes (under the simplifying assumption that all L_t becomes phosphorus of oxidative origin, P_{ox}) during the stagnation period, T_{st} (days), an approximate oxygen consumption, $\Delta(O_2)$ (mg m^{-3}) of the hypolimnion assumed to be homogeneously mixed of depth z_H(m) that is given by

$$\Delta[O_2] = 140 \frac{T_{st} L_t}{365 z_H} \tag{i}$$

Correspondingly, a maximum P loading L_{max} could be estimated for a tolerable oxygen consumption $[O_2]_{max}$:

$$L_{max} = \Delta[O_2]_{max} \times 7 \times 10^{-3} \times \frac{365}{T_{st}} z_H \tag{ii}$$

Many complicating factors, however, must not be overlooked. The simple stoichiometric relations may be too schematic, and they may change from lake to lake [24, 25].

As shown schematically in Figure 11.11, 1 mg of P can synthesize (assuming P to be the limiting factor) approximately 0.1 g of algal biomass (dry weight) in a single cycle of the limnological transformation. After settling to the deeper layers, this biomass exerts a biochemical oxygen demand of approximately 140 mg for its mineralization.

From this simple calculation it is obvious that the organic material introduced into the lake by domestic wastes (20 to 100 mg organic matter liter^{-1}) is small in comparison to the organic material that can be biosynthesized from introduced fertilizing constituents (3 to 8 mg P liter^{-1}, which can yield 300 to 800 mg organic matter per liter). Aerobic biological waste treatment with a heterotrophic enrichment culture mineralizes substantial fractions of bacterially oxidizable organic substances but is not capable of eliminating more than 20 to 50% of nitrogen and phosphate constituents.

The same type of calculation can also be made in assessing the value of secondary (i.e., aerobic biological) sewage treatment with respect to ocean outfalls. With regard to the balance of dissolved oxygen, the oxygen demands for oxidation of an untreated municipal effluent may be compared with the oxygen demand caused ultimately by the degradation of phytoplankton photosynthesized in response to algal nutrients (nitrogen more likely will limit in ocean outfalls). Such an analysis [26] shows that the BOD potential of the latter is greater than that for direct waste oxidation, particularly at large values of residence time in the receiving water. But, as emphasized earlier, dissolved oxygen must not be the only criterion for judging the value of waste treatment. Inorganic and organic contaminants may adversely affect the ecology of the receiving waters. The inefficiency of biological treatment with respect to removal of algal nutrients is a consequence of the elemental composition of domestic sewage as compared to the stoichiometric relation among C, N, and P in bacterial sludges. Most municipal wastes are nutritionally unbalanced for heterotrophic growth in the sense that they are deficient in organic carbon (see Figure 11.12).

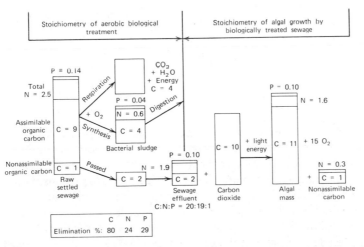

Figure 11.12 Stoichiometry of aerobic biological waste treatment. Comparing the relative composition of average domestic sewage with the mean stoichiometric relations among C, N, and P in bacterial organisms shows in a schematic way that only a fraction of the phosphate and nitrogenous material can become incorporated into the sludge. The inorganic nutrients as released by biologically treated sewage effluents can be converted into algal cell material. CO_2 (or HCO_3^-) and other essential elements are usually available in sufficient quantities relative to nitrogen and phosphorus. [Modified from W. Stumm, in *Advances in Water Research, Proceedings of the First International Conference*, 1962, W. W. Eckenfelder, Ed. Pergamon, Elmsford, N.Y., 1964, p. 216.]

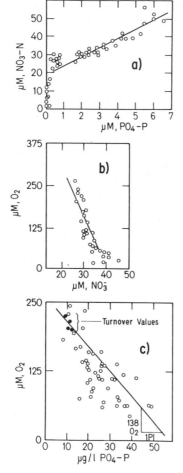

Figure 11.13 Stoichiometric correlations among concentrations of Nitrate, Phosphate, and Oxygen. (*a* and *b*) Lake Zürich. [Data from P. Zimmermann, *Schweiz. Z. Hydrol.*, **23**, 342 (1962) and I. Ahlgren, *Schweiz. Z. Hydrol.*, **29**, 53 (1967).] (*c*) Lake Gersau. [From H. Ambühl, *Schweiz. Z. Hydrol*, **37**, 35 (1975).] For (*b*) and (*c*) only results from the deeper water layers were considered.

To what extent it is possible to represent and interpret complicated natural processes in terms of simplified schematic stoichiometric equations, since we know that physiological conditions of growth differ from organism to organism, that there are algae capable of storing excess (luxury) phosphate, and that special growth factors, for example, biotin and niacin, can influence algal growth? Analytical data for individual lakes, however, show that, as in the oceans (cf. Figure 8.14), simple stoichiometric relations indeed hold. As shown in Figure 11.13, simple correlations between ΔP, ΔN, and ΔO_2 exist in these lakes. The observed molar ratios may differ somewhat from the average ratios.

Differential Distribution of Nutrients in Estuaries, Fjords, Coastal Waters, and the Mediterranean

We have illustrated how the characteristics of the circulation in lakes can markedly influence the distribution of biochemically important elements.

In estuaries and most fjords the circulation creates a trap in which nutrients tend to accumulate. Algae grown from nutrients carried seaward in the surface outflow eventually settle and become mineralized. The nutrients are then carried landward by the countercurrent of more dense seawater moving in from the outer sea to replace those entrained in the surface outflow. As pointed out by Redfield, Ketchum, and Richards [27, 28], the amount of circulation of nutrients varies greatly in different estuaries. It may be expected to increase with the rate of production of organic matter and with the length of the basin, to decrease with turbulence and the velocity of flow, and to vary with the relative depth and velocity of the surface and deep layers [29]. Several Norwegian fjords [30] and parts of the Baltic Sea are known to have become anoxic [31].

A reversal of the currents characteristic of estuarine circulation, that is, *antiestuarine circulation*, leads to an impoverishment in nutrients of a natural body of water. An example is the Mediterranean Sea which is the most impoverished large body of water known. Its antiestuarine circulation results from the fact that evaporation exceeds the accession of fresh water by about 4%.

A much larger volume of water flows in through the Strait of Gibraltar than is required to replace the loss by evaporation. The excess escapes through the strait as a countercurrent moving toward the Atlantic below the inflowing surface layer. Nutrients that tend to accumulate in the deeper layers are continuously being pumped out toward the Atlantic and toward the Black Sea.

As these cases illustrate, countercurrent systems are particularly effective in producing changes in the distribution of nutrients along the direction of flow. They lead to nutrient accumulation in the direction from which the surface current is flowing.

Phosphorus Models of Lake Eutrophication

Vollenweider [32] has related external sources of nutrients such as natural and agricultural runoff and municipal and industrial contributions to the enrichment of lakes. He has shown convincingly on the basis of data for 20 lakes that a valid correlation can be established between areal limiting nutrient loading (g m^{-2} year^{-1}) and mean lake depth on the one hand and the degree of enrichment on the other hand (Figure 11.14). The demarcation line indicated in Figure 11.14 gives a relevant reference value for permissive P loadings. Numerous models in which productivity is related to various lake variables have been reviewed by Shannon and Brezonik [33].

Source Functions. Obviously the quantity of nutrients introduced into a surface water from a given drainage area is dependent on the density of population and livestock, on the methods or intensity of fertilization, on the type of cultivation (e.g., forests, grassland, cropland), on the pedological characteristics of the soil, on topography, on climate and rainfall, and on the type of waste treatment system involved.

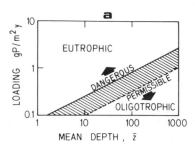 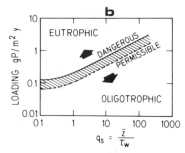

Figure 11.14 Vollenweider's model. (*a*) The relationship between areal phosphorus loading and mean depth provides estimates of dangerous or permissible loadings. (*b*) Vollenweider [*Schweiz. Z. Hydrol.*, **37**, 53 (1975)] later modified his loading concept to consider hydraulic residence time, τ_w (yv).

Chapra [34] has generated a historical simulation with variables indicative of human development (population density, land use, etc.) that compares favorably with present measurements. The historical profiles (Figure 11.15) indicate that lakes have been subject to two major periods of increased phosphorus load; first, during the latter part of the nineteenth century, resulting from a change from forested to agricultural land use; second, since 1945, because of increased sewering, population growth, and the introduction of phosphate detergents.

Conceptual models attempt to explain interactions between different system components and thus are potentially capable of providing a sensitivity analysis and insight into the dynamics of the system. Vollenweider [35] has summarized one-box nutrient mass balance models. Other more refined models [25, 35–38] have also considered (1) physical partitioning in lakes (e.g., division into epi-

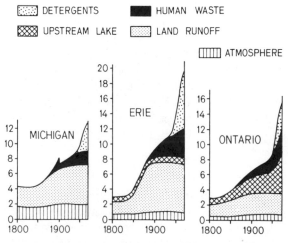

Figure 11.15 Historical loadings of total phosphorus in some Great Lakes, in thousands of tons per year, as calculated by a model. (From St. C. Chapra [34].)

limnion and hypolimnion or into trophogenic and tropholytic zones), (2) chemical-biological transformations (e.g., uptake of nutrients by the biota and its mineralization), and (3) hydraulic loading.

11.5   THE RESPONSE OF AQUATIC ECOSYSTEMS TO XENOBIOTIC SUBSTANCES; SOME ASPECTS OF ECOTOXICOLOGY

The harmfulness of a substance depends on its interaction with organisms or entire communities. The intensity of this interaction depends on the specific structure and on the activity of the substance under consideration; but other factors such as temperature, turbulence, and the presence of other substances are also important.

In evaluating toxicity, we need to distinguish between (1) substances that endanger animals and humans, that is, impair their health, or poison aquatic organisms [one often speaks of acute toxicity, especially if effects are observed within short time periods ($\leq$ days)]; and (2) substances that affect primarily the organization and structure of aquatic ecosystems. In this interaction contaminants may impair the self-regulatory functions of the system or interfere with food chains.

While there is some knowledge about the impact of chemical substances on individual organisms, less is known about their impact on ecosystems. The natural distribution of organisms depends on their ability to compete under given conditions, not merely their ability to survive the physical and chemical environment; a population will be eliminated when its competitive power is so much reduced that it can be replaced by another species. As shown by F. Taub [39], the competitive abilities of an organism are an interplay of its reproductive rate, which is related to food and physiological potential, and its mortality rate from all sources, including predation and imposed toxicity. Thus, in an ecosystem, a population may be eliminated even at apparently trivial toxicity levels if its competitive ability is marginal or if it is the most sensitive of the competitors. Similarly, a population may be eliminated if its food source is affected even if the population of interest is completely resistant to the toxicant [39]. It is thus often not possible to predict the ecological effects of substances from bioassays with individual organisms. Even if no acutely damaging effects on individual organisms are observed, the ecological consequences of such impairment may over long time spans often be more detrimental to aquatic ecosystems than acute toxic effects.

It has been postulated, for example [40], that altering a phytoplankton community may profoundly affect the healthy distribution and abundance of many animal populations higher in the food web. Thus, in eutrophic environments, alterations in algal communities can further reduce an already decreased species diversity, aggravating problems of algal blooms and contributing to general degradation of the ecosystem [41].

Refractory Substances

The number of synthesized chemicals now totals 1.8 million and is growing at a rate of 250,000 new formulations annually of which 300 to 500 reach the stage of commercial production [42] with a global production of 100 to 200 million tons [30]. It is estimated that up to one-third of the total production of these synthetic organic chemicals finds its way into the environment.

Persistence. Many of these synthetic chemicals survive long enough in the environment—most often because they are not readily biodegradable—to accumulate in the water; in the biosphere, concentrations at higher trophic levels may often exceed the relatively low levels in solution. Not all refractory compounds found in the environment are of synthetic origin. Alexander [43] has surveyed the long-lived natural organics occurring in the environment, including components of humus, kerogens, petroleum hydrocarbons, lignins, tannins, and the like. Many substances are altered readily to the extent that they lose their original structural entity, but the metabolites so formed are non- or slowly degradable. For example, so-called biodegradable detergents yield refractory degradation intermediates that may persist much longer than the parent compound [44]. Because, as mentioned before, the half-rate concentrations for the uptake of organic substrates by microorganisms (equation 3) lie in the range of 10^{-5} to 10^{-10} M, the rate of uptake and biodegradation at lower concentrations can be very slow.

Figure 11.16 gives gas chromatographic results for the analysis of volatile substances found in ground water and river and drinking waters within the same polluted watershed [45]. Some of the more refractory substances (some hydrocarbons and chlorinated hydrocarbons) detected in the river water may also typically occur in ground water and drinking water.

Even in such a detailed analysis, as exemplified by Figure 11.16, only a small fraction of the organic substances present in the water is detected. The sensitive gas chromatographic techniques are restricted to relatively volatile and nonpolar substances. These substances may be especially important, because they are lipophilic; that is, they tend to become concentrated in organisms. The tendency to accumulate refractory substances in the food chain is greater in aquatic than in terrestrial ecosystems. Bioaccumulation or biomagnification may also be controlled by absorption into the biota, that is, partition of lipophilic substances between water and organisms. Algae, invertebrates, and fish—like solvents in a separatory funnel—continuously exchange lipophilic substances with their aqueous environment. Biological uptake, lipophilic storage, and biomagnification and adsorption tendency may be related to the *n*-octanol–water partition coefficient [46]:

$$K = \frac{a_o}{a_w} \simeq \frac{c_o}{c_w} \tag{5}$$

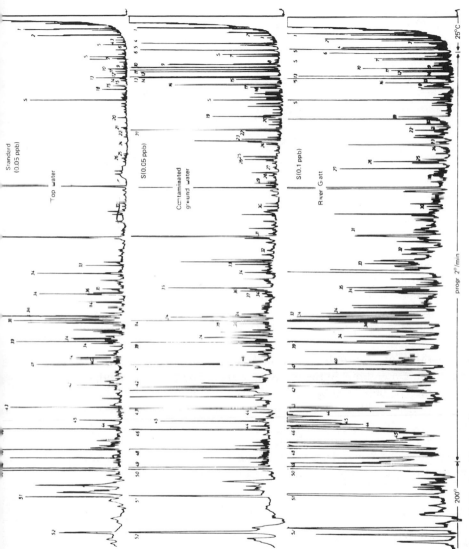

Figure 11.16 Specific detection of organic constituents in river water, ground water and drinking water. Each peak represents (at least) one compound. Peaks with a number represents substances identified with mass spectrometry. Some of the more refractory pollutants, especially some hydrocarbons and chlorinated hydrocarbons, occur also in groundwater and drinking water. (Analysis by E. Roman.) (From K. Grob et al. [45].)

713

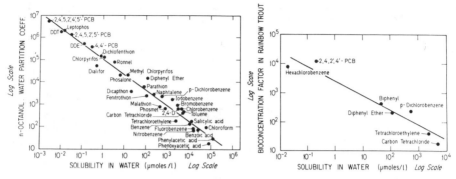

Figure 11.17 Lipophilicity and bioaccumulation. Both the *n*-octanol–water partition coefficient and the bioconcentration factor in rainbow trout are inversely proportional in aqueous solubility. (From Chiou et al. [46].)

where *a* and *c* refer to activity and concentration, respectively and the subscripts *o* and *w* refer to the octanol and water phases. It has been shown [46] that the octanol–water partition coefficient of a wide variety of chemicals is inversely related to the aqueous solubility and that a correlation is observed between the bioconcentration factors in rainbow trout and the aqueous solubility (Figure 11.17).†

Biodegradability is another important factor that determines the residence time or oganic pollutants in the receiving waters. Sometimes degradability of a molecule may be improved by modifying its structure; for example, biodegradability is enhanced by replacing a sulfonate group by a carboxylic group or more generally by incorporating *degradophores*, that is, molecular groupings that are substrates for attack by microbial oxidases. These enzymes effect hydroxylation and convert persistent molecules into more water-partitioning derivatives which are excreted rather than bioconcentrated [47, 48] (see Table 11.5).

Interference of Pollutants with Interorganismic Communication

Even if they exhibit neither acute nor chronic toxicity with regard to individual organisms chemicals may nevertheless disturb selectively the self-regulation of aquatic ecosystems. All organisms produce chemicals; some of these are dispersed in the aquatic environment, for example, humic acids, hydrocarbons, alkaloids, terpenes, vitamins, hormones, antibiotics, and insecticides. These substances may have various inter- and intraspecific effects [49] (Table 11.6). They are usually produced to enhance the dominance of certain organisms in

† As shown by Matter et al. [*Prog. Water Technol.*, **12**, 299 (1980)], the lipophilicity of a substance, as characterized by the octanol water partition coefficient, can also be used to predict the extent of absorption of such substances into the activated sludge in biological waste treatment.

TABLE 11.5 EFFECTS OF DEGRADOPHORES ON BIODEGRADATION AND IN TURN ON BIOCONCENTRATION OF DDT-TYPE ANALOGS IN THE FISH GAMBUSIA AFFINIS[a]

R^1	R^2	Ecological Magnification (ppm fish/ppm H_2O)
Cl	Cl	84,500
CH_3O	CH_3O	1,545
C_2H_5O	C_2H_5O	1,536
CH_3	CH_3	140
CH_3S	CH_3S	5.5
CH_3O	CH_3S	310
CH_3	C_2H_5O	400
Cl	CH_3	1,400

[a] From Kapoor et al. [47]. An enhanced biogradability is reflected in a smaller biomagnification.

their competition with other organisms. Insecticides and antibiotics are excreted by organisms to protect them from other organisms. Many substances exuded by algae or other organisms are chemotactically active. Behavioral patterns such as food finding, avoidance of injury, choosing of a habitat or host, social communication, sexual behavior, and migration or recognition of territory appear to be controlled sensitively and specifically by chemical cues. The production and exudation of chemicals is thus of adaptive value for the species and the community (enhancement of information transfer).

Humans also compete with other species for food; they enhance their dominance by destroying pathogens and pests. Industrial production of chemicals may be considered part of human biological activity. There is, however, a significant difference in tolerance toward interspecific chemicals in humans and organisms. Humans have learned in a long evolution to tolerate or to avoid most toxins of organisms. Organisms and ecosystems on the other hand have had little time to adapt to synthetic chemicals.

Some of the objectionable contaminants, even though they differ structurally from signaling substances (pheromones, allelochemical substances, telemediators), may blur chemotactic stimuli or may mislead by mimicking signals and thus cause sociological changes. Some of them have been found to be effective at very low concentrations. For example, 10^{-10} M morpholine (10^{-5}

TABLE 11.6   INTERORGANISMIC CHEMICAL EFFECTS (FUNCTIONS)

Intraspecific Effects	Interspecific Effects
Toxic or inhibitory repellents (autotoxins or autoinhibitors)	Repellents, suppressants, inductants, counteractants
Pheromones—chemical messages; signals for reproductive behavior, social regulation, alarm and defense, territory and trail marking, food location	Venoms, attractants, chemical lures, stimulants (hormones), olfactants
Roles	
Adaptation of species and organization of communities	
Niche differentiation	
Succession of communities	
Increase of information and of rate of information transfer in chemical signaling	

mg liter^{-1}) in river water can influence the homing of salmon [50]. Trace concentrations of petroleum hydrocarbons have been shown [51] to inhibit the decomposition of organic matter. The ecological aspects of microbial chemotactic behavior have been reviewed by Chet and Mitchell [51].

Dynamic Model for Residue Accumulation

The residual concentration of a lipophilic substance in an organism depends on the rate of uptake and the rate of clearance (elimination and degradation) by the organism [52]. The uptake of residue from an environmental reservoir with concentration c_{out} (assumed to remain constant) by an organism of mass W_1 occurs with a constant R (mass time^{-1}), while the rate of clearance occurs at a rate proportional to the internal concentration c_{in} with a first-order rate constant k (time^{-1}).

The rate of change of c_{in} with time can be written as

$$\frac{dc_{in}}{dt} = R \frac{c_{out}}{W_1} - kc_{in} \tag{i}$$

which upon integration gives

$$c_{in} = \frac{c_{out}R}{kW_1}(1 - e^{-kt}) \tag{ii}$$

Equation ii can be used to model the extent of bioaccumulation with time. Ultimately at steady state ($e^{-kt} = 0$) the residual concentration will be given by the distribution equilibrium (cf. equation 5)

$$\frac{c_{in}}{c_{out}} = \frac{R}{kW_1} \tag{iii}$$

Shifts in Species Abundance

Any change in chemical or physical variables will affect different species differently and thus will shift the species abundance.

Figure 11.18 gives an example of how a community of organisms may become perturbed as a result of pollution [53]; the frequency distribution of diatoms is modified: The number of species that occur with low frequency (few individuals per species) is reduced, while a few species become very abundant. As a consequence of the presence of many microhabitats in a "healthy" water, many species can survive. Because of interspecies competition, most species are present in a low population density. Pollution destroys microhabitats, diminishes the chance of survival for some of the species, and thus in turn reduces the competition; the more tolerant species become more numerous.

This shift in the frequency distribution of species—also observed typically with other organisms—is a general consequence of the chemical impact on waters by substances nonindigenous to nature.

Aquatic ecosystems may be perturbed in different ways (emissions of degradable or nondegradable chemicals, toxins, heat shocks, etc.). Some of the gross changes in communities resulting from many different types of disturbances are similar and predictable: The structure of the ecosystem is simplified, for example, by elimination of some of the organisms either through toxic or inhibitory effects or through competitive displacement by more tolerant

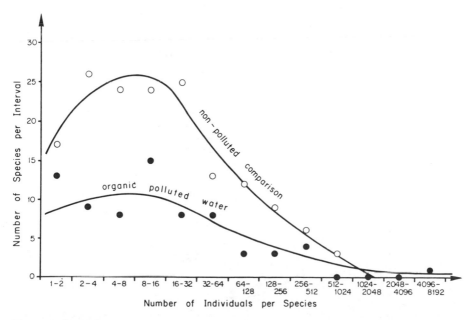

Figure 11.18   Shift in the frequency distribution of diatoms as a result of subtle organic pollution. (From R. Patrick et al. [53].)

organisms, by perturbation of regulatory factors, such as masking or disturbing chemotactic signals, and by acceleration of nutrient cycles.

Toxicological Effects on Humans

Of particular interest in ecotoxicology are the effects of xenobiotic substances on human health. We will not discuss acute effects and, furthermore, will restrict ourselves to a brief discussion of some aspects of health hazards resulting from the long-term effects of trace concentrations of pollutants in drinking water [54–58].

Epidemiological studies have shown significant correlations between cancer morbidity and origins of potable water resources. (Although such a correlation does not establish a cause- and-effect relationship, it appears difficult to explain the correlations by other factors.) Furthermore, substances contained in drinking water (concentrates from inverse osmosis or eluates from adsorbent columns) have been shown to produce mutations in bacteria.

Mutagenic and Carcinogenic Effects.   Cancer can occur in all organisms. Mutations (changes occurring in genes—or gene nucleotides—giving rise to hereditary changes) occur spontaneously in all organisms; they can also be induced by external agents such as ionizing radiation and certain specific chemicals. Such mutagens increase the mutation rate. There is increasing evidence that mutations can (but need not) induce carcinogenic effects. As illustrated in Figure 11.19 the substance entering the cell is either already reactive or—more typically for environmental pollutants—converted enzymatically into a reactive compound. Reactive compounds are often electrophilic and react (often by biochemical oxidation or hydroxylation) with nucleophilic macro-

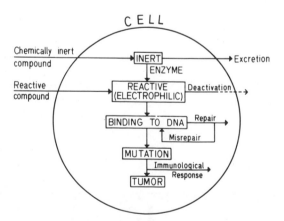

Figure 11.19 Induction of a carcinogenic effect by a mutation. The mutation may result from the chemical interaction of a reactive (often electrophilic) substance with deoxyribonucleic acid (or ribonucleic acid). [Modified from Lutz, W. K., *Mutation Res.*, in press (1979).]

molecules, especially with the proteins and nucleic acids of the cell. The nucleo-tide bases of DNA (and RNA) and tryptophan, tyrosine, methionine, and histidine of the proteins are particularly sensitive. As with ionizing radiation, one assumes that the effect does not require exceeding a threshold in the dose. On the other hand, often a very long induction period exists between initiation of the reaction and the generation of a tumor.

Since many carcinogenic substances are also mutagenic (not a necessity), tests for mutagenic effects in bacteria (e.g., the Ames test [59]) are often used as screening tests for recognizing potential carcinogens.

How Useful are Toxicity Tests?

The ultimate objective is to be able to develop sufficient information about a given compound to allow prediction of its environmental behavior and its ecotoxicological effects. Dozens of standard tests have been developed for predicting such effects. If one considers, however, that such short- and long-term tests have to be carried out for each substance (and its degradation intermedi-ates) with different organisms, that synergistic and antagonistic effects with other substances present may occur, and that each negative test result does not necessarily mean "zero toxicity," one realizes that the predictive value of such standard tests is limited. New chemicals may be produced and enter the environ-ment faster than they can be tested satisfactorily.

The pollution potential of individual compounds may be assessed on the basis of a few physicochemical parameters which—in addition to the biodegrad-ability—permit a forecast of the manner in which the compounds are distributed in the environment (cf. Figure 9.18) and thus in turn their fate and approximate residence time and their ecological impact [58].

Table 11.7 summarizes some of the parameters pertinent for assessment of the pollution potential of individual substances. The information on biode-gradability of a pollutant is of particular importance in estimating its residence time in water. The latter is also affected by the tendency of a substance to escape into the atmosphere as characterized by its vapor pressure, the distribution equilibrium between water and atmosphere, and the gas transfer coefficient (cf. Section 11.7). As we have discussed before, the lipophility—as measured by the n-octanol–water distribution coefficient—is a good measure of the tendency of a substance to accumulate in the food chain and to become biomagnified.

The best way to safeguard aquatic ecosystems and humans against un-anticipated effects of organic pollutants is to require that all substances that are mass produced and may become dispersed in the environment, especially substances used in households and in agriculture (e.g., detergents, cleaning fluids, pesticides), be fully† biodegradable.

---

† Full biodegradation is the conversion into CO_2, H_2O, and other oxides (With many so-called biodegradable detergents, the degradation goes merely to persistent biodegradation intermediates (cf. H. Leidner, R. Gloor, and K. Wuhrmann, *Tenside Deterg.*, **13**, 122 (1976)).

TABLE 11.7   PERTINENT INFORMATION FOR ASSESSING THE FATE
AND RESIDENCE TIME OF POLLUTANTS

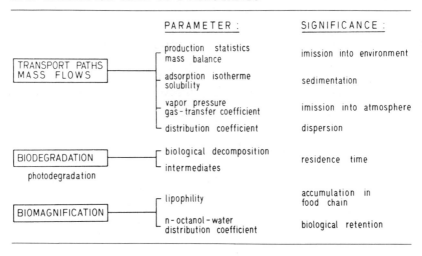

	PARAMETER :	SIGNIFICANCE :
TRANSPORT PATHS MASS FLOWS	production statistics mass balance	imission into environment
	adsorption isotherme solubility	sedimentation
	vapor pressure gas-transfer coefficient	imission into atmosphere
	distribution coefficient	dispersion
BIODEGRADATION photodegradation	biological decomposition intermediates	residence time
BIOMAGNIFICATION	lipophility	accumulation in food chain
	n-octanol-water distribution coefficient	biological retention

11.6   PERTURBATION AND ECOSYSTEM STRUCTURE

Every aquatic ecosystem (receiving waters plus drainage area) can be characterized by a flow of energy and nutrients through a network of interlocking
cycles. Figure 11.20 generalizes in a concise, oversimplified way how chemical
and physical perturbations affect an ecosystem in general. Every stress, that is,
a change in a chemical or physical variable that cannot be anticipated†—
occurring as a by-product of direct or indirect technological energy inputs into
the system—increases, macroscopically speaking, the biotic energy flux/biomass
ratio (Margalef [56]). Correspondingly, nature simplifies its ecosystem structure; by impairing negative feedback mechanisms, by enhancing competitive
exclusions, and by accelerating nutrient cycles, the randomness of the system
is increased. This model permits in a gross way the generalization of many
aspects of human action on aquatic ecosystems.

A number of scholars, especially E. Odum [1], Margalef [60], and Woodwell
[61] have tried to abstract from a large number of observations and to generalize
the temporal development of ecological communities. Based on the ideas of
Lotka (1924), the bioenergetic basis of the succession of ecosystems has been
recognized. Like every other closed (but energy-transparent) system, an ecosystem has a tendency to develop toward a macroscopic steady state. Actually,
natural nonperturbed ecosystems exposed to a solar input tend to change into
a climax community. Such a climax community can maintain a macroscopically
constant composition if cybernetic regulatory and symbiotic functions prevail.
At steady state an optimum in metabolic efficiency is attained. In a diverse

† Predictable variations such as tides and diurnal changes are not considered stressful.

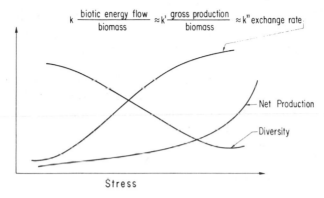

Figure 11.20   Effect of stress upon an ecosystem. Qualitative and schematic illustration indicating how every stress causes a simplification in the organization of the ecosystem and an acceleration of its nutrient cycles. Stress is interpreted as a change in chemical and physical variables that cannot be anticipated by the organisms.

system the flow of energy is shared by many kinds of species; since there is little energy for each species, productivity may not be very great. The complex system may support a great diversity of specialists that are better able to utilize and degrade the various chemical substances from wastes. The changes schematically depicted in Figure 11.20 that occur with increasing stress may be interpreted as a reversal of the ecological succession.

The interdependence between the acceleration of nutrient cycles (e.g., as indicated by an increase in the specific growth rate of certain organisms) and the simplification of ecosystem structure is typically observed as a consequence of water pollution with organic or inorganic substances. The disturbance of the ecological balance between photosynthesis P and respiration R $(P \gtrless R)$, caused by either organic pollution or pollution by fertilizing nutrients, leads to an increase in the specific growth rate of certain heterotrophic or phototrophic organisms, that is, to an increase in the energy flux/biomass ratio (corresponding to the metabolic rate per heterotrophic cell or rate of energy fixation). This increase is accompanied by a decrease in the organization of the ecosystem, for example, narrowing of the food chain and a decrease in the number of trophic levels.

Lake eutrophication demonstrates the dramatic change in the entire lake community as a consequence of the supply of a fertilizing nutrient to the lake: the phytoplankton increase and change their composition; despite a corresponding increase in zooplankton density, a significant part of the phytoplankton biomass settles into the deeper water layers where the resulting oxygen consumption and ultimate anaerobiosis drastically change the fauna at the sediment–water interface. As long as the phosphorus loading of the lake is small, and as long as dissolved oxygen prevails at the sediment–water interface, sediments are a sink for phosphates [which become bound to iron(III) oxides].

As soon as the loading exceeds a critical limit, a significant lowering of the redox intensity at the sediment–water interface results in a change in sign of the feedback mechanism of the P regulation; under anaerobic conditions, sediments release P accumulated in earlier years. As a further consequence of the nutrient supply, dramatic changes in species distribution occur at all trophic levels; pronounced shifts in fish populations have been observed [62]. Obviously an increase in primary production affects an aquatic ecosystem and its food web more than a terrestrial system.

Sanders [63] has shown that in marine systems the number of species per square meter decreases in the decreasing temperature and salinity gradients. His measurements on benthic diversity in coastal regions of southwest Africa and of Peru and Chile are particularly interesting; they show that stresses resulting from lowered oxygen concentrations are reflected in less benthic diversity, while the density of organisms is more a function of available food. Stress may reduce the capability of a community of organisms for self-regulation and thus in turn decrease its chance for survival. Sanders [59] postulates that, if environmentally stable (predictable) conditions persist, organism speciation and immigration will cause species diversity to increase gradually as species in the community become biologically accommodated to each other. On the other hand, if as a consequence of physical fluctuations or changes in chemical variables the gradient of physiological stresses increases, the community of organisms will adjust from being a physiologically accommodated community to one that is controlled by chemical and physical factors. As a consequence the number of species will decrease with the increasing stress gradient.

With many kinds of pollutants, for example, chemical or petrochemical substances or metals, one finds similar reductions in diversity. King and Ball [64], for example, illustrate that two entirely different types of pollutions (galvanic wastes in one case and untreated sewage in another case) cause—despite different mechanisms of ecological impairment—similar marked reductions in the ratio of insects to oligochaetes in different stretches of the same river.

In order to expand the food output for the increasing world population, a higher energy flow has to be diverted to agriculture (monocultures). The minimal food needs of the present world population are about two orders of magnitude smaller than the net biomass production (photosynthesis) per year. Supplying food—compatible with present European energy consumption—to an ultimate steady-state population of 15×10^9 would bring the biomass necessary for human food production to within one order of magnitude of present net photosynthesis. Because humans are not the only animals, such—probably technologically attainable—an increase in food production could be achieved only by a dramatic increase in man's dominance, that is, by reducing species diversity and by simplifying ecosystem structure and perhaps impairing the biosphere's resilience and buffering capacity. Do we know the consequences of such massive intervention in the pathways of photosynthetic energy flow?

11.7 GLOBAL CHEMICAL CYCLES AND POLLUTION

Pollution used to be considered a local or regional problem. But the effects of technology and the consequences of energy dissipation are now being felt over increasingly larger distances—at times becoming global in character. Humans in their social and cultural evolution continue to be successful in diverting energy to the advancement of their own civilization. As we have seen (Figure 11.3), anthropogenic energy use already exceeds biotic energy flux (photosynthesis) in most countries of the Northern Hemisphere. Receiving waters reflect not just the activities within their drainage area but also the impact of emissions carried over large distances through the atmosphere. The rapid changes observed in the last decades in chemical and biological properties of many coastal and fresh waters reflect the human influence on the environment and are a by-product of progressive energy dissipation. The burning of fossil fuels and the generation of nuclear energy produce a variety of gaseous oxides, compounds of heavy metals, acidity, radionuclides, and solid particles. The redistribution of these substances between water and land is influenced by their residence time in the atmosphere. Of special interest here are substances that are dispersed over large distances and may adversely affect the ecology of natural waters.

As we have already mentioned, of increasing importance is also the fact that, in order to expand the food output for an increasing population, in many drainage areas of streams and lakes, an ever-increasing amount of energy flow has to be diverted to agriculture (by deforestation, conversion of grassland to cropland, excessive application of fertilizers, herbicides, and pesticides), to the benefit of ever fewer species (approaching monocultures). This promotes the transport of fertilizers and xenobiotic substances to the waters and enhances increased rates of erosion and chemical denudation, thus potentially causing a disturbance of the ecological balance in natural water systems. Despite the ecosphere's remarkable resilience and feedback mechanisms, humans have now become sufficiently powerful to influence global chemical cycles.

Of prime concern are (1) human alteration of the natural CO_2 cycle, resulting from fossil fuel burning and perhaps from deforestation which may lead to climate changes which in turn could trigger profound effects on human society; (2) enhanced production of NO_2, a by-product of man's nitrogen fixation (20 to 30 % of that occurring by natural processes), which may interact with the stratosphere's ozone; (3) increased erosion rates resulting mainly from deforestation and modern agriculture; and (4) the release of oxides of N, H, and S and of trace metals and ash into the atmosphere, which may cause production of acid rain and fallout (oxidation of S and N oxides to H_2SO_4 and HNO_3), nutrient enrichment of inland and coastal waters, and perturbations of natural biological cycles by heavy metals.

As Figure 11.21 suggests, the "civilization engine" is driven by the political society. Scientists and engineers must assess more quantitatively human influences on the earth's natural metabolism and on the various interlocking

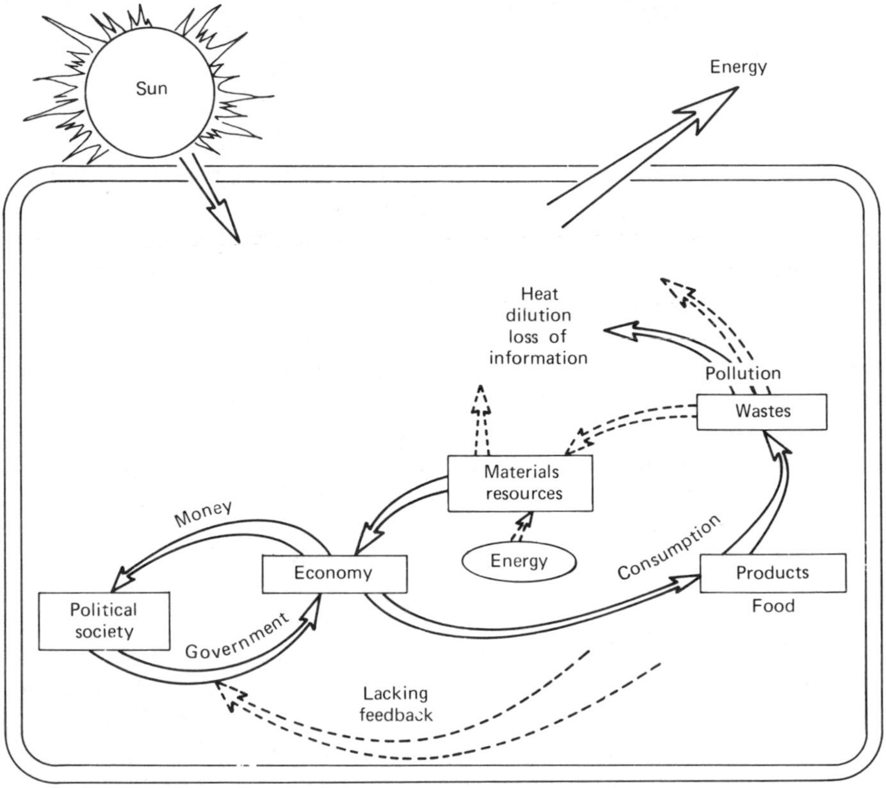

Figure 11.21 The "civilization engine." Superimposed upon the ecosphere, this engine converts, with the help of energy, resources into products which ultimately become wastes. Some of the wastes may be recycled, but most of them will become dispersed into the air, water, or soil environment. The faster the engine runs, the larger the potential extent of waste dispersion (pollution). The engine is driven by the political society; our economy with its capital-intensive production appears to function satisfactorily only under the condition of capital flow increase which leads to further enhancement of production, to an acceleration of hydrogeochemical cycles, and in turn to a progressive increase in potential pollution.

hydrogeochemical cycles. By evaluating the strength of various emission sources and by comparing natural and pollutant fluxes we will be able to understand better the present and future distribution of pollutants and their ecological consequences and effects on human health.

Simple Thermodynamic Implications

Every flux of energy and material is accompanied by degradation of material and energy. According to the first law of thermodynamics, wastes are merely resources at the wrong place. Energy cannot be destroyed and resources cannot

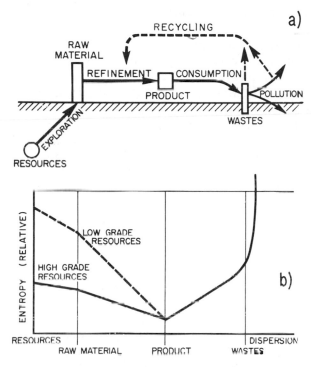

Figure 11.22 Conversion of resources into raw materials, consumer goods, and eventually into wastes.

be lost. Exhaustion of nature, depletion of resources, and every kind of pollution are in principle a consequence of the second law of thermodynamics: entropy of dilution and mixing, heat production, destruction of information (e.g., species depletion, interference with chemical communication). According to the first law we can close our cycles and recycle wastes into raw materials. Figure 11.22, however, illustrates in a schematic way that, because of entropy production in waste generation and in waste dispersion, energy requirements for recycling dispersed wastes may become prohibitive.

In exploiting resources, materials of relatively low entropy content (e.g., ores containing metals well ordered and in great concentration) are converted into consumer goods (materials of even lower entropy content, that is, of a higher degree of order). Figure 11.22b shows the relative entropy per unit consumer good, for example, per ton of Cu. The wastes ultimately formed, on the other hand, have a higher entropy content. The dispersion of wastes into the air, water, and soil is accompanied by a further very large production of entropy.

Limits to Waste Treatment. There are definite limits to the extent of waste treatment and water purification, because pollution control measures themselves

cause environmental loadings (secondary pollution). Recycling is a commend-
able means toward environmental protection only as long as it results in less
pollution loading than would arise from further extraction of the original resource. The limit to recycling is thus set by the extent of dissipative use of
materials. For example, ca. 40% of Cu and 95% of Cd are used in dissipative
ways (e.g., in alloys, pigments, coatings, etc.). To recycle more than ca. 60%
of the Cu and ca. 5% of the Cd would thus result in exorbitant energy demands
and secondary pollution. *Water*, on the other hand, can often be recycled (waste
treatment, purification, distillation, reverse osmosis) without excessive energy
requirements and cost. Our waste treatment practices, however, often ignore
the thermodynamic implications regarding entropy of dilution and mixing.
They rely on "end of the pipe" therapy rather than on prevention of waste
generation and waste dilution at the source. It is more expensive to remove (e.g.,
by biological treatment or adsorption) 1 mol of phenol or of a chlorinated
hydrocarbon from a highly dilute waste than from a concentrated one.

The Atmosphere as a Conveyor Belt of Pollutants

The concentrations of many constituents (gases and particulate matter) of the
atmosphere are determined by the dynamics of their cycles between sources and
sinks [65,66].

The residence time can be estimated if the source function or sink function is
known: If Q and R represent the input and output rates of a constituent in the
atmosphere, respectively, and M is the total mass of this constituent in the
atmosphere, then

$$\frac{dM}{dt} = Q - R \tag{6}$$

If one assumes that the removal rate is proportional to the mass of a constituent
in the atmosphere, then

$$\frac{dM}{dt} = Q - kM \tag{7}$$

At steady state, the residence time τ is given by

$$\tau = \frac{M}{Q} = \frac{M}{R} = \frac{1}{k} \tag{8}$$

The mean residence time provides a relative measure of the reactivity of a
substance in the atmosphere and is, as a first approximation, related to the
removal rate constant of this constituent [67].

A pollutant that has a very short residence time will be removed very efficiently. CO for example has a τ of ca. 3 weeks. If no CO were removed or converted to other compounds, the atmospheric content would double in about

3 weeks. Because the global level has not changed measurably for a number of years, it can be concluded that the removal of CO (microbial oxidation in soils and oxidation by OH radicals in the atmosphere) must be keeping pace with its production (from automobile exhaust and from microbial activity in surface waters) [65].

Example 11.3. Are Human Activities Changing the O_2 and CO_2 Content of the Atmosphere?

So far we have burned ca. 10^{16} mol of fossil carbon. To what extent could this cumulative combustion affect the atmospheric contents of O_2 and CO_2?

The weight of gas above each square centimeter is about 10^3 g cm^{-2} (~ 1 atm); with an average molecular weight of air of 29 g this corresponds to 34.5 mol. Of this, 20% is O_2 or 7.5 mol cm^{-2}. Furthermore, 1.1×10^{-2} mol of CO_2 (0.03% of air) are above each square centimeter. For the total earth surface (5.1 $\times$ 10^{18} cm^2), the atmospheric reservoir of O_2 and CO_2 are 3.8×10^{19} mol O_2 and 5.6×10^{16} mol CO_2. Combustion of all the fossil carbon exploited thus far would lead to a production of 10^{16} mol CO_2 (ca. 20% of the CO_2 contained presently in the atmosphere) and a depletion of O_2 corresponding to 2×10^{16} mol of O_2 (0.05%) of the O_2 contained in the atmosphere. Obviously the oxygen reservoir cannot seriously be depleted by human actions. Quite obviously, however, the CO_2 content of the atmosphere has increased markedly. (About 50% of the fossil CO_2 added to the atmosphere has remained in the atmosphere.)

This problem is taken from a lucid paper entitled, "Man's Oxygen Reserves," by Broecker [68]. Broecker also showed that in the extreme case of all fossil fuel reserves being burned in one moment, the percentage change of oxygen in the atmospheric reservoir would still be trivial.

Persistent Pollutants Transported in the Atmosphere

The vapor phase is the principal state for many persistent pollutants during transport from land to fresh water and from continents to the oceans. Even substances with small vapor pressures, as low as 10^{-10} atm, are released into the atmosphere. These substances include many pesticides, DDT, lindane, dieldrin, PCBs aromatic and aliphatic hydrocarbons, and more volatile metals and metalloids or their compounds. The widespread occurrence of some of these pollutants in plankton, in marine organisms in every ocean, and in arctic seals and antarctic penguins is evidence for the efficiency of the atmosphere as a conveyor belt [69].

These substances evaporate from polluted fresh waters and from soils. Their removal from the atmosphere can occur by the following processes: (1) Chemical destruction (oxidation); (2) destruction within the stratosphere, that is, escape through the tropopause; (3) dissolution of the gas phase in clouds and raindrops and subsequent removal from the atmosphere by precipitation; (4) attachment of the trace gas to atmospheric aerosols and subsequent removal by washout or rainout; (5) adsorption at land and ocean surfaces.

Distribution Equilibria Atmosphere–Water

The transport of a voltatile substance from water into the atmosphere depends on a Henry's distribution equilibrium of the substance between the gas and water phases. This distribution equilibrium is defined by a Henry's law-type distribution coefficient H[†]:

$$H = \frac{c_{sg}^0}{C_{sl}^0} = \frac{p_{sA}^0}{C_{sl}^0 RT} = \frac{c_g}{C_A} = \frac{p_A}{C_A RT} \tag{9}$$

where p_A = partial pressure of compound A in the gas phase; p_w = partial pressure of water vapor; p_{sA}^0 = vapor pressure of compound A in pure (solid or liquid) form, or in equilibrium with a saturated aqueous solution of A; C_A = concentration of A in water; c_g = concentration of A in gas phase C_{sl}^0 = solubility of A in water, c_{sg}^0 = concentration of A in the gas phase in equilibrium with an aqueous saturated solution of A; $c_{sg}^0 = p_{sA}^0/RT$.

One may note from equation 9 that the partial pressure of A, p_A, can be obtained for a given solute concentration if the solubility of A in water and the vapor pressure of A in pure form are known:

$$p_A = \frac{C_A p_{sA}^0}{C_{sl}^0} \tag{10}$$

Table 11.8 gives evaporation parameters for some representative low-water-solubility contaminants [70].

It is seen from these data that at equilibrium the ratio of contaminants to water in the gas phase is larger than in the liquid phase; $p_A/p_w > C_A/C_{H_2O}$. Thus these contaminants, if introduced into the water, will tend to gas out, and the concentration of the contaminant in the water will fall. Note that even pesticides with a very low vapor pressure, because of their low water solubility and their high activity coefficients, tend to escape from an aqueous solution[‡].

Example 11.4. Distribution between Water and Atmosphere

1. One liter of water initially contains 1 μg liter^{-1} of the contaminants n-octane, toluene, DDT, and Hg. Placed in a closed 10-liter bottle, what fractions of these contaminants are lost from the liquid at equilibrium?

[†] Note that H is defined differently from K_H used in Chapters 4 and 5. See Section 2.16 and note c of Table 11.8.
[‡] Examples of how to calculate the equilibrium distribution of pollutants between different reservoirs (atmosphere, water, sediments) have been given by D. McKay ("Finding Fugacity Feasible"), *Environ. Sci. Technol.*, **13**, 1218 (1979).

TABLE 11.8 EVAPORATION PARAMETERS FOR VARIOUS COMPOUNDS AT 25°C

Compound	Molecular Weight	Water Solubility, C_{sl}^0 (mg liter^{-1})	Vapor Pressure, p_{sA}^0 (mm Hg)	Concentration in Gas Phase, c_{sg}^0 (mg liter^{-1})[b]	H, c_{sg}^0/C_{sl}^0[c]	K_L (m hr^{-1})[e]
Alkanes						
n-Octane	114	0.66	14.1	85.5	1.4×10^2	0.124
2,2,4-Trimethylpentane	114	2.44	49.3	299	1.2×10^2	0.124
Aromatics						
Benzene	78	1780	95.2	395	0.22	0.144
Toluene	92	515	28.4	139	0.27	0.133
o-Xylene	106	175	6.6	37	0.21	0.123
Cumene	120	50	4.6	29	0.58	0.119
Naphthalene	128	33	0.23	1.6	0.048	0.096
Biphenyl	154	7.48	0.057	0.47	0.063	0.092
Pesticides						
DDT ($C_{14}H_9Cl_5$)	355	0.0012	1×10^{-7}	1.9×10^{-6}	1.6×10^{-3}	9.34×10^{-3}
Lindane	291	7.3	9.4×10^{-6}	1.4×10^{-4}	1.9×10^{-5}	1.5×10^{-4}
Dieldrin	381	0.25	1×10^{-7}	2.0×10^{-6}	8×10^{-6}	5.33×10^{-5}
Aldrin	365	0.2	6×10^{-6}	1.2×10^{-4}	6×10^{-4}	3.72×10^{-3}
Polychlorinated biphenyls (PCBs)						
Aroclor 1242	258	0.24	4.06×10^{-4}	5.6×10^{-3}	0.023	0.057
Aroclor 1248 ($C_{12}H_6Cl_4$)	—	5.4×10^{-2}	4.94×10^{-4}	6.8×10^{-3}	0.13	0.072
Aroclor 1254 ($C_{12}H_5Cl_5$)	—	1.2×10^{-2}	7.71×10^{-5}	1.1×10^{-3}	0.09	0.067
Aroclor 1260 ($C_{12}H_4Cl_6$)	—	2.7×10^{-3}	4.05×10^{-5}	5.5×10^{-4}	0.20	0.067
Other						
Mercury	201	3×10^{-2} [d]	1.3×10^{-3} [d]	1.4×10^{-2}	0.47	0.092
Water	18	—	23.8	23.0	—	—

[a] Mostly from Mackay and Leinonen [70].

[b] $C_{sg}^0 = (m_w p_{sA}^0/760)/RT$.

[c] c_g/C_i at equilibrium; can be converted into a Henry's law constant [mol/atm l] by dividing by $RT/(\rho_{sA}^0/760)$.

[d] D. N. Glew and D. H. Hames, [*Can. J. Chem.*, **49**, 3114 (1971)] give $C_s = 6 \times 10^{-2}$ mg liter^{-1}.

[e] K_L is a physical chemical characteristic of the system and is influenced by mixing within the system. Data are given here for relative comparison.

729

Table 11.8 gives the necessary data: Considering that $C_1 + 9c_g = 1$ μg liter^{-1} and $H = c_g/C_l$, we obtain for the fractions lost from the water: n-octane, 0.999; toluene, 0.71; DDT, 0.014; Hg, 0.81.

2. What is the efficiency of the removal of these pollutants from the atmosphere in the dissolved phase by rain? Assume a homogeneous atmosphere of 5-km height and 2.5 cm of rain water; assume that we can expect fairly rapid equilibration between the gas and the dissolved phase in cloud and rain droplets.

Essentially we equilibrate 5×10^5 cm^3 air (above 1 cm^2) with 2.5 cm^3 rainwater. The calculation shows that the fraction removed by rainwater in the dissolved phase is very small: n-octane, 3.6×10^{-8}; toluene, 2×10^{-5}; DDT, 3×10^{-3}; Hg, 1×10^{-5}.

Because the equilibrium distribution of these substances between the water and the atmosphere is so much in favor of the latter, their removal by dissolution into rain is not very efficient (as shown in Example 11.4). Some pollutants, especially those with saturation vapor pressures lower than 10^{-6} atm (e.g., DDT and dieldrin) may be removed by adsorption to atmospheric aerosols and subsequently washed and rained out.

Elements such as Hg, As, Se, and Pb are introduced into the atmosphere also by natural processes. These trace elements with low oxide boiling points (a measure of volatility) are more concentrated in the atmosphere, as compared to sedimentary rocks, than elements with high boiling points (see Section 11.8). Mackenzie and Wollast [71] suggest that elements highly enriched in the atmosphere fall into two groups: (1) elements such as Hg, Se, and As which are emitted into the atmosphere in vapor form and later removed as dissolved gases in rain; and (2) elements such as Pb, Zn, and V which may also be released in part to the atmosphere in vapor form, but they condense in the atmosphere and are removed principally as solid particles in rain or dry fallout.

As the vapor pressure is reduced by increased dilution in water or in other liquids, so is the vapor pressure influenced by adsorption on surfaces such as soils; adsorption reduces the chemical activity (fugacity) below that of the pure compound. Most of the more volatile pesticides are not strongly sorbed on clays but have an affinity for organic materials (the humus type). The higher the organic content of the soil, the smaller the vapor pressure or the potential volatility of a pesticide.

Rate of Escape. As discussed earlier, the rate of mass transfer of a substance, F, across a water–gas phase boundary can be expressed [70,72] by

$$F = K_L(C_l - C_{sl}) \tag{11}$$

where K_L is the overall liquid transfer coefficient (m hr^{-1}), F is in g m^{-2} hr^{-1} if concentrations are expressed in g m^{-3}, C_l is the concentration of the substance in the water, C_{sl} is its concentration in the liquid film on the gas side assumed to be in equilibrium with the concentration in the gas phase c_{sg}; the latter can be calculated from the distribution coefficient $H = c_g/C_{sl}$

$$F = K_L\left(C_l - \frac{c_g}{H}\right) \tag{12}$$

which leads to the differential equation

$$-\frac{dC_l}{dt} = \frac{K_L(C_l - c_g/H)}{L} \tag{13}$$

where L is the depth in meters (unit volume per interfacial area). (A water column of 1-m^2 cross section and a depth of L meters of water contains $C_l L$ grams.) Other possible rate-limiting steps, for example, eddy diffusion from deep waters, are neglected.

Upon integration equation 13 becomes [70]

$$C_l = \frac{c_g}{H} + \left(C_{l_0} - \frac{c_g}{H}\right)\exp\left(\frac{-K_L t}{L}\right) \tag{14}$$

c_g/H may be small in comparison to C_{l0} if the background atmospheric level of the contaminant is low compared to the local level; then equation 14 simplifies to

$$C_l = C_{l_0}\exp\left(\frac{-K_L t}{L}\right) \tag{15}$$

Representative K_L values evaluated by Mackay and Leinonen [70] on the basis of criteria developed for the air–sea system by Liss and Slater [72] are given in Table 11.8.

As discussed in Section 2.16 and as shown by Liss and Slater [72], the reciprocal of the transfer coefficient (time/length) is a measure of the resistance to the gas transfer and is composed of two resistances in series, the liquid-phase resistance r_l and the gas-phase resistance r_g:

$$\frac{1}{K_L} = R = r_l + r_g \tag{16}$$

For substances with relatively high values of H, the rate of evaporation is controlled by the liquid-phase resistance. Conversely for substances with low values of H (i.e., pesticides), the evaporation rate is controlled by the concentration gradient in the vapor.

Example 11.5. Freon Flux into the Oceans

Estimate the flux of Freon 11 (CCl$_3$F) into the oceans from the following data (Lovelock et al., *Nature*, **241**, 194 (1973)); mean concentration of CCl$_3$F in the marine atmosphere over the Atlantic, $c_g = 50 \times 10^{-6}$ ppm (by volume); mean surface water concentration, $C_l = 7.6 \times 10^{-12}$ cm^3 CCl$_3$F cm^{-3} water. H and K_L for CCl$_3$F are given by Liss and Slater [72] as 5 and 11.3 cm hr^{-1} respectively [cf. 70].

The concentration gradient in the liquid phase is given by $(c_g/H) - C_l$. The concentration, $c_g = 50 \times 10^{-6}$ ppm (by volume) is equal to 50×10^{-12} cm^3 CCl$_3$F cm^{-3} air. The corresponding liquid-phase value, $C_{sl} = 10 \times 10^{-12}$ cm^3 CCl$_3$F cm^{-3} water. $C_{sl} - C_1 = 2.4 \times 10^{-12}$ cm^3 Freon cm^{-3} water $= 1.5 \times 10^{-14}$ g cm^{-3}. Then F becomes 1.5×10^{-9} g cm^{-2} year^{-1}. The flux of CCl$_3$F from the atmosphere to the entire ocean (3.6×10^{18} cm^2) is 5.4×10^9 g year^{-1} (Liss and Slater estimate that this is about 2% of the world production).

Example 11.6. The Effect of Increase in Atmospheric CO$_2$ on the Extent of CaCO$_3$ Saturation of Ocean Water

At present the surface waters of the ocean are oversaturated with respect to calcite and aragonite. Some concern has been expressed that the increase in atmospheric CO$_2$ might lead to calcite undersaturation and that existing biotic CaCO$_3$ structures will begin to dissolve. What is the effect of a twofold (most likely to be reached early in the next century) and a fivefold (maximum possible) increase in the CO$_2$ partial pressure? The following information is available: alkalinity of present seawater due to HCO$_3^-$ and CO$_3^{2-}$ is 2.94×10^{-3} eq liter^{-1}; present $p_{CO_2} = 3.3 \times 10^{-4}$ atm. On an ionic medium scale, we have the following equilibrium constants (25°C):

$$\frac{[H^+][CO_{3_{tot}}^{2-}]}{[HCO_{3_{tot}}^-]} = 10^{-8.95} \, M$$

$$\frac{[H^+][HCO_{3_{tot}}^-]}{[H_2CO_3^*]} = 10^{-5.86} \, M$$

$$\frac{[H_2CO_3^*]}{p_{CO_2}} = 2.89 \times 10^{-2} \, M \, \text{atm}^{-1}$$

solubility of calcite: $[Ca_{tot}][CO_{3_{tot}}^{2-}] = 5 \times 10^{-7} \, M^2$

solubility of aragonite: $[Ca_{tot}][CO_{3_{tot}}^{2-}] = 9 \times 10^{-7} \, M^2$

The ion product of CaCO$_3$ of surface seawater is $[Ca^{2+}][CO_{3_{tot}}^{2-}]$
$$= 2 \times 10^{-6} \, M^2$$

In order to assess the possible effect, we evaluate first the effect of an increase in p_{CO_2} on seawater (25°C) in equilibrium with the atmosphere. For different p_{CO_2} pressures we can compute pH and $[CO_{3_{tot}}^{2-}]$. This permits us to compare effective CaCO$_3$ ion products with the solubilities of calcite and aragonite. An increase in CO_2 has no effect on alkalinity as long as $CaCO_3$ does not dissolve (cf. Section 4.3):

$$[\text{Alk}] = [HCO_{3\text{tot}}^-] + 2[CO_{3\text{tot}}^{2-}] + [B(OH)_4]^- + [OH^-] - [H^+]$$
$$= \text{constant} \tag{i}$$

$B(OH)_4^-$, OH^-, and H^+ in equation i may be negligible in comparison to HCO_3^- and CO_3^{2-}.

$$[\text{Alk}] = 2.94 \times 10^{-3} = [HCO_{3\text{tot}}^-] + 2[CO_{3\text{tot}}^{2-}] \tag{ii}$$

and (cf. equation 11 in Chapter 4)

$$[\text{Alk}] = C_T(\alpha_1 + 2\alpha_2) \tag{iii}$$

$$= \frac{K_H p_{CO_2}}{\alpha_0}(\alpha_1 + 2\alpha_2) \tag{iv}$$

Any change in p_{CO_2} will affect the $[H^+]$-dependent ratio $(\alpha_1 + 2\alpha_2)/\alpha_0$. This ratio can be computed with the equilibrium constants given for every $[H^+]$. For every p_{CO_2} assumed we can compute $(\alpha_1 + 2\alpha_2)/\alpha_0$ and obtain the corresponding $[H^+]$ (either from a graph that plots $(\alpha_1 + 2\alpha_2)/\alpha_0$ versus pH or by trial and error, for example, most conveniently with a programmable calculator). For the present situation, $p_{CO_2} = 3.3 \times 10^{-4}$ atm, we obtain $(\alpha_1 + 2\alpha_2)/\alpha_0 = 3.09 \times 10^2$. The corresponding pH is 8.21_4. Correspondingly we have $C_T = 2.56 \times 10^{-3}$ M, $[H_2CO_3^*] = 9.54 \times 10^{-6}$ M, $[HCO_{3\text{tot}}^-] = 2.15 \times 10^{-3}$ M, and $[CO_{3\text{tot}}^{2-}] = 3.96 \times 10^{-4}$ M.

Doubling the CO_2 partial pressure leads to $(\alpha_1 + 2\alpha_2)/\alpha_0 = 1.54 \times 10^2$, pH $= 7.96_6$, $C_T = 2.706 \times 10^{-3}$ M, and $[CO_{3\text{tot}}^{2-}] = 2.53 \times 10^{-4}$ M. A fivefold increase in p_{CO_2} gives pH $= 7.61_2$, $C_T = 2.95 \times 10^{-3}$ M, and $[CO_{3\text{tot}}^{2-}] = 1.28 \times 10^{-4}$ M.

Since the carbonate concentration decreases in comparison to that in present seawater by a factor of 1.56 (twofold p_{CO_2}), and since present seawater is ca. 4 and 2 times supersaturated with respect to calcite and aragonite, respectively, this calculation shows that at higher p_{CO_2} pressures the water would remain supersaturated; in the case of aragonite, the equilibrium system would just about attain saturation equilibrium. The calculated effects would be too small, since the surface seawater mixes with deep ocean water. An increase in the CO_2 of the atmosphere is thus not likely to affect significantly the extent of surface seawater supersaturation with respect to $CaCO_3$. [Cf. G. Skirrow and M. Whitfield, *Limnol, Oceanogr.*, **20**, 103 (1975).]

Indirect effects of increased p_{CO_2}, for example, increased chemical weathering or enhanced release of phosphate and metal ions from the soil, and shifts in the the extent of adsorption and ion-exchange equilibrium on sedimenting materials in lakes and oceans, cannot be assessed yet with any certainty. A doubling of p_{CO_2} will approximately double the $[H^+]$ of a natural water system in equilibrium with the atmosphere (even if it is in equilibrium with $CaCO_3$); since the concentration of free metal ions is regulated often by equilibria at the solid–solution interface [$XOH + Me^{2+} \rightleftharpoons XOMe + H^+$ or $2XOH + Me^{2+} \rightleftharpoons (XO)_2Me + 2H^+$], the free metal-ion concentration will increase with an increase in [H^+]: $d[Me^{2+}]/d[H^+] \sim 1$ to 2.

Petroleum Hydrocarbons

The U.S. National Academy of Sciences workshop (1975) estimates that ca. 6 million tons of petroleum hydrocarbons enter the oceans yearly. This flux is composed of natural seeps (10%), inputs from the atmosphere (10%), urban and river runoff and coastal industrial wastes (40%), and tanker operations (especially wasting of cargo tankers) (40%). The entry of petroleum hydrocarbons into the aquatic food web has been clearly demonstrated [73]. Experiences with oil spills have been particularly revealing in regard to the vulnerability of the marine environment to petroleum. Certain petroleum hydrocarbons have been shown to interfere with the processes of chemoreception through the blocking of receptive organs. Reproductive processes may become impaired as a result of the making of the presence of pheromones.

In assessing the effects of accidental hydrocarbon spills it is important to know the solubilities of the individual components of a hydrocarbon mixture in water. Few data are available on this solubility under environmental conditions.†

Example 11.7. Multicomponent Solubility of Hydrocarbons in Water

What is the water solubility of benzene if the latter is present at equilibrium in oil droplets at a concentration of 0.1% (by weight). The solubility of pure benzene in water (25°C) is 1765 mg liter^{-1}.

Without additional information on activity coefficients of benzene in water and benzene in this particular oil, an exact answer cannot be given. We would expect that at equilibrium the concentration of benzene in water $C_{B(H_2O)}$ is proportional to the mole fraction of benzene in the oil phase and the saturation concentration of pure benzene in water, $C_{sB(H_2O)}$:

$$f_{B(H_2O)} C_{B(H_2O)} = X_{B(oil)} \gamma_{B(oil)} C_{sB(H_2O)} \tag{17}$$

where $f_{B(H_2O)}$ and $\gamma_{B(oil)}$ are the activity coefficients of benzene in the water and oil phase, respectively. If it is assumed as a first and crude approximation that

---

† Experimental data on the evaporation and solution of low molecular weight hydrocarbons from four ocean spills of two different crude oils have been reported by C. D. McAuliffe in *Fate and Effects of Petroleum Hydrocarbons in Marine Ecosystems*, Pergamon, New York, 363–372, 1977.

these constants are 1 and that 0.1 % by weight corresponds to ca. 0.25 mol %, $X_{B(oil)}$ is ca. 2.5×10^{-3}. Thus we would expect an equilibrium water solubility of 4 to 5 mg liter^{-1}. The activity coefficients of hydrocarbons in the aqueous phase appear to be reduced by the presence of other hydrocarbons, while the activity coefficients of hydrocarbons in the oil phase may be assumed to be close to unity; thus the presence of other hydrocarbons in the water enhances the solubility of the hydrocarbon under consideration more than would be predicted from a molal average solubility. Leinonen and Mackey [*Can. J. Chem. Eng.*, **51**, 230 (1973)] have measured the aqueous solubility of six binary hydrocarbon systems. They found that the solubility of each hydrocarbon component was greater by 6 to 35 % than the value expected by assuming the solubility of the component to be proportional to its mole fraction in the hydrocarbon phase.

Radionuclides

A great number of radionuclides are released from nuclear plants into the environment. With respect to global contamination, however, only nuclides with a long radioactive half-life and enough mobility to achieve widespread distribution in the environment are important. ^3H, ^{14}C, ^{85}Kr, and ^{121}I belong to this group. Actinides such as ^{239}Pu, ^{141}Am, and ^{244}Cm become important on a global scale if they are directly released into seawater. This release thus is highly undesirable.

The global distribution models for radionuclides generally need improvement. Especially a model for stable iodine is required in order to obtain a more realistic estimation of the global impact of ^{129}I. Up to the year 2000 the global exposure to radioactive effluents from nuclear facilities will be small compared with the natural radiation exposure of the population.

Recent reviews on the possible impact of synthetic radioactivity on the hydrosphere are available [74–77].

Increase in Erosion Rates

A significant part of the fluxes of dissolved and suspended materials mediated by rivers results from erosion and chemical denudation. These fluxes may have increased substantially because of changes in land use and acidification of rain. The transformation of forests to croplands and grazing areas, as well as road building, increase erosion rates by as much as one order of magnitude. Degens et al. [78] have estimated from Black Sea sediment cores rates of soil erosion in the drainage area of the Black Sea, a region characterized by a diversity comprising all transitions from arid to humid climates and from lowlands to mountainous areas. The present denudation rate is ca. 100 tons km^{-2}. The high levels in sedimentation rates from A.D. 200 to the present time were accounted for by agricultural activities and deforestation. It appears from these data that agricultural activities have accelerated soil erosion by a factor of about 3.

11.8   GLOBAL MOBILIZATION OF METAL IONS

The dispersion of metals into the atmosphere appears to rival, and sometimes exceed, natural mobilizations. Metals are released into the atmosphere, both as particles and as vapors, as a result not only of fossil fuel (coal, oil, natural gas) combustion but also of cement production and extractive metallurgy.

Computations of anthropogenic fluxes have been provided by Bertine and Goldberg [79], and more extensively and more recently, by Mackenzie et al. [80, 81]. Table 11.9 (from Lantzy and Mackenzie [81] shows the magnitude of fluxes derived for 20 trace metals. The importance of the anthropogenic flux for any metal is given by the interference factor, IF, which is calculated as total (anthropogenic emissions ÷ total natural emissions) × 100. It can be seen from Table 11.9 that, for the so-called atmophile elements† Sn, Cu, Cd, Zn, As, Se, Sb, Mo, Ag, Hg, and Pb, IF values are considerably higher than 100%.

Lantzy and Mackenzie [81], however, point out that further documentation is needed to interpret the data and to establish global balances for atmophile elements. Indeed, there are arguments inferring that for Hg, As, and Se, and perhaps other atmophile metals, there are significant fluxes from the sea surface to the atmosphere. If such fluxes are considered in the inference calculations for these metals, the factors may be reduced to levels of no interference.

That pollution of coastal and inland waters by trace elements has increased markedly over the last decades is evident in the memory record of sediments (Figure 11.23). Increases in Cd, Pb, Zn, and Cu are seen to have started about 1820. This observation is in accordance with the notation that the industrial revolution could be detected in the Greenland ice sheet layers formed about 1850.

Sediments are good indicators of the input of metals into receiving waters, because settling biogenic and other particles act as transporting carriers for heavy-metal cations (Figure 11.21). Metals are also efficiently sequestered by soil surfaces, during stream transport, in estuaries and in lakes [82–84]. Humic colloids carrying Fe and other elements become coagulated at the stream-ocean boundary and especially in estuaries. In the latter the trace metal cycle—as

---

† Elements are termed atmophile when their mass transport through the atmosphere is greater than that in streams. Many atmophile elements are volatile and have metal oxides of relatively low boiling points. It is also known that the metals Hg, As, Se, Sn, and Pb can be methylated and released into the atmosphere as vapors and that Hg and probably As and Se are released as inorganic vapor from the burning of coal. In contrast, the elements Al, Ti, Sn, Mn, Co, Cr, V, and Ni are termed lithophile because their mass transport to the oceans by streams exceeds their transport through the atmosphere.

TABLE 11.9   NATURAL AND ANTHROPOGENIC SOURCES OF ATMOSPHERIC EMISSIONS[a]

Element	Continental Dust Flux	Volcanic Dust Flux	Volcanic Gas Flux	Industrial Particulate Emissions	Fossil Fuel Flux	Total Emissions, Industrial Plus Fossil Fuel	Atmospheric Interference Factor (%)[b]
Al	356,500	132,750	8.4	40,000	32,000	72,000	15
Ti	23,000	12,000	—	3,600	1,600	5,200	15
Sm	32	9	—	7	5	12	29
Fe	190,000	87,750	3.7	75,000	32,000	107,000	39
Mn	4,250	1,800	2.1	3,000	160	3,160	52
Co	40	30	0.04	24	20	44	63
Cr	500	84	0.005	650	290	940	161
V	500	150	0.05	1,000	1,100	2,100	323
Ni	200	83	0.0009	600	380	980	346
Sn	50	2.4	0.005	400	30	430	821
Cu	100	93	0.012	2,200	430	2,630	1,363
Cd	2.5	0.4	0.001	40	15	55	1,897
Zn	250	108	0.14	7,000	1,400	8,400	2,346
As	25	3	0.1	620	160	780	2,786
Se	3	1	0.13	50	90	140	3,390
Sb	9.5	0.3	0.013	200	180	380	3,878
Mo	10	1.4	0.02	100	410	510	4,474
Ag	0.5	0.1	0.0006	40	10	50	8,333
Hg	0.3	0.1	0.001	50	60	110	27,500
Pb	50	8.7	0.012	16,000	4,300	20,300	34,583

[a] From Lantzy and Mackenzie [81]. See original paper for assumptions and discussion of sources of data. All fluxes are in units of 10^8 g year^{-1}.

[b] Atmospheric Interference factor = (total emissions ÷ (continental + volcanic fluxes) × 100

TABLE 11.10   MASS-BALANCE MODEL EXEMPLIFIED FOR HEAVY METALS IN GREIFENSEE (EXAMPLE 11.8)

Accessible data:

Lake volume $V = 1.25 \times 10^8$ m³; $Q = 8.9 \times 10^7$ m³ year⁻¹; $P = 3.7 \times 10^7$ kg year⁻¹

Metal	c_{in} (mg m⁻³)	D (m³ kg⁻¹)	J_2 (rain) (kg year⁻¹)
Zn	19.8	25	1316
Pb	3.2	120	1376
Cu	3.8	35	224
Cd	0.1	65	6

where J_i = inputs (kg year⁻¹); F_i = fluxes out (kg year⁻¹; V = volume of lake (m³); Q = inflow or outflow rate (m³ year⁻¹); c = concentration of soluble species; C = concentration of species in suspended matter (mg kg⁻¹); D = distribution coefficient = C/c (m³ kg⁻¹); P = sedimentation rate (kg year⁻¹)

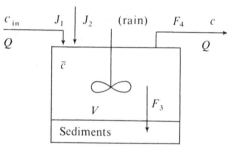

Mass Balance:

$$J_1 = Qc_{in} \qquad F_3 = CP = DcP \qquad F_4 = Qc \qquad \text{(i)}$$

$$V\frac{dc}{dt} = J_1 + J_2 - F_3 - F_4 \qquad \frac{dc}{dt} = \frac{Qc_{in} + J_2}{V} - c\left(\frac{Q + DP}{V}\right) \qquad \text{(ii)}$$

$$c(t) = c_0 \exp\left(-\frac{Q + DP}{V}t\right) + \frac{Qc_{in} + J_2}{Q + DP}\left[1 - \exp\left(-\frac{Q + DP}{V}t\right)\right] \qquad \text{(iii)}$$

At steady state:

$$t \to \infty, \, c(t) = \bar{c}, \qquad \bar{c} = \frac{Qc_{in} + J_2}{Q + DP} \qquad \text{(iv)}$$

Comparison with Field Data for Greifensee:

	Zn	Pb	Cu	Cd
Calculated $\bar{c}$ (mg m⁻³)	3.1	0.4	0.4	0.01
Measured $\bar{c}$ (mg m⁻³)	4.1	0.6	1.0	0.06

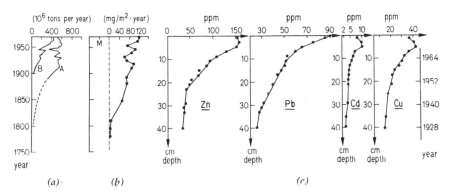

Figure 11.23   Sediments as indicators of heavy-metal pollution. (a) European coal production during the last 170 years; A: coal, B: lignite. (b) The anthropogenic input of heavy metals to the Baltic Sea [Erlenkeuser et al., *Geochim. Cosmochim. Acta*, **38**, 823 (1974)]. (c) Zn, Pb, Cd, and Cu in Greifensee. [(a) and (b) reproduced from E. D. Goldberg [69], (c) from Imboden et al. [85] (dating by Pb-210).]

shown by Boyle et al. [82] and by Turekian [84] is mainly self-contained; release and deposition occur virtually entirely within the system.

Example 11.8.   Mass-Balance Model for Heavy Metals in a Lake

For Greifensee, Switzerland, Imboden et al. [85] determined the inputs of Zn, Pb, Cu, and Cd both by the main tributary and by rainwater; they also determined the mean sedimentation rate P (kg year^{-1}) and the coefficient D for the distribution of metals between solution and suspended matter. Table 11.10 gives the characteristics of the lake and the experimental data.

Develop a mass-balance model and estimate the concentration of soluble metals in the lake, assuming attainment of steady state. (The concentrations of metal contaminants in this lake—see Figure 11.23—are not time-invariant. Nevertheless, steady-state models give useful limits for comparison purposes, i.e., concentrations obtained when the rate of input was kept constant for a time period of approximately three times the residence time of the substance under consideration.) Example 11.8 illustrates that the residence time of heavy-metal ions in a lake is drastically reduced because of the scavenging action of sediments. It may also be noted in this case that the input of metals through rain and aerial fallout exceeds that by tributaries.

11.9   ACID PRECIPITATION RESULTING FROM INTERACTIONS AMONG THE BIOGEOCHEMICAL CYCLES

The cycles for C, N, P, S, and O are interlocked in the biosphere [86]. The release of fuel combustion products may alter these cycles either by affecting

TABLE 11.11 INORGANIC COMPOSITION OF LAKES ON THE WEST COAST OF SWEDEN[a]

	pH	H	Na	K	Ca + Mg	Alk	Cl	SO$_4$	NO$_3$—N mg liter^{-1}	NH$_3$—N mg liter^{-1}
Precipitation at Plönninge, 1967–1969	4.6	0.024	0.096	0.016	0.122	−0.024	0.105	0.145	0.61	0.79
Precipitation at Plönninge, 1967–1969, concentrated 1.5 times	4.4	0.04	0.14	0.02	0.18	−0.04	0.16	0.22	0.91	1.19
85 lakes 1970 to 71, pH Interval					Mean Values					
<4.9	4.4	0.04	0.30	0.02	0.26	0	0.35	0.22	—	—
5.0–5.9	5.6	—	0.30	0.03	0.33	0.04	0.36	0.26	—	—
6.0–6.9	6.4	—	0.37	0.04	0.48	0.15	0.39	0.35	—	—
7.0	7.2	—	0.38	0.05	0.82	0.39	0.45	0.43	—	—

[a] From Almer et al. [88]. In this region the precipitation is concentrated about 1.5 times because of evaporation and the transpiration of plants before it reaches the lakes. Concentrations of ions and Alk are given as meq liter^{-1}.

740

the dynamics of processes or by uncoupling interactions among these elements. Two effects are of concern: (1) the production of acid rain and fallout resulting from the oxidation of S and N to sulfuric and nitric acid, respectively; (2) nutrient enrichment of aquatic ecosystems by products of fossil fuel combustion.

The harmful effects of S are due either to high concentrations of SO_2 or H_2SO_4 droplets in the air (local effects) or to deposition of acid on the ground. Acid precipitation has been falling on much of the eastern United States, portions of Canada, and Scandinavia. Long-distance nonpoint source emissions from one or more countries contribute a major portion of the atmospheric loading and of the ecological damage done.

As discussed in Section 3.3, acid rain results from an excess of mineral acidity (mostly H_2SO_4 and HNO_3, occasionally HCl) over reactive bases of dust particles (Ca and Mg carbonates and less reactive aluminum silicates, iron oxides, and other H^+ ion buffers). In crystalline rock areas acidity more likely remains in excess of reactive bases [3,87–89].

In order to assess the impact of acid precipitation, the atmospheric pathways to water bodies must be considered. The extent of evapotranspiration, type of surface canopy, type of lithology, interaction with organisms, and so on, affect the resulting water chemistry and the resulting $[H^+]$. The latter is the key parameter in assessing the ecological impact of acid precipitation because $[H^+]$ affects the chemical species distribution of metal ions and nutrients, the trace-metal migration, the extent of metal-ion adsorption at interfaces, and primary and secondary aquatic production in water and on land. Because metal binding to particles is inversely proportional to $[H^+]$ or $[H^+]^2$ (cf. Section 10.4), waters of high $[H^+]$ are also high in heavy-metal ions.

There has been a slight upward trend in H^+ ion input both in Scandinavia and in the eastern United States. Although H_2SO_4 dominates the acidity in these regions, the increase in input of H^+ during the past decade apparently has been caused by an increase in the HNO_3 content of precipitation.

Table 11.11 gives the chemical composition of lakes on the west coast of Sweden. During 1970 to 1972 more than half of these lakes had pH values below 6. Zooplankton and phytoplankton diversity has also markedly decreased; diatoms are nonexistant in lakes with pH < 5. Fish extinction in acidified waters has been well documented in Scandinavia and Canada. A study on 2083 lakes in Norway [89] showed that 36 % had no fish and that 23 % had been devoid of fish since 1940. Acidification usually does not result in immediate mortality of the stock but in reproductive failure.

11.10   THE BIOSPHERE AND THE GLOBAL ENVIRONMENT

In dealing with global chemical cycles, one copes by necessity with a broad range of mutually interacting systems. The *sensitivity of the system to disturbance* needs to be examined. For equilibrium systems, buffer intensities, buffer factors, and interaction intensities have been put forward as sensitivity parameters for each constituent in the face of variation in any other component. In a similar sense one needs to elaborate a response theory of cycle interactions in a system of different fluxes.

There are many inadequacies in our understanding of the interactions of global cycles. The types of questions for which we lack answers are: What is the effect of human perturbation of the N cycle on the magnitude of the atmospheric CO_2 increase? What interactions of the C cycle produce significant positive or negative effects on the N cycle? What is the effect of further acceleration of the S cycle on chemical erosion?

Effective feedback mechanisms appear to have protected the earth's surface environment from severe perturbations [90]. Does life merely have a passive role in cycling chemical elements or does the biosphere play an active role in maintaining the earth in a homeostatic state? As discussed in 7.5, the major and minor components of the earth's atmosphere go through *mostly biospheric redox cycles*; because their mass transfer through the atmosphere has been many times the current content of this reservoir, a steady-state composition of redox components of the atmosphere and of the ocean is implied. The earth indeed is alive and not in equilibrium. Earth without life would have an atmosphere with a chemical composition somewhere between those of Mars and Venus [91].

Photosynthesis imposes an electron flux of ca. 1.3 μeq m^{-2} sec^{-1}. As we have seen, the atmospheric concentration of CO_2 in the atmosphere is regulated by biological (photosynthesis and respiration) and geochemical (weathering) processes and by chemical equilibria in the sea. N_2 has the longest residence time in the atmosphere (ca. 50 million years); it is chemically remarkably inert. Nevertheless the output (microbial fixation of N_2) and input (microbial reduction of NO_3^- and NO_2^- to N_2) of N_2 into the atmospheric reservoir is regulated to a large extent by biota. The minor gas components (less than 0.01 %) are most likely also "buffered" by biological processes in the ocean and the soil.

Lovelock and Margulis [92] in their stimulating paper, "Atmospheric Homeostasis by and for the Biosphere: The Gaia Hypothesis," suggest that life at an early stage of evolution acquired the capacity to control the global environment to suit its needs. In this article "the biosphere" (the authors give this hypothetical creature, a personification of Mother Earth, the name "Gaia") is an entity with properties different from the simple sum of its parts; it interacts actively with the larger environment and has a powerful capacity to achieve homeostasis.

Lovelock and Margulis point out that, during the time ($> 10^9$ years) that life has been present on earth, the physical and chemical conditions of most of the planetary surface have never deviated from those most favorable for life. Humans have become a more and more powerful component of the biosphere. Will their influence on the control of the environment preserve the conditions best suited for life and human needs?

REFERENCES

1   E. P. Odum, *Science*, **164**, 262 (1969).

2   H. J. Morowitz, *Energy Flow in Biology*, Academic, New York, 1968.

3   G. E. Likens, F. H. Borman, R. S. Pierce, J. S. Eaton, and W. M. Johnson, *The Biogeochemistry of a Forested Ecosystem*, Springer, New York, 1977.

4   Y. H. Li, *Environ. Conserv.*, **3**, 171 (1976).

5   J. Ceasar, R. Collier, J. Edmond, F. Frey, G. Matisoff, A. Ng, and R. Stallard, *Environ. Sci. Technol.*, **20**, 697 (1976).

6   R. Eliassen, *Sci. Amer.*, **186**(3), 17 (1952).

7   K. Wuhrmann, *Int. Ver. Limnol.*, **20**, 324 (1974).

8   M. Blumer, *Angew. Chem. Int. Ed.*, **14**, 507 (1975).

9   J. H. Steele, *The Structure of Marine Ecosystems*, Harvard University Press, Cambridge, Mass., 1974.

10 J. E. Hobbie and R. T. Wright, *Limnol. Oceanogr.*, **10**, 471 (1965).

11 H. U. Jannasch, *Limnol. Oceanogr.*, **12**, 264 (1967).

12 E. Stumm-Zollinger and R. H. Harris, in *Organic Compounds in Aquatic Environments*, S. Faust and J. Hunter, Eds., Dekker, New York, 1971, p. 555.

13 K. Wuhrmann, in *Water Pollution Microbiology*, R. Mitchell, Ed., Wiley-Interscience, New York, 1972, p. 119.

14 R. T. Barber, in *Trace Metals and Metal-Organic Interactions in Natural Waters*, P. Singer, Ed., Ann Arbor Science Publications, 1973, p. 321.

15 G. A. Jackson and J. J. Morgan, *Limnol. Oceanogr.*, **23**, 268 (1978).

16 W. Sunda and R. L. Guillard, *J. Mar. Res.*, **34**, 511 (1976).

17 S. E. Manahan and M. J. Smith, *Environ. Sci. Technol.*, **7**, 829 (1973).

18 Environmental Protection Agency, *Water Quality Criteria*, Washington, D.C., 1972.

19 J. M. Schurr and J. Ruchti, *Limnol. Oceanogr.*, **22**, 208 (1977).

20 J. H. Ryther and W. M. Dunstan, *Science*, **171**, 1008 (1971).

21 D. W. Schindler, *Science*, **184**, 897 (1974); **195**, 260 (1977).

22 S. T. Emerson, *Limnol. Oceanogr.*, **20**, 743 (1975).

23 J. Shapiro, *Science*, **179**, 382 (1973).

24 W. Stumm and J. J. Morgan, *Transactions of the 12th Annual Conference on Sanitary Engineering*, University of Kansas, Lawrence, 1962, p. 16.

25 D. Imboden, *Limnol. Oceanogr.*, **19**, 297 (1974).

26 C. B. Officer and J. H. Ryther, *Science*, **197**, 1056 (1977).

27 F. A. Richards, in *Chemical Oceanography*, J. P. Riley and G. Skirrow, Eds., Academic, New York, 1965, Chapter 13.

28 A. C. Redfield, B. H. Ketchum, and F. A. Richards, in *The Sea*, Vol. 2, M. N. Hill, Ed., Wiley-Interscience, New York, 1963, p. 26.

29 B. H. Ketchum, in *Eutrophication: Causes, Consequences, Correctives*, National Academy of Sciences, Washington, D.C., 1969, p. 197.

30 E. Føyn, in *Eutrophication in Large Lakes and Impoundments*, Report by the Organization for Economic Cooperation and Development, Paris, 1970.

31   S. H. Fonselius, *Environment*, **12**(6), 2 (1970).

32   R. A. Vollenweider, *The Scientific Basis of Lake and Stream Eutrophication*, Technical Report DAS/CSI/68, The Organization for Economic Cooperation and Development, Paris, 1968, p. 27.

33   E. E. Shannon and P. L. Brezonik, *Environ. Sci. Technol.*, **6**, 719 (1972).

34   St. C. Chapra, *J. Environ. Eng. Div. Amer. Soc. Civ. Eng.*, **103**, 147 (1977).

35   R. A. Vollenweider, *Schweiz. Z. Hydrol.*, **37**, 53 (1975).

36   W. J. Snodgrass and C. R. O'Melia, *Environ. Sci. Technol.*, **9**, 937 (1975).

37   D. Imboden and R. Gächter, *Int. J. Ecol. Modelling*, **4**, 77 (1978).

38   J. G. Yeasted and F. M. Morel, *Environ. Sci. Technol.*, **12**, 195 (1978).

39   F. Taub, *Int. J. Environ. Stud.*, **10**, 23 (1976); personal communication, 1977.

40   J. L. Mosser, N. S. Fisher, C. F. Wurster, *Science*, **176**, 533 (1972); J. L. Mosser et al., *Science*, **175**, 191 (1972).

41   G. M. Woodwell, *Science*, **168**, 429 (1970).

42   R. Gillette, *Science*, **186**, 242 (1974).

43   M. Alexander, *Microbiol. Ecol.*, **2**, 17 (1975).

44   H. Leidner, R. Gloor, and K. Wuhrmann, *Tenside Deterg.*, **13**, 122 (1976).

45   K. Grob, G. Grob, and K. Grob, *J. Chromatogr.*, **106**, 299 (1975).

46   C. T. Chiou, V. H. Freed, D. W. Schmedding, and R. L. Kohnert, *Environ. Sci. Technol.*, 475 (1977).

47   I. P. Kapoor et al., *J. Agr. Food Chem.*, **21**, 310 (1973).

48   R. L. Metcalf, in *Fate of Pollutants in the Air and Water Environments*, Vol. 2, I. H. Suffet, Ed., Wiley-Interscience, 1977.

49   E. Sondheimer and J. B. Simeone, Eds., *Chemical Ecology*, Academic, New York, 1970.

50   A. T. Scholz, R. M. Horall, J. C. Cooper, and A. D. Hasler, *Science*, **192**, 1247 (1976).

51   I. Chet and R. Mitchell, *Ann. Rev. Microbiol.*, **30**, 221 (1976).

52   S. R. Kerr and W. P. Vass in *Environmental Pollution by Pesticides*, C. A. Edwards, Ed., Plenum, New York, 1973.

53   R. M. Patrick, H. Hohn, and J. H. Wallace, *Proc. Natl. Acad. Sci. U.S.A.*, **259**, 1 (1954).

54   World Health Organization, *Health Hazards from New Environmental Pollutants*, Technical Report 586, Geneva, 1976.

55   R. L. Jolley, H. Gorchev, and D. H. Hamilton, Eds., *Water Chlorination*, Vol. II; *Environmental Impact and Health Effects*, Ann Arbor Science Publications, Ann Arbor, Mich., 1978.

56   K. B. Cantor and L. J. McCabe, in *Water Chlorination*, Vol. II, *Environmental Impact and Health Effects*, R. L. Jolley, H. Gorchev, and D. H. Hamilton, Eds., Ann Arbor Science Publications, Ann Arbor Mich., 1978, p. 379.

57   R. H. Harris, T. Page, and N. A. Reiches, *Cold Spring Harbor Symp.* "Origins of Human Cancer," p. 309 (1977).

58   O. Hutzinger, I. H. Van Lelyveld, and B. C. J. Zoeteman, Eds., *Aquatic Pollutants: Transformation and Biological Effects*, Pergamon, Elmsford, N.Y., 1977.

59   B. N. Ames, J. McCann, and E. Yamasaki, *Mutat. Res.*, **31**, 347 (1975).

60   R. Margalef, *Perspectives in Ecological Theory*, University of Chicago Press, Chicago, 1968.

61 G. M. Woodwell, *Science*, **168**, 429 (1970).

62 A. M. Beeton, in *Eutrophication, Causes, Consequences and Correctives*, National Academy of Sciences, Washington, D.C., 1969, p. 150.

63 H. L. Sanders, *Brookhaven Symp. Biol.*, **22**, 1969, p. 63.

64 D. L. King and R. C. Ball, *Limnol. Oceanogr.*, **12**, 27 (1967).

65 R. M. Garrels, F. T. Mackenzie, and C. Hunt, *Chemical Cycles and the Global Environment: Assessing Human Influences*, W. Kaufmann, Los Altos, Calif., 1975.

66 C. E. Junge, *J. Roy. Meteorol. Soc.*, **98**, 711 (1972); in *Fate of Pollutants in the Air and Water Environments*, I. H. Suffet, Ed., Wiley-Interscience, 1977, p. 7.

67 Y. H. Li, *Geochim. Cosmochim Acta*, **41**, 555 (1977).

68 W. S. Broecker, *Science*, **168**, 1537 (1970).

69 E. D. Goldberg, *The Health of the Oceans*, UNESCO, Paris, 1976.

70 D. Mackay and A. W. Wolkoff, *Environ. Sci. Technol*, **7**, 611 (1973); D. Mackay and P. J. Leinonen, *Environ. Sci. Technol.*, **9**, 1178 (1975).

71 F. T. Mackenzie and R. Wollast, in *The Sea*, Vol. 6, E. D. Goldberg, Ed., Wiley-Interscience, New York, 1977, p. 739.

72 P. S. Liss and P. G. Slater, *Nature*, **247**, 181 (1974).

73 M. Blumer and J. Sass, *Science*, **176**, 1120 (1972); *Mar Pollut Bull*, **3**, 92 (1972).

74 E. D. Goldberg, in *The Health of the Oceans*, UNESCO, Paris, 1976, Chapter 4.

75 S. Hauri and W. Schikarsky, in *Global Chemical Cycles and Their Alterations by Man*, W. Stumm, Ed., Abakon, Berlin, 1977.

76 International Atomic Energy Agency, *Radioactive Contamination of the Environment*, Vienna, 1973.

77 D. Preston, in *The Sea*, Vol. 5, E. D. Goldberg, Ed., Wiley-Interscience, New York, 1974, p. 817.

78 E. T. Degens, A. Paluska, and E. Eriksson, in *Nitrogen, Phosphorus, and Sulfur–Global Cycles*, B. H. Svensson and R. Söderlund, Eds., Scope Report 7, Swedish National Science Research Council, Stockholm, Sweden, 1976.

79 K. K. Bertine and E. D. Goldberg, *Science*, **173**, 233 (1971).

80 F. T. Mackenzie and R. Wollast, in *The Sea*, Vol. 6, E. D. Goldberg, Ed., Wiley-Interscience, New York, 1977.

81 R. J. Lantzy and F. T. Mackenzie, *Geochim. Cosmochim. Acta*, **43**, 511 (1979).

82 E. A. Boyle, J. M. Edmond, and E. R. Sholkovitz, *Geochim. Cosmochim. Acta*, **41**, 1313 (1977).

83 E. R. Sholkovitz, *Geochim. Cosmochim. Acta*, **40**, 831 (1976).

84 K. K. Turekian, *Geochim. Cosmochim. Acta*, **41**, 1139 (1977).

85 D. Imboden, J. Tschopp, and W. Stumm, *Schweiz. Z. Hydrol.*, in press.

86 B. H. Svensson and R. Söderlund, Eds., N, P, S Global Cycles. Scope Report 7, Ecological Bulletin, Swedish National Science Research Council, 1976.

87 J. R. Kramer, in *Sulfur*, J. Nriagu, Ed., Wiley, New York, 1978, p. 325.

88 C. Brosset, *Ambio*, **5**, 157 (1976).

89 R. F. Wright and E. T. Gjessing, *Ambio*, **5**, 219 (1976).

90 R. M. Garrels, A. Lerman, and R. T. Mackenzie, *Amer. Sci.*, **63**, 306 (1976).

91   J. E. Lovelock, *Atmos. Environ.* **6**, 579 (1972).

92   J. E. Lovelock and L. Margulis, *Tellus*, **26**, 1 (1974).

READING SUGGESTIONS

Pollution and Water Quality

Blumer, M., and J. Sass, "Oil Pollution: Persistence and Degradation of Spilled Fuel Oil," *Science*, **167**, 1120 (1972).

Dagley, S., "Microbial Degradation of Organic Compounds in the Biosphere," *Amer. Sci.*, **63**, 681 (1975).

Environmental Protection Agency (EPA), *Water Quality Criteria*, Washington, D.C., 1972, 594 pp. [This report, prepared by a committee of the National Academy of Science and Engineering, critically reviews literature data and develops scientific criteria for major beneficial water use categories (recreation and esthetics, public water supplies, freshwater aquatic life and wildlife, marine aquatic life and wildlife, agricultural uses of water, industrial water supplies).]

Hutzinger, O., I. H. van Lelyveld, and B. C. J. Zoetman, Eds., *Aquatic Pollutants: Transformation and Biological Effects*, Pergamon, Elmsford, N.Y., 1978.

Suffet, I. H., Ed., *Fate of Pollutants in the Air and Water Environments*, Vols. 1 and 2, Wiley-Interscience, New York, 1977.

Tinsley, I. J., *Chemical Concepts in Pollutant Behavior*, Wiley-Interscience, New York, 1979.

Warren, C. W., *Biology and Water Pollution Control*, Saunders, Philadelphia, 1971.

Woodwell, G. M., "Toxic Substances and Ecological Cycles," *Sci. Amer.*, **216**, 23 (1967).

Disturbance of Balance of Photosynthesis and Respiration

Hutchinson, G. E., "Eutrophication," *Amer. Sci.*, **61**, 269–279 (1973).

Lerman, A., Ed., *Lakes: Chemistry, Geology, Physics*. Springer, New York, 1978.

Schindler, D. W., "Eutrophication and Recovery in Environmental Lakes; Implications for Lake Management," *Science*, **184**, 897–899 (1974).

Schindler, D. W., "Evolution of Phosphorus Limitation in Lakes," *Science*, **195**, 260–262 (1977).

Vallentyne, J. R., *The Algal Bowl; Algae and Man*, Environment Canada, 1974. (A most readable and interesting account for nonspecialists. Copies may be obtained from Information Canada, Ottawa, Canada K1A 0S9.)

Pollution of the Oceans

Dyrssen, D., and O. Jagner, Eds., *The Changing Chemistry of the Oceans*, Almqvist and Wiksell, Stockholm, 1972.

Goldberg, E. D., *The Health of the Oceans*, UNESCO, Paris, 1976.

Hood, D. W., Ed., *Impingement of Man on the Oceans*, Wiley-Interscience, New York, 1971.

International Atomic Energy Association (IAEA), *Radioactive Contamination of the Marine Environment*, Vienna, 1973.

Odum, H. T., B. J. Copeland, and E. A. McMahan, *Coastal Ecological Systems of the United States*, Vols. 1–4, The Conservation Foundation, Washington, D.C., 1974.

Global Chemical Cycles and Pollution

Broecker, W. S., T. Takahashi, H. J. Simpson, T. H. Peng, "Fate of Fossil Fuel Carbon Dioxide and the Global Carbon Budget," *Science*, **206**, 409 (1979).

Garrels, R. M., A. Lerman, and F. T. Mackenzie, "Controls of Atmospheric O_2 and CO_2: Past, Present and Future," *Amer. Sci.*, **63**, 306–315 (1976).

Garrels, R. M., F. T. Mackenzie, and C. Hunt, *Chemical Cycles and the Global Environment; Assessing Human Influences*, W. Kaufmann, Los Altos, Calif., 1975. (After considering the pre-human global and local fluxes of elements in the natural environment, the authors assess present fluxes, including human contributions.)

Georgescu-Roegen, N., "Economics and Entropy," *Ecologist*, **2**, 13–18 (1972).

Lantzy, R. L., and F. T. Mackenzie, "Global Cycles and Assessment of Man's Impact," *Geochim. Cosmochim. Acta*, **43**, 511–515 (1979).

Likens, G. E., F. H. Borman, R. S. Pierce, J. S. Eaton, N. M. Johnson, *Biogeochemistry of a Forested Ecosystem*, Springer, New York, 1977.

Singer, S. F., Ed., *The Changing Global Environment*, Reidel, Dordrecht, The Netherlands, 1975.

Stumm, W., Ed., *Global Chemical Cycles and Their Alterations by Man*, Report of the DAHLEM Workshop, Dahlem Konferenzen, Berlin, 1977.

Svensson, B. H., and R. Söderlund, Eds., *Nitrogen, Phosphorus and Sulphur—Global Cycles*, Ecological Bulletin No. 22, SCOPE Report No. 7, Swedish Natural Science Research Council, Stockholm, 1976.

Pollution and Ecosystem Structure

Cloud, P., Evolution of Ecosystems, *Amer. Sci.*, **62**, 54–66 (1974).

Ehrenfeld, D. W., "The Conservation of Non-Resources," *Amer. Sci.*, **64**, 648–656 (1973).

Lotka, A. J., *Elements of Mathematical Biology*, reprinted by Dover, New York, 1956.

Odum, E., "The Strategy of Ecosystem Development," *Science*, **164**, 262 (1969).

Odum, H. T., *Environment, Power and Society*, Wiley, New York, 1970.

Tyler Miller, G., Jr., *Energetics, Kinetics and Life; An Ecological Approach*, Wadsworth, Belmont, Calif., 1971. (An introductory text on chemical thermodynamics and kinetics and their application to life processes.)

Woodwell, G. M., "Effects of Pollution on the Structure and Physiology of Ecosystems," *Science*, **168**, 429 (1970).

Woodwell, G. M., and H. H. Smith, *Diversity and Stability in Ecological Systems*, Brookhaven Symp. Biol., **22**, (1969).

Chemical Ecology

H. J. M. Bowen, *Environmental Chemistry of the Elements*, Academic, New York, 1979.

Sondheimer, E., and J. B. Simeone, Eds., *Chemical Ecology*, Academic, New York, 1970. (Of particular interest are the chapters by E. O. Wilson ("Chemical Communication within Animal Species") and A. T. Hasler ("Chemical Ecology of Fish.").

Todd, J. H., "The Chemical Languages of Fishes," *Sci. Amer.*, **224**, 98–108 (1971).

Whittaker, R. H., and P. P. Feeny, "Allelochemics: Chemical Interaction between Species, *Science*, **171**, 757–770 (1971).

Appendix: Thermodynamic Properties; Table of G_f°, H_f°, and $\bar{S}^\circ$ Values for Common Chemical Species in Aquatic Systems[a]

Valid at 25°C, 1 atm Pressure and Standard States[b]

Species	Formation from the Elements		Entropy	Reference[c]
	G_f^0 (kJ mol^{-1})	H_f^0 (kJ mol^{-1})	$\bar{S}^0$ J mol^{-1} K^{-1}	
Ag (Silver)				
Ag (metal)	0	0	42.6	NBS
Ag$^+$(aq)	77.12	105.6	73.4	NBS
AgBr	−96.9	−100.6	107	NBS
AgCl	−109.8	−127.1	96	NBS
AgI	−66.2	−61.84	115	NBS
Ag$_2$S(α)	−40.7	−29.4	14	NBS
AgOH(aq)	−92	—	—	NBS
Ag(OH)$_2^-$ (aq)	−260.2	—	—	NBS
AgCl(aq)	−72.8	−72.8	154	NBS
AgCl$_2^-$ (aq)	−215.5	−245.2	231	NBS
Al (Aluminum)				
Al	0	0	28.3	R
Al^{3+}(aq)	−489.4	−531.0	−308	R
AlOH^{2+}(aq)	−698	—	—	S
Al(OH)$_2^+$(aq)	−911	—	—	S
Al(OH)$_3$(aq)	−1115	—	—	S
Al(OH)$_4^-$(aq)	−1325	—	—	S
Al(OH)$_3$ (amorph)	−1139	—	—	R
Al$_2$O$_3$ (corundum)	−1582	−1676	50.9	R
AlOOH (boehmite)	−922	−1000	17.8	R
Al(OH)$_3$ (gibbsite)	−1155	−1293	68.4	R
Al$_2$Si$_2$(OH)$_4$ (kaolinite)	−3799	−4120	203	R
KAl$_3$Si$_3$O$_{10}$(OH)$_2$ (muscovite)	−1341	—	—	G

$Mg_5Al_2Si_3O_{10}(OH)_8$ (chlorite)	-1962	—	—	R
$CaAl_2Si_2O_8$ (anorthite)	-4017.3	-4243.0	199	R
$NaAlSiO_3O_8$ (albite)	-3711.7	-3935.1	—	R
As (Arsenic)				
As (α metal)	0	0	35.1	NBS
H_3AsO_4(aq)	-766.0	-898.7	206	NBS
$H_2AsO_4^-$(aq)	-748.5	-904.5	117	NBS
$HAsO_4^{2-}$(aq)	-707.1	-898.7	3.8	NBS
AsO_4^{3-}(aq)	-636.0	-870.3	-145	NBS
$H_2AsO_3^-$(aq)	-587.4			NBS
Ba (Barium)				
Ba^{2+}(aq)	-560.7	-537.6	9.6	R
$BaSO_4$ (barite)	-1362	-1473	132	R
$BaCO_3$ (witherite)	-1132	-1211	112	R
Be (Beryllium)				
Be^{2+}(aq)	-380	-382	-130	NBS
$Be(OH)_2(\alpha)$	-815.0	-902	51.9	NBS
$Be_3(OH)_3^{3+}$	-1802	—	—	NBS
B (Boron)				
H_3BO_3(aq)	-968.7	-1072	162	NBS
$B(OH)_4^-$(aq)	-1153.3	-1344	102	NBS
Br (Bromide)				
Br_2(l)	0	0	152	NBS
Br_2(aq)	3.93	-2.59	130.5	NBS
Br^-(aq)	-104.0	-121.5	82.4	NBS
HBrO(aq)	-82.2	-113.0	147	NBS
BrO^-(aq)	-33.5	-94.1	42	NBS
C (Carbon)				
C (graphite)	0	0	152	NBS
C(diamond)	3.93	-2.59	130.5	NBS
CO_2(g)	-394.37	-393.5	213.6	NBS
$H_2CO_3^*$(aq)	-623.2	-699.7	187.0	R
H_2CO_3(aq) ("true")	~ -607.1	—	—	S
HCO_3^-(aq)	-586.8	-692.0	91.2	S
CO_3^{2-}(aq)	-527.9	-677.1	-56.9	NBS
CH_4(g)	-50.75	-74.80	186	NBS
CH_4(aq)	-34.39	-89.04	83.7	NBS
CH_3OH(aq)	-175.4	-245.9	133	NBS
HCOOH(aq)	-372.3	-425.4	163	NBS
$HCOO^-$(aq)	-351.0	-425.6	92	NBS
HCN(aq)	119.7	107.1	124.6	NBS
CN^-(aq)	172.4	150.6	94.1	NBS

CH$_3$COOH(aq)	−396.6	−485.8	179	NBS
CH$_3$COO$^-$(aq)	−369.4	−486.0	86.6	NBS
C$_2$H$_5$OH(aq)	−181.8	−288.3	149	NBS
NH$_2$CH$_2$COOH(aq)	−370.8	−514.0	158	NBS
NH$_2$CH$_2$COO$^-$(aq)	−315.0	−469.8	119	NBS

Ca   (Calcium)

Ca^{2+}(aq)	−553.54	−542.83	−53	R
CaOH$^+$(aq)	−718.4	—	—	NBS
Ca(OH)$_2$(aq)	868.1	1003	−74.5	NBS
Ca(OH)$_2$ (portlandite)	−898.4	−986.0	83	R
CaCO$_3$ (calcite)	−1128.8	1207.4	91.7	R
CaCO$_3$ (aragonite)	−1127.8	−1207.4	88.0	R
CaMg(CO$_3$)$_2$ (dolomite)	−2161.7	−2324.5	155.2	R
CaSiO$_3$ (wollastonite)	−1549.9	−1635.2	82.0	R
CaSO$_4$ (anhydrite)	−1321.7	−1434.1	106.7	R
CaSO$_4 \cdot 2$H$_2$O (gypsum)	−1797.2	−2022.6	194.1	R
Ca$_5$(PO$_4$)$_3$OH (hydroxyapatite)	−6338.4	−6721.6	390.4	R

Cd (Cadmium)

Cd (γ metal)				
Cd^{2+}(aq)	−77.58	−75.90	−73.2	R
CdOH$^+$(aq)	284.5			R
Cd(OH)$_3^-$(aq)	−600.8			R
Cd(OH)$_4^{2-}$(aq)	−758.5			R
Cd(OH)$_2$(aq)	−392.2			R
CdO (s)	−228.4	−258.1	54.8	
Cd(OH)$_2$ (precip.)	−473.6	−560.6	96.2	R
CdCl$^+$(aq)	−224.4	−240.6	43.5	R
CdCl$_2$(aq)	−340.1	−410.2	39.8	R
CdCl$_3^-$(aq)	−487.0	−561.0	203	R
CdCO$_3$ (s)	−669.4	−750.6	92.5	R

Cl (Chlorine)

Cl$^-$(aq)	−131.3	−167.2	56.5	NBS
Cl$_2$(g)	0	0	223.0	NBS
Cl$_2$(aq)	6.90	−23.4	121	NBS
HClO(aq)	−79.9	−120.9	142	NBS
ClO$^-$(aq)	−36.8	−107.1	42	NBS
ClO$_2$(aq)	117.6	74.9	173	NBS
ClO$_2^-$(aq)	17.1	−66.5	101	NBS
ClO$_3^-$(aq)	−3.35	−99.2	162	NBS
ClO$_4$(aq)	−8.62	−129.3	182	NBS

Co   (Cobalt)

Co (metal)	0	0	30.04	R
Co^{2+}(aq)	−54.4	−58.2	−113	R
Co^{3+}	−134	−92	−305	R

$HCoO_2^-$(aq)	−407.5	—	—	NBS
$Co(OH)_2$(aq)	−369	−518	134	NBS
$Co(OH)_2$ (blue precip.)	−450			NBS
CoO	−214.2	−237.9	53.0	R
Co_3O_4 (cobalt spinel)	−725.5	−891.2	102.5	R

Cr (Chromium)

Cr (metal)	0	0	23.8	NBS
Cr^{2+}(aq)	—	−143.5	—	NBS
Cr^{3+}(aq)	−215.5	−256.0	308	NBS
Cr_2O_3 (eskolaite)	−1053	−1135	81	R
$HCrO_4^-$(aq)	−764.8	−878.2	184	R
CrO_4^{2-}(aq)	−727.9	−881.1	50	R
$Cr_2O_7^{2-}$(aq)	−1301	−1490	262	R

Cu (Copper)

Cu(metal)	0	0	33.1	NBS
Cu^+(aq)	50.0	71.7	40.6	NBS
Cu^{2+}(aq)	65.5	64.8	−99.6	NBS
$Cu(OH)_2$(aq)	−249.1	−395.2	−121	NBS
$HCuO_2^-$(aq)	−258	—	—	
CuS (covellite)	−53.6	−53.1	66.5	NBS
Cu_2S (α)	−86.2	−79.5	121	NBS
CuO (tenorite)	−129.7	−157.3	43	NBS
$CuCO_3 \cdot Cu(OH)_2$ (malachite)	−893.7	−1051.4	186	NBS
$2CuCO_3 \cdot Cu(OH)_2$ (azurite)		−1632		NBS

F (Fluorine)

F_2(g)	0	0	202	NBS
F^-(aq)	−278.8	−332.6	−13.8	NBS
HF(aq)	−296.8	320.0	88.7	NBS
HF_2^-(aq)	−578.1	−650	92.5	NBS

Fe (Iron)

Fe (metal)	0	0	27.3	NBS
Fe^{2+}(aq)	−78.87	−89.10	−138	NBS
$FeOH^+$(aq)	−277.3	—	—	NBS
Fe^{3+}(aq)	−4.60	−48.5	−316	NBS
$FeOH^{2+}$(aq)	−229.4	−324.7	−29.2	NBS
$Fe(OH)_2^+$(aq)	−438	—	—	NBS
$Fe(OH)_2^-$(aq)	−659	—	—	NBS
$Fe_2(OH)_2^{4+}$(aq)	−467.3	—	—	NBS
FeS_2 (pyrite)	−160.2	−171.5	52.9	R
FeS_2 (marcasite)	−158.4	−169.4	53.9	R
FeO(s)	−251.1	−272.0	59.8	R
$Fe(OH)_2$ (precip.)	−486.6	−569	87.9	NBS

α-Fe$_2$O$_3$ (hematite)^e	−742.7	−824.6	87.4	R
Fe$_3$O$_4$ (magnetite)	−1012.6	−1115.7	146	R
α-FeOOH (goethite)^e	−488.6	−559.3	60.5	R
FeOOH (amorph)^e	−462	—	—	S
Fe(OH)$_3$ (amorph)^e	−699(−712)			S
FeCO$_3$ (siderite)	−666.7	−737.0	105	R
Fe$_2$SiO$_4$ (fayalite)	−1379.4	−1479.3	148	R

H (Hydrogen)

H$_2$(g)	0	0	130.6	NBS
H$_2$(aq)	17.57	−4.18	57.7	NBS
H$^+$(aq)	0	0	0	NBS
H$_2$O(l)	−237.18	−285.83	69.91	NBS
H$_2$O$_2$(aq)	−134.1	−191.1	144	NBS
HO$_2^-$(aq)	−67.4	−160.3	23.8	NBS
H$_2$O(g)	−228.57	−241.8	188.72	R

Hg (Mercury)

Hg(l)	0	0	76.0	NBS
Hg$_2^{2+}$(aq)	153.6	172.4	84.5	NBS
Hg^{2+}(aq)	164.4	171.0	−32.2	NBS
Hg$_2$Cl$_2$ (calomel)	−210.8	265.2	192.4	NBS
HgO (red)	−58.5	−90.8	70.3	NBS
HgS (metacinnabar)	−43.3	−46.7	96.2	NBS
HgI$_2$ (red)	101.7	−105.4	180	NBS
HgCl$^+$(aq)	−5.44	−18.8	75.3	NBS
HgCl$_2$(aq)	−173.2	−216.3	155	NBS
HgCl$_3$ (aq)	−309.2	−388.7	209	NBS
HgCl$_4^{2-}$ (aq)	−446.8	−554.0	293	NBS
HgOH$^+$(aq)	−52.3	−84.5	71	NBS
Hg(OH)$_2$(aq)	−274.9	−355.2	142	NBS
HgO$_2^-$(aq)	−190.3	—	—	NBS

I (Iodine)

I$_2$ (crystal)	0	0	116	NBS
I$_2$(aq)	16.4	22.6	137	NBS
I$^-$(aq)	−51.59	−55.19	111	NBS
I$_3^-$(aq)	−51.5	−51.5	239	NBS
HIO(aq)	−99.2	−138	95.4	NBS
IO$^-$(aq)	−38.5	−107.5	−5.4	NBS
HIO$_3$(aq)	−132.6	−211.3	167	NBS
IO$_3^-$	−128.0	−221.3	118	NBS

Mg (Magnesium)

Mg (metal)	0	0	32.7	R
Mg^{2+}(aq)	−454.8	−466.8	−138	R
MgOH$^+$(aq)	−626.8	—	—	S
Mg(OH)$_2$(aq)	−769.4	−926.8	−149	NBS
Mg(OH)$_2$ (brucite)	−833.5	−924.5	63.2	R

Mn (Manganese)

Mn (meta)	0	0	32.0	R
Mn^{2+}(aq)	−228.0	−220.7	−73.6	R
$Mn(OH)_2$ (precip.)	−616			S
Mn_3O_4 (hausmannite)	−1281			S
MnOOH (α manganite)	−557.7			S
MnO_2 (manganate) (IV)				
($MnO_{1.7}$ − MnO_2)	−453.1			S
MnO_2 (pyrolusite)	−465.1	−520.0	53	R
$MnCO_3$ (rhodochrosite)	−816.0	−889.3	100	R
MnS (albandite)	−218.1	−213.8	87	R
$MnSiO_3$ (rhodonite)	−1243	−1319	131	R

N (Nitrogen)

N_2(g)	0	0	191.5	NBS
N_2O(g)	104.2	82.0	220	NBS
NH_3(g)	−16.48	−46.1	192	NBS
NH_3(aq)	−26.57	−80.29	111	NBS
NH_4^+(aq)	−79.37	−132.5	113.4	NBS
HNO_2(aq)	−42.97	−119.2	153	NBS
NO_2^-(aq)	−37.2	−104.6	140	NBS
HNO_3(aq)	−111.3	−207.3	146.	NBS
NO_3^-(aq)	−111.3	−207.3	146.4	NBS

Ni (Nickel)

Ni^{2+}(aq)	−45.6	−54.0	−129	R
NiO (bunsenite)	−211.6	−239.7	38	R
NiS (millerite)	−86.2	−84.9	66	R

O (Oxygen)

O_2(g)	0	0	205	NBS
O_2(aq)	16.32	−11.71	111	NBS
O_3(g)	163.2	142.7	239	NBS
OH^-(aq)	−157.3	−230.0	−10.75	NBS

P (Phosphorus)

P (α, white)	0	0	41.1	
PO_4^{3-}(aq)	−1018.8	−1277.4	−222	NBS
HPO_4^{2-}(aq)	−1089.3	−1292.1	−33.4	NBS
$H_2PO_4^-$(aq)	−1130.4	−1296.3	90.4	NBS
H_3PO_4(aq)	−1142.6	−1288.3	158	NBS

Pb (Lead)

Pb (metal)	0	0	64.8	NBS
Pb^{2+}(aq)	−24.39	−1.67	10.5	NBS
$PbOH^+$(aq)	−226.3	—	—	NBS
$Pb(OH)_3^-$(aq)	−575.7			NBS
$Pb(OH)_2$ (precip.)	−452.2			NBS
PbO (yellow)	−187.9	−217.3	68.7	NBS

PbO_2	-217.4	-277.4	68.6	NBS
Pb_3O_4	-601.2	-718.4	211	NBS
PbS	-98.7	-100.4	91.2	NBS
$PbSO_4$	-813.2	-920.0	149	NBS
$PbCO_3$ (cerussite)	-625.5	-699.1	131	NBS

S (Sulfur)

S (rhombic)	0	0	31.8	NBS
$SO_2(g)$	-300.2	-296.8	248	NBS
$SO_3(g)$	-371.1	-395.7	257	NBS
$H_2S(g)$	-33.56	-20.63	205.7	NBS
$H_2S(aq)$	-27.87	-39.75	121.3	NBS
$S^{2-}(aq)$	85.8	33.0	-14.6	NBS
$HS^-(aq)$	12.05	-17.6	62.8	NBS
$SO_3^{2-}(aq)$	-486.6	-635.5	29	NBS
$HSO_3^-(aq)$	-527.8	-626.2	140	NBS
$H_2SO_3^*(aq)$	-537.9	-608.8	232	NBS^f
$H_2SO_3(aq)$ ("true")	~ -534.5			S
$SO_4^{2-}(aq)$	-744.6	-909.2	20.1	NBS
$HSO_4^-(aq)$	-756.0	-887.3	132	NBS

Se (Selenium)

Se (black)	0	0	42.4	NBS
$SeO_3^{2-}(aq)$	-369.9	-509.2	12.6	NBS
$HSeO_3^-(aq)$	431.5	-514.5	135	NBS
$H_2SeO_3(aq)$	-426.2	-507.5	208	NBS
$SeO_4^{2-}(aq)$	-441.4	-599.1	54.0	NBS
$HScO_4^-(aq)$	-452.3	-581.6	149	NBS

Si (Silicon)

Si (metal)	0	0	18.8	NBS
SiO_2 (α, quartz)	-856.67	-910.94	41.8	NBS
SiO_2 (α, cristobalite)	-855.88	-909.48	42.7	NBS
SiO_2 (α, tridymite)	-855.29	-909.06	43.5	NBS
SiO_2 (amorph)	-850.73	-903.49	46.9	NBS
$H_4SiO_4(aq)$	-1316.7	-1468.6	180	NBS

Sr (Strontium)

$Sr^{2+}(aq)$	-559.4	-545.8	-33	R
$SrOH^+(aq)$	-721	—	—	NBS
$SrCO_3$ (strontianite)	-1137.6	-1218.7	97	R
$SrSO_4$ (celestite)	-1341.0	-1453.2	118	R

Zn (Zinc)

Zn, metal	0	0	29.3	NBS
$Zn^{2+}(aq)$	-147.0	-153.9	112	NBS
$ZnOH^+(aq)$	-330.1			NBS
$Zn(OH)_2(aq)$	-522.3			NBS
$Zn(OH)_3^-(aq)$	-694.3			NBS
$Zn(OH)_4^{2-}(aq)$	-858.7			NBS

$Zn(OH)_2$ (solid β)	−553.2	−641.9	81.2	R
$ZnCl^+$(aq)	−275.3			NBS
$ZnCl_2$(aq)	−403.8			NBS
$ZnCl_3^-$(aq)	−540.6			NBS
$ZnCl_4^{2-}$(aq)	−666.1			S
$ZnCO_3$ (smithsonite)	−731.6	−812.8	82.4	NBS

[a] The quality of the data is highly variable; the authors do not claim to have critically selected the "best" data. For information on precision of the data and for a more complete compendium which includes less common substances, the reader is referred to the references. For research work, the original literature should be consulted.

[b] Thermodynamic properties taken from Robie, Hemingway, and Fisher are based on a reference state of the elements in their standard states at 1 bar (10^5 P = 0.987 atm). This change in reference pressure has a negligible effect upon the tabulated values for the condensed phases. [For gas phases only data from NBS (reference state = 1 atm) are given].

[c] NBS: D. D. Wagman et al., Selected Values of Chemical Thermodynamic Properties, U.S. National Bureau of Standards, Technical Notes 270–3 (1968), 270–4 (1969), 270–5 (1971). R: R. A. Robie, B. S. Hemingway, and J. R. Fisher, Thermodynamic Properties of Minerals and Related Substances at 298.15 K and 1 Bar (10^5 Pascals) Pressure and at Higher Temperatures, Geological Survey Bulletin No. 1452, Washington D.C., 1978. S: Other sources (e.g., computed from data in Stability Constants).

[d] $[H_2CO_3^*]$ = CO_2(aq) + "true" $[H_2CO_3]$.

[e] The thermodynamic stability of oxides, hydroxides, or oxyhydroxides of Fe(III) depends on mode of preparation, age, and molar surface. Reported solubility products (K_{so} = $\{Fe^{3+}\}$ $\{OH^-\}^3$) range from $10^{-37.3}$ to $10^{-43.7}$. Correspondingly FeOOH may have G_f^0 values between −452 J mol^{-1} (freshly precipitated amorphous FeOOH) and −489 J mol^{-1} (aged goethite). If the precipitate is written as Fe(OH)$_3$, its G_f^0 values vary from −692 to −729 J mol^{-1}. (See also Figure 5.23b).

[f] $[H_2SO_3^*]$ = $[SO_2$(aq)] + "true" $[H_2SO_3]$.

Author Index

Subject Index